Names and Formulas of Common Ions

Positive Ions		Negative Ions	
Ammonium	NH_4^+	Acetate	$C_2H_3O_2^-$
Copper (I)	Cu^+	Bromate	BrO_3^-
(Cuprous)		Bromide	Br^-
Hydrogen	H^+	Chlorate	ClO_3^-
Potassium	K^+	Chloride	Cl^-
Silver	Ag^+	Cyanide	CN^-
Sodium	Na^+	Fluoride	F^-
Barium	Ba^{2+}	Hydride	H^-
Cadmium	Cd^{2+}	Hydrogen carbonate	HCO_3^-
Calcium	Ca^{2+}	(Bicarbonate)	
Cobalt (II)	Co^{2+}	Hydrogen sulfate	HSO_4^-
Copper (II)	Cu^{2+}	(Bisulfate)	
(Cupric)		Hydroxide	OH^-
Iron (II)	Fe^{2+}	Iodide	I^-
(Ferrous)		Nitrate	NO_3^-
Lead (II)	Pb^{2+}	Nitrite	NO_2^-
Magnesium	Mg^{2+}	Permanganate	MnO_4^-
Manganese (II)	Mn^{2+}	Thiocyanate	SCN^-
Mercury (II)	Hg^{2+}	Carbonate	CO_3^{2-}
(Mercuric)		Chromate	CrO_4^{2-}
Nickel (II)	Ni^{2+}	Dichromate	$Cr_2O_7^{2-}$
Tin (II)	Sn^{2+}	Oxalate	$C_2O_4^{2-}$
(Stannous)		Oxide	O^{2-}
Zinc	Zn^{2+}	Peroxide	O_2^{2-}
Aluminum	Al^{3+}	Silicate	SiO_3^{2-}
Arsenic (III)	As^{3+}	Sulfate	SO_4^{2-}
Bismuth (III)	Bi^{3+}	Sulfide	S^{2-}
Chromium (III)	Cr^{3+}	Sulfite	SO_3^{2-}
Iron (III)	Fe^{3+}	Arsenate	AsO_4^{3-}
(Ferric)		Phosphate	PO_4^{3-}
Manganese (IV)	Mn^{4+}		
Tin (IV)	Sn^{4+}		
(Stannic)			
Arsenic (V)	As^{5+}		

COLLEGE CHEMISTRY

An Introduction to Inorganic, Organic, and Biochemistry (Second Edition)

The Brooks/Cole Series in Chemistry

FOUNDATIONS OF COLLEGE CHEMISTRY, Fourth Edition, by Morris Hein

FOUNDATIONS OF COLLEGE CHEMISTRY IN THE LAB, Fourth Edition, by Morris Hein, Leo R. Best, and Robert L. Miner

Study Guide for FOUNDATIONS OF COLLEGE CHEMISTRY, Fourth Edition, by Peter Scott

COLLEGE CHEMISTRY: AN INTRODUCTION TO INORGANIC, ORGANIC, AND BIOCHEMISTRY, Second Edition, by Morris Hein and Leo R. Best

COLLEGE CHEMISTRY IN THE LABORATORY, by Morris Hein, Leo R. Best, and Robert L. Miner

Study Guide for COLLEGE CHEMISTRY: AN INTRODUCTION TO INORGANIC, ORGANIC, AND BIOCHEMISTRY, Second Edition, by Peter Scott

FOUNDATIONS OF COLLEGE CHEMISTRY: THE ALTERNATE EDITION, by Morris Hein

Study Guide for FOUNDATIONS OF COLLEGE CHEMISTRY: THE ALTERNATE EDITION, by Peter Scott

*Note: FOUNDATIONS OF COLLEGE CHEMISTRY IN THE LAB, Fourth Edition, is compatible with FOUNDATIONS OF COLLEGE CHEMISTRY: THE ALTERNATE EDITION

COLLEGE CHEMISTRY

An Introduction to Inorganic, Organic, and Biochemistry (Second Edition)

Morris Hein
Leo R. Best
Mount San Antonio College

Brooks/Cole Publishing Company
Monterey, California
A Division of Wadsworth, Inc.

TO EDNA AND LOUISE

© 1980 by Wadsworth, Inc., Belmont, California 94002.
All rights reserved. No part of this book may be reproduced, stored in a retrieval system, or transcribed, in any form or by any means—electronic, mechanical, photocopying, recording, or otherwise—without the prior written permission of the publisher, Brooks/Cole Publishing Company, Monterey, California 93940, a division of Wadsworth, Inc.

Printed in the United States of America
10 9 8 7 6 5 4 3 2

Library of Congress Cataloging in Publication Data

Hein, Morris.
  College chemistry.

  Includes index.
  1. Chemistry. I. Best, Leo R., joint author.
II. Title.
QD31.2.H43   1980   540   80-257
ISBN 0-8185-0349-1

Acquisition Editor: *James F. Leisy, Jr.*
Manuscript Editor: *Phyllis Niklas*
Production Staff: *Joan Marsh, Marilu Uland, Vicki Bamman*
Interior and Cover Design: *Jamie Sue Brooks*
Illustrations: *Cyndie Jo Clark*
Typesetting: *Syntax*
Cover Photograph: © *Batista Moon Studio*
Cover Assistance: *Gordon Williams, Monterey Peninsula College*

Preface

College Chemistry evolved over a considerable period of time. After the appearance of the senior author's *Foundations of College Chemistry* in 1967, various colleagues expressed a desire for a more comprehensive text, written at about the same level, and containing an introduction to organic and biochemistry. *College Chemistry* was published in 1976 in response to this need. Many users of the first edition have asked us to include more material on organic and biochemistry in the next edition. The second edition was developed in response to these requests.

Like the first edition, this book was written for students who, while not majoring in chemistry, require a basic course in the subject including the fundamentals of organic and biochemistry. It is written with the assumption that the student will not have had a previous course in chemistry. We are aware that many students who will use this text will not have a strong background in science and mathematics and may even be somewhat apprehensive about undertaking the study of chemistry. Accordingly one of our major goals has been to present the subject matter in a well organized and easily understandable fashion. Each chapter was written with this thought in mind: Is our language clear—will this material be comprehensible to the student?

We believe a fair amount of quantitative reasoning is essential in the study of chemistry. We also know from experience that many students need considerable practice in order to develop computational skills and quantitative reasoning. Consequently, we have carefully explained each topic that requires mathematical reasoning. In nearly every instance, the explanation is illustrated with one or more problems set up and solved by a straightforward step-by-step procedure. These problems are, with very few exceptions, solved by the conversion-factor dimensional analysis method. This method has proven to be an effective tool and requires no mathematics beyond arithmetic and elementary algebra. In addition to the many solved problems in the text, a mathematical review section covering the pertinent material on handling numbers, algebraic equations, exponents, dimensional analysis, and graphical representation of data is given in Appendix I.

The authors remain convinced that a sound foundation in general chemistry is highly desirable for an understanding of organic and biochemistry. The general chemistry sections in our first edition were patterned after those in the various editions of Hein's *Foundations of College Chemistry*. Hence this material has, for the most part, been retained with relatively minor additions and deletions in content. Some sections dealing with general chemistry have been rewritten, and new or modified problems have been used wherever possible throughout the book. These questions and problems are numerous and varied to give the instructor flexibility in making assignments.

The chapters dealing with organic and biochemistry have been rewritten, updated, and expanded to cover many additional topics. Twelve chapters of organic and biochemistry are included in this edition instead of the nine chapters in the first edition. In courses where more time is desired for organic and biochemistry, Chapters 18, 19, 20, and 21 may be omitted without loss of continuity.

The following features are intended to serve both as learning aids and to make the book convenient to use.

- A list of achievement goals is given at the beginning of each chapter to serve as a guide to the student.
- Each new term that is defined is identified by bold faced type and is also printed in color in the margin.
- A second color is used to point out and emphasize noteworthy aspects of the figures, tables, equations, and summaries given in the text.
- Illustrative problems and examples are set up and worked by logical step-by-step procedures. The conversion-factor dimensional analysis approach is used for nearly all numerical problems.
- Questions and problems are given in several groups at the end of each chapter. The questions and problems within each group are arranged in the order of the related material in the chapter.
- Answers to all numerical problems are given in Appendix V. These answers should be used by students to verify their own calculations.
- Nearly all chapters have a self evaluation question containing a fairly large number of "correct" or "not correct" statements. Answers to these questions are also in Appendix V.
- Material on inorganic nomenclature appears in several chapters, but for convenient reference, the basic rules for naming inorganic compounds are brought together in Chapter 8.
- For quick reference a list of the names and formulas of common inorganic ions is given on the inside front cover.

We wish to acknowledge the contributions of the following for their professional and critical reviews and suggestions: Professors David Adams of North Shore Community College, Eleanor Behrman of the University of Cincinnati, Keith Biever of Bellevue Community College, Walter Brooks of Santa Ana College, Tom Cramblet of Cabrillo College, W. L. Felty of Pennsylvania State University, Wilkes-Barre Campus, Andrew Glaid of Duquesne University, Stewart Karp of Long Island University, Thomas L. McCarley of Iowa Western Community College, Daniel Meloon of State University of New York at Buffalo, Beatrice Paige of Antelope Valley College, Jim Ritchey of California State University, Sacramento, Sally Solomon of Drexel University, and C. M. Wilkerson of Western Kentucky University.

We express our gratitude to Robert L. Miner of Mount San Antonio College, who prepared the Solutions Manual for the text, and we also thank Peter C. Scott of Linn-Benton Community College, who prepared the Study Guide.

We also wish to express our gratitude to all our colleagues in the Mount San Antonio College Chemistry Department for their many helpful suggestions. We would be remiss if we did not mention the cooperation and help that we received from the editorial and production staff of Brooks/Cole Publishing Company.

Last, but certainly not least, we are forever grateful to our wives, Edna and Louise, for their continued support and understanding and to whom we fondly dedicate this volume.

Morris Hein
Leo R. Best

Contents

1 Introduction 1

- 1.1 The Nature of Chemistry 1
- 1.2 History of Chemistry 2
- 1.3 The Branches of Chemistry 4
- 1.4 Relationship of Chemistry to Other Sciences and Industry 4
- 1.5 Scientific Method 6
- 1.6 How to Study Chemistry 7

2 Standards for Measurement 9

- 2.1 Mass and Weight 9
- 2.2 Measurement, Significant Figures, and Calculations 10
- 2.3 The Metric System 11
- 2.4 Measurement of Length 12
- 2.5 Problem Solving 13
- 2.6 Measurement of Mass 16
- 2.7 Measurement of Volume 17
- 2.8 Temperature Scales 19
- 2.9 Heat and Temperature 21
- 2.10 Tools for Measurement 23
- 2.11 Density 24
- 2.12 Specific Gravity 27

3 Properties of Matter 33

- 3.1 Matter Defined 33
- 3.2 Physical States of Matter 33
- 3.3 Substances and Mixtures 36
- 3.4 Properties of Substances 37
- 3.5 Physical Changes 38
- 3.6 Chemical Changes 39
- 3.7 Conservation of Mass 41
- 3.8 Energy 42
- 3.9 Energy in Chemical Changes 43
- 3.10 Conservation of Energy 43
- 3.11 Interchangeability of Matter and Energy 44

4 Elements and Compounds 48

- 4.1 Elements 48
- 4.2 Distribution of Elements 49
- 4.3 Names of the Elements 50
- 4.4 Symbols of the Elements 50
- 4.5 Compounds 52
- 4.6 Law of Definite Composition of Compounds 54
- 4.7 Chemical Formulas 54
- 4.8 Mixtures 56
- 4.9 Metals, Nonmetals, and Metalloids 57
- 4.10 Nomenclature and Chemical Equations 59

5 Atomic Theory and Structure 64

- 5.1 Early Thoughts 64
- 5.2 Dalton's Atomic Theory 65
- 5.3 Subatomic Parts of the Atom 65
- 5.4 The Nuclear Atom 67
- 5.5 General Arrangement of Subatomic Particles 69
- 5.6 The Bohr Atom 69
- 5.7 The Quantum Mechanical Atom 70
- 5.8 Energy Levels of Electrons 72
- 5.9 Energy Sublevels 73
- 5.10 Atomic Numbers of the Elements 75
- 5.11 The Simplest Atom—Hydrogen 75
- 5.12 Isotopes of the Elements 76
- 5.13 Atomic Structure of the First Twenty Elements 77
- 5.14 Electron Structure of the Elements Beyond Calcium 80
- 5.15 Diagramming Atomic Structures 82
- 5.16 Electron-Dot Representation of Atoms 84
- 5.17 The Noble Gases 85
- 5.18 Atomic Weight 87
- 5.19 The Mole 88

6 The Periodic Arrangement of the Elements 96

- 6.1 Early Attempts to Classify the Elements 96
- 6.2 The Periodic Law 97
- 6.3 Arrangement of the Periodic Table 100
- 6.4 Periods of Elements 101
- 6.5 Groups or Families of Elements 102
- 6.6 Predicting Formulas by Use of the Periodic Table 104
- 6.7 Transition Elements 105
- 6.8 New Elements 105
- 6.9 Summary 105

7 The Formation of Compounds from Atoms 109

- 7.1 Ionization Energy 110
- 7.2 Electrons in the Outer Shell 112

7.3 Transfer of Electrons from One Atom to Another 112
7.4 Sharing of Electrons 118
7.5 Chemical Bonds 121
7.6 The Electrovalent Bond 121
7.7 The Covalent Bond 121
7.8 Polar Covalent Bonds 123
7.9 Coordinate Covalent Bonds 124
7.10 Polyatomic Ions 125
7.11 Oxidation Numbers of Atoms 126
7.12 Oxidation Number Tables 128
7.13 Formulas of Electrovalent Compounds 129
7.14 Determining Oxidation Numbers and Ionic Charges from a Formula 131
7.15 Oxidation—Reduction 133

8 Nomenclature of Inorganic Compounds 137

8.1 Common, or Trivial, Names 137
8.2 Systematic Chemical Nomenclature 139
8.3 Binary Compounds 139
8.4 Ternary Compounds 143
8.5 Salts 145
8.6 Salts with More than one Positive Ion 147
8.7 Bases 148

9 Quantitative Composition of Compounds 151

9.1 Formula Weight or Molecular Weight 151
9.2 Determination of Molecular Weights from Formulas 152
9.3 Gram-Molecular Weight; Gram-Formula Weight; the Mole 153
9.4 Percentage Composition of Compounds 155
9.5 Empirical Formula versus Molecular Formula 156
9.6 Calculation of Empirical Formula 157
9.7 Calculation of the Molecular Formula from the Empirical Formula 160

10 Chemical Equations 164

10.1 The Chemical Equation 164
10.2 Format for Writing Chemical Equations 165
10.3 Writing and Balancing Equations 165
10.4 What Information Does an Equation Tell Us? 169
10.5 Types of Chemical Equations 170
10.6 Heat in Chemical Reactions 173

11 Calculations from Chemical Equations 178

11.1 A Short Review 178
11.2 Calculations from Chemical Equations: The Mole Method 179

x Contents

    11.3   Mole-Mole Calculations   181
    11.4   Mole-Weight and Weight-Weight Calculations   184
    11.5   Limiting Reagent and Yield Calculations   185

12 The Gaseous State of Matter   191

    12.1   General Properties of Gases   191
    12.2   The Kinetic-Molecular Theory   192
    12.3   Measurement of Pressure of Gases   194
    12.4   Dependence of Pressure on Number of Molecules   197
    12.5   Boyle's Law—The Relationship of the Volume and Pressure of a Gas   198
    12.6   Charles' Law—The Effect of Temperature on the Volume of a Gas   201
    12.7   Standard Temperature and Pressure   205
    12.8   Simultaneous Changes in Pressure, Volume, and Temperature (Combined Gas Laws)   205
    12.9   Dalton's Law of Partial Pressures   207
    12.10 Avogadro's Hypothesis   209
    12.11 Weight-Volume Relationship of Gases   210
    12.12 Density and Specific Gravity of Gases   212
    12.13 Calculations from Chemical Equations Involving Gases   214
    12.14 Ideal Gas Equation   217

13 Water and the Properties of Liquids   223

    13.1   Occurrence of Water   224
    13.2   Physical Properties of Water   224
    13.3   Structure of the Water Molecule   226
    13.4   The Hydrogen Bond   227
    13.5   Formation of Water and Chemical Properties of Water   229
    13.6   Hydrates   231
    13.7   Hygroscopic Substances: Deliquescence; Efflorescence   232
    13.8   Hydrogen Peroxide   233
    13.9   Natural Waters   235
    13.10 Water Pollution   237
    13.11 Evaporation   239
    13.12 Vapor Pressure   240
    13.13 Boiling Point   241
    13.14 Freezing Point or Melting Point   243

14 Solutions   250

    14.1   Components of a Solution   251
    14.2   Types of Solutions   251
    14.3   General Properties of Solutions   251
    14.4   Solubility   252
    14.5   Factors Related to Solubility   254
    14.6   Solutions: A Reaction Zone   258

14.7 Concentration of Solutions   259
14.8 Colligative Properties of Solutions   271
14.9 Osmosis and Osmotic Pressure   274

15 Acids, Bases, Salts   282

15.1 Acids and Bases   282
15.2 Reactions of Acids   285
15.3 Reactions of Bases   286
15.4 Salts   287
15.5 Electrolytes and Nonelectrolytes   288
15.6 Dissociation and Ionization of Electrolytes   289
15.7 Strong and Weak Electrolytes   292
15.8 Ionization of Water   294
15.9 Introduction to pH   294
15.10 Neutralization   298
15.11 Writing Ionic Equations   301
15.12 Colloids: Introduction   302
15.13 Preparation of Colloids   304
15.14 Properties of Colloids   305
15.15 Stability of Colloids   306
15.16 Applications of Colloidal Properties   307

16 Chemical Equilibrium   313

16.1 Reversible Reactions   313
16.2 Rates of Reaction   315
16.3 Chemical Equilibrium   316
16.4 Principle of Le Chatelier   317
16.5 Effect of Concentration on Reaction Rate and Equilibrium   318
16.6 Effect of Pressure on Reaction Rate and Equilibrium   320
16.7 Effect of Temperature on Reaction Rate and Equilibrium   322
16.8 Effect of Catalysts on Reaction Rate and Equilibrium   323
16.9 Equilibrium Constants   324
16.10 Ionization Constants   325
16.11 Ion Product Constant for Water   327
16.12 Solubility Product Constant   329
16.13 Buffer Solutions   332
16.14 Mechanism of Reactions   333

17 Oxidation-Reduction   341

17.1 Oxidation Number   341
17.2 Oxidation-Reduction   343
17.3 Balancing Oxidation-Reduction Equations   345
17.4 Balancing Ionic Redox Equations   348
17.5 Activity Series of Metals   350
17.6 Electrolytic and Voltaic Cells   352

18 Radioactivity and Nuclear Chemistry 360

18.1 Discovery of Radioactivity 361
18.2 Natural Radioactivity 362
18.3 Properties of Alpha, Beta, and Gamma Rays 363
18.4 Radioactive Disintegration Series 367
18.5 Transmutation of Elements 367
18.6 Artificial Radioactivity 368
18.7 Measurement of Radioactivity 369
18.8 Nuclear Fission 370
18.9 Nuclear Power 371
18.10 The Atomic Bomb 372
18.11 Nuclear Fusion 374
18.12 Mass-Energy Relationship in Nuclear Reactions 375
18.13 Transuranium Elements 375
18.14 Biological Effects of Radiation 377
18.15 Applications of Radioisotopes 378

19 Chemistry of Some Selected Metals 384

19.1 The Alkali Metals 385
19.2 Alloys 390
19.3 The Alkaline Earth Metals 391
19.4 Flame Tests and Spectroscopy 398
19.5 Iron 399
19.6 Aluminum 405

20 Chemistry of Some Selected Nonmetals 412

20.1 The Halogen Family 413
20.2 The Sulfur Family 427
20.3 The Nitrogen Family 437

21 Air Pollution 448

21.1 Introduction 449
21.2 Major Air Pollution Episodes 450
21.3 Atmospheric Pollutants 451
21.4 Gaseous Pollutants 452
21.5 Ozone in the Stratosphere 458
21.6 Photochemical Smog 459
21.7 Atmospheric Temperature Inversion 461
21.8 Air Pollution Control 463

22 Organic Chemistry; Saturated Hydrocarbons 469

22.1 Organic Chemistry: History and Scope 469
22.2 The Need for Classification of Organic Compounds 470

22.3   The Carbon Atom: Tetrahedral Structure   471
22.4   Carbon-Carbon Bonds   472
22.5   Hydrocarbons   473
22.6   Saturated Hydrocarbons: Alkanes   473
22.7   Carbon Bonding in Alkanes   474
22.8   Structural Formulas and Isomerism   476
22.9   Naming Organic Compounds   480
22.10 Reactions of Alkanes   485
22.11 Cycloalkanes   489

23 Unsaturated Hydrocarbons: Alkenes, Alkynes, and Aromatic Hydrocarbons   494

23.1   Alkenes and Alkynes   495
23.2   Bond Formation in Alkenes and Alkynes   496
23.3   Naming Alkenes and Alkynes   499
23.4   Geometric Isomerism   501
23.5   Reactions of Alkenes   503
23.6   Octane Rating of Gasoline   507
23.7   Acetylenes: Preparation and Properties   508
23.8   Aromatic Hydrocarbons: Structure   510
23.9   Naming Aromatic Compounds   513
23.10 Fused Aromatic Rings   517
23.11 Sources and Properties of Aromatic Hydrocarbons   518
23.12 Reactions of Aromatic Hydrocarbons   518

24 Alcohols, Phenols, and Ethers   525

24.1   Functional Groups   526
24.2   Classification of Alcohols   528
24.3   Naming Alcohols   530
24.4   Physical Properties of Alcohols   531
24.5   Reactions of Alcohols   533
24.6   Preparation and Properties of Common Alcohols   536
24.7   Phenols   540
24.8   Selected Phenolic Compounds   541
24.9   Properties of Phenols   542
24.10 Production of Phenol   543
24.11 Ethers   544
24.12 Structures and Properties of Ethers   545
24.13 Preparation of Ethers   547
24.14 Ether as an Anesthetic   547

25 Aldehydes and Ketones   552

25.1   Structure of Aldehydes and Ketones   553
25.2   Naming Aldehydes and Ketones   553
25.3   Bonding and Physical Properties   556
25.4   Reactions of Aldehydes and Ketones   558

xiv   Contents

    25.5   Preparation and Properties of Common Aldehydes and Ketones   563
    25.6   Iodoform Test   565
    25.7   Grignard Reactions   566

26 Carboxylic Acids, Esters, and Amines   573

    26.1   Carboxylic Acids   574
    26.2   Nomenclature and Sources of Aliphatic Carboxylic Acids   574
    26.3   Physical Properties of Carboxylic Acids   577
    26.4   Classification of Carboxylic Acids   578
    26.5   Preparation of Carboxylic Acids   581
    26.6   Reactions of Carboxylic Acids   584
    26.7   Esters   587
    26.8   Glycerol Esters   589
    26.9   Soaps and Synthetic Detergents   592
    26.10   Amines   596

27 Polymers—Macromolecules   607

    27.1   Introduction   607
    27.2   Synthetic Polymers   608
    27.3   Polymer Types   609
    27.4   Addition Polymerization   609
    27.5   Butadiene Polymers   611
    27.6   Stereochemistry of Polymers   613
    27.7   Condensation Polymers   614
    27.8   Silicone Polymers   617

28 Optical Isomerism   621

    28.1   Review of Isomerism   621
    28.2   Plane-Polarized Light   622
    28.3   Optical Activity   624
    28.4   Projection Formulas   625
    28.5   Enantiomers   627
    28.6   Racemic Mixtures   632
    28.7   Diastereomers and *Meso* Compounds   633

29 Carbohydrates   640

    29.1   Introduction   641
    29.2   Classification   641
    29.3   Monosaccharides   644
    29.4   Structure of Glucose and Other Aldoses   645
    29.5   Cyclic Structure of Glucose; Mutarotation   647
    29.6   Hemiacetals and Acetals   649
    29.7   Structure of Galactose and Fructose   651
    29.8   Pentoses   652

29.9   Disaccharides   653
29.10 Structure and Properties of Disaccharides   653
29.11 Reducing Sugars   656
29.12 Reactions of Monosaccharides   658
29.13 Polysaccharides   661

30 Amino Acids, Polypeptides, and Proteins   667

30.1   Introduction   668
30.2   The Nature of Amino Acids   669
30.3   Essential Amino Acids   669
30.4   D-Amino Acids and L-Amino Acids   671
30.5   Amphoterism   672
30.6   Formation of Polypeptides   674
30.7   Protein Structure   676
30.8   Enzymes   681
30.9   Hydrolysis of Proteins   685
30.10 Denaturation of Proteins   686
30.11 Tests for Proteins and Amino Acids   687
30.12 Determination of the Primary Structure of Polypeptides   688
30.13 Synthesis of Peptides and Proteins   690

31 Nucleic Acids and Heredity   695

31.1   Introduction   695
31.2   Nucleosides   696
31.3   Nucleotides—Phosphate Esters   698
31.4   High-Energy Nucleotides   699
31.5   Polynucleotides; Nucleic Acids   700
31.6   Structure of DNA   701
31.7   DNA—The Genetic Substance   705
31.8   The Genetic Code   707
31.9   Genetic Transcription, RNA   708
31.10 Biosynthesis of Proteins   710
31.11 Mutations   713

32 Digestion; Carbohydrate Metabolism; Hormones; Vitamins   717

32.1   Introduction   718
32.2   Human Digestion   718
32.3   Absorption   723
32.4   NAD, NADP, and FAD   723
32.5   Metabolism   726
32.6   Carbohydrate Metabolism   726
32.7   Overview of Human Carbohydrate Metabolism   727
32.8   Anaerobic Sequence   729
32.9   Citric Acid Cycle (Aerobic Sequence)   730
32.10 Gluconeogenesis   733

32.11 Hormones   733
32.12 Vitamins   735
32.13 Glucose Concentration in the Blood   738

33 Lipids; Metabolism of Lipids and Proteins   743

33.1 Classification of Lipids   743
33.2 Fats and Oils   744
33.3 Compound Lipids   746
33.4 Steroids   748
33.5 Fat Absorption and Distribution   750
33.6 Fatty Acid Oxidation   751
33.7 Fat Storage and Utilization   754
33.8 Biosynthesis of Fatty Acids (Lipogenesis)   755
33.9 Amino Acid Absorption and Distribution   757
33.10 Amino Acid Utilization   759

Appendix I   Mathematical Review   A-1
Appendix II   Vapor Pressure of Water at Various Temperatures   A-18
Appendix III   Units of Measurements   A-19
Appendix IV   Solubility Table   A-20
Appendix V   Answers to Problems   A-21

Index   A-29

1 Introduction

1.1 The Nature of Chemistry

What is chemistry? A popular dictionary gives this definition: Chemistry is the science of the composition, structure, properties, and reactions of matter, especially of atomic and molecular systems. Another and somewhat simpler dictionary definition is: Chemistry is the science dealing with the composition of substances and the transformations that they undergo. Neither of these definitions is entirely adequate. Chemistry, along with the closely related science of physics, is a very fundamental branch of science. The scope of chemistry, as implied by the definitions just quoted, is extremely broad—it includes the whole universe and everything, animate and inanimate, to be found in the universe. Chemistry is concerned not only with the composition and changes of composition of matter, but equally importantly, it deals with the energy and energy changes associated with matter. Through chemistry we seek to learn and to understand the general principles that govern the behavior of all matter.

The chemist, like other scientists, observes nature at work and attempts to unlock its secrets: What makes a rose red? Why is sugar sweet? Why is water wet? Why is carbon monoxide poisonous? Why do people wither with age? Problems such as these—some of which have been solved, some of which are still to be solved—are part of what we call chemistry.

A chemist may interpret natural phenomena, devise experiments that will reveal the composition and structure of complex substances, study methods for improving natural processes, or, sometimes, synthesize substances unknown in nature. Ultimately, the efforts of successful chemists advance the frontiers of knowledge and at the same time contribute to the well-being of humanity. Chemistry helps us to understand nature; however, one need not be a professional chemist or scientist to enjoy natural phenomena. Nature and its beauty, its simplicity within complexity, is for all to appreciate.

The body of chemical knowledge is so vast that no one can hope to master it all even in a lifetime of study. However, many of the basic concepts can be learned in a relatively short period of time. These basic concepts have become part of the education required for many professionals, including agriculturists, biologists, dental hygienists, dentists, medical technologists, microbiologists, nurses, nutritionists, pharmacists, physicians, and veterinarians to name a few.

1.2 History of Chemistry

From the earliest times, people have practiced empirical chemistry. Ancient civilizations were practicing the art of chemistry in such processes as wine-making, glass-making, pottery, dyeing, and elementary metallurgy. The early Egyptians, for example, had considerable knowledge of certain chemical processes. Excavations into ancient tombs dated about 3000 B.C. have uncovered workings of gold, silver, copper, iron, pottery from clay, glass beads, beautiful dyes and paints, as well as bodies of Egyptian kings in unbelievably well-preserved states. Many other cultures made significant developments in chemistry. However, all these developments were empirical; that is, they were achieved by trial and error and did not rest upon any valid theory of matter.

Philosophical ideas relating to the properties of matter (chemistry) did not develop as early as those relating to astronomy and mathematics. The ancient Greek philosophers made great strides in philosophical speculation concerning materialistic ideas about chemistry. They led the way to placing chemistry on a highly intellectual, scientific basis. They first introduced the concepts of elements, atoms, shapes of atoms, chemical combinations, and so on. The Greeks believed that there were four elements—earth, air, fire, and water—and that all matter was derived from these elements. The Greek philosophers had very keen minds and perhaps came very close to establishing chemistry on a sound basis similar to the one that was to develop about 2000 years later. The main shortcoming of the Greek approach to scientific work was a failure to carry out systematic experimentation.

The Greek civilization declined and was succeeded by the Roman civilization. The Romans were outstanding in military, political, and economic affairs. They continued to practice empirical chemical arts such as metallurgy, enameling, glass-making, and pottery-making, but they did very little to advance new and theoretical knowledge. Eventually, the Roman civilization declined and was succeeded in Europe by the Dark Ages. During this period, European civilization and learning were at a very low ebb.

In the Middle East and in North Africa knowledge did not decline during the Dark Ages as it did in Western Europe. During this period Arabic cultures made contributions that were of great value to the later development of modern chemistry. In particular, the Arabic number system, including the use of zero, gained acceptance; the branch of mathematics known as *algebra* was developed; and alchemy, a sort of pseudochemistry, was practiced extensively.

One of the more interesting periods in the history of chemistry was that of the alchemists (500–1600 A.D.). People have long had a lust for gold, and in those days gold was considered the ultimate, most perfect metal formed in nature. The principal goals of the alchemists were to find a method of prolonging human life indefinitely and to change the base metals—such as iron, zinc, and copper—into gold. They searched for a universal solvent to transmute base metals into gold and for the "philosopher's stone" to rid the body of all diseases and to renew life. In the course of their labors, they learned a great deal of chemistry. Unfortunately, much of their work was done secretly because of the mysticism that shrouded their activity, and very few records remain.

Although the alchemists were not guided by sound theoretical reasoning and were clearly not in the intellectual class of the Greek philosophers, they did something that the philosophers had not considered worthwhile. They subjected various materials to prescribed treatments under what might be loosely described as laboratory methods. These manipulations, carried out in alchemical laboratories, not only uncovered many facts of nature but paved the way for the systematic experimentation that is characteristic of modern science.

Alchemy began to decline in the 16th century when Paracelsus (1493–1541), a Swiss physician and outspoken revolutionary leader in chemistry, strongly advocated that the objectives of chemistry be directed toward the needs of medicine and the curing of human ailments. He openly condemned the mercenary efforts of alchemists to convert cheaper metals to gold.

But the real beginning of modern science can be traced to astronomy during the Renaissance. Nicolaus Copernicus (1473–1543), a Polish astronomer, succeeded in upsetting the generally accepted belief in a geocentric universe. Although not all the Greek philosophers had believed that the sun and the stars revolved about the earth, the geocentric concept had come to be accepted without question. The heliocentric (sun-centered) universe concept of Copernicus was based on direct astronomical observation and represented a radical departure from the concepts handed down from Greek and Roman times. The ideas of Copernicus and the invention of the telescope stimulated additional work in astronomy. This work, especially that of Galileo Galilei (1564–1642) and Johannes Kepler (1571–1630), led directly to a rational explanation of the general laws of motion by Sir Isaac Newton (1642–1727) from about 1665 to 1685.

Modern chemistry was slower to develop than astronomy and physics; it began in the 17th and 18th centuries when Joseph Priestley (1733–1804), who discovered oxygen in 1774, and Robert Boyle (1627–1691) began to record and publish the results of their experiments and to discuss their theories openly. Boyle, who has been called the founder of modern chemistry, was one of the first to practice chemistry as a true science. He believed in the experimental method. In his most important book, *The Skyptical Chemist*, he clearly distinguished between an element and a compound or mixture. Boyle is best known today for the gas law that bears his name. A French chemist, Antoine Lavoisier (1743–1794), placed the science on a firm foundation with experiments in which he used a chemical balance to make quantitative measurements of the weights of substances involved in chemical reactions.

The use of the chemical balance by Lavoisier and others later in the 18th century was almost as revolutionary in chemistry as the use of the telescope had been in astronomy. Thereafter, chemistry was a highly quantitative experimental science. Lavoisier also contributed greatly to the organization of chemical data, to chemical nomenclature, and to the establishment of the Law of Conservation of Mass in chemical changes. During the period from 1803 to 1810, John Dalton (1766–1844), an English schoolteacher, advanced his atomic theory. This theory (see Section 5.2) placed the atomistic concept of matter on a valid rational basis. It remains today as a tremendously important **general concept of modern science**.

Since the time of Dalton, knowledge of chemistry has advanced in great strides, with the most rapid advancement occurring at the end of the 19th century and during the 20th century. Especially outstanding achievements have been made in determining the structure of the atom, understanding the biochemical fundamentals of life, developing chemical technology, and the mass production of chemicals and related products.

1.3 The Branches of Chemistry

Chemistry may be broadly classified into two main branches: *organic* chemistry and *inorganic* chemistry. Organic chemistry is concerned with compounds containing the element carbon. The term *organic* was originally derived from the chemistry of living organisms—plants and animals. Inorganic chemistry deals with all the other elements as well as with some carbon compounds. Substances classified as inorganic are derived mainly from mineral sources rather than from animal or vegetable sources.

Other subdivisions of chemistry, such as analytical chemistry, physical chemistry, biochemistry, electrochemistry, geochemistry, and radiochemistry, may be considered specialized fields of, or auxiliary fields to, the two main branches.

Chemical engineering is the branch of engineering that deals with the development, design, and operation of chemical processes. A chemical engineer generally begins with a chemist's laboratory-scale process and develops it into an industrial-scale operation.

1.4 Relationship of Chemistry to Other Sciences and Industry

Besides being a science in its own right, chemistry is the servant of other sciences and industry. Chemical principles contribute to the study of physics, biology, agriculture, engineering, medicine, space research, oceanography, and many other sciences. Chemistry and physics are overlapping sciences, since both are based on the properties and behavior of matter. Biological processes are chemical in nature. The metabolism of food to provide energy to living organisms is a chemical process. Knowledge of molecular structure of proteins, hormones, enzymes, and the nucleic acids is assisting biologists in their investigations of the composition, development, and reproduction of living cells.

Chemistry is playing an important role in alleviating the growing shortage of food in the world. Agricultural production has been increased with the use of chemical fertilizers, pesticides, and improved varieties of seeds. Chemical refrigerants make possible the frozen food industry that preserves large amounts of food that might otherwise spoil. Chemistry is also producing synthetic nutrients, but much remains to be done as the world population multiplies with respect to the land available for cultivation. Expanding energy needs have

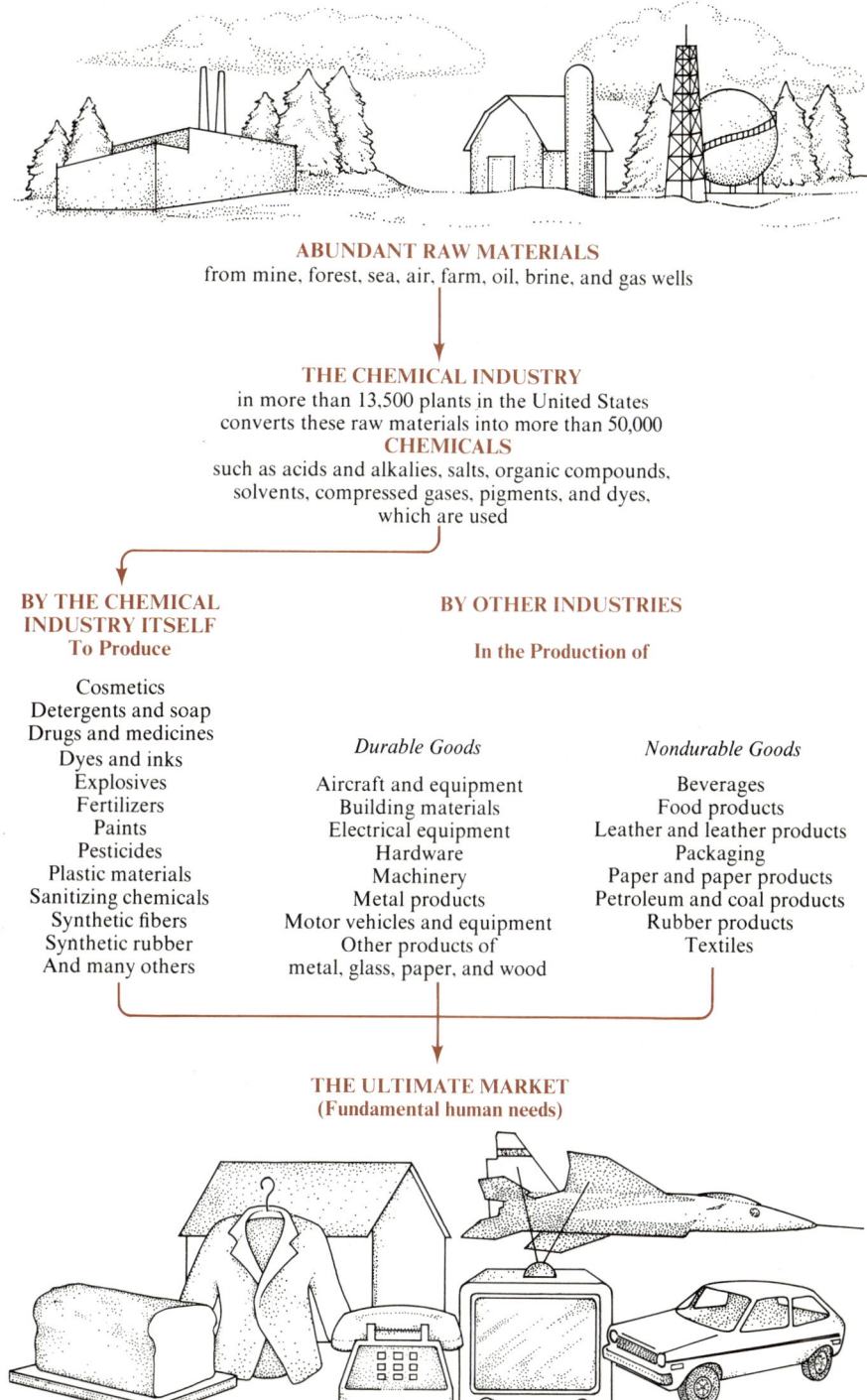

Figure 1.1. Broad scope of the chemical industry today. (Courtesy Chemical Manufacturers Association.)

brought about difficult environmental problems in the form of air and water pollution. Chemists as well as other scientists are working diligently to alleviate these problems.

Advances in medicine and chemotherapy, through the development of new drugs, have contributed to prolonged life and the relief of human suffering. More than 90% of the drugs and pharmaceuticals being used in the United States today have been developed commercially within the past 40 years. The entire plastics and polymer industry, unknown 50 years ago, has revolutionized the packaging and textile industries and is producing more durable and useful construction materials. Energy derived from chemical processes is used universally for heating, lighting, and transportation. There is virtually no industry that is not dependent on chemicals—for example, the petroleum, steel, rubber, pharmaceutical, electronic, transportation, cosmetic, garment, aircraft, and television industries (the list could go on and on). Figure 1.1 illustrates the conversion of natural resources by the chemical industry into useful products for commerce, industry, and human needs.

1.5 Scientific Method

Chemistry as a science or field of knowledge is concerned with ideas and concepts relating to the behavior of matter. Although these concepts are abstract, their application has had an extraordinarily concrete impact on human culture. This impact is due to modern technology, which may be said to have begun about 200 years ago and has continued to grow at an accelerating rate ever since.

There is a very important difference between the science of chemistry and technology. The science represents a basic body of knowledge; technology represents the physical application of this knowledge to the real world in which we live.

Why has the science of chemistry and its associated technology flourished so abundantly in the last two centuries? Is it because we are growing more intelligent? No, there is absolutely no reason to believe that the general level of human intelligence is any higher today than it was 1000 years ago in the Dark Ages. The use of the scientific method is usually credited with being the most important single factor in the amazing development of chemistry and technology. Although complete agreement is lacking on exactly what is meant by "using the scientific method," the general approach is as follows:

1. Collect facts or data that are relevant to the problem or question at hand. This is usually done by planned experimentation.
2. Analyze the data to find trends (regularities) that are pertinent to the problem. Formulate a hypothesis that will account for the data that have been accumulated and that can be tested by further experimentation.
3. Plan and do additional experiments to test the hypothesis. Such experiments extend beyond the range that is covered in Step 1.
4. Modify the hypothesis as necessary so that it is compatible with all the pertinent experimental data.

Confusion sometimes arises regarding the exact meanings of the words *hypothesis*, *theory*, and *law*. A well-established hypothesis is often called a theory. Hypotheses and theories explain natural phenomena, whereas scientific laws are simple statements of natural phenomena to which no exceptions are known.

While the four steps listed above are a broad outline of the general procedure that is followed in much scientific work, they do not provide a recipe for doing chemistry or any other science. But chemistry is an experimental science; and much of its progress has been due to application of the scientific method through systematic research. Occasionally, a great discovery is made by accident, but the majority of scientific achievements are accomplished by well-planned experiments.

Many theories and laws are studied in chemistry. They make the study of chemistry or any science easier, because they summarize a particular aspect of the science. Although the student will see that theories advanced by great thinkers have been subject to change, this change does not mean that their contributions are of lesser significance than the discoveries of today. Change is the natural evolution of scientific knowledge.

1.6 How to Study Chemistry

How do you as a student approach a subject such as chemistry, with its unfamiliar terminology, symbols, formulas, theories, and laws? All the normally accepted habits of good study are applicable to the study of chemistry. Budget your study time and spend it wisely. In particular, you can spend your study time more profitably in regular, relatively short periods rather than in one prolonged cram session.

Chemistry has its own language, and learning this language is of prime importance to the successful study of chemistry. Chemistry is a subject of many facts. At first you will simply have to memorize some of them. However, you will also learn these facts by referring to them frequently in your studies and by repetitive use. For example, you must learn the symbols of 30 or 40 common elements in order to be able to write chemical formulas and equations. As with the alphabet, repetitive use of these symbols will soon make them part of your vocabulary.

Careful reading of assigned material cannot be overemphasized. You should read each chapter at least twice. The first time, read the chapter rapidly, noting especially topic headings, diagrams, and other outstanding features. Then read more thoroughly and deliberately for better understanding. It may be profitable to underline and abstract material during the second reading. Isolated reading may be sufficient for learning some subjects, but it is not sufficient for learning chemistry. During the lectures, become an active mental participant and try to think along with your instructor—do not just occupy a seat. Lecture and laboratory sessions will be much more meaningful if you have already read the assigned material.

Your studies must include a good deal of written chemistry. Chemical symbolism, equations, problem solving, and so on, require much written practice for proficiency. One does not become an accomplished pianist by merely reading or listening to music—it takes practice. One does not become a good baseball player by reading the rules and watching baseball games—it takes practice. So it is with chemistry. One does not become proficient in chemistry by only reading about it—it takes practice.

In solving a numerical problem, you should read the problem carefully to determine what is being asked. Then develop a plan for solving the problem. It is a good idea to start by writing down something—a formula, a diagram, an equation, the data given in the problem. This will give you something to work with, to think about, to modify—and finally to expand into an answer. When you have arrived at an answer, consider it carefully to make sure that it is a reasonable one. The solutions to problems should be recorded in a neat, orderly, stepwise fashion. Fewer errors and time saved are the rewards of a neat and orderly approach to problem solving. If you need to read and study still further for complete understanding, do it!

2 Standards for Measurement

After studying Chapter 2 you should be able to:

1. Understand the terms listed in Question A at the end of the chapter.
2. Differentiate clearly between mass and weight.
3. Know the basic metric units of mass, length, and volume.
4. Give the numerical equivalents of the metric prefixes deci, centi, milli, micro, deka, hecto, kilo, and mega.
5. Express any number in exponential notation form.
6. Express the results of arithmetic operations to the proper number of significant figures.
7. Set up and solve problems by the dimensional analysis, or factor-label, method.
8. Convert any measurement of mass, length, or volume in American units to metric units and vice versa.
9. Make conversions between Fahrenheit, Celsius, and Kelvin temperatures.
10. Differentiate clearly between temperature and heat.
11. Make calculations using the equation

    calories = (Grams of substance) × (Heat capacity of substance) × (Δt)

12. Calculate density, mass, or volume of an object (or substance) from appropriate data.
13. Calculate the specific gravity when given the density of a substance and vice versa.
14. Recognize the common laboratory measuring instruments illustrated in this chapter.

2.1 Mass and Weight

Chemistry is an experimental science. The results of experiments are usually determined by making measurements. In elementary experiments, the quantities that are commonly measured are mass (weight), length, volume, and time. Measurements of electrical and optical quantities may also be needed in more sophisticated experimental work.

mass

Although mass and weight are often used interchangeably, the two words have quite different meanings. The **mass** of a body is defined as the amount of matter in that body. The amount of mass in an object is a fixed and unvarying quantity that is independent of the object's location.

weight

The **weight** of a body is the measure of the earth's gravitational attraction for that body. Unlike its mass, the weight of an object varies in relation to (1) its position on or its distance from the earth and (2) whether the rate of motion of the object is changing with respect to the motion of the earth. Consider an astronaut of mass 70.0 kilograms (154 pounds) who is being shot into a space orbit. At the instant before blastoff the weight of the astronaut is also 70.0 kilograms. As the distance from the earth increases and the rocket turns into an orbiting course, the gravitational pull on the astronaut's body decreases until a state of "weightlessness" is attained. However, the mass of the astronaut's body has remained constant at 70.0 kilograms during the entire process of lift-off and going into orbit.

The mass of an object may be measured on a chemical balance by comparing it with other known masses. Two objects of equal mass will also have equal weights if they are measured in the same place. Thus, under these conditions the terms *mass* and *weight* are used interchangeably. Although the chemical balance is used to determine mass, it is said to *weigh* objects, and we often speak of the *weight* of an object when we really mean its mass.

2.2 Measurement, Significant Figures, and Calculations

To understand certain phases of chemistry it is necessary to set up and solve problems. Problem solving requires an understanding of the elementary mathematical operations used to manipulate numbers. Numerical values or data are obtained from measurements made in an experiment. A chemist may use these data to calculate the extent of the physical and chemical changes occurring in the substances that are being studied. By appropriate calculations, the results of an experiment may be compared with those of other experiments and summarized in ways that are meaningful.

When expressing a quantity of something, one must state both the numerical value and the units in which the quantity is expressed. In the statement "1 kilometre contains 1000 metres," kilometre and metre are the units in which this length is expressed.

There is some degree of uncertainty in every experimental measurement—due to inherent limitations of the measuring instrument and in the skill of the experimenter. The value recorded for a measurement should give some indication of its reliability (precision). To express maximum precision, this value should contain all the digits that are known plus one digit that is estimated. This last estimated digit introduces some degree of uncertainty. Because of this uncertainty, every number that expresses a measurement can have only a limited number of digits. These digits, used to express a measured quantity, are known as **significant figures** or **significant digits**.

significant figures

A detailed discussion of significant figures in calculations, exponents, powers of 10, rounding off numbers, large and small numbers, the dimensional analysis method of calculation, and other mathematical operations is given in the Mathematical Review in Appendix I. You are urged to review Appendix I and to study carefully any portions that are not familiar to you. This study may be done at various times during the course as the need for additional knowledge of certain mathematical operations arises.

2.3 The Metric System

metric system or SI

The **metric system**, or **International System (SI)**, is a decimal system of units for measurements of mass, length, time, and other physical constants. It is built around a set of basic units and uses factors of 10 to express larger or smaller quantities of these units. To express larger and smaller quantities, prefixes are added to the names of the units. These prefixes represent multiples of 10, making the metric system a total decimal system of measurements. Table 2.1 shows the names, symbols, and numerical values of the prefixes. These are also shown in Appendix III. Some of the more commonly used prefixes are

kilo	One thousand (1000) times the unit expressed
deci	One-tenth (0.1) of the unit expressed
centi	One-hundredth (0.01) of the unit expressed
milli	One-thousandth (0.001) of the unit expressed
micro	One-millionth (0.000001) of the unit expressed

Examples are

1 kilometre = 1000 metres
1 kilogram = 1000 grams
1 microsecond = 0.000001 second

Table 2.1. Prefixes used in the metric system and their numerical values.

Prefix	Symbol	Numerical value		
tera	T	1,000,000,000,000	or	10^{12}
giga	G	1,000,000,000	or	10^{9}
mega	M	1,000,000	or	10^{6}
kilo	k	1,000	or	10^{3}
hecto	h	100	or	10^{2}
deka	da	10	or	10^{1}
deci	d	0.1	or	10^{-1}
centi	c	0.01	or	10^{-2}
milli	m	0.001	or	10^{-3}
micro	μ	0.000001	or	10^{-6}
nano	n	0.000000001	or	10^{-9}
pico	p	0.000000000001	or	10^{-12}
femto	f	0.000000000000001	or	10^{-15}
atto	a	0.000000000000000001	or	10^{-18}

The seven base units of measurement in the International System are given below. Other units are derived from these base units.

Quantity	Name of Unit	Symbol
Length	Metre	m
Mass	Kilogram	kg
Time	Second	s
Electric current	Ampere	A
Temperature	Kelvin	K
Luminous intensity	Candela	cd
Amount of substance	Mole	mol

The metric system, or International System, is currently used by most of the countries in the world, not only for scientific and technical work, but also in commerce and industry. The United States is currently in the process of changing to the metric system of weights and measurements.

2.4 Measurement of Length

Standards for the measurement of length have an interesting historical development. The Old Testament mentions such units as the cubit (the distance from a man's elbow to the tip of his outstretched hand). In ancient Scotland the inch was once defined as a distance equal to the width of a man's thumb.

Reference standards of measurements have undergone continuous improvements in precision. The standard unit of length in the metric system is the **metre**. When the metric system was first introduced in the 1790s, the metre was defined as one ten-millionth of the distance from the equator to the North Pole measured along the meridian passing through Dunkirk, France. In 1889, the metre was redefined as the distance between two engraved lines on a platinum–iridium alloy bar maintained at 0° Celsius. This international metre bar is stored in a vault at Sèvres near Paris. Duplicate metre bars have been made and used as standards by many nations.

By the 1950s, length could be measured with such precision that a new standard was needed. Accordingly, in 1960, by international agreement the metre was again redefined, this time as 1,650,763.73 wavelengths of a particular spectral emission line of krypton-86. This is the reference standard for length presently in use.

A metre is 39.37 inches, a little longer than 1 yard. One metre contains 10 decimetres, 100 centimetres, or 1000 millimetres (see Figure 2.1). A kilometre contains 1000 metres. Table 2.2 shows the relationships of these units.

The angstrom unit (10^{-8} cm) is used extensively in expressing the wavelength of light and in atomic dimensions. Other important relationships are

$1\ m = 100\ cm = 1000\ mm = 10^6\ \mu = 10^{10}\ Å$
$1\ cm = 10\ mm = 0.01\ m$
$1\ in. = 2.54\ cm$

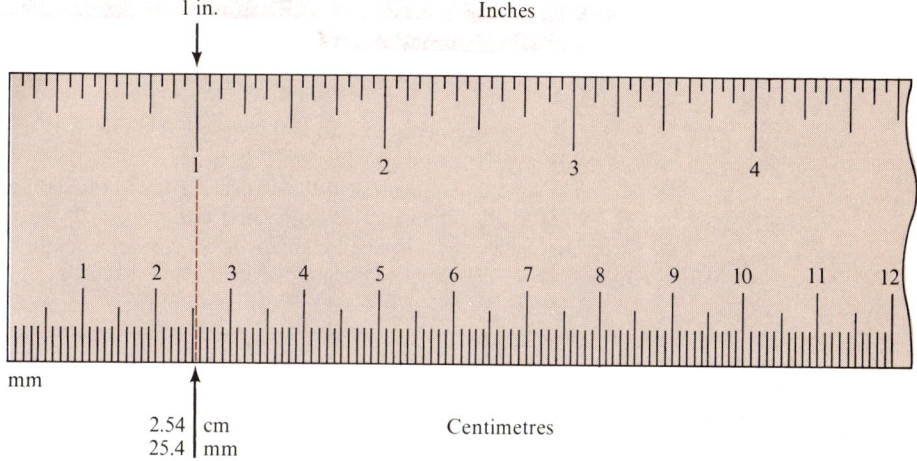

Figure 2.1. Comparison of the metric and American systems of length measurement: 2.54 cm = 1 in.

Table 2.2. Metric units of length.

Unit	Abbreviation	Metre equivalent	Exponential equivalent
Kilometre	km	1000 m	10^3 m
Metre	m	1 m	10^0 m
Decimetre	dm	0.1 m	10^{-1} m
Centimetre	cm	0.01 m	10^{-2} m
Millimetre	mm	0.001 m	10^{-3} m
Micrometre	μm	0.000001 m	10^{-6} m
Micron	μ	0.000001 m	10^{-6} m
Nanometre	nm	0.000000001 m	10^{-9} m
Angstrom	Å	0.0000000001 m	10^{-10} m

2.5 Problem Solving

One of the most consistently troublesome areas in chemistry involves the solving of mathematical problems. Since many chemical principles are illustrated by mathematical concepts, it is necessary to learn to solve problems dealing with these concepts.

There are usually several methods by which a problem can be solved. But in all methods it is best, especially for beginners, to use a systematic, orderly approach. The dimensional analysis, or factor-label, method (Appendix I, Section 16) is stressed in this book because:

1. It provides a systematic, straightforward way to set up problems.
2. It gives a clear understanding of the principles involved.
3. It helps in learning to organize and evaluate data.

14   Chapter 2

   4. It helps to identify errors because unwanted units are not eliminated if the setup of the problem is incorrect.

The basic steps for solving problems are:

1. Read the problem very carefully to determine what is to be solved for. Write down what you are solving for.
2. Tabulate the data given in the problem. Even in tabulating data, it is important to label all factors and measurements with the proper units.
3. Determine which principles are involved and which unit relationships are needed to solve the problem. It may be necessary to refer to certain tables to obtain other data needed. Use sample problems in the text to help set up and solve the problem.
4. Proceed with the necessary mathematical operations. Make certain that the answer contains the proper number of significant figures.
5. Check the answer to see if it is reasonable.
6. Do your work in a neat and organized form.

Label all factors with the proper units.

Just a few more words about problem solving. Don't allow any formal method of problem solving to limit your use of common sense and intuition. If the solution to a problem is obvious and seems simpler to you by another method, by all means use it. But in the long run you should be able to solve many otherwise difficult problems by using the dimensional analysis method. Here are some examples of problem solving by the dimensional analysis method.

Suppose you want to change 2.5 metres to an equivalent number of millimetres. Write

$$m \longrightarrow mm$$

To accomplish this you need to find a conversion unit that will change the given units (metres) to the desired units (millimetres). Write

$$1 \text{ m} = 1000 \text{ mm}$$

From this statement these two factors can be derived:

$$\frac{1 \text{ m}}{1000 \text{ mm}} = 1 \quad \text{and} \quad \frac{1000 \text{ mm}}{1 \text{ m}} = 1$$

These factors are read "1 metre per 1000 millimetres" and "1000 millimetres per 1 metre." They can also be written as 1 m/1000 mm and 1000 mm/1 m. Since either factor is equal to the number 1, any number of metres can be converted to the equivalent number of millimetres by multiplying by 1000 mm/m; and, any number of millimetres can be converted to the equivalent number of metres by multiplying by 1 m/1000 mm. (Any number multiplied by 1 has the same value as the number.) To convert metres to millimetres we choose the factor 1000 mm/1 m because in the multiplication, metres cancel, leaving the answer in the desired units, millimetres.

$$2.5 \text{ m} \times \frac{1000 \text{ mm}}{1 \text{ m}} = 2500 \text{ mm} \qquad (2.5 \times 10^3 \text{ mm, two significant figures})$$

Note that in making this calculation, units are treated the same as numbers—metres in the numerator are canceled by metres in the denominator.

$$\text{m} \times \frac{\text{mm}}{\text{m}} = \text{mm}$$

Now suppose you need to change 215 centimetres to metres. Write

$$\text{cm} \longrightarrow \text{m}$$

Then write

$$100 \text{ cm} = 1 \text{ m}$$

The two possible conversion factors are

$$\frac{100 \text{ cm}}{\text{m}} = 1 \qquad \text{and} \qquad \frac{1 \text{ m}}{100 \text{ cm}} = 1$$

Next, set up the calculation choosing the conversion factor that will give the answer in the desired units, metres.

$$215 \text{ cm} \times \frac{1 \text{ m}}{100 \text{ cm}} = \frac{215 \text{ m}}{100} = 2.15 \text{ m}$$

The dimensional analysis, or factor-label, method used in the preceding work shows how unit conversion factors are derived and used in calculations. After you become more proficient with the terms, you can save steps by writing the factors directly in the calculation. Here are some examples of the conversion from American to metric units.

Problem 2.1

How many centimetres are in 2.00 ft?
The stepwise conversion of units from feet to centimetres may be done in this manner: Convert feet to inches; then convert inches to centimetres.

$$\text{ft} \rightarrow \text{in.} \rightarrow \text{cm}$$

The needed conversion factors are

$$\frac{12 \text{ in.}}{1 \text{ ft}} \qquad \text{and} \qquad \frac{2.54 \text{ cm}}{1 \text{ in.}}$$

$$2.00 \text{ ft} \times \frac{12 \text{ in.}}{\text{ft}} = 24.0 \text{ in.}$$

$$24.0 \text{ in.} \times \frac{2.54 \text{ cm}}{\text{in.}} = 61.0 \text{ cm} \quad \text{(Answer)}$$

Since 1 ft and 12 in. are considered to be exact numbers, the number of significant figures allowed in the answer is three, based on the number 2.00.

16   Chapter 2

Problem 2.2 How many metres are there in a 100 yd football field? The stepwise conversion of units from yards to metres may be done in this manner, using the proper conversion factors.

$$yd \rightarrow ft \rightarrow in. \rightarrow cm \rightarrow m$$

$$100 \text{ yd} \times \frac{3 \text{ ft}}{\text{yd}} = 300 \text{ ft} \qquad (3 \text{ ft/yd})$$

$$300 \text{ ft} \times \frac{12 \text{ in.}}{\text{ft}} = 3600 \text{ in.} \qquad (12 \text{ in./ft})$$

$$3600 \text{ in.} \times \frac{2.54 \text{ cm}}{\text{in.}} = 9144 \text{ cm} \qquad (2.54 \text{ cm/in.})$$

$$9144 \text{ cm} \times \frac{1 \text{ m}}{100 \text{ cm}} = 91.4 \text{ m} \qquad (1 \text{ m}/100 \text{ cm}) \qquad \text{(three significant figures)}$$

Problems 2.1 and 2.2 may be solved using a running linear expression, writing down each conversion factor in succession. Very often this saves one or two calculation steps, and numerical values may be reduced to simpler terms leading to simpler calculations. The single linear expressions for Problems 2.1 and 2.2 are

$$2.00 \text{ ft} \times \frac{12 \text{ in.}}{\text{ft}} \times \frac{2.54 \text{ cm}}{\text{in.}} = 61.0 \text{ cm}$$

$$100 \text{ yd} \times \frac{3 \text{ ft}}{\text{yd}} \times \frac{12 \text{ in.}}{\text{ft}} \times \frac{2.54 \text{ cm}}{\text{in.}} \times \frac{1 \text{ m}}{100 \text{ cm}} = 91.4 \text{ m}$$

Using the units alone (Problem 2.2), we see that the stepwise cancellation proceeds in succession until the unit desired is reached.

$$yd \times \frac{ft}{yd} \times \frac{in.}{ft} \times \frac{cm}{in.} \times \frac{m}{cm} = \text{metres}$$

2.6   Measurement of Mass

kilogram

The standard unit of mass in the metric system is the **kilogram**. This amount of mass is defined by international agreement as being exactly equal to the mass of a platinum–iridium weight (*Kilogramme de Archive*) kept in a vault at Sèvres. A kilogram contains 1000 grams. Comparing this unit of mass to 1 pound (16 ounces), we find that a kilogram is equal to 2.2 pounds. A pound is equal to 454 grams (0.454 kilogram). The same prefixes used in length measurement are used to indicate larger and smaller gram units (see Table 2.3).

It is convenient to remember that

1 g = 1000 mg
1 kg = 1000 g
1 kg = 2.2 lb
1 lb = 454 g

Standards for Measurement

Table 2.3. Metric units of mass.

Unit	Abbreviation	Gram equivalent	Exponential equivalent
Kilogram	kg	1000 g	10^3 g
Gram	g	1 g	10^0 g
Decigram	dg	0.1 g	10^{-1} g
Centigram	cg	0.01 g	10^{-2} g
Milligram	mg	0.001 g	10^{-3} g
Microgram	μg	0.000001 g	10^{-6} g

To change grams to milligrams, multiply grams by the conversion factor 1000 mg/g. The setup for converting 25 g is

$$25 \text{ g} \times \frac{1000 \text{ mg}}{1 \text{ g}} = 25,000 \text{ mg} \quad (2.5 \times 10^4 \text{ mg}) \quad \text{(Answer)}$$

To change milligrams to grams, multiply milligrams by the conversion factor 1 g/1000 mg. For example, to convert 150 mg to grams:

$$150 \text{ mg} \times \frac{1 \text{ g}}{1000 \text{ mg}} = 0.150 \text{ g} \quad \text{(Answer)}$$

Examples of converting weights from American to metric units are shown below.

Problem 2.3

A 1.50 lb package of sodium bicarbonate costs 80 cents. How many grams of this substance are in this package?
We are solving for the number of grams equivalent to 1.50 lb. Since 1 lb = 454 g, the factor to convert pounds to grams is 454 g/lb.

$$1.50 \text{ lb} \times \frac{454 \text{ g}}{\text{lb}} = 681 \text{ g} \quad \text{(Answer)}$$

Note: The cost of the sodium bicarbonate has no bearing on the question asked in this problem.

Problem 2.4

Suppose four ostrich feathers weigh 1.00 lb. Assuming that each feather is equal in weight, how many milligrams does a single feather weigh? The unit conversion in this problem is from 1 lb/4 feathers to milligrams per feather. Since the unit feathers occurs in the denominator of both the starting unit and the desired unit, the unit conversions needed are

lb → g → mg

$$\frac{1.00 \text{ lb}}{4 \text{ feathers}} \times \frac{454 \text{ g}}{\text{lb}} \times \frac{1000 \text{ mg}}{\text{g}} = \frac{113,500 \text{ mg}}{\text{feather}} \quad (1.14 \times 10^5 \text{ mg/feather})$$

2.7 Measurement of Volume

litre

The metric system unit of volume is the **litre**. The litre is a little larger than a U.S. quart; 1.000 litre equals 1.057 quarts. The most commonly used fractional unit of a litre is the millilitre: 1 litre = 1000 ml, and 946 ml = 1.00 qt. This small unit of volume is also commonly referred to as a cubic centimetre, abbreviated cm³ (or sometimes cc). This is because a litre corresponds to the

18 Chapter 2

volume enclosed in a cube measuring exactly 10 cm on an edge. The volume of this cube is determined by multiplying length times width times height—that is, 10 cm × 10 cm × 10 cm = 1000 cm³ (see Figure 2.2). The relationship between a millilitre and a cubic centimetre is that 1 ml equals 1 cm³ exactly (see Table 2.4).

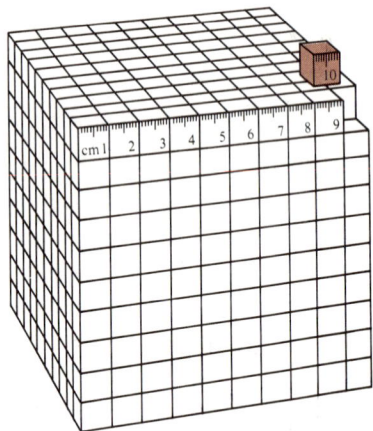

Figure 2.2. The large cube, 10.0 cm on a side, has a volume of 1000 cm³, or 1.0 litre. The small cube on top is 1.0 cm³. The large block contains 1000 of these small cubes.

Table 2.4. Metric units of volume.

Unit	Abbreviation	Litre equivalent
Litre	1	1.0 litre
Millilitre	ml	0.001 litre
Cubic centimetre	cm³ or cc	0.001 litre

Examples of volume conversions are shown below.

Problem 2.5

How many millilitres are contained in 3.5 litres?
The conversion factor to change litres to millilitres is 1000 ml/litre.

$$3.5 \text{ litres} \times \frac{1000 \text{ ml}}{\text{litre}} = 3500 \text{ ml} \quad (3.5 \times 10^3 \text{ ml})$$

Litres may be changed to millilitres by moving the decimal point three places to the right and changing the units to millilitres.

$$1.500 \text{ litres} = 1500 \text{ ml}$$

Problem 2.6

How many cubic centimetres are in a cube that is 11.1 inches on a side?
First change inches to centimetres.

$$11.1 \text{ in.} \times \frac{2.54 \text{ cm}}{\text{in.}} = 28.2 \text{ cm on a side}$$

Then change to cubic volume (Length × Width × Height).

28.2 cm × 28.2 cm × 28.2 cm = 22,426 cm³   (2.24 × 10⁴ cm³)

Table 2.5 summarizes the units and conversion factors that are used most often.

Table 2.5. Most often used units and their equivalents.

$$1 \text{ m} = 1000 \text{ mm}$$
$$1 \text{ m} = 39.37 \text{ in.}$$
$$1 \text{ cm} = 10 \text{ mm}$$
$$2.54 \text{ cm} = 1 \text{ in.}$$
$$1 \text{ g} = 1000 \text{ mg}$$
$$1 \text{ kg} = 1000 \text{ g}$$
$$454 \text{ g} = 1 \text{ lb}$$
$$1 \text{ litre} = 1000 \text{ ml}$$
$$1 \text{ ml} = 1 \text{ cm}^3 \text{ or cc}$$
$$946 \text{ ml} = 1 \text{ qt}$$

2.8   Temperature Scales

The *temperature* of a system measures how hot or cold that system is and can be expressed by several different temperature scales. Three commonly used temperature scales are the Celsius (centigrade) scale, the Kelvin (absolute) scale, and the Fahrenheit scale. A unit of temperature on each of these scales is called a *degree,* although the size of the degree varies. The symbol for the degree is ° and it is placed as a superscript after the number and before the temperature scales for Celsius and Fahrenheit. Thus, 100°C means 100 *degrees Celsius.* The degree sign is not used with Kelvin temperatures.

$$\text{Degrees Celsius (centigrade)} = °C$$
$$\text{Degrees Kelvin (absolute)} = K$$
$$\text{Degrees Fahrenheit} = °F$$

The Celsius scale is based on dividing the interval between the freezing and boiling temperatures of water into 100 equal parts, or degrees. The freezing point of water is assigned a temperature of 0°C and the boiling point of water a temperature of 100°C. The Kelvin temperature scale is also known as the absolute temperature scale because 0 K is the lowest possible temperature theoretically attainable. The Kelvin zero is 273.16 degrees below the Celsius zero. Kelvin degrees are equal in size to Celsius degrees. The freezing point of water on the Kelvin scale is 273.16 K (usually rounded to 273 K). On the Fahrenheit scale there are 180 degrees between the freezing and boiling temperatures of water. On this scale, the freezing point of water is 32°F and the boiling point is 212°F.

$$0°C = 273 \text{ K} = 32°F$$

20 Chapter 2

The three scales are compared in Figure 2.3. Although absolute zero is the lower limit of temperature on these scales, there is no known upper limit to temperature. (Temperatures of several million degrees are known to exist in the sun and in other stars.)

By examining Figure 2.3 we can see that there are 100 Celsius degrees and 100 Kelvin degrees between the freezing and boiling points of water. But there are 180 Fahrenheit degrees between these two temperatures. Hence, the size of a degree on the Celsius scale is the same as the size of a degree on the Kelvin scale, but the Celsius degree corresponds to 1.8 degrees on the Fahrenheit scale. From these data, mathematical formulas have been derived to convert a temperature on one scale to the corresponding temperature on another scale. These formulas are

$$K = °C + 273 \tag{1}$$

$$°F = (1.8 \times °C) + 32 \tag{2}$$

$$°C = \frac{(°F - 32)}{1.8} \tag{3}$$

Interpretation: Formula (1) states that the addition of 273 to the degrees Celsius converts the temperature to degrees Kelvin. Formula (2) states that to obtain the Fahrenheit temperature corresponding to a given Celsius temperature we multiply the degrees Celsius by 1.8 and then add 32. Formula (3) states that to

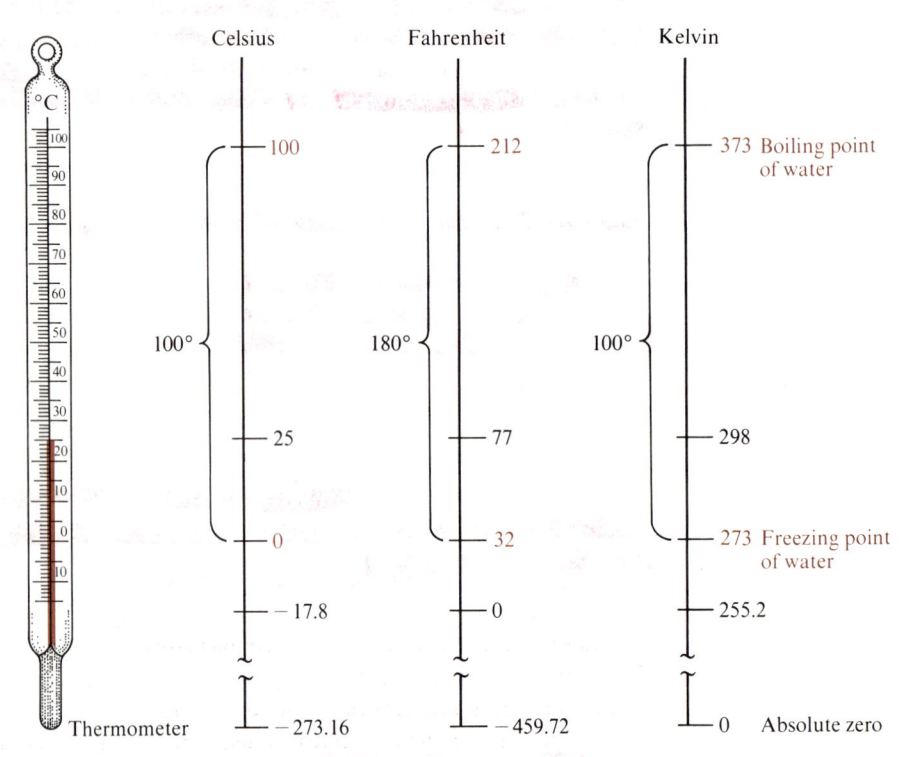

Figure 2.3. Comparison of Celsius, Kelvin, and Fahrenheit temperature scales.

obtain the corresponding Celsius temperature we subtract 32 from the degrees Fahrenheit and then divide this figure by 1.8.

Examples of temperature conversions follow.

Problem 2.7

The temperature at which salt (sodium chloride) melts is 800°C. What is this temperature on the Kelvin and Fahrenheit scales?

We need to calculate K from °C, so we use formula (1) above. We also need to calculate °F from °C; for this we use formula (2).

$$K = °C + 273$$
$$K = 800°C + 273 = 1073 \text{ K}$$

$$°F = (1.8 \times °C) + 32$$
$$°F = (1.8 \times 800°C) + 32$$
$$°F = 1440 + 32 = 1472°F$$

$$800°C = 1073 \text{ K} = 1472°F$$

Problem 2.8

The temperature for December 1 was 110°F, a new record. Calculate this temperature in °C.

Formula (3) applies here.

$$°C = \frac{(°F - 32)}{1.8}$$

$$°C = \frac{(110 - 32)}{1.8} = \frac{78}{1.8} = 43°C$$

Problem 2.9

What temperature on the Fahrenheit scale corresponds to −8.0°C? (Be alert to the presence of the minus sign in this problem.)

$$°F = (1.8 \times °C) + 32$$
$$°F = [1.8 \times (-8.0)] + 32 = -14.4 + 32$$
$$°F = 17.6$$

Temperatures used in this book are in degrees Celsius (°C) unless specified otherwise.

2.9 Heat and Temperature

heat

temperature

Heat is a form of energy associated with the motion of small particles of matter. Heat is associated with a quantity of energy within a system or a quantity of energy supplied to a system. **Temperature** is a measure of the intensity of heat, or how hot a system is, regardless of its size. Heat always flows from a region of high temperature to one of lower temperature.

calorie
kilocalorie

The unit of heat commonly used in chemical systems is the calorie. A **calorie** (cal) is the quantity of heat required to change the temperature of 1 g water 1°C, usually measured from 14.5 to 15.5°C. The **kilocalorie** (kcal), also

22   Chapter 2

known as the nutritional or large Calorie (spelled with a capital C and abbreviated Cal), is equal to 1000 small calories. Temperature is measured in degrees and heat quantity is measured in calories.

The difference in the meanings of the terms *heat* and *temperature* can be seen by this example: Visualize two beakers, A and B. Beaker A contains 100 g water at 20°C, and beaker B contains 200 g water also at 20°C. The beakers are now heated until the temperature of the water in each reaches 30°C. The *temperature* of the water in the beakers was raised by exactly the same amount, 10°C. Yet twice as much *heat* (2000 cal) was required by the water in beaker B than was required by the water in beaker A (1000 cal).

heat capacity

The **heat capacity**, also known as *specific heat*, of any substance is the quantity of heat (in calories) required to change the temperature of 1 g of that substance by 1°C. It follows from the definition of the calorie that the heat capacity of water is 1 cal/g°C. The heat capacity of water is high compared to most substances. Aluminum and copper, for example, have heat capacities of 0.215 and 0.0921 cal/g°C, respectively (see Table 2.6). The relation of mass, heat capacity, temperature change (Δt), and quantity of heat lost or gained by a system is expressed by this general equation:

$$\begin{pmatrix} \text{Grams of} \\ \text{substance} \end{pmatrix} \times \begin{pmatrix} \text{Heat capacity} \\ \text{of substance} \end{pmatrix} \times \Delta t = \text{calories} \qquad (1)$$

Thus, the amount of heat needed to warm 200 g water 10°C can be calculated as follows:

$$200 \text{ g} \times \frac{1.00 \text{ cal}}{\text{g°C}} \times 10°C = 2000 \text{ cal}$$

Table 2.6. Heat capacity of selected substances.

Substance	Heat capacity cal/g°C
Water	1.00
Ethyl alcohol	0.511
Ice	0.492
Aluminum	0.215
Iron	0.113
Copper	0.0921
Gold	0.0312
Lead	0.0305

The application of equation (1) to the solution of a more complicated example is illustrated by Problem 2.10.

Problem 2.10

The heat capacity of aluminum (Al) is 0.215 cal/g°C. How many grams of water (H_2O) can be raised from 24.0°C to 33.0°C when a 10.0 g ingot of Al at 143°C is dropped into the water?

The heat lost or gained by a system is given by equation (1). For our problem, we have

Δt = Change in temperature (°C) of Al

Heat lost by Al = Heat gained by H_2O

The Al ingot is dropped into water and is allowed to cool to 33.0°C. Therefore, its temperature change (Δt) is 110°C (143 − 33.0).

Heat lost by Al = 10.0 g Al × 0.215 cal/g°C × 110°C = 236 cal

Heat gained by H_2O = 236 cal

Solving equation (1) for grams of water, we obtain

$$g\ H_2O = \frac{cal}{Heat\ capacity \times \Delta t}$$

Δt = 9.0°C (change in water temperature from 24.0°C to 33.0°C)

$$g\ H_2O = \frac{236\ cal}{1\ cal/g°C \times 9.0°C}$$

$g\ H_2O$ = 26 g

2.10 Tools for Measurement

Common measuring instruments used in chemical laboratories are illustrated in Figures 2.4 and 2.5. A balance is used to measure mass. Balances are obtainable that will weigh objects to the nearest microgram. The choice of the balance depends on the accuracy required and the amount of material being

Figure 2.4. At left is a Cent-o-Gram R311. This balance has four calibrated horizontal beams, each fitted with a specific movable weight. The weight of the object placed on the pan is determined by moving the weights along the beams until the swinging beam is in balance, as shown by the indicator on the right. (Courtesy Ohaus Scale Corporation.) In the center is a single-pan, top-loading, rapid-weighing balance with direct read-out to the nearest milligram (for example, 125.456 g). (Courtesy Sartorius Balance Div., Brinkmann Instruments Inc.) At right is a single-pan analytical balance for high-precision weighing. The precision of this balance is 0.1 mg (0.0001 g). (Courtesy Mettler Instrument Corporation, Highstown, N.J.)

24 Chapter 2

Graduated cylinder   Volumetric flask   Buret   Pipet

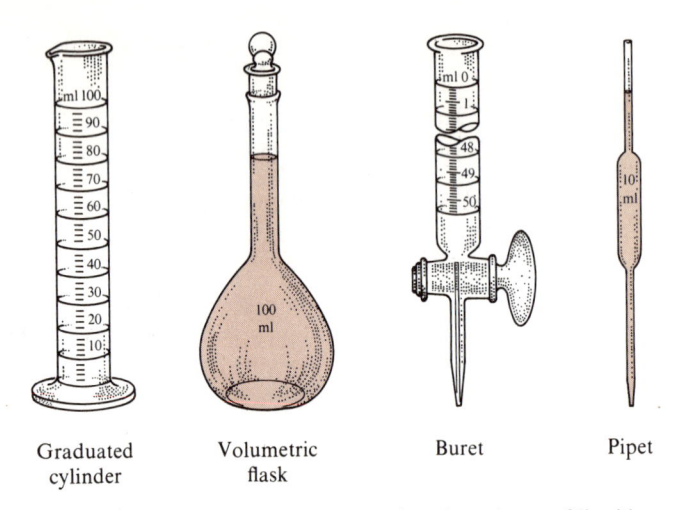

Figure 2.5. Calibrated glassware for measuring the volume of liquids.

weighed. Three standard balances are shown in Figure 2.4: a triple-beam balance with precision up to 0.01 g; a single-pan, top-loading balance with a precision of 0.001 g (1 mg); a single-pan analytical balance with a precision up to 0.0001 g. Automatic-recording balances are also available.

The most common instruments for measuring liquids are the graduated cylinder, volumetric flask, buret, and pipet, which are shown in Figure 2.5. These calibrated pieces are usually made of glass and are available in various sizes.

The common laboratory tool for measuring temperature is a thermometer (see Figure 2.3).

2.11 Density

density

Density (d) is the ratio of the mass of a substance to the volume occupied by that mass; it is the mass per unit of volume and is given by the equation

$$d = \frac{\text{Mass}}{\text{Volume}}$$

Density is a physical characteristic of a substance and may be used as an aid to its identification. When the density of a solid or a liquid is given, the mass is usually expressed in grams and the volume in millilitres or cubic centimetres.

$$d = \frac{\text{Mass}}{\text{Volume}} = \frac{g}{ml} \quad \text{or} \quad d = \frac{g}{cm^3}$$

Since the volume of a substance (especially liquids and gases) varies with temperature, it is important to state the temperature along with the density. For example, the volume of 1.0000 g water at 4°C is 1.0000 ml, while at 20°C it is 1.0018 ml, and at 80°C it is 1.0290 ml. Density, therefore, also varies with temperature.

The density of water at 4°C is 1.0000 g/ml but at 80°C the density of water is 0.9718 g/ml.

$$d^{4°C} = \frac{1.0000 \text{ g}}{1.0000 \text{ ml}} = 1.0000 \text{ g/ml}$$

$$d^{80°C} = \frac{1.0000 \text{ g}}{1.0290 \text{ ml}} = 0.9718 \text{ g/ml}$$

Densities for liquids and solids are usually represented in terms of grams per millilitre or grams per cubic centimetre. The density of gases, however, is normally expressed in terms of grams per litre (g/litre). Unless otherwise stated, gas densities are given for 0°C and 1 atmosphere pressure (discussed further in Chapter 12). Table 2.7 lists the densities of a number of common materials.

Table 2.7. Densities of some selected materials. For comparing densities, the density of water is the reference for solids and liquids; air is the reference for gases.

Liquids and solids		Gases	
Substance	Density (g/ml at 20°C)	Substance	Density (g/litre at 0°C)
Wood (Douglas fir)	0.512	Hydrogen	0.090
Ethyl alcohol	0.79	Helium	0.178
Cottonseed oil	0.926	Methane	0.714
Water (4°C)	**1.0000**	Ammonia	0.771
Sugar	1.59	Neon	0.90
Carbon tetrachloride	1.595	Carbon monoxide	1.25
Magnesium	1.74	Nitrogen	1.251
Sulfuric acid	1.84	**Air**	**1.293**
Sulfur	2.07	Oxygen	1.429
Salt	2.16	Hydrogen chloride	1.63
Aluminum	2.70	Argon	1.78
Silver	10.5	Carbon dioxide	1.963
Lead	11.34	Chlorine	3.17
Mercury	13.55		
Gold	19.3		

Suppose that water, carbon tetrachloride, and cottonseed oil are successively poured into a graduated cylinder. The result is a layered three-liquid system (Figure 2.6). Can we predict the order of the liquid layers? Yes, by looking up the liquid densities in Table 2.7. Carbon tetrachloride has the greatest density (1.595 g/ml) and cottonseed oil has the lowest density (0.926 g/ml). Carbon tetrachloride will, therefore, form the bottom layer and cottonseed oil the top layer. Water, with a density between the other two liquids, will, of course, form the middle layer. This information can also be determined readily by experiment. Add a few millilitres of carbon tetrachloride to a beaker of water. The carbon tetrachloride, being more dense than the water, will sink. Cottonseed oil, being less dense than water, will float when added to water.

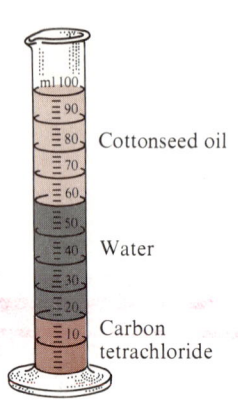

Figure 2.6. Relative density of liquids. When three immiscible (not capable of mixing) liquids are poured together, the liquid with the highest density will be the bottom layer. In the case of cottonseed oil, water, and carbon tetrachloride, cottonseed oil is the top layer.

Direct comparisons of density in this manner can be made only with liquids that are *immiscible* (do not dissolve in one another).

The density of air at 0°C is approximately 1.293 g/litre. Gases with densities less than this value are said to be "lighter than air." A helium-filled balloon will rise rapidly in air because the density of helium is only 0.178 g/litre.

When an insoluble solid object is dropped into water, the object will sink or float—depending on its density. If the object is less dense than water, it will float, displacing a *mass* of water equal to the mass of the object. If the object is more dense than water, it will sink, displacing a *volume* of water equal to the volume of the object. This information can be utilized to determine the volume (and density) of irregularly shaped objects.

Sample calculations of density problems follow.

Problem 2.11

What is the density of a mineral if 427 g of the mineral occupy a volume of 35.0 ml? We need to solve for density, so we start by writing the formula for calculating density.

$$d = \frac{\text{Mass}}{\text{Volume}}$$

Then we substitute the data given in the problem into the equation and solve.

Mass = 427 g   Volume = 35.0 ml

$$d = \frac{\text{Mass}}{\text{Volume}} = \frac{427 \text{ g}}{35.0 \text{ ml}} = 12.2 \text{ g/ml} \quad \text{(Answer)}$$

Problem 2.12

The density of gold is 19.3 g/ml. What is the mass of 25.0 ml of gold?
(a) First write the formula for calculating density.

$$d = \frac{\text{Mass}}{\text{Volume}}$$

(b) We need to solve for mass; therefore, we solve the density equation to obtain mass on one side by itself.

$$\text{Mass} = \text{Volume} \times d$$

(c) Substitute data given in the problem and calculate.

$$\text{Mass} = 25.0 \, \cancel{\text{ml}} \times \frac{19.3 \, \text{g}}{\cancel{\text{ml}}} = 482 \, \text{g}$$

Problem 2.13   The water level in a graduated cylinder stands at 20.0 ml before and at 26.2 ml after a 16.74 g metal bolt is submerged in the water. (a) What is the volume of the bolt? (b) What is the density of the bolt?

(a)   26.2 ml = Volume of water plus bolt
     − 20.0 ml = Volume of water
     ─────────
      6.2 ml = Volume of bolt   (Answer)

(b) $d = \dfrac{\text{Mass of bolt}}{\text{Volume of bolt}} = \dfrac{16.74 \, \text{g}}{6.2 \, \text{ml}} = 2.7 \, \text{g/ml}$ (Answer)

2.12   Specific Gravity

specific gravity

The **specific gravity** (sp gr) of a substance is a ratio of the density of that substance to the density of another substance. Water is usually used as the reference standard for solids and liquids.

$$\text{sp gr} = \frac{\text{Density of a liquid or solid}}{\text{Density of water}} \quad \text{or} \quad \frac{\text{Density of a gas}}{\text{Density of air}}$$

Specific gravity has no units but is a number that compares the density of a liquid or a solid with that of water, or the density of a gas with that of air.

Problem 2.14   What is the specific gravity of mercury with respect to water at 4°C? (Density of water at 4°C is 1.000 g/ml.)

$$\text{sp gr} = \frac{\text{Density of mercury}}{\text{Density of water}} = \frac{13.55 \, \text{g/ml}}{1.000 \, \text{g/ml}}$$

sp gr of mercury = 13.55

The value for the specific gravity of mercury (13.55) tells us that, per unit volume, mercury is 13.55 times as heavy as water. Do you think that you could readily lift a litre (approximately 1 quart) of mercury?

hydrometer

A **hydrometer** consists of a weighted bulb at the end of a sealed, calibrated tube. This instrument is used to measure the specific gravity of a liquid (see Figure 2.7). When a hydrometer is floated in a liquid, the specific gravity is indicated on the scale at the surface of the liquid.

(a) Water   (b) Sulfuric acid

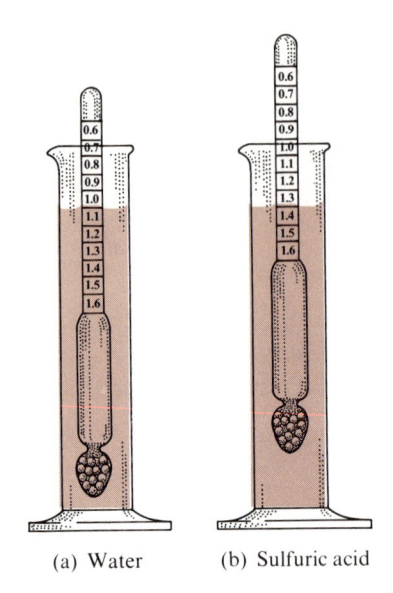

Figure 2.7. Specific gravity determination using hydrometers. The hydrometer in (a) is floating in water, showing a specific gravity of 1.0. The hydrometer in (b) is floating in dilute sulfuric acid (battery acid), showing a specific gravity of 1.3.

Questions

A. Review the meanings of the new terms introduced in this chapter. The terms listed in Section A of each set of Questions are new terms defined in the chapter. They appear in boldface type and occur in the chapter in the order listed in Question A.

1. Mass
2. Weight
3. Significant figures
4. Metric system, or SI
5. Metre
6. Kilogram
7. Litre
8. Heat
9. Temperature
10. A calorie
11. Kilocalorie
12. Heat capacity
13. Density
14. Specific gravity
15. Hydrometer

B. Answers to the following questions will be found in tables and figures.
1. Use Table 2.2 to determine how many centimetres make 1 km.
2. What is the temperature difference in Celsius degrees between 0°C and 0°F?
3. Why do you suppose the top ends of the pipet and the volumetric flask are narrower than the bulk of these volumetric instruments? (See Figure 2.5.)
4. Refer to Table 2.7 and describe the arrangement you would see when these three immiscible materials are placed in a 100 ml graduated cylinder: 135.5 g mercury, 25 ml cottonseed oil, and a cube of sulfur measuring 2.0 cm on an edge. Mercury and cottonseed oil are liquids.
5. Arrange the following materials in order of increasing density: lead, sulfur, gold, wood, and water.
6. Would argon be a satisfactory lifting gas for a balloon? Explain.

C. Review questions.
1. Why is the metric system of weights and measurements more desirable than the American system?

2. What are the abbreviated symbols for the following?
   (a) Gram   (d) Megagram   (g) Cubic centimetre   (i) Millilitre
   (b) Milligram   (e) Decimetre   (h) Nanometre   (j) Kilolitre
   (c) Kilogram   (f) Centimetre
3. In a number, when is zero significant and when is it not significant? See Appendix I.
4. What are the three rules for rounding off a number? See Appendix I.
5. Suppose you had a litre of water in a flask, a quart bottle, and a balance with a capacity of 200 g. Describe how you would go about determining the volume of the water required to fill the quart bottle. Assume the density of water is 1.00 g/ml.
6. Distinguish between heat and temperature.
7. Will aluminum or iron become hotter when 100 cal of energy are added to 10 g samples of each of these metals?
8. Ice floats in water and sinks in ethyl alcohol. What information does this give you about the density of ice?
9. Which of the following statements are correct?
   (a) The prefix *milli* indicates one-millionth of the unit expressed.
   (b) The quantity 10 cm is equal to 100 mm.
   (c) The number 383.263 reduced to four significant figures becomes 383.3.
   (d) The number of significant figures in the number 29,004 is three.
   (e) The sum of 24.928 g + 2.126 g should contain five significant figures.
   (f) The product of 14.63 cm × 2.50 cm should contain three significant figures.
   (g) One microsecond is 10^{-6} second.
   (h) One thousand metres is a shorter distance than 1000 yards.
   (i) The number 0.002894 expressed in scientific notation is 2.894×10^{-3}.
   (j) $2.0 \times 10^4 \times 6.0 \times 10^6 = 1.2 \times 10^{11}$
   (k) $5.0 \times 10^5 \times 5.0 \times 10^{-3} = 2.5 \times 10^9$
   (l) One degree on the Celsius scale is equal to one degree on the Kelvin scale and to 1.8 degrees on the Fahrenheit scale.
   (m) The direction of heat flow is from cold to hot.
   (n) A calorie is a unit of temperature.
   (o) Temperature is a form of energy.
   (p) The density of water at 4°C is 1.00 g/ml.
   (q) A graduated cylinder would be a more accurate instrument for measuring 10.0 ml water than would a pipet.
   (r) A hydrometer is an instrument for measuring the specific gravity of liquids.

D. Review problems.
   1. How many significant figures are in each of the following numbers? (See Mathematical Review in Appendix I.)
      (a) 0.007   (c) 0.1002   (e) 345,409   (g) 0.0283
      (b) 22.2   (d) 300.0   (f) 82.060   (h) 0.0720
   2. Round off the following numbers to four significant figures. (See Mathematical Review in Appendix I.)
      (a) 3.00051 (c) 41.127   (e) 2144.4   (g) 19.995
      (b) 9.3775   (d) 25.5555 (f) 82.365
   3. Express each of the following numbers in exponential notation (as a power of 10). (See Mathematical Review in Appendix I.)
      (a) 847   (c) 22,400   (e) 0.0000611 (g) 0.0650
      (b) 0.000586 (d) 0.088   (f) 4286
   4. Solve the following mathematical problems. (See Mathematical Review in Appendix I.)

30 Chapter 2

(a) $23.89 + 13.0 + 1.3 =$
(b) $33.04 + 9.009 + 106.8 =$
(c) $15.3 \times 6.82 =$
(d) $2.90 \times 29.0 \times 290 =$
(e) $\dfrac{5}{6} \times \dfrac{2}{3} =$
(f) Change to decimal fractions: $\dfrac{3}{7}, \dfrac{11}{15}, \dfrac{6}{9}, \dfrac{98}{125}$
(g) $\dfrac{1}{3} + \dfrac{5}{9} =$
(h) $2.5(3.0X + 12) =$
(i) $\dfrac{(°F - 32)}{1.8}$, where $°F = 200$
(j) $124.36 \div 6.40$
(k) $\dfrac{0.2386}{0.2550}$

5. Show calculations for the following conversions:
   (a) 12.0 cm to m   (g) 2.10 m to cm
   (b) 142 m to km   (h) 3.0 km to m
   (c) 2.5 cm to Å   (i) 10 Å to cm
   (d) 42.4 cm to mm   (j) 400 mm to cm
   (e) 12.0 in. to cm   (k) 22.0 cm to in.
   (f) 5.00 miles to km   (l) 70.0 km to miles

6. An automobile travelling at 55 miles per hour is moving at what speed in kilometres per hour?

7. The speed of light in a vacuum is 3.00×10^{10} cm per second. Calculate the speed of light in miles per second.

8. The sun is approximately 92 million miles from the earth. How many seconds will it take light to travel from the sun to the earth if the velocity of light is 3.00×10^{10} cm per second?

9. Oil spreads in a thin layer on water and is commonly called an "oil slick." How much area in square metres (m^2) will 100 cm^3 of oil cover if it forms a layer 5 Å in thickness?

10. Show calculations for the following conversions:
    (a) 1.200 g to mg   (e) 50 mg to g
    (b) 454 g to kg   (f) 2.2 kg to g
    (c) 1000 mg to kg   (g) 0.350 kg to mg
    (d) 2.55 lb to g   (h) 25.6 g to lb

11. How many kilograms does a 170 lb man weigh?

12. The usual aspirin tablet contains 5.0 grains of aspirin. How many grams of aspirin are in one tablet (1 grain = 1/7000 lb)?

13. The price of gold varies greatly and has been almost as high as $400 per ounce. What is the value of 250 g of gold at $325 per ounce?

14. The largest nugget of gold on record was found in 1872 in New South Wales, Australia, and weighed 93.3 kg. What was the volume of this nugget in cubic centimetres? In litres?

15. At 35 cents per litre, how much will it cost to fill a 15 gal tank with gasoline?

16. A French automobile manufacturer claims that its sedan uses only 6.0 litres of

gasoline per 100 km. How many miles per gallon of gasoline could be expected from the car?

17. An adult ruby-throated hummingbird has an average weight of 3.2 g, whereas an adult California condor may attain a weight of 21 lb. How many times heavier than the hummingbird is the condor?
18. The average weight of the heart of a human baby is about 1 oz. What is this weight in milligrams?
19. More sulfuric acid is manufactured in the United States than any other chemical; the annual production is 6.4×10^{10} lb. What is the average daily production of sulfuric acid in tons? In kilograms?
20. Control of automobile exhaust emission of oxides of nitrogen began with the 1971 car models. The Clean Air Act of 1970 states that these emissions must be reduced from 6.2 g per mile to 0.62 g per mile. What will be the average daily emission (in grams) of oxides of nitrogen in a city having 6 million automobiles, each driving an average of 10,000 miles per year, when this goal is reached?
21. Show calculations for the following conversions:
    (a) 145 ml to litres
    (b) 5.00 in.^{3} to cubic centimetres
    (c) 150 gal to litres
    (d) 2.50 litres to millilitres
    (e) 6.00 litres to millilitres
    (f) 58.0 cm^3 to cubic inches
    (g) 2.50 litres to gallons
    (h) 22.4 litres to millilitres
22. Calculate the volume, in litres, of a box 1.8 m long, 16 cm wide, and 55 mm deep.
23. At a price of $1.15 per gallon, what will it cost to fill a 50 litre tank with gasoline?
24. An aquarium has the following dimensions: 30 in. long by 10 in. wide by 20 in. high. How much water will it take to fill the aquarium three-fourths full? Express your answer in both litres and gallons.
25. Show calculations for the following conversions:
    (a) 140°F to °C
    (b) 0°F to °C
    (c) 0°F to K
    (d) −12°F to °C
    (e) 25°C to °F
    (f) −12°C to °F
    (g) 273°C to K
    (h) 0°C to K
26. Normal body temperature for humans is 37.0°C. What is this temperature on the Fahrenheit scale?
27. Which is colder, −90°C or −135°F?
28. (a) At what temperature are the Fahrenheit and Celsius scales exactly equal?
    (b) At what temperature are they numerically equal but opposite in sign?
29. How many calories of heat are required to raise the temperature of 100 g water from 20°C to 50°C?
30. How many calories are required to raise the temperature of 100 g iron from 20°C to 50°C?
31. A 20.0 g piece of copper at 203°C is dropped into 80.0 g water at 25.0°C. The water temperature rises to 29.0°C. Calculate the heat capacity (specific heat) of copper. Assume all the heat lost by the copper is transferred to the water.
32. Assuming no heat losses by the system, what will be the final temperature when 50 g water at 10°C is mixed with 10 g water at 50°C?
33. Calculate the density of a liquid if 16.60 ml of the liquid weighs 17.25 g.
34. A 25.0 ml sample of bromine weighs 78.0 g. What is the density of bromine?
35. When a 15.6 g piece of chromium metal was placed into a graduated cylinder containing 25.0 ml water, the water level rose to 27.2 ml. Calculate the density of the chromium.
36. Concentrated hydrochloric acid has a density of 1.19 g/ml. Calculate the weight, in grams, of 1.00 litre of this acid.

37. Thirty-five millilitres of ethyl alcohol (density 0.79 g/ml) is added to a graduated cylinder that weighs 44.28 g. What will be the weight of the cylinder plus the alcohol?
38. What weight of mercury (density 13.6 g/ml) will occupy a volume of 50.0 ml?
39. You are given three cubes, A, B, and C; one is magnesium, one is aluminum, and the other is silver. All three cubes weigh the same, but cube A has a volume of 29.0 ml, cube B a volume of 18.8 ml, and cube C a volume of 4.81 ml. Identify cubes A, B, and C.
40. A cube of aluminum weighs 500 g. What will be the weight of a cube of silver of the same dimensions?
41. Twenty-five millilitres of water at 90°C weighs 24.12 g. Calculate the density of water at this temperature.
42. Calculate (a) the density and (b) the specific gravity of a solid that weighs 160 g and has a volume of 50.0 ml.
43. Which liquid will occupy the greater volume, 100 g water or 100 g ethyl alcohol? Explain!
44. A solution made by adding 143 g sulfuric acid to 500 ml water had a volume of 554 ml.
    (a) What value will a hydrometer read when placed in this solution?
    (b) What volume of concentrated sulfuric acid ($d = 1.84$ g/ml) was added?

3 Properties of Matter

After studying Chapter 3 you should be able to:

1. Understand the new terms listed in Question A at the end of the chapter.
2. Identify the three physical states of matter and list the physical properties that characterize each state.
3. Distinguish between the physical and chemical properties of matter.
4. Classify changes undergone by matter as being either physical or chemical changes.
5. Distinguish between substances and mixtures.
6. Distinguish between kinetic and potential energy.
7. State the Law of Conservation of Mass.
8. State the Law of Conservation of Energy.
9. Explain why the laws dealing with the conservation of mass and of energy may be combined into a single more accurate general statement.
10. Calculate the percent composition of compounds from the weights of the elements involved in a chemical reaction or vice versa.

3.1 Matter Defined

matter

The entire universe consists of matter and energy. Every day we come into contact with countless kinds of matter. Air, food, water, rocks, soil, glass, this book—all are different types of matter. Broadly defined, **matter** is *anything that has mass and occupies space.*

Matter may be quite invisible. If an apparently empty test tube is submerged mouth downward in a beaker of water, the water rises only slightly into the tube. The water cannot rise further because the tube is filled with invisible matter—air (see Figure 3.1).

To the eye matter appears to be continuous and unbroken. However, it is actually discontinuous and is composed of discrete, tiny particles called *atoms*.

3.2 Physical States of Matter

solid

Matter exists in three physical states: solid, liquid, and gas. A **solid** is characterized by having a definite shape and volume, with particles that cohere rigidly to one another. The shape of a solid may be independent of its container.

33

Figure 3.1. An apparently empty test tube is submerged, mouth downward, in water. Only a small volume of water rises into the tube, which is actually filled with air. This experiment proves that air, which is matter, occupies space.

For example, a crystal of sulfur has the same shape and volume whether it is placed in a beaker or simply laid on a glass plate.

amorphous

Most commonly occurring solids, such as salt, sugar, quartz, and metals, are *crystalline*. Crystalline materials exist in regular, recurring geometric patterns. Solids such as plastics, glass, and gels, because they do not have any particular regular internal geometric form, are called **amorphous** solids. ("Amorphous" means without shape or form.) Figure 3.2 illustrates three crystalline solids—salt, quartz, and gypsum.

liquid

A **liquid** is characterized by having a definite volume, but not a definite shape, with particles that cohere firmly but not rigidly. Although held together by strong attractive forces, the particles are able to move freely but remain in close contact with each other. Particle mobility gives the liquid fluidity and causes it to take the shape of the container in which it is stored. Figure 3.3 shows the same amount of liquid in differently shaped containers.

gas

A **gas** is characterized by having no fixed shape, with particles that are moving independently of each other. The particles in the gaseous state have gained enough energy to overcome the attractive forces holding them together as liquids or solids. A gas presses continuously and in all directions upon the walls of any container. Because of this quality, a gas completely fills a container. The particles of a gas are relatively far apart, compared to those of solids and liquids. The actual volume of the gas particles is usually very small in comparison to the volume of space occupied by the gas. A gas therefore may be compressed into a very small volume or expanded practically indefinitely. Liquids cannot be compressed to any great extent, and solids are even less compressible than liquids.

When a bottle of ammonia solution is opened in one corner of the laboratory, you can soon smell its familiar odor in all parts of the room. The ammonia gas escaping from the solution demonstrates that gaseous particles move freely and rapidly, and tend to permeate the entire area into which they are released.

Although matter is discontinuous, attractive forces exist that hold the particles together and give matter its appearance of continuity. These attractive

Figure 3.2. These three naturally occurring substances are examples of regular geometric formations that are characteristic of crystalline solids: (a) salt; (b) quartz; (c) gypsum.

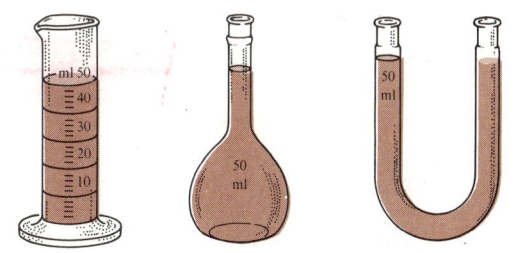

Figure 3.3. Liquids have the property of fluidity and assume the shape of their container, as illustrated in each of the three different calibrated containers.

forces are strongest in solids, giving them rigidity; they are weaker in liquids, but still strong enough to hold liquids to definite volumes. In gases the attractive forces are so weak that the particles of a gas are practically independent of each other. Table 3.1 lists a number of common materials that exist as solids, liquids, and gases. Table 3.2 summarizes comparative properties of solids, liquids, and gases.

36 Chapter 3

Table 3.1. Common materials in the solid, liquid, and gaseous states of matter.

Solids	Liquids	Gases
Aluminum	Alcohol	Acetylene
Copper	Blood	Air
Gold	Gasoline	Butane
Polyethylene	Honey	Carbon dioxide
Salt	Mercury	Chlorine
Sand	Oil	Helium
Steel	Vinegar	Methane
Sulfur	Water	Oxygen

Table 3.2. Physical properties of solids, liquids, and gases.

State	Shape	Volume	Particles	Compressibility
Solid	Definite	Definite	Rigidly cohering; tightly packed	Very slight
Liquid	Indefinite	Definite	Mobile; cohering	Slight
Gas	Indefinite	Indefinite	Independent of each other and relatively far apart	High

3.3 Substances and Mixtures

The term *matter* refers to the total concept of material things. There are thousands of distinct and different kinds of matter. Upon closely examining different samples of matter, we can observe them to be either homogeneous or heterogeneous. By homogeneous we mean uniform in appearance when observed by the unaided eye or through a microscope. Matter that has identical properties throughout is **homogeneous**. Matter consisting of two or more physically distinct phases is **heterogeneous**. A **phase** is a homogeneous part of a system separated from other parts by physical boundaries. A system of ice and water is heterogeneous, containing both solid and liquid phases, although each physical state of water is uniform in composition and is homogeneous. Whenever we have a system in which definite boundaries exist between the components, no matter whether they are in the solid, liquid, or gaseous states, the system has more than one phase and is heterogeneous. Thus, when we first put a spoonful of sugar into water, there exist a solid and a liquid phase, and the system is heterogeneous. After the sugar has been stirred and dissolved, the system has only one phase and is homogeneous.

A **substance** is a particular kind of matter that is homogeneous and has a definite, fixed composition. Substances, sometimes known as pure substances, occur in two forms: elements and compounds. Several examples of elements and compounds are copper, gold, oxygen, salt, sugar, and water. Elements and compounds are discussed in more detail in Chapter 4.

mixture

Matter that contains two or more substances mixed together is known as a **mixture.** Mixtures are variable in composition and may be either homogeneous or heterogeneous. When sugar is dissolved in water, a sugar solution is formed. All parts of this solution are sweet and contain both substances, sugar and water, uniformly mixed. Solutions are homogeneous mixtures. Air is a homogeneous mixture (solution) of several gases. If we examine ordinary concrete, granite, iron ore, or other naturally occurring mineral deposits, we observe them to be heterogeneous mixtures of several different substances. Of course, it is very easy to prepare a heterogeneous mixture simply by physically mixing two or more substances, such as sugar and salt. We will consider mixtures again in Chapter 4. Figure 3.4 illustrates the relationship between homogeneous and heterogeneous matter.

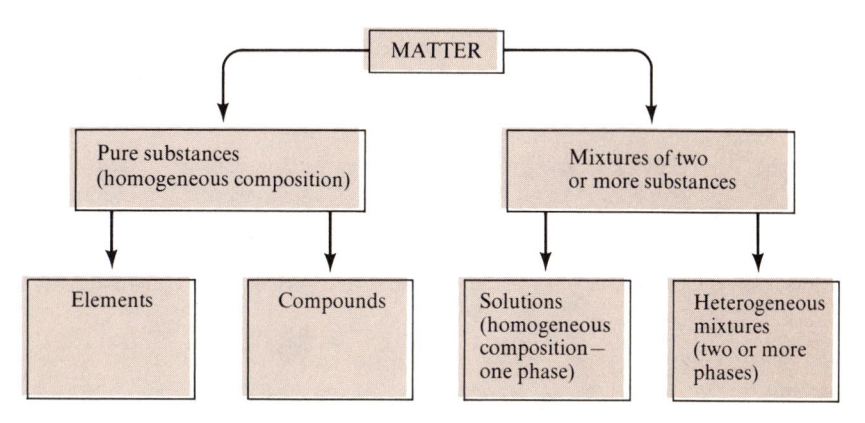

Figure 3.4. Forms of matter: A pure substance is always homogeneous. A mixture always contains two or more substances and may be either homogeneous (a solution) or heterogeneous.

3.4 Properties of Substances

properties

physical properties

chemical properties

How do we recognize substances? Each substance has a set of **properties** that is characteristic of that substance and gives it a unique identity. Properties are the personality traits of substances and are classified as either physical or chemical. **Physical properties** are the inherent characteristics of a substance that may be determined without altering the composition of that substance; they are associated with its physical existence. Common physical properties are color, taste, odor, state of matter (solid, liquid, or gas), density, melting point, and boiling point. **Chemical properties** describe the ability of a substance to form new substances, either by reaction with other substances or by decomposition.

We can select a few of the physical and chemical properties of chlorine as an example. Physically, chlorine is a gas about 2.4 times heavier than air. It is yellowish-green in color and has a disagreeable odor. Chemically, chlorine will not burn but will support the combustion of certain other substances. It can be used as a bleaching agent, as a disinfectant for water, and in many chlorinated substances such as refrigerants and insecticides. When chlorine combines with the metal sodium, it forms a salt, sodium chloride. These properties, among others, help to characterize and identify chlorine.

Substances, then, are recognized and differentiated by their properties. Table 3.3 lists four substances and tabulates several of their common physical properties. Information about common physical properties, such as given in Table 3.3, is readily available in handbooks of chemistry and physics. Scientists don't pretend to know all the answers or to remember voluminous amounts of data, but it is important for them to know where to look for data in the literature. Handbooks are one of the most widely used resources for scientific data.

Table 3.3. Physical properties of chlorine, water, sugar, and acetic acid.

Substance	Color	Odor	Taste	Physical state	Boiling point (°C)	Freezing point (°C)
Chlorine	Yellowish-green	Sharp, suffocating	Sharp, sour	Gas	−34.6	−101.6
Water	Colorless	Odorless	Tasteless	Liquid	100.0	0.0
Sugar	White	Odorless	Sweet	Solid	Decomposes 170–186	—
Acetic acid	Colorless	Like vinegar	Sour	Liquid	118.0	16.7

3.5 Physical Changes

physical change

Matter can undergo two types of changes, physical and chemical. **Physical changes** are mainly changes in physical properties (such as size, shape, density) or state of matter without an accompanying change in composition. The changing of ice into water and water into steam are physical changes from one state of matter into another. No new substances are formed in these physical changes (see Figure 3.5).

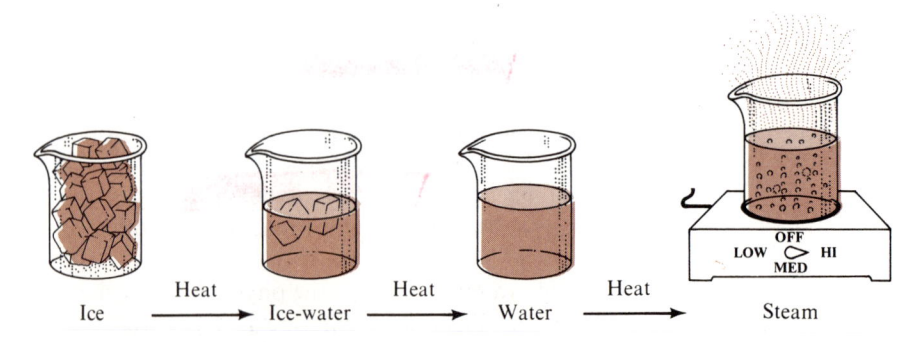

Ice $\xrightarrow{\text{Heat}}$ Ice-water $\xrightarrow{\text{Heat}}$ Water $\xrightarrow{\text{Heat}}$ Steam

Figure 3.5. Physical changes in the appearance and state of water.

If we heat a platinum wire in a burner flame, the wire will become red hot. It returns to its original silvery, metallic form after cooling. The platinum undergoes a physical change in appearance while in the flame, but its composition remains the same.

3.6 Chemical Changes

chemical change

In a **chemical change**, new substances are formed that have different properties and composition from the original material. The new substances need not in any way resemble the initial material.

If a clean copper wire is heated in a burner flame, a change in the appearance of the wire is readily noted after it cools. The copper no longer has its characteristic color, but now appears black. The black material is copper(II) oxide, a new substance formed when copper is combined chemically with oxygen in the air during the heating process. The wire before heating was essentially 100% copper, whereas the black copper(II) oxide contains only 79.9% copper, the rest being oxygen (see Figure 3.6). When both platinum and copper are heated under the conditions described, platinum, which does not readily combine with oxygen, changes only physically, but copper changes chemically as well as physically.

Before heating (copper-colored)

Copper and oxygen from the air combine chemically on heating

After heating (black)

Copper wire: 1.00 g (100% Copper)

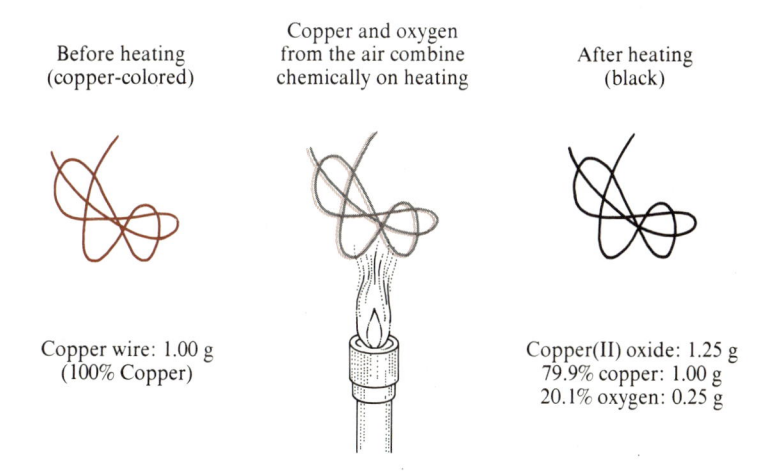

Copper(II) oxide: 1.25 g
79.9% copper: 1.00 g
20.1% oxygen: 0.25 g

Figure 3.6. Chemical change: formation of copper(II) oxide from copper and oxygen.

Mercuric oxide is an orange-red powder which, when subjected to high temperature (500–600°C), decomposes into a colorless gas (oxygen) and a silvery, liquid metal (mercury). The composition of both of these products, as well as their physical appearances, is noticeably different from that of the starting compound. When mercuric oxide is heated in a test tube (see Figure 3.7), small globules of mercury are observed collecting on the cooler part of the tube. Evidence of the oxygen formed is observed when a glowing wood splint,

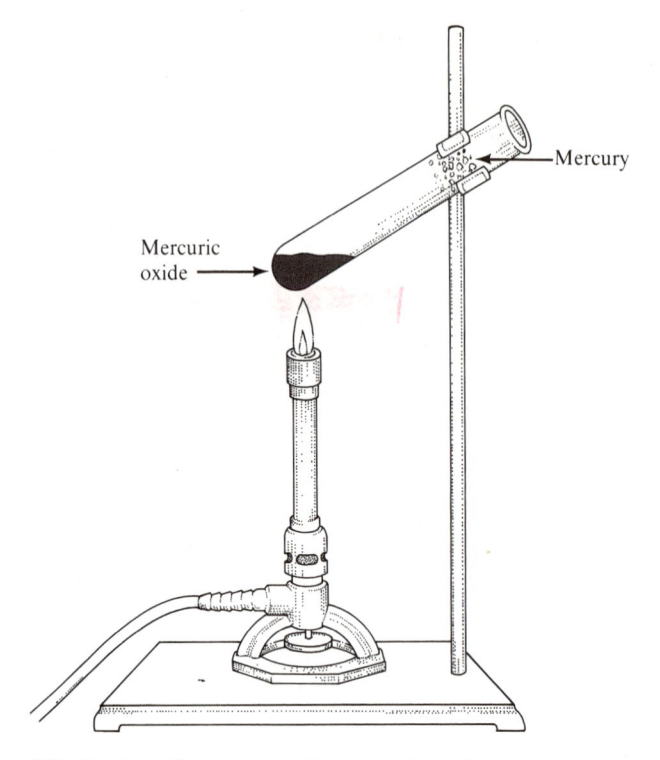

Figure 3.7. Heating of mercuric oxide causes it to decompose into mercury and oxygen. Observation of the mercury and oxygen with properties different from mercuric oxide is evidence that a chemical change has occurred.

lowered into the tube, bursts into flame. Oxygen supports and intensifies the combustion of the wood. From these observations, we can conclude that a chemical change has taken place.

Chemists have devised *chemical equations* as a shorthand method of expressing chemical changes. The two examples of chemical changes presented above may be represented by the following word equations:

$$\text{Copper} + \text{Oxygen} \xrightarrow{\Delta} \text{Copper(II) oxide} \quad \text{(Cupric oxide)} \tag{1}$$

$$\text{Mercuric oxide} \xrightarrow{\Delta} \text{Mercury} + \text{Oxygen} \tag{2}$$

Equation (1) states: copper plus oxygen when heated produce copper(II) oxide. Equation (2) states: mercuric oxide when heated produces mercury plus oxygen. The arrow means "produces"; it points to the products. The delta sign (Δ) represents heat. The starting substances (copper, oxygen, and mercuric oxide) are called the *reactants*; and the substances produced (copper(II) oxide, mercury, and oxygen) are called the *products.* In later chapters, equations are presented in a still more abbreviated form, with symbols used for each substance.

Physical change inevitably accompanies a chemical change. Table 3.4 lists common physical and chemical changes. In the examples given in the

Properties of Matter 41

Table 3.4. Examples of processes involving physical or chemical changes.

Process taking place	Type of change	Accompanying physical changes
Rusting of iron	Chemical	Shiny, bright metal changes to reddish-brown rust
Boiling of water	Physical	Liquid changes to vapor
Burning of sulfur in air	Chemical	Yellow solid sulfur changes to gaseous, choking sulfur dioxide
Melting of lead	Physical	Solid changes to liquid
Combustion of gasoline	Chemical	Liquid gasoline burns to gaseous carbon monoxide, carbon dioxide, and water
Cutting of a diamond	Physical	Small diamonds are made from a larger diamond
Sawing of wood	Physical	Smaller pieces of wood plus sawdust are made from a larger piece of wood
Burning of wood	Chemical	Wood burns to ashes and gaseous carbon dioxide and water
Heating of glass	Physical	Solid becomes pliable during heating and the glass may change its shape

table, you will note that wherever a chemical change occurs, a physical change occurs also. However, wherever a physical change is listed, only a physical change occurs.

3.7 Conservation of Mass

Law of Conservation of Mass

The **Law of Conservation of Mass** states that there is no detectable change in the total mass of the substances involved in a chemical change. This law, tested by extensive laboratory experimentation, is the basis for the quantitative weight relationships among reactants and products.

The decomposition of mercuric oxide into mercury and oxygen illustrates this law. One hundred grams of mercuric oxide decomposes into 92.6 g of mercury and 7.4 g of oxygen.

Mercuric oxide $\longrightarrow$ Magnesium + Oxygen
100.0 g   92.6 g   7.4 g

| 100 g Reactant | | 100 g Products |

Sealed within the ordinary photographic flashbulb are fine wires of magnesium (a metal) and oxygen (a gas). When these reactants are energized, they combine chemically, producing magnesium oxide, together with a blinding white light and considerable heat. The chemical change may be represented by this equation:

Magnesium + Oxygen ⟶ Magnesium oxide + Heat + Light

When weighed before and after the chemical change, as illustrated in Figure 3.8, the bulb shows no increase or decrease in weight.

Figure 3.8. The flashbulb, containing magnesium and oxygen, weighs the same (a) before and (b) after the bulb is flashed. When the bulb is flashed, a chemical change occurs. The original substances are changed into the white powder, magnesium oxide.

⊙ *Mass of reactants = Mass of products*

3.8 Energy

energy

From the prehistoric discovery that fire could be used to warm shelters and cook food to the modern-day discovery that nuclear reactors can be used to produce vast amounts of controlled energy, man's progress has been directed by the ability to harness, produce, and utilize energy. **Energy** is the capacity of matter to do work. Energy exists in several forms; some of the more common forms are mechanical, chemical, electrical, heat, nuclear, and radiant or light energy. Matter can have both potential and kinetic energy.

potential energy

Potential energy is stored energy, or energy an object possesses due to its relative position. For example, a ball located 20 feet above the ground has more potential energy than when located 10 feet above the ground, and will bounce higher when allowed to fall. Water backed up behind a dam represents potential energy that can be converted into useful work in the form of electrical energy. Gasoline represents a source of stored chemical potential energy that can be released during combustion.

kinetic energy

Kinetic energy is the energy that matter possesses due to its motion. When the water behind the dam is released and allowed to flow, its potential energy is changed into kinetic energy, which may be used to drive generators and produce electricity. All moving bodies possess kinetic energy. The pressure exerted by a confined gas is due to the kinetic energy of rapidly moving gas particles. We all know the results when two moving vehicles collide—their kinetic energy is expended in the crash that occurs.

Energy may be converted from one form to another form. Some kinds of energy can be converted to other forms easily and efficiently. For example, mechanical energy can be converted to electrical energy with an electric generator at better than 90% efficiency. On the other hand, solar energy has thus far been directly converted to electrical energy at an efficiency of about 15%.

3.9 Energy in Chemical Changes

In all chemical changes, matter either absorbs or releases energy. Chemical changes can be used to produce different forms of energy. Electrical energy to start automobiles is produced by chemical changes in the lead storage battery. Light energy for photographic purposes occurs as a flash during the chemical change in the magnesium flashbulb. Heat and light energies are released from the combustion of fuels. All the energy needed for our life processes—breathing, muscle contraction, blood circulation, and so on—is produced by chemical changes occurring within the cells of the body.

Conversely, energy is used to cause chemical changes. For example, a chemical change occurs in the electroplating of metals when electrical energy is passed through a salt solution in which the metal is submerged. A chemical change also occurs when radiant energy from the sun is utilized by plants in the process of photosynthesis. And, as we saw, a chemical change occurs when heat causes mercuric oxide to decompose into mercury and oxygen. Chemical changes are often used primarily to produce energy rather than to produce new substances. The heat or thrust generated during the combustion of fuels is more important than the new substances formed.

3.10 Conservation of Energy

An energy transformation occurs whenever there is a chemical change. If energy is absorbed during the change, the products will have more chemical or potential energy than the reactants. Conversely, if energy is given off in a chemical change, the products will have less chemical or potential energy than the reactants. Water, for example, can be decomposed in an electrolytic cell. Electrical energy is absorbed in the decomposition, and the products—hydrogen and oxygen—have a greater chemical or potential energy level than that of water. This potential energy is released in the form of heat and light when the hydrogen and oxygen are burned to form water again (see Figure 3.9). Thus, energy can be changed from one form to another or from one substance to another and therefore is not lost.

The energy changes occurring in many systems have been thoroughly studied by many investigators. No system has been found to acquire energy

44 Chapter 3

[Diagram: Electrical energy is converted to chemical energy → ELECTROLYSIS (Chemical change): Water → Hydrogen + Oxygen; COMBUSTION (Chemical change): Hydrogen + Oxygen → Water → Chemical energy is converted to heat and light energy]

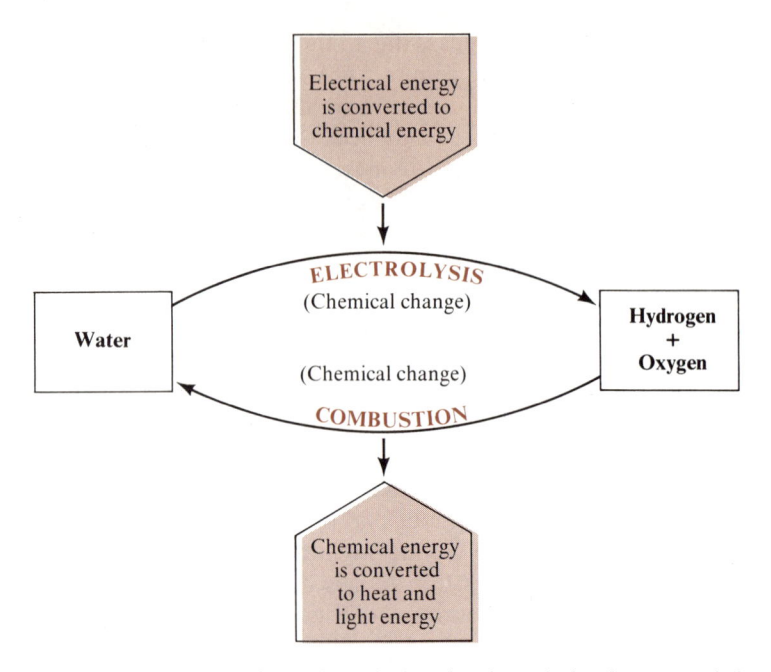

Figure 3.9. Energy transformations during the electrolysis of water and the combustion of hydrogen and oxygen. Electrical energy is converted to chemical energy in the electrolysis, and chemical energy is converted to heat and light energy in the combustion.

Law of Conservation of Energy

except at the expense of energy possessed by another system. This is stated in other words as the **Law of Conservation of Energy**: Energy can be neither created nor destroyed, though it may be transformed from one form to another.

3.11 Interchangeability of Matter and Energy

Sections 3.7–3.10 dealt with matter and energy. The two are clearly related; any attempt to deal with one inevitably involves the other. The nature of this relationship eluded the most able scientists until the beginning of the 20th century. Then, in 1905, Albert Einstein (Figure 3.10) presented one of the most original scientific concepts ever devised.

Einstein stated that the quantity of energy (E) equivalent to the mass (m) could be calculated by the equation $E = mc^2$, where m is in grams and c is the velocity of light (3.0×10^{10} cm/s). According to Einstein's equation, whenever energy is absorbed or released by a substance, there must be a loss or gain of mass. Although the energy changes in chemical reactions are measurable and may appear to be large, the amounts are relatively small. The accompanying difference in mass between reactants and products in chemical changes is so small that it cannot be detected by available measuring instruments. According to Einstein's equation, 2.2×10^7 cal (9×10^{14} ergs) of energy are equivalent to 0.000001 g (1 microgram) of mass. In a more practical sense, when 2.8×10^3 g

Figure 3.10. Albert Einstein (1879–1955), world-renowned physicist and author of the theory of relativity and the interrelationship between matter and energy: $E = mc^2$. (Courtesy of The Bettmann Archive.)

of carbon are burned to carbon dioxide, 2.2×10^7 cal of energy are released. Of this very large amount of carbon only about one millionth of a gram, which is 3.6×10^{-8} % of the starting mass, is converted to energy. Therefore, in actual practice we may treat the reactants and products of chemical changes as having constant mass. However, because mass and energy are interchangeable, the two laws dealing with the conservation of matter may be combined into a single and generally more accurate statement:

> *The total amount of mass and energy remains constant during chemical change.*

Questions

A. Review the meanings of the new terms introduced in this chapter.
   1. Matter
   2. Solid
   3. Amorphous
   4. Liquid
   5. Gas
   6. Homogeneous
   7. Heterogeneous
   8. Phase
   9. Substance
   10. Mixture
   11. Properties
   12. Physical properties
   13. Chemical properties
   14. Physical change
   15. Chemical change
   16. Law of Conservation of Mass
   17. Energy
   18. Potential energy
   19. Kinetic energy
   20. Law of Conservation of Energy

B. Answers to the following questions will be found in tables and figures.
   1. Name three liquids listed in Table 3.1 that are mixtures.
   2. Which of the gases listed in Table 3.1 is not a pure substance?
   3. What physical properties do solids and liquids have in common? (See Table 3.2.)
   4. In what physical state will acetic acid exist at 150°C? (See Table 3.3.)

5. In what physical state will water exist at 293 K? (See Table 3.3.)
6. From Figure 3.1 what evidence can you find that gases occupy space?
7. What effect does the absorption of heat energy have on mercuric oxide? (See Figure 3.7.)
8. What physical changes occur to the matter in the flashbulb of Figure 3.8 when the bulb is flashed?

C. *Review questions.*
1. List three substances in each of the three physical states of matter.
2. Explain why a gas can be compressed and why a liquid cannot be compressed appreciably.
3. In terms of the properties of the ultimate particles explain why a liquid can be poured but a solid cannot be poured.
4. When the stopper is removed from a partly filled bottle containing solid and liquid acetic acid at 16.7°C, a strong vinegar-like odor is noticeable immediately. How many acetic acid phases must be present in the bottle? Explain.
5. Is the system enclosed in the bottle of Question 4 homogeneous or heterogeneous? Explain.
6. Distinguish between physical and chemical properties of matter.
7. Is a system containing only water necessarily homogeneous? Explain.
8. Is a system containing only one substance necessarily homogeneous? Explain.
9. Distinguish between physical and chemical changes.
10. Classify the following as primarily physical or chemical changes:
    (a) Boiling water   (c) Boiling an egg   (e) Souring milk
    (b) Freezing ice cream   (d) Homogenizing milk
11. Reread Section 3.4 and list those properties given for chlorine that are physical and those that are chemical.
12. Cite the evidence demonstrating that the heating of mercuric oxide brings about a chemical change.
13. Distinguish between potential and kinetic energy.
14. Is chemical energy potential or kinetic?
15. In an ordinary chemical change, why can we consider that mass is neither lost nor gained (for practical purposes)?
16. When the flashbulb of Figure 3.8 is flashed, energy is given off to the surroundings. Explain why the apparent mass of the bulb was the same after flashing as it was before, although according to Einstein, energy is equivalent to mass.
17. Which of the following statements are correct? (Try to answer this question without referring to the text.)
    (a) Liquids are the most compact state of matter.
    (b) Matter in the solid state is discontinuous—that is, it is made up of discrete particles.
    (c) Seawater, although homogeneous, is considered to be a mixture.
    (d) Any system that consists of two or more phases is heterogeneous.
    (e) A solution, although it contains dissolved material, is considered to be homogeneous.
    (f) Boiling water represents a chemical change because no change in composition occurs.
    (g) All of the following represent chemical change: baking a cake, frying an egg, leaves changing color, iron changing to rust.
    (h) All of the following represent physical change: breaking a stick, melting wax, folding a napkin, burning hydrogen to form water.

(i) A stretched rubber band possesses kinetic energy.
(j) An automobile rolling down a hill possesses both kinetic and potential energy.

D. *Review problems.*
  1. Calculate the boiling point of chlorine in degrees Fahrenheit (see Table 3.3).
  2. What weight of mercury can be obtained from 75.0 g of mercuric oxide?
  3. Given the chemical reaction

  Magnesium + Oxygen → Magnesium oxide

  When 9.50 g of magnesium is heated in air, 15.75 g of magnesium oxide is produced.
  (a) What weight of oxygen has combined with the magnesium?
  (b) What percentage of the magnesium oxide is magnesium?
  4. If a nickel weighs about 5.0 g:
  (a) How many calories would be released by the complete conversion of a nickel to energy?
  (b) If 3.03×10^5 cal are needed to heat a gallon of water from room temperature (20°C) to boiling temperature (100°C), how many gallons of water could be heated to the boiling point by the energy from part (a)?

4 Elements and Compounds

After studying Chapter 4 you should be able to:

1. Understand the terms listed in Question A at the end of the chapter.
2. List in order of abundance the five most abundant elements in the earth's crust, seawater, and atmosphere.
3. List in order of abundance the six most abundant elements in the human body.
4. Classify common materials as elements, compounds, or mixtures.
5. Write the symbols when given the names or write the names when given the symbols of the common elements listed in Table 4.3.
6. State the Law of Definite Composition.
7. Understand how symbols, including subscripts and parentheses, are used to write chemical formulas.
8. Differentiate among atoms, molecules, and ions.
9. List the characteristics of metals and nonmetals.
10. Name binary compounds from their formulas.
11. Balance simple chemical equations when the formulas are given.
12. List the elements that occur as diatomic molecules.

4.1 Elements

element

All the words in the English dictionary are formed from an alphabet consisting of only 26 letters. All known substances on earth—and most probably in the universe, too—are formed from a sort of "chemical alphabet" consisting of 106 known elements. An **element** is a fundamental or elementary substance that cannot be broken down, by chemical means, to simpler substances. Elements are the basic building blocks of all substances. The elements are numbered in order of increasing complexity beginning with hydrogen, number 1. Of the first 92 elements, 88 are known to occur in nature. The other four—technetium (43), promethium (61), astatine (85), and francium (87)—either do not occur in nature or have only transitory existences resulting from radioactive decay. With the exception of number 94, elements 93–106 are not known to occur naturally, but have been synthesized—usually in very small quantities—in laboratories. The discovery of trace amounts of element 94 (plutonium) in nature has been reported recently. Element 106 was reported to have been synthesized in 1974. No elements other than those on the earth have been detected on other bodies in the universe.

Most substances can be decomposed into two or more other simpler substances. We have seen that mercuric oxide can be decomposed into mercury and oxygen and that water can be decomposed into hydrogen and oxygen. Sugar may be decomposed into carbon, hydrogen, and oxygen. Table salt is easily decomposed into sodium and chlorine. An element, however, cannot be decomposed into simpler substances by ordinary chemical changes.

atom

If we could take a small piece of an element, say copper, and divide it and subdivide it into smaller and smaller particles, we finally would come to a single unit of copper that we could no longer divide and still have copper. This ultimate particle, the smallest particle of an element that can exist, is called an **atom**. An atom is also the smallest unit of an element that can enter into a chemical reaction. Chapter 5 describes the smaller subatomic particles that make up atoms, but these particles no longer have the properties of elements.

4.2 Distribution of Elements

Elements are distributed very unequally in nature. Table 4.1 shows that ten of the elements make up about 99% of the weight of the earth's crust, seawater, and the atmosphere (see Figure 4.1). Oxygen, the most abundant of these, constitutes about 50% of this mass. This list does not include the mantle

Table 4.1. Distribution of the elements in the earth's crust, seawater, and atmosphere.

Element	Weight percent	Element	Weight percent
Oxygen	49.20	Chlorine	0.19
Silicon	25.67	Phosphorus	0.11
Aluminum	7.50	Manganese	0.09
Iron	4.71	Carbon	0.08
Calcium	3.39	Sulfur	0.06
Sodium	2.63	Barium	0.04
Potassium	2.40	Nitrogen	0.03
Magnesium	1.93	Fluorine	0.03
Hydrogen	0.87		
Titanium	0.58	All others	0.47

and the core of the earth, which are believed to be composed of metallic iron and nickel. The average distribution of the elements in the human body is shown in Table 4.2. Note again the high percentage of oxygen.

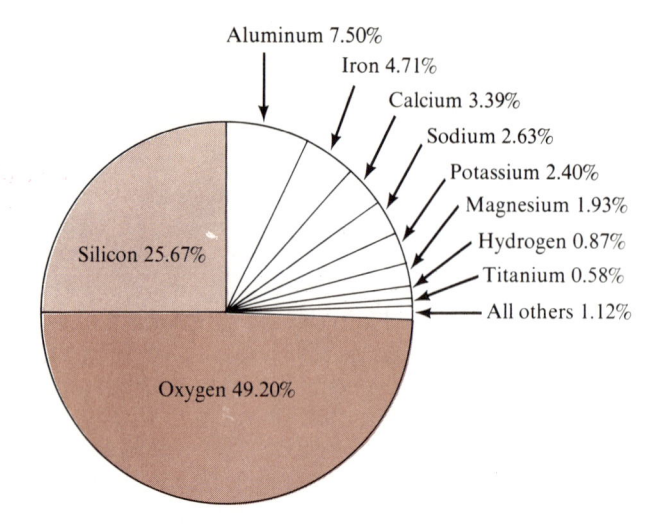

Figure 4.1. Weight percent of the elements in the earth's crust, seawater, and atmosphere.

Table 4.2. Average elemental composition of the human body.

Element	Weight percent
Oxygen	65.0
Carbon	18.0
Hydrogen	10.0
Nitrogen	3.0
Calcium	2.0
Phosphorus	1.0
Traces of several other elements	1.0

4.3 Names of the Elements

The names of the elements came to us from various sources. Many are derived from early Greek, Latin, or German words that generally described some property of the element. For example, iodine is taken from the Greek word *iodes*, meaning violet-like. Iodine, indeed, is violet in the vapor state. The name of the metal bismuth had its origin from the German words *weisse masse*, which means white mass. Miners called it *wismat*; it was later changed to *bismat*, and finally to bismuth. Some elements are named for the location of their discovery—for example, germanium, discovered in 1886 by Winkler, a German chemist. Others are named in commemoration of famous scientists, such as einsteinium and curium, for Albert Einstein and Marie Curie, respectively.

4.4 Symbols of the Elements

symbol

We all recognize Mr., N.Y., and St. as abbreviations for "mister," "New York," and "street." In like manner, chemists have assigned specific abbreviations to each element; these are called **symbols** of the elements. Fourteen of the elements have a single letter as their symbol, and all the others have two letters. The symbol stands for the element itself, for one atom of the element, and, as we shall see later, for a particular quantity of the element.

Rules governing symbols of elements are as follows:

1. Symbols are composed of one or two letters.
2. If one letter is used, it is capitalized.
3. If two letters are used, the first is capitalized and the second is a lowercase letter.

Examples: Sulfur S
   Barium Ba

The symbols and names of all the elements are given in the table on the inside back cover of this book. Table 4.3 lists the more commonly used symbols. If we examine this table carefully, we note that most of the symbols start with the same letter as the name of the element that is represented. A number of symbols, however, appear to have no connection with the names of the elements they represent (see Table 4.4). These symbols have been carried over from earlier names (usually in Latin) of the elements and are so firmly implanted in the literature that their use is continued today.

Special care must be used in writing symbols. Begin each with a capital letter and use a lowercase second letter if needed. For example, consider Co, the symbol for the element cobalt. If through error CO (capital C and capital O) is written, the two elements carbon and oxygen (the *formula* for carbon monoxide) are represented instead of the single element cobalt. Another example of

Table 4.3. Symbols of the most common elements.

Element	Symbol	Element	Symbol	Element	Symbol
Aluminum	Al	Fluorine	F	Phosphorus	P
Antimony	Sb	Gold	Au	Platinum	Pt
Argon	Ar	Helium	He	Potassium	K
Arsenic	As	Hydrogen	H	Radium	Ra
Barium	Ba	Iodine	I	Silicon	Si
Bismuth	Bi	Iron	Fe	Silver	Ag
Boron	B	Lead	Pb	Sodium	Na
Bromine	Br	Lithium	Li	Strontium	Sr
Cadmium	Cd	Magnesium	Mg	Sulfur	S
Calcium	Ca	Manganese	Mn	Tin	Sn
Carbon	C	Mercury	Hg	Titanium	Ti
Chlorine	Cl	Neon	Ne	Tungsten	W
Chromium	Cr	Nickel	Ni	Uranium	U
Cobalt	Co	Nitrogen	N	Zinc	Zn
Copper	Cu	Oxygen	O		

Table 4.4. Symbols of the elements derived from early names. These symbols are in use today, even though they do not appear to correspond to the current name of the element.

Present name	Symbol	Former name
Antimony	Sb	Stibium
Copper	Cu	Cuprum
Gold	Au	Aurum
Iron	Fe	Ferrum
Lead	Pb	Plumbum
Mercury	Hg	Hydrargyrum
Potassium	K	Kalium
Silver	Ag	Argentum
Sodium	Na	Natrium
Tin	Sn	Stannum
Tungsten	W	Wolfram

the need for care in writing symbols is the symbol Ca for calcium versus Co for cobalt.

A knowledge of symbols is essential for writing chemical formulas and equations. You should begin to learn symbols immediately since they will be used extensively in the remainder of this book and in any future chemistry courses you may take. One way to learn the symbols is to practice a few minutes a day by making side-by-side lists of names and symbols and then covering each list alternately and writing the corresponding name or symbol. Initially, it is a good plan to learn the symbols of the most common elements shown in Table 4.3.

The experiments of alchemists paved the way for the development of chemistry. Alchemists surrounded their work in mysticism partly by devising a system of symbols known only to practitioners of alchemy (see Figure 4.2). In the early 1800s the Swedish chemist J. J. Berzelius (1779–1848) made a great contribution to chemistry by devising the present system of symbols using letters of the alphabet.

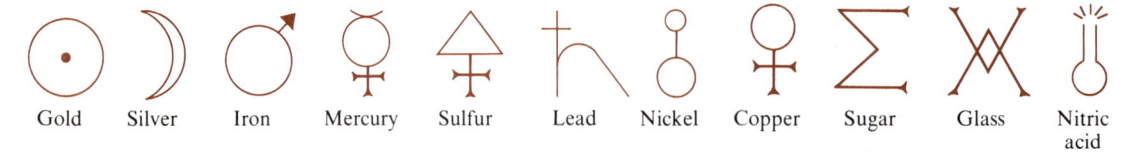

Gold   Silver   Iron   Mercury   Sulfur   Lead   Nickel   Copper   Sugar   Glass   Nitric acid

Figure 4.2. Some typical alchemists' symbols.

4.5 Compounds

compound **Compounds**, unlike elements, can be decomposed chemically into simpler substances—that is, into other compounds and/or elements.

The atoms of the elements that form a compound are combined in whole-number ratios, never as fractional atoms. Compounds exist either as molecules

molecule

or as ions. A **molecule** is the smallest uncharged individual unit of a compound formed by the union of two or more atoms. If we subdivide a drop of water into smaller and smaller particles, we ultimately obtain a single molecule of water consisting of two hydrogen atoms bonded to one oxygen atom. This molecule cannot be further subdivided without destroying the water and forming the elements hydrogen and oxygen. Thus, a molecule of water is the smallest unit of the substance water.

ion

An **ion** is a positive or negative electrically charged atom or group of atoms. The ions in a compound are held together in a crystalline structure by the attractive forces that exist between their positive and negative charges. Compounds consisting of ions do not exist as molecules. The formula of such a compound usually represents the simplest ratio of the charged atoms or ions that exist in the substance. Sodium chloride (table salt) is such a substance. It consists of sodium ions (positively charged sodium atoms) and chloride ions (negatively charged chlorine atoms). The two types of compounds, *molecular* and *ionic*, are illustrated in Figure 4.3.

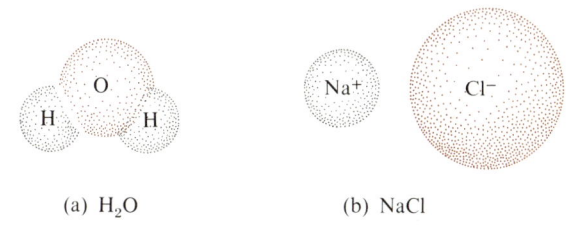

(a) H_2O (b) NaCl

Figure 4.3. Representation of molecular and ionic (nonmolecular) compounds. (a) Two hydrogen atoms combined with an oxygen atom to form a molecule of water. (b) A positively charged sodium ion and a negatively charged chloride ion form the compound sodium chloride.

The compound carbon monoxide (CO) is composed of carbon and oxygen in the ratio of one atom of carbon to one atom of oxygen. Hydrogen chloride (HCl) contains a ratio of one atom of hydrogen to one atom of chlorine. Compounds may contain more than one atom of the same element. Methane (natural gas, CH_4) is composed of a ratio of one atom of carbon to four atoms of hydrogen; ordinary table sugar (sucrose, $C_{12}H_{22}O_{11}$) contains a ratio of 12 atoms of carbon to 22 atoms of hydrogen to 11 atoms of oxygen. These atoms are held together in the compound by *chemical bonds.*

Substance	Each molecule composed of	Formula
Carbon monoxide	1 carbon atom + 1 oxygen atom	CO
Hydrogen chloride	1 hydrogen atom + 1 chlorine atom	HCl
Methane	1 carbon atom + 4 hydrogen atoms	CH_4
Sugar (sucrose)	12 carbon atoms + 22 hydrogen atoms + 11 oxygen atoms	$C_{12}H_{22}O_{11}$
Water	2 hydrogen atoms + 1 oxygen atom	H_2O

There are about 4 million known compounds, with no end in sight as to the number that can and will be prepared in the future. Each compound is unique and has characteristic physical and chemical properties. Let us consider in some detail two compounds—water and mercuric oxide. Water is a colorless, odorless, tasteless liquid that can be changed to a solid (ice) at 0°C and to a gas (steam) at 100°C. It is composed of two atoms of hydrogen and one atom of oxygen per molecule, which represents 11.2% hydrogen and 88.8% oxygen by weight. Water reacts chemically with sodium to produce hydrogen gas and sodium hydroxide, with lime to produce calcium hydroxide, and with sulfur trioxide to produce sulfuric acid. No other compound has all these exact physical and chemical properties; they are characteristic of water alone.

Mercuric oxide is a dense, orange-red powder composed of a ratio of one atom of mercury to one atom of oxygen. Its composition by weight is 92.6% mercury and 7.4% oxygen. When it is heated to temperatures greater than 360°C, a colorless gas, oxygen, and a silvery liquid metal, mercury, are produced. These are specific physical and chemical properties belonging to mercuric oxide and to no other substance. Thus, a compound may be identified and distinguished from all other compounds by its characteristic properties.

4.6 Law of Definite Composition of Compounds

Many experiments extending over a long period of time have established the fact that a specific compound always contains the same elements in a fixed proportion by weight. For example, water contains 11.2% hydrogen and 88.8% oxygen by weight. Water will always contain hydrogen and oxygen in this fixed weight ratio. The fact that water contains hydrogen and oxygen in this particular ratio does not mean that hydrogen and oxygen cannot combine in some other ratio. However, the resulting compound will not be water. In fact, hydrogen peroxide is made up of two atoms of hydrogen and two atoms of oxygen per molecule and contains 5.9% hydrogen and 94.1% oxygen by weight; its properties are markedly different from those of water.

	Water	Hydrogen peroxide
Percent H	11.2	5.9
Percent O	88.8	94.1
Atomic composition	2H + 1O	2H + 2O

Law of Definite Composition

The **Law of Definite Composition** states: A compound always contains two or more elements combined in a definite proportion by weight. The reliability of this law, which in essence states that the composition of a substance will always be the same no matter what its origin or how it is formed, is the cornerstone of chemical science.

4.7 Chemical Formulas

chemical formulas

In a manner similar to the use of symbols for elements, chemists use **chemical formulas** as abbreviations for compounds. The chemical formula represents two or more elements that are in chemical combination. Sodium chloride contains one atom of sodium per atom of chlorine; its formula is NaCl. The formula for water is H_2O; it shows that a molecule of water contains two atoms of hydrogen and one atom of oxygen.

The formula of a compound tells us which elements it is composed of and how many atoms of each element are present in a formula unit. For example, a molecule of sulfuric acid is composed of two atoms of hydrogen, one atom of sulfur, and four atoms of oxygen. We can express this compound as HHSOOOO, but the usual formula for writing sulfuric acid is H_2SO_4. The formula may be expressed verbally as "H-two-S-O-four." Characteristics of chemical formulas are summarized below.

1. The formula of a compound contains the symbols of all the elements in the compound.
2. When the formula contains one atom of an element, the symbol of that element represents that one atom. The number one (1) is not used as a subscript to indicate one atom of an element.
3. When the formula contains more than one atom of an element, the number of atoms is indicated by a subscript written to the right of the symbol of that atom. For example, the two (2) in H_2O indicates two atoms of H in the formula.
4. When the formula contains more than one of a group of atoms that occurs as a unit, parentheses are placed around the group and the number of units of the group are indicated by a subscript placed to the right of the parentheses. Consider

$$H_2SO_4$$

Indicates the element hydrogen

Indicates two atoms of hydrogen

Indicates the element sulfur (one atom)

Indicates the element oxygen

Indicates four atoms of oxygen

$$Ca(NO_3)_2$$

Indicates the element calcium (one atom)

Indicates the nitrate group composed of one nitrogen atom and three oxygen atoms

Indicates two nitrate (NO_3) groups

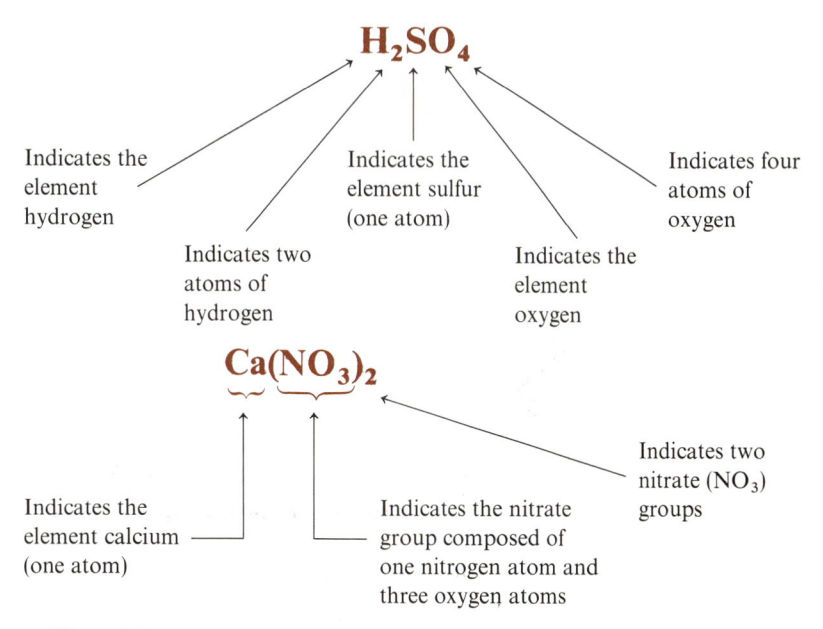

Figure 4.4. Explanation of the formulas H_2SO_4 and $Ca(NO_3)_2$.

the nitrate group, NO_3^-. In the formula for sodium nitrate, $NaNO_3$, there is only one nitrate group; therefore, no parentheses are needed. In calcium nitrate, $Ca(NO_3)_2$, there are two nitrate groups, and parentheses and the subscript 2 are used to indicate this. There are a total of nine atoms in $Ca(NO_3)_2$—one Ca, two N, and six O atoms.

5. Formulas written as H_2O, H_2SO_4, $Ca(NO_3)_2$, $C_{12}H_{22}O_{11}$ show only the number and kind of each atom contained in the compound; they do not show the arrangement of the atoms in the compound or how they are chemically bonded to one another.

Figure 4.4 illustrates how symbols and numbers are used in chemical formulas. There is more extensive use of formulas in later chapters.

4.8 Mixtures

Single substances—elements or compounds—seldom occur naturally in the pure state. Air is a mixture of gases; seawater is a mixture containing a variety of dissolved minerals; ordinary soil is a complex mixture of minerals and various organic materials.

How is a mixture distinguished from a pure substance? A mixture (see Section 3.3) always contains two or more substances that can be present in varying concentrations. Let us consider an example of a homogeneous mixture and an example of a heterogeneous mixture. Homogeneous mixtures (solutions) containing either 5% or 10% salt in water can be prepared by simply mixing the correct amounts of salt and water. These mixtures can be separated by boiling away the water, leaving the salt as a residue. The composition of a heterogeneous mixture of sulfur crystals and iron filings can be varied by merely blending in either more sulfur or more iron filings. This mixture can be separated physically, either by using a magnet to attract the iron or by adding carbon disulfide to dissolve the sulfur.

Iron(II) sulfide (FeS) contains 63.5% Fe and 36.5% S by weight. If we mix iron and sulfur in this proportion, do we have iron(II) sulfide? No, it is still a

Table 4.5. Comparison of mixtures and compounds.

	Mixture	Compound
Composition	May be composed of elements, compounds, or both in variable composition	Composed of two or more elements in a definite, fixed proportion by weight
Separation of components	Separation may be made by simple physical or mechanical means	Elements can be separated by chemical changes only
Identification of components	Components do not lose their identity	A compound does not resemble the elements from which it is formed

mixture; the iron is still attracted by a magnet. But if this mixture is heated strongly, a chemical change (reaction) occurs, forming iron(II) sulfide. This is a substance with properties that are different from either those of iron or of sulfur—FeS is neither attracted by a magnet nor dissolved by carbon disulfide. Thus, the properties of the reactants are lost and a compound (a pure substance) is formed. Key differences between mixtures and compounds are summarized in Table 4.5.

4.9 Metals, Nonmetals, and Metalloids

metal

nonmetal

metalloid

Three primary classifications of the elements are **metals, nonmetals**, and **metalloids**. Most of the elements are metals. We are familiar with metals because of their widespread use in tools, materials of construction, automobiles, and so on. But nonmetals are equally useful in our everyday life as major components of such items as clothing, food, fuel, glass, plastics, and wood.

The metallic elements are solids at room temperature (mercury is an exception). They have a high luster, are good conductors of heat and electricity, can be rolled or hammered into sheets (they are *malleable*), and can be drawn into wires (they are *ductile*). In addition, most metals have a high melting point and a high density. Metals familiar to most of us are aluminum, chromium, copper, gold, iron, lead, magnesium, mercury, nickel, platinum, silver, tin, and zinc. Other less familiar but still important metals are calcium, cobalt, potassium, sodium, uranium, and titanium.

Metals have little tendency to combine with each other to form compounds. But many metals readily combine with nonmetals such as chlorine, oxygen, and sulfur to form mainly ionic compounds such as metallic chlorides, oxides, and sulfides, respectively. The more active metals are found in nature combined with other elements as minerals. A few of the less active ones—such as copper, gold, and silver—are sometimes found in a native or free state as well.

Nonmetals, unlike metals, are not lustrous, have relatively low melting points and densities, and are generally poor conductors of heat and electricity. Carbon, phosphorus, sulfur, selenium, and iodine are solids; bromine is a liquid; the rest of the nonmetals are gases. Common nonmetals found uncombined in nature are carbon (graphite and diamond), nitrogen, oxygen, sulfur, and the noble gases (helium, neon, argon, krypton, xenon, and radon).

Nonmetals combine with one another to form molecular compounds such as carbon dioxide (CO_2), methane (CH_4), butane (C_4H_{10}), and sulfur dioxide (SO_2). Fluorine, the most reactive nonmetal, combines readily with almost all the other elements.

Several elements (boron, silicon, germanium, arsenic, antimony, tellurium, and polonium) are classified as *metalloids* and have properties that are intermediate between those of metals and those of nonmetals. The intermediate position for these elements is shown in Table 4.6, which classifies all the elements as metals, nonmetals, or metalloids. Certain of the metalloids are the raw materials for the semiconductor devices that make our modern electronics industry possible.

Table 4.6. Classification of the elements into metals, metalloids, and nonmetals.

1 H																1 H	2 He
3 Li	4 Be		Key									5 B	6 C	7 N	8 O	9 F	10 Ne
11 Na	12 Mg		☐ Metals ▓ Metalloids ▨ Nonmetals									13 Al	14 Si	15 P	16 S	17 Cl	18 Ar
19 K	20 Ca	21 Sc	22 Ti	23 V	24 Cr	25 Mn	26 Fe	27 Co	28 Ni	29 Cu	30 Zn	31 Ga	32 Ge	33 As	34 Se	35 Br	36 Kr
37 Rb	38 Sr	39 Y	40 Zr	41 Nb	42 Mo	43 Tc	44 Ru	45 Rh	46 Pd	47 Ag	48 Cd	49 In	50 Sn	51 Sb	52 Te	53 I	54 Xe
55 Cs	56 Ba	57 La	72 Hf	73 Ta	74 W	75 Re	76 Os	77 Ir	78 Pt	79 Au	80 Hg	81 Tl	82 Pb	83 Bi	84 Po	85 At	86 Rn
87 Fr	88 Ra	89 Ac	104 Ku	105 Ha	106 —												

58 Ce	59 Pr	60 Nd	61 Pm	62 Sm	63 Eu	64 Gd	65 Tb	66 Dy	67 Ho	68 Er	69 Tm	70 Yb	71 Lu
90 Th	91 Pa	92 U	93 Np	94 Pu	95 Am	96 Cm	97 Bk	98 Cf	99 Es	100 Fm	101 Md	102 No	103 Lr

Seven of the elements (all nonmetals) occur as *diatomic molecules* (consisting of two atoms). These seven elements, together with their formulas and brief descriptions, are listed below.

Element	Molecular formula	Normal state
Hydrogen	H_2	Colorless gas
Nitrogen	N_2	Colorless gas
Oxygen	O_2	Colorless gas
Fluorine	F_2	Pale yellow gas
Chlorine	Cl_2	Yellow-green gas
Bromine	Br_2	Reddish-brown liquid
Iodine	I_2	Greyish-black solid

Whether found free in nature or prepared in the laboratory, the molecules of each of these elements contain two atoms. Their formulas, therefore, are always written to show the molecular composition of the element—that is, H_2, N_2, O_2, F_2, Cl_2, Br_2, and I_2.

It is important to understand how symbols are used to designate either an atom or a molecule of an element. Consider the elements hydrogen and oxygen. Hydrogen is found in gases coming from volcanoes and in certain natural gas supplies; it can also be prepared by many different chemical reactions. Regardless of the source, all samples of hydrogen are identical and consist of diatomic molecules. The composition of this molecular hydrogen is expressed by the formula H_2. Free oxygen makes up about 21% (by volume) of the air that we breathe. Whether obtained from the air or prepared by chemical reaction, free oxygen also exists as diatomic molecules; its composition is expressed by the formula O_2. Now consider hydrogen in the compound water, which has the composition expressed by the formula H_2O (or HOH). Water contains neither free hydrogen (H_2) nor free oxygen (O_2); the H_2 part of the formula H_2O simply tells us that each molecule of water contains two hydrogen atoms combined chemically with one oxygen atom. Symbols and subscripts are used in this way to show the molecular composition of elements and to show the composition of compounds.

4.10 Nomenclature and Chemical Equations

Nomenclature

Knowledge of chemical names and of the writing and balancing of chemical equations is vital to the study of chemistry. This section serves only as an introduction to the naming of compounds and writing equations. More complete details of the systematic methods of naming inorganic compounds are given in Chapter 8, and a more detailed explanation of chemical equations is given in Chapter 10. Refer to these two chapters often, as needed. Neither chapter is intended to be studied only in the sequence given in the text; rather, they are common depositories of information on chemical nomenclature and equations.

We have already used such names as hydrogen chloride (HCl), mercuric oxide (HgO), magnesium oxide (MgO), and carbon dioxide (CO_2). Note that all four names end in *ide*. This *ide* ending is characteristic of the names of compounds composed of atoms of two different elements. Compounds composed of two elements are called *binary* compounds. Some compounds contain several atoms of the same element (for example, CCl_4, carbon tetrachloride), but as long as there are only two different kinds of atoms, the compound is considered to be binary.

When naming a compound consisting of a metal and a nonmetal, the name of the metal is given first, followed by the name of the nonmetal, which is modified to end in *ide*. There are, of course, exceptions to this rule, and the names of some compounds containing more than two elements end in *ide* (for example, NH_4Cl, ammonium chloride; NaOH, sodium hydroxide). Refer to Section 8.3 for more details on naming binary compounds. Examples of binary compounds with names ending in *ide* are given below.

NaCl	Sodium chlor *ide*	H_2S	Hydrogen sulf *ide*
CO_2	Carbon diox *ide*	$AlBr_3$	Aluminum brom *ide*
NaI	Sodium iod *ide*	K_2S	Potassium sulf *ide*
CaF_2	Calcium fluor *ide*	Mg_3N_2	Magnesium nitr *ide*

Chemical Equations

Chemical changes or reactions result in the formation of substances with compositions that are different from the starting substances. A chemical equation is a shorthand expression for a chemical reaction. Substances in the reaction are represented by their symbols or formulas in the equation. The equation indicates both the reactants (starting substances) and the products, and often shows the conditions necessary to facilitate the chemical change. The reactants are written on the left side and the products on the right side of the equation. An arrow (→) pointing to the products separates the reactants from the products. A plus sign (+) is used to separate one reactant (or product) from another.

Reactants ⟶ Products

We will see in Chapter 5 that every atom has a specific mass. In an equation, the symbols or formulas that represent a substance also represent a specific mass of that substance. Since no detectable change in mass results from a chemical change, the mass of the products must equal the mass of the reactants. In representing a chemical change by an equation, this conservation of mass is attained by balancing the equation. After establishing the correct formulas for the reactants and products, an equation is balanced by placing integral numbers (as needed) in front of the formulas of the substances in the equation. We use these numbers to obtain an equation with the same number of atoms of each kinds of element on each side of the equation.

Consider, again, the reaction of metallic copper heated in air. The chemical change may be represented by the following equations:

$$\text{Copper} + \text{Oxygen} \xrightarrow{\Delta} \text{Copper(II) oxide} \quad (1)$$

$$\text{Cu} + \text{O}_2 \xrightarrow{\Delta} \text{CuO} \quad \text{(Unbalanced)} \quad (2)$$

Copper and oxygen are the reactants, and copper(II) oxide is the product. Equation (2) as written is not balanced because there are two oxygen atoms on the left side and only one on the right side. We place a 2 in front of Cu and a 2 in front of CuO to obtain the balanced equation (3):

$$2\,\text{Cu} + \text{O}_2 \xrightarrow{\Delta} 2\,\text{CuO} \quad \text{(Balanced)} \quad (3)$$

This balanced equation contains 2 Cu atoms and 2 O atoms on both sides of the equation.

A very important factor to remember when balancing equations is that a correct formula of a substance may not be changed for the convenience of balancing the equation. In the unbalanced equation (2) above, we cannot change the formula of CuO to CuO_2 to balance the equation, even though by so doing we balance the number of atoms of each element on each side of the equation. The formula CuO_2 is not the correct formula for the product. It is also important to be aware that a number in front of a formula multiplies every atom in that formula by that number. Thus,

2 CuO   means 2 Cu atoms and 2 O atoms
3 H_2O means 6 H atoms and 3 O atoms
4 H_2SO_4 means 8 H atoms, 4 S atoms, and 16 O atoms

Questions

A. Review the meanings of the new terms introduced in this chapter.
1. Element
2. Atom
3. Symbol
4. Compound
5. Molecule
6. Ion
7. Law of Definite Composition
8. Chemical formula
9. Metal
10. Nonmetal
11. Metalloid

B. Answers to the following questions will be found in tables and figures.
1. List, in decreasing order of abundance, the six most abundant elements in the human body.
2. Are there more atoms of silicon or hydrogen in the earth's crust, seawater, and atmosphere? Use Table 4.1 and the fact that the mass of a silicon atom is about 28 times that of a hydrogen atom.
3. Why is the symbol for lead Pb instead of Le?
4. Make a list of the names of the elements in Table 4.3. Now see how many of the symbols you know by writing the correct symbol after each name.
5. How many metals are there? Nonmetals? Metalloids? (See Table 4.6.)

C. Review questions.
1. What does the symbol of an element stand for?
2. Write down what you believe to be the symbols for the elements argon, lithium, manganese, nickel, nitrogen, platinum, plutonium, and uranium. Now look up the correct symbols and rewrite them, comparing the two sets.
3. Interpret the difference in meanings for each of these pairs:
   (a) Pb and PB   (b) Co and CO
4. Distinguish between an element and a compound.
5. Explain why the Law of Definite Composition does not pertain to mixtures.
6. Does the Law of Definite Composition pertain to an element? Discuss.
7. Distinguish between a chemical formula and a symbol.
8. Given the following list of compounds and their formulas, what elements are present in each compound?

(a)	Potassium iodide	KI
(b)	Sodium carbonate	Na_2CO_3
(c)	Aluminum oxide	Al_2O_3
(d)	Calcium bromide	$CaBr_2$
(e)	Carbon tetrachloride	CCl_4
(f)	Magnesium bromide	$MgBr_2$
(g)	Nitric acid	HNO_3
(h)	Barium sulfate	$BaSO_4$
(i)	Aluminum phosphate	$AlPO_4$
(j)	Acetic acid	$HC_2H_3O_2$

9. Write the formula for each of the following compounds, the composition of which is given after each name:

(a)	Zinc oxide	1 atom Zn, 1 atom O
(b)	Potassium chlorate	1 atom K, 1 atom Cl, 3 atoms O
(c)	Sodium hydroxide	1 atom Na, 1 atom O, 1 atom H
(d)	Aluminum bromide	1 atom Al, 3 atoms Br
(e)	Calcium fluoride	1 atom Ca, 2 atoms F
(f)	Lead(II) chromate	1 atom Pb, 1 atom Cr, 4 atoms O
(g)	Ethyl alcohol	2 atoms C, 6 atoms H, 1 atom O
(h)	Benzene	6 atoms C, 6 atoms H

10. Explain the meaning of each symbol and number in the following formulas:
    (a) H₂O   (b) CCl₄   (c) Cd(NO₃)₂   (d) Cu₂S   (e) C₆H₁₂O₆ (glucose)
11. How many atoms are represented in each of these formulas?
    (a) AgI   (d) K₂Cr₂O₇   (g) CCl₂F₂ (freon)
    (b) H₃PO₄   (e) Cl₂   (h) C₆H₈N₂O₂S (sulfanilamide)
    (c) LiNO₃   (f) Mg(NO₃)₂
12. How many atoms of oxygen are contained in one molecule of oxygen?
13. How many atoms of hydrogen and oxygen are contained in one molecule of water? In one molecule of hydrogen peroxide?
14. Write the names and formulas of the elements that exist as diatomic molecules.
15. How many atoms of oxygen are represented in each expression?
    (a) 4 H₂O   (b) 3 CuSO₄   (c) H₂O₂   (d) 3 Fe(OH)₃   (e) Al(ClO₃)₃
16. Are all mixtures heterogeneous? Explain.
17. Classify each of the following materials as an element, compound, or mixture:
    Air   Coal   Oil   Magnesium
    Brass   Water   Silver   Wine
    Cement   Milk   Sugar   Sodium chloride
18. A white solid, on heating, formed a colorless gas and a yellow solid. Assuming there was no reaction with the air, is the original solid an element or a compound? Explain.
19. Tabulate the properties that characterize metals and nonmetals.
20. Which of the following are diatomic molecules?
    (a) H₂   (c) HCl   (e) NO   (g) MgCl₂
    (b) SO₂   (d) H₂O   (f) NO₂
21. Name the following binary compounds. Refer to Chapter 8 if necessary.
    (a) CaCl₂   (c) AgCl   (e) RaBr₂   (g) H₂S   (i) ZnCl₂   (k) CaC₂
    (b) HBr   (d) Al₂S₃   (f) CdO   (h) BN   (j) BaF₂   (l) KI
22. Which of the compounds listed in Questions 8 and 9 are binary compounds? What is common to the names of these binary compounds?
23. An atom of silver is represented by the symbol Ag; a hydrogen molecule by the formula H₂; a water molecule by H₂O. Write expressions to represent:
    (a) Five silver atoms
    (b) Four hydrogen molecules
    (c) Three water molecules
24. Balance these equations (all formulas are correct as written):
    (a) H₂ + Cl₂ → HCl
    (b) Zn + CuSO₄ → Cu + ZnSO₄
    (c) HCl + NaOH → NaCl + H₂O
    (d) Ca + O₂ → CaO
    (e) Fe + HCl → FeCl₂ + H₂
    (f) P + I₂ → PI₃
    (g) MgO + HCl → MgCl₂ + H₂O
    (h) HNO₃ + Ba(OH)₂ → Ba(NO₃)₂ + H₂O
    (i) BiCl₃ + H₂S → Bi₂S₃ + HCl
    (j) Mg₃N₂ + H₂O → Mg(OH)₂ + NH₃
25. Balance the following equations, each of which represents a method of preparing oxygen gas:
    (a) H₂O₂ → H₂O + O₂
    (b) KClO₃ $\xrightarrow{\Delta}$ KCl + O₂
    (c) KNO₃ $\xrightarrow{\Delta}$ KNO₂ + O₂

(d) $Na_2O_2 + H_2O \rightarrow NaOH + O_2$

(e) $H_2O \xrightarrow[H_2SO_4]{\text{Electrical energy}} H_2 + O_2$

26. Balance the following equations, each of which represents a method of preparing hydrogen gas:
    (a) $Zn + HCl \rightarrow ZnCl_2 + H_2$
    (b) $Al + H_2SO_4 \rightarrow Al_2(SO_4)_3 + H_2$
    (c) $Na + H_2O \rightarrow NaOH + H_2$
    (d) $C + H_2O \text{ (steam)} \rightarrow CO + H_2$
    (e) $Fe + H_2O \text{ (steam)} \rightarrow Fe_3O_4 + H_2$

D. Review problems.
   1. Common table salt, NaCl, contains 39.3% sodium and 60.7% chlorine. What weight of sodium is present in 25.0 g of salt?
   2. Yellow brass is a homogeneous mixture of 67% copper and 33% zinc. If 50 g of copper is added to 100 g of yellow brass, what will be the new composition?
   3. Calcium oxide, CaO, contains 71.5% calcium. What size sample of CaO would contain 12.0 g of calcium?
   4. What would be the density of a solution made by mixing 2.50 ml of carbon tetrachloride (CCl_4, $d = 1.595$ g/ml) and 3.50 ml of carbon tetrabromide (CBr_4, $d = 3.420$ g/ml)? Assume that the volume of the mixed liquids is the sum of the two volumes used.
   5. When 4.00 g of calcium and 4.00 g of sulfur were mixed and reacted to give the compound calcium sulfide (CaS), 0.80 g of sulfur remained unreacted.
      (a) What percentage of the compound is sulfur?
      (b) An atom of which element, Ca or S, has the greater mass? Explain.
      (c) How many grams of sulfur will combine with 20.0 g of calcium?
   6. Pure gold is too soft a metal for many uses, so it is alloyed to give it more mechanical strength. One particular alloy is made by mixing 60 g of gold, 8.0 g of silver, and 12 g of copper. What carat gold is this alloy if pure gold is considered to be 24 carat?
   7. Methane, the chief component of natural gas, has the formula CH_4. Each atom of carbon weighs 12 times as much as an atom of hydrogen. Calculate the weight percent of hydrogen in methane.

5 Atomic Theory and Structure

After studying Chapter 5 you should be able to:

1. Understand the terms listed in Question A at the end of the chapter.
2. State the major provisions of Dalton's atomic theory.
3. Give the names, symbols, charges, and relative masses of the three principal subatomic particles.
4. Describe the atom as conceived by Ernest Rutherford after his alpha scattering experiments.
5. Describe the atom as conceived by Niels Bohr.
6. Discuss the contributions to atomic theory made by Dalton, Thomson, Rutherford, Bohr, Moseley, Chadwick, and Schrödinger.
7. State the Pauli exclusion principle.
8. Determine the maximum number of electrons that can exist in a given main energy level.
9. Determine the atomic number, atomic mass, or number of neutrons of any isotope when given any two of these three terms.
10. Draw the diagram of any isotope of the first 38 elements, showing the composition of the nucleus and the numbers of electrons in the main energy levels.
11. Give the electron structure ($1s^2 2s^2 2p^6$, etc.) for any of the first 38 elements.
12. Explain what is represented by the electron-dot (Lewis-dot) structure of an element.
13. Draw the electron-dot (Lewis-dot) diagrams for any of the first 20 elements.
14. Name and distinguish among the three isotopes of hydrogen.
15. Convert grams or atoms of an element to moles of that element, and vice versa.
16. Understand the relationships among a mole, Avogadro's number, and a gram-atomic weight of an element.

5.1 Early Thoughts

The structure of matter has long intrigued and engaged the minds of people. The seed of modern atomic theory was sown during the time of the ancient Greek philosophers. About 440 B.C. Empedocles stated that all matter was composed of four "elements"—earth, air, water, and fire. Democritus (about 470–370 B.C.), one of the early atomistic philosophers, thought that all forms of matter were finitely divisible into invisible particles, which he called atoms. He held that atoms were in constant motion and that they combined with one another in various ways. This purely speculative hypothesis was not based on scientific observations. Shortly thereafter, Aristotle (384–322 B.C.) opposed the theory of Democritus and endorsed and advanced the Empedoclean theory. So strong was the influence of Aristotle that his theory dominated the thinking of scientists and philosophers until the beginning of the 17th century. The term *atom* is derived from the Greek word *atomos*, meaning "indivisible."

5.2 Dalton's Atomic Theory

More than 2000 years after Democritus, the English schoolmaster John Dalton (1766–1844) revived the concept of atoms and proposed an atomic theory based on facts and experimental evidence. This theory, described in a series of papers published during the period 1803–1810, rested on the idea of a different kind of atom for each element. The essence of **Dalton's atomic theory** may be summed up as follows:

Dalton's atomic theory

1. Elements are composed of minute, indivisible particles called atoms.
2. Atoms of the same element are alike in mass and size.
3. Atoms of different elements have different masses and sizes.
4. Chemical compounds are formed by the union of two or more atoms of different elements.
5. When atoms combine to form compounds, they do so in simple numerical ratios, such as one to one, two to one, two to three, and so on.
6. Atoms of two elements may combine in different ratios to form more than one compound.

Dalton's atomic theory stands as a landmark in the development of chemistry. The major premises of his theory are still valid today. However, some of the statements must be modified or qualified because investigations since Dalton's time have shown that (1) atoms are composed of subatomic particles; (2) all the atoms of a specific element do not have the same mass; and (3) atoms, under special circumstances, can be decomposed.

5.3 Subatomic Parts of the Atom

The concept of the atom—a particle so small that it cannot be seen even with the most powerful microscope—and the subsequent determination of its structure stand among the very greatest creative intellectual human achievements.

When we refer to any visible quantity of an element, we are considering a vast number of identical atoms of that element. But when we refer to an atom of an element, we isolate a single atom from the multitude in order to present that element in its simplest form. Figure 5.1 illustrates the hypothetical isolation of a single copper atom from its crystal lattice.

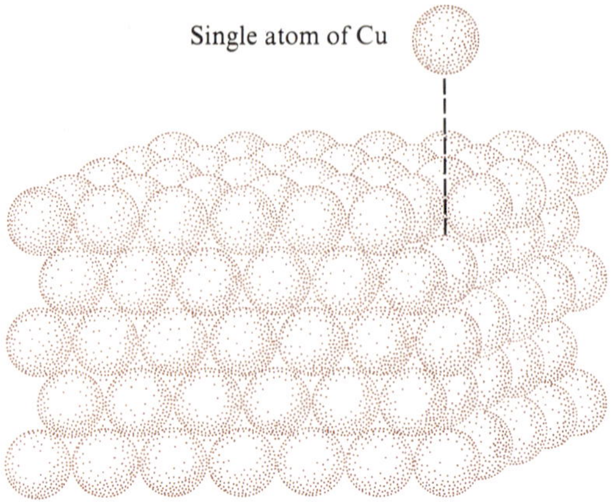

Single atom of Cu

Crystalline structure of copper atoms

Figure 5.1. A single atom of copper compared with copper as it occurs in its regular crystalline lattice structure. Billions of atoms are present in even the smallest strand of copper wire.

Let us examine this tiny particle we call the atom. The diameter of a single atom ranges from 1 to 5 angstroms (1 Å = 1×10^{-8} cm). Hydrogen, the smallest atom, has a diameter of about 1 Å. To arrive at some idea of how small an atom is, consider this dot (•), which has a diameter of about 1 mm, or 1×10^7 Å. It would take 10 million hydrogen atoms to form a line stretching across this dot. To carry this size illustration a bit further, 10 million of the 1 millimetre dots laid edge to edge would extend for 10,000 metres, or more than 6 miles! As inconceivably small as atoms are, they contain even more minute particles, the **subatomic particles**, such as electrons, protons, and neutrons.

The experimental discovery of the electron (e^-) was made in 1897 by J. J. Thomson (1856–1940). The **electron** is a particle with a negative electrical charge and has a mass of 9.107×10^{-28} gram. This mass is 1/1837 the mass of a hydrogen atom and corresponds to 0.0005486 atomic mass unit (amu). One atomic mass unit has a mass of 1.660×10^{-24} gram. Although the actual electrical charge of an electron is known, its value is too cumbersome for practical use. The electron, therefore, has been assigned a relative electrical charge of -1. The size of an electron has not been determined exactly, but its diameter is believed to be less than 10^{-12} cm.

Protons were first observed by E. Goldstein (1850–1930) in 1886. However, it was J. J. Thomson who discovered the nature of the proton. He showed that the proton is a particle and he calculated its mass to be about 1837 times that of an electron. The **proton** (p) is a particle with a relative mass of 1 amu and an actual mass of 1.672×10^{-24} g. Its charge ($+1$) is equal in magnitude but of opposite sign to the charge on the electron. The mass of a proton is only very slightly less than that of a hydrogen atom.

The third major subatomic particle was discovered in 1932 by James Chadwick (1891–1974). This particle, the **neutron** (n) bears neither a positive nor

Table 5.1. Electrical charge and relative mass of electrons, protons, and neutrons.

Particle	Symbol	Electrical charge	Relative mass (amu)	Actual mass (g)
Electron	e^-	-1	$\frac{1}{1837}$	9.107×10^{-28}
Proton	p	$+1$	1	1.672×10^{-24}
Neutron	n	0	1	1.675×10^{-24}

a negative charge and has a relative mass of about 1 amu. Its actual mass (1.675×10^{-24} g) is only very slightly greater than that of a proton. The properties of these three subatomic particles are summarized in Table 5.1.

Nearly all the ordinary chemical properties of matter can be explained in terms of atoms consisting of electrons, protons, and neutrons. The discussion of atomic structure that follows is based on the assumption that atoms contain only these principal subatomic particles. Many other subatomic particles such as mesons, positrons, neutrinos, and antiprotons have been discovered. At this time it is not clear whether all these particles are originally present in the atom or whether they are produced by reactions occurring within the nucleus. The field of atomic physics is fascinating and has attracted many young scientists in recent years. This interest has resulted in a great deal of research that is producing a long list of additional subatomic particles. Descriptions of the properties of many of these particles are to be found in recent textbooks on atomic physics and in various articles appearing in *Scientific American* over the past several years.

5.4 The Nuclear Atom

The discovery that positively charged particles were present in atoms came soon after the discovery of radioactivity by Henri Becquerel in 1896.

Ernest Rutherford (Figure 5.2) had, by 1907, established that the positively charged alpha particles emitted by certain radioactive elements were ions of the element helium. Rutherford used these alpha particles to establish the nuclear nature of atoms. In some experiments performed in 1911, he directed a stream of positively charged helium ions (alpha particles) at a very thin sheet of gold foil (about 1000 atoms thick). He observed that most of the alpha particles passed through the foil with little or no deflection; but a few of the particles were deflected at large angles, and occasionally one even bounced back from the foil (see Figure 5.3). It was known that like charges repel each other and that an electron with a mass of 1/1837 amu could not have an appreciable effect upon the path of a far more massive (4 amu) alpha particle. Rutherford therefore reasoned that each gold atom must contain a positively charged mass occupying a relatively tiny volume and that when an alpha particle approached close enough to this positive mass, it was deflected. Rutherford spoke of this positively

Figure 5.2. Ernest Rutherford (1871–1937), British physicist, who identified two of the three principal rays emanating from radioactive substances. His experiments with alpha particles led to the first laboratory transmutation of an element and to his formulation of the nuclear atom. Rutherford was awarded the Nobel prize in 1908 for his work on transmutation. (Courtesy Rutherford Museum, McGill University.)

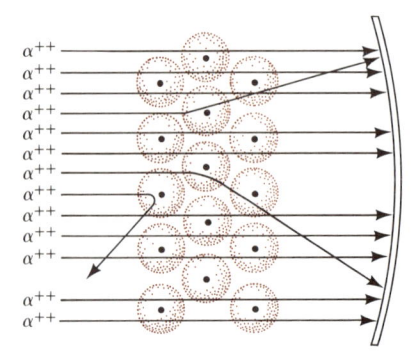

Figure 5.3. Diagram representing Rutherford's experiment on alpha particle scattering. Positive alpha particles, emanating from a radioactive source, were directed at a thin metal foil. Diagram illustrates the repulsion of the positive alpha particles by the positive nucleus of the metal atom.

charged mass as the *nucleus* of the atom. Since alpha particles are relatively high in mass, the extent of the deflections—remember some actually bounced back—indicated to Rutherford that the nucleus is relatively very heavy and dense. Since most of the alpha particles passed through the thousand or so gold atoms without any apparent deflection, he further concluded that the bulk of an atom consists of empty space.

When we speak of the mass of an atom, we are, for practical purposes, referring primarily to the mass of the nucleus. This is because the nucleus contains all the protons and neutrons, and these represent more than 99.9% of the total mass of any atom (see Table 5.1). By way of illustration, the largest number of electrons known to exist in an atom is 106. The mass of even 106 electrons represents only about 1/17 of the mass of a single proton or neutron. The mass of an atom, therefore, is primarily determined by the mass of its protons and neutrons.

5.5 General Arrangement of Subatomic Particles

The alpha particle scattering experiments of Rutherford established that the atom contains a dense, positively charged nucleus. The later work of Chadwick demonstrated that the atom contains neutrons, which are particles with mass but no charge. Light negatively charged electrons are also present and offset the positive charges in the nucleus. Based on this experimental evidence, a general description of the atom and location of its subatomic particles was devised. Each atom consists of a nucleus surrounded by electrons. The nucleus contains protons and neutrons, but electrons are not found in the nucleus. In a neutral atom, the positive charge of the nucleus (due to protons) is exactly offset by the negative electrons. Since the charge on an electron is equal to but of opposite sign to the charge on a proton, a neutral atom must contain exactly the same number of electrons as protons. However, this generalized picture of atomic structure provides no information on the arrangement of electrons within the atom.

nucleus

5.6 The Bohr Atom

At high temperatures or when subjected to high voltages, elements in the gaseous state give off colored light. Neon signs illustrate this property of matter very well. When passed through the prism or grating of a spectroscope, the light emitted by a gas appears as a set of bright colored lines (band spectra). These colored lines indicate that the light is being emitted only at certain wavelengths or frequencies that correspond to specific colors. Each element possesses a unique set of these spectral lines that is different from the sets of all the other elements.

In 1912–1913, while studying the line spectra of hydrogen, Niels Bohr (1885–1962), a Danish physicist, made a significant contribution to the rapidly growing knowledge of atomic structure. His research led him to believe that electrons in an atom exist in specific regions at various distances from the nucleus. He also visualized the electrons as rotating in orbits around the nucleus like planets rotating around the sun.

Bohr's first paper in this field dealt with the hydrogen atom, which he described as a single electron rotating in an orbit about a relatively heavy nucleus (see Figure 5.5). He applied the concept of energy quanta, proposed in 1900 by the German physicist Max Planck (1858–1947), to the observed line spectra of hydrogen. Planck stated that energy is never emitted in a continuous stream, but only in small discrete packets called quanta (Latin, *quantus*, how much). Bohr theorized that there are several possible orbits for electrons at different distances from the nucleus. But an electron had to be in one specific orbit or another; it could not exist between orbits. Bohr also stated that when a hydrogen atom absorbed one or more quanta of energy, its electron would "jump" to another orbit at a greater distance from the nucleus.

Bohr was able to account for spectral lines this way. Each orbit corresponds to a different energy level, the one closest to the nucleus representing the lowest or ground-state energy level. Orbits at increasing distances from the nucleus represent the second, third, fourth energy levels. When an electron "falls" from a high-energy orbit to one of lower energy, a quantum of energy in the form of light is emitted, thus giving rise to a spectral line at a specific frequency. A number of orbits exist, and when electrons "fall" different distances, correspondingly different quanta of energy are emitted, producing the several lines visible in the hydrogen spectrum.

Bohr contributed greatly to the advancement of our knowledge of atomic structure by (1) suggesting quantized energy levels for electrons and (2) showing that spectral lines result from the radiation of small increments of energy (Planck's quanta) when electrons shift from one energy level to another.

Much of Bohr's work related to the simplest atom, hydrogen. Difficulties arose when his energy calculations were applied to atoms containing many electrons. Bohr's concept of the atom has been replaced by the quantum mechanics theory. One of the chief differences between these two theories is that in the quantum mechanics theory electrons are not considered to be revolving around the nucleus in orbits, but to occupy *orbitals*—somewhat cloudlike regions surrounding the nucleus and corresponding to energy levels. The concept of electrons being in specific energy levels is still retained in the modern theory.

5.7 The Quantum Mechanical Atom

The discussion in this section and Sections 5.8–5.17 describes the ordered system by which the electrons are distributed within the atoms of all the elements. The classifications given in the periodic table (Chapter 6), as well as the chemical properties of the elements, are dependent on their electronic arrangements.

The most important feature of the Bohr atom is the concept of definite energy levels for electrons in atoms. In 1924, the French physicist Louis de Broglie (1892–   ) suggested that moving electrons had the properties of

Atomic Theory and Structure   71

waves as well as mass. In 1926, the Austrian physicist Erwin Schrödinger (1887–1961) introduced a new method of calculation—quantum mechanics, or wave mechanics. Schrödinger's equation, which is a complex mathematical expression, describes an electron as simultaneously having properties of a wave and a particle. Thus, the electron was given dual characteristics—some of its properties are best described in terms of waves (like light), and other properties are described in terms of a particle having mass.

The solution of the Schrödinger equation is complex and we will not concern ourselves with it here. However, the solution leads to the introduction of three quantum numbers that define the probabilities of location and spatial properties of electrons in atoms. An extension of this theory shows that a fourth quantum number (the spin quantum number) is also necessary to define an electron completely.

quantum numbers

The four **quantum numbers—n, l, m, and s**—specify the energy and probable location of each electron in an atom:

1. Electrons exist in energy levels at different distances from the nucleus. The principal quantum number **n** indicates the energy levels of the electrons relative to their distance from the nucleus. The number **n** may have any positive integral value up to infinity ($n = 1, 2, 3, 4, \ldots, \infty$), but only values of 1–7 have be established for atoms of known elements in their ground state (lowest energy state). Energy level $n = 1$ is closest to the nucleus and is the lowest principal energy level. (See Figure 5.7.)

2. Electrons exist in orbitals that have specific shapes. The principal energy levels (except the first) contain closely grouped sublevels. These sublevels consist of orbitals of specific shape. In quantum mechanics the term **orbital** (not orbit) is used and refers to the region around the nucleus in which we may expect to find a particular electron. The orbital quantum number **l** (ell) specifies the shape of the electron cloud about the nucleus. The four common sublevels (orbitals) normally encountered are designated by the lowercase italic letters s, p, d, and f.

orbital

3. Electron orbitals have specific orientation in space. The magnetic quantum number **m** designates this orientation. This quantum number accounts for the number of s, p, d, and f orbitals that can be present in the principal energy levels. There can be at most one s orbital, three p orbitals, five d orbitals, and seven f orbitals in any given principal energy level.

4. An electron spins about its own axis in either a clockwise or counterclockwise direction. The spin quantum number **s** relates to the direction of spin of an electron. Because there are only two possible directions of spin, each orbital, no matter what its designation, can contain a maximum of two electrons. When two electrons occupy the same orbital, they must have opposite spins. When an orbital contains two electrons, the electrons are said to be *paired*.

Thus, the quantum numbers tell us relatively how far the electrons are located from the nucleus, the shapes of the electron orbitals, the orientation of the orbitals in space, and whether the electrons are paired within a given orbital.

The basic rules and limitations regarding the state of electrons in atoms are the following:

1. In the ground state (lowest energy state) of an atom, electrons tend to occupy orbitals of the lowest possible energy. Thus, an electron will occupy an s orbital in the $n = 1$ level before it occupies an s orbital in the $n = 2$ level.

72 Chapter 5

2. Each orbital may contain a maximum of two electrons (with opposite spins).
3. No two electrons in an atom can have the same four quantum numbers. This is a statement of the *Pauli exclusion principle*.

The use of quantum numbers in atomic structures is illustrated later in the chapter.

5.8 Energy Levels of Electrons

energy levels of electrons

electron shells

All the electrons in an atom are not located the same distance from the nucleus. As pointed out by both the Bohr theory and quantum mechanics, the probability of finding the electrons is greatest at certain specified distances—called **energy levels**—from the nucleus. Energy levels are also referred to as **electron shells** and may contain only a limited number of electrons. Energy levels are numbered, starting with $n = 1$ as the shell nearest the nucleus and going to $n = 7$. They are also identified by the letters K, L, M, N, O, P, Q, with K equivalent to the first energy level, L to the second level, and so on, as follows:

Energy level	n	Letter designation
First	1	K
Second	2	L
Third	3	M
Fourth	4	N
Fifth	5	O
Sixth	6	P
Seventh	7	Q

Each succeeding energy level is located farther away from the nucleus.

The maximum number of electrons that can occupy a specific energy level can be calculated from the formula $2n^2$, where n is the number of the principal energy level. For example, shell K, or energy level 1, can have a maximum of two electrons ($2 \times 1^2 = 2$); shell L, or energy level 2, can have a maximum

Table 5.2. Maximum number of electrons that can occupy each principal energy level.

Energy level, n	Letter designation	Maximum number of electrons in each energy level, $2n^2$
1	K	$2 \times 1^2 = 2$
2	L	$2 \times 2^2 = 8$
3	M	$2 \times 3^2 = 18$
4	N	$2 \times 4^2 = 32$
5	O	$2 \times 5^2 = 50$[a]

[a] The theoretical value of 50 electrons in energy level 5 has never been attained in any element known to date.

of eight electrons ($2 \times 2^2 = 8$), and so on. Table 5.2 shows the maximum number of electrons that can exist in each of the first five energy levels.

5.9 Energy Sublevels

The principal energy levels contain sublevels designated by the letters s, p, d, f. The s sublevel consists of one orbital; the p sublevel consists of three orbitals; the d sublevel consists of five orbitals; and the f sublevel consists of seven orbitals. Since no more than two electrons can exist in an orbital, the maximum numbers of electrons that can exist in the sublevels are 2 in the s orbital, 6 in the three p orbitals, 10 in the five d orbitals, and 14 in the seven f orbitals.

Type of sublevel	Number of orbitals possible	Number of electrons possible
s	1	2
p	3	6
d	5	10
f	7	14

The order of energy of the sublevels within a specific principal energy level is the following: s electrons are lower in energy than p electrons, which are lower than d electrons, which are lower than f electrons. This may be expressed in the following manner:

Sublevel energy: $s < p < d < f$

The order of energy of the principal energy levels, **n**, is

Principal energy levels: $1 < 2 < 3 < 4 < 5 < 6 < 7$

We can determine what type of sublevels occur in any energy level from the maximum number of electrons that can exist in that energy level. To make this determination, we need to know the maximum number of electrons possible in an energy level and to use two of the rules set forth in Section 5.7: (1) no more than two electrons can occupy one orbital, and (2) an electron will occupy the lowest possible sublevel. The maximum number of electrons in the first energy level is two; both of these are s electrons. They are designated as $1s^2$, indicating two s electrons in the first energy level. The s orbital in the second energy level is written as $2s$, in the third energy level as $3s$, and so on. The second energy level, with a maximum of eight electrons, will contain only s and p electrons—namely, a maximum of two s and six p electrons. They are designated as $2s^2 2p^6$.

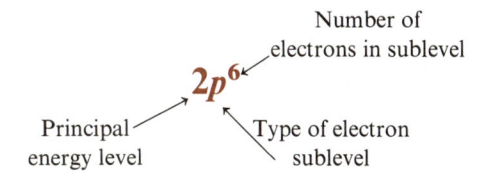

$$2p^6$$

Principal energy level — Type of electron sublevel — Number of electrons in sublevel

If each orbital contains two electrons, the second energy level can have four orbitals: one s orbital and three individual p orbitals. These three p orbitals are energetically equivalent to each other and are labeled $2p_x$, $2p_y$, and $2p_z$ to indicate their orientation in space (see Figure 5.4). The symbols $3s^2$, $3p^6$, and $3d^{10}$ illustrate the sublevel breakdown of electrons in the third energy level. From this line of reasoning, we can see that if there are sufficient electrons, f electrons first appear in the fourth energy level. Table 5.3 shows the type of sublevel electrons and maximum number of orbitals and electrons in each energy level.

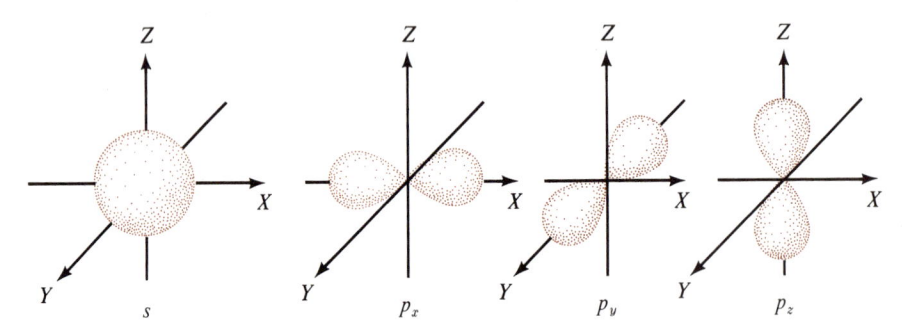

Figure 5.4. Perspective representation of the s, p_x, p_y, and p_z atomic orbitals.

Table 5.3. Sublevel electrons in each principal energy level and the maximum number of orbitals and electrons in each energy level.

Energy level	Sublevel electrons	Maximum number of orbitals	Total number of electrons
1 (K)	s	1	2
2 (L)	s, p	4	8
3 (M)	s, p, d	9	18
4 (N)	s, p, d, f	16	32
5[a] (O)	s, p, d, f	Incomplete	—
6[a] (P)	s, p, d	Incomplete	—
7[a] (Q)	s	Incomplete	—

[a] Insufficient electrons to complete the shell.

Since the *spdf* atomic orbitals have definite distribution in space, they are represented by particular spatial shapes. At this time we will consider only the s and p orbitals. The s orbitals are spherically symmetrical about the nucleus, as illustrated in Figure 5.4. A $2s$ orbital is a larger sphere than a $1s$ orbital. The p orbitals (p_x, p_y, p_z) are dumbbell-shaped and are oriented at right angles to each other along the x, y, and z axes in space (see Figure 5.4). An electron has equal probability of being located in either lobe of the p orbital. In illustrations such as Figure 5.4, the boundaries of the orbitals enclose the region of the greatest probability (about 90% chance) of finding an electron.

In the ground state of a hydrogen atom, this falls within a sphere having a radius of 0.53 Å.

5.10 Atomic Numbers of the Elements

atomic number

The **atomic number** of an element is the number of protons in the nucleus of an atom of that element. The atomic number is a fundamental characteristic and identifies an atom as being a particular element. The presently known elements are numbered consecutively from 1 to 106 to coincide with the number of protons in their nuclei. Thus, hydrogen, element number 1, has one proton; calcium, element number 20, has 20 protons; and uranium, element number 92, has 92 protons in the nucleus. The atomic number tells us not only the amount of positive charge in the nucleus, but also the number of electrons in the neutral atom.

5.11 The Simplest Atom—Hydrogen

The hydrogen atom, consisting of a nucleus containing one proton and an electron orbital containing one electron, is the simplest known atom. (Some hydrogen atoms are known to contain one or two neutrons in their nucleus—see Section 5.12 on isotopes.) The electron in hydrogen occupies an s orbital in the first energy level. This electron does not move in any definite path but rather in a random motion within its orbital, forming an electron cloud about the nucleus. The diameter of the nucleus is believed to be about 10^{-13} cm, and the diameter of the electron orbital to be about 10^{-8} cm. Hence, the diameter of the electron orbital of a hydrogen atom is about 100,000 times greater than the diameter of the nucleus.

What we have, then, is a positive nucleus surrounded by an electron cloud formed by an electron in an s orbital. The net electrical charge on the hydrogen atom is zero; it is called a *neutral atom*. Figure 5.5 shows two methods of representing a hydrogen atom.

(a)   (b)

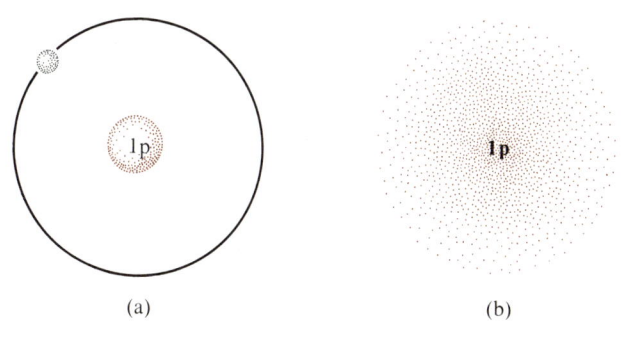

Figure 5.5. The hydrogen atom. (a) Illustration of the Bohr description, indicating a discrete electron moving around its nucleus of one proton. (b) Illustration of the quantum mechanical concept, showing the electron orbital as a cloud surrounding the nucleus.

5.12 Isotopes of the Elements

Shortly after Rutherford's conception of the nuclear atom, experiments were performed to determine the masses of individual atoms. These experiments showed that the masses of nearly all atoms were greater than could be accounted for by simply adding up the masses of all the protons and electrons that were known to be present. This fact led to the concept of the neutron, a particle with no charge but with a mass about the same as that of a proton. Since this particle has no charge, it was very difficult to detect, and the existence of the neutron was not proven experimentally until 1932. All atomic nuclei except that of the simplest hydrogen atom are now believed to contain neutrons.

All atoms of a given element have the same number of protons, but experimental evidence has shown that, in most cases, all atoms of a given element do not have identical masses. This is because atoms of the same element may have different numbers of neutrons in their nuclei.

isotopes

Atoms of an element having the same atomic number but different atomic masses are called **isotopes** of that element. Atoms of the isotopes of an element, therefore, have the same number of protons and electrons but different numbers of neutrons.

Three isotopes of hydrogen (atomic number 1) are known. Each has one proton in the nucleus and one electron in the first energy level. The first isotope (protium) without a neutron has a mass of 1; the second isotope (deuterium) with one neutron in the nucleus has a mass of 2; the third isotope (tritium) with two neutrons has a mass of 3 (see Figure 5.6).

$^{1}_{1}H$ $^{2}_{1}H$, D $^{3}_{1}H$, T
Protium   Deuterium   Tritium

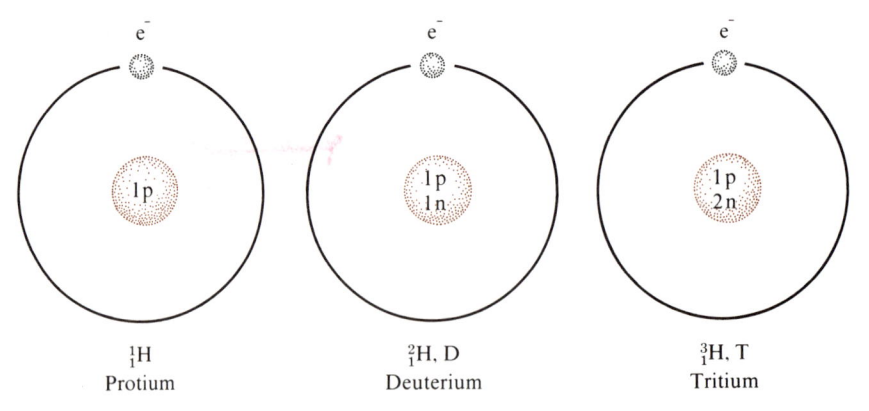

Figure 5.6. Diagram of the isotopes of hydrogen. The number of protons (p) and neutrons (n) are shown within the nucleus.

The three isotopes of hydrogen may be represented by the symbols $^{1}_{1}H$, $^{2}_{1}H$, $^{3}_{1}H$, indicating an atomic number of 1 and mass numbers of 1, 2, and 3, respectively. This method of representing atoms is called *isotopic notation*. The subscript number is the atomic number; the superscript number is the **mass number** (total number of protons and neutrons). The hydrogen isotopes may also be referred to as hydrogen-1, hydrogen-2, and hydrogen-3

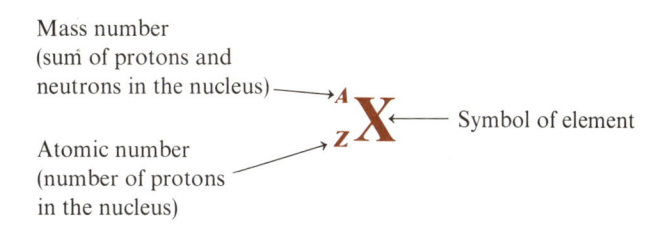

Mass number
(sum of protons and
neutrons in the nucleus) $\longrightarrow {}_{Z}^{A}X \longleftarrow$ Symbol of element

Atomic number
(number of protons
in the nucleus)

Two or more isotopes are known for all elements. However, not all isotopes are stable; some are radioactive and are continually decomposing to form other elements. For example, of the seven known isotopes of carbon, only two, carbon-12 and carbon-13, are stable. Of the seven known isotopes of oxygen, only three, $^{16}_{8}O$, $^{17}_{8}O$, and $^{18}_{8}O$, are stable. Of the fifteen known isotopes of arsenic, $^{75}_{33}As$ is the only one that is stable.

5.13 Atomic Structure of the First Twenty Elements

Starting with hydrogen, and progressing in order of increasing atomic number to helium, lithium, beryllium, and so on, the atoms of each successive element contain one more proton and one more electron than do the atoms of the preceding element. This sequence continues, without exception, throughout the entire list of known elements and is one of the most impressive examples of order in nature.

The number of neutrons also increases as we progress through the elements. But this number, unlike the number of protons and electrons, does not increase in a perfectly uniform manner as we go from elements of low atomic numbers to those of higher atomic numbers. Furthermore, atoms of the same element may contain different numbers of neutrons (see Section 5.12). For example, the predominant isotope of hydrogen contains no neutrons, but two other hydrogen isotopes, containing one and two neutrons, respectively, are known. The predominant isotope of helium (element number 2) has two neutrons, but helium isotopes containing one and four neutrons are known.

The ground-state electronic structures of the first 20 elements fall into a regular pattern. The one hydrogen electron is in the first energy level, and both helium electrons are in the first energy level. The electron structures for hydrogen and helium are written $1s^1$ and $1s^2$, respectively. The maximum number of electrons in the first energy level is two ($2 \times 1^2 = 2$). In lithium (atomic number 3), the third electron is in the $2s$ sublevel of the second energy level. Lithium has the electron structure $1s^2 2s^1$.

In succession, the atoms of beryllium (4), boron (5), carbon (6), nitrogen (7), oxygen (8), fluorine (9), and neon (10) have one more proton and one more electron than the preceding element. Both the first and second energy levels are filled to capacity by the ten electrons of neon, which has two electrons in the first and eight electrons in the second energy level.

H	$1s^1$
He	$1s^2$
Li	$1s^2 2s^1$
Be	$1s^2 2s^2$
B	$1s^2 2s^2 2p^1$
C	$1s^2 2s^2 2p^2$
N	$1s^2 2s^2 2p^3$
O	$1s^2 2s^2 2p^4$
F	$1s^2 2s^2 2p^5$
Ne	$1s^2 2s^2 2p^6$

Element 11, sodium (Na), has two electrons in the first energy level and eight electrons in the second energy level, with the remaining electron occu-

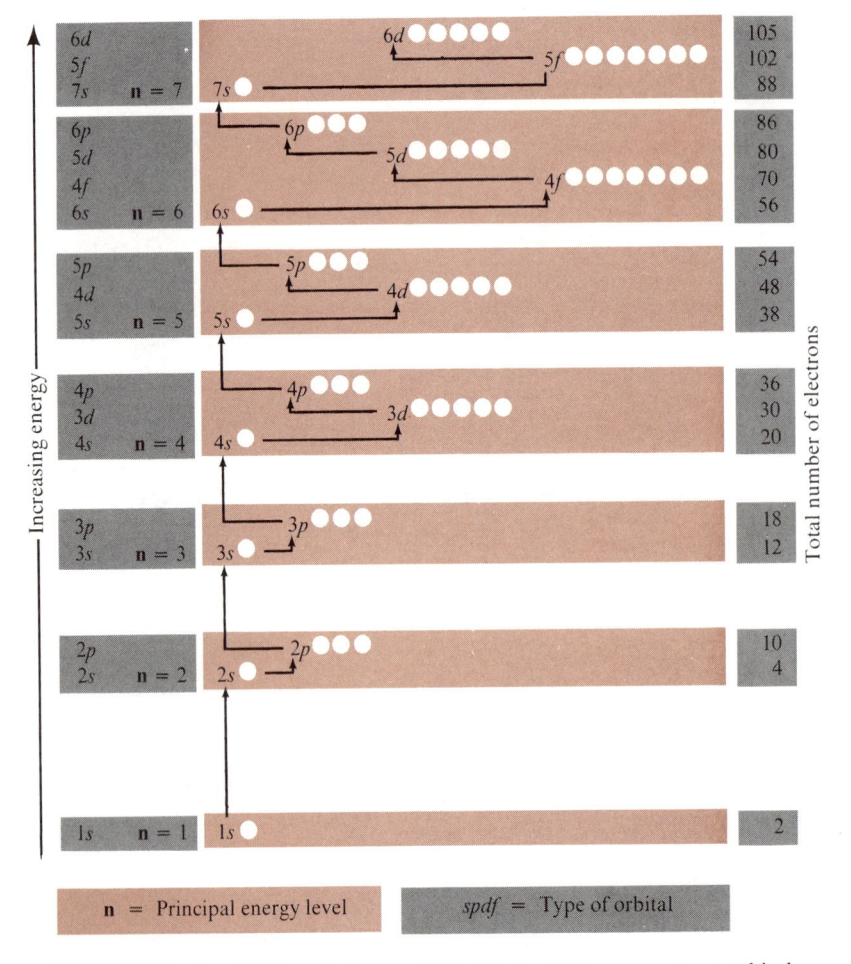

Figure 5.7. Order of filling electron orbitals. Each circle represents an orbital, which can contain two electrons. (Some exceptions to this order are known.)

pying the 3s orbital in the third energy level. The electron structure of sodium is $1s^22s^22p^63s^1$. Magnesium (12), aluminum (13), silicon (14), phosphorus (15), sulfur (16), chlorine (17), and argon (18) follow in order, each adding one electron to the third energy level up to argon, which has eight electrons in the M shell.

The placement of the last electron in potassium and calcium, elements numbers 19 and 20, departs somewhat from the expected order. One might expect that if the third energy level can contain a maximum of 18 electrons (see Table 5.2), electrons would continue to fill this shell until the maximum capacity was reached. However, this is not the case. The 4s sublevel is at a lower energy state than the 3d sublevel (see Figure 5.7). Hence, in elements 19 and 20, the last electron is found in the 4s level. The electron structure for potassium is $1s^22s^22p^63s^23p^64s^1$. Calcium has an electron structure similar to potassium, except that it has two 4s electrons. This break in sequence does not invalidate the formula $2n^2$, which merely prescribes the maximum number of electrons that each shell may contain, but does not state the order in which the shells are filled. Table 5.4 shows the electron arrangement of the first 20 elements.

Table 5.4. Electron arrangement of the first 20 elements.

Element	Symbol	Number of protons (atomic number)	Total number of electrons	Arrangement of electrons $n=$ 1 2 3 4
Hydrogen	H	1	1	1
Helium	He	2	2	2
Lithium	Li	3	3	2 1
Beryllium	Be	4	4	2 2
Boron	B	5	5	2 3
Carbon	C	6	6	2 4
Nitrogen	N	7	7	2 5
Oxygen	O	8	8	2 6
Fluorine	F	9	9	2 7
Neon	Ne	10	10	2 8
Sodium	Na	11	11	2 8 1
Magnesium	Mg	12	12	2 8 2
Aluminum	Al	13	13	2 8 3
Silicon	Si	14	14	2 8 4
Phosphorus	P	15	15	2 8 5
Sulfur	S	16	16	2 8 6
Chlorine	Cl	17	17	2 8 7
Argon	Ar	18	18	2 8 8
Potassium	K	19	19	2 8 8 1
Calcium	Ca	20	20	2 8 8 2

The relative energies of the electron orbitals are shown in Figure 5.7. The order given can be used to determine the electron distribution in the atoms of the elements, although some exceptions to the pattern are known. Suppose we wish to determine the electron structure of a chlorine atom, which has 17 electrons. Following the order in Figure 5.7, we begin by placing two electrons in the 1s orbital, then two electrons in the 2s orbital, and six electrons in the 2p orbitals. We now have used ten electrons. Finally, we place the next two electrons in the 3s orbital and the remaining five electrons in the 3p orbitals, which uses all 17 electrons, giving the electron structure for a chlorine atom as $1s^2 2s^2 2p^6 3s^2 3p^5$. The sum of the superscripts equals 17, the number of electrons in the atom. This procedure is summarized below.

Order of orbitals to be filled: $1s2s2p3s3p$
Distribution of the 17 electrons in a chlorine atom: $1s^2 2s^2 2p^6 3s^2 3p^5$

Problem 5.1

What is the electron distribution in a phosphorus atom?

First determine the number of electrons contained in a phosphorus atom. The atomic number of phosphorus is 15; therefore, each atom contains 15 protons and 15 electrons. Now tabulate the number of electrons in each principal and subenergy level until all 15 electrons are assigned.

Sublevel	Number of e^-	Total e^-
1s orbital	2 e^-	2
2s orbital	2 e^-	4
2p orbital	6 e^-	10
3s orbital	2 e^-	12
3p orbital	3 e^-	15

Therefore, the electron distribution in phosphorus is $1s^2 2s^2 2p^6 3s^2 3p^3$.

5.14 Electron Structure of the Elements Beyond Calcium

The elements following calcium have a less regular pattern of adding electrons. The lowest energy level available for the twenty-first electron is the 3d level. Thus, scandium (21) has the following electron arrangement: first energy level, two electrons; second energy level, eight electrons; third energy level, nine electrons; fourth energy level, two electrons. The last electron is located in the 3d level. The structure for scandium is $1s^2 2s^2 2p^6 3s^2 3p^6 3d^1 4s^2$. The elements following scandium, titanium (22) through copper (29), continue to add d electrons until the third energy level has its maximum of 18. Two exceptions in the orderly electron addition are chromium (24) and copper (29), the structures of which are given in Table 5.5. The third energy level of the electrons is first completed in the element copper. Table 5.5 shows the order of filling of the electron orbitals and the electron configuration of all the known elements.

Table 5.5. Electron structure of the elements. (For simplicity of expression, symbols of the chemically stable noble gases are used as a portion of the electron structure for the elements beyond neon. For example, the electron structure of a sodium (Na) atom consists of ten electrons, as in neon [Ne], plus a $3s^1$ electron.) Detailed electron structures for the noble gases are given in Table 5.6.

Element	Atomic number	Electron structure	Element	Atomic number	Electron structure
H	1	$1s^1$	Ru	44	$[Kr]\,4d^75s^1$
He	2	$1s^2$	Rh	45	$[Kr]\,4d^85s^1$
Li	3	$1s^22s^1$	Pd	46	$[Kr]\,4d^{10}$
Be	4	$1s^22s^2$	Ag	47	$[Kr]\,4d^{10}5s^1$
B	5	$1s^22s^22p^1$	Cd	48	$[Kr]\,4d^{10}5s^2$
C	6	$1s^22s^22p^2$	In	49	$[Kr]\,4d^{10}5s^25p^1$
N	7	$1s^22s^22p^3$	Sn	50	$[Kr]\,4d^{10}5s^25p^2$
O	8	$1s^22s^22p^4$	Sb	51	$[Kr]\,4d^{10}5s^25p^3$
F	9	$1s^22s^22p^5$	Te	52	$[Kr]\,4d^{10}5s^25p^4$
Ne	10	$1s^22s^22p^6$	I	53	$[Kr]\,4d^{10}5s^25p^5$
Na	11	$[Ne]\,3s^1$	Xe	54	$[Kr]\,4d^{10}5s^25p^6$
Mg	12	$[Ne]\,3s^2$	Cs	55	$[Xe]\,6s^1$
Al	13	$[Ne]\,3s^23p^1$	Ba	56	$[Xe]\,6s^2$
Si	14	$[Ne]\,3s^23p^2$	La	57	$[Xe]\,5d^16s^2$
P	15	$[Ne]\,3s^23p^3$	Ce	58	$[Xe]\,4f^15d^16s^2$
S	16	$[Ne]\,3s^23p^4$	Pr	59	$[Xe]\,4f^36s^2$
Cl	17	$[Ne]\,3s^23p^5$	Nd	60	$[Xe]\,4f^46s^2$
Ar	18	$[Ne]\,3s^23p^6$	Pm	61	$[Xe]\,4f^56s^2$
K	19	$[Ar]\,4s^1$	Sm	62	$[Xe]\,4f^66s^2$
Ca	20	$[Ar]\,4s^2$	Eu	63	$[Xe]\,4f^76s^2$
Sc	21	$[Ar]\,3d^14s^2$	Gd	64	$[Xe]\,4f^75d^16s^2$
Ti	22	$[Ar]\,3d^24s^2$	Tb	65	$[Xe]\,4f^96s^2$
V	23	$[Ar]\,3d^34s^2$	Dy	66	$[Xe]\,4f^{10}6s^2$
Cr	24	$[Ar]\,3d^54s^1$	Ho	67	$[Xe]\,4f^{11}6s^2$
Mn	25	$[Ar]\,3d^54s^2$	Er	68	$[Xe]\,4f^{12}6s^2$
Fe	26	$[Ar]\,3d^64s^2$	Tm	69	$[Xe]\,4f^{13}6s^2$
Co	27	$[Ar]\,3d^74s^2$	Yb	70	$[Xe]\,4f^{14}6s^2$
Ni	28	$[Ar]\,3d^84s^2$	Lu	71	$[Xe]\,4f^{14}5d^16s^2$
Cu	29	$[Ar]\,3d^{10}4s^1$	Hf	72	$[Xe]\,4f^{14}5d^26s^2$
Zn	30	$[Ar]\,3d^{10}4s^2$	Ta	73	$[Xe]\,4f^{14}5d^36s^2$
Ga	31	$[Ar]\,3d^{10}4s^24p^1$	W	74	$[Xe]\,4f^{14}5d^46s^2$
Ge	32	$[Ar]\,3d^{10}4s^24p^2$	Re	75	$[Xe]\,4f^{14}5d^56s^2$
As	33	$[Ar]\,3d^{10}4s^24p^3$	Os	76	$[Xe]\,4f^{14}5d^66s^2$
Se	34	$[Ar]\,3d^{10}4s^24p^4$	Ir	77	$[Xe]\,4f^{14}5d^76s^2$
Br	35	$[Ar]\,3d^{10}4s^24p^5$	Pt	78	$[Xe]\,4f^{14}5d^96s^1$
Kr	36	$[Ar]\,3d^{10}4s^24p^6$	Au	79	$[Xe]\,4f^{14}5d^{10}6s^1$
Rb	37	$[Kr]\,5s^1$	Hg	80	$[Xe]\,4f^{14}5d^{10}6s^2$
Sr	38	$[Kr]\,5s^2$	Tl	81	$[Xe]\,4f^{14}5d^{10}6s^26p^1$
Y	39	$[Kr]\,4d^15s^2$	Pb	82	$[Xe]\,4f^{14}5d^{10}6s^26p^2$
Zr	40	$[Kr]\,4d^25s^2$	Bi	83	$[Xe]\,4f^{14}5d^{10}6s^26p^3$
Nb	41	$[Kr]\,4d^45s^1$	Po	84	$[Xe]\,4f^{14}5d^{10}6s^26p^4$
Mo	42	$[Kr]\,4d^55s^1$	At	85	$[Xe]\,4f^{14}5d^{10}6s^26p^5$
Tc	43	$[Kr]\,4d^55s^2$	Rn	86	$[Xe]\,4f^{14}5d^{10}6s^26p^6$

Table 5.5 (continued)

Element	Atomic number	Electron structure	Element	Atomic number	Electron structure
Fr	87	[Rn] $7s^1$	Bk	97	[Rn] $5f^9 7s^2$
Ra	88	[Rn] $7s^2$	Cf	98	[Rn] $5f^{10} 7s^2$
Ac	89	[Rn] $6d^1 7s^2$	Es	99	[Rn] $5f^{11} 7s^2$
Th	90	[Rn] $6d^2 7s^2$	Fm	100	[Rn] $5f^{12} 7s^2$
Pa	91	[Rn] $5f^2 6d^1 7s^2$	Md	101	[Rn] $5f^{13} 7s^2$
U	92	[Rn] $5f^3 6d^1 7s^2$	No	102	[Rn] $5f^{14} 7s^2$
Np	93	[Rn] $5f^4 6d^1 7s^2$	Lr	103	[Rn] $5f^{14} 6d^1 7s^2$
Pu	94	[Rn] $5f^6 7s^2$	Ku	104	[Rn] $5f^{14} 6d^2 7s^2$
Am	95	[Rn] $5f^7 7s^2$	Ha	105	[Rn] $5f^{14} 6d^3 7s^2$
Cm	96	[Rn] $5f^7 6d^1 7s^2$	—	106	[Rn] $5f^{14} 6d^4 7s^2$

5.15 Diagramming Atomic Structures

Several methods can be used to diagram atomic structures of atoms, depending on what we are trying to illustrate. When we want to show both the nuclear makeup and the total electron structure of each energy level (without orbital detail), we can use a diagram such as shown in Figure 5.8.

A method of diagramming subenergy levels is shown in Figure 5.9. Each orbital is represented by a circle ◯. When the orbital contains one electron, an arrow (↑) is placed in the circle. A second arrow, pointing downward (↓), indicates the second electron in that orbital.

The diagram for hydrogen is ⓛ. Helium, with two electrons, is drawn as ⓝ; both electrons are 1s electrons. The diagram for lithium shows three electrons in two energy levels, $1s^2 2s^1$. All four electrons of beryllium are s electrons, $1s^2 2s^2$. Boron has the first p electron, which is located in the $2p_x$ orbital. Since it is energetically more difficult for the next p electron to pair up with the electron in the p_x orbital than to occupy a second p orbital, the second p electron in carbon is located in the $2p_y$ orbital. The third p electron in nitrogen is still unpaired and is found in the $2p_z$ orbital. The next three electrons pair with each of the 2p electrons up to the element neon. Also shown in Figure 5.9 are the equivalent linear expressions for these orbital electron structures.

The electrons in successive elements are found in sublevels of increasing energy. The general sequence of increasing energy of subshells is $1s 2s 2p 3s 3p 4s 3d 4p 5s 4d 5p 6s 4f 5d 6p 7s 5f 6d 7p$. Figure 5.10 is a useful mnemonic device for

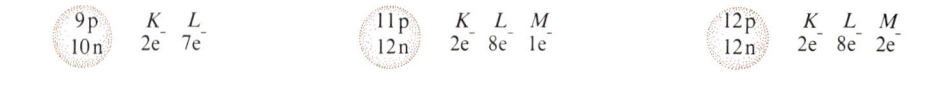

	K	L			K	L	M			K	L	M
9p 10n	2e⁻	7e⁻		11p 12n	2e⁻	8e⁻	1e⁻		12p 12n	2e⁻	8e⁻	2e⁻

Fluorine atom   Sodium atom   Magnesium atom

Figure 5.8. Atomic structure diagrams of fluorine, sodium, and magnesium atoms. The number of protons and neutrons are shown in the nucleus; outside the nucleus are shown the number of electrons in each principal energy level.

Element	Orbital electron structure						Linear expression of electron structure
	1s	2s	$2p_x$	$2p_y$	$2p_z$	3s	
H	↑						$1s^1$
He	↑↓						$1s^2$
Li	↑↓	↑					$1s^2 2s^1$
Be	↑↓	↑↓					$1s^2 2s^2$
B	↑↓	↑↓	↑				$1s^2 2s^2 2p_x^1$
C	↑↓	↑↓	↑	↑			$1s^2 2s^2 2p_x^1 2p_y^1$
N	↑↓	↑↓	↑	↑	↑		$1s^2 2s^2 2p_x^1 2p_y^1 2p_z^1$
O	↑↓	↑↓	↑↓	↑	↑		$1s^2 2s^2 2p_x^2 2p_y^1 2p_z^1$
F	↑↓	↑↓	↑↓	↑↓	↑		$1s^2 2s^2 2p_x^2 2p_y^2 2p_z^1$
Ne	↑↓	↑↓	↑↓	↑↓	↑↓		$1s^2 2s^2 2p_x^2 2p_y^2 2p_z^2$
Na	↑↓	↑↓	↑↓	↑↓	↑↓	↑	$1s^2 2s^2 2p_x^2 2p_y^2 2p_z^2 3s^1$

Figure 5.9. Subenergy-level electron structure of hydrogen through sodium atoms. Each electron is indicated by an arrow placed in the circle, which represents the orbitals.

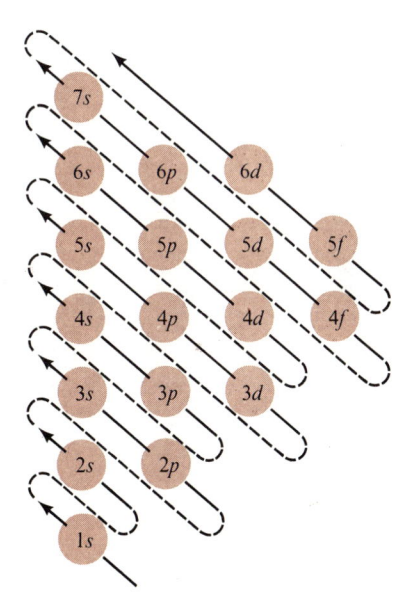

Figure 5.10. Approximate order for placing electrons in subenergy levels. Follow the arrows as indicated. Fill each successive subenergy level with the proper number of electrons, starting with 1s ($s = 2, p = 6, d = 10, f = 14$ electrons), until all the electrons of an atom have been assigned. (There are a few exceptions to this order—for example, chromium and copper among the first 30 elements.)

writing electron structures. Minor variations from the electron structure predicted by the foregoing general sequence or mnemonic device of Figure 5.10 are found in a number of atoms. Table 5.5 shows the accepted ground-state electron structures for all the elements.

Problem 5.2   Diagram the electron structure of a zinc atom and a rubidium atom. Use the $1s^22s^22p^6$, etc., method.

The atomic number of zinc is 30; therefore, it has 30 protons and 30 electrons in a neutral atom. Using Figure 5.7 or Figure 5.10, tabulate the 30 electrons as follows:

Orbital	Number of e^-	Total e^-
1s	2 e^-	2
2s	2 e^-	4
2p	6 e^-	10
3s	2 e^-	12
3p	6 e^-	18
4s	2 e^-	20
3d	10 e^-	30

The electron distribution in Zn is $1s^22s^22p^63s^23p^63d^{10}4s^2$. Check by adding the superscripts, which should equal 30.

The atomic number of rubidium is 37; therefore, it has 37 protons and 37 electrons in a neutral atom. With a little practice, and using either Figure 5.7 or Figure 5.10, the electron structure may be written directly in the linear form. The structure for rubidium is $1s^22s^22p^63s^23p^63d^{10}4s^24p^65s^1$. Check by adding the superscripts, which should equal 37.

5.16 Electron-Dot Representation of Atoms

The electron-dot (or Lewis-dot) method of representing atoms uses the symbol of the element, together with dots as electrons. The number of dots, which are shown around the symbol, are equal to the number of electrons in the outermost energy level of the atom. Paired dots represent paired electrons; unpaired dots represent unpaired electrons.

The nucleus and all the electrons other than those in the outermost energy level are called the **kernel** of the atom and are represented by the symbol of the element. For example, :Ḃ represents a boron atom and tells us that boron has three electrons in its outermost energy level. The kernel of the boron atom includes the nucleus and its $1s^2$ electrons.

:Ḃ

Dots represent electrons in the outer energy level

Symbol represents the kernel of the atom

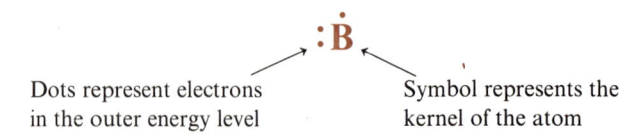

:Ï· indicates an iodine atom, which has seven electrons in its outermost principal energy level. The electron-dot system is used a great deal, not only because of its simplicity of expression, but also because much of the chemistry of the atom is directly associated with the electrons in the outermost energy level. This association is especially true for the first 20 elements and the remaining Group A elements of the periodic table (see Chapter 6). Figure 5.11 shows electron-dot diagrams for the elements hydrogen through calcium.

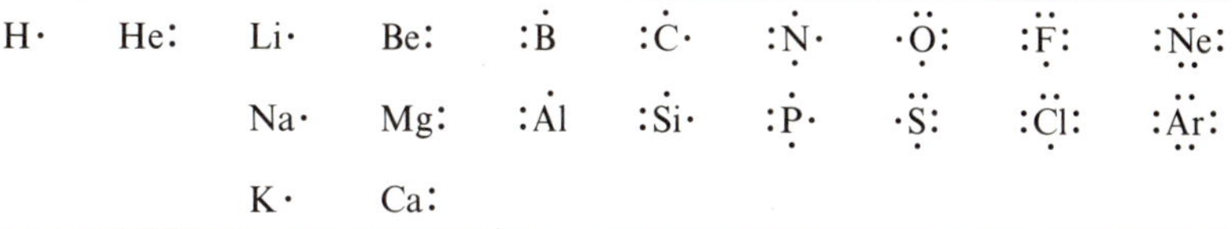

H·	He:	Li·	Be:	:B̈	:C̈·	:N̈·	·Ö·	:F̈:	:N̈ë:
		Na·	Mg:	:Äl	:S̈i·	:P̈·	·S̈:	:C̈l:	:Är:
		K·	Ca:						

Figure 5.11. Electron-dot diagrams of the first 20 elements. Dots represent electrons in the outermost energy level only. Symbol represents the kernel of the atom.

Problem 5.3

Write the electron-dot structure for a phosphorus atom.
First establish the electron structure for a phosphorus atom. It is $1s^2 2s^2 2p^6 3s^2 3p^3$. Note that there are five electrons in the outermost principal energy level; they are $3s^2 3p^3$. Write the symbol for phosphorus and place the five electrons as dots around it.

$$:\overset{\cdot}{\text{P}}\cdot$$

The $3s^2$ electrons are paired and are represented by the paired dots. The $3p^3$ electrons, which are unpaired, are represented by the single dots.

5.17 The Noble Gases

noble gases

The family of elements consisting of helium, neon, argon, krypton, xenon, and radon is known as the **noble gases**. The electron structure of these gases has particular interest for the chemist. Until 1961, attempts to prepare compounds of these elements met with failure. Because of their supposed inability to enter into chemical combinations, these elements were formerly called *inert* gases. They have also been called the *rare* gases. Because of their chemical inactivity, the electron structure of the noble gases is considered to be extraordinarily stable.

Each of these elements, except helium, has eight electrons in its outer shell (see Table 5.6). This electron structure is such that the s and p orbitals in the outer shell are filled with paired electrons, an arrangement that is very stable and chemically unreactive.

Recognition of this structure led to the *rule of eight* principle: The elements, through chemical changes, attempt to attain an electron structure of eight electrons in the outermost energy level identical with that of the chemically stable noble gases. Although the rule of eight principle applies to the chemical behavior of many elements and compounds, it is not universally applicable; some elements do not follow this rule.

All the noble gases are present in the atmosphere. Argon is the most abundant and is found at a concentration of about 1% by volume. The others are present in only trace amounts. Argon was discovered in 1894 by Lord

Table 5.6. Arrangement of electrons in the noble gases. Each gas except helium has eight electrons in its outermost energy level.

Noble gas	Symbol	n = 1	2	3	4	5	6
Helium	He	$1s^2$					
Neon	Ne	$1s^2$	$2s^22p^6$				
Argon	Ar	$1s^2$	$2s^22p^6$	$3s^23p^6$			
Krypton	Kr	$1s^2$	$2s^22p^6$	$3s^23p^63d^{10}$	$4s^24p^6$		
Xenon	Xe	$1s^2$	$2s^22p^6$	$3s^23p^63d^{10}$	$4s^24p^64d^{10}$	$5s^25p^6$	
Radon	Rn	$1s^2$	$2s^22p^6$	$3s^23p^63d^{10}$	$4s^24p^64d^{10}4f^{14}$	$5s^25p^65d^{10}$	$6s^26p^6$

Raleigh (1842–1919) and Sir William Ramsay (1852–1916). Helium was first observed in the spectrum of the sun during an eclipse in 1868. It was not until 1894 that Ramsay recognized that helium exists in the earth's atmosphere. In 1898 he and his coworker, Morris W. Travers (1872–1961), announced the discovery of neon, krypton, and xenon, having isolated them from liquid air. Friedrich E. Dorn (1848–1916) first identified radon, the heaviest member of the noble gases, as a radioactive gas emanating from the element radium.

Because of its low density and nonflammability, helium has been used for filling balloons and dirigibles. Only hydrogen surpasses helium in lifting power. Helium mixed with oxygen is used by deep-sea divers for breathing. This mixture reduces the danger of acquiring the "bends" (caisson disease), pains and paralysis suffered by divers on returning from the ocean depths to normal atmospheric pressures. Helium is also used in "heliarc welding," where it supplies an inert atmosphere for the welding of active metals such as magnesium. Helium is found in some natural gas wells in the southwestern United States. As a liquid, it is used to study the properties of substances at very low temperatures. The boiling point of liquid helium, 4.2 K, is not far above absolute zero.

We are all familiar with the neon sign, in which a characteristic red color is produced when an electric discharge is passed through a tube filled with neon. This color may be modified by mixing the neon with other gases or by changing the color of the glass. Argon is used primarily in gas-filled electric light bulbs and other types of electronic tubes to provide an inert atmosphere for prolonging tube life. Argon is also used in some welding applications where an inert atmosphere is needed. Krypton and xenon have not been used extensively because of their limited availability. Radon is radioactive; it has been used medicinally in the treatment of cancer.

For many years it was believed that the noble gases could not be made to combine chemically with any other element. Then, in 1962, Neil Bartlett at the University of British Columbia, Vancouver, synthesized the first noble gas compound, xenon hexafluoroplatinate, $XePtF_6$. This outstanding discovery opened a new field in the techniques of preparing noble gas compounds and investigating their chemical bonding and properties. Other compounds of xenon as well as compounds of krypton and radon have been prepared. Some of these are XeF_2, XeF_4, XeF_6, $XeOF_4$, $Xe(OH)_6$, and KrF_2.

5.18 Atomic Weight

By means of an instrument called *mass spectrometer*, it is possible to make fairly precise physical determinations of the masses of individual atoms. In the ordinary sense of weighing, single atoms are far too tiny actually to be weighed (the mass of a single hydrogen atom is 1.67×10^{-24} g). If we could magnify the mass of such an atom by a factor of 100 billion, it would still require about 300 million of these magnified hydrogen atoms to equal the weight of a single drop of water (0.05 g)!

Because it is not only inconvenient but impractical to express the actual weights of atoms in grams, chemists have devised a useful table of relative atomic weights. The carbon isotope having six protons and six neutrons and designated carbon-12, or $^{12}_{6}C$, was chosen as the standard of reference for atomic weights. (Until 1961, oxygen had been the chemical standard of reference.) The mass of this isotope is assigned a value of exactly 12 atomic mass units (amu). Thus, **1 atomic mass unit** is defined as 1/12 the mass of a carbon-12 atom. In the table of atomic weights all other elements are assigned values proportional to the arbitrary mass assigned to the reference isotope carbon-12. A table of atomic weights is given on the inside back cover of this book. Hydrogen atoms, with a mass of about 1/12 that of a carbon atom, have an average atomic mass of 1.00797 amu on this relative scale. Magnesium atoms, which are about twice as heavy as carbon, have an average mass of 24.312 amu. The average atomic mass of oxygen is 15.9994 amu (usually rounded off to 16.0 for calculations).

Since all elements exist as isotopes having different masses, the atomic weight of an element represents the average relative mass of all the naturally occurring isotopes of that element. The atomic weights of the individual isotopes are approximately whole numbers, because the relative masses of the protons and neutrons are approximately 1.0 amu each. Yet we find that the atomic weights given for many of the elements deviate considerably from whole numbers. For example, the atomic weight of rubidium is 85.47 amu, that of copper is 63.54 amu, and that of magnesium is 24.312 amu. The deviation of an atomic weight from a whole number is due mainly to the unequal occurrence of the various isotopes of an element. It is also due partly to the difference between the mass of a free proton or neutron and the mass of these same particles in the nucleus. For example, the two principal isotopes of chlorine are $^{35}_{17}Cl$ and $^{37}_{17}Cl$. It is apparent that chlorine-35 atoms are the more abundant isotope, since the atomic weight of chlorine, 35.453 amu, is closer to 35 than to 37 amu. The actual values of the chlorine isotopes observed by mass spectra determination are shown below.

atomic mass unit

Isotope	Isotopic mass (amu)	Abundance (%)	Average atomic mass (amu)
$^{35}_{17}Cl$	34.969	75.53	35.453
$^{37}_{17}Cl$	36.966	24.47	

atomic weight

The **atomic weight** of an element is the average relative mass of the isotopes of that element referred to the atomic mass of carbon-12 (exactly 12.0000 amu).

The relationship of mass number and atomic number is such that if we subtract the atomic number from the mass number, we obtain the number of neutrons in the nucleus of the atom. Table 5.7 shows the application of this method of determining the number of neutrons. For example, the fluorine atom ($^{19}_{9}F$), atomic number 9, having a mass of 19 amu, contains 10 neutrons:

Mass number $-$ Atomic number $=$ Number of neutrons
19 $-$ 9 $=$ 10

The atomic weights given in the table on the inside back cover of this book are values accepted by international agreement. There is no need to memorize atomic weights. In most of the calculations needed in this book, their use to the first decimal place will give results of sufficient accuracy.

Table 5.7. Calculation of the number of neutrons in an atom by subtracting the atomic number from the mass number.

	Hydrogen ($^{1}_{1}H$)	Oxygen ($^{16}_{8}O$)	Sulfur ($^{32}_{16}S$)	Fluorine ($^{19}_{9}F$)	Iron ($^{56}_{26}Fe$)
Mass number	1	16	32	19	56
Atomic number	1	8	16	9	26
Number of neutrons	0	8	16	10	30

5.19 The Mole

According to the atomic theory set forth by John Dalton, atoms always combine in whole-number ratios. Since individual atoms certainly cannot be weighed, there was a need to establish some weighable unit for comparing the quantities of elements involved in chemical reactions. The working unit devised for this purpose is the **gram-atomic weight** (g-at. wt), which is defined as the number of grams of an element *numerically* equal to its atomic weight. Be very careful to note this distinction: The atomic weight of magnesium, for example, as given in the table, is 24.312. This value indicates the average mass of magnesium atoms relative to the carbon-12 isotope. But the gram-atomic weight of magnesium is 24.312 *grams* of magnesium.

gram-atomic weight

It is known that 1 g-at. wt of magnesium reacts chemically with 1 g-at. wt of sulfur to form magnesium sulfide:

$$Mg + S \longrightarrow MgS$$

Thus, magnesium atoms and sulfur atoms react in a 1:1 atom ratio. It follows, then, that a gram-atomic weight of magnesium must contain the same number of atoms as a gram-atomic weight of sulfur.

Since the actual weight of an atom is very minute, 1 g-at. wt of atoms will contain a very large number of individual atoms. The number of atoms in 1 g-at.

Avogadro's number

wt of any element has been experimentally determined to be 6.02×10^{23} (that is, 602,000,000,000,000,000,000,000 atoms). This number is known as **Avogadro's number**. Thus, 1 g-at. wt (24.312 g) of magnesium contains 6.02×10^{23} magnesium atoms. A gram-atomic weight of any element—for example, 32.064 g of sulfur, 15.9994 g of oxygen, 1.00797 g of hydrogen, and so on—therefore contains Avogadro's number of atoms. Table 5.8 summarizes these relationships.

Table 5.8. Avogadro's number related to the gram-atomic weights of oxygen, hydrogen, sulfur, and magnesium.

Element	Gram-atomic weight	Avogadro's number (Number of atoms/g-at. wt)
Oxygen	15.9994 g	6.02×10^{23}
Hydrogen	1.00797 g	6.02×10^{23}
Sulfur	32.064 g	6.02×10^{23}
Magnesium	24.312 g	6.02×10^{23}

Avogadro's number $= 6.02 \times 10^{23}$

It is difficult to imagine how large Avogadro's number really is, but perhaps the following analogy will help express it: If 10,000 people started to count Avogadro's number and each counted at the rate of 100 numbers per minute each minute of the day, it would take them over 1 trillion (10^{12}) years to count the total number. So, even the minutest amount of matter contains extremely large numbers of atoms.

Avogadro's number is the basis for an additional very important quantity used to express a particular number of chemical species—such as atoms, molecules, formula units, ions, or electrons. This quantity is the mole. We define a **mole** as an amount of a substance containing the same number of formula units as there are atoms in exactly 12 g of carbon-12. Other definitions are used, but they all relate to a mole being Avogadro's number of formula units of a substance. A formula unit is whatever is indicated by the formula of the substance under consideration—for example, Mg, MgS, H_2O, O_2, $^{75}_{33}As$. The gram-atomic weight of any element contains a mole of atoms of that element.

mole

1 gram-atomic weight = 1 mole of atoms = Avogadro's number (6.02×10^{23}) of atoms

Thus, for example, 1 mole of hydrogen atoms represents Avogadro's number of hydrogen atoms and has a mass of 1.008 g (the gram-atomic weight).

The term *mole* is so commonplace in chemical jargon that chemists use it as freely as the words *atom* and *molecule*. The mole is used in conjunction with many different particles, such as atoms, molecules, ions, and electrons, to represent Avogadro's number of these particles. If we can speak of a mole of atoms, we can also speak of a mole of molecules, a mole of electrons, a mole of ions, understanding that in each case we mean 6.02×10^{23} formula units of these particles.

90   Chapter 5

Avogadro's number of formula units = 1 mole of formula units

We frequently encounter problems that require interconversions involving the quantities of mass, numbers, and moles of atoms of an element. Conversion factors that may be used for this purpose are

(a) Grams to atoms: $\dfrac{6.02 \times 10^{23} \text{ atoms of the element}}{\text{g-at. wt of the element}}$

(b) Atoms to grams: $\dfrac{\text{g-at. wt of the element}}{6.02 \times 10^{23} \text{ atoms of the element}}$

(c) Grams to moles: (monatomic elements) $\dfrac{1 \text{ mole of the element}}{1 \text{ g-at. wt of the element}}$

(d) Moles to grams: (monatomic elements) $\dfrac{1 \text{ g-at. wt of the element}}{1 \text{ mole of the element}}$

Sample problems follow.

Problem 5.4

How many moles of iron does 25.0 g of Fe represent?
The problem requires that we change grams of Fe to moles of Fe. We look up the atomic weight of Fe in the atomic weight table and find it to be 55.8. Then we use the proper conversion factor to obtain moles. The conversion factor is (c) above.

$$\dfrac{1 \text{ mole Fe}}{55.8 \text{ g Fe}}$$

$$25.0 \text{ g Fe} \times \dfrac{1 \text{ mole Fe}}{55.8 \text{ g Fe}} = 0.448 \text{ mole Fe} \quad \text{(Answer)}$$

Problem 5.5

How many magnesium atoms does 5.00 g of Mg represent?
The problem requires that we change grams of magnesium to atoms of magnesium through the following sequence: Grams Mg → Moles Mg → Atoms Mg. We find the atomic weight of magnesium to be 24.3 and set up the sequence using conversion factors (c) and (a) above.

$$\dfrac{1 \text{ mole Mg}}{24.3 \text{ g Mg}} \quad \text{and} \quad \dfrac{6.02 \times 10^{23} \text{ Mg atoms}}{1 \text{ mole Mg}}$$

$$5.00 \text{ g Mg} \times \dfrac{1 \text{ mole Mg}}{24.3 \text{ g Mg}} \times \dfrac{6.02 \times 10^{23} \text{ Mg atoms}}{1 \text{ mole Mg}}$$

$$= 1.24 \times 10^{23} \text{ Mg atoms} \quad \text{(Answer)}$$

Thus, there are 1.24×10^{23} atoms of Mg in 5.00 g of Mg.

Problem 5.6

What is the mass (in grams) of one atom of carbon?
From the table of atomic weights, we see that 1 g-at. wt of carbon is 12.0 g. The factor to

Atomic Theory and Structure 91

convert atoms to grams therefore is

$$\frac{12.0 \text{ g C}}{6.02 \times 10^{23} \text{ atoms C}}$$

Then

$$1 \text{ atom C} \times \frac{12.0 \text{ g C}}{6.02 \times 10^{23} \text{ atoms C}} = 1.99 \times 10^{-23} \text{ g C} \quad \text{(Answer)}$$

Problem 5.7

How much do 3.01×10^{23} atoms of sodium weigh?

The information needed to solve this problem is the gram-atomic weight of Na (23.0 g) and the conversion factors

$$\frac{1 \text{ mole}}{6.02 \times 10^{23} \text{ atoms}} \quad \text{and} \quad \frac{1 \text{ g-at. wt}}{1 \text{ mole}}$$

$$3.01 \times 10^{23} \text{ atoms Na} \times \frac{1 \text{ mole Na}}{6.02 \times 10^{23} \text{ atoms Na}} \times \frac{23.0 \text{ g Na}}{1 \text{ mole Na}} = 11.5 \text{ g Na} \quad \text{(Answer)}$$

Problem 5.8

How much does Avogadro's number of copper atoms weigh?

Avogadro's number of Cu atoms = 1 mole of Cu atoms

One mole of Cu atoms by definition is the gram-atomic weight of Cu. We therefore look up the atomic weight of Cu in the atomic weight table to find the answer.

Avogadro's number of Cu atoms = 1 mole Cu = 63.5 g Cu (Answer)

Or, by conversion factors,

$$6.02 \times 10^{23} \text{ atoms Cu} \times \frac{1 \text{ mole Cu}}{6.02 \times 10^{23} \text{ atoms Cu}} \times \frac{63.5 \text{ g Cu}}{1 \text{ mole Cu}} = 63.5 \text{ g Cu} \quad \text{(Answer)}$$

Questions

A. Review the meanings of the new terms introduced in this chapter.
  1. Dalton's atomic theory
  2. Subatomic particles
  3. Electron
  4. Proton
  5. Neutron
  6. Nucleus
  7. Quantum number
  8. Orbital
  9. Energy level of electrons
  10. Electron shell
  11. Atomic number
  12. Isotopes
  13. Kernel of an atom
  14. Noble gases
  15. Atomic mass unit
  16. Atomic weight
  17. Gram-atomic weight
  18. Avogadro's number
  19. Mole

B. Answers to the following questions will be found in tables and figures.
  1. What are the atomic numbers of aluminum, platinum, radium, and krypton?
  2. How many electron orbitals can be present in the third energy level? What are they?
  3. Show the sublevel electron structure ($1s^2 2s^2 2p^6$, etc.) for elements of atomic numbers 13, 15, 19, 24, and 36.

4. Using only Table 5.6, write electron structures ($1s^2 2s^2 2p^6$, etc.) for elements containing the following number of electrons:
   (a) 21 (b) 33 (c) 52 (d) 55
5. Diagram the atomic structure of the following atoms: F, Ca, Li, K, Br, and Mg (see Figure 5.8).
6. Show electron-dot structures for K, O, Mg, Kr, Br, Co, Si, and Al.
7. Explain the meaning of the following symbols. $^{131}_{53}I$, $^{235}_{92}U$.

C. Review questions.
   1. From the point of view of a chemist, what are the essential differences among a proton, a neutron, and an electron?
   2. Describe the general arrangement of particles in the atom.
   3. What part of the atom contains practically all its mass?
   4. What experimental evidence led Rutherford to conclude that:
      (a) The nucleus of the atom contains most of the atomic mass?
      (b) The nucleus of the atom is positively charged?
      (c) The atom consists of mostly empty space?
   5. (a) What is an atomic orbital?
      (b) What are the shapes of an s orbital and a p orbital?
   6. What is the Pauli exclusion principle?
   7. Under which conditions can a second electron enter an orbital already containing one electron?
   8. What is meant when we say that the electron structure of an atom is in its ground state?
   9. List the following electron sublevels in order of increasing energy: $2s$, $2p$, $4s$, $1s$, $3d$, $3p$, $4p$, $3s$.
   10. How many s electrons, p electrons, and d electrons are possible in any electron shell?
   11. How many protons are in the nucleus of an atom of each of these elements: H, N, B, Sc, Hg, U, Br, Sn, and Pb?
   12. Why is the eleventh electron of the sodium atom located in the third energy level rather than in the second energy level?
   13. Why are the last two electrons in calcium located in the fourth energy level rather than in the third energy level?
   14. Which atoms have the following structures?
      (a) $1s^2 2s^2 2p^6 3s^2$
      (b) $1s^2 2s^2 2p^5$
      (c) $1s^2 2s^2 2p^6 3s^2 3p^6 3d^8 4s^2$
      (d) $1s^2 2s^2 2p^6 3s^2 3p^6 3d^5 4s^2$
      (e) $1s^2 2s^2 2p^6 3s^2 3p^6 3d^{10} 4s^2 4p^6 4d^5 5s^1$
   15. What does the atomic number of an atom represent?
   16. Which of the following statements are correct?
      (a) The maximum number of p electrons in the first energy level is six.
      (b) A $2s$ electron is in a lower energy state than a $2p$ electron.
      (c) The energy level of a $3d$ electron is higher than a $4s$ electron.
      (d) The electron structure for a carbon atom is $1s^2 2s^2 2p_x^2$.
      (e) The $2p_x$, $2p_y$, and $2p_z$ electron orbitals are in the same energy state.
   17. In what ways are isotopes alike? In what ways are they different?
   18. An atom of an element has a mass number of 108 and has 62 neutrons in its nucleus.
      (a) What is the symbol and name of the element?
      (b) What is its nuclear charge?
   19. List the similarities and differences in the three isotopes of hydrogen.

20. What is the nuclear and electron composition of the six naturally occurring isotopes of calcium having mass numbers of 40, 42, 43, 44, 46, and 48?
21. How many electrons are not shown in each electron-dot structure in Si, N, O, Ca, and He atoms?
22. What characterizes the special chemical stability of the noble gases?
23. Which of the following statements about the neutral atoms $^{23}_{11}Na$ and $^{24}_{11}Na$ are true?
    (a) $^{23}_{11}Na$ and $^{24}_{11}Na$ are isotopes.
    (b) $^{24}_{11}Na$ has one more electron than $^{23}_{11}Na$.
    (c) $^{24}_{11}Na$ has one more proton than $^{23}_{11}Na$.
    (d) $^{24}_{11}Na$ has one more neutron than $^{23}_{11}Na$.
    (e) A mole of $^{24}_{11}Na$ contains more atoms than does a mole of $^{23}_{11}Na$.
24. Explain why the atomic weights of elements are not whole numbers.
25. Distinguish between atomic weights expressed in atomic mass units and in grams.
26. What information is needed to calculate the approximate atomic weight of an atom?
27. Which of the isotopes of calcium in Question 20 is the most abundant isotope? Explain your choice.
28. What is the significance of Avogadro's number?
29. Complete the following statements, supplying the proper quantity:
    (a) A mole of oxygen atoms (O) contains _____ atoms.
    (b) A mole of oxygen molecules (O_2) contains _____ molecules.
    (c) A mole of oxygen molecules (O_2) contains _____ atoms.
    (d) A mole of oxygen atoms (O) weighs _____ grams.
    (e) A mole of oxygen molecules (O_2) weighs _____ grams.
30. What contribution did each of the following scientists make to the atomic theory?
    (a) Dalton   (c) Rutherford   (e) Bohr
    (b) Thomson   (d) Chadwick   (f) Schrödinger
31. Which of the following statements are correct?
    (a) Hydrogen is the smallest atom.
    (b) A proton is about 1837 times as heavy as an electron.
    (c) The nucleus of an atom contains protons, neutrons, and electrons.
    (d) The Bohr structure of the atom had one electron in each of several orbits surrounding the nucleus.
    (e) There are seven principal electron energy levels.
    (f) The second principal energy level can have four subenergy levels and contain a maximum of eight electrons.
    (g) The M energy level can have a maximum of 32 electrons.
    (h) The number of possible d electrons in the third energy level is ten.
    (i) The first f electron occurs in the fourth principal energy level.
    (j) The $4s$ subenergy level is at a higher energy than the $3d$ subenergy level.
    (k) An element with an atomic number of 29 has 29 protons, 29 neutrons, and 29 electrons.
    (l) An atom of the isotope $^{60}_{26}Fe$ has 34 neutrons in its nucleus.
    (m) ^2_1H is a symbol for the isotope deuterium.
    (n) The electron-dot structure, P·, is used for potassium.
    (o) Atoms of all the noble gases (except helium) have eight electrons in their outermost energy level.
    (p) One gram-atomic weight of any element contains 6.02×10^{23} atoms.

(q) The mass of one atom of chlorine is $\dfrac{35.5 \text{ g}}{6.02 \times 10^{23} \text{ atoms}}$.

(r) A mole of magnesium atoms (24.3 g) contains the same number of atoms as a mole of sodium atoms (23.0 g).

(s) A mole of bromine atoms contains 6.02×10^{23} atoms of bromine.

(t) A mole of chlorine molecules (Cl_2) contains 6.02×10^{23} atoms of chlorine.

D. Review problems.

1. Change the following to powers of 10:
   (a) 640,000   (b) 0.000568   (c) 0.01^2 (d) 5^3
   (See Mathematical Review in Appendix I.)

2. Express the following as numbers without using powers of 10:
   (a) 4.2×10^6 (b) 9×10^{-5} (c) 10^{-4} (d) 35×10^3
   (See Mathematical Review in Appendix I.)

3. Using the formula $2n^2$, calculate the maximum number of electrons that can exist in electron shells K, L, M, N, O, and P. Energy levels: $n = 1, 2, 3, 4, 5,$ and 6.

4. How many neutrons are in an atom of $^{40}_{20}Ca$, $^{59}_{28}Ni$, $^{119}_{50}Sn$, $^{207}_{82}Pb$, $^{235}_{92}U$, $^{254}_{102}No$?

5. How many individual atoms are contained in the following?
   (a) 12.0 g Na   (c) 160 g C   (e) 56.2 g Cd
   (b) 0.75 g P   (d) 25.0 g Cu   (f) 1.00×10^{-4} g H

6. Complete the following table with the appropriate data for each element given:

Atomic number	Symbol of element	Number of protons	Atomic weight	Gram-atomic weight	Mass (g) of 1 atom
9					
33					
82					

7. Calculate the weight, in grams, of one atom of:
   (a) Sulfur   (b) Tin   (c) Mercury   (d) Helium

8. How many grams are represented by each of the following?
   (a) 1,000,000 atoms of Ar   (c) 3.00 g-at. wt of Cl
   (b) 4.50 moles of Al   (d) 0.00100 mole of Ag

9. Make the following conversions:
   (a) 2.62 moles of C to grams of C
   (b) 0.150 mole of Cu to grams of Cu
   (c) 22.5 moles of Ag to kilograms of Ag
   (d) 3.0 moles of Cl_2 to grams of Cl_2
   (e) 125 g of Fe to moles of Fe
   (f) 2.00 g of Sn to moles of Sn
   (g) 28.0 g of N_2 to moles of N_2
   (h) 25.0 ml Hg ($d = 13.6$ g/ml) to moles of Hg

10. One mole of water contains:
    (a) How many water molecules?
    (b) How many oxygen atoms?
    (c) How many hydrogen atoms?

11. White phosphorus is one of several forms of phosphorus and exists as a waxy solid consisting of P_4 molecules. How many atoms are present in 0.50 mole of this phosphorus?
12. How many grams of magnesium contain the same number of atoms as 6.00 g of calcium?
13. One atom of an unknown element is found to have a mass of 2.28×10^{-22} g. What is the gram-atomic weight of the element?
14. There are about four billion (4×10^9) people on the earth. If 1 mole of dollars were distributed equally among these people, how many dollars would each person receive?
15. A ping-pong ball is about 4 cm in diameter. If a mole of ping-pong balls were laid end to end, what distance, in kilometres, would they cover? How many miles is this?

E. Review exercises.
1. Show the atomic structure of the most abundant isotope of cobalt. Of sulfur.
2. Which atom would you expect to have the larger volume, chlorine or bromine? Why?
3. Draw the atomic structure of an atom with an atomic weight of 34 and an atomic number of 16. What element is this?
4. In your own words, describe a chlorine atom.
5. Would you expect the density of the nucleus of an atom to be relatively large or small? Explain.
6. What mass of sodium contains the same number of atoms as 5.00 g of calcium?
7. One atom of an unknown element is found to have a mass of 9.75×10^{-23} g. What is the gram-atomic weight of the element?

6 The Periodic Arrangement of the Elements

After studying Chapter 6 you should be able to:

1. Understand the terms listed under Question A at the end of the chapter.
2. Describe briefly the contributions of Döbereiner, Newlands, Mendeleev, Meyer, and Moseley to the development of the periodic law.
3. State the periodic law in its modern form.
4. Explain why there were blank spaces in Mendeleev's periodic table and how he was able to predict the properties of the elements that belonged in those spaces.
5. Indicate the locations of the metals, nonmetals, metalloids, and noble gases in the periodic table.
6. Indicate in the periodic table the areas where the *s, p, d,* and *f* orbitals are being filled.
7. Describe how atomic radii vary (a) from left to right in a period and (b) from top to bottom in a group.
8. Describe the changes in outer-level electron structure when (a) moving from left to right in a period and (b) going from top to bottom in a group.
9. List the general characteristics of group properties.
10. Predict formulas of simple compounds formed between Group A elements using the periodic table.
11. Point out how the change in electron structure in going from one transition element to the next differs from that in nontransition elements.

6.1 Early Attempts to Classify the Elements

Chemists of the early 19th century had sufficient knowledge of the properties of elements to recognize similarities among groups of elements. J. W. Döbereiner (1780–1849), professor at the University of Jena in Germany, observed the existence of "triads" of similarly behaving elements, in which the middle element had an atomic weight approximating the average of the other two elements. He also noted that many other properties of the central element were approximately the average of the other two elements. Table 6.1 presents comparative data on atomic weight and density for two sets of Döbereiner's triads.

Table 6.1. Döbereiner's triads.

Triads	Atomic weight	Density (g/ml at 4°C)
Chlorine	35.5	1.56[a]
Bromine	79.9	3.12
Iodine	126.9	4.95
Average of chlorine and iodine	81.2	3.26
Calcium	40.1	1.55
Strontium	87.6	2.6
Barium	137.4	3.5
Average of calcium and barium	88.8	2.52

[a] Density at −34°C (liquid).

In 1864, J. A. R. Newlands (1837–1898), an English chemist, reported his *Law of Octaves*. In his studies, Newlands observed that when the elements were arranged according to increasing atomic weights, every eighth element had similar properties. (The noble gases were not yet discovered at that time.) But Newlands' theory was ridiculed by his contemporaries in the Royal Chemical Society, and they refused to publish his work. Many years later, however, Newlands was awarded the highest honor of the society for this important contribution to the development of the periodic law.

In 1869, Dmitri Ivanovitch Mendeleev (1834–1907) of Russia and Lothar Meyer (1830–1895) of Germany independently published their periodic arrangements of the elements that were based on increasing atomic weights. Because his arrangement was published slightly earlier and was in a somewhat more useful form than that of Meyer, Mendeleev's name is usually associated with the modern periodic table.

6.2 The Periodic Law

Only about 63 elements were known when Mendeleev constructed his table. These elements were arranged so that those with similar chemical properties fitted into columns to form family groups. The arrangement left many gaps between elements. Mendeleev predicted that these spaces would be filled as new elements were discovered. For example, spaces for undiscovered elements were left after calcium, under aluminum, and under silicon. He called these unknown elements eka-boron, eka-aluminum, and eka-silicon. The term *eka* comes from Sanskrit meaning "one" and was used to indicate that the missing element was one place away in the table from the element indicated. Mendeleev even went so far as to predict with high accuracy the physical and chemical properties of these elements yet to be discovered. The three elements, scandium (atomic number 21), gallium (31), and germanium (32) were in fact discovered during Mendeleev's lifetime and were found to have properties agreeing very closely with the predictions that he had made for eka-boron,

Table 6.2. Comparison of the properties of eka-silicon predicted by Mendeleev with the properties of germanium.

Property	Mendeleev's predictions in 1871 for eka-silicon (Es)	Observed properties for germanium (Ge)
Atomic weight	72	72.6
Color of metal	Dirty gray	Grayish-white
Density	5.5 g/ml	5.47 g/ml
Oxide formula	EsO_2	GeO_2
Oxide density	4.7 g/ml	4.70 g/ml
Chloride formula	$EsCl_4$	$GeCl_4$
Chloride density	1.9 g/ml	1.89 g/ml
Boiling temperature of chloride	Under 100°C	86°C

eka-aluminum, and eka-silicon. The amazing way in which Mendeleev's predictions were fulfilled is illustrated in Table 6.2, which compares the predicted properties of eka-silicon with those of germanium, discovered by the German chemist C. Winkler in 1886.

Two major additions have been made to the periodic table since Mendeleev's time: (1) a new family of elements, the noble gases, was discovered and added; and (2) elements having atomic numbers greater than 92 have been discovered and fitted into the table.

The original table was based on the premise that the properties of the elements are a periodic function of their atomic weights. However, there were some disturbing discrepancies to this basic premise. For example, the atomic weight for argon is greater than that of potassium. Yet potassium had to be placed after argon since argon is certainly one of the noble gases and potassium behaves like the other alkali metals. These discrepancies were resolved by the work of the British physicist, H. G. J. Moseley (1887–1915) and by the discovery of the existence of isotopes. Moseley noted that the x-ray emission frequencies of the elements increased in a regular, stepwise fashion each time the nuclear charge (atomic number) increased by one unit. This showed that the basis for placing an element in the periodic table should be dependent on atomic number rather than on atomic weight. Atomic weights are the average masses of the naturally occurring mixtures of isotopes of each element. The atomic weight for argon is greater than that of potassium (the next element of higher atomic number) because the average mass of the argon isotopes is greater than the average mass of the potassium isotopes. The current statement of the **periodic law** is:

periodic law

The properties of the elements are a periodic function of their atomic numbers.

As the format of the periodic table is studied, it becomes evident that the periodicity of the properties of the elements is due to the recurring similarities of their electron structures.

Table 6.3. Periodic table of the elements.

6.3 Arrangement of the Periodic Table

periodic table

periods of elements

groups or families of elements

The most commonly used **periodic table** is the long form shown in Table 6.3. In this table the elements are arranged horizontally in numerical sequence, according to their atomic numbers; the result is seven horizontal **periods**. Each period, with the exception of the first, starts with an alkali metal and ends with a noble gas. By this arrangement, vertical columns of elements are formed, having identical or similar outer-shell electron structures and thus similar chemical properties. These columns are known as **groups** or **families of elements**.

The heavy zigzag line starting at boron and running diagonally down the table separates the elements into metals and nonmetals. The elements to the right of the line are nonmetallic, and those to the left are metallic. The elements bordering the zigzag line are the metalloids, which show both metallic and nonmetallic properties. With some exceptions, the characteristic electronic arrangement of metals is that their atoms have one, two, or three electrons in the outer energy level, while nonmetals have five, six, or seven electrons in the outer energy level.

With this periodic arrangement, the elements fall into blocks according to the sublevel of electrons that is being filled. The grouping of the elements into *spdf* blocks is shown in Table 6.4. The *s* block comprising Groups IA and IIA has one or two *s* electrons in its outer energy level. The *p* block includes Groups IIIA through VIIA and the noble gases (except helium). In these elements, electrons are filling the *p* sublevel orbitals. The *d* block includes the

Table 6.4. Arrangement of the elements into blocks according to the sublevel of electrons being filled in their atomic structure.

s block | *d* block: Transition elements | *p* block | Noble gases

IA	IIA	IIIB	IVB	VB	VIB	VIIB	VIII	VIII	VIII	IB	IIB	IIIA	IVA	VA	VIA	VIIA	Noble gases
1 H																	2 He
3 Li	4 Be											5 B	6 C	7 N	8 O	9 F	10 Ne
11 Na	12 Mg											13 Al	14 Si	15 P	16 S	17 Cl	18 Ar
19 K	20 Ca	21 Sc	22 Ti	23 V	24 Cr	25 Mn	26 Fe	27 Co	28 Ni	29 Cu	30 Zn	31 Ga	32 Ge	33 As	34 Se	35 Br	36 Kr
37 Rb	38 Sr	39 Y	40 Zr	41 Nb	42 Mo	43 Tc	44 Ru	45 Rh	46 Pd	47 Ag	48 Cd	49 In	50 Sn	51 Sb	52 Te	53 I	54 Xe
55 Cs	56 Ba	57* La	72 Hf	73 Ta	74 W	75 Re	76 Os	77 Ir	78 Pt	79 Au	80 Hg	81 Tl	82 Pb	83 Bi	84 Po	85 At	86 Rn
87 Fr	88 Ra	89** Ac	104 Ku	105 Ha	106												

f block: Inner transition elements

*58 Ce	59 Pr	60 Nd	61 Pm	62 Sm	63 Eu	64 Gd	65 Tb	66 Dy	67 Ho	68 Er	69 Tm	70 Yb	71 Lu
**90 Th	91 Pa	92 U	93 Np	94 Pu	95 Am	96 Cm	97 Bk	98 Cf	99 Es	100 Fm	101 Md	102 No	103 Lr

s block
p block
d block
f block

transition elements of Groups IB through VIIB and Group VIII. The d sublevels of electrons are being filled in these elements. The f block of elements includes the inner transition series. In the lanthanide series, electrons are filling the $4f$ sublevel. In the actinide series, electrons are filling the $5f$ sublevel.

6.4 Periods of Elements

The number of elements in each period is shown in Table 6.5. The first period contains 2 elements, hydrogen and helium, and coincides with the full K shell of electrons. Period 2 contains 8 elements, starting with lithium and ending with neon. Period 3 also contains 8 elements, sodium to argon. Periods 4 and 5 each contain 18 elements; period 6 has 32 elements; and period 7, which is incomplete, contains the remaining 20 elements.

Table 6.5. The number of elements in each period.

Period number	Number of elements	Electron orbitals in each period being filled
1	2	$1s$
2	8	$2s2p$
3	8	$3s3p$
4	18	$4s3d4p$
5	18	$5s4d5p$
6	32	$6s4f5d6p$
7	20	$7s5f6d$

The first three periods are known as *short periods*; the others, as *long periods*. The number of each period corresponds to the outermost energy level in which electrons are located in the neutral atom. For example, the elements in period 1 contain electrons in the first energy level only; period 2 elements contain electrons in the first and second levels; period 3 elements contain electrons in the first, second, and third levels; and so on. Moving horizontally across periods 2 through 6, we find that the properties of the elements vary from strongly metallic at the beginning to nonmetallic at the end of the period. Starting with the third element of the long periods 4, 5, 6, and 7 (scandium, Sc; yttrium, Y; lanthanum, La; actinium, Ac), the inner shells of d and f orbital electrons begin to fill in.

In general, the atomic radii of the elements within a period decrease with increasing nuclear charge. Therefore, the size of the atoms becomes progressively smaller from left to right within each period. Because the noble gases do not readily combine with other elements to form compounds, the radii of their atoms are not determined in the same comparative manner as the other elements. However, calculations have shown that the radii of the noble gas atoms are about the same as, or slightly smaller than, the element immediately preceding them. Slight deviations in atomic radii occur in the middle of the long periods of the elements. The elements of period 3 serve to illustrate this principle. The radii are given in angstrom units:

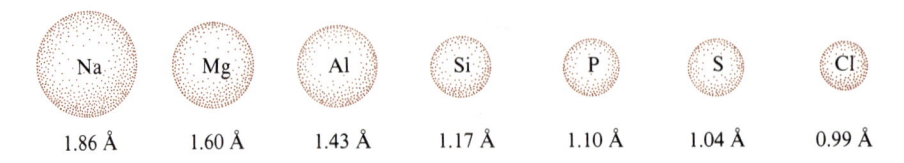

Na	Mg	Al	Si	P	S	Cl
1.86 Å	1.60 Å	1.43 Å	1.17 Å	1.10 Å	1.04 Å	0.99 Å

6.5 Groups or Families of Elements

The groups, or families, of elements are numbered IA through VIIA, IB through VIIB, VIII, and noble gases.

The elements comprising each family have similar outer energy-level electron structures. In Group A elements, the number of electrons in the outer energy level is identical to the group number. Group IA is known as the *alkali metal* family. Each atom of this family of elements has one *s* electron in its outer energy level. Group IIA atoms, the *alkaline earth metals*, each have two *s* electrons in their outer energy level. All atoms of the *halogen* family, Group VIIA, have an outer energy-level electron structure of ns^2np^5. The noble gases (except helium) have an outer energy-level structure of ns^2np^6.

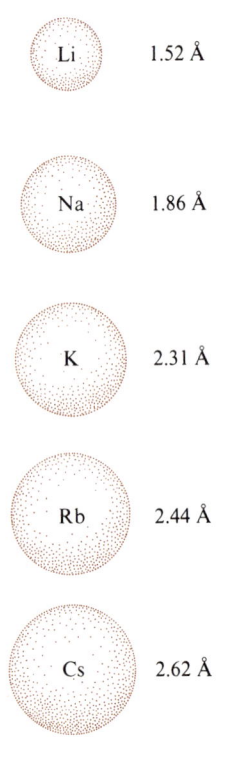

Li	1.52 Å
Na	1.86 Å
K	2.31 Å
Rb	2.44 Å
Cs	2.62 Å

Each alkali metal starts a new period of elements in which the *s* electron occupies a principal energy level one greater than in the previous period. As a result, the size of the atoms of this and other families of elements increases from the top to the bottom of the family. The relative size of the alkali metals is illustrated at the left. The radii are given in angstrom units.

One major distinction between groups lies in the energy level to which the last electron is added. In the elements of Groups IA through VIIA, IB, IIB, and the noble gases, the last electron is added either to an *s* or to a *p* orbital located in the outermost energy level (copper is an exception; see Table 5.5). In the elements of Groups IIIB through VIIB and VIII, the last electron goes to a *d* or to an *f* orbital located in an inner energy level. For example, the last electron in potassium (a fourth-period, Group IA element) is a 4*s* electron located in the fourth energy level (an outer shell); in scandium (a fourth-period Group IIIB element) the last electron added is a 3*d* electron located in the third energy level (an inner shell). Figure 6.1 compares the locations of the last electron added in Group A and B types of elements.

Group	Noble gas	IA	IIA	IIIB	IVB
Element	Ar	K	Ca	Sc	Ti
Electron structure	2,8,8	2,8,8,1	2,8,8,2	2,8,9,2	2,8,10,2
Energy level	1,2,3	1,2,3,4	1,2,3,4	1,2,3,4	1,2,3,4
Last electron added	↑	↑	↑	↑	↑

Figure 6.1. Comparison of the placement of the last electron in Group A and Group B elements. The energy level to which the last electron is added is indicated by the arrow—an outer level for Group A elements, an inner level for Group B elements.

The general characteristics of group properties are as follows:

1. The number of electrons in the outer energy level of Groups IA through VIIA, IB, and IIB elements is the same as the group number. The other B groups and Group VIII do not show this characteristic. Each noble gas except helium has eight electrons in its outer energy level.
2. The group number in which an element is located will indicate one of its possible oxidation states, with some exceptions, notably in Group VIII.
3. The groups on the left and in the middle sections of the table tend to be metallic in nature. The groups on the right tend to be nonmetallic.
4. The radii of the elements increase from top to bottom within a particular group (for example, from lithium to francium).
5. Elements at the bottom of a group tend to be more metallic in their properties than those at the top. This tendency is especially noticeable in Groups IVA through VIIA.
6. Elements within an A group have the same number of electrons in their outer shell and show closely related chemical properties.
7. Elements within a B group have some similarity in electron structure and also show some similarities in chemical properties.

6.6 Predicting Formulas by Use of the Periodic Table

The periodic table can be used to predict the formulas of simple compounds. As we shall see in Chapter 7, the chemical properties of the elements are dependent on their electrons. Group A elements ordinarily form compounds using only the electrons in their outer energy level. If we examine Group IA, the alkali metals, we see that all of them have one electron in their outer energy level, all follow a noble gas in the table, and all, except lithium, have eight electrons in their next inner shell. These likenesses suggest that there should be a great deal of similarity in the chemistry of these metals, since their chemical properties are vested primarily in their outer-shell electron. And there is similarity—all readily lose their outer electron and attain a noble gas electron structure. In doing this, they form compounds with similar atomic compositions. For example, all the monoxides of Group IA contain two atoms of the alkali metal to one atom of oxygen. Their formulas are Li_2O, Na_2O, K_2O, Rb_2O, Cs_2O, and Fr_2O.

How can we use the table to predict formulas of other compounds? Because of similar electron structures, the elements in a family generally form compounds with the same atomic ratios. This was shown for the oxides of the Group IA metals above. In general, if we know the atomic ratio of a particular compound, say sodium chloride (NaCl), we can predict the atomic ratios and formulas of the other alkali metal chlorides. These formulas are LiCl, KCl, RbCl, CsCl, and FrCl.

In a similar way, if we know that the formula of the oxide of hydrogen is H_2O, we predict that the formula of the sulfide will be H_2S because sulfur has the same outer-shell electron structure as oxygen. It must be recognized, however, that these are only predictions; it does not necessarily follow that every element in a group will behave like the others, or even that a predicted compound will actually exist. Knowing the formulas for potassium chlorate, bromate, and iodate to be $KClO_3$, $KBrO_3$, and KIO_3, we can correctly predict the corresponding sodium compounds to have the formulas $NaClO_3$, $NaBrO_3$, and $NaIO_3$. Fluorine belongs to the same family of elements (Group VIIA) as chlorine, bromine, and iodine. Therefore, we can predict that the formulas for potassium and sodium fluorates will be KFO_3 and $NaFO_3$. These compounds are not known to exist; however, if they did exist, the formulas could very well be correct, for these predictions are based on comparisons with known formulas and/or similar electron structures.

Problem 6.1

The formula for magnesium sulfate is $MgSO_4$ and that for potassium sulfate is K_2SO_4. Predict the formulas for:
(a) Barium sulfate   (b) Lithium sulfate

(a) Look in the periodic table for the locations of magnesium and barium. They are both in Group IIA. Since Mg and Ba are in the same group, we can predict that the ratio of Ba^{2+} to SO_4^{2-} in barium sulfate will be the same as the ratio of Mg^{2+} to SO_4^{2-} in magnesium sulfate—namely 1:1. Therefore, the formula of barium sulfate will be $BaSO_4$.

(b) Check the periodic table and locate potassium and lithium in Group IA. Since both elements are in the same group and the formula of potassium sulfate is K_2SO_4, the formula for lithium sulfate will be Li_2SO_4.

6.7 Transition Elements

transition elements

Elements in Groups IB, IIIB through VIIB, and VIII are known as the **transition elements**. There are four series of transition elements, one in each of the periods 4, 5, 6, and 7. The transition elements are characterized by an increasing number of d or f electrons in an inner shell; they all have either one or two electrons in their outer shell. In period 4, electrons enter the $3d$ sublevel. In period 5, electrons enter the $4d$ sublevel. The transition elements in period 6 include the lanthanide series (or rare earth elements, La to Lu), in which electrons are entering the $4f$ sublevel. The $5d$ sublevel also fills up in the sixth period. The seventh period of elements is an incomplete period. It includes the actinide series (Ac to Lr) in which electrons are entering the $5f$ sublevel.

In the formation of compounds of the transition metals, electrons may come from more than one energy level. For this reason these metals form multiple series of compounds.

6.8 New Elements

Mendeleev allowed gaps in his orderly periodic table for elements whose discovery he predicted. These were actually discovered, as were all the elements up to atomic number 92 that occur naturally on the earth. Fourteen elements beyond uranium (atomic numbers 93–106) have been discovered or synthesized since 1939. All these elements have unstable nuclei and are radioactive. Beyond element 101, the isotopes synthesized thus far have such short lives that chemical identification has not been accomplished.

Intensive research is continuing on the synthesis of still heavier elements. Extending the periodic table beyond the presently known elements, it is predicted that elements 110 to 118 will be very stable, but still radioactive. Element 114 will lie below lead (82) and should be exceptionally stable. Element 118 should be a member of the noble gas family. Elements 119 and 120 should be in Groups IA and IIA, respectively, and have electrons in the $8s$ sublevel.

6.9 Summary

The periodic table has been used for studying the relationships of many properties of the elements. Ionization energies, densities, melting points, atomic radii, atomic volumes, oxidation states, electrical conductance, and electronegativity are just a few of these properties. However, a detailed discussion of all these properties is not practical at this time.

The periodic table is still used as a guide in predicting the synthesis of possible new elements. It presents a very large amount of chemical information in compact form and correlates the properties and relationships of all the elements. The table is so useful that a copy hangs in nearly every chemistry lecture hall and laboratory in the world. Refer to it often.

Questions

A. Review the meanings of the new terms introduced in this chapter.
   1. Periodic law
   2. Periodic table
   3. Periods of elements
   4. Groups or families of elements
   5. Transition elements

B. Answers to the following questions are to be found in tables and figures.
   1. How many elements are present in each period?
   2. Write the symbols of the alkali metal family in the order of increasing size of their atoms.
   3. What similarities do you observe in the elements of Group IIA?
   4. Write the symbols for the elements with atomic numbers 18, 36, 54, and 86. What do these elements have in common?
   5. Write the names and symbols of the halogens.
   6. Write the symbols for the family of elements that have two electrons in their outer energy level.
   7. Point out similarities and differences between Group IA and Group IB elements.
   8. What similarities do the elements of the lanthanide series possess?
   9. Where are the elements with the most metallic characteristics located in the periodic table?

C. Review questions.
   1. Write a paragraph describing the general features of the periodic table.
   2. What do you feel is the basis for Newlands' Law of Octaves?
   3. How are elements in a period related to one another?
   4. How are elements in a group related to one another?
   5. Classify the following elements as metals, nonmetals, or metalloids:
      (a) Potassium   (c) Sulfur   (e) Iodine   (g) Molybdenum
      (b) Plutonium   (d) Antimony   (f) Radium
   6. What is common about the electron structures of the alkali metals?
   7. Draw the electron-dot diagrams for Cs, Ba, Tl, Pb, Po, At, and Rn. How do these structures correlate with the group in which each element occurs?
   8. Pick the electron structures below that represent elements in the same chemical family.
      (a) $1s^2 2s^1$
      (b) $1s^2 2s^2 2p^4$
      (c) $1s^2 2s^2 2p^2$
      (d) $1s^2 2s^2 2p^6 3s^2 3p^4$
      (e) $1s^2 2s^2 2p^6 3s^2 3p^6$
      (f) $1s^2 2s^2 2p^6 3s^2 3p^6 4s^2$
      (g) $1s^2 2s^2 2p^6 3s^2 3p^6 4s^1$
      (h) $1s^2 2s^2 2p^6 3s^2 3p^6 3d^{14} 4s^2$
   9. In how many different principal energy levels do electrons occur in period 1, period 3, and period 5?
   10. In which period and group does an electron first appear in a *d* orbital?
   11. How many electrons occur in the outer shell of Groups IIIA and IIIB elements? Why are they different?

12. In which groups are transition elements located?
13. How do transition elements differ from other elements?
14. Which element in each of the following pairs has the larger atomic radius?
    (a) Na or K   (b) Na or Mg   (c) O or F   (d) Br or I   (e) Ti or Zr
15. Which element in each of Groups IA–VIIA has the smallest atomic radius?
16. Why does the atomic size increase in going down any group of the periodic table?
17. Letting E be an element in any group, the table represents the possible formulas

Group	IA	IIA	IIIA	IVA	VA	VIA	VIIA
	EH	EH$_2$	EH$_3$	EH$_4$	EH$_3$	H$_2$E	HE
	E$_2$O	EO	E$_2$O$_3$	EO$_2$	E$_2$O$_5$	EO$_3$	E$_2$O$_7$

of such compounds. Following the pattern in the table, write the formulas for the hydrogen and oxygen compounds of:
   (a) Na   (b) Ca   (c) Al   (d) Sn   (e) Sb   (f) Se   (g) Cl
18. Group IB elements have one electron in their outer shell, as do group IA elements. Would you expect them to form compounds such as CuCl, AgCl, and AuCl? Explain.
19. The formula for lead(II) bromide is PbBr$_2$; predict formulas for tin(II) and germanium(II) bromides. (See Section 8.3b for the use of Roman numerals in naming compounds.)
20. The formula for sodium sulfate is Na$_2$SO$_4$. Write the names and formulas for the other alkali metal sulfates.
21. All the atoms within each Group A family of elements can be represented by the same electron-dot structure. Complete the table, expressing the electron-dot structure for each group. Use E to represent the elements.

Group	IA	IIA	IIIA	IVA	VA	VIA	VIIA
	E·						

22. Oxygen and sulfur are extremely different elements in that one is a colorless gas and the other, a yellow crystalline solid. Why, then, are both located in Group VIA?
23. Why should the discovery of the existence of isotopes have bearing on the fact that the periodicity of the elements is a function of their atomic numbers and not their atomic weights?
24. Hydrogen can fit into Group VIIA as well as Group IA. Explain why it would not be wrong to place hydrogen in Group VIIA.

Try to answer Questions 25 through 28 without referring to the periodic table.

25. The atomic numbers of the noble gases are 10, 18, 36, 54, and 86. What are the atomic numbers for the elements with six electrons in their outer electron shells?
26. Element number 87 is in Group IA, period 7. Describe its outermost energy level. How many energy levels of electrons does it have?
27. If element 36 is a noble gas, in which group would you expect elements 35 and 37 to occur?
28. Which of the following statements about the elements in the periodic table are correct?
    (a) Properties of the elements are periodic functions of their atomic numbers.

(b) There are more nonmetallic elements than metallic elements.
(c) Metallic properties of the elements increase as you go from left to right across a period.
(d) Metallic properties of the elements increase as you go from top to bottom in a family of elements.
(e) Calcium is a member of the alkaline earth family.
(f) Group A elements do not contain any d or f electrons.
(g) An atom of element 37 has a larger volume than an atom of element 19.
(h) An atom of element 13 has a larger volume than an atom of element 11.
(i) An atom of germanium (Group IVA) has six electrons in its outer shell.
(j) If the formula for calcium iodide is CaI_2, then the formula for cesium iodide is CsI_2.
(k) Uranium is a transition element.
(l) If the electron-dot structure for barium (Group IIA) is Ba:, then the electron-dot structure for thallium (Group IIIA) is Tl:.
(m) The element with the electron configuration $1s^2 2s^2 2p^6 3s^2 3p^6 3d^{10} 4s^2 4p^3$ belongs to Group VB.

7 The Formation of Compounds from Atoms

After studying Chapter 7 you should be able to:

1. Understand the terms listed in Question A at the end of the chapter.
2. Describe how the ionization energies of the elements vary with respect to (1) position in the periodic table and (2) the removal of successive electrons.
3. Describe (1) the formation of ions by electron transfer and (2) the nature of the chemical bond formed by electron transfer.
4. Show by means of electron-dot structures the formation of an ionic compound from atoms.
5. Describe a crystal of sodium chloride.
6. Predict the formulas of the monatomic ions of Group A metals and nonmetals.
7. Predict the relative sizes of an atom and a monatomic ion for a given element.
8. Describe the covalent bond and predict whether a given covalent bond would be polar or nonpolar.
9. Determine from its electron dot structure whether a molecule is a dipole.
10. Describe the changes in electronegativity in (1) moving across a period and (2) moving down a group in the periodic table.
11. Distinguish clearly between ionic and molecular substances.
12. Predict whether the bonding in a compound is primarily ionic or covalent.
13. Distinguish coordinate covalent from covalent bonds in an electron-dot structure.
14. Draw electron-dot structures for monatomic and simple polyatomic ions.
15. Write the formulas of compounds formed by combining the ions from Tables 7.6 and 7.7 (or from the inside front cover of this book) in the correct ratios.
16. Assign the oxidation number to each element in a compound or ion.
17. Distinguish between oxidation number and ionic charge.
18. Distinguish between oxidation and reduction.
19. Identify in an equation the element that has been oxidized and the element that has been reduced.

7.1 Ionization Energy

Niels Bohr's description of the atom showed that electrons can exist at various energy levels when an atom absorbs energy. If sufficient energy is applied to an atom, it is possible to remove completely (or "knock out") one or more electrons from its structure, thereby forming a positive ion.

Atom + Energy ⟶ Positive ion + Electron (e^-)

ionization energy

The amount of energy required to completely remove one electron from an atom is known as the **ionization energy**. This energy may be expressed in units of kilocalories per mole (kcal/mole), indicating the number of kilocalories required to remove an electron from each atom in a mole of atoms. Thus, 314 kcal are required to remove one electron from a mole of hydrogen atoms; 567 kcal are required to remove the first and 1254 kcal to remove the second electron from a mole of helium atoms. Other units are often used to express ionization energy. The unit of energy in the International System of measurements is the joule (pronounced *jool*), where 1 calorie = 4.184 joules.

Table 7.1 gives ionization energies for the removal of five electrons (where available) from several selected elements. The table shows that it requires increasingly higher amounts of energy to remove the second, third, fourth, and fifth electrons. This is a logical sequence because as electrons are removed, the nuclear charge remains the same, and thereby holds the remaining electrons more tightly. The data also show an extra large increase in the ionization energy when an electron is removed from a noble gas electron structure, indicating the high stability of this structure.

Table 7.1. Ionization energies for selected elements. Values are expressed in kilocalories per mole, showing energies required to remove up to five electrons per atom. The values shown in color indicate the energy needed to remove an electron from a noble gas electron structure.

Element	1st e^-	2nd e^-	3rd e^-	4th e^-	5th e^-
H	314				
He	567	1254			
Li	124	1744	2823		
Be	215	420	3548	5020	
B	191	580	874	5980	7843
C	260	562	1104	1487	9034
Ne	497	947	1500	2241	2913
Na	118	1091	1652	2280	3192

Ionization energies have been experimentally determined for most of the elements. First ionization energies are given in Table 7.2. Figure 7.1 is a graphic plot of the first ionization energies of the first 56 elements, H through Ba.

The Formation of Compounds from Atoms 111

Table 7.2. First ionization energies of the elements (in kilocalories per mole).

1 H 314					Key												2 He 567
3 Li 124	4 Be 215			Atomic number — Symbol — Ionization energy (kcal/mole)		1 H 314						5 B 191	6 C 260	7 N 335	8 O 314	9 F 402	10 Ne 497
11 Na 118	12 Mg 176											13 Al 138	14 Si 188	15 P 242	16 S 239	17 Cl 300	18 Ar 363
19 K 100	20 Ca 141	21 Sc 151	22 Ti 158	23 V 155	24 Cr 156	25 Mn 171	26 Fe 182	27 Co 181	28 Ni 176	29 Cu 178	30 Zn 217	31 Ga 138	32 Ge 182	33 As 226	34 Se 225	35 Br 273	36 Kr 322
37 Rb 96.3	38 Sr 131	39 Y 147	40 Zr 158	41 Nb 159	42 Mo 164	43 Tc 168	44 Ru 170	45 Rh 172	46 Pd 192	47 Ag 175	48 Cd 207	49 In 133	50 Sn 169	51 Sb 199	52 Te 208	53 I 241	54 Xe 280
55 Cs 89.8	56 Ba 120	57 La 129	72 Hf 127	73 Ta 182	74 W 184	75 Re 181	76 Os 201	77 Ir 207	78 Pt 207	79 Au 213	80 Hg 240	81 Tl 141	82 Pb 171	83 Bi 184	84 Po 194	85 At 219	86 Rn 248
87 Fr 88.3	88 Ra 122	89 Ac 159	104 Ku	105 Ha	106 —												

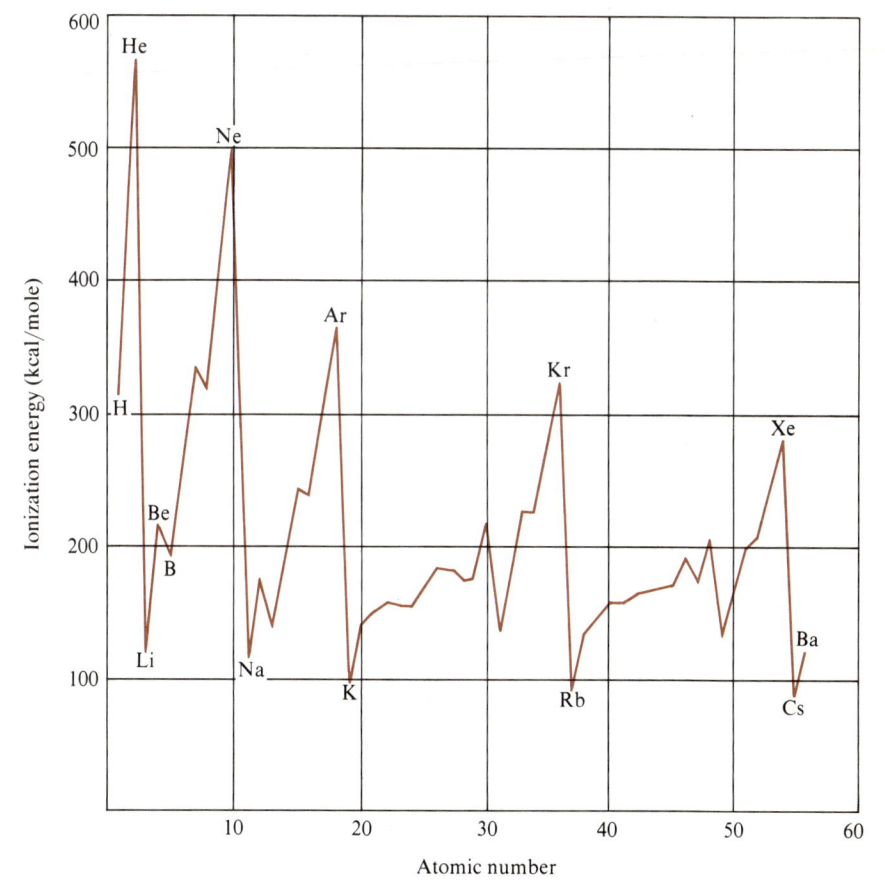

Figure 7.1. Periodic relationship of the first ionization energies to the atomic number of the elements.

Certain periodic relationships are noted in Table 7.2 and Figure 7.1. All the alkali metals have relatively low ionization energies, indicating that they each have one electron that is easily removed. Furthermore, the ionization energy decreases from Li to Cs, showing that Cs loses an electron more easily than the other elements. There are two main reasons for this family trend: the electron being removed (1) is farther away from its nucleus and (2) is shielded from its nucleus by more shells of electrons as one goes down the family from Li to Cs.

From left to right within a period, the ionization energy, despite some irregularities, gradually increases. The noble gases have relatively high values, confirming the nonreactivity of these elements and the stability of an electron structure containing eight electrons in the outer energy level.

7.2 Electrons in the Outer Shell

Two outstanding properties of the elements are their tendencies (1) to have two electrons in each atomic orbital and (2) to form a stable outer-shell electron structure. For many elements this stable outer shell contains eight electrons and is similar to the outer-shell electron structure of the noble gases. Atoms undergo electron structure rearrangements to attain a state of greater stability. These rearrangements are accomplished by losing, gaining, or sharing electrons with other atoms. For example, a hydrogen atom has a tendency to accept another electron and thus attain an electron structure like that of the stable noble gas helium; a fluorine atom can accommodate one more electron and attain a stable electron structure like neon; a sodium atom tends to lose one electron to attain a stable electron structure like neon. This process requires energy.

$$Na + Energy \longrightarrow Na^+ + 1\,e^-$$

valence electrons

The electrons in the outermost shell of an atom are responsible for most of this electron activity and are called the **valence electrons**. In electron-dot formulas of atoms the dots represent the outer-shell electrons and thus also represent the valence electrons. For example, hydrogen has one valence electron; sodium, one; aluminum, three; and oxygen, six. When a rearrangement of these electrons takes place between atoms, a chemical change occurs.

7.3 Transfer of Electrons from One Atom to Another

Let us look at the electron structures of sodium and chlorine to see how each element may attain a structure of 8 electrons in its outer shell. A sodium atom has 11 electrons: 2 in the first energy level, 8 in the second energy level, and 1 in the third energy level. Chlorine has 17 electrons: 2 in the first energy level, 8 in the second energy level, and 7 in the third energy level. If a sodium atom transfers or loses its 3s electron, its third energy level becomes vacant and it becomes a sodium ion with an electron configuration identical to that of the noble gas neon.

$$\underset{\text{Na atom }(1s^22s^22p^63s^1)}{\boxed{\begin{array}{c}11\,p\\12\,n\end{array}}\;2\,e^-8\,e^-1\,e^-} \longrightarrow \underset{\text{Na}^+\text{ ion }(1s^22s^22p^6)}{\boxed{\begin{array}{c}11\,p\\12\,n\end{array}}^{1+}\;2\,e^-8\,e^- + 1\,e^-}$$

An atom that has lost or gained electrons will have a plus or minus electrical charge, depending on which charged particles, protons or electrons, are in excess. A charged atom or group of atoms is called an *ion*. A positively charged ion is called a *cation* (pronounced cat-ion); a negative ion is called an *anion* (pronounced an-ion).

By losing a negatively charged electron, the sodium atom becomes a positively charged particle known as a sodium ion. The charge, +1, occurs because the nucleus still contains 11 positively charged protons but the electron orbitals now contain only 10 negatively charged electrons. The charge is indicated by a plus sign (+) and is written as a superscript after the symbol of the element (Na$^+$).

Determination of the charge of a sodium ion:

Na atom	11 p	Na$^+$ ion	11 p
	11 e$^-$		10 e$^-$
Charge	0	Charge	+1

A chlorine atom with 7 electrons in the third energy level needs 1 electron to pair up with its one unpaired 3p electron to attain the stable outer-shell electron structure of argon. By gaining 1 electron, the chlorine atom becomes a chloride ion (Cl$^-$), a negatively charged particle containing 17 protons and 18 electrons.

$$\underset{\text{Cl atom }(1s^22s^22p^63s^23p^5)}{\boxed{\begin{array}{c}17\,p\\18\,n\end{array}}\;2\,e^-8\,e^-7\,e^- + 1\,e^-} \longrightarrow \underset{\text{Cl}^-\text{ ion }(1s^22s^22p^63s^23p^6)}{\boxed{\begin{array}{c}17\,p\\18\,n\end{array}}^{1-}\;2\,e^-8\,e^-8\,e^-}$$

Determination of the charge of a chloride ion:

Cl atom	17 p	Cl$^-$ ion	17 p
	17 e$^-$		18 e$^-$
Charge	0	Charge	−1

Consider the case in which sodium and chlorine atoms react with each other. The 3s electron from the sodium atom transfers to the vacant 3p orbital in the chlorine atom to form a positive sodium ion and a negative chloride ion.

114   Chapter 7

The compound sodium chloride results because the Na$^+$ and Cl$^-$ ions are strongly attracted to one another by their opposite electrostatic charges:

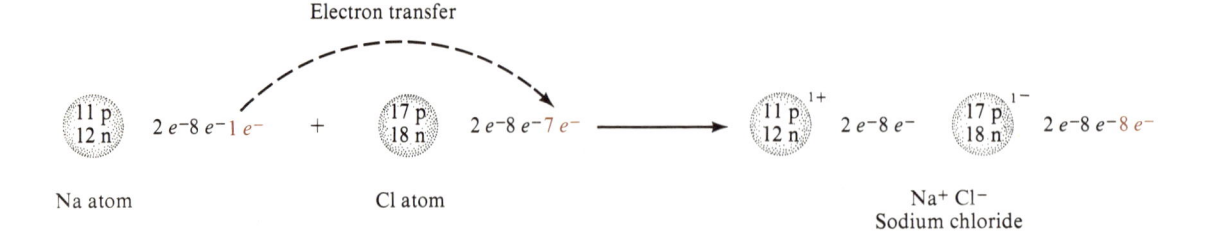

The electron-dot representation of sodium chloride formation is shown below.

$$Na\cdot + :\!\ddot{\underset{..}{Cl}}\!: \longrightarrow Na^+ \; :\!\ddot{\underset{..}{Cl}}\!:^-$$

The chemical reaction between sodium and chlorine is a very vigorous one. It is highly *exothermic* (evolving heat), liberating 90,200 cal when 1 gram-atomic weight of sodium (23.0 g) combines with 1 gram-atomic weight of chlorine (35.5 g).

Sodium chloride is actually made up of cubic crystals, in which each sodium ion is surrounded by six chloride ions and each chloride ion by six sodium ions, except at the crystal surface. A visible crystal is a regularly arranged aggregate of millions of these ions, but the ratio of sodium to chloride ions is one-to-one. The cubic crystalline lattice arrangement of sodium chloride is shown in Figure 7.2.

Figure 7.3 contrasts the relative sizes of sodium and chlorine atoms with that of their ions. The sodium ion is smaller than the atom due primarily to

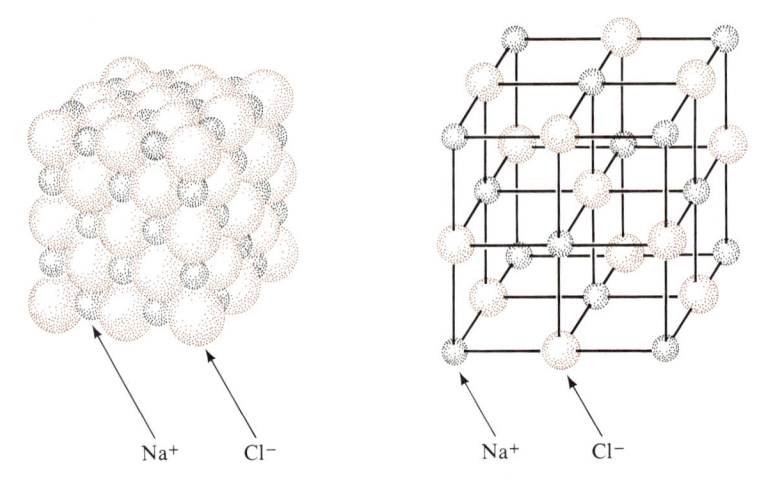

Figure 7.2. Sodium chloride crystal. Diagram represents a small fragment of sodium chloride, which forms cubic crystals. Each sodium ion is surrounded by six chloride ions, and each chloride ion is surrounded by six sodium ions.

The Formation of Compounds from Atoms 115

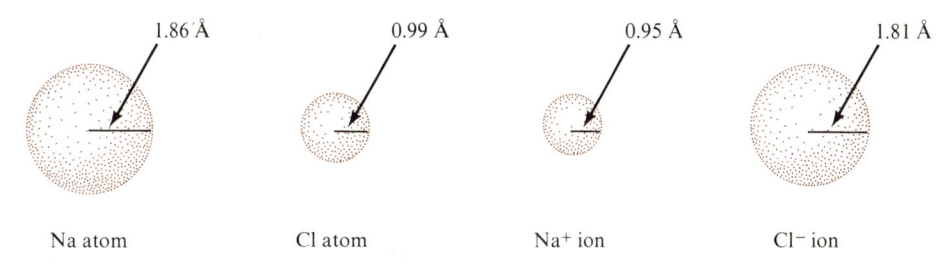

1.86 Å   0.99 Å   0.95 Å   1.81 Å

Na atom   Cl atom   Na⁺ ion   Cl⁻ ion

Figure 7.3. Relative sizes of sodium and chlorine atoms and their ions.

two factors: (1) the sodium atom has lost its outer shell, consisting of 1 electron, thereby reducing its size; and (2) the 10 remaining electrons are now attracted by 11 protons and are thus drawn closer to the nucleus. Conversely, the chloride ion is larger than the atom because it has 18 electrons but only 17 protons. The nuclear attraction on each electron is thereby decreased, allowing the chlorine atom to expand as it forms an ion.

A metal will ordinarily have one, two, or three electrons in its outer energy level. In reacting, metal atoms characteristically lose these electrons, attain the electron structure of a noble gas, and become positive ions. A nonmetal, on the other hand, lacks a small number of electrons of having a complete octet in its outer energy level and thus has a tendency to gain electrons (electron affinity). In reacting with metals, nonmetal atoms characteristically gain one, two, or three electrons, attain the electron structure of a noble gas, and become negative ions. The ions formed by loss of electrons are much smaller than the corresponding metal atoms; the ions formed by gaining electrons are larger than the corresponding nonmetal atoms. The actual dimensions of the atomic and ionic radii of several metals and nonmetals are given in Table 7.3.

Table 7.3. Changes in atomic size of selected metals and nonmetals. The metals shown lose electrons to become positive ions. The nonmetals gain electrons to become negative ions.

Atomic radius (Å)	Ionic radius (Å)	Atomic radius (Å)	Ionic radius (Å)
Li 1.52	Li⁺ 0.60	F 0.71	F⁻ 1.36
Na 1.86	Na⁺ 0.95	Cl 0.99	Cl⁻ 1.81
K 2.27	K⁺ 1.33	Br 1.14	Br⁻ 1.95
Mg 1.60	Mg²⁺ 0.65	O 0.74	O²⁻ 1.40
Al 1.43	Al³⁺ 0.50	S 1.03	S²⁻ 1.84

A magnesium atom of electron structure $1s^2 2s^2 2p^6 3s^2$ must lose two electrons or gain six electrons to reach a stable electron structure. If magnesium reacts with chlorine and each chlorine atom has room for only one electron, two chlorine atoms will be needed to accept the two electrons from one magnesium atom. The compound formed will contain one magnesium ion and two chloride ions. The magnesium ion will have a +2 charge, having lost two electrons. Each chloride ion will have a −1 charge. The transfer of electrons from a magnesium atom to two chlorine atoms is shown in the following illustration.

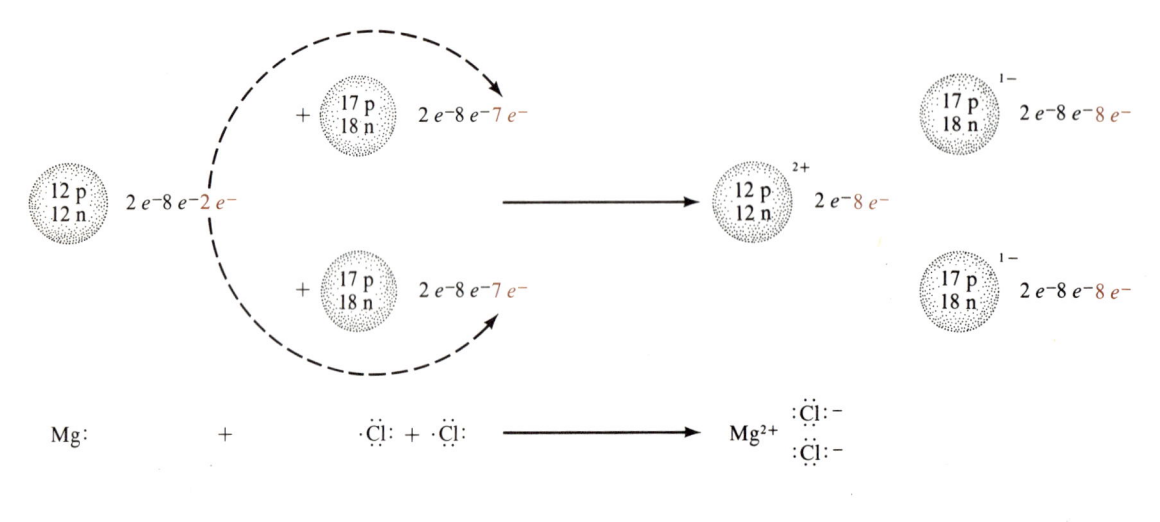

| Mg: | + | ·C̈l: + ·C̈l: | ⟶ | Mg²⁺ :C̈l:⁻ :C̈l:⁻ |

Mg atom   2 Cl atoms   Magnesium chloride

Study the following examples. Note the loss and gain of electrons between atoms; also note that the ions in each compound have a stable noble gas electron structure.

(a) Sodium fluoride, NaF

Na·   +   ·F̈:   ⟶   Na⁺ :F̈:⁻

Sodium atom   Fluorine atom   Sodium fluoride

The fluorine atom, with seven electrons in its outer shell, behaves similarly to a chlorine atom.

(b) Aluminum chloride, AlCl₃

$$Al· + \begin{array}{c} ·\ddot{C}l: \\ ·\ddot{C}l: \\ ·\ddot{C}l: \end{array} \longrightarrow Al^{3+} \begin{array}{c} :\ddot{C}l:^- \\ :\ddot{C}l:^- \\ :\ddot{C}l:^- \end{array}$$

Aluminum atom   Chlorine atoms   Aluminum chloride

Each chlorine atom can accept only one electron. Therefore, three chlorine atoms are needed to combine with the three outer-shell electrons of one aluminum atom. The aluminum atom has lost three electrons to become an aluminum ion, Al^{3+}, with a $+3$ charge.

(c) Magnesium oxide, MgO

$$\begin{pmatrix}12\ p\\12\ n\end{pmatrix}\ 2e^{-}8e^{-}2e^{-} + \begin{pmatrix}8\ p\\8\ n\end{pmatrix}\ 2e^{-}6e^{-} \longrightarrow \begin{pmatrix}12\ p\\12\ n\end{pmatrix}^{2+}\ 2e^{-}8e^{-}\quad \begin{pmatrix}8\ p\\8\ n\end{pmatrix}^{2-}\ 2e^{-}8e^{-}$$

$$Mg\!: \quad + \quad \cdot\ddot{O}\!: \quad \longrightarrow \quad Mg^{2+}\!:\!\ddot{O}\!:^{2-}$$

Magnesium atom   Oxygen atom   Magnesium oxide

The magnesium atom, with two electrons in the outer energy level, exactly fills the need of two electrons of one oxygen atom. The resulting compound has a ratio of one atom of magnesium to one atom of oxygen. The oxygen (oxide) ion has a -2 charge, having gained two electrons. In combining with oxygen, magnesium behaves the same way as when combining with chlorine—it loses two electrons.

(d) Sodium sulfide, Na_2S

$$\begin{matrix}Na\cdot\\ \\Na\cdot\end{matrix}\quad +\quad \cdot\ddot{S}\!:\quad \longrightarrow \quad \begin{matrix}Na^+\\ \\ \!:\!\ddot{S}\!:^{2-}\\ \\Na^+\end{matrix}$$

Sodium atoms   Sulfur atom   Sodium sulfide

Two sodium atoms supply the electrons that one sulfur atom needs to make eight in its outer shell.

(e) Aluminum oxide, Al_2O_3

$$\begin{matrix}\ddot{Al}\cdot\\ \\ \\ \ddot{Al}\cdot\end{matrix}\quad +\quad \begin{matrix}\cdot\ddot{O}\!:\\ \\ \cdot\ddot{O}\!:\\ \\ \cdot\ddot{O}\!:\end{matrix}\quad \longrightarrow \quad \begin{matrix}\!:\!\ddot{O}\!:^{2-}\\ Al^{3+}\\ \!:\!\ddot{O}\!:^{2-}\\ Al^{3+}\\ \!:\!\ddot{O}\!:^{2-}\end{matrix}$$

Aluminum atoms   Oxygen atoms   Aluminum oxide

One oxygen atom, needing two electrons, cannot accommodate the three electrons from one aluminum atom. One aluminum atom falls one electron short of the four electrons needed by two oxygen atoms. A ratio of two atoms of aluminum to three atoms of oxygen, involving the transfer of six electrons (two to each oxygen atom), gives each atom a stable electron configuration.

Note that in each of the examples above outer shells containing eight electrons were formed in all the negative ions. This formation resulted in the pairing of all the *s* and *p* electrons in these outer shells.

118 Chapter 7

Chemistry would be considerably simpler if all compounds were made by the direct formation of ions as outlined in the examples just given. Unfortunately, this is only one of the two general methods of compound formation. The second general method will be outlined in the sections that follow.

7.4 Sharing of Electrons

The formula of chlorine gas is Cl$_2$. When the two atoms of chlorine combine to form this molecule, the electrons must interact by a method that is different from that shown in the preceding examples. Each chlorine atom would be more stable with eight electrons in its outer shell. But if an electron transfers from one chlorine atom to the other, the first chlorine atom, with only six electrons remaining in its outer shell, would be highly unstable. What actually happens when the two chlorine atoms join together is this: the unpaired 3p electron orbital of one chlorine atom overlaps the unpaired 3p electron orbital of the other atom, resulting in a pair of electrons that are mutually shared between the two atoms. Each atom furnishes one of the pair of shared electrons. Thus, each atom attains a stable structure of eight electrons by sharing an electron pair with the other atom. The pairing of the p electrons and formation of a chlorine molecule is illustrated in the following diagrams.

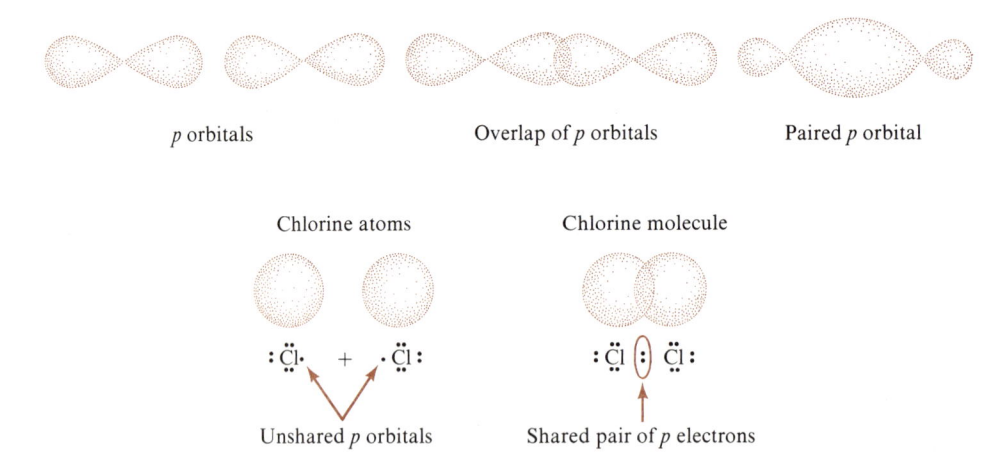

p orbitals   Overlap of *p* orbitals   Paired *p* orbital

Chlorine atoms   Chlorine molecule

:Cl· + ·Cl: :Cl(:)Cl:

Unshared *p* orbitals   Shared pair of *p* electrons

Neither chlorine atom has a positive or negative charge, since both contain the same number of protons and have equal attraction for the pair of electrons being shared. Other examples of molecules in which there is equal sharing of electrons between two atoms are hydrogen, H$_2$; oxygen, O$_2$; nitrogen, N$_2$; fluorine, F$_2$; bromine, Br$_2$; and iodine, I$_2$. Note that two or even three pairs of electrons may be shared between atoms.

H:H   :F:F: :Br:Br: :I:I: :O::O: :N:::N:

Hydrogen   Fluorine   Bromine   Iodine   Oxygen   Nitrogen

The Formation of Compounds from Atoms 119

electronegativity

When two different kinds of atoms share a pair of electrons, one atom assumes a partial positive charge and the other a partial negative charge (in respect to each other). This is because the two atoms exert unequal attraction for the pair of shared electrons. The attractive force that an atom of an element has for shared electrons in a molecule is known as its **electronegativity**. Elements differ in their electronegativities. For example, both hydrogen and chlorine need one electron to form stable electron configurations. They share a pair of electrons in the substance hydrogen chloride, HCl. Chlorine is more electronegative and therefore has a greater attraction for the shared electrons than does hydrogen. As a result, the pair of electrons is displaced toward the chlorine atom, giving it a partial negative charge and leaving the hydrogen atom with a partial positive charge. It should be understood that the electron is not transferred entirely to the chlorine atom, as in the case of sodium chloride, and no ions are formed. The entire molecule, HCl, is electrically neutral.

H:Cl:
Hydrogen chloride

The pair of shared electrons in HCl is closer to the more electronegative chlorine atom than to the hydrogen atom, giving chlorine a partial negative charge with respect to the hydrogen atom

The electronegativity, or ability of an atom to attract an electron, depends on several factors: (1) the charge on the nucleus, (2) the distance of the outer electrons from the nucleus, and (3) the amount of shielding of the nucleus by intervening shells of electrons between the outer-shell electrons and the nucleus. A scale of relative electronegativities, in which the most electronegative element, fluorine, is assigned a value of 4.0, was developed by Linus Pauling (1901–) and is given in Table 7.4. This table shows that the relative electronegativity of the nonmetals is high and that of the metals is low. These electronegativities

Table 7.4. Relative electronegativity of the elements. The electronegativity value is given below the symbol of each element.

1 H 2.1																	2 He
3 Li 1.0	4 Be 1.5											5 B 2.0	6 C 2.5	7 N 3.0	8 O 3.5	9 F 4.0	10 Ne —
11 Na 0.9	12 Mg 1.2											13 Al 1.5	14 Si 1.8	15 P 2.1	16 S 2.5	17 Cl 3.0	18 Ar —
19 K 0.8	20 Ca 1.0	21 Sc 1.3	22 Ti 1.4	23 V 1.6	24 Cr 1.6	25 Mn 1.5	26 Fe 1.8	27 Co 1.8	28 Ni 1.8	29 Cu 1.9	30 Zn 1.6	31 Ga 1.6	32 Ge 1.8	33 As 2.0	34 Se 2.4	35 Br 2.8	36 Kr —
37 Rb 0.8	38 Sr 1.0	39 Y 1.2	40 Zr 1.4	41 Nb 1.6	42 Mo 1.8	43 Tc 1.9	44 Ru 2.2	45 Rh 2.2	46 Pd 2.2	47 Ag 1.9	48 Cd 1.7	49 In 1.1	50 Sn 1.8	51 Sb 1.9	52 Te 2.1	53 I 2.5	54 Xe —
55 Cs 0.7	56 Ba 0.9	57–71 La-Lu 1.1–1.2	72 Hf 1.3	73 Ta 1.5	74 W 1.7	75 Re 1.9	76 Os 2.2	77 Ir 2.2	78 Pt 2.2	79 Au 2.4	80 Hg 1.9	81 Tl 1.8	82 Pb 1.8	83 Bi 1.9	84 Po 2.0	85 At 2.2	86 Rn —
87 Fr 0.7	88 Ra 0.9	89– Ac– 1.1–1.7	104 Ku	105 Ha	106 —												

9 — Atomic number
F — Symbol
4.0 — Electronegativity

indicate that atoms of metals have a greater tendency to lose electrons than do atoms of nonmetals, and that nonmetals have a greater tendency to gain electrons. The higher the electronegativity value, the greater the attraction for electrons.

The following examples further illustrate electrons shared between atoms to form molecules. The electron-dot structure is used, showing the individual atoms and then the molecule formed from them. In each case, an electron structure is formed to give each atom a noble gas electron structure.

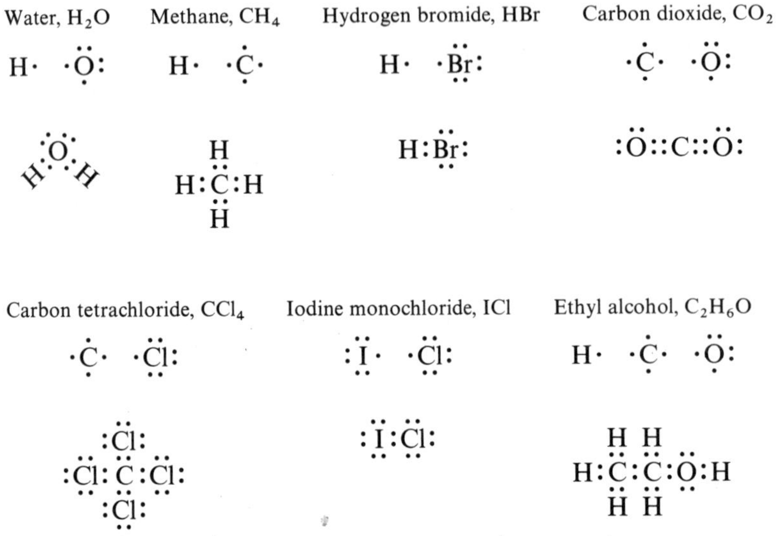

Water, H₂O   Methane, CH₄   Hydrogen bromide, HBr   Carbon dioxide, CO₂

H· ·Ö: H· ·C·   H· ·B̈r: ·C· ·Ö:

H:Ö:H   H   H:B̈r: :Ö::C::Ö:
              H:C:H
              H

Carbon tetrachloride, CCl₄   Iodine monochloride, ICl   Ethyl alcohol, C₂H₆O

·C· ·C̈l: :Ï· ·C̈l: H· ·C· ·Ö:

    :C̈l: :Ï:C̈l: H H
:C̈l:C:C̈l: H:C:C:Ö:H
    :C̈l: H H

A dash (—) written between two atoms is often used to represent the pair of shared electrons. These same structures would then appear as follows:

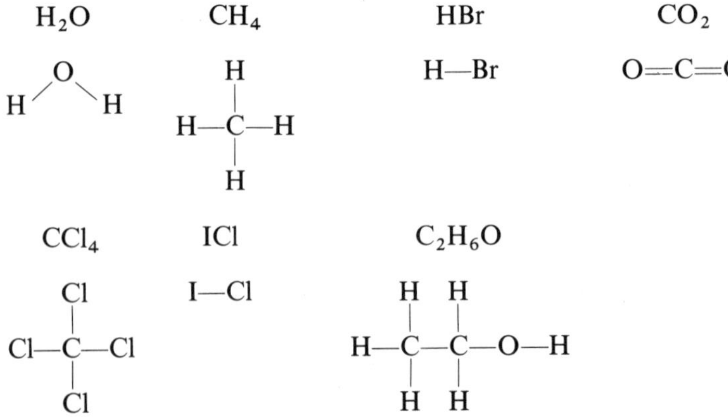

H₂O   CH₄   HBr   CO₂

              H   H—Br   O=C=O
  O
H   H   H—C—H
              |
              H

CCl₄   ICl   C₂H₆O

  Cl   I—Cl   H H
  | | |
Cl—C—Cl   H—C—C—O—H
  | | |
  Cl   H H

One should not get the impression that these shared electrons are in a fixed position between their respective atoms. The placement of the dots in these diagrams is merely a convenient method of showing the shared pairs of electrons. Both atoms use these shared electrons in their orbitals to complete their octets of electrons.

7.5 Chemical Bonds

chemical bond

Except in very rare instances matter does not fly apart spontaneously. It is prevented from doing so by forces acting at the ionic and molecular levels. Through chemical reactions, atoms tend to attain more stable states at lower chemical potential energy levels. Atoms react chemically by losing, gaining, or sharing electrons. Forces arise from electron transferring and sharing interactions. These forces that hold oppositely charged ions together or that bind atoms together in molecules or in polyatomic ions are called **chemical bonds**. There are two principal types of bonds: the electrovalent, or ionic, bond and the covalent bond. We will study these two bond types and their modifications in more detail in the next four sections of this chapter.

7.6 The Electrovalent Bond

electrovalent bond

ionic bond

When sodium reacts with chlorine, each atom becomes an electrically charged ion. Sodium chloride, and indeed all ionic substances, is bonded together by the attraction existing between positive and negative charges. An **electrovalent** or **ionic bond** is the electrostatic attraction existing between oppositely charged ions.

Electrovalent, or ionic, bonds are formed whenever there is a complete transfer of one or more electrons from one atom to another. The metals, which have comparatively low electronegativities and little attraction for their valence electrons, tend to form ionic bonds when they combine with nonmetals. Section 7.3 gives several examples of electron transfer reactions that result in the formation of electrovalent bonds. Restudy these examples. Substances that contain polyatomic ions (see Section 7.10) are also electrostatically bonded.

It is important to recognize that electrovalently bonded substances do not exist as molecules. In sodium chloride, for example, the bond does not exist solely between a single sodium ion and a single chloride ion. Each sodium ion in the crystal attracts six near-neighbor negative chloride ions; in turn, each negative chloride ion attracts six near-neighbor positive sodium ions (see Figure 7.2).

7.7 The Covalent Bond

covalent bond

A pair of electrons shared between two atoms constitutes a **covalent bond**. It is the most predominant chemical bond in nature. The concept of the covalent bond was introduced in 1916 by Gilbert N. Lewis (1875–1946) of the University of California at Berkeley.

True molecules exist in compounds that are held together by covalent bonds. It is not correct to refer to "a molecule" of sodium chloride or other ionic compounds, since these compounds exist as large aggregates of positive and negative ions. But we can refer to a molecule of hydrogen, chlorine, hydrogen chloride, carbon tetrachloride, sugar, or carbon dioxide, because these compounds contain only covalent bonds and exist in molecular aggregates.

A study of the hydrogen molecule will give us a better insight into the nature of the covalent bond and its formation. The formation of a hydrogen molecule, H_2, involves the overlapping and pairing of $1s$ electron orbitals from two hydrogen atoms. This overlapping and pairing is shown in Figure 7.4. Each atom contributes one electron of the pair that is shared jointly by two hydrogen nuclei. The orbital of the electrons now includes both hydrogen nuclei, but probability factors show that the most likely place to find the electrons (the point of highest electron density) is between the two nuclei.

The tendency for hydrogen atoms to form a molecule is very strong. In the molecule, each electron is attracted by two positive nuclei. This attraction gives the hydrogen molecule a more stable structure than the individual hydrogen atoms had. Experimental evidence of stability is shown by the fact that 104.2 kcal are needed to break the bonds between the hydrogen atoms in one mole of hydrogen (2.0 g). The strength of a bond may be determined by the energy required to break it. The energy required to break a covalent bond is known as the *bond dissociation energy*. The following bond dissociation energies illustrate relative bond strengths. (All substances are considered to be in the gaseous state and to form neutral atoms.)

Reaction	Bond dissociation energy (kcal/mole)
$H_2 \rightarrow 2\,H$	104.2
$N_2 \rightarrow 2\,N$	226.0
$O_2 \rightarrow 2\,O$	118.3
$F_2 \rightarrow 2\,F$	36.6
$Cl_2 \rightarrow 2\,Cl$	58.0
$Br_2 \rightarrow 2\,Br$	46.1
$I_2 \rightarrow 2\,I$	36.1

The covalent bond is designated by a dash (—) between the two atoms (for example, H–H, H–Cl, C=O). A single dash means one pair of electrons; two dashes mean two pairs, or four electrons, are shared between two atoms. All the examples in Section 7.4 contain covalent bonds between atoms. Study each

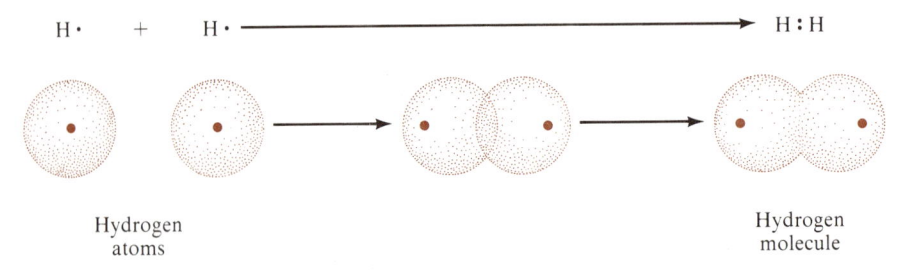

Figure 7.4. The formation of a hydrogen molecule from two hydrogen atoms.

example carefully, noting especially the pair of shared electrons that form each covalent bond.

7.8 Polar Covalent Bonds

We have considered bonds to be either covalent or electrovalent, according to whether electrons are shared between atoms or are transferred from one atom to another. In most covalent bonds, the pairs of electrons are not shared equally between the atoms. Such bonds are known as polar covalent bonds.

nonpolar covalent bond

In a **nonpolar covalent bond** the shared pair of electrons is attracted equally by the two atoms. A bond between the same kind of atoms, such as that in a hydrogen molecule, is nonpolar because the electronegativity difference between identical atoms is zero.

When a covalent bond is formed between two atoms of different electronegativities, the more electronegative atom attracts the shared electron pair toward itself. As a result, the atom with the higher electronegativity acquires a partial negative charge and the other atom, a partial positive charge. However, the overall molecule is still neutral. Due to this greater attraction of the electron pair, the bond formed between the two atoms has partial ionic character and is known as a **polar covalent bond**. The resulting molecule is said to be *polar*.

polar covalent bond

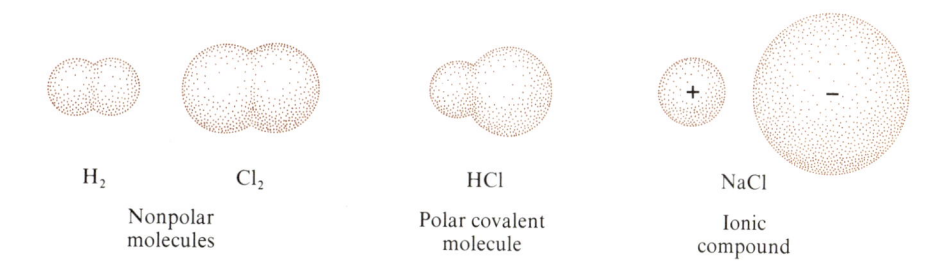

H_2	Cl_2	HCl	NaCl
Nonpolar molecules		Polar covalent molecule	Ionic compound

dipole

A **dipole** is a molecule that is electrically unsymmetrical, causing it to be oppositely charged at two points. A dipole is often written as $(+\ -)$. A hydrogen chloride molecule is polar and behaves as a small dipole. The HCl dipole may be written as H ⟶ Cl. The arrow points toward the negative end of the dipole. Molecules of H_2O, HBr, and ICl are polar; CH_4, CCl_4, and CO_2 are nonpolar.

$$H \longrightarrow Cl \qquad H \longrightarrow Br \qquad I \longrightarrow Cl \qquad \overset{O}{\underset{H\quad H}{}}$$

The greater the difference in electronegativity between two atoms, the more polar is the bond between them. When this difference is sufficiently large (greater than 1.7–1.9 electronegativity units), the bond between the two atoms will be essentially ionic (with some exceptions, of course). If the difference is less than 1.7 units, the bond will be essentially covalent. The difference in electronegativity between atoms can also give us a guide to the relative strength of covalent bonds. The greater the difference, the stronger the bond—that is, the more energy required to break the bond. For example, HF has the strongest bond in the series HF, HCl, HBr, and HI, as seen by the bond dissociation data in the following table.

Compound	Electronegativity difference	Bond dissociation energy (kcal/mole)
HF	1.9	134.6
HCl	0.9	103.2
HBr	0.7	87.5
HI	0.4	71.4

Care must be taken to distinguish between polar bonds and polar molecules. A covalent bond between different kinds of atoms is always polar. But a molecule containing different kinds of atoms may or may not be polar, depending on its shape or geometry. The HF, HCl, HBr, HI, and ICl molecules just mentioned are all polar because each contains a single polar bond. However, CO_2, CH_4, and CCl_4 are nonpolar molecules despite the fact that all three contain polar bonds. The carbon dioxide molecule, O=C=O, is nonpolar because the carbon–oxygen dipoles cancel each other by acting in opposite directions.

$$\overleftarrow{O=C}\overrightarrow{=O}$$

Dipoles in opposite directions

Methane (CH_4) and carbon tetrachloride (CCl_4) are nonpolar because the C–H and C–Cl dipoles form tetrahedral angles (see Section 22.3) and thereby cancel one another.

7.9 Coordinate Covalent Bonds

In Section 7.7, we saw that a covalent bond was formed by the overlapping of electron orbitals between two atoms. The two atoms each furnish an electron to make a pair that is shared between them.

Covalent bonds can also be formed by a single atom furnishing both electrons that are shared between the two atoms. The bond so formed is called a **coordinate covalent**, or **semipolar**, **bond**. This bond is often designated by an arrow pointing away from the electron donor (for example, A → B). Once formed, a coordinate covalent bond has the same properties as any other covalent bond—it simply is a pair of electrons shared between two atoms.

The electron-dot structures of sulfurous and sulfuric acids show a coordinate covalent bond between the sulfur and the oxygen atoms that are not

coordinate covalent bond

bonded to hydrogen atoms. The colored dots indicate the electrons of the sulfur atom.

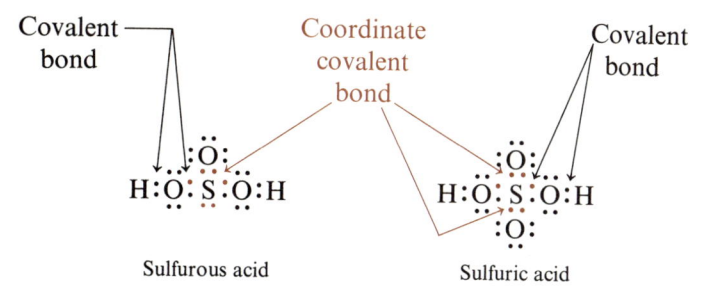

$$\underset{\text{Sulfurous acid}}{H:\overset{..}{\underset{..}{O}}:\overset{:\overset{..}{O}:}{\underset{..}{S}}:\overset{..}{\underset{..}{O}}:H} \qquad \underset{\text{Sulfuric acid}}{H:\overset{..}{\underset{..}{O}}:\overset{:\overset{..}{O}:}{\underset{:\underset{..}{\overset{..}{O}}:}{S}}:\overset{..}{\underset{..}{O}}:H}$$

Covalent bond — Coordinate covalent bond — Covalent bond

The open (unbonded) pair of electrons on the sulfur atom in sulfurous acid allows room for another oxygen atom with six electrons to fit perfectly into its structure and form sulfuric acid. Other atoms with six electrons in their outer shell, such as sulfur, could also fit into this pattern. The coordinate covalent bond explains the formation of many complex molecules.

7.10 Polyatomic Ions

polyatomic ion

A **polyatomic ion** is a stable group of atoms that has either a positive or a negative charge and behaves as a single unit in many chemical reactions. Sodium sulfate, Na_2SO_4, contains two sodium ions and a sulfate ion. The sulfate ion, SO_4^{2-}, is a polyatomic ion composed of one sulfur atom and four oxygen atoms, and has a charge of -2. One sulfur and four oxygen atoms have a total of 30 electrons in their outer shells. The sulfate ion contains 32 outer-shell electrons and therefore has a charge of -2. In this case, the two additional electrons come from the two sodium atoms, which are now sodium ions.

$$Na^+ \left[:\overset{:\overset{..}{O}:}{\underset{:\underset{..}{\overset{..}{O}}:}{\overset{..}{\underset{..}{O}}:S:\overset{..}{\underset{..}{O}}:}} \right]^{2-} Na^+ \qquad \left[:\overset{:\overset{..}{O}:}{\underset{:\underset{..}{\overset{..}{O}}:}{\overset{..}{\underset{..}{O}}:S:\overset{..}{\underset{..}{O}}:}} \right]^{2-}$$

Sodium sulfate   Sulfate ion

Sodium sulfate has both ionic and covalent bonds. Ionic bonds exist between each of the sodium ions and the sulfate ion. Covalent bonds are present between the sulfur and oxygen atoms within the sulfate ion. One important difference between the ionic and covalent bonds in this compound may be demonstrated by dissolving sodium sulfate in water. It dissolves in water, forming three charged particles—two sodium ions and one sulfate ion per formula unit of sodium sulfate:

$$\underset{\text{Sodium sulfate}}{Na_2SO_4} \xrightarrow{\text{Water}} \underset{\text{Sodium ions}}{2\,Na^+} + \underset{\text{Sulfate ion}}{SO_4^{2-}}$$

The ion SO_4^{2-} remains as a unit, held together by covalent bonds; whereas, where there were ionic bonds, dissociation of the ions took place. Do not think, however, that polyatomic ions are so stable that they cannot be altered. They may indeed be changed into other compounds or ions in certain chemical changes.

The electron-dot formulas for several common polyatomic ions are shown below.

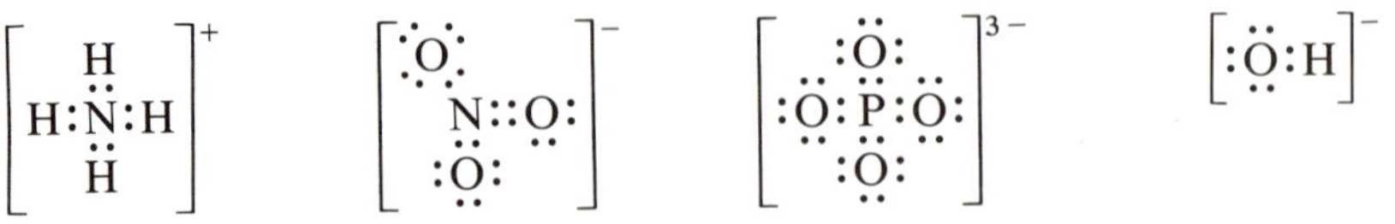

$$\left[\begin{matrix} & H & \\ H:&\overset{..}{N}:&H \\ & H & \end{matrix}\right]^+ \quad \left[\begin{matrix} & :\overset{..}{O}: & \\ & N::O: & \\ & :\overset{..}{O}: & \end{matrix}\right]^- \quad \left[\begin{matrix} & :\overset{..}{O}: & \\ :\overset{..}{O}:&P:&\overset{..}{O}: \\ & :\overset{..}{O}: & \end{matrix}\right]^{3-} \quad \left[:\overset{..}{\underset{..}{O}}:H\right]^-$$

Ammonium ion, NH_4^+ Nitrate ion, NO_3^- Phosphate ion, PO_4^{3-} Hydroxide ion, OH^-

7.11 Oxidation Numbers of Atoms

oxidation number

We have seen that atoms can combine to form compounds by losing, gaining, or sharing electrons. The **oxidation number** of an element is a number having a positive, a negative, or a zero value that may be assigned to an atom of that element in a compound. These positive and negative numbers are directly related to the positive and negative charges that result from the transfer of electrons from one atom to another in ionic compounds or from an unequal sharing of electrons between atoms forming covalent bonds. Oxidation numbers are assigned by a somewhat arbitrary system of rules. They are useful for writing formulas, naming compounds, and balancing chemical equations.

In a compound having ionic bonds, the oxidation number of an atom or group of atoms existing as an ion is the same as the *charge of the ion*. Thus, in sodium chloride, NaCl, the oxidation number of sodium is $+1$ and that of chlorine is -1; in magnesium oxide, MgO, the oxidation number of magnesium is $+2$ and that of oxygen is -2; in calcium chloride, $CaCl_2$, the oxidation number of calcium is $+2$ and that of chlorine is -1. The sum of the oxidation numbers of all the atoms in a compound is numerically equal to zero, since a compound is electrically neutral.

For practical purposes it is also convenient to assign oxidation numbers to the individual atoms comprising molecules and polyatomic ions. Here the electrons have not been completely transferred from one atom to another and the assignment cannot be done solely on the basis of ionic charges. However, oxidation numbers can be readily assigned to the atoms in either molecules or polyatomic ions by this general method: For each covalent bond, first assign the shared pair of electrons to the more electronegative atom. Then assign an oxidation number to each atom corresponding to its apparent net charge based (1) on the number of electrons gained or lost and (2) on the fact that the sum of the oxidation numbers must equal zero for a compound or must equal the charge on a polyatomic ion. Consider these substances, H_2, H_2O, CH_4, and CCl_4:

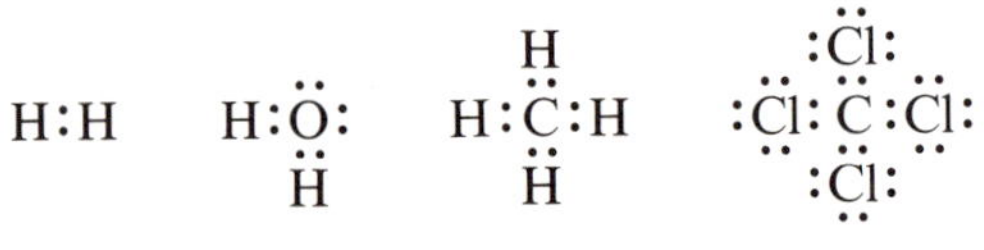

$$H:H \quad H:\overset{..}{\underset{..}{O}}:\!\!\! \atop \phantom{H:\underset{..}{O}:}H \quad H:\overset{H}{\underset{H}{C}}:H \quad :\overset{..}{\underset{..}{Cl}}:\overset{:\overset{..}{Cl}:}{\underset{:\overset{..}{Cl}:}{C}}:\overset{..}{\underset{..}{Cl}}:$$

In H_2, the pair of electrons is shared equally between the two atoms; therefore, each H is assigned an oxidation number of zero. In H_2O, oxygen is the more electronegative atom and is assigned the two pairs of shared electrons. The oxygen atom now has two additional electrons over the neutral atom, and therefore is assigned an oxidation number of -2. Each hydrogen atom in H_2O has one less electron than the neutral atom and is assigned an oxidation number of $+1$. In CH_4, all four shared pairs of electrons are assigned to the more electronegative carbon atom. The carbon atom then has an additional four electrons and is assigned an oxidation number of -4. Each hydrogen atom has one less electron than the neutral atom and is assigned an oxidation number of $+1$. In CCl_4, one pair of electrons is assigned to each of the four more electronegative chlorine atoms. The carbon atom therefore has four less electrons than the neutral atom and is assigned an oxidation number of $+4$. Each chlorine atom has one additional electron and is assigned an oxidation number of -1.

The following rules govern the assignment of oxidation numbers:

1. The oxidation number of any free element is zero, even when the atoms are combined with themselves. (*Examples*: Na, Mg, H_2, O_2)
2. Metals generally have positive oxidation numbers in compounds.
3. The oxidation number of hydrogen in a compound or an ion is $+1$ except in metal hydrides.
4. The oxidation number of oxygen in a compound or an ion is -2 except in peroxides.
5. The oxidation number of a monatomic ion is the same as the charge on the ion.
6. The oxidation number of an atom in a covalent compound is equal to the net apparent charge on the atom after each pair of shared electrons is assigned to the more electronegative element sharing the pair of electrons.
7. The algebraic sum of the oxidation numbers for all the atoms in a compound must equal zero.
8. The algebraic sum of the oxidation numbers for all the atoms in a polyatomic ion must equal the charge on the ion.

The oxidation numbers of many elements are predictable from their position in the periodic table. This is especially true of the Group A elements because the number of electrons in their outer shells corresponds to the group number. Remember that metals lose electrons, becoming positively charged ions. Nonmetals tend to gain electrons, and become negatively charged ions, but they can also share electrons with other atoms to assume a positive or negative oxidation number. Hydrogen can have a $+1$ or -1 oxidation number, depending on the relative electronegativity of the element with which it is combined.

The predictable oxidation numbers of the Group A elements are given in the following table:

	IA	IIA	IIIA	IVA	VA	VIA	VIIA
Oxidation number	$+1$	$+2$	$+3$	$+4$ to -4	-3 to $+5$	-2 to $+6$	-1 to $+7$

Table 7.5 illustrates the use of oxidation numbers to predict formulas of binary compounds from representative members of these groups.

Table 7.5. Selected binary hydrogen, oxygen, and chlorine compounds of Group A elements.

	IA	IIA	IIIA	IVA	VA	VIA	VIIA
Hydrogen compound	NaH	CaH_2	AlH_3	CH_4	NH_3	H_2S	HCl
Oxygen compound	Na_2O	CaO	Al_2O_3	CO_2	N_2O_5	SO_3	Cl_2O
Chlorine compound	NaCl	$CaCl_2$	$AlCl_3$	CCl_4	NCl_3	SCl_2	Cl_2

7.12 Oxidation Number Tables

The writing of formulas of compounds and chemical equations is facilitated by a knowledge of oxidation numbers and ionic charges. Table 7.6 lists the names and ionic charges of common monatomic ions. Monatomic ions of Group A elements are not given in Table 7.6 because the charges and oxidation numbers of these ions are readily determined from the periodic table. The charges and oxidation numbers of the Groups IA, IIA, and IIIA metal ions are positive and correspond to the group number (for example, Na^+, Ca^{2+}, Al^{3+}). The negative charges and oxidation numbers of Groups VA, VIA, and VIIA monatomic ions can be determined by subtracting eight from the group number. Sulfur, for example, is in Group VIA, and $6 - 8 = -2$. Therefore, the oxidation number of the sulfide ion (S^{2-}) is -2. All the halogens (F, Cl, Br, I) in binary compounds with metals or hydrogen have an oxidation number of -1.

Table 7.6. Names and charges of selected monatomic ions. Ions of Group A elements are not shown.

Name	Formula	Charge	Name	Formula	Charge
Arsenic(III)	As^{3+}	3+	Manganese(II)	Mn^{2+}	2+
Cadmium	Cd^{2+}	2+	Mercury(I)	Hg^+	1+
Chromium(III)	Cr^{3+}	3+	Mercury(II)	Hg^{2+}	2+
Copper(I)	Cu^+	1+	Nickel(II)	Ni^{2+}	2+
Copper(II)	Cu^{2+}	2+	Silver	Ag^+	1+
Iron(II)	Fe^{2+}	2+	Tin(II)	Sn^{2+}	2+
Iron(III)	Fe^{3+}	3+	Tin(IV)	Sn^{4+}	4+
Lead(II)	Pb^{2+}	2+	Zinc	Zn^{2+}	2+

The names, formulas, and ionic charges of some common polyatomic ions are given in Table 7.7. A more comprehensive list of both monatomic and polyatomic ions is given on the inside front cover of this book. Table 7.8 lists the principal oxidation numbers of common elements that have variable oxidation states.

Table 7.7. Names, formulas, and charges of some common polyatomic ions.

Name	Formula	Charge	Name	Formula	Charge
Acetate	$C_2H_3O_2^-$	1−	Cyanide	CN^-	1−
Ammonium	NH_4^+	1+	Dichromate	$Cr_2O_7^{2-}$	2−
Arsenate	AsO_4^{3-}	3−	Hydroxide	OH^-	1−
Bicarbonate	HCO_3^-	1−	Nitrate	NO_3^-	1−
Bisulfate	HSO_4^-	1−	Nitrite	NO_2^-	1−
Bromate	BrO_3^-	1−	Permanganate	MnO_4^-	1−
Carbonate	CO_3^{2-}	2−	Phosphate	PO_4^{3-}	3−
Chlorate	ClO_3^-	1−	Sulfate	SO_4^{2-}	2−
Chromate	CrO_4^{2-}	2−	Sulfite	SO_3^{2-}	2−

Table 7.8. Principal oxidation numbers of some common elements that have variable oxidation numbers.

Element	Oxidation number	Element	Oxidation number
Cu	+1, +2	Cl	−1, +1, +3, +5, +7
Hg	+1, +2	Br	−1, +1, +3, +5, +7
Sn	+2, +4	I	−1, +1, +3, +5, +7
Pb	+2, +4	S	−2, +4, +6
Fe	+2, +3	N	−3, +1, +2, +3, +4, +5
Au	+1, +3	P	−3, +3, +5
Ni	+2, +3	C	−4, +4
Co	+2, +3		
As	+3, +5		
Bi	+3, +5		
Cr	+2, +3, +6		

7.13 Formulas of Electrovalent Compounds

The sum of the oxidation numbers of all the atoms in a compound is zero. This statement applies to all substances, regardless of whether they are electrovalently or covalently bonded. For electrovalently bonded compounds the sum of the charges on all the ions in the compound must also be zero. Hence the formulas of ionic (electrovalently bonded) substances can be determined and written readily. Simply combine the ions in the simplest proportion so that the sum of the ionic charges adds up to zero.

To illustrate: Sodium chloride consists of Na^+ and Cl^- ions. Since $(1+) + (1-) = 0$, these ions combine in a one-to-one ratio, and the formula is written NaCl. Calcium fluoride is made up of Ca^{2+} and F^- ions; one Ca^{2+} and two F^- ions are needed, so the formula is CaF_2. Aluminum oxide is a bit more complicated, because it consists of Al^{3+} and O^{2-} ions. Since 6 is the lowest common multiple of 3 and 2, we have $2(3+) + 3(2-) = 0$; that is, two Al^{3+} ions and three O^{2-} ions are needed; therefore, the formula is Al_2O_3.

The foregoing compounds all are made up of monatomic ions. The same procedure is used for polyatomic ions. Consider calcium hydroxide, which is made up of Ca^{2+} and OH^- ions. Since $(2+) + 2(1-) = 0$, one Ca^{2+} and two OH^- ions are needed, the formula is $Ca(OH)_2$. The parentheses are used to indicate that the formula has two hydroxide ions. It is not correct to write CaO_2H_2 in place of $Ca(OH)_2$ because the identity of the compound would be lost by so doing. Note that the positive ion is written first in formulas. The following table provides examples of formula writing for ionic compounds (see Tables 7.6 and 7.7 or inside the front cover for formulas of common ions).

> *The sum of the charges on the ions of an electrovalently bonded compound must equal zero.*

Name of compound	Ions	Lowest common multiple	Sum of charges on ions	Formula
Sodium bromide	Na^+, Br^-	1	$(1+) + (1-) = 0$	$NaBr$
Potassium sulfide	K^+, S^{2-}	2	$2(1+) + (2-) = 0$	K_2S
Zinc sulfate	Zn^{2+}, SO_4^{2-}	2	$(2+) + (2-) = 0$	$ZnSO_4$
Ammonium phosphate	NH_4^+, PO_4^{3-}	3	$3(1+) + (3-) = 0$	$(NH_4)_3PO_4$
Aluminum chromate	Al^{3+}, CrO_4^{2-}	6	$2(3+) + 3(2-) = 0$	$Al_2(CrO_4)_3$

It is not always easy to distinguish between the terms *oxidation number* and *charge*. Oxidation numbers are assigned to atoms according to a set of rules. The charge on an ion is the actual electron excess or deficiency when compared to the neutral atom—or group of atoms in the case of a polyatomic ion. Oxidation numbers may be assigned to all the atoms, including monatomic ions, in any compound; but only ions have charges. There is no problem with covalently bonded molecules such as methane (CH_4) because they contain no ions. The oxidation number of the hydrogen is $+1$, and that of carbon is -4. Electrovalently bonded compounds are apt to be troublesome because the oxidation number and the charge on a monatomic ion have the same numerical value. In sodium chloride (NaCl), the oxidation number of the sodium ion is $+1$ and the charge is $1+$; the oxidation number of the chloride ion is -1 and the charge is $1-$.

In a compound composed of monatomic ions, such as sodium chloride, there is no practical difference in writing the formula regardless of whether oxidation numbers or ionic charges are used. But there is a difference with compounds containing polyatomic ions. Sodium sulfate (Na_2SO_4), for example, consists of two sodium ions (Na^+) and a polyatomic sulfate ion (SO_4^{2-}). The sum of the ionic charges is zero: $2(1+) + (2-) = 0$; and it is convenient to make use of this fact in writing the formula. The charge on the polyatomic sulfate ion is not an oxidation number. However, the sum of the oxidation numbers of all the atoms in sodium sulfate is zero: $2(Na^+) + (S^{6+}) + 4(O^{2-}) = 0$; and the sum of the oxidation numbers of all the atoms in the sulfate ion is -2: $(S^{6+}) + 4(O^{2-}) = -2$ (see Section 7.14).

Problem 7.1   Write formulas for (a) calcium chloride; (b) iron(III) sulfide; (c) aluminum sulfate. Refer to Tables 7.6 and 7.7 as needed.

(a) *Step 1.* From the name we know that calcium chloride is composed of calcium and chloride ions. First write down the formulas of these ions.

Ca^{2+} and Cl^-

Step 2. To write the formula of the compound, combine the smallest numbers of Ca^{2+} and Cl^- ions to give a charge sum equal to zero. In this case, the lowest common multiple of the charges is 2:

$(Ca^{2+}) + 2(Cl^-) = 0$

$(2+) + 2(1-) = 0$

Therefore, the formula is $CaCl_2$.

(b) Use the same procedure for iron(III) sulfide.

Step 1. Write down the formulas for the iron(III) and sulfide ions.

Fe^{3+} and S^{2-}

Step 2. Use the smallest numbers of these ions required to give a charge sum equal to zero. The lowest common multiple of the charges is 6:

$2(Fe^{3+}) + 3(S^{2-}) = 0$

$2(3+) + 3(2-) = 0$

Therefore, the formula is Fe_2S_3.

(c) Use the same procedure for aluminum sulfate.

Step 1. Write down the formulas for the aluminum and sulfate ions.

Al^{3+} and SO_4^{2-}

Step 2. Use the smallest numbers of these ions required to given a charge sum equal to zero. The lowest common multiple of the charges is 6:

$2(Al^{3+}) + 3(SO_4^{2-}) = 0$

$2(3+) + 3(2-) = 0$

Therefore, the formula is $Al_2(SO_4)_3$. Note the use of parentheses around the SO_4^{2-} ion.

7.14 Determining Oxidation Numbers and Ionic Charges from a Formula

If the formula of a compound is known, the oxidation number of an element or the charge on a polyatomic ion in the formula can often be determined by algebraic difference. To begin, you must know the oxidation numbers of a few elements. Excellent ones with which to work are hydrogen, H^+, always a $+1$ except in hydrides (a hydride is a compound of hydrogen combined with a metal); oxygen, O^{2-}, always a -2 except in peroxides; and sodium, Na^+, always a $+1$. Using the compound sulfuric acid, H_2SO_4, as an example, let us determine the charge of the sulfate ion and oxidation number of the sulfur atom. The sulfate ion is combined with two hydrogen atoms, each with a $+1$ oxidation number. The sulfate ion must then have a -2 charge in order for the net charge in the compound to be zero:

$$
\begin{array}{ll}
H^+ & +1 \\
H^+ & +1 \\
SO_4^{2-} & -2 \\
\hline
 & 0
\end{array}
$$

To find the oxidation number of sulfur, we proceed as follows:

Step 1. Write the oxidation number of a single atom of hydrogen and a single atom of oxygen below the atoms in the formula.

Step 2. Below this, write the sums of the oxidation numbers of all the H and O atoms: $2(+1) = +2$ and $4(-2) = -8$.

Step 3. Then, add together the total oxidation numbers of all the atoms, including the sulfur atom, and set them equal to zero: $+2 + S + (-8) = 0$. Solving the equation for S, we determine that the oxidation number of sulfur is $+6$, the value needed to give the sum of zero.

$$
\begin{array}{lll}
 & H_2 \quad S \quad O_4 & \\
\text{Step 1.} & \quad +1 \qquad\quad -2 & (1) \\
\text{Step 2.} \quad 2(+1) = +2 \qquad 4(-2) = -8 & & (2) \\
\text{Step 3.} \qquad\qquad +2 + S + (-8) = 0 & & (3) \\
 & \qquad\qquad S = +6 &
\end{array}
$$

The oxidation number of sulfur in H_2SO_4 is $+6$.

What is the oxidation number of chromium in sodium dichromate, $Na_2Cr_2O_7$? Using the same method as for H_2SO_4, we have

$$
\begin{array}{ll}
 & Na_2 \quad Cr_2 \quad O_7 \\
\text{Step 1.} & \quad +1 \qquad\qquad -2 \\
\text{Step 2.} \quad 2(+1) = +2 \qquad\quad 7(-2) = -14 \\
\text{Step 3.} \qquad\qquad +2 + 2\,Cr + (-14) = 0 \\
 & \qquad\qquad\quad 2\,Cr = +12 \\
 & \qquad\qquad\quad\; Cr = +6
\end{array}
$$

The oxidation number of chromium in $Na_2Cr_2O_7$ is $+6$.

The formula of radium chloride is $RaCl_2$. What is the oxidation number of radium? If you remember that the oxidation number of chloride is -1, then the value for radium is $+2$, since one radium ion is combined with two Cl^- ions. If you do not remember the oxidation number of chloride, then you should try to recall the formula of another chloride. One that might come to mind is sodium chloride, NaCl, in which the chloride is -1 because of its combination with one sodium ion of $+1$. This recollection establishes the oxidation number of chloride, which then enables you to calculate the value for radium.

What is the oxidation number of phosphorus in the phosphate ion, PO_4^{3-}? First of all, note that this is a polyatomic ion with a charge of -3. The sum of the oxidation numbers of phosphorus and oxygen must equal -3 and not zero.

Four oxygen atoms, each with a -2, give a total of -8. The oxidation number of the phosphorus atom must then be $+5$:

$$P\ O_4^{3-}$$
$$-2$$
$$P + 4(-2) = -3$$
$$P = +5$$

The sum of the oxidation numbers of the atoms in a polyatomic ion must equal the charge of the polyatomic ion.

7.15 Oxidation–Reduction

Magnesium burns brilliantly in air, forming magnesium oxide:

$$2\ Mg^0 + O_2^0 \longrightarrow 2\ Mg^{2+}O^{2-}$$

In this reaction, two electrons are transferred from each magnesium atom to the oxygen atoms, resulting in an increase in the oxidation number of magnesium from 0 to $+2$. At the same time, the oxidation number of oxygen has decreased from 0 to -2. Oxidation and reduction have occurred in this reaction. The Mg^0 was oxidized and the O_2^0 was reduced. **Oxidation** is defined as an increase in the oxidation number or oxidation state of an element. **Reduction** is defined as a decrease in the oxidation number or oxidation state of an element. *Oxidation–reduction* occurs as a result of a loss and gain of electrons. Oxidation and reduction occur simultaneously. The element that loses electrons (increases in oxidation number) is oxidized, and the element that gains electrons (decreases in oxidation number) is reduced.

oxidation

reduction

$$2\ Hg^{2+}O^{2-} \xrightarrow{\Delta} 2\ Hg^0 + O_2^0$$

In the decomposition of mercury(II) oxide, the oxidation number of mercury(II) changes from $+2$ to 0; the oxidation number of oxygen changes from -2 to 0. Therefore, oxidation–reduction occurs; oxygen (O^{2-}) is oxidized and mercury (Hg^{2+}) is reduced.

The process of oxidation and reduction involves the transfer of electrons and results in changes of oxidation numbers. The element oxygen is not necessarily involved in this process. For example, in the chemical reaction between sodium and chlorine to form sodium chloride, we saw that electrons were transferred from sodium atoms to chlorine atoms:

$$2\ Na^0 + Cl_2^0 \longrightarrow 2\ Na^+Cl^-$$

In this reaction, the oxidation number of sodium increases from 0 to $+1$, and therefore sodium is oxidized. The oxidation number of chlorine decreases from 0 to -1, and consequently, chlorine is reduced.

A more detailed discussion of oxidation–reduction is given in Chapter 17.

Questions

A. Review the meanings of the new terms introduced in this chapter.
1. Ionization energy
2. Valence electrons
3. Electronegativity
4. Chemical bond
5. Electrovalent bond
6. Ionic bond
7. Covalent bond
8. Nonpolar covalent bond
9. Polar covalent bond
10. Dipole
11. Coordinate covalent bond
12. Polyatomic ion
13. Oxidation number
14. Ionic charge
15. Oxidation
16. Reduction

B. Answers to the following questions will be found in tables and figures.
1. Explain the large increase in ionization energy required to remove the second electron from a lithium atom compared to the first electron removed. (See Table 7.1.)
2. Arrange the following elements in the order of increasing attraction by which their valence electrons are held in the atom: aluminum, sulfur, silicon, magnesium, chlorine, phosphorus, argon, and sodium.
3. In which general areas of the periodic table are the elements with the lowest and the highest ionization energies and electronegativities located?
4. Using the table of electronegativities (Table 7.4), indicate which element is positive and which is negative in the following compounds:
   (a) MgH_2
   (b) NaCl
   (c) CO_2
   (d) PCl_3
   (e) NH_3
   (f) ICl
   (g) Br_2
   (h) Cl_2O
   (i) OF_2
   (j) NO_2
   (k) CF_4
   (l) KBr
   (m) CuS
5. Classify the bond between the following pairs of atoms as either principally ionic or covalent (use Table 7.4):
   (a) Phosphorus and hydrogen
   (b) Sodium and fluorine
   (c) Chlorine and carbon
   (d) Hydrogen and chlorine
   (e) Magnesium and oxygen
   (f) Hydrogen and sulfur
6. Using the principle employed in Table 7.5, write formulas for:
   (a) The hydrogen compounds of Li, Ca, Sb, Br, and S
   (b) The oxygen compounds of K, Si, N, Sr, and Ga
   (c) The bromine compounds of Na, Mg, Al, C, P, and S
7. Use the oxidation number tables and determine the formulas for compounds composed of the following ions:
   (a) Hydrogen and cyanide
   (b) Silver and nitrate
   (c) Potassium and carbonate
   (d) Mercury(I) and oxide
   (e) Aluminum and acetate
   (f) Iron(III) and sulfate

C. Review questions.
1. Write an equation representing the change of a fluorine atom to a fluoride ion (F^-).
2. Write an equation representing the change of a calcium atom to a calcium ion (Ca^{2+}).
3. Why does barium (Ba) have a lower ionization energy than beryllium (Be)?
4. Why is there such a large increase in the ionization energy required to remove the second electron from a sodium atom?

The Formation of Compounds from Atoms 135

5. How many electrons must be gained or lost for each of the following to achieve a noble gas electron structure?
   (a) A calcium atom (d) A chloride ion
   (b) A sulfur atom (e) A nitrogen atom
   (c) A helium atom
6. Explain why potassium forms a K^+ ion but not a K^{2+} ion.
7. What portion of an atom is represented by the kernel?
8. Why does an aluminum ion have a $+3$ charge?
9. Which would be larger, a potassium ion or a potassium atom? Explain.
10. Which would be smaller, a bromine atom or a bromide ion? Explain.
11. Which would be larger, a magnesium ion or an aluminum ion? Explain.
12. What causes a bond to be polar?
13. How does a coordinate covalent bond differ from an ordinary covalent bond?
14. How does a covalent bond differ from an electrovalent bond?
15. Draw electron-dot structures for:
    (a) Mg (c) H_2O (e) CO_2 (g) H_2S (i) CaF_2
    (b) F_2 (d) NH_3 (f) HCl (h) MgO (j) ZnI_2
16. Draw electron-dot structures for:
    (a) Mg^{2+} (c) Cl^- (e) SO_4^{2-} (g) CN^- (i) ClO_3^-
    (b) Al^{3+} (d) S^{2-} (f) SO_3^{2-} (h) CO_3^{2-} (j) NO_3^-
17. The electron-dot structure for chloric acid is

    $$H:\overset{..}{\underset{..}{O}}:\overset{..}{\underset{..}{Cl}}:\overset{..}{\underset{..}{O}}:$$
    $$:\overset{..}{\underset{..}{O}}:$$

    Point out the covalent and coordinate covalent bonds in this structure.
18. Draw the electron-dot structure for ammonia (NH_3). What type of bonds are present? Can this molecule form coordinate covalent bonds? How many?
19. Classify the following molecules as polar or nonpolar:
    (a) NH_3 (b) HBr (c) CF_4 (d) F_2 (e) CO_2
20. Is it possible for a molecule to be nonpolar even though it contains polar covalent bonds? Explain.
21. Determine the oxidation number of each element in the following:
    (a) HBr (c) MgO (e) F_2 (g) H_2SO_4 (i) K_2CrO_4
    (b) NH_3 (d) $FeCl_2$ (f) SiH_4 (h) KNO_3 (j) C_2H_6
22. Determine the oxidation number of the element in italic in each formula:
    (a) *Ba*SO_4 (c) Li*H* (e) *As*$_2O_5$ (g) *Cu*Cl_2 (i) *C*O
    (b) *Cl*$_2O_7$ (d) H*N*O_3 (f) *Na*$_2O$ (h) Na_2*Cr*$_2O_7$ (j) H*C*$_2H_3O_2$
23. Write the formula of the compound that would be formed between the given elements:
    (a) Na and I (d) H and N (g) Ca and P (i) C and F
    (b) Ba and Br (e) Rb and Cl (h) Li and Se (j) I and Cl
    (c) H and S (f) Al and S
24. Write the formula of the compound formed from the given ions:
    (a) Na^+ and F^- (f) PO_4^{3-} and Fe^{3+}
    (b) Mg^{2+} and O^{2-} (g) NH_4^+ and S^{2-}
    (c) Cl^- and Al^{3+} (h) Zn^{2+} and AsO_4^{3-}
    (d) Fe^{2+} and OH^- (i) NO_3^- and Ni^{2+}
    (e) SO_4^{2-} and K^+ (j) NH_4^+ and CrO_4^{2-}
25. Which of these statements are correct? (Try to answer this question without referring to your book.)

(a) A chlorine atom has less electrons than a chloride ion.
(b) The noble gases have a tendency to lose one electron to become a positively charged ion.
(c) The chemical bonds in a water molecule are ionic.
(d) The chemical bonds in a water molecule are polar.
(e) Valence electrons are those electrons in the highest occupied energy level, **n**, of an atom.
(f) An atom with eight electrons in its outer shell has all its *s* and *p* orbitals filled.
(g) Fluorine has the lowest electronegativity of all the elements.
(h) A neutral atom with eight electrons in its valence shell must be an atom of a noble gas.
(i) A nitrogen atom has four valence electrons.
(j) An aluminum atom must lose three electrons to become an aluminum ion, Al^{3+}.
(k) In an ethylene molecule, C_2H_4,

$$\begin{array}{c} H \\ \diagdown \\ C=C \\ \diagup \\ H \end{array} \begin{array}{c} \\ \diagup \\ H \\ \diagdown \\ H \end{array}$$

two pairs of electrons are shared between the carbon atoms.
(l) The octet rule is mainly useful for atoms where only *s* and *p* electrons enter into bonding.
(m) When electrons are transferred from one atom to another, the resulting compound contains ionic bonds.
(n) A phosphorus atom, ·P̈·, needs three additional electrons to attain a stable octet of electrons.
(o) The simplest compound between oxygen, ·Ö·, and fluorine, :F̈·, atoms is FO_2.
(p) In the molecule H:C̈l:, there are three unshared pairs of electrons.
(q) The bonds in a water molecule are formed by overlapping of *s* and *p* electron orbitals.
(r) The smaller the difference in electronegativity between two atoms, the more ionic will be the bond between them.
(s) In the reaction $2\ H_2O \rightarrow 2\ H_2 + O_2$, hydrogen is oxidized and oxygen is reduced.
(t) Oxidation occurs when an atom loses electrons.
(u) In the oxide WO_3, the oxidation number of tungsten is +6 and that of oxygen is −2.

D. *Review exercises.*
1. (a) In terms of electron structure, why is the oxidation number of nitrogen never higher than +5 or lower than −3?
   (b) What are the highest and lowest possible oxidation states for bromine?
2. Why do chemical bonds form?

8 Nomenclature of Inorganic Compounds

After studying Chapter 8 you should be able to:

1. Give the name or formula for inorganic binary compounds in which the metal has only one common oxidation state.
2. Give the name or formula for inorganic binary compounds that contain metals of variable oxidation state, using either the Stock System or classical nomenclature.
3. Give the name or formula for inorganic binary compounds that contain two nonmetals.
4. Give the name or formula for binary acids.
5. Give the name or formula for ternary inorganic acids.
6. Give the name or formula for ternary salts.
7. Given the formula of a salt, write the name and formula of the acid from which the salt may be derived.
8. Give the name or formula for salts that contain more than one positive ion.
9. Give the name or formula for inorganic bases.

8.1 Common, or Trivial, Names

Chemical nomenclature is the system of names that chemists use to identify compounds. When a new substance is formulated, it must be named in order to distinguish it from all other substances. Before chemistry was systematized, a substance was given a name that generally associated it with one of its outstanding physical or chemical properties. For example, *quicksilver* is a common name for mercury; it describes two properties of mercury—a silvery appearance and quick, liquidlike movement. Nitrous oxide, N_2O, used as an anesthetic in dentistry, has been called *laughing gas*, because it induces laughter when inhaled. The name *nitrous oxide* is now giving way to the more systematic name *dinitrogen oxide*. Nonsystematic names are called *common*, or *trivial*, names.

Common names for chemicals are widely used in many industries, since the systematic name frequently is too long or too technical for everyday use.

Chapter 8

For example, CaO is called *lime*, not *calcium oxide*, by plasterers; photographers refer to $Na_2S_2O_3$ as *hypo*, rather than *sodium thiosulfate*; gardeners refer to $CCl_3CH(C_6H_4Cl)_2$ by the abbreviation *DDT*, not as *dichlorodiphenyltrichloroethane*. These common names are chemical nicknames, and, as the DDT example shows, there is a practical need for short, common names. Table 8.1 lists the common names, formulas, and chemical names of some familiar substances.

Table 8.1. Common names, formulas, and chemical names of some familiar substances.

Common name	Formula	Chemical name
Acetylene	C_2H_2	Ethyne
Lime	CaO	Calcium oxide
Slaked lime	$Ca(OH)_2$	Calcium hydroxide
Water	H_2O	Water
Galena	PbS	Lead(II) sulfide
Alumina	Al_2O_3	Aluminum oxide
Baking soda	$NaHCO_3$	Sodium hydrogen carbonate
Cane or beet sugar	$C_{12}H_{22}O_{11}$	Sucrose
Blue stone, blue vitriol	$CuSO_4 \cdot 5H_2O$	Copper(II) sulfate pentahydrate
Borax	$Na_2B_4O_7 \cdot 10H_2O$	Sodium tetraborate decahydrate
Brimstone	S	Sulfur
Calcite, marble, limestone	$CaCO_3$	Calcium carbonate
Cream of tartar	$KHC_4H_4O_6$	Potassium hydrogen tartrate
Epsom salts	$MgSO_4 \cdot 7H_2O$	Magnesium sulfate heptahydrate
Gypsum	$CaSO_4 \cdot 2H_2O$	Calcium sulfate dihydrate
Grain alcohol	C_2H_5OH	Ethyl alcohol, ethanol
Hypo	$Na_2S_2O_3$	Sodium thiosulfate
Laughing gas	N_2O	Dinitrogen oxide
Litharge	PbO	Lead(II) oxide
Lye, caustic soda	NaOH	Sodium hydroxide
Milk of magnesia	$Mg(OH)_2$	Magnesium hydroxide
Muriatic acid	HCl	Hydrochloric acid
Oil of vitriol	H_2SO_4	Sulfuric acid
Plaster of paris	$CaSO_4 \cdot \frac{1}{2} H_2O$	Calcium sulfate hemihydrate
Potash	K_2CO_3	Potassium carbonate
Pyrites (fool's gold)	FeS_2	Iron disulfide
Quicksilver	Hg	Mercury
Sal ammoniac	NH_4Cl	Ammonium chloride
Saltpeter (chile)	$NaNO_3$	Sodium nitrate
Table salt	NaCl	Sodium chloride
Washing soda	$Na_2CO_3 \cdot 10H_2O$	Sodium carbonate decahydrate
Wood alcohol	CH_3OH	Methyl alcohol, methanol

8.2 Systematic Chemical Nomenclature

The trivial name is not entirely satisfactory to the chemist, who requires a name that will identify precisely the composition of each substance. Therefore, as the number of known compounds increased, it became more and more necessary to develop a scientific, systematic method of identifying compounds by name. The systematic method of naming inorganic compounds considers the compound to be composed of two parts, one positive and one negative. The positive part, which is either a metal, hydrogen or other positively charged group, is named and written first. The negative part, generally nonmetallic, follows. The names of the elements are modified with suffixes and prefixes to identify the different types or classes of compounds. Thus, the compound composed of sodium ions and chloride ions is named sodium chloride; the compound composed of calcium ions and bromide ions is named calcium bromide; the compound composed of iron(II) ions and chloride ions is named iron(II) chloride (read as "iron-two chloride").

We will consider the naming of acids, bases, salts, and oxides. Refer to Tables 7.6, 7.7, and 7.8 for the names, formulas, and oxidation numbers of ions. For handy, quick reference, the names and formulas of some common ions are given on the inside front cover of this book.

8.3 Binary Compounds

Binary compounds contain only two different elements. Their names consist of two parts: the name of the more electropositive element followed by the name of the electronegative element, which is modified to end in *ide*. [The names of nonbinary compounds that use the *ide* ending but are exceptions to the rule are discussed in part (d) of this section.]

(a) Binary compounds in which the electropositive element has a fixed oxidation state. The majority of these compounds contain a metal and a nonmetal. The chemical name is composed of the name of the metal, which is written first, followed by the name of the nonmetal, which has been modified to an identifying stem plus the suffix *ide*. For example, sodium chloride, NaCl, is composed of one atom each of sodium and chlorine. The name of the metal, sodium, is written first and is not modified. The second part of the name is derived from the nonmetal, chlorine, by using the stem *chlor* and adding the ending *ide*; it is named *chloride*. The compound name is sodium chloride.

NaCl

Elements: Sodium (metal)
Chlorine (nonmetal)
name modified to the stem *chlor* + *ide*
Name of compound: Sodium chloride

Stems of the more common negative-ion forming elements are shown in the following table.

Symbol	Element	Stem	Binary name ending
B	Boron	Bor	Boride
Br	Bromine	Brom	Bromide
Cl	Chlorine	Chlor	Chloride
F	Fluorine	Fluor	Fluoride
H	Hydrogen	Hydr	Hydride
I	Iodine	Iod	Iodide
N	Nitrogen	Nitr	Nitride
O	Oxygen	Ox	Oxide
P	Phosphorus	Phosph	Phosphide
S	Sulfur	Sulf	Sulfide

Compounds may contain more than one atom of the same element, but as long as they contain only two different elements and if only one compound of these two elements exists, the name follows the rule for binary compounds:

Examples: $CaBr_2$ Mg_3N_2 KI

Calcium bromide Magnesium nitride Potassium iodide

Table 8.2 shows more examples of compounds with names ending in *ide*.

Table 8.2. Examples of compounds with names ending in *ide*.

Formula	Name
$MgBr_2$	Magnesium bromide
Na_2O	Sodium oxide
NaH	Sodium hydride
HCl	Hydrogen chloride
HI	Hydrogen iodide
CaC_2	Calcium carbide
$AlCl_3$	Aluminum chloride
PbS	Lead(II) sulfide
LiI	Lithium iodide
Al_2O_3	Aluminum oxide

(b) **Binary compounds containing metals of variable oxidation numbers.**
Two systems are commonly used for compounds in this category. The official system, designated by the International Union of Pure and Applied Chemistry (IUPAC), is known as the *Stock System*. In the Stock System, when a compound contains a metal that can have more than one oxidation number, the oxidation number of the metal in the compound is designated by a roman numeral written immediately after the name of the metal. The negative element is treated in the usual manner for binary compounds.

Examples: $FeCl_2$ Iron(II) chloride Fe^{2+}
             $FeCl_3$ Iron(III) chloride Fe^{3+}
             $CuCl$ Copper(I) chloride Cu^{+}
             $CuCl_2$ Copper(II) chloride Cu^{2+}

When a metal has only one possible oxidation state, there is no need to distinguish one oxidation state from another, so roman numerals are not needed. Thus, we do not say calcium(II) chloride for $CaCl_2$, but rather calcium chloride, since the oxidation number of calcium is understood to be $+2$.

In classical nomenclature, when the metallic ion has only two oxidation numbers, the name of the metal is modified with the suffixes *ous* and *ic* to distinguish between the two. The lower oxidation state is given the *ous* ending and the higher one, the *ic* ending.

Examples:

$FeCl_2$	Ferrous chloride	Fe^{2+}
$FeCl_3$	Ferric chloride	Fe^{3+}
$CuCl$	Cuprous chloride	Cu^+
$CuCl_2$	Cupric chloride	Cu^{2+}

Table 8.3 lists some common metals with more than one oxidation number.

Table 8.3. Names and oxidation numbers of some common metal ions that have more than one oxidation number.

Formula	Stock System name	Classical name
Cu^{1+}	Copper(I)	Cuprous
Cu^{2+}	Copper(II)	Cupric
$Hg^{1+}(Hg_2)^{2+}$	Mercury(I)	Mercurous
Hg^{2+}	Mercury(II)	Mercuric
Fe^{2+}	Iron(II)	Ferrous
Fe^{3+}	Iron(III)	Ferric
Sn^{2+}	Tin(II)	Stannous
Sn^{4+}	Tin(IV)	Stannic
As^{3+}	Arsenic(III)	Arsenous
As^{5+}	Arsenic(V)	Arsenic
Sb^{3+}	Antimony(III)	Stibnous
Sb^{5+}	Antimony(V)	Stibnic
Ti^{3+}	Titanium(III)	Titanous
Ti^{4+}	Titanium(IV)	Titanic

Notice that the *ous–ic* naming system does not give the oxidation state of an element but merely indicates that at least two oxidation states exist. The Stock System avoids any possible uncertainty by clearly stating the oxidation number.

(c) **Binary compounds containing two nonmetals.** The chemical bond that exists between two nonmetals is predominantly covalent. In a covalent compound, positive and negative oxidation numbers are assigned to the elements according to their electronegativities. The most electropositive element is named first. In a compound between two nonmetals, the element that occurs earlier in the following sequence is written and named first.

B, Si, C, P, N, H, S, I, Br, Cl, O, F.

A Latin or Greek prefix is attached to each element to indicate the number of atoms of that element in the molecule. The second element still retains the modified binary ending. The prefix *mono* is generally omitted except when needed to distinguish between two or more compounds, such as carbon monoxide, CO, and carbon dioxide, CO_2. Some common prefixes and their numerical equivalences are the following:

Mono = 1   Hexa = 6
Di = 2   Hepta = 7
Tri = 3   Octa = 8
Tetra = 4   Nona = 9
Penta = 5   Deca = 10

Here are some examples of compounds that illustrate this system:

CO	Carbon monoxide
CO_2	Carbon dioxide
PCl_3	Phosphorus trichloride
PCl_5	Phosphorus pentachloride
P_2O_5	Diphosphorus pentoxide
CCl_4	Carbon tetrachloride
N_2O	Dinitrogen oxide
S_2Cl_2	Disulfur dichloride
N_2O_4	Dinitrogen tetroxide
NO	Nitrogen oxide
N_2O_3	Dinitrogen trioxide

(d) Exceptions that use *ide* endings. Three notable exceptions that use the *ide* ending are hydroxides (OH^-), cyanides (CN^-), and ammonium (NH_4^+) compounds. These polyatomic ions, when combined with another element, take the ending *ide*, even though more than two elements are present in the compound.

NH_4I	Ammonium iodide
$Ca(OH)_2$	Calcium hydroxide
KCN	Potassium cyanide

(e) Acids derived from binary compounds. Certain binary hydrogen compounds, when dissolved in water, form solutions that have **acid** properties. Because of this property, these compounds are given acid names in addition to their regular *ide* names. For example, HCl is a gas and is called *hydrogen chloride*, but its water solution is known as *hydrochloric acid*. Binary acids are composed of hydrogen and one other nonmetallic element. However, not all binary hydrogen compounds are acids. To express the formula of a binary acid, it is customary to write the symbol of hydrogen first, followed by the symbol of the second element (for example, HCl, HBr, H_2S). When we see formulas such as CH_4 or NH_3, we understand that these compounds are not normally considered to be acids.

To name a binary acid, place the prefix *hydro* in front of, and the suffix *ic* after, the stem of the nonmetal. Then add the word *acid*.

	HCl	H$_2$S
Examples:	Hydro chlor/ic acid	Hydro sulfur/ic acid
	(hydrochloric acid)	(hydrosulfuric acid)

Acids are hydrogen-containing substances that liberate hydrogen ions when dissolved in water. The same formula is often used to express binary hydrogen compounds such as HCl, regardless of whether they are dissolved in water. Table 8.4 shows examples of other binary acids.

Table 8.4. Names and formulas of selected binary acids.

Formula	Acid name
HF	Hydrofluoric acid
HCl	Hydrochloric acid
HBr	Hydrobromic acid
HI	Hydriodic acid
H$_2$S	Hydrosulfuric acid
H$_2$Se	Hydroselenic acid

8.4 Ternary Compounds

Ternary compounds contain three elements: an electropositive group, which is either a metal or hydrogen, combined with a polyatomic negative ion. We will consider the naming of compounds in which one of the three elements is oxygen.

In general, in naming ternary compounds the positive group is given first, followed by the name of the negative ion. Rules for naming the positive groups are identical to those used in naming binary compounds. The negative group usually contains two elements: oxygen and a metal or a nonmetal. To name the polyatomic negative ion, add the endings *ite* or *ate* to the stem of the element other than oxygen. Thus, SO$_4^{2-}$ is called *sulfate*, and SO$_3^{2-}$ is called *sulfite*. Note that oxygen is not specifically included in the name, but is understood to be present when the endings *ite* and *ate* are used. The suffixes *ite* and *ate* repre-

ite	*ate*
+4	+6
CaSO$_3$	CaSO$_4$
Calcium sulfite	Calcium sulfate
ite ending indicates lower oxidation state of sulfur	*ate* ending indicates higher oxidation state of sulfur

sent different oxidation states of the element other than oxygen in the polyatomic ion. The *ite* ending represents the lower and the *ate* the higher oxidation state. When an element has only one oxidation state, such as C in carbonate,

Table 8.5. Names and formulas of selected ternary compounds.

Formula	Name
Na$_2$SO$_3$	Sodium sulfite
Na$_2$SO$_4$	Sodium sulfate
K$_2$CO$_3$	Potassium carbonate
CaCO$_3$	Calcium carbonate
Al$_2$(SO$_4$)$_3$	Aluminum sulfate
KClO$_3$	Potassium chlorate
AlPO$_4$	Aluminum phosphate
FeSO$_4$	Iron(II) sulfate or ferrous sulfate
Fe$_2$(SO$_4$)$_3$	Iron(III) sulfate or ferric sulfate
PbCrO$_4$	Lead(II) chromate
H$_2$SO$_4$	Hydrogen sulfate
HNO$_3$	Hydrogen nitrate
Cu$_2$SO$_4$	Copper(I) sulfate or cuprous sulfate
CuSO$_4$	Copper(II) sulfate or cupric sulfate
Li$_3$AsO$_4$	Lithium arsenate
NaNO$_2$	Sodium nitrite
ZnMoO$_4$	Zinc molybdate

the *ate* ending is used. Examples of ternary compounds and their names are given in Table 8.5.

Ternary oxy-acids.   Inorganic ternary compounds containing hydrogen, oxygen, and one other element are called *oxy-acids*. The element other than hydrogen or oxygen in these acids is usually a nonmetal, but in some cases it can be a metal. The *ous–ic* system is used in naming ternary acids. The suffixes *ous* and *ic* are used to indicate different oxidation states of the element other than hydrogen and oxygen. The *ous* ending again indicates the lower oxidation state and the *ic* ending, the higher oxidation state.

To name these acids, we place the ending *ic* or *ous* after the stem of the element other than hydrogen and oxygen, and add the word *acid*. If an element has only one usual oxidation state, the *ic* ending is used. Hydrogen in a ternary oxy-acid is not specifically designated in the acid name but its presence is implied by use of the word *acid*.

Examples: H$_2$SO$_3$ H$_2$SO$_4$
Sulfur/*ous* acid   Sulfur/*ic* acid

Once again, the acid name is associated with the water solution of the pure compound. In the pure state, the usual ternary name may be used. Thus, H$_2$SO$_4$ is called both hydrogen sulfate and sulfuric acid.

In cases where there are more than two oxy-acids in a series, the *ous–ic* names are further modified with the prefixes *per* and *hypo*. *Per* is placed before the stem of the element other than hydrogen and oxygen when the element has a higher oxidation number than in the *ic* acid. *Hypo* is used as a prefix before the stem when the element has a lower oxidation number than in the *ous* acid. The use of *per* and *hypo* is illustrated in the oxy-acids of chlorine.

Formula	Name	Oxidation number of chlorine
HClO	Hypochlorous acid	+1
HClO$_2$	Chlorous acid	+3
HClO$_3$	Chloric acid	+5
HClO$_4$	Perchloric acid	+7

The electron-dot structures of the oxy-acids of chlorine are

H:Ö:Cl: H:Ö:Cl: H:Ö:Cl:Ö: H:Ö:Cl:Ö:
                :Ö: :Ö:
Hypochlorous acid   Chlorous acid   Chloric acid   Perchloric acid

Check the oxidation number of chlorine in each of these oxy-acids using the method for assigning oxidation numbers.

Examples of other ternary oxy-acids and their names are shown in Table 8.6.

Table 8.6. Names and formulas of selected ternary oxy-acids.

Formula	Acid name	Formula	Acid name
H$_2$SO$_3$	Sulfurous acid	HNO$_2$	Nitrous acid
H$_2$SO$_4$	Sulfuric acid	HNO$_3$	Nitric acid
H$_3$PO$_2$	Hypophosphorous acid	HBrO$_3$	Bromic acid
H$_3$PO$_3$	Phosphorous acid	HIO$_3$	Iodic acid
H$_3$PO$_4$	Phosphoric acid	H$_3$BO$_3$	Boric acid
HClO	Hypochlorous acid	H$_2$C$_2$O$_4$	Oxalic acid
HClO$_2$	Chlorous acid	HC$_2$H$_3$O$_2$	Acetic acid
HClO$_3$	Chloric acid	H$_2$CO$_3$	Carbonic acid
HClO$_4$	Perchloric acid		

The endings *ous*, *ic*, *ite*, and *ate* are part of classical nomenclature; they are not used in the Stock System to indicate different oxidation states of the elements. These endings are still used, however, in naming many common compounds. The Stock name for H$_2$SO$_4$ is tetraoxosulfuric(VI) acid, and that for H$_2$SO$_3$ is trioxosulfuric(IV) acid. These Stock names are awkward and are not commonly used.

8.5 Salts

When the hydrogen of an acid is replaced by a metal ion or an ammonium (NH$_4^+$) ion, the compound formed is classified as a *salt*. Therefore, we can have a series of metal chlorides, bromides, sulfides, sulfates, sulfites, nitrates, phosphites, borates, and so on.

We have already considered the rules for naming salts, but a comparison of the salt and acid names will reveal definite patterns that are used. The same rules given above are used in naming the positive part of a salt. In binary compounds, the usual *ide* ending is given to the negative part of the salt name. In ternary compounds, the *ous* and *ic* endings of the acids become *ite* and *ate*, respectively, in the salt names, but the names of the stems remain the same.

Ternary oxy-acid
  ous ending of acid   becomes
  ic ending of acid   becomes

Ternary oxy-salt
  ite ending in salt
  ate ending in salt

Acid
H_2SO_4 Sulfur/*ic* acid
H_2SO_3 Sulfur/*ous* acid
HClO   Hypochlor/*ous* acid
$HClO_4$ Perchlor/*ic* acid

Salt
Na_2SO_4 Sodium sulf/*ate*
$CaSO_3$ Calcium sulf/*ite*
LiClO   Lithium hypochlor/*ite*
$NaClO_4$ Sodium perchlor/*ate*

Other examples of ternary acids and salts are given in Table 8.7.

Table 8.7. Comparison of acid and salt names in ternary oxy-compounds.

Acid	Salt	Name of salt
H_2SO_4 Sulfuric acid	Na_2SO_4	Sodium sulfate
	$CuSO_4$	Copper(II) sulfate or cupric sulfate
	$CaSO_4$	Calcium sulfate
	$Fe_2(SO_4)_3$	Iron(III) sulfate or ferric sulfate
HNO_3 Nitric acid	KNO_3	Potassium nitrate
	$HgNO_3$	Mercury(I) nitrate or mercurous nitrate
	$Hg(NO_3)_2$	Mercury(II) nitrate or mercuric nitrate
	$Fe(NO_3)_2$	Iron(II) nitrate or ferrous nitrate
	$Al(NO_3)_3$	Aluminum nitrate
HNO_2 Nitrous acid	KNO_2	Potassium nitrite
	$Co(NO_2)_2$	Cobalt(II) nitrite or cobaltous nitrite
	$Mg(NO_2)_2$	Magnesium nitrite
HClO	NaClO	Sodium hypochlorite
$HClO_2$	$NaClO_2$	Sodium chlorite
$HClO_3$	$NaClO_3$	Sodium chlorate
$HClO_4$	$NaClO_4$	Sodium perchlorate
H_2CO_3 Carbonic acid	Li_2CO_3	Lithium carbonate
	$CaCO_3$	Calcium carbonate

8.6 Salts with More than one Positive Ion

Salts may be formed from acids that contain two or more acid hydrogen atoms by replacing only one of the hydrogen atoms with a metal or by replacing both hydrogen atoms with different metals. Each positive group is named first and then the appropriate salt ending is added.

Acid	Salt	Name of salt
H_2CO_3	$NaHCO_3$	Sodium hydrogen carbonate or sodium bicarbonate
H_2S	$NaHS$	Sodium hydrogen sulfide or sodium bisulfide
H_3PO_4	$MgNH_4PO_4$	Magnesium ammonium phosphate
H_2SO_4	$NaKSO_4$	Sodium potassium sulfate

Note the name *sodium bicarbonate* given in the table. The prefix *bi* is commonly used to indicate a compound in which one of two acid hydrogen atoms has been replaced by a metal. Another example is sodium bisulfate, which has the formula $NaHSO_4$. Table 8.8 shows examples of other salts that contain more than one positive ion.

Table 8.8. Names of selected salts that contain more than one positive ion.

Acid	Salt	Name of salt
H_2SO_4	$KHSO_4$	Potassium hydrogen sulfate or potassium bisulfate
H_2SO_3	$Ca(HSO_3)_2$	Calcium hydrogen sulfite or calcium bisulfite
H_2S	NH_4HS	Ammonium hydrogen sulfide or ammonium bisulfide
H_3PO_4	$MgNH_4PO_4$	Magnesium ammonium phosphate
H_3PO_4	NaH_2PO_4	Sodium dihydrogen phosphate
H_3PO_4	Na_2HPO_4	Disodium hydrogen phosphate
$H_2C_2O_4$	KHC_2O_4	Potassium hydrogen oxalate or potassium binoxalate
H_2SO_4	$KAl(SO_4)_2$	Potassium aluminum sulfate
H_2CO_3	$Al(HCO_3)_3$	Aluminum hydrogen carbonate or aluminum bicarbonate

Note that prefixes are also used in chemical nomenclature to give special clarity or emphasis to certain compounds as well as to distinguish between two or more compounds.

Examples: Na_3PO_4 Trisodium phosphate
Na_2HPO_4 Disodium hydrogen phosphate
NaH_2PO_4 Sodium dihydrogen phosphate

8.7 Bases

Inorganic bases contain the hydroxyl group, OH⁻, in chemical combination with a metal ion. These compounds are called *hydroxides*. The OH⁻ group is named as a single ion and is given the ending *ide*. Several common bases are listed below.

NaOH	Sodium hydroxide
KOH	Potassium hydroxide
NH₄OH	Ammonium hydroxide
Ca(OH)₂	Calcium hydroxide
Ba(OH)₂	Barium hydroxide

We have now looked at methods of naming inorganic acids, bases, salts, and oxides. These four classes are just a handful of the classified chemical compounds. Most other classes fall under the broad field of organic chemistry. A few of these are alcohols, hydrocarbons, ethers, aldehydes, ketones, phenols, and carboxylic acids.

Problem 8.1   Name the compound CaS.

Step 1. From the formula, it is a two-element compound and follows the rules for binary compounds.

Step 2. The compound is composed of Ca, a metal, and S, a nonmetal. From oxidation number tables, determine whether the metal has a single or a variable oxidation number. We find that Ca has only one oxidation state; therefore, we name the positive part of the compound *calcium*.

Step 3. Modify the name of the second element to the identifying stem *sulf* and add the binary ending *ide* to form the name of the negative part, *sulfide*.

Step 4. The name of the compound, therefore, is *calcium sulfide*.

Problem 8.2   Name the compound FeS.

Step 1. This compound follows the rules for a binary compound and, like CaS, must be a sulfide.

Step 2. It is a compound of Fe, a metal, and S, a nonmetal. From the oxidation number tables, we see that Fe has variable oxidation numbers. In sulfides, the oxidation number of S is −2. Therefore, the oxidation number of Fe must be +2. Thus, the name of the positive part of the compound is *iron(II)* or *ferrous*.

Step 3. We have already determined that the name of the negative part of the compound will be *sulfide*.

Step 4. The name of FeS is *iron(II) sulfide* or *ferrous sulfide*.

Problem 8.3   (a) Name the salt KNO₃; and (b) name the acid HNO₃, from which this salt can be derived.

Nomenclature of Inorganic Compounds   149

(a) *Step 1.* From the formula, the compound contains three elements and follows the rules for ternary compounds.

Step 2. The salt is composed of a K^+ ion and a NO_3^- ion. The name of the positive part of the compound is *potassium*.

Step 3. Since it is a ternary salt, the name will end in *ite* or *ate*. From the oxidation number tables, we see that the name of the NO_3^- ion is *nitrate*.

Step 4. The name of the compound is *potassium nitrate*.

(b) The name of the acid follows the rules for ternary oxy-acids. Since the name of the salt KNO_3 ends in *ate*, the name of the corresponding acid will end in *ic acid*. Change the *ate* ending of nitrate to *ic*. Thus, *nitrate* becomes *nitric*, and the name of the acid is *nitric acid*.

Questions

In naming compounds, be careful to use correct spelling. For additional assistance in naming compounds refer to Tables 7.6, 7.7, and 7.8.

1. Write formulas for the following cations (do not forget to include the charges): sodium, magnesium, aluminum, copper(II), iron(II), ferric, lead(II), silver, barium, hydrogen, mercury(II), tin(II), chromium(III).

2. Write formulas for the following anions (do not forget to include the charges): chloride, bromide, fluoride, iodide, cyanide, oxide, hydroxide, sulfide, sulfate, bisulfate, bisulfite, chromate, carbonate, bicarbonate, acetate, chlorate, permanganate, oxalate.

3. Complete the table, filling in each box with the proper formula.

	Br^-	CO_3^{2-}	NO_3^-	PO_4^{3-}	OH^-	$C_2H_3O_2^-$
K^+	KBr	K_2CO_3				
Mg^{2+}						
Al^{3+}						
Fe^{3+}						
Zn^{2+}						
Ag^+						

4. State how each of the following terms are used in naming inorganic compounds: ide, ous, ic, hypo, per, ite, ate, roman numerals.

5. Name the following binary compounds, all of which are composed of nonmetals:
   (a) CO (c) NF_3 (e) SO_2 (g) CCl_4 (i) N_2O_5 (k) NH_3
   (b) CO_2 (d) PBr_5 (f) SO_3 (h) Cl_2O_7 (j) H_2S (l) OF_2

6. Name each compound listed by (1) the Stock (IUPAC) System and (2) the *ous–ic* system:
   (a) $CuCl_2$ (c) $TiCl_4$ (e) $Fe(OH)_3$ (g) $SnCl_4$ (i) Hg_2S
   (b) SnF_2 (d) Ti_2O_3 (f) FeO (h) $MnCO_3$ (j) $HgSO_4$

7. Provide two names for each compound listed: (1) the name of the pure substance and (2) the name of the substance when dissolved in water to form an acid or a base.
   (a) HCl (d) NH_3 (g) H_2Se (j) HBr (l) H_2SO_3
   (b) H_2SO_4 (e) H_2S (h) NaOH (k) H_3PO_4 (m) $Ba(OH)_2$
   (c) HNO_3 (f) $HC_2H_3O_2$ (i) $HClO_4$

150   Chapter 8

8. Name each compound.
   - (a) AgCl
   - (b) Na$_2$SO$_3$
   - (c) CuI$_2$
   - (d) MgBr$_2$
   - (e) Cu$_2$S
   - (f) Fe(NO$_3$)$_2$
   - (g) AlPO$_4$
   - (h) KHCO$_3$
   - (i) KOCl
   - (j) Na$_2$C$_2$O$_4$
   - (k) NH$_4$SCN
   - (l) KAl(SO$_4$)$_2$
   - (m) BaCr$_2$O$_7$
   - (n) Cd(CN)$_2$
   - (o) NaMnO$_4$
   - (p) HgF$_2$
   - (q) ZnCO$_3$
   - (r) PbCrO$_4$
   - (s) Na$_2$O$_2$
   - (t) CO
   - (u) Co(NO$_2$)$_2$
   - (v) SO$_3$
   - (w) K$_2$SiO$_3$
   - (x) LiH
   - (y) Mn(C$_2$H$_3$O$_2$)$_2$
   - (z) Si$_3$N$_4$

9. Name each compound.
   - (a) NH$_4$I
   - (b) PF$_3$
   - (c) KOH
   - (d) Ca(ClO$_3$)$_2$
   - (e) NaC$_2$H$_3$O$_2$
   - (f) CaF$_2$
   - (g) Sn$_3$(PO$_4$)$_2$
   - (h) SbCl$_5$
   - (i) MnSO$_4$
   - (j) Hg$_2$O
   - (k) Co(HCO$_3$)$_2$
   - (l) BaSO$_3$
   - (m) HOCl
   - (n) N$_2$O$_5$
   - (o) NH$_4$NO$_2$
   - (p) Na$_2$Cr$_2$O$_7$
   - (q) Li$_2$CO$_3$
   - (r) NiCl$_2$
   - (s) Na$_2$SiO$_3$
   - (t) CaC$_2$O$_4$
   - (u) BF$_3$
   - (v) CaH$_2$
   - (w) HgBr$_2$
   - (x) Sn(CN)$_4$
   - (y) KSCN
   - (z) BaO$_2$

10. Write formulas for each substance.
    - (a) Ammonium iodide
    - (b) Silver oxide
    - (c) Sulfurous acid
    - (d) Chlorous acid
    - (e) Stannous fluoride
    - (f) Copper(I) oxide
    - (g) Beryllium nitride
    - (h) Potassium chlorate
    - (i) Calcium carbonate
    - (j) Magnesium oxalate
    - (k) Bismuth(III) sulfide
    - (l) Aluminum carbide
    - (m) Silicon carbide
    - (n) Sodium permanganate
    - (o) Zinc hypochlorite
    - (p) Carbonic acid
    - (q) Zinc phosphate
    - (r) Hydriodic acid
    - (s) Mercuric oxide
    - (t) Nickel(II) carbonate
    - (u) Ammonium hydroxide
    - (v) Strontium chloride
    - (w) Cadmium nitrate
    - (x) Lead(II) nitrate
    - (y) Iodine
    - (z) Acetic acid
    - (aa) Iron(III) sulfate
    - (bb) Zinc oxide
    - (cc) Ferric acetate
    - (dd) Oxalic acid

11. Write the name of each salt and the formula and name of the acid from which the salt may be derived.
    - (a) Mg(NO$_2$)$_2$
    - (b) NiCl$_2$
    - (c) KNO$_3$
    - (d) CaCO$_3$
    - (e) KMnO$_4$
    - (f) NaNO$_3$
    - (g) NaC$_2$H$_3$O$_2$
    - (h) KNO$_2$
    - (i) HgI$_2$
    - (j) KHSO$_4$
    - (k) Na$_3$BO$_3$
    - (l) KOCl
    - (m) NaF
    - (n) BaBr$_2$
    - (o) Fe$_2$(SO$_4$)$_3$
    - (p) Na$_2$HPO$_4$
    - (q) Zn(HSO$_3$)$_2$
    - (r) RaCl$_2$
    - (s) SnS$_2$
    - (t) Mn(ClO$_3$)$_2$
    - (u) Na$_2$C$_2$O$_4$
    - (v) LiBrO$_3$
    - (w) Fe(ClO$_2$)$_2$
    - (x) Ca(ClO$_4$)$_2$

12. Refer to an outside reference (chemical handbook, encyclopedia, dictionary) and write the chemical formulas for each of the following substances.
    - (a) Baking soda
    - (b) Calomel
    - (c) Carbolic acid
    - (d) Epsom salts
    - (e) Fool's gold
    - (f) Glauber's salt
    - (g) Litharge
    - (h) Lunar caustic
    - (i) Muriatic acid
    - (j) Prussian blue
    - (k) Prussic acid
    - (l) Quicksilver
    - (m) Sal soda
    - (n) Vinegar
    - (o) Vitriolic acid

9 Quantitative Composition of Compounds

After studying Chapter 9 you should be able to:

1. Understand the new terms listed in Question A at the end of the chapter.
2. Determine the formula weight or molecular weight of a compound from the formula.
3. Convert moles (gram-molecular weights or gram-formula weights) to grams, to molecules, or to formula units, and vice versa.
4. Calculate the percentage composition by weight of a compound from its formula.
5. Explain the relationship between an empirical formula and a molecular formula.
6. Determine the empirical formula of a compound from its percentage composition.
7. Calculate the molecular formula of a compound from its percentage composition and molecular weight.

9.1 Formula Weight or Molecular Weight

The quantitative composition of a compound can be determined from its formula; and the formula of a compound can be determined from its quantitative composition. In order to make these determinations, chemists have established a scale of relative masses for atoms known as *atomic weights*. This scale is based on the carbon-12 isotope having a mass of exactly 12 amu (see Section 5.18).

Because compounds are composed of atoms, they may be represented by a mass known as the formula weight or the molecular weight. The **formula weight** of a substance is the total mass of all the atoms in the chemical formula of that substance. The **molecular weight** of a substance is the total mass of all the atoms in a molecule of that substance. Formula weight and molecular weight are used interchangeably. However, the term *formula weight* is more inclusive, since it includes both molecular and ionic substances.

formula weight

molecular weight

152   Chapter 9

9.2 Determination of Molecular Weights from Formulas

If the formula of a substance is known, its formula weight or molecular weight may be determined by adding together the atomic weights of all the atoms in the formula. If more than one atom of any element is present, it must be added as many times as it is used in the formula.

Problem 9.1

The molecular formula for water is H_2O. What is its molecular weight?
Proceed by looking up the atomic weights of H (1.008 amu) and O (15.999 amu) and adding together the masses of all the atoms in the formula unit. Water contains two atoms of H and one atom of O. Thus,

$$2\,H \text{ atoms} = 2 \times 1.008 = 2.016 \text{ amu}$$
$$1\,O \text{ atom} = 1 \times 15.999 = 15.999 \text{ amu}$$
$$\overline{18.015 \text{ amu}} = \text{Molecular or formula weight}$$

Problem 9.2

Calculate the formula weight of calcium hydroxide, $Ca(OH)_2$.
The formula of this substance contains one atom of Ca and two atoms each of O and H. Thus,

$$1\,Ca \text{ atom} = 1 \times 40.08 = 40.08 \text{ amu}$$
$$2\,O \text{ atoms} = 2 \times 15.999 = 31.998 \text{ amu}$$
$$2\,H \text{ atoms} = 2 \times 1.008 = 2.016 \text{ amu}$$
$$\overline{74.094 \text{ amu}} = \text{Formula weight}$$

The atomic weights of elements are often rounded off to one decimal place to simplify calculations. (However, this simplification cannot be made in the most exacting chemical work.) If we calculate the formula weight of $Ca(OH)_2$ on the basis of one decimal place, we find the value to be 74.1 amu instead of 74.094 amu. The formula weight of $Ca(OH)_2$ would then be calculated as follows:

$$1\,Ca = 1 \times 40.1 = 40.1 \text{ amu}$$
$$2\,O = 2 \times 16.0 = 32.0 \text{ amu}$$
$$2\,H = 2 \times 1.0 = 2.0 \text{ amu}$$
$$\overline{74.1 \text{ amu}} = \text{Formula weight}$$

Problem 9.3

The formula for barium chloride dihydrate is $BaCl_2 \cdot 2H_2O$. What is its formula weight?
This formula contains one atom of Ba, two atoms of Cl, and two molecules of H_2O. Thus,

$$1\,Ba = 1 \times 137.3 = 137.3 \text{ amu}$$
$$2\,Cl = 2 \times 35.5 = 71.0 \text{ amu}$$
$$2\,H_2O = 2 \times 18.0 = 36.0 \text{ amu}$$
$$\overline{244.3 \text{ amu}} = \text{Formula weight}$$

9.3 Gram-Molecular Weight; Gram-Formula Weight; The Mole

gram-formula weight

gram-molecular weight

The quantity of any substance having a mass in grams that is numerically equal to its formula weight is the **gram-formula weight (g-form. wt)** or **gram-molecular weight (g-mol. wt)** of that substance. This quantity represents the weight of 1 mole (6.02×10^{23} formula units or molecules) of the substance. As an illustration, consider the compound hydrogen chloride, HCl. When 1 gram-atomic weight of H (1.00 gram representing 6.02×10^{23}, or 1 mole of, H atoms) and 1 gram-atomic weight of Cl (35.5 grams representing 6.02×10^{23}, or 1 mole of, Cl atoms) combine, they produce 1 gram-molecular weight of HCl (36.5 grams representing 6.02×10^{23}, or 1 mole of, HCl molecules). Since 36.5 grams of HCl contains 6.02×10^{23} molecules, we may refer to this quantity as a gram-molecular weight, a gram-formula weight, or simply as a mole of HCl. These relationships are summarized in tabular form for hydrogen chloride.

H	Cl	HCl
6.02×10^{23} H *atoms*	6.02×10^{23} Cl *atoms*	6.02×10^{23} HCl *molecules*
1 mole H *atoms*	1 mole Cl *atoms*	1 mole HCl *molecules*
1.00 g H	35.5 g Cl	36.5 g HCl
1 g-at. wt H	1 g-at. wt Cl	1 g-mol. wt HCl or
		1 g-form. wt HCl

In dealing with diatomic elements (H_2, O_2, N_2, F_2, Cl_2, Br_2, I_2), special care must be taken to distinguish between a mole of atoms (gram-atomic weight) and a mole of molecules (gram-molecular weight). For example, consider *one* mole of oxygen molecules, which weighs 32.0 g. This quantity is equal to *two* gram-atomic weights of the element oxygen and thus represents *two* moles of oxygen atoms. It is important, therefore, in the case of diatomic elements, to be certain which form we are considering, the atom or the molecule.

1 mole $= 6.02 \times 10^{23}$ *molecules or formula units*

1 mole = *1 gram-molecular weight of a compound*

1 mole = *1 gram-atomic weight of a monatomic element*

The conversion factors for changing grams of a compound to moles and vice versa are

Grams to moles: $\dfrac{1 \text{ mole of a compound}}{1 \text{ g-mol. wt of the compound}}$

Moles to grams: $\dfrac{1 \text{ g-mol. wt of a compound}}{1 \text{ mole of the compound}}$

154   Chapter 9

Problem 9.4   What is the weight of 1 mole (gram-molecular weight) of sulfuric acid, H_2SO_4? This problem is solved in a similar manner to Problems 9.1 through 9.3, using atomic weights of the elements in gram units instead of amu. Look up the atomic weights of H, S, and O, and solve.

$$2H = 2 \times 1.0 = 2.0 \text{ g}$$
$$1S = 1 \times 32.1 = 32.1 \text{ g}$$
$$4O = 4 \times 16.0 = 64.0 \text{ g}$$
$$\overline{98.1 \text{ g}} = \text{Weight of 1 mole (1 g-mol. wt) of } H_2SO_4$$

Problem 9.5   How many moles of NaOH are there in 1 kg of sodium hydroxide?
First, we know that
$$1 \text{ mole} = 1 \text{ g-mol. wt} = 40.0 \text{ g } (23.0 + 16.0 + 1.0 \text{ g}) \text{ NaOH}$$
$$1 \text{ kg} = 1000 \text{ g}$$

To convert grams to moles we use the conversion factor

$$\frac{1 \text{ mole}}{1 \text{ g-mol. wt}} = \frac{1 \text{ mole NaOH}}{40.0 \text{ g NaOH}}$$

The calculation is

$$1000 \text{ g NaOH} \times \frac{1 \text{ mole NaOH}}{40.0 \text{ g NaOH}} = 25.0 \text{ moles NaOH}$$

$$1 \text{ kg NaOH} = 25.0 \text{ moles NaOH}$$

Problem 9.6   What is the weight in grams of 5.00 moles of water?
First, we know that
$$1 \text{ mole } H_2O = 18.0 \text{ g} \quad \text{(Problem 9.1)}$$

To convert moles to grams, use the conversion factor

$$\frac{1 \text{ g-mol. wt } H_2O}{1 \text{ mole } H_2O} = \frac{18.0 \text{ g } H_2O}{1 \text{ mole } H_2O}$$

The calculation is

$$5.00 \text{ moles } H_2O \times \frac{18.0 \text{ g } H_2O}{1 \text{ mole } H_2O} = 90.0 \text{ g } H_2O \quad \text{(Answer)}$$

Problem 9.7   How many molecules of HCl are there in 25.0 g of hydrogen chloride?
From the formula we find that the gram-molecular weight (1 mole) of HCl is 36.5 g. The sequence of conversions is from

$$\text{grams HCl} \longrightarrow \text{moles HCl} \longrightarrow \text{molecules HCl}$$

using the conversion factors

$$\frac{1 \text{ mole HCl}}{36.5 \text{ g HCl}} \quad \text{and} \quad \frac{6.02 \times 10^{23} \text{ molecules HCl}}{1 \text{ mole HCl}}$$

$$25.0 \text{ g HCl} \times \frac{1 \text{ mole HCl}}{36.5 \text{ g HCl}} \times \frac{6.02 \times 10^{23} \text{ molecules HCl}}{1 \text{ mole HCl}}$$

$$= 4.12 \times 10^{23} \text{ molecules HCl} \quad \text{(Answer)}$$

Quantitative Composition of Compounds   155

9.4 Percentage Composition of Compounds

percentage composition of a compound

Just as each piece of pie represents a percentage of the whole pie, so the mass of each element in a compound represents a percentage of the total mass of that compound. The formula weight may be used to represent the total mass, or 100%, of a compound. If the *weight-percent* of each element in a compound is known, we have the **percentage composition of the compound**. The composition of water, H_2O, is two atoms of H and one atom of O per molecule, or 11.1% H and 88.9% O by weight.

The percentage composition of a compound can be determined if its formula is known or if the weights of two or more elements that have combined with each other are known or are experimentally determined.

If the formula is known, it is essentially a two-step process to determine the percentage composition.

Step 1. Determine the gram-formula weight.

Step 2. Determine the total weight of each element in the gram-formula weight. Divide each of these weights by the gram-formula weight and multiply by 100%. This gives the percentage.

Problem 9.8   Calculate the percentage composition of sodium chloride, NaCl.
Step 1.   Gram-formula weight of NaCl:

1 Na = 23.0 g
1 Cl = 35.5 g
―――――――
          58.5 g

Step 2. $Na = \dfrac{23.0 \text{ g}}{58.5 \text{ g}} \times 100\% = $ 39.3% Na

$Cl = \dfrac{35.5 \text{ g}}{58.5 \text{ g}} \times 100\% = $ 60.7% Cl
―――――――――
                                              100.0% Total

In any two-component system, if one percentage is known, the other is automatically defined by difference; that is, if Na = 39.3%, then Cl = 100 − 39.3 = 60.7%. However, the calculation of the percentage of each component should be carried out, since this provides a check against possible error. The percentage composition data should add up to 100 ± 0.5%.

Problem 9.9   Calculate the percentage composition of potassium sulfate, K_2SO_4.
Step 1.   Gram-formula weight of K_2SO_4:

2 K = 2 × 39.1 = 78.2 g
1 S = 1 × 32.1 = 32.1 g
4 O = 4 × 16.0 = 64.0 g
                ―――――――
                 174.3 g

156 Chapter 9

Step 2. $K = \dfrac{78.2 \text{ g}}{174.3 \text{ g}} \times 100\% = \quad 44.9\% \text{ K}$

$S = \dfrac{32.1 \text{ g}}{174.3 \text{ g}} \times 100\% = \quad 18.4\% \text{ S}$

$O = \dfrac{64.0 \text{ g}}{174.3 \text{ g}} \times 100\% = \quad \underline{36.7\% \text{ O}}$

$\qquad\qquad\qquad\qquad\qquad\qquad 100.0\% \quad \text{Total}$

Problem 9.10 When heated in the air, 1.63 g of zinc, Zn, combine with 0.40 g of oxygen, O_2, to form an oxide of zinc. Calculate the percentage composition of the compound formed.

The percentage composition may be calculated on the basis of the individual elements as parts or percentages of the total weight of the compound formed. First, calculate the total weight of the compound formed.

$\quad$ 1.63 g Zn
$\quad$ 0.40 g O
$\quad$ ──────
$\quad$ 2.03 g $\quad$ = Total weight of product

Then divide the weight of each element by the total weight (Step 1) and multiply by 100%.

$\dfrac{1.63 \text{ g}}{2.03 \text{ g}} \times 100\% = \quad 80.3\% \text{ Zn}$

$\dfrac{0.40 \text{ g}}{2.03 \text{ g}} \times 100\% = \quad \underline{19.7\% \text{ O}}$

$\qquad\qquad\qquad\qquad\qquad 100.0\% \quad \text{Total}$

The compound formed contains 80.3% Zn and 19.7% O.

9.5 Empirical Formula versus Molecular Formula

empirical formula

The **empirical formula**, or **simplest formula**, gives the smallest ratio of the atoms that are present in a compound. This formula gives the relative number of atoms of each element in the compound. The empirical formula contains the smallest whole-number ratio that can be derived from the percentages of the different elements in the compound.

molecular formula

The **molecular formula** is the true formula, representing the total number of atoms of each element present in one molecule of a compound. It is entirely possible that two or more substances will have the same percentage composition, yet be distinctly different compounds. For example, acetylene, C_2H_2, is a common gas used in welding; benzene, C_6H_6, is an important solvent obtained from coal tar and is used in the synthesis of styrene and nylon. Both acetylene and benzene contain 92.3% C and 7.7% H. The smallest ratio of C and H corresponding to these percentages is CH (1:1). Therefore, the empirical formula for both acetylene and benzene is CH—even though it is known that the molecular formulas are C_2H_2 and C_6H_6, respectively. It is not uncommon for the molecular formula to be the same as the empirical formula. If the molecular formula is not the same, it will be an integral multiple of the empirical formula.

CH = Empirical formula
(CH)$_2$ = C$_2$H$_2$ = Acetylene (molecular formula)
(CH)$_6$ = C$_6$H$_6$ = Benzene (molecular formula)

Table 9.1 summarizes the data concerning these CH formulas. Table 9.2 shows empirical and molecular formula relationships of other compounds.

Table 9.1. Molecular formulas of two compounds having an empirical formula with a 1:1 ratio of carbon and hydrogen atoms.

Formula	%C	%H	Molecular weight
CH (empirical)	92.3	7.7	13.0 (empirical)
C$_2$H$_2$ (acetylene)	92.3	7.7	26.0 (2 × 13.0)
C$_6$H$_6$ (benzene)	92.3	7.7	78.0 (6 × 13.0)

Table 9.2. Some empirical and molecular formulas.

Compound	Empirical formula	Molecular formula	Compound	Empirical formula	Molecular formula
Acetylene	CH	C$_2$H$_2$	Diborane	BH$_3$	B$_2$H$_6$
Benzene	CH	C$_6$H$_6$	Hydrazine	NH$_2$	N$_2$H$_4$
Ethylene	CH$_2$	C$_2$H$_4$	Hydrogen	H	H$_2$
Formaldehyde	CH$_2$O	CH$_2$O	Chlorine	Cl	Cl$_2$
Acetic acid	CH$_2$O	C$_2$H$_4$O$_2$	Bromine	Br	Br$_2$
Dextrose	CH$_2$O	C$_6$H$_{12}$O$_6$	Oxygen	O	O$_2$
Hydrogen chloride	HCl	HCl	Nitrogen	N	N$_2$
Carbon dioxide	CO$_2$	CO$_2$			

9.6 Calculation of Empirical Formula

It is possible to establish an empirical formula because (1) the individual atoms in a compound are combined in whole-number ratios and (2) each element has a specific atomic weight.

In order to calculate the empirical formula, we need to know (1) the elements that are combined, (2) their atomic weights, and (3) the ratio by weight or percentage in which they are combined. If elements A and B form a compound, we may represent the empirical formula as A$_x$B$_y$, where x and y are small whole numbers that represent the number of atoms of A and B. To write the empirical formula, we must determine x and y.

The solution to this problem requires three or four steps.

Step 1. Assume a definite quantity (usually 100 g) of the compound, if not given, and express the weight of each element in grams.

158 Chapter 9

Step 2. Multiply the weight (grams) of each element by the factor 1 mole/1 g-at. wt to convert grams to moles. This conversion gives the number of moles of atoms of each element in the quantity assumed. At this point, these numbers will usually not be whole numbers.

Step 3. Divide each of the values obtained in Step 2 by the smallest of these values. If the numbers obtained by this procedure are whole numbers, use them as subscripts in writing the empirical formula. If the numbers obtained are not whole numbers, go on to Step 4.

Step 4. Multiply the values obtained in Step 3 by the smallest number that will convert them to whole numbers. Use these whole numbers as the subscripts in the empirical formula. For example, if the ratio of A to B is 1.0:1.5, multiply both numbers by 2 to obtain a ratio of 2:3. The empirical formula would then be A_2B_3.

Problem 9.11

Calculate the empirical formula of a compound containing 11.19% hydrogen, H, and 88.89% oxygen, O.

Step 1. Express each element in grams; if we assume that there are 100 g of material, then the percentage of each element is equal to the grams of each element in 100 g.

$$H = 11.19\% = \frac{11.19 \text{ g}}{100 \text{ g}}$$

$$O = 88.89\% = \frac{88.89 \text{ g}}{100 \text{ g}}$$

Step 2. Multiply the grams of each element by the proper factor to obtain the relative number of moles of atoms:

$$H: \quad 11.19 \text{ g H} \times \frac{1 \text{ mole H atoms}}{1.01 \text{ g H}} = 11.1 \text{ moles H atoms}$$

$$O: \quad 88.89 \text{ g O} \times \frac{1 \text{ mole O atoms}}{16.0 \text{ g O}} = 5.55 \text{ moles O atoms}$$

The formula could be expressed as $H_{11.1}O_{5.55}$. However, it is customary to use the smallest whole-number ratio of atoms. This ratio is calculated in Step 3.

Step 3. Change these numbers to whole numbers by dividing each by the smallest.

$$H = \frac{11.1 \text{ moles}}{5.55 \text{ moles}} = 2 \quad O = \frac{5.55 \text{ moles}}{5.55 \text{ moles}} = 1$$

In this step, the ratio of atoms has not changed, because we divided the number of moles of each element by the same number.

The simplest ratio of H to O is 2:1.

Empirical formula = H_2O

Problem 9.12

The analysis of a salt showed that it contained 56.58% potassium, K, 8.68% carbon, C, and 34.73% oxygen, O. Calculate the empirical formula for this substance.

Steps 1 and 2. After changing the percentage of each element to grams, find the relative number of moles of each element by multiplying by the proper mole/g-at. wt factor.

Quantitative Composition of Compounds 159

$$\text{K:} \quad 56.58 \text{ g K} \times \frac{1 \text{ mole K atoms}}{39.1 \text{ g K}} = 1.45 \text{ moles K atoms}$$

$$\text{C:} \quad 8.68 \text{ g C} \times \frac{1 \text{ mole C atoms}}{12.0 \text{ g C}} = 0.720 \text{ mole C atoms}$$

$$\text{O:} \quad 34.73 \text{ g O} \times \frac{1 \text{ mole O atoms}}{16.0 \text{ g O}} = 2.17 \text{ moles O atoms}$$

Step 3. Divide each number of moles by the smallest.

$$\text{K} = \frac{1.45 \text{ moles}}{0.720 \text{ mole}} = 2.01$$

$$\text{C} = \frac{0.720 \text{ mole}}{0.720 \text{ mole}} = 1.00$$

$$\text{O} = \frac{2.17 \text{ moles}}{0.720 \text{ mole}} = 3.01$$

The simplest ratio of K:C:O is therefore 2:1:3.

Empirical formula = K_2CO_3

Problem 9.13   A sulfide of iron was formed by combining 2.233 g of iron, Fe, with 1.926 g of sulfur, S. What is the empirical formula of the compound?

Steps 1 and 2.   Find the relative number of moles of each element by multiplying grams of each element by the proper mole/g-at. wt factor.

$$\text{Fe:} \quad 2.233 \text{ g Fe} \times \frac{1 \text{ mole Fe atoms}}{55.8 \text{ g Fe}} = 0.0400 \text{ mole Fe atoms}$$

$$\text{S:} \quad 1.926 \text{ g S} \times \frac{1 \text{ mole S atoms}}{32.1 \text{ g S}} = 0.0600 \text{ mole S atoms}$$

Step 3.   Divide each number of moles by the smaller of the two numbers.

$$\text{Fe} = \frac{0.0400 \text{ mole}}{0.0400 \text{ mole}} = 1.00$$

$$\text{S} = \frac{0.0600 \text{ mole}}{0.0400 \text{ mole}} = 1.50$$

Step 4.   Since we still have not reached a ratio that will give a formula containing whole numbers of atoms, we must double each value to obtain a ratio of 2.00 atoms of Fe to 3.00 atoms of S. Doubling both values does not change the ratio of Fe and S atoms.

Fe: $1.00 \times 2 = 2.00$

S: $1.50 \times 2 = 3.00$

Empirical formula = Fe_2S_3

In many of these calculations, results may vary somewhat from an exact whole number. This can be due to experimental errors in obtaining the data.

Calculations that vary by no more than ±0.1 from a whole number can usually be rounded off to the nearest whole number. Deviations greater than about 0.1 unit usually mean that the calculated ratios need to be multiplied by a factor to make them all whole numbers.

9.7 Calculation of the Molecular Formula from the Empirical Formula

The molecular formula can be calculated from the empirical formula if the molecular weight, in addition to data for calculating the empirical formula, is known. The molecular formula, as stated in Section 9.5, will be equal to, or some multiple of, the empirical formula. For example, if the empirical formula of a compound between hydrogen and fluorine is HF, the molecular formula can be expressed as $(HF)n$, where $n = 1, 2, 3, 4, \ldots$. This n means that the molecular formula could be HF, H_2F_2, H_3F_3, H_4F_4, and so on. To determine the molecular formula, we must evaluate n.

$$n = \frac{\text{Molecular weight}}{\text{Empirical formula weight}} = \text{Number of empirical formula units}$$

Problem 9.14 A compound of nitrogen and oxygen with a molecular weight of 92.0 was found to have an empirical formula of NO_2. What is its molecular formula?

Step 1. Let n be the number of (NO_2) units in a molecule; then the molecular formula is $(NO_2)n$.

Step 2. Each (NO_2) unit weighs 46.0 g [14 + (2)(16)]. The gram-molecular weight of $(NO_2)n = 92.0$ g and the number of (46.0) units in 92.0 is 2.

$$n = \frac{92.0 \text{ g}}{46.0 \text{ g}} = 2 \quad \text{(Empirical formula units)}$$

Step 3. The molecular formula is $(NO_2)_2$, or N_2O_4.

Problem 9.15 The hydrocarbon propylene has a gram-molecular weight of 42.0 g/mole and contains 14.3% H and 85.7% C. What is its molecular formula?

Step 1. First find the empirical formula:

$$\text{C:} \quad 85.7 \text{ g C} \times \frac{1 \text{ mole C atoms}}{12.0 \text{ g C}} = 7.14 \text{ moles C atoms}$$

$$\text{H:} \quad 14.3 \text{ g H} \times \frac{1 \text{ mole H atoms}}{1.0 \text{ g H}} = 14.3 \text{ moles H atoms}$$

Divide each value by the smallest number of moles.

$$C = \frac{7.14 \text{ moles}}{7.14 \text{ moles}} = 1.0$$

$$H = \frac{14.3 \text{ moles}}{7.14 \text{ moles}} = 2.0$$

Empirical formula = CH_2

Step 2. Determine the molecular formula from the empirical formula and molecular weight.

Molecular formula = $(CH_2)_n$
Molecular weight = 42.0

Each CH_2 unit weighs 14.0 (12 + 2). The number of CH_2 units in 42.0 is 3.

$$n = \frac{42.0}{14.0} = 3 \text{ (Empirical formula units)}$$

The molecular formula is $(CH_2)_3$, or C_3H_6.

Questions

A. Review the meanings of the new terms introduced in this chapter.
   1. Formula weight
   2. Molecular weight
   3. Gram-formula weight
   4. Gram-molecular weight
   5. Percentage composition of a compound
   6. Empirical formula
   7. Molecular formula

B. Review questions.
   1. How are formula weight and molecular weight related to each other? In what respects are they different?
   2. How many molecules are present in 1 g-mol. wt of nitric acid, HNO_3? How many atoms are present?
   3. What is the relationship between the following?
      (a) Mole and molecular weight   (b) Mole and formula weight
   4. Why is it correct to refer to the weight of 1 mole of sodium chloride, but incorrect to refer to a molecular weight of sodium chloride?
   5. In calculating the empirical formula of a compound from its percentage composition, why do we choose to start with 100 g of the compound?
   6. Which of the following statements are correct?
      (a) A mole of sodium and a mole of sodium chloride contain the same number of sodium atoms.
      (b) A compound such as NaCl has a formula weight but no true molecular weight.
      (c) One mole of nitrogen gas weighs 14.0 g.
      (d) The percentage of oxygen is higher in K_2CrO_4 than it is in Na_2CrO_4.
      (e) The number of Cr atoms is the same in a mole of K_2CrO_4 as it is in a mole of Na_2CrO_4.
      (f) Both K_2CrO_4 and Na_2CrO_4 contain the same percentage by weight of Cr.
      (g) A gram-molecular weight of sucrose, $C_{12}H_{22}O_{11}$, contains 1 mole of sucrose molecules.
      (h) Two moles of sulfuric acid, H_2SO_4, contain 8 moles of oxygen atoms.
      (i) The empirical formula of sucrose, $C_{12}H_{22}O_{11}$, is CH_2O.
      (j) A hydrocarbon that has a molecular weight of 280 and an empirical formula of CH_2 has a molecular formula of $C_{22}H_{44}$.

C. Problems.
   1. Determine the molecular weight of the following compounds:
      (a) NaI
      (b) $Fe(OH)_3$
      (c) K_2SO_4
      (d) Mn_3O_4
      (e) $Al_2(SO_4)_3$
      (f) C_2H_5Cl
      (g) $C_6H_{12}O_6$
      (h) Br_2
      (i) KH_2PO_4

162   Chapter 9

2. Determine the gram-molecular weight of the following compounds:
   (a) $HC_2H_3O_2$ (d) HNO_3 (g) UF_6
   (b) $Pb(NO_3)_2$ (e) $Ca(HCO_3)_2$ (h) $K_3Fe(CN)_6$
   (c) $C_6H_5NO_2$ (f) $ZnCl_2$ (i) $CoCl_2 \cdot 6\,H_2O$
3. How many moles are contained in the following?
   (a) 32.0 g NaOH   (d) 40.0 g $Ca(NO_3)_2$
   (b) 18.0 g N_2 (e) 0.953 g $MgCl_2$
   (c) 50.0 g CH_3OH (f) 1.0 lb KCl
4. How many moles of atoms are contained in each of the following?
   (a) 7.2 g Mg   (c) 12.0 g Cl_2
   (b) 39.9 g Ar   (d) 3.01×10^{23} atoms F
5. Calculate the number of grams contained in each of the following:
   (a) 50.0 moles H_2O (d) 0.500 mole O_2
   (b) 0.100 mole $SnCl_2$ (e) 4.25×10^{-4} mole NH_4Br
   (c) 1.21 moles H_3PO_4 (f) 4.50 moles CH_4
6. How many molecules are contained in each of the following?
   (a) 1.0 mole F_2 (c) 3.0 g C_2H_6
   (b) 0.35 mole N_2 (d) 50.0 g SO_3
   How many atoms are present in each of these amounts?
7. What is the weight in grams of each of the following?
   (a) 1 atom of mercury   (c) 1 atom of helium
   (b) 1 molecule of water   (d) 1 molecule of CO_2
8. How many moles are contained in each of the following?
   (a) 1000 molecules C_6H_6 (d) 6000 molecules NO_2
   (b) 1×10^{12} atoms Zn   (e) 1 atom Mg
   (c) 1000 molecules CH_4 (f) 2×10^6 molecules H_2O
9. How many atoms of carbon are contained in each of the following?
   (a) 1.00 mole C_4H_{10} (d) 2.00 moles C_2H_5Br
   (b) 11.5 g $CaCO_3$ (e) 5.5 g CO_2
   (c) 6.00×10^{20} molecules $C_3H_8O_3$ (f) 3.00×10^{10} molecules CH_4
10. Calculate the number of:
    (a) Grams of silver in 40.0 g AgCl
    (b) Grams of bromine in 10.0 g $CaBr_2$
    (c) Grams of sulfur in 500 g $Na_2S_2O_7$
    (d) Grams of chromium in 25.0 g K_2CrO_4
11. A solution was made by dissolving 12.0 g of potassium chromate, K_2CrO_4, in 200 g of water. How many moles of each compound were used?
12. A sulfuric acid solution contains 65.0% H_2SO_4 by weight and has a density of 1.55 g/ml. How many moles of the acid are present in 1.00 litre of the solution?
13. A nitric acid solution containing 72.0% HNO_3 by weight has a density of 1.42 g/ml. How many moles of HNO_3 are present in 100 ml of the solution?
14. Calculate the percentage composition of the following compounds:
    (a) MgO   (c) $CaSO_4$ (e) AlN   (g) $AgNO_3$
    (b) CCl_4 (d) KNO_3 (f) HCl   (h) $Fe(OH)_3$
15. What is the oxidation number of the first element in each of the compounds in Problem 14?
16. Calculate the percentage of iron, Fe, in the following compounds:
    (a) FeO   (b) Fe_2O_3 (c) Fe_3O_4 (d) $K_4Fe(CN)_6$
17. Which one of the following oxides has the highest and which has the lowest percentage of oxygen, O, by weight, in its formula?
    (a) Li_2O (b) MgO   (c) Bi_2O_3 (d) TiO_2

18. A 6.23 g sample of silver, Ag, was converted to the oxide, which was found to weigh 6.69 g. Calculate the percentage composition of the compound formed.
19. Calculate the empirical formula of each compound from the percentage compositions given below.
    (a) 66.4% Cu,   33.6% S
    (b) 79.8% Cu,   20.2% S
    (c) 62.6% Ca,   37.4% C
    (d) 36.8% N,   63.2% O
    (e) 38.9% Cl,   61.2% O
    (f) 39.8% K,   27.8% Mn,   32.5% O
    (g) 32.4% Na,   22.6% S,   45.0% O
    (h) 52.0% Zn,   9.60% C,   38.4% O
    (i) 1.90% H,   67.6% Cl,   30.5% O
    (j) 60.0% C,   13.3% H,   26.7% O
20. Calculate the percentage of:
    (a) Cadmium in $CdCO_3$
    (b) Carbon in $C_5H_{11}Cl$
    (c) Manganese in $KMnO_4$
    (d) Nitrogen in NH_4NO_3
21. Answer the following by consideration of the formulas. Check your answers by calculation if you wish. Which compound has the:
    (a) Higher percent by weight of hydrogen, H_2O or H_2O_2?
    (b) Lower percent by weight of manganese, $NaMnO_4$ or Na_2MnO_4?
    (c) Higher percent by weight of chromium, K_2CrO_4 or $K_2Cr_2O_7$?
    (d) Higher percent by weight of nitrogen, NO_2 or N_2O_4?
    (e) Lower percent by weight of sulfur, $NaHSO_4$ or Na_2SO_4?
22. A 7.615 g sample of gallium, Ga, was found to react with 2.622 g of oxygen, O_2. What is the empirical formula of the compound formed?
23. Magnesium reacts with nitrogen to form the compound Mg_3N_2. Will a mixture of 20.0 g Mg and 10.0 g N_2 have sufficient magnesium atoms to react with all the nitrogen atoms? Show evidence for your answer.
24. Hydroquinone is an organic compound commonly used as a photographic developer. It has a molecular weight of 110 g/mole and a composition of 65.45% C, 5.45% H, and 29.09% O. Calculate the molecular formula of hydroquinone.
25. Fructose is a very sweet natural sugar that is present in honey, fruits, and fruit juices. It has a molecular weight of 180 g/mole and a composition of 40.0% C, 6.7% H, and 53.3% O. Calculate the molecular formula of fructose.
26. Listed below are the compositions of four different compounds of carbon, C, and chlorine, Cl. Derive the empirical and molecular formulas for each.

	Percent C	Percent Cl	Molecular weight
(a)	7.79	92.21	154
(b)	10.13	89.87	237
(c)	25.26	74.74	285
(d)	11.25	88.75	320

10 Chemical Equations

After studying Chapter 10 you should be able to:

1. Understand the terms listed in Question A at the end of the chapter.
2. Know the format used in setting up chemical equations.
3. Recognize the various symbols commonly used in writing chemical equations.
4. Balance simple chemical equations.
5. Interpret a balanced equation in terms of the relative numbers or amounts of molecules, atoms, grams, or moles of each substance represented.
6. Classify equations as representing combination, decomposition, single-replacement, or double-replacement reactions.
7. Complete and balance equations for simple combination, decomposition, single-replacement, and double-replacement reactions when given the reactants.
8. Distinguish between exothermic and endothermic reactions, and relate the quantity of heat to the amounts of substances involved in the reaction.

10.1 The Chemical Equation

chemical equation

word equation

A **chemical equation** is a shorthand expression for a chemical change or reaction. It shows, among other things, the rearrangement of atoms that are involved in the reaction. A **word equation** states in words, in equation form, the substances involved in a chemical reaction. For example, when mercuric oxide is heated, it decomposes to form mercury and oxygen. The word equation for this decomposition is

$$\text{Mercuric oxide} + \text{Heat} \longrightarrow \text{Mercury} + \text{Oxygen}$$

From the chemist's point of view, this method of describing a chemical change is very inadequate. It is bulky and cumbersome to use and does not give quantitative information. The chemical equation, using symbols and formulas, is a far better way to describe the decomposition of mercuric oxide:

$$2\,\text{HgO} \xrightarrow{\Delta} 2\,\text{Hg} + \text{O}_2\uparrow$$

This equation gives all the information from the word equation plus formulas, composition, reactive amounts of all the substances involved in the reaction, and much additional information (see Section 10.4). Even though a chemical equation provides much quantitative information, it is still not a complete description; it does not tell us how much heat is needed to cause decomposition, what we observe during the reaction, or anything about the rate of reaction. This information must be obtained from other sources or from experimentation.

10.2 Format for Writing Chemical Equations

reactant

product

The **reactants** are the substances that enter into a chemical change or reaction. The **products** are the substances produced by the reaction. A chemical equation uses the chemical symbols and formulas of the reactants and products and other symbolic terms to represent a chemical reaction. Equations are written according to this general format:

1. The reactants and products are separated by an arrow or other sign indicating equality between reactants and products ($\longrightarrow$, =, $\rightleftharpoons$).
2. The reactants are placed to the left and the products to the right of the arrow or equality sign. A plus sign (+) is placed between reactants and between products when needed.
3. Conditions required to carry out the reaction may, if desired, be placed above or below the arrow or equality sign. For example, a delta sign placed over the arrow ($\xrightarrow{\Delta}$) indicates that heat is supplied to the reaction.
4. Small integral numbers in front of substances (for example, 2 H_2O) are used to balance the equation and to indicate the number of formula units (atoms, molecules, moles, ions) of each substance reacting or being produced. When no number is shown, it is understood that one formula unit of the substance is indicated.

Symbols commonly used in equations are given in Table 10.1.

Table 10.1. Symbols commonly used in chemical equations.

Symbol	Meaning
$\rightarrow$	Yields, produces (points to products)
=	Equals; equilibrium between reactants and products
$\rightleftarrows$	Reversible reaction; equilibrium between reactants and products
↑	Gas evolved (written after a substance)
↓	Solid or precipitate formed (written after a substance)
(s)	Solid (written after a substance)
(l)	Liquid (written after a substance)
(g)	Gas (written after a substance)
Δ	Heat
+	Plus or added to
(aq)	Aqueous solution (substance dissolved in water)

10.3 Writing and Balancing Equations

balanced equation

To represent the quantitative relationships of a reaction, the chemical equation must be balanced. A **balanced equation** is one that contains the same number of each kind of atom on each side of the equation. The balanced equation, therefore, obeys the Law of Conservation of Mass.

The ability to balance equations must be acquired by every chemistry student. Simple equations are easy to balance, but some care and attention to detail are required. Clearly, the way to balance an equation is to adjust the number of atoms of each element so that it is the same on each side of the equation. But we must not change a correct formula in order to achieve a balanced equation. Each equation must be treated on its own merits; there is no simple "plug in" formula for balancing equations. The following outline gives a general procedure for balancing equations. Study this outline and refer to it as needed when working examples. There is no substitute for practice in learning to write and balance equations.

1. Identify the reaction for which the equation is to be written. Formulate a description or word equation for this reaction (for example, mercuric oxide decomposes yielding mercury and oxygen). This, of course, need not be done when the reactants and products are identified and their formulas are given.
2. Write the unbalanced, or skeleton, equation. Make sure that the formula for each substance is correct and that the reactants are written to the left and the products to the right of the arrow (for example, $HgO \rightarrow Hg + O_2$). The correct formulas must be known or ascertained from the periodic table, oxidation numbers, lists of ions, or experimental data.
3. Balance the equation. Use the following steps as necessary:
    (a) Count and compare the number of atoms of each element on each side of the equation and determine those that must be balanced.
    (b) Balance each element, one at a time, by placing small whole numbers (coefficients) in front of the formulas containing the unbalanced element. It is usually best to balance first metals, then nonmetals, then hydrogen and oxygen. Select the smallest coefficients that will give the same number of atoms of the element on each side. A coefficient placed before a formula multiplies every atom in the formula by that number (for example, $2\,H_2SO_4$ means two molecules of sulfuric acid and also means four H atoms, two S atoms, and eight O atoms).
    (c) Check all other elements after each individual element is balanced to see if, in balancing one, other elements have become unbalanced. Make adjustments as needed.
    (d) Balance polyatomic ions such as SO_4^{2-}, which remain unchanged from one side of the equation to the other, in the same way as individual atoms.
    (e) Do a final check, making sure that each element and/or polyatomic ion is balanced and that the smallest possible set of whole-number coefficients has been used.

$$4\,HgO \longrightarrow 4\,Hg + 2\,O_2 \quad \text{(Incorrect form)}$$
$$2\,HgO \longrightarrow 2\,Hg + O_2 \quad \text{(Correct form)}$$

The following examples show stepwise sequences leading to balanced equations. Study each example carefully.

Example 10.1   Write the balanced equation for the reaction that takes place when magnesium metal is burned in air to produce magnesium oxide.
1. *Word equation*:

    Magnesium + Oxygen $\longrightarrow$ Magnesium oxide

2. *Skeleton equation*:

$$Mg + O_2 \longrightarrow MgO \quad \text{(Unbalanced)}$$

3. *Balance*:
   (a) Oxygen is not balanced. There are two O atoms on the left side and one on the right side.
   (b) Place the coefficient 2 before MgO.

   $$Mg + O_2 \longrightarrow 2\,MgO \quad \text{(Unbalanced)}$$

   (c) Now Mg is not balanced. There is one Mg atom on the left side and two on the right side. Place a 2 before Mg.

   $$2\,Mg + O_2 \longrightarrow 2\,MgO \quad \text{(Balanced)}$$

   (d) *Check*: Each side has two Mg and two O atoms.

Example 10.2 Methane, CH_4, undergoes complete combustion to produce carbon dioxide and water. Write the balanced equation for this reaction.

1. *Word equation*:

$$\text{Methane} + \text{Oxygen} \longrightarrow \text{Carbon dioxide} + \text{Water}$$

2. *Skeleton equation*:

$$CH_4 + O_2 \longrightarrow CO_2 + H_2O \quad \text{(Unbalanced)}$$

3. *Balance*:
   (a) Hydrogen and oxygen are not balanced.
   (b) Balance H atoms by placing a 2 before H_2O

   $$CH_4 + O_2 \longrightarrow CO_2 + 2\,H_2O \quad \text{(Unbalanced)}$$

   Each side of the equation has four H atoms; oxygen is still not balanced. Place a 2 before O_2 to balance the oxygen atoms.

   $$CH_4 + 2\,O_2 \longrightarrow CO_2 + 2\,H_2O \quad \text{(Balanced)}$$

   (c) *Check*: The equation is correctly balanced; it has one C atom, four O atoms, and four H atoms on each side.

Example 10.3 Oxygen is prepared by heating potassium chlorate.

1. *Word equation*:

$$\text{Potassium chlorate} \xrightarrow{\Delta} \text{Potassium chloride} + \text{Oxygen}$$

2. *Skeleton equation*:

$$KClO_3 \xrightarrow{\Delta} KCl + O_2 \quad \text{(Unbalanced)}$$

3. *Balance*:
   (a) Oxygen is unbalanced (three O atoms on the left and two on the right side).
   (b) Balance by placing a 2 before $KClO_3$ and a 3 before O_2 to give six O atoms on each side.

   $$2\,KClO_3 \xrightarrow{\Delta} KCl + 3\,O_2 \quad \text{(Unbalanced)}$$

168 Chapter 10

Now K and Cl are not balanced. Place a 2 before KCl, which balances both K and Cl at the same time.

$$2\ KClO_3 \xrightarrow{\Delta} 2\ KCl + 3\ O_2 \quad \text{(Balanced)}$$

(c) *Check:* Each side contains two K, two Cl, and six O atoms.

Example 10.4 Balance by starting with the word equation given.
1. *Word equation:*

Silver nitrate + Hydrogen sulfide ⟶ Silver sulfide + Nitric acid

2. *Skeleton equation:*

$$AgNO_3 + H_2S \longrightarrow Ag_2S + HNO_3 \quad \text{(Unbalanced)}$$

3. *Balance:*
   (a) Ag and H are unbalanced.
   (b) Place a 2 in front of $AgNO_3$.

$$2\ AgNO_3 + H_2S \longrightarrow Ag_2S + HNO_3 \quad \text{(Unbalanced)}$$

   (c) This leaves H and NO_3^- unbalanced. Balance by placing a 2 in front of HNO_3.

$$2\ AgNO_3 + H_2S \longrightarrow Ag_2S + 2\ HNO_3 \quad \text{(Balanced)}$$

   (d) In this example, N and O atoms are balanced by balancing the NO_3^- ion as a unit.
   (e) *Check:* Each side has two Ag, two H, and one S atom. Also, each side has two NO_3^- ions.

Example 10.5 Balance by starting with the word equation given.
1. *Word equation:*

Aluminum hydroxide + Sulfuric acid ⟶ Aluminum sulfate + Water

2. *Skeleton equation:*

$$Al(OH)_3 + H_2SO_4 \longrightarrow Al_2(SO_4)_3 + HOH \quad \text{(Unbalanced)}$$

3. *Balance:*
   (a) All elements are unbalanced.
   (b) Balance Al by placing a 2 in front of $Al(OH)_3$. Treat the unbalanced SO_4^{2-} ion as a unit and balance by placing a 3 before H_2SO_4. Note that Step (d) may sometimes be combined with Step (b).

$$2\ Al(OH)_3 + 3\ H_2SO_4 \longrightarrow Al_2(SO_4)_3 + HOH \quad \text{(Unbalanced)}$$

Balance the unbalanced H and O by placing a 6 in front of HOH.

$$2\ Al(OH)_3 + 3\ H_2SO_4 \longrightarrow Al_2(SO_4)_3 + 6\ HOH \quad \text{(Balanced)}$$

   (c) *Check:* Each side has two Al, twelve H, and six O atoms; and also three SO_4^{2-} ions.

Example 10.6 The fuel in a butane gas stove undergoes complete combustion to carbon dioxide and water.
1. *Word equation:*

Butane + Oxygen ⟶ Carbon dioxide + Water

2. *Skeleton equation*:

$$C_4H_{10} + O_2 \longrightarrow CO_2 + H_2O \quad \text{(Unbalanced)}$$

3. *Balance*:
   (a) All elements are unbalanced.
   (b) Balance C by placing a 4 in front of CO_2.

   $$C_4H_{10} + O_2 \longrightarrow 4\,CO_2 + H_2O \quad \text{(Unbalanced)}$$

   Balance H by placing a 5 in front of H_2O.

   $$C_4H_{10} + O_2 \longrightarrow 4\,CO_2 + 5\,H_2O \quad \text{(Unbalanced)}$$

   Oxygen remains unbalanced. When we try to balance the O atoms, we find that there is no integer (whole number) that can be placed in front of O_2 to bring about a balance. The equation can be balanced if we use $6\frac{1}{2}\,O_2$ and then double the coefficients of each substance, including the $6\frac{1}{2}\,O_2$, to obtain the balanced equation.

   $$2\,C_4H_{10} + 13\,O_2 \longrightarrow 8\,CO_2 + 10\,H_2O \quad \text{(Balanced)}$$

   An alternate procedure is to rewrite the last unbalanced equation, doubling all the coefficients except that of O_2:

   $$2\,C_4H_{10} + O_2 \longrightarrow 8\,CO_2 + 10\,H_2O$$

   Now balance the O_2—the result is the same balanced equation as above.
   (c) *Check*: Each side now has eight C, twenty H, and twenty-six O atoms.

10.4 What Information Does an Equation Tell Us?

Interpreting the information given in an equation is important if we are to gain the full benefit of its use in evaluating a chemical reaction. The balanced equation tells us

1. what the reactants are and what the products are,
2. the formulas of the reactants and products,
3. the number of molecules or formula units of reactants and products in the reaction,
4. the number of atoms of each element involved in the reaction,
5. the number of molecular weights or formula weights of each substance used or produced,
6. the number of moles of each substance,
7. the number of gram-molecular weights or gram-formula weights of each substance used or produced,
8. the number of grams of each substance used or produced.

Consider the equation

$$H_2(g) + Cl_2(g) \longrightarrow 2\,HCl(g)$$

This equation states that hydrogen gas reacts with chlorine gas to produce hydrogen chloride, also a gas. Let us summarize all the information relating to the equation. The information that can be stated about the relative amount of each substance, with respect to all other substances in the balanced equation, is written below its formula in the following equation:

$H_2(g)$ +	$Cl_2(g)$	$\longrightarrow$ 2 HCl(g)
Hydrogen	Chlorine	Hydrogen chloride
1 molecule	1 molecule	2 molecules
2 atoms	2 atoms	2 atoms H + 2 atoms Cl
1 mol. wt	1 mol. wt	2 mol. wt
1 mole	1 mole	2 moles
1 form. wt	1 form. wt	2 form. wt
1 g-mol. wt	1 g-mol. wt	2 g-mol. wt
2.0 g	71.0 g	2 × 36.5 g (73.0 g)

These data are very useful in calculating quantitative relationships that exist among substances in a chemical reaction. For example, if we react 2 moles of hydrogen (twice as much as is indicated by the equation) with 2 moles of chlorine, we can expect to obtain 4 moles, or 146 g of hydrogen chloride, as a product. We will study this phase of using equations in more detail in the next chapter.

10.5 Types of Chemical Equations

Chemical equations represent chemical changes or reactions. To be of any significance, an equation must represent an actual or possible reaction. Part of the problem of writing equations is determining the products formed. There is no sure method of predicting products, nor do we have time to carry out experimentally all the reactions we may wish to consider. Therefore, we must use data reported in the writings of other workers, certain rules to aid in our predictions, and the atomic structure and combining capacities of the elements to help us predict the formulas of the products of a chemical reaction. The final proof of the existence of any reaction, of course, is in the actual observation of the reaction in the laboratory (or elsewhere).

Reactions are classified into types to assist in writing equations and to aid in predicting other reactions. Many chemical reactions fit one or another of the four principal reaction types that are discussed in the following paragraphs. Reactions are also classified as oxidation–reduction. Special methods are used to balance complex oxidation–reduction equations (see Chapter 17).

1. Combination or synthesis reaction. In this type of reaction, direct union or combination of two substances produces one new substance. Oxidation–reduction is involved in some, but not all, combination reactions. The general form of the equation is

combination or synthesis reaction

$A + B \longrightarrow AB$

Examples:
$S(s) + O_2(g) \longrightarrow SO_2(g)$
$2\,Mg(s) + O_2(g) \longrightarrow 2\,MgO(s)$
$2\,Na(s) + Cl_2(g) \longrightarrow 2\,NaCl(s)$
$CaO(s) + H_2O \longrightarrow Ca(OH)_2(aq)$
$SO_3(g) + H_2O \longrightarrow H_2SO_4(aq)$

decomposition reaction

2. Decomposition reaction. In this type of reaction a single substance is decomposed or broken down into two or more different substances. The reaction may be considered the reverse of combination. The starting material must be a compound, and the products may be elements or compounds. Oxidation–reduction is involved in some, but not all, decomposition reactions. The general form of the equation is

$$AB \longrightarrow A + B$$

Examples:
$$2\,HgO(s) \longrightarrow 2\,Hg(l) + O_2(g)$$
$$2\,H_2O \longrightarrow 2\,H_2(g) + O_2(g)$$
$$CaCO_3(s) \longrightarrow CaO(s) + CO_2(g)$$
$$2\,KClO_3(s) \longrightarrow 2\,KCl(s) + 3\,O_2(g)$$

single-replacement or substitution reaction

3. Single-replacement or substitution reaction. In this type of reaction one simple substance (element) reacts with a compound substance to form a new simple substance and a new compound, one element displacing another in a compound. Oxidation–reduction is always present in single-replacement reactions. The general form of the equation is

$$A + BC \longrightarrow B + AC$$

If A is a metal, it will replace B to form AC; if A is a nonmetal, it will replace C to form BA. Some reactions that fall into this category are the following:

(a) Active metal + Acid ⟶ Hydrogen + Salt
$$Zn(s) + 2\,HCl(aq) \longrightarrow H_2(g) + ZnCl_2(aq)$$
$$2\,Al(s) + 3\,H_2SO_4(aq) \longrightarrow 3\,H_2(g) + Al_2(SO_4)_3(aq)$$

(b) Active metal + Water ⟶ Hydrogen + Metal hydroxide or Metal oxide
$$2\,Na(s) + 2\,H_2O \longrightarrow H_2(g) + 2\,NaOH(aq)$$
$$Ca(s) + 2\,H_2O \longrightarrow H_2(g) + Ca(OH)_2(aq)$$
$$3\,Fe(s) + 4\,H_2O(g) \longrightarrow 4\,H_2(g) + Fe_3O_4(s)$$
(steam)

(c) Metal + Salt ⟶ Metal + Salt
$$Fe(s) + CuSO_4(aq) \longrightarrow Cu(s) + FeSO_4(aq)$$
$$Cu(s) + 2\,AgNO_3(aq) \longrightarrow 2\,Ag(s) + Cu(NO_3)_2(aq)$$

In the reactions in (c), the starting free metal must be more reactive than the metal ion in the salt that it displaces.

(d) Nonmetal + Salt ⟶ Nonmetal + Salt
$$Cl_2(g) + 2\,NaBr(aq) \longrightarrow Br_2(l) + 2\,NaCl(aq)$$
$$Cl_2(g) + 2\,KI(aq) \longrightarrow I_2(s) + 2\,KCl(aq)$$

In the reactions in (d), the starting free nonmetal must be more reactive than the nonmetal ion in the salt that it displaces.

double-replacement or metathesis reaction

4. Double-replacement or metathesis reaction. In this type of reaction, two compounds react with each other to produce two different compounds. Oxidation–reduction does not occur in double-replacement reactions. The general form of the equation is

$$AB + CD \longrightarrow AD + CB$$

This reaction may be thought of as an exchange of positive and negative groups, where the positive group (A) of the first reactant combines with the negative group (D) of the second reactant, and the positive group (C) of the second reactant combines with the negative group (B) of the first reactant. In writing the formulas of the products, we must take into account the oxidation numbers of the combining groups. Some reactions that fall into this category are the following:

(a) Neutralization of an acid and a base

Acid + Base ⟶ Salt + Water

$HCl(aq) + NaOH(aq) \longrightarrow NaCl(aq) + H_2O$

$H_2SO_4(aq) + Ba(OH)_2(aq) \longrightarrow BaSO_4\downarrow + 2\,H_2O$

(b) Formation of an insoluble precipitate

$BaCl_2(aq) + 2\,AgNO_3(aq) \longrightarrow 2\,AgCl\downarrow + Ba(NO_3)_2(aq)$

$FeCl_3(aq) + 3\,NH_4OH(aq) \longrightarrow Fe(OH)_3\downarrow + 3\,NH_4Cl(aq)$

(c) Metal oxide + Acid ⟶ Salt + Water

$CuO(s) + 2\,HNO_3(aq) \longrightarrow Cu(NO_3)_2(aq) + H_2O$

$CaO(s) + 2\,HCl(aq) \longrightarrow CaCl_2(aq) + H_2O$

(d) Formation of a gas

$H_2SO_4(l) + NaCl(s) \longrightarrow NaHSO_4(s) + HCl\uparrow$

$2\,HCl(aq) + ZnS(s) \longrightarrow ZnCl_2(aq) + H_2S\uparrow$

$2\,HCl(aq) + Na_2CO_3(s) \longrightarrow 2\,NaCl(aq) + H_2O(l) + CO_2\uparrow$

All substances that we attempt to react may not react, or the conditions under which they react may not be present. For example, mercuric oxide does not decompose until it is heated; magnesium does not burn in air or oxygen until the temperature is raised to the point at which it begins to react. When silver is placed in a solution of copper(II) sulfate, no reaction takes place; however, when a strip of copper is placed in a solution of silver nitrate, the single replacement reaction as given in 3 (c) above takes place because copper is a more reactive metal than silver. The successful prediction of the products of a reaction is not always easy. The ability to predict products correctly comes with knowledge and experience. Although you may not be able to predict many reactions at this point, as you continue, you will find that reactions can be categorized, as above, and that prediction of the products thereby becomes easier—if not always certain.

Consider the reaction between aqueous solutions of potassium hydroxide and hydrobromic acid. First, write the formula for each reactant, KOH and HBr. Then, after examining these formulas, ascertain that one is a base and the other is an acid. From the type reaction (Acid + Base → Salt + Water), begin to write the equation by putting down the formulas for the known substances.

$$HBr(aq) + KOH(aq) \longrightarrow Salt + H_2O$$

In this double-replacement reaction, the H^+ from the acid combines with the OH^- from the base to form water. The salt must be composed of the other two ions, K^+ and Br^-. We determine the formula of the salt to be KBr from the fact that K is a + ion and Br is a − ion. The final balanced equation is

$$HBr(aq) + KOH(aq) \longrightarrow KBr(aq) + H_2O$$

There is a great deal yet to learn about which substances react with each other, how they react, and what conditions are necessary to bring about their reaction. It is possible to make accurate predictions concerning the occurrence of proposed reactions. Such predictions require, in addition to appropriate data, a good knowledge of thermodynamics. But the study of this subject is usually reserved for advanced courses in chemistry and physics. Even without the formal use of thermodynamics, your knowledge of such generalities as the four reaction types just cited, the periodic table, atomic structure, oxidation numbers, and so on, can be put to good use in predicting reactions and in writing equations. Indeed, such applications serve to make chemistry an interesting and fascinating study.

10.6 Heat in Chemical Reactions

Energy changes always accompany chemical reactions. One reason why reactions may occur is that the products attain a lower, more stable energy state than the reactants. For the products to attain this more stable state, energy must be liberated and given off to the surroundings as heat (or as heat and work). When a solution of a base is neutralized by the addition of an acid, the liberation of heat energy is signaled by an immediate rise in the temperature of the solution. When an automobile engine burns gasoline, heat is certainly liberated, and, at the same time, part of the liberated energy does the work of moving the automobile.

exothermic reaction

endothermic reaction

Reactions are either exothermic or endothermic. **Exothermic reactions** liberate heat; **endothermic reactions** absorb heat. In an exothermic reaction, heat is a product and may be written on the right side of the equation for the reaction. In an endothermic reaction, heat can be regarded as a reactant and is written on the left side of the equation. Examples indicating heat in an exothermic and an endothermic reaction are shown below.

$$H_2(g) + Cl_2(g) \longrightarrow 2\,HCl(g) + 44.2 \text{ kcal} \quad \text{(Exothermic)}$$
$$N_2(g) + O_2(g) + 43.2 \text{ kcal} \longrightarrow 2\,NO(g) \quad \text{(Endothermic)}$$

heat of reaction

The quantity of heat produced by a reaction is known as the **heat of reaction**. The units used are calories or kilocalories. Consider the reaction represented by this equation:

$$C(s) + O_2(g) \longrightarrow CO_2(g) + 94.0 \text{ kcal}$$

When the heat liberated is expressed as part of the equation, the substances are expressed in units of moles. Thus, when 1 mole (12.0 g) of C combines with 1 mole (32.0 g) of O_2, 1 mole (44.0 g) of CO_2 is formed and 94.0 kcal of heat are liberated. Assuming that coal is 90% C and the combustion product is only CO_2, 6.4×10^9 cal would be released when 1 ton of coal is burned. In this reaction, as in many others, the heat or energy is more useful than the chemical products. At 1 cal per gram per degree, 6.4×10^9 cal is sufficient energy to heat about 20,000 gal

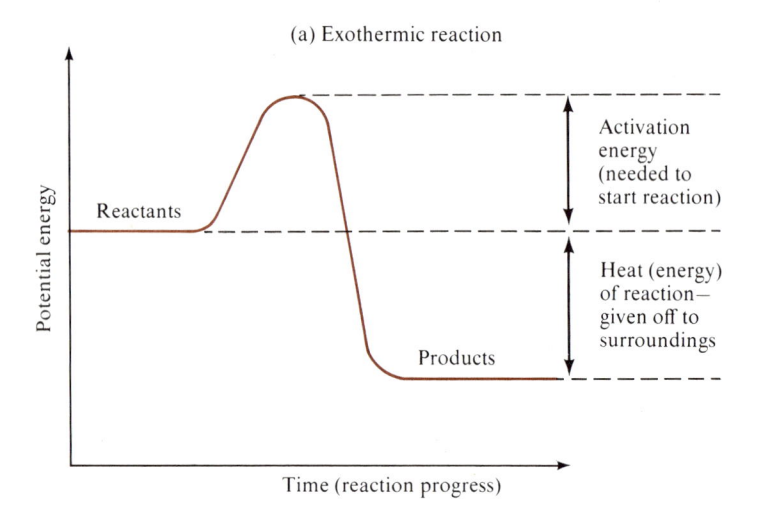

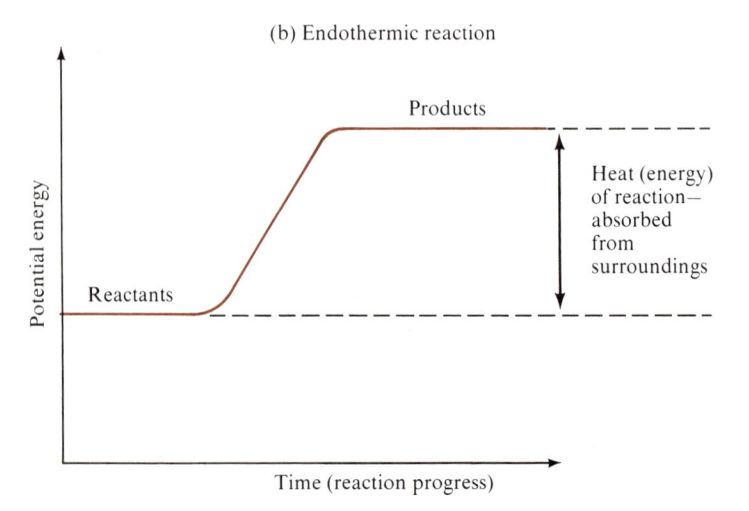

Figure 10.1. Energy states in exothermic and endothermic reactions.

of water from room temperature (20°C) to 100°C.

As another example, aluminum metal is produced by electrolyzing aluminum oxide.

$$2\,Al_2O_3 + 779\,kcal \longrightarrow 4\,Al + 3\,O_2$$

For each mole of Al_2O_3 the equivalent of 389.5 kcal must be supplied as electrical energy. Since only 54 g of aluminum are obtained from 1 mole of Al_2O_3, this means that each ton of aluminum produced requires more than 6.56×10^9 kcal of energy.

Be careful not to confuse an exothermic reaction that merely requires heat (activation energy) to get it started with a truly endothermic process. The combustion of magnesium is highly exothermic, yet magnesium must be heated to a fairly high temperature in air before combustion begins. Once started, however, the combustion reaction goes very vigorously until either the magnesium or the available supply of oxygen is exhausted. The electrolytic decomposition of water to hydrogen and oxygen is highly endothermic. If the electric current is shut off when this process is going on, the reaction stops instantly. The relative energy levels of reactants and products in exothermic and in endothermic processes are presented graphically in Figure 10.1.

In reaction (a) of Figure 10.1, the products are at a lower energy level than the reactants. Energy (heat) was given off to the surroundings and the reaction is exothermic. In reaction (b), the products are at a higher energy level than the reactants. Energy has therefore been absorbed and the reaction is endothermic.

Questions

A. Review the meanings of the new terms introduced in this chapter.
  1. Chemical equation
  2. Word equation
  3. Reactant
  4. Product
  5. Balanced equation
  6. Combination reaction
  7. Decomposition reaction
  8. Single-replacement reaction
  9. Double-replacement or metathesis reaction
  10. Exothermic reaction
  11. Endothermic reaction
  12. Heat of reaction

B. Study Table 10.1 so that you will be familiar with the more common symbols used in equations.

C. Review questions.
  1. Balance the following equations:
     (a) $H_2 + I_2 \rightarrow HI$
     (b) $Mg + N_2 \rightarrow Mg_3N_2$
     (c) $NH_4NO_2 \xrightarrow{\Delta} N_2\uparrow + H_2O$
     (d) $Ca(ClO_3)_2 \xrightarrow{\Delta} CaCl_2 + O_2\uparrow$
     (e) $Ca + HCl \rightarrow CaCl_2 + H_2\uparrow$
     (f) $K + H_2O \rightarrow KOH + H_2\uparrow$
     (g) $HCl + Ba(OH)_2 \rightarrow BaCl_2 + H_2O$
     (h) $Pb(NO_3)_2 + NaCl \rightarrow PbCl_2\downarrow + NaNO_3$
     (i) $Cl_2 + KI \rightarrow KCl + I_2$
     (j) $As_2S_3 + HCl \rightarrow AsCl_3 + H_2S\uparrow$
  2. Balance the following equations:
     (a) $SO_2 + O_2 \rightarrow SO_3$

(b) $Al + Br_2 \rightarrow AlBr_3$
(c) $NH_4NO_3 \xrightarrow{\Delta} N_2O + H_2O$
(d) $CuSO_4 \cdot 5 H_2O \rightarrow CuSO_4 + H_2O$
(e) $Al + H_2SO_4 \rightarrow Al_2(SO_4)_3 + H_2\uparrow$
(f) $Zn + HC_2H_3O_2 \rightarrow Zn(C_2H_3O_2)_2 + H_2\uparrow$
(g) $C_5H_{12} + O_2 \rightarrow CO_2 + H_2O$
(h) $C_6H_{14} + O_2 \rightarrow CO_2 + H_2O$
(i) $Fe_2O_3 + HCl \rightarrow FeCl_3 + H_2O$
(j) $BaCl_2 + (NH_4)_2CO_3 \rightarrow BaCO_3\downarrow + NH_4Cl$

3. Balance the following equations:
   (a) $Bi + O_2 \rightarrow Bi_2O_3$
   (b) $LiAlH_4 \xrightarrow{\Delta} LiH + Al + H_2$
   (c) $Mg + B_2O_3 \xrightarrow{\Delta} MgO + B$
   (d) $Al + MnO_2 \xrightarrow{\Delta} Mn + Al_2O_3$
   (e) $Na_2CO_3 + HCl \rightarrow NaCl + CO_2\uparrow + H_2O$
   (f) $Al_2S_3 + H_2O \rightarrow Al(OH)_3 + H_2S\uparrow$
   (g) $B_2O_3 + C \rightarrow B_4C + CO$
   (h) $C_5H_{11}OH + O_2 \rightarrow CO_2 + H_2O$
   (i) $Ca_3(PO_4)_2 + H_3PO_4 \rightarrow Ca(H_2PO_4)_2$
   (j) $C_3H_5(NO_3)_3 \xrightarrow{\Delta} N_2 + O_2 + CO_2 + H_2O$
       Nitroglycerin

4. Change the following word equations into formula equations and balance them:
   (a) Copper + Sulfur → Copper(I) sulfide
   (b) Acetic acid + Sodium hydroxide → Sodium acetate + Water
   (c) Iron(III) chloride + Potassium hydroxide →
       Iron(III) hydroxide + Potassium chloride
   (d) Aluminum + Copper(II) sulfate → Copper + Aluminum sulfate
   (e) Calcium hydroxide + Phosphoric acid → Calcium phosphate + Water
   (f) Iron(II) sulfide + Hydrobromic acid → Iron(II) bromide + Hydrogen sulfide
   (g) Ammonium sulfate + Barium chloride →
       Ammonium chloride + Barium sulfate
   (h) Sodium hydroxide + Sulfuric acid → Sodium hydrogen sulfate + Water
   (i) Silver nitrate + Aluminum chloride → Silver chloride + Aluminum nitrate
   (j) Sulfur tetrafluoride + Water → Sulfur dioxide + Hydrogen fluoride

5. Complete and balance the following equations: (Combination, a–d; decomposition, e–h; single replacement, i–m; double replacement, n–s.)
   (a) $Li + O_2 \rightarrow$
   (b) $Al + Cl_2 \rightarrow$
   (c) $SO_2 + H_2O \rightarrow$
   (d) $MgO + H_2O \rightarrow$
   (e) $CuSO_4 \xrightarrow{\Delta} SO_3 +$
   (f) $H_2O_2 \xrightarrow{\Delta}$
   (g) $Ca(HCO_3)_2 \xrightarrow{\Delta} CaO +$
   (h) $MnO_2 \xrightarrow{\Delta} Mn_3O_4 +$
   (i) $Al + Fe_2O_3 \rightarrow$
   (j) $Pb + AgNO_3 \rightarrow$
   (k) $Al + SnCl_2 \rightarrow$
   (l) $Br_2 + NaI \rightarrow$
   (m) $Ca + H_2O \rightarrow$
   (n) $Ba(OH)_2 + HNO_3 \rightarrow$
   (o) $(NH_4)_2S + HCl \rightarrow$
   (p) $Na_2O + H_2O \rightarrow$
   (q) $FeBr_2 + NH_4OH \rightarrow$
   (r) $Bi(NO_3)_3 + H_2S \rightarrow$
   (s) $CdO + HCl \rightarrow$

6. What is the purpose of balancing equations?
7. What is represented by the numbers that are placed in front of the formulas in a balanced equation?
8. Interpret the meaning of each of the following equations in terms of a chemical

reaction. Give the relative number of moles of each substance involved and indicate whether the reaction is exothermic or endothermic.
(a) $2\,Na + Cl_2 \rightarrow 2\,NaCl + 196.4$ kcal
(b) $PCl_5 + 22.2$ kcal $\rightarrow PCl_3 + Cl_2$

9. Why is an equation balanced when we have the same number of atoms of each kind of element on both sides of the equation?

10. Interpret the following chemical reactions from the point of view of the number of moles of reactants and products:
(a) $HBr + KOH \rightarrow KBr + H_2O$
(b) $MgBr_2 + 2\,AgNO_3 \rightarrow Mg(NO_3)_2 + 2\,AgBr\downarrow$
(c) $N_2 + 3\,H_2 \rightarrow 2\,NH_3$
(d) $3\,Ag + 4\,HNO_3 \rightarrow 3\,AgNO_3 + NO + 2\,H_2O$
(e) $2\,CH_3CH(OH)CH_3 + 9\,O_2 \rightarrow 6\,CO_2 + 8\,H_2O$
   Isopropyl alcohol

11. In the reaction $H_2(g) + Br_2(g) \rightarrow 2\,HBr(g) + 24.7$ kcal, the net heat liberated is a result of breaking the bonds of H_2 and Br_2 molecules and forming the bonds of 2 HBr molecules. Energy is absorbed in breaking bonds and energy is released in forming bonds. Use the bond dissociation energy data give in Section 7.7 and 7.8 to verify that this reaction is exothermic by 24.7 kcal.

12. Which of the following statements are correct?
(a) The coefficients in front of the formulas in a balanced chemical equation give the relative number of moles of the reactants and products in the reaction.
(b) A balanced chemical equation is one that has the same number of moles on each side of the equation.
(c) In a chemical equation, the symbol $\triangle$ indicates that the reaction is exothermic.
(d) A chemical change that absorbs heat energy is said to be endothermic.
(e) In the reaction $H_2 + Cl_2 \rightarrow 2\,HCl$, 100 molecules of HCl are produced for every 50 molecules of H_2 reacted.
(f) The symbol (aq) after a substance in an equation means that the substance is in a water solution.
(g) The equation $H_2O \rightarrow H_2 + O_2$ can be balanced by placing a 2 in front of H_2O.
(h) In the equation $3\,H_2 + N_2 \rightarrow 2\,NH_3$ there are fewer moles of product than there are moles of reactants.
(i) The total number of moles of reactants and products represented by this equation is 5 moles.

$Mg + 2\,HCl \rightarrow MgCl_2 + H_2$

(j) One mole of glucose, $C_6H_{12}O_6$, contains 6 moles of carbon atoms.

11 Calculations from Chemical Equations

After studying Chapter 11 you should be able to:

1. Understand the new terms listed in Question A at the end of the chapter.
2. Give mole ratios for any two substances involved in a chemical reaction.
3. Outline the mole or mole-ratio method for making stoichiometric calculations.
4. Calculate the number of moles of a desired substance obtainable from a given number of moles of a starting substance in a chemical reaction (mole to mole calculations).
5. Calculate the weight of a desired substance obtainable from a given number of moles of a starting substance in a chemical reaction and vice versa (mole to weight and weight to mole calculations).
6. Calculate the weight of a desired substance involved in a chemical reaction from a given weight of a starting substance (weight to weight calculations).
7. Deduce the limiting reagent or reactant when given the amounts of starting substances and then calculate the moles or weight of desired substance obtainable from a given chemical reaction (limiting reagent calculations).
8. Apply theoretical yield or actual yield to any of the foregoing types of problems, or calculate theoretical and actual yields of a chemical reaction.

This chapter shows the quantitative relationship between reactants and products in chemical reactions and also reviews and correlates such relationships as molecular weight, the molecule, the mole concept, and balancing equations.

11.1 A Short Review

(a) *Molecular weight or formula weight.* The molecular weight is the sum of the atomic weights of all the atoms in a molecule. The formula weight is the sum of the atomic weights of all the atoms in a given formula of a compound or an ion. The terms *molecular weight* and *formula weight* are commonly used interchangeably.

(b) *Relationship between molecule and mole.* A molecule is the smallest unit of a molecular substance (e.g., Cl_2) and a mole is Avogadro's Number,

6.02×10^{23}, of molecules of that substance. A mole of chlorine (Cl_2) has the same number of molecules as a mole of carbon dioxide, a mole of water, or a mole of any other molecular substance. When we relate molecules to gram-molecular weight, 1 g-mol. wt = 1 mole, or 6.02×10^{23} molecules.

In addition to molecular substances, the term *mole* may refer to any chemical species. It represents a quantity in grams equal to the formula weight (1 g-form. wt) and may be applied to atoms, ions, electrons, and formula units of nonmolecular substances. For example, a mole of water consists of 18.0 grams of water, or 6.02×10^{23} molecules; a mole of sodium is 23.0 grams of sodium, or 6.02×10^{23} atoms; a mole of chloride ions is 35.5 grams of chloride ions, or 6.02×10^{23} of these ions; a mole of electrons is 5.48×10^{-4} grams of electrons, or 6.02×10^{23} electrons, and a mole of sodium chloride (a nonmolecular substance) is 58.5 grams of sodium chloride, or 6.02×10^{23} formula units.

$$1 \text{ mole} = \begin{cases} 1 \text{ g-mol. wt} = 6.02 \times 10^{23} \text{ molecules} \\ 1 \text{ g-form. wt} = 6.02 \times 10^{23} \text{ formula units} \\ 1 \text{ g-atomic wt} = 6.02 \times 10^{23} \text{ atoms} \\ 1 \text{ g-ionic wt} = 6.02 \times 10^{23} \text{ ions} \end{cases}$$

Other useful mole relationships are

$$\text{Number of moles} = \frac{\text{Grams of a substance}}{\text{Gram-molecular weight of the substance}}$$

$$\text{Number of moles} = \frac{\text{Grams of a monatomic element}}{\text{Gram-atomic weight of the element}}$$

$$\text{Number of moles} = \frac{\text{Number of molecules}}{6.02 \times 10^{23} \text{ molecules/mole}}$$

Two other useful equalities may be derived algebraically from each of these mole relationships. What are they?

(c) **Balanced equations.** In using equations for calculations of mole–weight–volume relationships between reactants and products, the equations must be balanced. Remember that the number in front of a formula in a balanced equation represents the number of moles of that substance reacting in the chemical change.

11.2 Calculations from Chemical Equations: The Mole Method

In chemical work it is often necessary to calculate the amount of a substance that is produced from, or needed to react with, a given quantity of another substance. The area of chemistry that deals with the quantitative relationships among reactants and products is known as **stoichiometry**. Although

stoichiometry

several methods are known, we believe that the *mole* or *mole-ratio* method is best for solving problems in stoichiometry. This method is straightforward and, in our opinion, makes it easy to see and understand the relationships of the reacting species.

A **mole ratio** is a ratio between the number of moles of any two species involved in a chemical reaction. For example in the reaction

$$2H_2 + O_2 \longrightarrow 2H_2O$$
2 moles   1 mole   2 moles

there are six mole ratios that apply only to this reaction:

$$\frac{2 \text{ moles } H_2}{1 \text{ mole } O_2}; \quad \frac{2 \text{ moles } H_2}{2 \text{ moles } H_2O}; \quad \frac{1 \text{ mole } O_2}{2 \text{ moles } H_2};$$

$$\frac{1 \text{ mole } O_2}{2 \text{ moles } H_2O}; \quad \frac{2 \text{ moles } H_2O}{2 \text{ moles } H_2}; \quad \frac{2 \text{ moles } H_2O}{1 \text{ mole } O_2}$$

Since stoichiometric problems are encountered throughout the entire field of chemistry, it is profitable to master this general method for their solution. The mole method makes use of three simple basic operations.

A. Conversion of the starting substance to moles (if it is not given in moles).
B. Calculation of the moles of desired substance obtainable from the available moles of starting substance. This is done by multiplying the moles of starting substance (from A) by the mole ratio. This mole ratio is taken from the balanced equation and is the number of moles of the desired substance over the number of moles of the starting substance.
C. Conversion of the moles of desired substance (from B) to whatever units are required.

Like learning to balance chemical equations, learning to make stoichiometric calculations requires some practice. A detailed step-by-step description of the general method, together with a variety of worked examples, is given in the following paragraphs. Study this material and apply the method to the problems at the end of this chapter.

Step 1. Use a balanced equation. Write a balanced equation for the chemical reaction in question or check to see that the equation given is balanced.

Step 2. Determine the number of moles of starting substance. Identify the starting substance from the data given in the statement of the problem. When the starting substance is given in moles, use it in that form; if it is not in moles, convert the quantity of the starting substance to moles.

Step 3. Determine the mole ratio of the desired substance to the starting substance. The number of moles of each substance in the balanced equation is indicated by the coefficient in front of each substance. Use these coefficients to set up the mole ratio:

$$\text{Mole ratio} = \frac{\text{Moles of desired substance in the equation}}{\text{Moles of starting substance in the equation}}$$

Calculations from Chemical Equations

Step 4. Calculate the number of moles of the desired substance. Multiply the number of moles of starting substance (from Step 2) by the mole ratio (from Step 3) to obtain the number of moles of desired substance:

$$\left(\begin{array}{c}\text{Moles of desired}\\ \text{substance}\end{array}\right) = \underbrace{\left(\begin{array}{c}\text{Moles of starting}\\ \text{substance}\end{array}\right)}_{\text{From Step 2}} \times \underbrace{\frac{\left(\begin{array}{c}\text{Moles of desired}\\ \text{substance in}\\ \text{the equation}\end{array}\right)}{\left(\begin{array}{c}\text{Moles of starting}\\ \text{substance in}\\ \text{the equation}\end{array}\right)}}_{\text{Mole ratio from Step 3}}$$

Note that the units of moles of starting substance cancel out in the numerator and the denominator.

Step 5. Calculate the desired substance in the units asked for in the problem. If the answer is requested in moles, the problem is finished in Step 4. If units other than moles are wanted, multiply the moles of the desired substance from Step 4 by the appropriate factor to convert moles to the units required.

For example, if grams of the desired substance are wanted,

$$\left(\begin{array}{c}\text{Grams of desired}\\ \text{substance}\end{array}\right) = \underbrace{\left(\begin{array}{c}\text{Moles of desired}\\ \text{substance}\end{array}\right)}_{\text{From Step 4}} \times \left(\begin{array}{c}\text{Gram-molecular weight}\\ \text{of desired substance}\end{array}\right)$$

These steps are summarized in Figure 11.1.

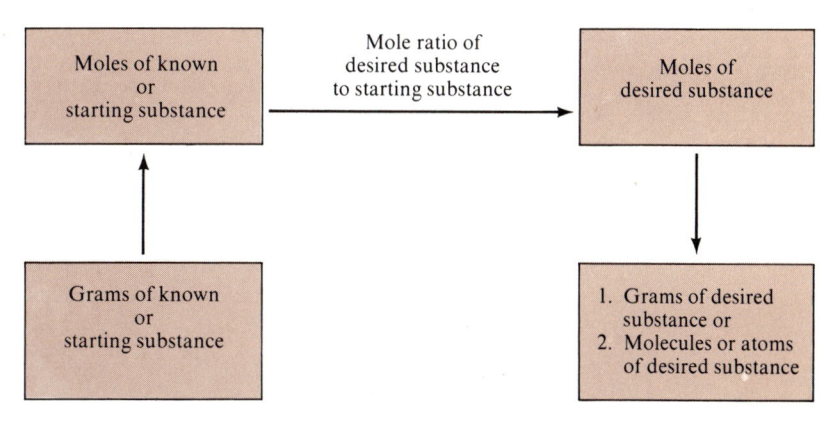

Figure 11.1. Basic steps in using the mole ratio to convert moles of one substance to moles of another substance in a chemical reaction.

11.3 Mole–Mole Calculations

The object of this type of problem is to calculate the moles of one substance that react with, or are produced from, a given number of moles of another substance. Illustrative problems follow.

182   Chapter 11

Problem 11.1

How many moles of water will be produced from 3.5 moles of magnesium hydroxide, $Mg(OH)_2$, according to the following reaction?

$$Mg(OH)_2 + H_2CO_3 \longrightarrow MgCO_3 + 2H_2O$$
$$\text{1 mole} \quad \text{1 mole} \quad \text{1 mole} \quad \text{2 moles}$$

When the equation is balanced, it states that 2 moles of water will be produced from 1 mole of $Mg(OH)_2$. Even though we can quickly ascertain that 7.0 moles of water will be formed from 3.5 moles of $Mg(OH)_2$, the mole method for solving the problem is shown below. This method of working with mole ratios will be very helpful in solving later problems.

Step 1. The equation given is balanced.
Step 2. The moles of starting substance are 3.5 moles of $Mg(OH)_2$.
Step 3. From the balanced equation, set up the mole ratio between the two substances in question, placing the moles of the substance being sought in the numerator and the moles of the starting substance in the denominator. The number of moles, in each case, is the same as the coefficient in front of the substance in the balanced equation.

$$\text{Mole ratio} = \frac{2 \text{ moles } H_2O}{1 \text{ mole } Mg(OH)_2} \quad \text{(From equation)}$$

Step 4. Multiply the 3.5 moles of $Mg(OH)_2$ given in the problem by this mole ratio.

$$3.5 \text{ moles } Mg(OH)_2 \times \frac{2 \text{ moles } H_2O}{1 \text{ mole } Mg(OH)_2} = 7.0 \text{ moles } H_2O \quad \text{(Answer)}$$

Again note the use of units. The moles of $Mg(OH)_2$ cancel, leaving the answer in units of moles of H_2O.

Problem 11.2

How many moles of ammonia can be produced from 8.00 moles of hydrogen reacting with nitrogen?

Step 1. First, we need the balanced equation

$$3H_2 + N_2 \longrightarrow 2NH_3$$

Step 2. The moles of starting substance are 8.00 moles of hydrogen.
Step 3. The balanced equation states that we get 2 moles of NH_3 for every 3 moles of H_2 that react. Set up the mole ratio of desired substance (NH_3) to starting substance (H_2):

$$\text{Mole ratio} = \frac{2 \text{ moles } NH_3}{3 \text{ moles } H_2}$$

Step 4. Multiplying the 8.00 moles of starting H_2 by this mole ratio, we get

$$8.00 \text{ moles } H_2 \times \frac{2 \text{ moles } NH_3}{3 \text{ moles } H_2} = 5.33 \text{ moles } NH_3 \quad \text{(Answer)}$$

Problem 11.3

Given the balanced equation

$$K_2Cr_2O_7 + 6KI + 7H_2SO_4 \longrightarrow Cr_2(SO_4)_3 + 4K_2SO_4 + 3I_2 + 7H_2O$$
$$\text{1 mole} \quad \text{6 moles} \quad \quad\quad\quad\quad\quad\quad\quad\quad\quad\quad\quad\quad\quad \text{3 moles}$$

Calculate (a) the number of moles of potassium dichromate ($K_2Cr_2O_7$) that will react with 2.0 moles of potassium iodide (KI); (b) the number of moles of iodine (I_2) that will be produced from 2.0 moles of potassium iodide.

After the equation is balanced, we are concerned only with $K_2Cr_2O_7$, KI, and I_2, and we can ignore all the other substances. The equation states that 1 mole of $K_2Cr_2O_7$ will react with 6 moles of KI to produce 3 moles of I_2.

(a) Calculate the moles of $K_2Cr_2O_7$.

Step 1. The equation given is balanced.
Step 2. The moles of starting substance are 2.0 moles of KI.
Step 3. Set up the mole ratio of desired substance to starting substance:

$$\text{Mole ratio} = \frac{1 \text{ mole } K_2Cr_2O_7}{6 \text{ moles KI}} \quad \text{(From equation)}$$

Step 4. Multiply the moles of starting material by this ratio to obtain the answer.

$$2.0 \text{ moles KI} \times \frac{1 \text{ mole } K_2Cr_2O_7}{6 \text{ moles KI}} = 0.33 \text{ mole } K_2Cr_2O_7 \quad \text{(Answer)}$$

(b) Calculate the moles of I_2.

Steps 1 and 2. The equation given is balanced and the moles of starting substance are 2.0 moles KI as in part (a).
Step 3. Set up the mole ratio of desired substance to starting substance:

$$\text{Mole ratio} = \frac{3 \text{ moles } I_2}{6 \text{ moles KI}} \quad \text{(From equation)}$$

Step 4. Multiply the moles of starting material by this ratio to obtain the answer.

$$2.0 \text{ moles KI} \times \frac{3 \text{ moles } I_2}{6 \text{ moles KI}} = 1.0 \text{ mole } I_2 \quad \text{(Answer)}$$

Thus, 2.0 moles KI will react with 0.33 moles $K_2Cr_2O_7$ to produce 1.0 mole I_2.

Problem 11.4

How many molecules of water can be produced by reacting 0.010 mole of oxygen with hydrogen?

The sequence of conversions needed in the calculation is

$$\text{Moles } O_2 \longrightarrow \text{Moles } H_2O \longrightarrow \text{Molecules } H_2O$$

Step 1. First, we write the balanced equation:

$$\underset{1 \text{ mole}}{2 H_2} + O_2 \longrightarrow \underset{2 \text{ moles}}{2 H_2O}$$

Step 2. The moles of starting substance is 0.010 mole O_2.
Step 3. Set up the mole ratio of desired substance to starting substance:

$$\text{Mole ratio} = \frac{2 \text{ moles } H_2O}{1 \text{ mole } O_2} \quad \text{(From equation)}$$

Step 4. Multiplying the 0.010 mole of oxygen by this ratio, we obtain

$$0.010 \text{ mole } O_2 \times \frac{2 \text{ moles } H_2O}{1 \text{ mole } O_2} = 0.020 \text{ mole } H_2O$$

184   Chapter 11

Step 5. Since the problem asks for molecules instead of moles of H$_2$O, we must convert moles to molecules. Use the conversion factor (6.02 × 10^{23} molecules)/mole.

$$0.020 \text{ mole H}_2\text{O} \times \frac{6.02 \times 10^{23} \text{ molecules}}{\text{mole}} = 1.2 \times 10^{22} \text{ molecules H}_2\text{O}$$

We should note that 0.020 mole is still quite a large number of water molecules

11.4   Mole–Weight and Weight–Weight Calculations

The object of these types of problems is to calculate the weight of one substance that reacts with, or is produced from, a given number of moles or a given weight of another substance in a chemical reaction. The mole ratio is used to convert from moles of starting substance to moles of desired substance.

Problem 11.5   What weight of hydrogen can be produced by reacting 6.0 moles of aluminum with hydrochloric acid?
First calculate the moles of hydrogen produced, using the mole-ratio method, and then calculate the weight of hydrogen by multiplying the moles of hydrogen by its weight per mole. The sequence of conversions in the calculation is

Moles Al ⟶ Moles H$_2$ ⟶ Grams H$_2$

Step 1. The balanced equation is

$$2 \text{ Al}(s) + 6 \text{ HCl}(aq) \longrightarrow 2 \text{ AlCl}_3(aq) + 3 \text{ H}_2(g)$$

2 moles   3 moles

Step 2. The moles of starting substance are 6.0 moles of aluminum.
Steps 3 and 4. Calculate moles of H$_2$.

$$6.0 \text{ moles Al} \times \frac{3 \text{ moles H}_2}{2 \text{ moles Al}} = 9.0 \text{ moles H}_2$$

Step 5. Convert moles of H$_2$ to grams [g = moles × (g/mole)]:

$$9.0 \text{ moles H}_2 \times \frac{2.0 \text{ g H}_2}{1 \text{ mole H}_2} = 18 \text{ g H}_2$$

We see that 18 g of H$_2$ can be produced by reacting 6.0 moles of Al with HCl. The following setup combines all the above steps into one continuous calculation:

$$6.0 \text{ moles Al} \times \frac{3 \text{ moles H}_2}{2 \text{ moles Al}} \times \frac{2.0 \text{ g H}_2}{1 \text{ mole H}_2} = 18 \text{ g H}_2$$

Problem 11.6   What weight of carbon dioxide is produced by the complete combustion of 100 g of propane gas, C$_3$H$_8$?
The sequence of conversions in the calculation is

Grams C$_3$H$_8$ ⟶ Moles C$_3$H$_8$ ⟶ Moles CO$_2$ ⟶ Grams CO$_2$

Step 1. The balanced equation is

$$\text{C}_3\text{H}_8 + 5 \text{ O}_2 \rightarrow 3 \text{ CO}_2 + 4 \text{ H}_2\text{O}$$

1 mole   3 moles

Step 2. The starting substance is 100 g of C_3H_8. Convert 100 g of C_3H_8 to moles [moles = g × (moles/g)]:

$$100 \text{ g } C_3H_8 \times \frac{1 \text{ mole } C_3H_8}{44.0 \text{ g } C_3H_8} = 2.27 \text{ moles } C_3H_8$$

Steps 3 and 4. Calculate the moles of CO_2 by the mole-ratio method.

$$2.27 \text{ moles } C_3H_8 \times \frac{3 \text{ moles } CO_2}{1 \text{ mole } C_3H_8} = 6.81 \text{ moles } CO_2$$

Step 5. Convert the moles of CO_2 to grams.

Moles CO_2 × Gram-molecular weight CO_2 = Grams CO_2

$$6.81 \text{ moles } CO_2 \times \frac{44.0 \text{ g } CO_2}{1 \text{ mole } CO_2} = 300 \text{ g } CO_2 \quad \text{(Answer)}$$

We see that 300 g of CO_2 are produced from the complete combustion of 100 g of C_3H_8. The calculation in a continuous setup is

$$100 \text{ g } C_3H_8 \times \frac{1 \text{ mole } C_3H_8}{44.0 \text{ g } C_3H_8} \times \frac{3 \text{ moles } CO_2}{1 \text{ mole } C_3H_8} \times \frac{44.0 \text{ g } CO_2}{1 \text{ mole } CO_2} = 300 \text{ g } CO_2$$

Grams C_3H_8 → Moles C_3H_8 → Moles CO_2 → Grams CO_2

Note that in the continuous setup, since both C_3H_8 and CO_2 have the same molecular weight, the numbers cancel, saving two steps in the calculation.

11.5 Limiting Reagent and Yield Calculations

In many chemical processes the quantities of the reactants used are such that the moles of one reactant are in excess of the moles of a second reactant in the reaction. The amount of the product(s) formed in such a case will be dependent on the reactant that is not in excess. Thus, the reactant that is not in excess is known as the **limiting reagent**, since it limits the amount of product that can be formed.

limiting reagent

As an example, consider the case where solutions containing 1.0 mole of sodium hydroxide and 1.5 moles of hydrochloric acid are mixed:

$$NaOH + HCl \longrightarrow NaCl + H_2O$$

1 mole   1 mole   1 mole   1 mole

According to the equation it is possible to obtain 1.0 mole of NaCl from 1.0 mole of NaOH, and 1.5 moles of NaCl from 1.5 moles of HCl. However, we cannot have two different yields of NaCl from the reaction. When 1.0 mole of NaOH and 1.5 moles of HCl are mixed, there is insufficient NaOH to react with all of the HCl. Therefore, HCl is the reagent in excess and NaOH is the limiting reagent. Since the NaCl formed is dependent on the limiting reagent, the amount of NaCl formed will be 1.0 mole. Since 1.0 mole of NaOH reacts with 1.0 mole of HCl, 0.5 mole of HCl remains unreacted.

Problems giving the amounts of two reactants are generally of the limiting reagent type and may be solved in the following manner:

1. Calculate the number of moles of each substance used.
2. Determine which substance is the limiting reagent.
   (a) Compare the ratio of the moles calculated with the mole ratio of the two substances in the balanced equation. Or
   (b) Determine the number of moles of the product that can be formed from each substance to see which produces the least amount.
3. Using the moles of the limiting reagent, calculate the amount of the product in the units asked for in the problem.

Problem 11.7

How many grams of silver bromide, AgBr, can be formed when solutions containing 50.0 g of $MgBr_2$ and 100 g of $AgNO_3$ are mixed together?

$$MgBr_2(aq) + 2\ AgNO_3(aq) \longrightarrow 2\ AgBr\downarrow + Mg(NO_3)_2(aq)$$

Step 1. Moles of $MgBr_2$ and $AgNO_3$:

$$50.0\ \text{g}\ MgBr_2 \times \frac{1\ \text{mole}\ MgBr_2}{184.1\ \text{g}\ MgBr_2} = 0.272\ \text{mole}\ MgBr_2$$

$$100\ \text{g}\ AgNO_3 \times \frac{1\ \text{mole}\ AgNO_3}{169.9\ \text{g}\ AgNO_3} = 0.589\ \text{mole}\ AgNO_3$$

Step 2. Determine which is the limiting reagent, $MgBr_2$ or $AgNO_3$.

$$\text{Ratio of moles calculated} = \frac{0.589\ \text{mole}\ AgNO_3}{0.272\ \text{mole}\ MgBr_2} = \frac{2.16\ \text{mole}\ AgNO_3}{1\ \text{mole}\ MgBr_2}$$

$$\text{Mole ratio in the equation} = \frac{2\ \text{moles}\ AgNO_3}{1\ \text{mole}\ MgBr_2}$$

Therefore, $AgNO_3$ is in excess, and $MgBr_2$ is the limiting reagent.

Step 3. Calculate the grams of AgBr using the moles of the limiting reagent.

Moles $MgBr_2 \longrightarrow$ Moles AgBr $\longrightarrow$ Grams AgBr

$$0.272\ \text{mole}\ MgBr_2 \times \frac{2\ \text{moles}\ AgBr}{1\ \text{mole}\ MgBr_2} \times \frac{187 \cdot 8\ \text{g}\ AgBr}{1\ \text{mole}\ AgBr} = 102\ \text{g}\ AgBr$$

Many reactions, especially those involving organic substances, fail to give a 100% yield of product. The main reasons for this failure are the side reactions that give products other than the main product and the fact that many reactions are reversible. In addition, some product may be lost in handling and transferring from one vessel to another. The yield calculated from the chemical equation is commonly known as the **theoretical yield**; this is the maximum amount of product that can be produced according to the equation. The **actual yield** is the amount of product that we finally obtain.

theoretical yield

actual yield

The percentage yield is calculated as follows:

$$\frac{\text{Actual yield}}{\text{Theoretical yield}} \times 100\% = \text{Percentage yield}$$

Problem 11.8

Carbon tetrachloride was prepared by reacting 100 g of carbon disulfide and 100 g of chlorine. Calculate the percentage yield if 65.0 g of CCl_4 was obtained from the reaction.

$$CS_2 + 3\,Cl_2 \longrightarrow CCl_4 + S_2Cl_2$$

In this problem we need to determine the limiting reagent in order to calculate the quantity of CCl_4 (theoretical yield) that can be formed according to the equation for the reaction. Then we can compare this amount with the 65.0 g CCl_4 actual yield to calculate the percentage yield.

Step 1. Moles of CS_2 and Cl_2 and the limiting reagent:

$$100 \text{ g } CS_2 \times \frac{1 \text{ mole } CS_2}{76.2 \text{ g } CS_2} = 1.31 \text{ moles } CS_2$$

$$100 \text{ g } Cl_2 \times \frac{1 \text{ mole } Cl_2}{71.0 \text{ g } Cl_2} = 1.41 \text{ moles } Cl_2$$

From the equation we see that 3 moles of Cl_2 are needed to react with 1 mole of CS_2. But we have only

$$\frac{1.41 \text{ moles } Cl_2}{1.31 \text{ moles } CS_2} = 1.08 \text{ moles } Cl_2 \text{ per mole } CS_2.$$

Therefore Cl_2 is the limiting reagent and is used to calculate the theoretical yield of CCl_4.

Step 2. Grams of CCl_4:

Moles $Cl_2 \longrightarrow$ Moles $CCl_4 \longrightarrow$ Grams CCl_4

$$1.41 \text{ moles } Cl_2 \times \frac{1 \text{ mole } CCl_4}{3 \text{ moles } Cl_2} \times \frac{154 \text{ g } CCl_4}{1 \text{ mole } CCl_4} = 72.4 \text{ g } CCl_4$$

Step 3. Percentage yield. According to the equation, 72.4 g CCl_4 is the maximum amount or theoretical yield of CCl_4 possible from 100 g Cl_2. Actual yield is 65.0 g CCl_4.

$$\text{Percentage yield} = \frac{65.0 \text{ g}}{72.4 \text{ g}} \times 100\% = 89.8\%$$

When solving problems, you will achieve better results if at first you do not try to take shortcuts. Write the data and numbers in a logical, orderly manner. Make certain that the equations are balanced and that the computations are accurate and expressed to the correct number of significant figures. Remember that units are very important; a number without units has little meaning. Finally, an electronic calculator can save you many hours of tedious computations.

Questions

A. Review the meaning of the new terms introduced in this chapter.
   1. Stoichiometry
   2. Mole ratio
   3. Limiting reagent
   4. Theoretical yield
   5. Actual yield

B. Review problems.
(In some of the following problems the equations shown are not balanced.)
1. Calculate the number of moles in each of the following quantities:
   (a) 25.0 g MnO_2
   (b) 350 g H_2SO_4
   (c) 45.0 g Br_2
   (d) 1.00 g CCl_4
   (e) 1.00 kg NaCl
   (f) 11.5 g C_2H_6O
   (g) 410.0 g CO_2
   (h) 60.0 g O_2
   (i) 100 millimoles HNO_3
   (j) 250 ml concentrated HCl ($d = 1.19$ g/ml, 37% HCl by weight)
2. Calculate the number of grams in each of the following quantities:
   (a) 0.400 mole C_3H_8
   (b) 1.50 moles Al
   (c) 7.50 moles H_2
   (d) 0.100 mole $AgNO_3$
   (e) 1.00 mole $FeSO_4$
   (f) 0.250 mole $AlCl_3$
   (g) 0.600 mole Au
   (h) 50 ml $Ni(NO_3)_2$ solution ($d = 1.107$ g/ml, 12.0% $Ni(NO_3)_2$ by weight)
3. Set up all the possible mole ratios between the reactants and the products in the following equations:
   (a) $2 Mg + O_2 \rightarrow 2 MgO$
   (b) $2 Al + 6 HCl \rightarrow 2 AlCl_3 + 3 H_2$
   (c) $3 Zn + N_2 \rightarrow Zn_3N_2$
   (d) $2 C_2H_6 + 7 O_2 \rightarrow 4 CO_2 + 6 H_2O$
4. Which contains the larger number of molecules?
   (a) 8.0 g of CH_4 or 30.0 g of SO_2 (b) 4.0 g of CO_2 or 4.0 g of CO
5. An early method of producing chlorine was by the reaction of pyrolusite, MnO_2, and hydrochloric acid. How many moles of HCl will react with 0.85 mole of MnO_2? (Balance the equation first.)

$$MnO_2(s) + HCl(aq) \longrightarrow Cl_2(g) + MnCl_2(aq) + H_2O$$

6. How many moles of carbon dioxide and how many moles of sulfur dioxide can be formed by the reaction of 4.0 moles of carbon disulfide with oxygen?

$$CS_2 + O_2 \longrightarrow CO_2 + SO_2$$

7. How many moles of oxygen are needed to react with 3.5 moles of isopropyl alcohol, C_3H_7OH?

$$2 C_3H_7OH + 9 O_2 \longrightarrow 6 CO_2 + 8 H_2O$$

8. How many moles of each reactant are required to produce 0.80 mole of sodium sulfate, Na_2SO_4, according to the following equation?

$$3 Na_2S_2O_3 + 8 KMnO_4 + H_2O \rightarrow 3 Na_2SO_4 + 3 K_2SO_4 + 8 MnO_2 + 2 KOH$$

9. What weight of nitric acid is needed to convert 0.125 mole of copper to copper(II) nitrate, $Cu(NO_3)_2$?

$$3 Cu(s) + 8 HNO_3(aq) \longrightarrow 3 Cu(NO_3)_2(aq) + 2 NO(g) + 4 H_2O$$

10. How many moles of oxygen are needed to completely burn 500 g of pentane, C_5H_{12}?

$$C_5H_{12}(l) + O_2(g) \longrightarrow CO_2(g) + H_2O(g)$$

11. Given the equation

$$6\ FeCl_2 + 14\ HCl + K_2Cr_2O_7 \longrightarrow 6\ FeCl_3 + 2\ KCl + 2\ CrCl_3 + 7\ H_2O$$

   (a) How many moles of $FeCl_3$ will be formed from 1.0 mole of $FeCl_2$?
   (b) How many moles of HCl will react with 2.5 moles of $FeCl_2$?
   (c) How many grams of $K_2Cr_2O_7$ will react with 0.50 mole of $FeCl_2$?
   (d) How many grams of $CrCl_3$ will be formed from 1.00 g of $K_2Cr_2O_7$?
   (e) How many moles of H_2O are formed when 2.0 moles of $FeCl_3$ are formed?
   (f) How many grams of $FeCl_3$ are formed from 6.00 g of $FeCl_2$?

12. Oyster shells are essentially pure limestone, $CaCO_3$. When heated, they form quicklime, CaO, and carbon dioxide. What weight of quicklime can be produced from 1.00 kg of oyster shells?

$$CaCO_3 \xrightarrow{\Delta} CaO + CO_2$$

13. The essential reactions in the production of steel (Fe) from iron ore in a blast furnace are

$$\underset{\text{Coke}}{2\ C} + O_2 \xrightarrow{\Delta} 2\ CO$$

$$3\ CO + Fe_2O_3 \longrightarrow 2\ Fe + 3\ CO_2$$

How many kilograms of coke are needed to produce a metric ton (1000 kg) of steel?

14. What weight of steam and iron must react in order to produce 250 g of magnetic iron oxide, Fe_3O_4?

$$Fe(s) + H_2O(g) \longrightarrow Fe_3O_4(s) + H_2\uparrow$$

15. Oxygen gas is obtained when potassium nitrate, KNO_3, is heated.

$$KNO_3(s) \xrightarrow{\Delta} KNO_2(s) + O_2(g)$$

What will be the weight loss when 22.4 g of KNO_3 are heated?

16. Calculate the weight of the *first product* listed in each equation that could be obtained by starting with 15.0 g of the *first reactant* listed in each equation.
   (a) $2\ Na_3PO_4 + 3\ H_2SO_4 \rightarrow 2\ H_3PO_4 + 3\ Na_2SO_4$
   (b) $4\ FeS_2 + 11\ O_2 \rightarrow 2\ Fe_2O_3 + 8\ SO_2$
   (c) $11\ Si + 4\ NpF_3 \rightarrow 3\ SiF_4 + 4\ NpSi_2$
   (d) $6\ B_2Cl_4 + 3\ O_2 \rightarrow 2\ B_2O_3 + 8\ BCl_3$

17. Both $CaCl_2$ and $MgCl_2$ react with $AgNO_3$ to precipitate AgCl. When solutions containing equal weights of $CaCl_2$ and $MgCl_2$ are reacted, which salt will produce the most AgCl? Show proof.

18. In the following equations, determine which reactant is the limiting reagent and which reactant is in excess. The amounts mixed together are shown below each reactant. Show evidence for your answers.
   (a) $KOH + HCl \rightarrow KCl + H_2O$
         3.00 g   2.20 g

   (b) $2\ Bi(NO_3)_3 + 3\ H_2S \rightarrow Bi_2S_3 + 6\ HNO_3$
         25.0 g   8.0 g

(c) $3\,Fe + 4\,H_2O \rightarrow Fe_3O_4 + 4\,H_2$
   55.8 g   18.0 g

(d) $2\,C_2H_6 + 7\,O_2 \rightarrow 4\,CO_2 + 6\,H_2O$
    100 g   400 g

19. Methyl alcohol, CH_3OH, is made by reacting carbon monoxide and hydrogen in the presence of certain metal oxide catalysts. How much alcohol can be obtained by reacting 500 g CO and 100 g H_2?

    $CO(g) + 2\,H_2(g) \longrightarrow CH_3OH(l)$

20. Iron was reacted with a solution containing 100 g of copper(II) sulfate. The reaction was stopped after 1 hour, and 37.4 g of copper were obtained. Calculate the percentage yield of copper obtained.

    $Fe(s) + CuSO_4(aq) \longrightarrow Cu(s) + FeSO_4(aq)$

21. Calcium carbide, CaC_2, is used for generating acetylene. It is made by reacting lime, CaO, and coke, C, in an electric furnace at 3000°C.

    $CaO + 3\,C \longrightarrow CaC_2 + CO$

    How many grams of CaC_2 can be prepared by reacting 1000 g of CaO and 500 g of C?

22. An astronaut excretes about 2500 g of water a day. If lithium oxide, Li_2O, is used in the spaceship to absorb this water, how many kilograms of Li_2O must be carried for a 30-day space trip for three astronauts?

    $Li_2O + H_2O \longrightarrow 2\,LiOH$

23. The equation representing the reaction used for the commercial preparation of hydrogen cyanide is

    $2\,CH_4 + 3\,O_2 + 2\,NH_3 \longrightarrow 2\,HCN + 6\,H_2O$

    Which of the following statements are correct?
    (a) Three moles of O_2 are required for 2 moles of NH_3.
    (b) Twelve moles of HCN are produced for every 16 moles of O_2 that react.
    (c) The mole ratio between H_2O and CH_4 is

    $$\frac{6 \text{ moles } H_2O}{2 \text{ moles } CH_4}$$

    (d) When 12 moles of HCN are produced, 4 moles of H_2O will also be formed.
    (e) When 10 moles CH_4, 10 moles O_2, and 10 moles NH_3 are mixed and reacted, O_2 is the limiting reagent.
    (f) When 3 moles each of CH_4, O_2, and NH_3 are mixed and reacted, 3 moles of HCN will be produced.

12 The Gaseous State of Matter

After studying Chapter 12 you should be able to:

1. Understand the terms listed in Question A at the end of the chapter.
2. State the principal assumptions of the Kinetic-Molecular Theory (KMT).
3. Estimate the relative rates of diffusion of two gases of known molecular weights.
4. Sketch and explain the operation of a mercury barometer.
5. Tell what two factors determine gas pressure in a vessel of fixed volume.
6. Work problems involving (a) Boyle's and (b) Charles' gas laws.
7. State what is meant by standard temperature and pressure (STP).
8. Give the equation for the combined gas law that deals with the pressure, volume, and temperature relationships expressed in Boyle's, Charles', and Gay-Lussac's gas laws.
9. Use Dalton's Law of Partial Pressures and the combined gas laws to calculate the dry STP volume of a gas collected over water.
10. State Avogadro's hypothesis.
11. Determine the density of any gas at STP.
12. Determine the molecular weight of a gas from its density at a known temperature and pressure.
13. Make mole-to-volume, weight-to-volume, and volume-to-volume stoichiometric calculations from a balanced chemical equation.
14. Define an ideal gas.
15. State two valid reasons why real gases may deviate from the behavior predicted for an ideal gas.
16. Solve problems involving the ideal gas equation.

12.1 General Properties of Gases

In Chapter 3, solids, liquids, and gases are described in a brief outline. In this chapter we will consider the behavior of gases in greater detail.

Of the three states of matter, gases are the least compact and most mobile. A solid has a rigid structure and its particles remain in essentially fixed positions. When a solid absorbs sufficient heat, it melts and changes into a liquid. Melting occurs because the molecules (or ions) have absorbed enough energy to break out of the rigid crystal lattice structure of the solid. The molecules or ions in the liquid are more energetic than they were in the solid, as shown by their increased mobility. Molecules in the liquid state are coherent—that is, they cling to one another. When the liquid absorbs additional heat, the more energetic molecules break away from the liquid surface and go into the gaseous state. Gases represent the most mobile state of matter. Gas molecules move with very high velocities and have high kinetic energy (KE). The average velocity of hydrogen molecules at 0°C is over 1600 metres (1 mile) per second. Because of the high velocities of their molecules, mixtures of gases are uniformly distributed within the container in which they are confined.

A quantity of a substance occupies a much greater volume as a gas than does a like quantity of the substance as a liquid or a solid. For example, 1 mole of water (18 g) has a volume of 18 ml at 4°C. This same amount of water would occupy about 22,400 ml in the gaseous state—more than a 1200-fold increase in volume. We may assume from this difference in volume that (1) gas molecules are relatively far apart, (2) gases are capable of being greatly compressed, and (3) the volume occupied by a gas is mostly empty space.

12.2 The Kinetic-Molecular Theory

Careful scientific studies of the behavior and properties of gases were begun in the 17th century by Robert Boyle (1627–1691). This work was carried forward by many investigators after Boyle. The accumulated data were used in the second half of the 19th century to formulate a general theory to explain the behavior and properties of gases, called the **Kinetic-Molecular Theory (KMT)**. The KMT has since been extended to cover, in part, the behavior of liquids and solids. It ranks today with the atomic theory as one of the greatest generalizations of modern science.

Kinetic-Molecular Theory (KMT)

The KMT is based on the motion of particles, particularly gas molecules. A gas that behaves exactly as outlined by the theory is known as an **ideal**, or **perfect**, **gas**. Actually, there are no ideal gases, but under certain conditions of temperature and pressure, gases approach ideal behavior, or at least show only small deviations from it. Under extreme conditions, such as very high pressure and low temperature, real gases may deviate greatly from ideal behavior. For example, at low temperature and high pressures many gases become liquids.

ideal, or perfect, gas

The principal assumptions of the Kinetic-Molecular Theory are

1. Gases consist of tiny (submicroscopic) molecules.
2. The distance between molecules is large compared to the size of the molecules themselves. The volume occupied by a gas consists mostly of empty space.
3. Gas molecules have no attraction for each other.
4. Gas molecules move in straight lines in all directions, colliding frequently with each other and with the walls of the container.
5. No energy is lost by the collision of a gas molecule with another gas molecule or with the walls of the container. All collisions are perfectly elastic.
6. The average kinetic energy for molecules is the same for all gases at the same temperature, and its value is directly proportional to the Kelvin temperature.

Let us consider the facts supporting the theory. Assumption 1 above is based on the size of atoms and molecules, already established in previous chapters. Assumptions 2 and 3 are based on the comparison of volumes occupied by equal masses of the solid, liquid, and gaseous states of a substance

diffusion

and the fact that gases continue to expand and completely fill any size container. Assumption 4, that gases are in constant motion, is shown by the fact that gases exert pressure, expand into larger containers, and diffuse.

The property of **diffusion**, the ability of two or more gases to spontaneously mix, also supports the assumption that gas molecules have very little attraction for each other. The diffusion of gases may be illustrated by use of the apparatus shown in Figure 12.1. Two large flasks, one containing reddish-brown bromine vapors and the other dry air, are connected by a side tube. When the stopcock between the flasks is opened, the bromine and air will diffuse into each other. After standing awhile, both flasks will contain bromine and air.

Figure 12.2 shows that a gas exerts the same pressure at all parts of a container. The three gauges, located at different parts of the cylinder, show the same pressure.

With billions of molecules present in even a very small mass of gas, it is safe to assume that there will be collisions between these molecules as well as collisions with the walls of the container. Assumption 5 above is borne out by the fact that gases do not change temperature upon standing (external causes excepted). This shows that the molecules do not suffer loss of energy by collisions. Although one molecule may transfer energy to another molecule in a collision, the average or total energy of the system remains the same. The kinetic energy (KE) of a molecule is one-half of its mass times its velocity

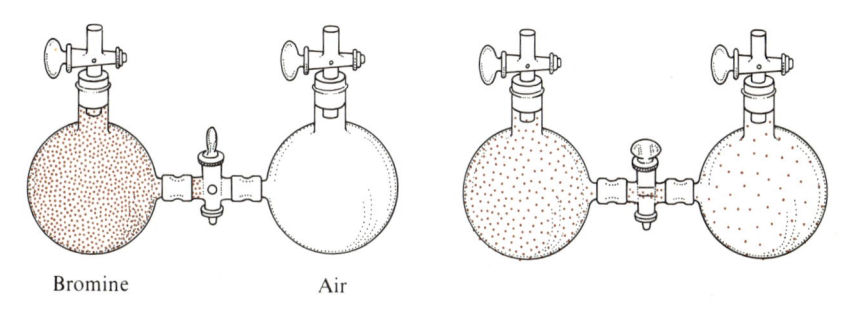

Bromine   Air

Figure 12.1. Diffusion of gases. When the stopcock between the two flasks is opened, colored bromine molecules can be seen diffusing into the flask containing air.

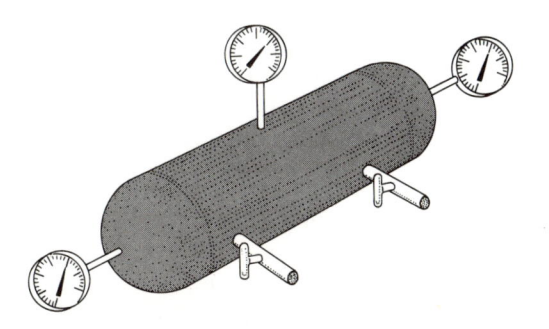

Figure 12.2. A gas moves in all directions and exerts the same pressure in all directions.

squared. It is expressed by the equation

$$KE = \frac{1}{2} mv^2$$

where m is the mass and v is the velocity of the molecule.

Experimental evidence shows that 2.0 g H_2 (1 mole) and 32.0 g O_2 (1 mole) in containers of equal volume at the same temperature exert the same pressure. This evidence supports assumption 6—that the kinetic energy for all gases is the same at the same temperature—and leads us to reason that the molecules of different gases, because of differing masses, will have different average velocities. The *relative* molecular velocities of different gases can be calculated from their kinetic energies. For example, the mass of any oxygen molecule is 32 amu and that of hydrogen is 2 amu. From the Kinetic-Molecular Theory, we have

$$KE \text{ of } H_2 = KE \text{ of } O_2$$

$$\frac{1}{2} m_{H_2} v^2_{H_2} = \frac{1}{2} m_{O_2} v^2_{O_2}$$

$$\frac{1}{2} \times 2 \times v^2_{H_2} = \frac{1}{2} \times 32 \times v^2_{O_2}$$

$$\frac{v^2_{H_2}}{v^2_{O_2}} = \frac{16}{1}$$

Taking the square root of both sides of the equation, we have

$$\frac{v_{H_2}}{v_{O_2}} = \frac{4}{1}$$

These calculations show that the average velocity of a hydrogen molecule is four times greater than that of an oxygen molecule.

The rates of diffusion of different gases are directly proportional to their molecular velocities. Inspection of the foregoing equations shows that molecular velocities—and therefore the rates of diffusion—of different gases are inversely proportional to the square roots of their molecular weights. This principle was first introduced by the Scottish chemist Thomas Graham (1805–1869) and is known as **Graham's law of diffusion**: The rates of diffusion of different gases are inversely proportional to the square roots of their molecular weights (or densities).

Graham's law of diffusion

These properties of an ideal gas are independent of the molecular constitution of the gas. Mixtures of gases also obey the Kinetic-Molecular Theory if the gases in the mixture do not enter into a chemical reaction with each other.

12.3 Measurement of Pressure of Gases

pressure

Pressure is defined as force per unit area. Do gases exert pressure? Yes. When a rubber balloon is inflated with air, it stretches and maintains an abnormally large size because the pressure on the inside is greater than that on the

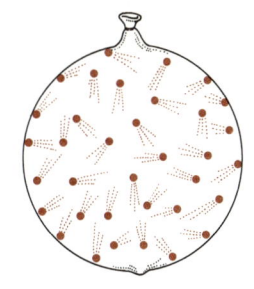

Figure 12.3. Pressure resulting from the collisions of gas molecules with the walls of the balloon keep the balloon inflated.

outside. Pressure results from the collisions of gas molecules with the walls of the balloon (see Figure 12.3). When the gas is released, the force or pressure of the air escaping from the small neck propels the balloon in a rapid, irregular path. If the balloon is inflated until it bursts, the gas escaping all at once causes a small explosive noise. This pressure that gases display can be measured; it can also be transformed into useful work. Steam under pressure, as used in the locomotive, played an important role in the early development of the United States. Compressed steam is used today to generate at least part of the electricity for many cities. Compressed air is used to operate many different kinds of mechanical equipment.

The mass of air surrounding the earth is called the *atmosphere*. It is composed of about 78% nitrogen, 21% oxygen, and 1% argon and other minor constituents (see Table 12.1). The outer boundary of the atmosphere is not known precisely, but more than 99% of the atmosphere is below an altitude of 20 miles (32 km). Thus, the concentration of gas molecules in the atmosphere decreases with altitude, and at about 4 miles, there is insufficient oxygen to sustain human life. The gases in the atmosphere exert a pressure known as **atmospheric pressure**. The pressure exerted by a gas depends on the number of molecules of gas present, the temperature, and the volume in which the gas is confined. Gravitational forces confine the atmosphere relatively close to the earth and act to prevent air molecules from flying off into outer space. Thus, the atmospheric pressure at any point is due to the weight of the atmosphere pressing downward at that point.

The pressure of a gas can be measured with a pressure gauge, a manometer, or a **barometer**. A mercury barometer is commonly used in the laboratory

Table 12.1. Average composition of normal dry air.

Gas	Percent by volume	Gas	Percent by volume
N_2	78.08	He	0.0005
O_2	20.95	CH_4	0.0002
Ar	0.93	Kr	0.0001
CO_2	0.033	Xe, H_2, and N_2O	Trace
Ne	0.0018		

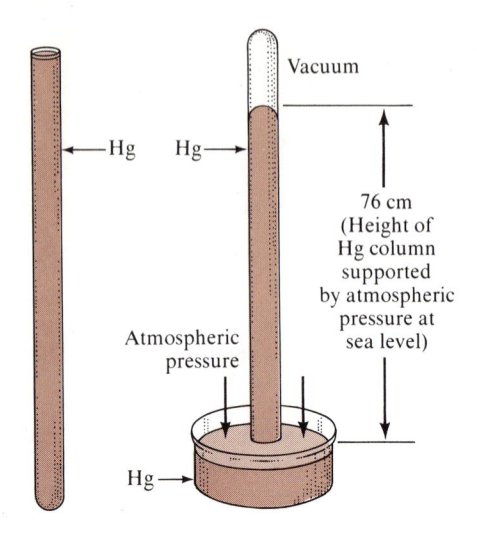

Figure 12.4. **Preparation of a mercury barometer.** The full tube of mercury at the left is inverted and placed in a dish of mercury.

to measure atmospheric pressure. A simple barometer of this type may be prepared by filling a long tube with pure, dry mercury and inverting the open end into an open dish of mercury. If the tube is longer than 76 cm, the mercury level will drop to a point at which the column of mercury in the tube is just supported by the pressure of the atmosphere. If the tube is properly prepared, a vacuum will exist above the mercury column. The weight of mercury, per unit area, is equal to the pressure of the atmosphere. The column of mercury is supported by the pressure of the atmosphere, and the height of the column is a measure of this pressure (see Figure 12.4). The mercury barometer was invented in 1643 by the Italian physicist E. Torricelli (1608–1647), for whom the unit of pressure *torr* was named.

1 atmosphere

The average pressure of the atmosphere at sea level is **1 atmosphere** (abbreviated as atm). This pressure is equivalent to that of a 76 cm column of mercury. Other units for expressing pressure are inches of mercury, millimeters of mercury, the torr, the millibar, and pounds per square inch (lb/in.^{2}). The meteorologist uses inches of mercury in reporting atmospheric pressure. The values of these units equivalent to 1 atm are summarized in Table 12.2. 1 atm ≡ 76 cm Hg ≡ 760 mm Hg ≡ 29.9 in. Hg ≡ 760 torr ≡ 1013 mbar ≡ 14.7 lb/in.^{2}

Table 12.2. Pressure units equivalent to 1 atmosphere.

1 atm
76 cm Hg
760 mm Hg
760 torr
1013 mbar
29.9 in. Hg
14.7 lb/in.^{2}

Atmospheric pressure varies with altitude. The average pressure at Denver, Colorado, 1 mile above sea level, is 63 cm Hg (0.83 atm). Other liquids besides mercury may be employed for barometers, but they are not as useful as mercury because of the difficulty of maintaining a vacuum above the liquid and because of impractical heights of the liquid column. For example, a pressure of 1 atm will support a column of water about 10,336 mm (33.9 ft) high.

12.4 Dependence of Pressure on Number of Molecules

Pressure is produced by gas molecules colliding with the walls of a container. At a specific temperature and volume, the number of collisions depends on the number of gas molecules present. The number of collisions may be increased by increasing the number of gas molecules present. If we double the number of molecules, the frequency of collisions and the pressure should double. We find, for an ideal gas, that this doubling is actually what happens. The pressure, therefore, when the temperature and volume are kept constant, is directly proportional to the number of moles or molecules of gas present. Figure 12.5 illustrates this.

A good example of this molecule–pressure relationship may be observed on an ordinary cylinder of compressed gas that is equipped with a pressure

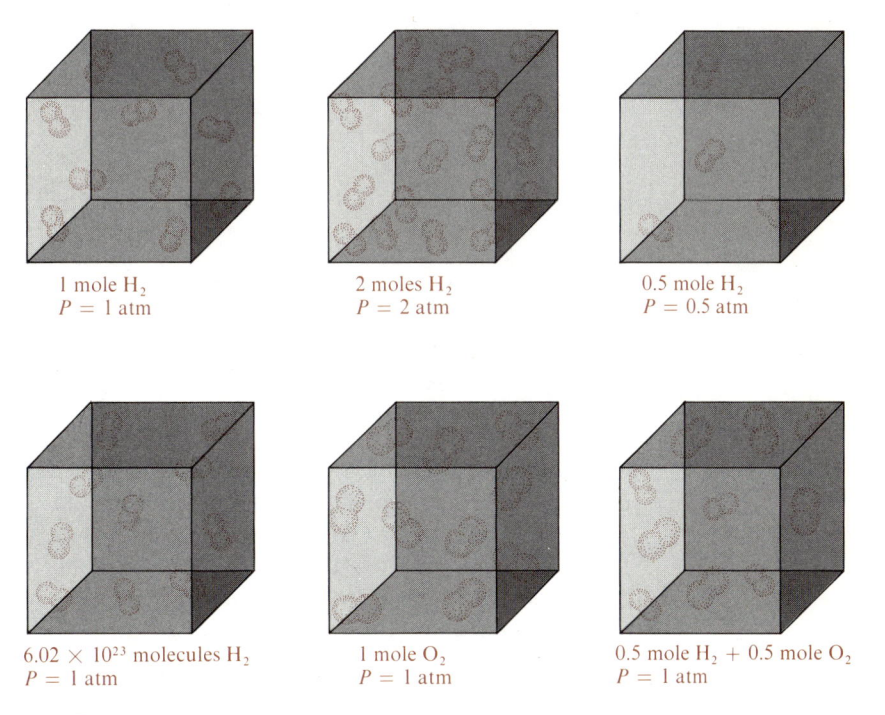

1 mole H_2
$P = 1$ atm

2 moles H_2
$P = 2$ atm

0.5 mole H_2
$P = 0.5$ atm

6.02×10^{23} molecules H_2
$P = 1$ atm

1 mole O_2
$P = 1$ atm

0.5 mole H_2 + 0.5 mole O_2
$P = 1$ atm

Figure 12.5. The pressure exerted by a gas is directly proportional to the number of molecules present. In each case, the volume is 22.4 litres and the temperature is 0°C.

198 Chapter 12

gauge. When the valve is opened, gas escapes from the cylinder. The volume of the cylinder is constant and the decrease in quantity of gas is registered by a drop in pressure indicated on the gauge.

12.5 Boyle's Law—The Relationship of the Volume and Pressure of a Gas

Boyle's law

Robert Boyle demonstrated experimentally that, at constant temperature, T, the volume, V, of a fixed mass of a gas is inversely proportional to the pressure, P. This relationship of P and V is known as **Boyle's law**. Mathematically, Boyle's law may be expressed as

$$V \propto \frac{1}{P} \quad \text{(Mass and Temperature are constant)}$$

This equation says that the volume varies inversely as the pressure, at constant mass and temperature. When the pressure on a gas is increased, its volume will decrease, and vice versa. The inverse relationship of pressure and volume is shown graphically in Figure 12.6.

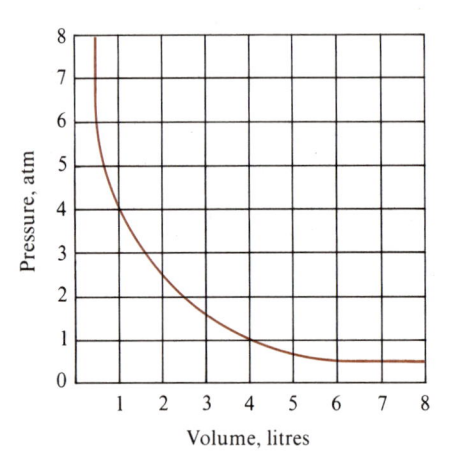

Figure 12.6. Graph of pressure versus volume showing inverse PV relationship of an ideal gas.

Boyle demonstrated that when he doubled the pressure on a specific quantity of a gas, keeping the temperature constant, the volume was reduced to one-half the original volume; when he tripled the pressure on the system, the new volume was one-third the original volume; and so on. His demonstration shows that the product of volume and pressure is constant if the temperature is not changed:

$$PV = \text{Constant} \quad \text{or} \quad PV = k \quad \text{(Mass and } T \text{ are constant)}$$

Let us demonstrate this law by taking a cylinder of gas with a movable piston, so that the volume may be varied by changing the external pressure (see Figure 12.7). We assume that there is no change in temperature or the

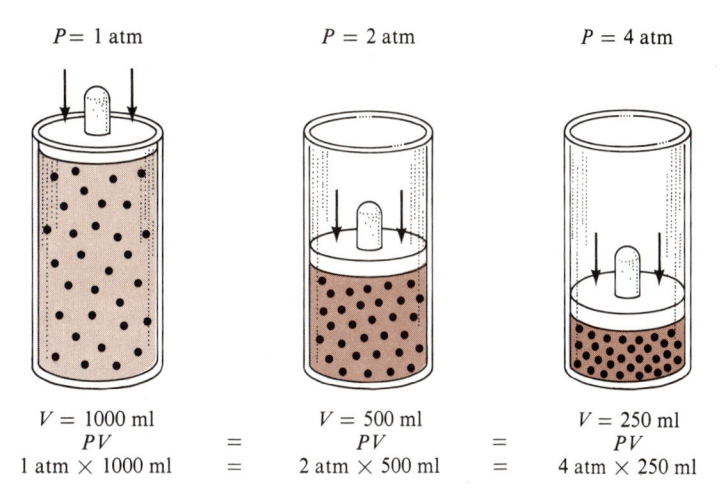

P = 1 atm		P = 2 atm		P = 4 atm
V = 1000 ml		V = 500 ml		V = 250 ml
PV	=	PV	=	PV
1 atm × 1000 ml	=	2 atm × 500 ml	=	4 atm × 250 ml

Figure 12.7. The effect of pressure on the volume of a gas.

number of molecules. Let us start with a volume of 1000 ml and a pressure of 1 atm. When we change the pressure to 2 atm, the gas molecules are crowded closer together and the volume is reduced to 500 ml. If we increase the pressure to 4 atm, the volume becomes 250 ml.

Note that the product of the pressure times the volume in each case is the same number, substantiating Boyle's law. We may then say that

$$P_1 V_1 = P_2 V_2$$

where $P_1 V_1$ is the pressure–volume product at one set of conditions and $P_2 V_2$, at another set of conditions. In each case, the new volume may be calculated by multiplying the starting volume by a ratio of the two pressures involved. Of course, the ratio of pressures used must reflect the direction in which the volume should change. When the pressure is changed from 1 atm to 2 atm, the ratio to be used is 1 atm/2 atm. Now we can verify the results given in Figure 12.7.

(a) Starting volume: 1000 ml; pressure change (1 atm → 2 atm)

$$1000 \text{ ml} \times \frac{1 \text{ atm}}{2 \text{ atm}} = 500 \text{ ml}$$

(b) Starting volume: 1000 ml; pressure change (1 atm → 4 atm)

$$1000 \text{ ml} \times \frac{1 \text{ atm}}{4 \text{ atm}} = 250 \text{ ml}$$

(c) Starting volume: 500 ml; pressure change (2 atm → 4 atm)

$$500 \text{ ml} \times \frac{2 \text{ atm}}{4 \text{ atm}} = 250 \text{ ml}$$

In summary, a change in the volume of a gas due to a change in pressure may be calculated by multiplying the original volume by a ratio of the two pressures. If the pressure is increased, the ratio should have the smaller pressure in the numerator and the larger pressure in the denominator. If the pressure

200   Chapter 12

is decreased, the larger pressure should be in the numerator and the smaller pressure in the denominator.

New volume = Original volume × Ratio of pressures

Examples of problems based on Boyle's law follow. If no mention is made of temperature, assume that it remains constant.

Problem 12.1   What volume will 2.50 litres of a gas occupy if the pressure is changed from 760 mm Hg to 630 mm Hg?

First we must determine whether the pressure is being increased or decreased. In this case it is being decreased. This decrease in pressure should result in an increase in the volume. Therefore, we need to multiply 2.50 litres by a ratio of the pressures, which will give us an increase in volume. This ratio is 760 mm Hg/630 mm Hg. The calculation is

$$V = 2.50 \text{ litres} \times \frac{760 \text{ mm Hg}}{630 \text{ mm Hg}} = 3.02 \text{ litres} \quad \text{(New volume)}$$

Alternatively, an algebraic approach may be used, solving $P_1 V_1 = P_2 V_2$ for V_2:

$$V_2 = V_1 \times \frac{P_1}{P_2} = 2.50 \text{ litres} \times \frac{760 \text{ mm Hg}}{630 \text{ mm Hg}} = 3.02 \text{ litres}$$

where $V_1 = 2.50$ litres, $P_1 = 760$ mm Hg, and $P_2 = 630$ mm Hg.

Problem 12.2   A given mass of hydrogen occupies 40.0 litres at 760 mm Hg pressure. What volume will it occupy at 5 atm pressure?

Since the units of the two pressures are not the same they must be made the same; otherwise, the units will not cancel in the final calculation. Since the pressure is increased, the volume should decrease. Therefore, we need to multiply 40.0 litres by a ratio of the pressures that will give us a decrease in volume.

First, convert 760 mm Hg to atmospheres by multiplying by the conversion factor 1 atm/760 mm Hg:

$$760 \text{ mm Hg} \times \frac{1 \text{ atm}}{760 \text{ mm Hg}} = 1 \text{ atm}$$

Second, set up a ratio of the pressures that will give a volume decrease:

$$\frac{1 \text{ atm}}{5 \text{ atm}}$$

Third, multiply the volume (40.0 litres) by this pressure ratio:

$$V = 40.0 \text{ litres} \times \frac{1 \text{ atm}}{5 \text{ atm}} = 8.00 \text{ litres} \quad \text{(Answer)}$$

Problem 12.3   A gas occupies a volume of 200 ml at 400 mm Hg pressure. To what pressure must the gas be subjected in order to change the volume to 75.0 ml?

In order to reduce the volume from 200 ml to 75.0 ml, it will be necessary to increase the pressure. In the same way we calculated volume change affected by a change in pressure, we must multiply the original pressure by a ratio of the two volumes. The volume ratio in this case should be 200 ml/75.0 ml. The calculation is

$$P = 400 \text{ mm Hg} \times \frac{200 \text{ ml}}{75.0 \text{ ml}} = 1067 \quad (1.07 \times 10^3 \text{mm Hg}) \quad \text{(New pressure)}$$

Algebraically, $P_1V_1 = P_2V_2$ may be solved for P_2:

$$P_2 = P_1 \times \frac{V_1}{V_2} = 400 \text{ mm Hg} \times \frac{200 \text{ ml}}{75.0 \text{ ml}} = 1.07 \times 10^3 \text{ mm Hg}$$

where $P_1 = 400$ mm Hg, $V_1 = 200$ ml, and $V_2 = 75.0$ ml.

In problems of this type, it is good practice to check the answers to see if they are consistent with the given facts. For example, if the data indicate that the pressure is increased, the final volume should be smaller than the initial volume.

12.6 Charles' Law—The Effect of Temperature on the Volume of a Gas

The effect of temperature on the volume of a gas was observed in about 1787 by the French physicist J. A. C. Charles (1746–1823). Charles found that various gases expanded by the same fractional amount when heated through the same temperature interval. Later it was found that if a given volume of any gas initially at 0°C was cooled by 1°C, the volume decreased by $\frac{1}{273}$; if cooled by 2°C, by $\frac{2}{273}$; if cooled by 20°C, by $\frac{20}{273}$; and so on. Since each degree of cooling reduced the volume by $\frac{1}{273}$, it was apparent that any quantity of any gas would have zero volume, if it could only be cooled to -273°C. Of course, no real gas can be cooled to -273°C for the simple reason that it liquefies before that temperature is reached. However, -273°C (more precisely -273.16°C) is referred to as *absolute zero*; this temperature is the zero point on the Kelvin (Absolute) temperature scale. It is the temperature at which the volume of an ideal or perfect gas would become zero.

The volume–temperature relationship for gases is shown graphically in Figure 12.8. Experimental data show the graph to be a straight line which when extrapolated crosses the temperature axis at -273.16°C, or absolute zero.

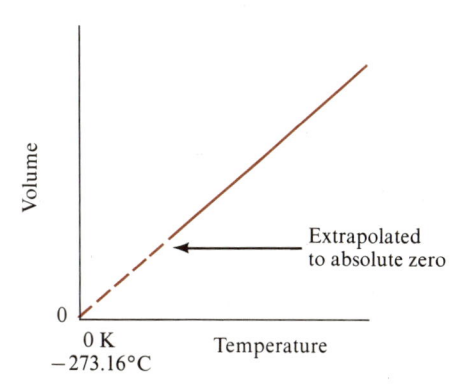

Figure 12.8. Volume–temperature relationship of gases. Extrapolated portion of the graph is shown by the broken line.

202 Chapter 12

Charles' law

In modern form, **Charles' law** states that at constant pressure the volume of a fixed weight of any gas is directly proportional to the absolute temperature. Mathematically, Charles' law may be expressed as

$$V \propto T \quad (P \text{ is constant})$$

which means that the volume of a gas varies directly with the absolute temperature when the pressure remains constant. In equation form Charles' law may also be written as

$$V = kT \quad (\text{At constant pressure})$$

where k is a constant for a fixed weight of the gas. If the absolute temperature of a gas is doubled, the volume will double. (A capital T is usually used for absolute temperature, K, and a small t for °C.)

To illustrate, let us return to the gas cylinder with the movable or free-floating piston (see Figure 12.9). Assume that the cylinder labeled (a) contains a quantity of gas and the pressure on it is 1 atm. When the gas is heated, the molecules move faster and their kinetic energy increases. This action should increase the number of collisions per unit of time and thereby the pressure. However, the increased internal pressure will cause the piston to rise to a level at which the internal and external pressures again equal 1 atm (cylinder b). The net result is an increase in volume due to an increase in temperature. Another equation relating the volume of a gas at two different temperatures is

$$\frac{V_1}{T_1} = \frac{V_2}{T_2} \quad (\text{Constant } P)$$

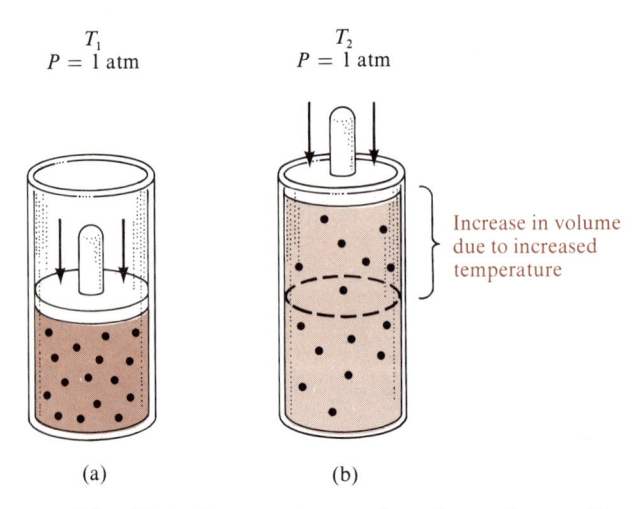

Figure 12.9. The effect of temperature on the volume of a gas. The gas in cylinder (a) is heated from T_1 to T_2. With the external pressure constant at 1 atm, the free-floating piston rises, resulting in an increased volume, as shown in cylinder (b).

where V_1 and T_1 are one set of conditions and V_2 and T_2 are another set of conditions.

A simple experiment showing the variation of the volume of a gas with temperature is illustrated in Figure 12.10. A small balloon attached to a bottle is immersed in either ice water or hot water. In ice water the volume is reduced, as shown by the collapse of the balloon; in hot water the gas expands and the balloon increases in size.

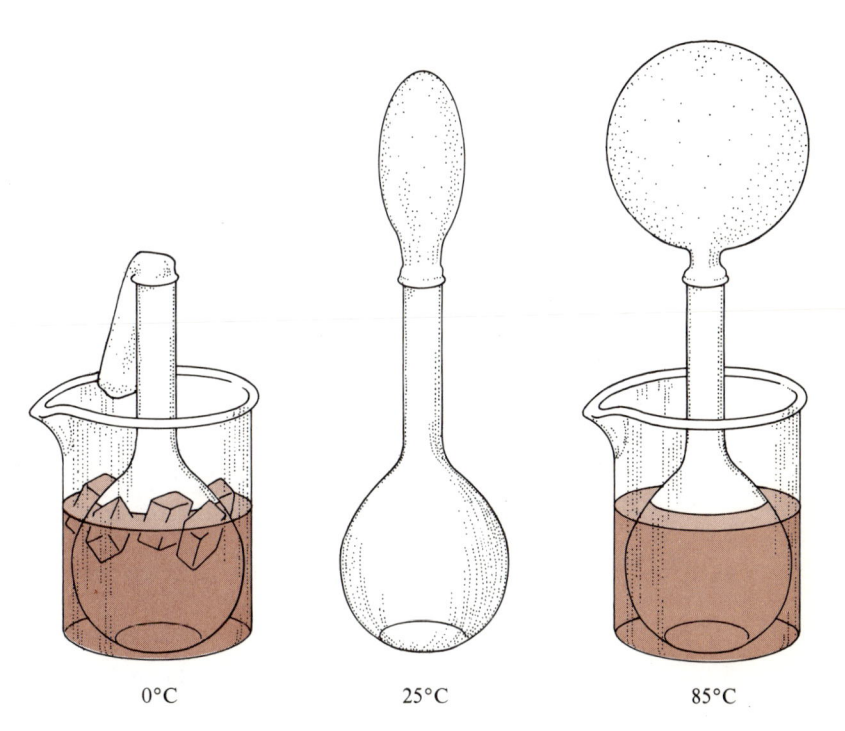

Figure 12.10. The effect of temperature on the volume of a gas. A volume decrease occurs when a balloon attached to a flask is immersed in ice water; the volume increases when the flask is immersed in hot water.

The calculation of changes in volume due to changes in temperature involves two basic steps: (1) changing the temperatures to K and (2) multiplying the original volume by a ratio of the initial and final temperatures. If the temperature is increased, the higher temperature is placed in the numerator of the ratio and the lower temperature in the denominator. If the temperature is decreased, the lower temperature is placed in the numerator of the ratio and the higher temperature in the denominator.

New volume = Original volume × Ratio of temperatures (K)

Problems based on Charles' law follow.

204 Chapter 12

Problem 12.4   Three litres of hydrogen at $-20°C$ are allowed to warm to a room temperature of $27°C$. What is the volume at room temperature if the pressure remains constant?
First change °C to K.

$$°C + 273 = K$$
$$-20°C + 273 = 253 \text{ K}$$
$$27°C + 273 = 300 \text{ K}$$

Since the temperature is increased, the volume should increase. The original volume should be multiplied by the temperature ratio of 300 K/253 K. The calculation is

$$V = 3.00 \text{ litres} \times \frac{300 \text{ K}}{253 \text{ K}} = 3.56 \text{ litres} \quad \text{(New volume)}$$

To obtain the answer by algebra, solve $V_1/T_1 = V_2/T_2$ for V_2:

$$V_2 = V_1 \times \frac{T_2}{T_1} = 3.00 \text{ litres} \times \frac{300 \text{ K}}{253 \text{ K}} = 3.56 \text{ litres}$$

where $V_1 = 3.00$ litres, $T_1 = 253$ K, and $T_2 = 300$ K.

Problem 12.5   If 20.0 litres of nitrogen are cooled from $100°C$ to $0°C$, what is the new volume? Since no mention is made of pressure, assume that there is no pressure change.
First change °C to K.

$$100°C + 273 = 373 \text{ K}$$
$$0°C + 273 = 273 \text{ K}$$

The ratio of temperature to be used is 273 K/373 K, since the final volume should be smaller than the original volume. The calculation is

$$V = 20.0 \text{ litres} \times \frac{273 \text{ K}}{373 \text{ K}} = 14.6 \text{ litres} \quad \text{(New volume)}$$

Three variables—pressure, P; volume, V; and temperature, T—are needed to describe a fixed amount of a gas. Boyle's law, $PV = k$, relates pressure and volume at constant temperature; Charles' law, $V = kT$, relates volume and temperature at constant pressure. A third relationship involving pressure and temperature at constant volume is also known and is stated: The pressure of a fixed weight of a gas, at constant volume, is directly proportional to the Kelvin temperature. In equation form, the relationship is

$$P = kT \quad \text{(At constant volume)}$$

This relationship is a modification of Charles' law and is sometimes called Gay-Lussac's law.

We may summarize the effect of changes in pressure, temperature, and quantity of a gas as follows:

1. In the case of a fixed or constant volume,
    (a) when the temperature is increased, the pressure increases.
    (b) when the quantity of a gas is increased, the pressure increases (T remaining constant).

2. In the case of a variable volume,
    (a) when the external pressure is increased, the volume decreases (T remaining constant).
    (b) when the temperature of a gas is increased, the volume increases (P remaining constant).
    (c) when the quantity of a gas is increased, the volume increases (P and T remaining constant).

12.7 Standard Temperature and Pressure

standard conditions

standard temperature and pressure (STP)

In order to compare volumes of gases, common reference points of temperature and pressure were selected and called **standard conditions** or **standard temperature and pressure** (abbreviated **STP**). Standard temperature is 273 K (0°C) and standard pressure is 1 atm, or 760 mm Hg. For purposes of comparison, volumes of gases are usually changed to STP conditions.

$$STP = 273 \text{ K } (0°C) \text{ and } 1 \text{ atm or } 760 \text{ mm Hg}$$

12.8 Simultaneous Changes in Pressure, Volume, and Temperature (Combined Gas Laws)

When both temperature and pressure change at the same time, the new volume may be calculated by multiplying the initial volume by the correct ratios of both pressure and temperature:

$$\text{Final volume} = \text{Initial volume} \times \left(\begin{array}{c}\text{Ratio of}\\ \text{pressures}\end{array}\right) \times \left(\begin{array}{c}\text{Ratio of}\\ \text{temperatures}\end{array}\right)$$

This equation combines both Boyle's and Charles' laws, and the same considerations for the pressure and the temperature ratios should be used in the calculation. There are four possible variations:

1. both T and P cause an increase in volume,
2. both T and P cause a decrease in volume,
3. T causes an increase and P a decrease in volume, and
4. T causes a decrease and P an increase in volume.

The P, V, and T relationships for a given weight of any gas, in fact, may be expressed as a single equation, $PV/T = k$. For problem solving, this equation is usually written

$$\frac{P_1 V_1}{T_1} = \frac{P_2 V_2}{T_2}$$

where $P_1 V_1/T_1$ are the initial conditions and $P_2 V_2/T_2$ are the final conditions.

This equation may be solved for any one of the six variables represented and is very generally useful in dealing with the pressure–volume–temperature relationships of gases. Note that when T is constant ($T_1 = T_2$), Boyle's law is represented; when P is constant ($P_1 = P_2$), Charles' law is represented; and when V is constant $V_1 = V_2$, the modified Charles' or Gay-Lussac's law is represented.

Problem 12.6

Given 20.0 litres of ammonia gas at 5°C and 730 mm Hg pressure, calculate the volume at 50°C and 800 mm Hg.

In order to get a better look at the data, tabulate the initial and final conditions:

	Initial	Final
V	20.0 litres	V_2
T	5°C	50°C
P	730 mm Hg	800 mm Hg

Change °C to K:

5°C + 273 = 278 K

50°C + 273 = 323 K

Set up ratios of T and P:

$$T \text{ ratio} = \frac{323 \text{ K}}{278 \text{ K}} \quad \text{(Increase in } T \text{ should increase } V\text{)}$$

$$P \text{ ratio} = \frac{730 \text{ mm Hg}}{800 \text{ mm Hg}} \quad \text{(Increase in } P \text{ should decrease } V\text{)}$$

The calculation is

$$V_2 = 20.0 \text{ litres} \times \frac{323 \text{ K}}{278 \text{ K}} \times \frac{730 \text{ mm Hg}}{800 \text{ mm Hg}} = 21.2 \text{ litres}$$

The algebraic solution is:

solve $\dfrac{P_1 V_1}{T_1} = \dfrac{P_2 V_2}{T_2}$ for V_2 by multiplying both sides of the equation by T_2/P_2 and rearranging to obtain

$$V_2 = V_1 \times \frac{P_1}{P_2} \times \frac{T_2}{T_1}$$

Tabulate the known values:

$V_1 = 20.0$ litres $\qquad V_2 = ?$

$T_1 = 5°C + 273 = 278$ K $\qquad T_2 = 50°C + 273 = 323$ K

$P_1 = 730$ mm Hg $\qquad P_2 = 800$ mm Hg

Substitute these values in the equation and calculate the value of V_2:

$$V_2 = 20.0 \text{ litres} \times \frac{730 \text{ mm Hg}}{800 \text{ mm Hg}} \times \frac{323 \text{ K}}{278 \text{ K}} = 21.2 \text{ litres}$$

Problem 12.7

To what temperature (°C) must 10.0 litres of nitrogen at 25°C and 700 mm Hg be heated in order to have a volume of 15.0 litres and a pressure of 760 mm Hg?

This problem is conveniently handled by an algebraic solution.

Solve $\dfrac{P_1 V_1}{T_1} = \dfrac{P_2 V_2}{T_2}$ for T_2 to obtain $T_2 = T_1 \times \dfrac{P_2}{P_1} \times \dfrac{V_2}{V_1}$

Tabulate the known values:

$P_1 = 700$ mm Hg $\qquad P_2 = 760$ mm Hg
$V_1 = 10.0$ litres $\qquad V_2 = 15.0$ litres
$T_1 = 25°C + 273 = 298$ K $\qquad T_2 = ?$

Substitute these known values in the equation and evaluate T_2:

$$T_2 = 298 \text{ K} \times \dfrac{760 \text{ mm Hg}}{700 \text{ mm Hg}} \times \dfrac{15.0 \text{ litres}}{10.0 \text{ litres}} = 485 \text{ K}$$

485 K − 273 = 212°C   (Answer)

Problem 12.8

The volume of a gas is 50.0 litres at 20°C and 742 mm Hg. What volume will it occupy at standard temperature and pressure (STP)? Tabulate the data.

	Initial	Final
V	50.0 litres	V_2
T	20°C	0°C
P	742 mm Hg	760 mm Hg

STP conditions are 0°C and 760 mm Hg. First change °C to K.

20°C + 273 = 293 K
0°C + 273 = 273 K

Then set up ratios of T and P.

T ratio $= \dfrac{273 \text{ K}}{293 \text{ K}}$ (Decrease in T should decrease V)

P ratio $= \dfrac{742 \text{ mm Hg}}{760 \text{ mm Hg}}$ (Increase in P should decrease V)

The calculation is

$$V_2 = 50.0 \text{ litres} \times \dfrac{273 \text{ K}}{293 \text{ K}} \times \dfrac{742 \text{ mm Hg}}{760 \text{ mm Hg}}$$

$V_2 = 45.5$ litres

12.9 Dalton's Law of Partial Pressures

If gases behave according to the Kinetic–Molecular Theory, there should be no difference in their pressure–volume–temperature relationships, whether the gas molecules are all the same or different from each other. This similarity

Dalton's Law of Partial Pressures

is the basis for an understanding of **Dalton's Law of Partial Pressures**, which states that in a mixture of gases, each gas exerts a pressure independent of the other gases present, and that the total pressure is the sum of the partial pressures exerted by each gas in the mixture. Thus, if we have a mixture of three gases, *A, B,* and *C,* exerting 50 mm, 150 mm, and 400 mm Hg pressure, respectively, the total pressure will be 600 mm Hg.

$$P_{Total} = P_A + P_B + P_C$$
$$P_{Total} = 50 \text{ mm Hg} + 150 \text{ mm Hg} + 400 \text{ mm Hg} = 600 \text{ mm Hg}$$

We can see an application of Dalton's law in the collection of gases over water. When oxygen is prepared in the laboratory, it is commonly collected over water (see Figure 12.11). The O_2, collected by the downward displacement of water, is not pure but contains water vapor mixed with it. When the water level is adjusted to be the same inside and outside the bottle, the pressure of the oxygen plus water vapor inside the bottle is equal to the atmospheric pressure:

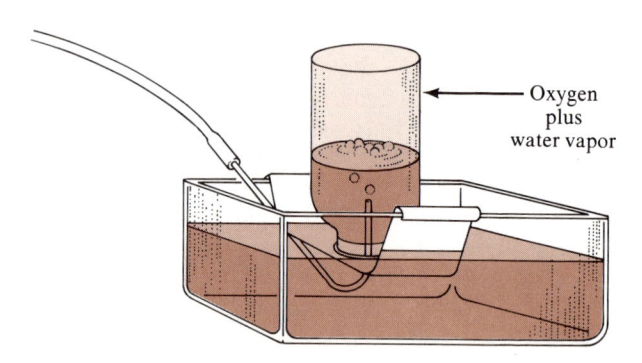

$$P_{atm} = P_{O_2} + P_{H_2O}$$
$$P_{O_2} = P_{atm} - P_{H_2O}$$

Figure 12.11. Oxygen collected over water.

To determine the amount of O_2 or any other gas collected over water, we must subtract the pressure of the water vapor from the total pressure of the gas. The vapor pressure of water at various temperatures is tabulated in Appendix II.
An illustrative problem follows.

Problem 12.9

A 500 ml sample of oxygen, O_2, was collected over water at 23°C and 760 mm Hg pressure. What volume will the dry O_2 occupy at 23°C and 760 mm Hg? The vapor pressure of water at 23°C is 21.0 mm Hg.
To solve this problem, we must first find the pressure of the O_2 alone by subtracting the pressure of the water vapor present.

$$P_{\text{total}} = 760 \text{ mm Hg} = P_{O_2} + P_{H_2O}$$
$$P_{O_2} = 760 \text{ mm Hg} - 21.0 \text{ mm Hg} = 739 \text{ mm Hg}$$

Thus, the pressure of dry O_2 is 739 mm Hg.

The problem is now of the Boyle's law type. It is treated as if we had 500 ml of dry O_2 at 739 mm Hg pressure, which is then changed to 760 mm Hg pressure, with the temperature remaining constant. The calculation is

$$V = 500 \text{ ml} \times \frac{739 \text{ mm Hg}}{760 \text{ mm Hg}} = 486 \text{ ml dry } O_2$$

This means that 486 ml of the 500 ml mixture of O_2 and water vapor is pure O_2. Figure 12.12 depicts the pressure and volume changes involved in this problem.

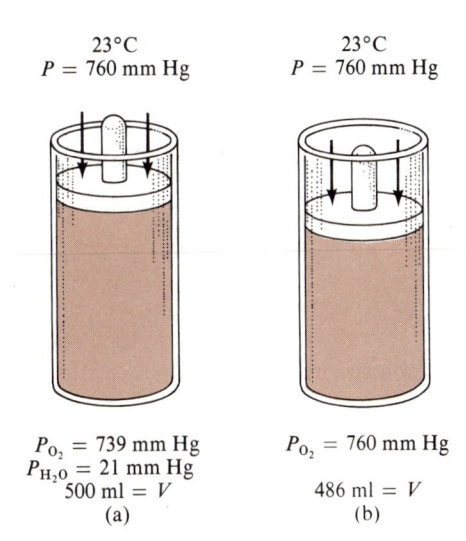

Figure 12.12. A 500 ml sample of oxygen was collected over water at 23°C and 760 mm Hg pressure. The original gas collected is shown in cylinder (a). When the water vapor is removed (cylinder b), the volume is reduced. The external pressure, being greater than the pressure of the oxygen, forces the cylinder lid downward until the pressure of the oxygen is 760 mm Hg. The volume of dry oxygen is 486 ml.

12.10 Avogadro's Hypothesis

Gay-Lussac's Law of Combining Volumes of Gases

Early in the 19th century, J. L. Gay-Lussac (1778–1850) of France studied the volume relationships of reacting gases. His results, published in 1809, were summarized in a statement known as **Gay-Lussac's Law of Combining Volumes of Gases**: *When measured at constant temperature and pressure, the ratios of the volumes of reacting gases are small whole numbers.* Thus, H_2 and O_2 combine in a volume ratio of 2:1; H_2 and Cl_2 react in a volume ratio of 1:1; H_2 and N_2 react in a volume ratio of 3:1; and so on.

Avogadro's hypothesis

Two years later, in 1811, Amadeo Avogadro of Italy used the Law of Combining Volumes of Gases to make a simple but very significant and far-reaching generalization concerning gases. **Avogadro's hypothesis** states:

Equal volumes of different gases at the same temperature and pressure contain the same number of molecules.

This hypothesis was a real breakthrough in understanding the nature of gases. (1) It offered a rational explanation of Gay-Lussac's Law of Combining Volumes of Gases and indicated the diatomic nature of such elemental gases as hydrogen, chlorine, and oxygen; (2) it provided a method for determining the molecular weights of gases and for comparing the densities of gases of known molecular weight (see Section 12.11); and (3) it afforded a firm foundation for the development of the Kinetic-Molecular Theory.

On a volume basis, hydrogen and chlorine react thus:

Hydrogen + Chlorine → Hydrogen chloride
1 volume   1 volume   2 volumes

By Avogadro's hypothesis, equal volumes of hydrogen and chlorine must contain the same number of molecules. Therefore, hydrogen molecules react with chlorine molecules in a 1:1 ratio. Since two volumes of hydrogen chloride are produced, one molecule of hydrogen and one of chlorine must produce two molecules of hydrogen chloride. Therefore, each hydrogen molecule and each chlorine molecule is made up of two atoms. The coefficients of the balanced equation for the reaction give the correct ratios for volumes, molecules, and moles of reactants and products:

$$H_2 + Cl_2 \longrightarrow 2\,HCl$$

1 volume   1 volume   2 volumes
1 molecule   1 molecule   2 molecules
1 mole   1 mole   2 moles

By like reasoning, oxygen molecules must contain at least two atoms because one volume of oxygen reacts with two volumes of hydrogen to produce two volumes of steam (H_2O).

12.11   Weight–Volume Relationship of Gases

gram molecular volume

A mole of any gas contains 6.02×10^{23} molecules (Avogadro's number). Therefore, a gram-molecular weight (1 mole) of any gas, at STP conditions, occupies about the same volume—namely, 22.4 litres. This volume, 22.4 litres, occupies a cube about 28.2 cm (11.1 in.) on a side (see Figure 12.13) and is called the molar volume or **gram-molecular volume**. The gram-molecular weights of several gases, each occupying 22.4 litres at STP, are also shown in Figure 12.13.

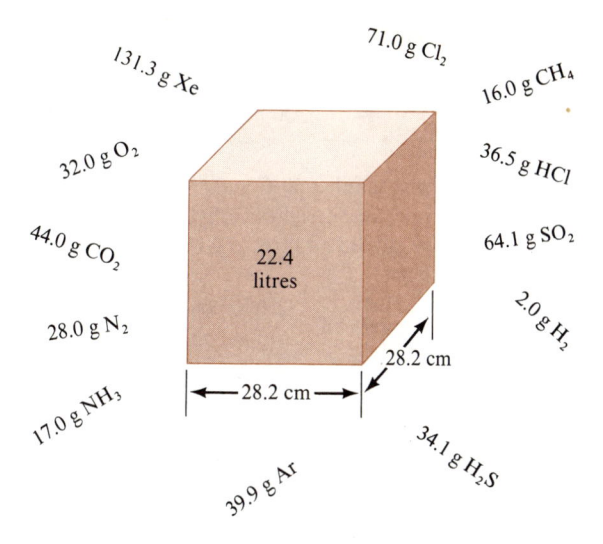

Figure 12.13. The gram-molecular weight of a gas at STP occupies 22.4 litres. The weight given for each gas is its gram-molecular weight.

One gram-molecular weight (one mole) of a gas occupies 22.4 litres at STP.

This relationship is useful for determining the molecular weight of a gas or of substances that can be easily vaporized into gases. If the weight and the volume of a gas at STP are known, we can calculate its molecular weight. One litre of pure oxygen at STP weighs 1.429 g. The molecular weight of oxygen may be calculated by multiplying the weight of 1 litre by 22.4 litres/mole.

$$\frac{1.429 \text{ g}}{\text{litre}} \times \frac{22.4 \text{ litres}}{\text{mole}} = 32.0 \text{ g/mole} \quad \text{(g-mol. wt)}$$

If the weight and volume are at other than standard conditions, we first change the volume to STP and then calculate the molecular weight. Note that we do not correct the weight to standard conditions—only the volume.

The gram-molecular volume, 22.4 litres/mole, is used as a conversion factor to convert g/litre to g/mole and also to convert litres to moles. The two conversion factors are

$$\frac{22.4 \text{ litres}}{\text{mole}} \quad \text{and} \quad \frac{1 \text{ mole}}{22.4 \text{ litres}}$$

These conversions must be done at STP conditions except under certain special circumstances. Examples of problems follow.

Problem 12.10 If 2.00 litres of a gas measured at STP weigh 3.23 g, what is the molecular weight of the gas?

g-mol. wt at STP = 22.4 litres/mole

212 Chapter 12

If 2.00 litres weigh 3.23 g then 22.4 litres will weigh 22.4/2.00 times as much as 3.23 g, or

$$\text{g-mol. wt} = \frac{3.23 \text{ g}}{2.00 \text{ litres}} \times \frac{22.4 \text{ litres}}{\text{mole}} = 36.2 \text{ g/mole}$$

Problem 12.11   Measured at 40°C and 630 mm Hg, 691 ml of ethyl ether weigh 1.65 g. Calculate the gram-molecular weight of ethyl ether.

In order to use 22.4 litres = g-mol. wt, we must first correct the volume to standard conditions. Thus,

$$V = 691 \text{ ml} \times \frac{273 \text{ K}}{313 \text{ K}} \times \frac{630 \text{ mm Hg}}{760 \text{ mm Hg}}$$

V at (STP) = 500 ml = 0.500 litre

The weight of the gas has not been altered by correcting the volume to STP, so that 500 ml at STP weigh 1.65 g. Then

$$\text{g-mol. wt} = \frac{1.65 \text{ g}}{0.500 \text{ litre}} \times \frac{22.4 \text{ litres}}{\text{mole}} = 73.9 \text{ g/mole}$$

12.12 Density and Specific Gravity of Gases

The density, d, of a gas is its mass per unit volume, which is generally expressed in grams per litre (g/litre):

$$d = \frac{\text{Mass}}{\text{Volume}} = \frac{\text{g}}{\text{litre}}$$

Because the volume of a gas depends on temperature and pressure, both of these should be given when stating the density of a gas. The volume of solids and liquids is hardly affected by changes in pressure and is changed only by a small degree when the temperature is varied. Increasing the temperature from 0°C to 50°C will reduce the density of a gas by about 18% if the gas is allowed to expand. In comparison, a 50°C rise in the temperature of water (0°C → 50°C) will change its density by less than 0.2%.

The density of a gas at any temperature and pressure may be determined by calculating the weight of gas present in 1 litre. At STP, in particular, the density may be calculated by dividing the gram-molecular weight of the gas by 22.4 litres/mole.

$$\text{Density (at STP)} = \frac{\text{g-mol. wt}}{22.4 \text{ litres/mole}} = \frac{\text{g/mole}}{\text{litre/mole}} = \frac{\text{g}}{\text{litre}}$$

Or

$$d(\text{at STP}) \times 22.4 \text{ litres/mole} = \text{g-mol. wt}$$

The density of Cl_2 at STP is calculated as follows:

g-mol. wt of Cl_2 = 71.0 g/mole

$$d = \frac{\text{g-mol. wt}}{22.4 \text{ litres/mole}} = \frac{71.0 \text{ g/mole}}{22.4 \text{ litres/mole}} = 3.17 \text{ g/litre}$$

The specific gravity (sp gr) of a gas is the ratio of the mass of any volume of the gas to the mass of an equal volume of some reference gas. Specific gravities of gases are commonly quoted in reference to air = 1.00. The actual mass of air at STP is 1.29 g/litre, which is the density of air.

The specific gravity can be calculated by dividing the density of gas by the density of air. Both gases must be the same temperature and pressure.

$$\text{sp gr} = \frac{\text{Density of a gas}}{\text{Density of air}}$$

The specific gravity of Cl_2, for example, is

$$\text{sp gr of } Cl_2 = \frac{\text{Density of } Cl_2}{\text{Density of air}} = \frac{3.17 \text{ g/litre}}{1.29 \text{ g/litre}} = 2.46$$

This indicates that Cl_2 is 2.46 times as heavy as air. Table 12.3 lists the density and specific gravity of some common gases.

Table 12.3. Density and specific gravity of common gases at STP.

Gas	Molecular weight	Density (g/litre at STP)	Specific gravity (air = 1.00)
H_2	2.0	0.090	0.070
CH_4	16.0	0.714	0.553
NH_3	17.0	0.760	0.589
C_2H_2	26.0	1.16	0.899
HCN	27.0	1.21	0.938
CO	28.0	1.25	0.969
N_2	28.0	1.25	0.969
O_2	32.0	1.43	1.11
H_2S	34.1	1.52	1.18
HCl	36.5	1.63	1.26
F_2	38.0	1.70	1.32
CO_2	44.0	1.96	1.52
C_3H_8	44.1	1.96	1.52
O_3	48.0	2.14	1.66
SO_2	64.1	2.86	2.22
Cl_2	71.0	3.17	2.46

The volume of a gas depends on the temperature, the pressure, and the number of gas molecules. Two or more gases at the same temperature have the same average kinetic energy. If these gases occupy the same volume, they will exhibit the same pressure. Such a system of identical *PVT* properties can only be produced by the same number of molecules having the same average kinetic energy.

12.13 Calculations from Chemical Equations Involving Gases

(a) Mole–volume (gas) and weight–volume (gas) calculations. Stoichiometric problems involving gas volumes can be solved by the general mole-ratio method outlined in Chapter 11. The factors 1 mole/22.4 litres and 22.4 litres/1 mole are used as needed for converting volume to moles and moles to volume, respectively. These conversion factors are used under the assumptions that the gases are at STP and behave as ideal gases. In actual practice, gases are measured at other than STP conditions, and the volumes are converted to STP for stoichiometric calculations.

In a balanced equation, the number in front of the formula of a gaseous substance represents the number of moles or molar volumes (22.4 litres at STP) of that substance.

The following are examples of typical problems involving gases and chemical equations:

Problem 12.12 What volume of oxygen (at STP) can be formed from 0.500 mole of potassium chlorate?

Step 1. Write the balanced equation:

$$2\,KClO_3 \rightarrow 2\,KCl + 3\,O_2\uparrow$$

2 moles   3 moles

Step 2. The moles of starting substance is 0.500 mole $KClO_3$.

Step 3. Calculate the moles of O_2, using the mole-ratio method:

$$0.500 \text{ mole } KClO_3 \times \frac{3 \text{ moles } O_2}{2 \text{ moles } KClO_3} = 0.750 \text{ mole } O_2$$

Step 4. Convert moles of O_2 to litres of O_2. The moles of a gas at STP are converted to litres by multiplying by the molar volume, 22.4 litres per mole:

$$0.750 \text{ mole } O_2 \times \frac{22.4 \text{ litres}}{\text{mole}} = 16.8 \text{ litres } O_2 \quad \text{(Answer)}$$

Setting this up in a continuous calculation, we obtain

$$0.500 \text{ mole } KClO_3 \times \frac{3 \text{ moles } O_2}{2 \text{ moles } KClO_3} \times \frac{22.4 \text{ litres } O_2}{\text{mole } O_2} = 16.8 \text{ litres } O_2$$

Problem 12.13 How many grams of zinc must react with sulfuric acid to produce 1000 ml of hydrogen gas at STP?

Step 1. The equation is

$$Zn + H_2SO_4 \longrightarrow ZnSO_4 + H_2\uparrow$$

1 mole   1 mole

Step 2. Moles of H_2. The equation states that 1 mole of H_2 is produced from 1 mole of Zn; 1000 ml of H_2 equals 1 litre of H_2 and represents a fraction of a mole.

$$1000 \text{ ml } H_2 \times \frac{1 \text{ litre } H_2}{1000 \text{ ml } H_2} \times \frac{1 \text{ mole}}{22.4 \text{ litres}} = 0.0446 \text{ mole } H_2$$

Step 3. Convert to moles of Zn:

$$0.0446 \text{ mole } H_2 \times \frac{1 \text{ mole } Zn}{1 \text{ mole } H_2} = 0.0446 \text{ mole } Zn$$

Step 4. Convert to grams of Zn:

$$0.0446 \text{ mole } Zn \times \frac{65.4 \text{ g } Zn}{\text{mole } Zn} = 2.92 \text{ g } Zn \quad \text{(Answer)}$$

The continous calculation setup is

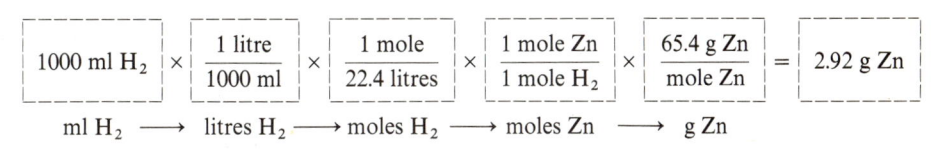

$$\boxed{1000 \text{ ml } H_2} \times \boxed{\frac{1 \text{ litre}}{1000 \text{ ml}}} \times \boxed{\frac{1 \text{ mole}}{22.4 \text{ litres}}} \times \boxed{\frac{1 \text{ mole } Zn}{1 \text{ mole } H_2}} \times \boxed{\frac{65.4 \text{ g } Zn}{\text{mole } Zn}} = \boxed{2.92 \text{ g } Zn}$$

ml H_2 ⟶ litres H_2 ⟶ moles H_2 ⟶ moles Zn ⟶ g Zn

Problem 12.14   What volume of hydrogen, collected at 30°C and 700 mm Hg pressure, will be formed by reacting 50.0 g of aluminum with hydrochloric acid?

$$2 \text{ Al} + 6 \text{ HCl} \longrightarrow 2 \text{ AlCl}_3 + 3 \text{ H}_2$$

2 moles   3 moles

In this problem, the volume of H_2 is first calculated from the equation, as we have done before. But, because the volume calculated by use of the equation is at STP, it must be changed to the conditions at which the gas is collected.

Step 1. Change grams of Al to moles of Al:

$$50.0 \text{ g Al} \times \frac{1 \text{ mole Al}}{27.0 \text{ g Al}} = 1.85 \text{ moles Al}$$

Step 2. Calculate litres of H_2 using the mole-ratio method:

$$1.85 \text{ moles Al} \times \frac{3 \text{ moles } H_2}{2 \text{ moles Al}} \times \frac{22.4 \text{ litres } H_2}{1 \text{ mole } H_2} = 62.2 \text{ litres } H_2 \quad \text{(at STP)}$$

Step 3. Calculate the volume of H_2 at 30°C and 700 mm Hg pressure:

$$\text{Volume} = 62.2 \text{ litres} \times \frac{303 \text{ K}}{273 \text{ K}} \times \frac{760 \text{ mm Hg}}{700 \text{ mm Hg}} = 75.0 \text{ litres } H_2 \quad \text{(Answer)}$$

(b) Volume–volume calculations. When all substances in a reaction are in the gaseous state, simplifications in the calculation can be made based on Avogadro's hypothesis that gases under identical conditions of temperature and pressure contain the same number of molecules and occupy the same volume. Using this same hypothesis, we can state that, under the same conditions of temperature and pressure, the volumes of gases reacting are proportional to the number of moles of the gases in the balanced equation. Consider the reaction

$$H_2(g) + Cl_2(g) \longrightarrow 2 \text{ HCl}(g)$$

1 mole	1 mole	2 moles
22.4 litres	22.4 litres	2 × 22.4 litres
1 volume	1 volume	2 volumes
Y volume	Y volume	2Y volumes

216   Chapter 12

In this reaction, 22.4 litres of hydrogen will react with 22.4 litres of chlorine to give 2 × 22.4, or 44.8, litres of hydrogen chloride gas. This is true because these volumes are equivalent to the number of reacting moles in the equation. Therefore, Y volume of H_2 will combine with Y volume of Cl_2 to give $2Y$ volume of HCl. For example, 100 litres of H_2 reacts with 100 litres of Cl_2 to give 200 litres of HCl; if the 100 litres of H_2 and Cl_2 are at 50°C, they will give 200 litres of HCl at 50°C. When the temperature and pressure before and after a reaction are the same, calculation of volumes can be done without correcting the volumes to STP.

⊙ *For reacting gases: Volume-volume relationships are the same as the mole-mole relationships.*

Problem 12.15

What volume of oxygen will react with 150 litres of hydrogen to form water vapor? What volume of water vapor will be formed? Assume that both reactants and products are measured at the same conditions. Let us compare the two methods for solving this problem, using the mole method first and then the principle of reacting volumes.

$$2H_2(g) + O_2(g) \longrightarrow 2H_2O(g)$$
$$\text{2 moles} \quad \text{1 mole} \quad \text{2 moles}$$

Mole method:

Step 1.   Moles of H_2 = 150 litres $H_2 \times \dfrac{1\ \text{mole}}{22.4\ \text{litres}}$ = 6.70 moles H_2

Step 2.   Moles of O_2 = 6.70 moles $H_2 \times \dfrac{1\ \text{mole}\ O_2}{2\ \text{moles}\ H_2}$ = 3.35 moles O_2

Step 3.   Volume of O_2 = 3.35 moles $O_2 \times \dfrac{22.4\ \text{litres}}{\text{mole}}$ = 75.0 litres O_2

Step 4.   Volume of $H_2O(g)$ = 150 litres $H_2 \times \dfrac{1\ \text{mole}\ H_2}{22.4\ \text{litres}\ H_2} \times \dfrac{2\ \text{moles}\ H_2O}{2\ \text{moles}\ H_2} \times \dfrac{22.4\ \text{litres}}{\text{mole}}$

= 150 litres $H_2O(g)$ (Answer)

Calculation by reacting volumes:

$$2H_2(g) \quad + \quad O_2(g) \quad \longrightarrow \quad 2H_2O(g)$$

2 moles	1 mole	2 moles
2 × 22.4 litres	22.4 litres	2 × 22.4 litres
2 volumes	1 volume	2 volumes
150 litres	1/2 × 150 = 75 litres	2/2 × 150 = 150 litres

Thus, 150 litres of H_2 will react with 75 litres of O_2 to produce 150 litres of $H_2O(g)$. The calculation by reacting volumes, which may be done by inspection, is certainly simpler and more direct.

Problem 12.16

The equation for the preparation of ammonia is

$$3H_2 + N_2 \xrightarrow{400°C} 2NH_3$$

Assuming that the reaction goes to completion,

(a) What volume of H_2 will react with 50 litres of N_2?
(b) What volume of NH_3 will be formed from 50 litres of N_2?
(c) What volume of N_2 will react with 100 ml of H_2?
(d) What volume of NH_3 will be produced from 100 ml of H_2?
(e) If 600 ml of H_2 and 400 ml of N_2 are sealed in a flask and allowed to react, what amounts of H_2, N_2, and NH_3 are in the flask at the end of the reaction?

The answers to parts (a)–(d), shown in the boxes below, can be determined from the equation by inspection, using the principle of reacting volumes.

$$3\,H_2 \;+\; N_2 \;\longrightarrow\; 2\,NH_3$$

3 volumes   1 volume   2 volumes

(a) $\boxed{150\ \text{litres}}$ 50 litres

(b)   50 litres $\boxed{100\ \text{litres}}$

(c) 100 ml $\boxed{33.3\ \text{ml}}$

(d) 100 ml $\boxed{66.7\ \text{ml}}$

(e) Volume ratio from the equation $= \dfrac{3\ \text{volumes}\ H_2}{1\ \text{volumes}\ N_2}$

Volume ratio used $= \dfrac{600\ \text{ml}\ H_2}{400\ \text{ml}\ N_2} = \dfrac{3\ \text{volumes}\ H_2}{2\ \text{volumes}\ N_2}$

Comparing these two ratios, we see that an excess of N_2 is present in the gas mixture. Therefore, the reagent limiting the amount of NH_3 that can be formed is H_2:

$$3\,H_2 \;+\; N_2 \;\longrightarrow\; 2\,NH_3$$

600 ml   200 ml   400 ml

In order to have a 3:1 ratio of volumes reacting, 600 ml of H_2 will react with 200 ml of N_2 to produce 400 ml of NH_3, leaving 200 ml of N_2 unreacted. At the end of the reaction, the flask will contain 400 ml of NH_3 and 200 ml of N_2.

12.14 Ideal Gas Equation

We have used four variables in calculations involving gases: the volume, V; the pressure, P; the absolute temperature, T; and the number of molecules or moles, n. Combining these variables into a single expression, we obtain

$$V \propto \frac{nT}{P} \quad \text{or} \quad V = \frac{nRT}{P}$$

where R is a proportionality constant known as the *ideal gas constant*. The equation is commonly written as

$$PV = nRT$$

ideal gas equation

and is known as the **ideal gas equation**. This equation states in a single expression what we have considered earlier in our discussions—that the volume of a gas varies directly with the number of gas molecules and the absolute temperature and varies inversely with the pressure. The value and units of R depend on the units of P, V, and T. We can calculate one value of R by taking 1 mole of a gas at STP conditions. Solve the equation for R:

$$R = \frac{PV}{nT} = \frac{1 \text{ atm} \times 22.4 \text{ litres}}{1 \text{ mole} \times 273 \text{ K}} = 0.0821 \frac{\text{litre-atm}}{\text{mole-K}}$$

The units of R in this case are litre-atmospheres per mole-K.

The ideal gas equation can be used to calculate any one of the four variables if the other three are known. When the value of R is 0.0821 litre-atm/mole-K, the other units must be as follows: P in atm, V in litres, n in moles, and T in K. Any problem that can be solved by the ideal gas equation can also be solved by direct application of the gas laws.

Problem 12.17

What pressure will be exerted by 0.400 mole of a gas in a 5.00-litre container at 17°C? First solve the ideal gas equation for P.

$$PV = nRT \quad \text{or} \quad P = \frac{nRT}{V}$$

Then substitute the data in the problem into the equation and solve.

$$P = \frac{0.400 \text{ mole} \times 0.0821 \text{ litre-atm} \times 290 \text{ K}}{5.00 \text{ litre} \times \text{mole-K}} = 1.90 \text{ atm} \quad \text{(Answer)}$$

Problem 12.18

How many moles of oxygen gas are in a 50.0 litre tank at 22°C if the pressure gauge reads 2000 lb/in^2?

First change to pressure in atmospheres. Then solve the ideal gas equation for n (moles), and then substitute the data in the equation to complete the calculation.

Step 1. Pressure in atmospheres:

$$\frac{2000 \text{ lb}}{\text{in.}^2} \times \frac{1 \text{ atm}}{14.7 \text{ lb/in.}^2} = 136 \text{ atm}$$

Step 2. Solve for moles using the ideal gas equation:

$$PV = nRT \quad \text{or} \quad n = \frac{PV}{RT}$$

$$n = \frac{136 \text{ atm} \times 50.0 \text{ litres}}{(0.0821 \text{ litre-atm/mole-K}) \times 295 \text{ K}} = 281 \text{ moles O}_2 \quad \text{(Answer)}$$

The ideal gas equation can also be used for problems involving a specific mass of gas by substituting the mass–mole relationship, n = g/g-mol. wt, into the equation.

All the gas laws are based on the behavior of an ideal gas—that is, a gas with a behavior that is described exactly by the gas laws for all possible values of P, V, and T. Most real gases actually do behave as predicted by the gas laws over a fairly wide range of temperatures and pressures. However, when conditions are such that the gas molecules are crowded closely together (high pressure and/or low temperature), they show marked deviations from ideal behavior. Deviations

Questions

occur because molecules have finite volumes and also exhibit intermolecular attractions. This results in less compressibility at high pressures and greater compressibility at low temperatures than predicted by the gas laws.

A. Review the meanings of the new terms introduced in this chapter.
1. Kinetic-Molecular Theory (KMT)
2. Ideal, or perfect gas
3. Diffusion
4. Graham's law of diffusion
5. Pressure
6. Atmospheric pressure
7. Barometer
8. 1 atmosphere
9. Boyle's law
10. Charles' law
11. Standard conditions
12. Standard temperature and pressure (STP)
13. Dalton's Law of Partial Pressures
14. Gay-Lussac's Law of Combining Volumes of Gases
15. Avogadro's hypothesis
16. Gram-molecular volume
17. Ideal gas equation

B. Answers to the following questions will be found in tables and figures.
1. What evidence is used to show diffusion in Figure 12.1? If methane and oxygen were substituted for bromine and air, how could we prove that diffusion had taken place?
2. Given the situation represented in Figure 12.3, are all the gas molecules traveling at the same speed? Explain.
3. In the preparation of a barometer at sea level, as in Figure 12.4, the height of the mercury column was found to be less than 76 cm. Suggest possible reasons for this discrepancy.
4. What is the weight of a column of mercury 76 cm long with a cross-sectional area of one square inch?
5. List in order of decreasing abundance the five most common gases found in normal dry air.
6. What would happen to the balloon in Figure 12.10 if the flask were immersed in a mixture of dry ice and acetone at $-78°C$?
7. List the gases in Table 12.3 that have densities at least 1.5 times greater than the density of air.
8. What volume would the box of Figure 12.13 have if it contained 14.0 g N_2, 8.0 g CH_4, and 66.0 g CO_2 all together at STP?
9. Explain why the pressure of O_2, P_{O_2}, has changed from 739 mm Hg to 760 mm Hg in Figure 12.12.

C. Review questions.
1. Outline the basic assumptions of the Kinetic-Molecular Theory.
2. What two factors determine the gas pressure in a container of fixed volume?
3. What determines the rate at which a gas diffuses?

220   Chapter 12

4. List the following gases in order of increasing relative molecular velocities: CH_4, SO_2, Ne, H_2, CO_2. What is your basis for determining this order? (Assume that all the gases are at the same temperature and pressure.)
5. At 100°C, which, if any, of the gases in Question 4 (above) has the highest kinetic energy? Explain.
6. What is an ideal gas?
7. Under what kind of conditions are real gases likely to deviate widely from the behavior of an ideal gas?
8. Some aerosol cans that are pressurized with a noncombustible gas bear a warning: "Caution: keep away from fire; do not incinerate empty container." Explain the logic of this warning in terms of the KMT.
9. What is the reason for referring gases to STP conditions?
10. Explain Dalton's Law of Partial Pressures in terms of the KMT.
11. What major exception can you visualize where mixtures of gases will not obey Dalton's Law of Partial Pressures?
12. Which of the following statements are correct? (Try to answer this question without the use of your text.)
    (a) When the pressure on a sample of gas is increased with the temperature kept constant, the gas will be compressed.
    (b) $V_1/V_2 = P_1/P_2$ is a statement of Boyle's law.
    (c) To calculate the volume of a gas resulting from a change in pressure, the original volume is multiplied by the ratio of the final pressure over the initial pressure.
    (d) If the pressure of a gas is kept constant, the volume can be changed by changing the number of molecules of the gas or by changing the temperature.
    (e) $PV = k$ is a statement of Charles' law.
    (f) If the temperature on a sample of gas is increased from 25°C to 50°C, the volume of the gas will increase by 100%.
    (g) According to Charles' law, the volume of a gas would be zero at $-273°C$.
    (h) Increasing the temperature of a fixed volume of a gas causes the pressure of the gas to decrease.
    (i) One mole of chlorine, Cl_2, at 20°C and 600 mm Hg pressure contains 6.02×10^{23} molecules.
    (j) At a given P and T, the volume of an ideal gas will be determined by the number of molecules in the sample.
    (k) A mixture of 0.5 mole CH_4 and 0.5 mole CO_2 will occupy 11.2 litres at STP.
    (l) Although a nitrogen molecule is 14 times as heavy as a hydrogen molecule, they both have the same kinetic energy at the same temperature and pressure.
    (m) The expression $n = PV/RT$ can be derived from the ideal gas equation.
    (n) When the pressure on a sample of gas is halved with the temperature kept constant, the density of the gas is also halved.
    (o) When the temperature of a sample of gas is increased at constant pressure, the density of the gas will decrease.
    (p) In a mixture containing an equal number of ammonia and oxygen molecules, the oxygen molecules, on the average, are moving faster than the ammonia molecules.

D. *Review problems.*
    1. The barometer reads 630 mm Hg. Calculate the corresponding atmospheric pressure in:

(a) Atmospheres
(b) Inches of Hg
(c) Pounds per square inch
(d) Torrs
(e) Millibars

2. The barometric pressure at the top of Pike's Peak in Colorado was recorded at 525 mm Hg. What is this pressure in atmospheres?
3. A gas occupies a volume of 200 ml at 600 mm Hg pressure. What will be its volume if the pressure is changed to the following, with the temperature remaining constant? (a) 800 mm Hg (b) 200 mm Hg
4. A 500 ml sample of gas is at a pressure of 720 mm Hg. What must be the pressure, with the temperature remaining constant, if the volume is changed to (a) 700 ml; (b) 350 ml?
5. What pressure would be required to compress 2000 litres of hydrogen at 1 atm into a 25 litre tank? (Assume constant temperature.)
6. Given 3.00 litres of nitrogen gas at $-20°C$, what volume will the nitrogen occupy at the following temperatures? (Assume constant pressure.)
   (a) 0°C   (b) $-80°C$ (c) 200 K   (d) 300 K
7. Given a 250 ml sample of oxygen at 22°C, at what temperature (°C) would the volume of oxygen be doubled if the pressure remains constant?
8. Early in the morning, the pressure in a tire is 25 pounds per square inch (psi) and its temperature is 18°C. At noon, after hard driving, the temperature of the tire is 46°C. What is the pressure in the tire? (Assume constant volume.)
9. The volume of a gas at STP is 650 ml. What volume will the gas occupy at 50°C and 380 mm Hg pressure?
10. Given 500 ml of a gas at STP. At what pressure will the gas volume be 200 ml at 30°C?
11. An expandable balloon contains 1000 litres of a gas at 1 atm pressure and 25°C. What will be the volume of the balloon when it rises to 22 miles altitude where the pressure is 4 mm Hg and the temperature is 2°C?
12. What volume would 1 mole of a gas occupy at 100°C and 630 mm Hg pressure?
13. What volume will a mixture of 4.00 moles of H_2 and 1.00 mole of CO_2 occupy at STP?
14. Four moles of O_2 in a small tank exert a pressure of 1520 lb/in.² How many moles are in the tank if the pressure reads 900 lb/in.²? (Assume constant T.)
15. How many moles of hydrogen are present in 2500 ml of pure H_2 at STP?
16. If 350 ml of a gas at STP weigh 0.726 g, what is its gram-molecular weight?
17. Calculate the gram-molecular weight of a gas if 1.52 g occupy 425 ml at 10°C and 720 mm Hg pressure.
18. Calculate the volume of 0.500 g of CO_2 at STP.
19. Calculate the density of the following gases at STP:
    (a) C_2H_6 (b) SO_3 (c) NF_3 (d) He
20. (a) Calculate the density of chlorine, Cl_2, at 100°C and 760 mm Hg pressure.
    [Hint: First change the volume of Cl_2 from STP to 100°C.]
    (b) At what temperature (°C) will the density of methane be 1.0 g/litre?
21. The density of a certain gas is 3.55 g/litre at STP. Calculate its gram-molecular weight.
22. Given the weight of the following gases, what volume will each occupy at STP?
    (a) 34.1 g H_2S (b) 0.525 g CO_2 (c) 8.50 g NH_3 (d) 6.00 g HCl
23. What volume will a mixture of 4.60 g of CH_4 and 7.00 g of N_2 occupy at 200°C and 3.00 atm pressure?

24. An equilibrium mixture is composed of hydrogen, nitrogen, and ammonia, where the pressure of each gas is as follows: H_2, 650 mm Hg; N_2, 250 mm Hg; NH_3, 450 mm Hg. What is the total pressure of the system?
25. How many total gas molecules are present in Problem 24 (above) if the volume of the mixture is 22.4 litres at 0°C?
26. Suppose that a sample of H_2 was collected over water at 23°C and 740 mm Hg pressure. What is the partial pressure of H_2 in this system? (Check Appendix II for the vapor pressure of water.)
27. A sample of O_2 collected over water at 30°C and 742 mm Hg pressure occupied a volume of 455 ml. Calculate the volume the dry O_2 would occupy at STP.
28. A 600 ml sample of methane, CH_4, was collected over water at 30°C and 750 mm Hg. What was the weight of the methane?
29. A mixture of noble gases at 760 mm Hg pressure consists of 50% He, 30% Ne, and 20% Ar. What is the partial pressure of each gas?
30. What volume of hydrogen at STP can be produced by reacting 3.60 moles of aluminum with sulfuric acid according to the following equation?

$$2\,Al(s) + 3\,H_2SO_4(aq) \rightarrow Al_2(SO_4)_3(aq) + 3\,H_2(g)$$

31. Given the equation: $4\,NH_3(g) + 5\,O_2(g) \rightarrow 4\,NO(g) + 6\,H_2O(g)$
    (a) How many moles of NH_3 must react to produce 5.0 moles of NO?
    (b) How many moles of O_2 must react to produce 5.0 moles of NO?
    (c) How many litres of NH_3 and O_2 must react to produce 100 litres of NO?
    (d) How many litres of O_2 will react with 100 grams of NH_3?
    (e) How many litres of NO are formed by reacting 10 moles of NH_3 with 10 moles of O_2?
32. Given the equation: $4\,FeS_2(s) + 11\,O_2(g) \xrightarrow{\Delta} 2\,Fe_2O_3(s) + 8\,SO_2(g)$:
    (a) How many litres of O_2 at STP will react with 1.00 kg of FeS_2?
    (b) How many litres of SO_2 will be produced from 1.00 kg of FeS_2?
33. Assume that the reaction $2\,CO(g) + O_2(g) \rightarrow 2\,CO_2(g)$ goes to completion. When 15 moles of carbon monoxide and 10 moles of oxygen are mixed and reacted in a closed flask, how many moles of CO, O_2, and CO_2 are present in flask at the end of the reaction?
34. In the preparation of hydrogen from methane, CH_4, and steam, what volume of H_2 is produced per cubic foot of CH_4 reacted? What volume of carbon monoxide is also produced as a by-product?

$$2\,CH_4(g) + 2\,H_2O(g) \xrightarrow{\Delta} 2\,CO(g) + 6\,H_2(g)$$

35. How many litres of air (21% oxygen) are needed to burn 5.00 litres of methane gas, CH_4, to carbon dioxide and water? Assume STP conditions.
36. A 25.00 g mixture of KCl and $KClO_3$ was heated, driving off all the oxygen. The volume of oxygen collected was 4.50 litres at STP. What is the percentage of $KClO_3$ in the mixture?
37. Calculate, using the ideal gas equation:
    (a) Volume of 0.820 mole of oxygen at 27°C and 720 mm Hg pressure
    (b) Weight of 16.0 litres of ethane, C_2H_6, at 20°C and 40.0 atm pressure
    (c) Weight of 10.2 litres of N_2 at 427°C and 2 atm pressure
    (d) Volume of 88.0 g of CO_2 at 700 mm Hg pressure and 27°C

13 Water and the Properties of Liquids

After studying Chapter 13 you should be able to:

1. Understand the terms listed in Question A at the end of the chapter.
2. Describe a water molecule with respect to electron-dot structure, bond angle, and polarity.
3. Make sketches showing hydrogen bonding (a) between water molecules and (b) between hydrogen fluoride molecules.
4. Explain the effect of hydrogen bonding on the physical properties of water.
5. Determine whether a compound will or will not form hydrogen bonds.
6. Complete and balance equations showing the formation of water (a) from hydrogen and oxygen, (b) by neutralization, and (c) by combustion of hydrogen-containing compounds.
7. Complete and balance equations for (a) the electrolysis of water, (b) the reactions of water with Groups IA and IIA metals, (c) the reactions of steam with other metals, (d) the reaction of steam with carbon, and (e) the reactions of water with halogens.
8. Identify metal oxides as basic anhydrides and write balanced equations for their reactions with water.
9. Identify nonmetal oxides as acid anhydrides and write balanced equations for their reactions with water.
10. Deduce the formula of the acid anhydride or of the basic anhydride when given the formula of the corresponding acid or base.
11. Identify, name, and write equations for the complete dehydration of hydrates.
12. Distinguish clearly between peroxides and ordinary oxides.
13. Discuss the occurrence of ozone and its effects on humans.
14. Outline the processes needed to prepare a potable water supply from a contaminated river source.
15. Describe how water may be softened by distillation, chemical precipitation, ion exchange, and demineralization—including chemical equations where appropriate.
16. Explain the process of evaporation from the standpoint of kinetic energy.
17. Relate vapor pressure data or vapor pressure curves of different substances to their relative rates of evaporation and to their relative boiling points.

18. Explain what is happening in the different segments of the time-temperature phase diagram of water.

13.1 Occurrence of Water

Water is our most common natural resource; it covers about three-fourths of the earth's surface. Not only is it found in the oceans and seas, in lakes, rivers, streams, and in glacial ice deposits; it is also always present in the atmosphere and in cloud formations. Moreover, water is an essential constituent of all living matter.

About 97% of the earth's water is in the oceans. This is saline water that contains vast amounts of dissolved minerals. The world's fresh water comprises the other 3%, of which about two-thirds is locked up in polar ice caps and glaciers. The remaining fresh water is found in ground water, lakes, and the atmosphere. More than 70 elements have been detected in the mineral content of seawater. Only four of these—chlorine, sodium, magnesium, and bromine—are now commercially obtained from the sea.

13.2 Physical Properties of Water

Water is a colorless, odorless, tasteless liquid with a melting point of 0°C and a boiling point of 100°C at 1 atm pressure. Two additional physical properties of matter are introduced with the study of water: heat of fusion and heat of vaporization. **Heat of fusion** is the amount of heat required to change one gram of a solid into a liquid at its melting point. The heat of fusion of water is 80 calories per gram. The temperature of the solid–liquid system does not change during the absorption of this heat. The heat of fusion is the energy used in breaking down the crystalline lattice of ice from a solid to a liquid. **Heat of vaporization** is the amount of heat required to change one gram of liquid to a vapor at its normal boiling point. The value for water is 540 calories per gram. Once again, there is no change in temperature during the absorption of this heat. The heat of vaporization is the energy needed to overcome the attractive forces between molecules in changing them from the liquid to the gaseous state. The values for water for both the heat of fusion and the heat of vaporization are relatively high compared to those for other substances; these high values indicate that strong attractive forces are acting between the molecules.

Ice and water exist together in equilibrium at 0°C, as shown in Figure 13.1. When ice at 0°C melts, it absorbs 80 cal/g in changing into a liquid; the temperature remains at 0°C. In order to refreeze the water, we have to remove 80 cal/g from the liquid at 0°C.

Both boiling water and steam are shown in Figure 13.2 to have a temperature of 100°C. It takes 100 cal to heat 1 g of water from 0°C to 100°C, but water at its boiling point absorbs 540 cal/g in changing to steam. Although boiling water and steam are both at the same temperature, steam contains

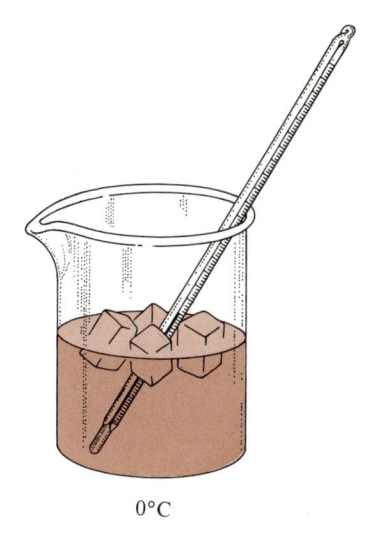

Figure 13.1. Ice and water in equilibrium at 0°C.

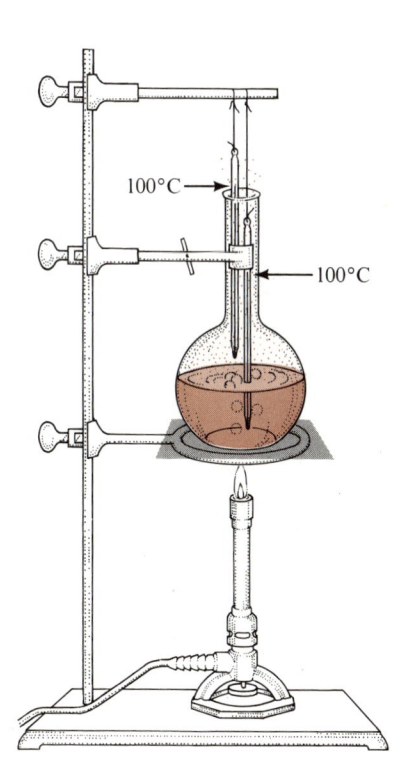

Figure 13.2. Boiling water and steam in equilibrium at 100°C.

considerably more heat per gram and can cause more severe burns than hot water. The physical properties of water are tabulated and compared with other hydrogen compounds of Group VIA elements in Table 13.1.

Table 13.1. Physical properties of water and other hydrogen compounds of Group VIA elements.

Formula	Color	Molecular weight	Melting point (°C)	Boiling point, 760 mm Hg (°C)	Heat of fusion (cal/g)	Heat of vaporization (cal/g)
H_2O	Colorless	18.0	0.00	100.0	80.0	540
H_2S	Colorless	34.1	−85.5	−60.3	16.7	131
H_2Se	Colorless	81.0	−65.7	−41.3	7.4	57.0
H_2Te	Colorless	129.6	−51	−2.3	—	42.8

The maximum density of water is 1.000 g/ml at 4°C. Water has the unusual property of contracting in volume as it is cooled to 4°C and then expanding when cooled from 4°C to 0°C. (Most liquids contract in volume all the way down to the point at which they solidify.) Therefore, 1 g of water occupies a volume greater than 1 ml at all temperatures above and below 4°C. Water, on the other hand, shows a large increase (about 9%) in volume when water is changed from a liquid at 0°C to a solid (ice) at 0°C. The density of ice at 0°C is 0.915 g/ml, which means that ice, being less dense than water, will float in water.

13.3 Structure of the Water Molecule

A single water molecule consists of two hydrogen atoms and one oxygen atom. Each H atom is attached to the O atom by a single covalent bond. This bond is formed by the overlap of the 1s orbital of hydrogen with an unpaired 2p electron orbital of oxygen. The average distance between the two nuclei is known as the *bond length*. The O–H bond length in water is 0.96 Å. The water molecule is nonlinear and has a bent structure with an angle of about 105° between the two bonds (see Figure 13.3).

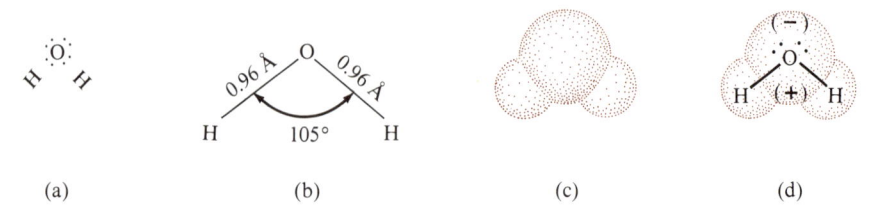

Figure 13.3. Diagrams of a water molecule: (a) electron distribution, (b) bond angle and O–H bond length, (c) molecular orbital structure, (d) dipole representation.

Oxygen is the second most electronegative element. As a result, the two covalent OH bonds in water are polar. If the three atoms in a water molecule were aligned in a linear structure such as H⊢→O←⊣H, the two polar bonds would be acting in equal and opposite directions and the molecule would be nonpolar. However, water is a highly polar molecule. Therefore, it does not have a linear structure; instead, it has a bent structure. When atoms are bonded together in a nonlinear fashion, the angle formed by the bonds is called the *bond angle*. In water the HOH bond angle is 105°. The two polar covalent bonds and the bent structure result in the oxygen atom having a partial negative charge and each hydrogen atom having a partial positive charge. The polar nature of water is responsible for many of its properties, including its behavior as a solvent.

13.4 The Hydrogen Bond

Table 13.1 compares the physical properties of H_2O, H_2S, H_2Se, and H_2Te. From this comparison it is apparent that four physical properties of water—melting point, boiling point, heat of fusion, and heat of vaporization—are extremely high and do not fit the trend for molecular weight. If water fitted the trend shown by the other compounds, we would expect the melting point of water to be below $-85°C$ and the boiling point to be below $-60°C$.

hydrogen bond

Why does water have these anomalous physical properties? The answer is that liquid water molecules are linked together by hydrogen bonds. A **hydrogen bond** is a chemical bond that is formed between polar molecules that contain hydrogen covalently bonded to a small, highly electronegative atom such as fluorine, oxygen, or nitrogen. The bond is actually the dipole–dipole attraction of polar molecules.

What is a hydrogen bond, or H-bond? Because a hydrogen atom has only one electron, it can form only one covalent bond. When it is attached to a strong electronegative atom such as oxygen, a hydrogen atom will also be attracted to an oxygen atom of another molecule, forming a bond (or bridge) between the two molecules. Water has two types of bonds: covalent bonds that exist between hydrogen and oxygen atoms within a molecule, and hydrogen bonds that exist between hydrogen and oxygen atoms in different water molecules.

Hydrogen bonds are *intermolecular* bonds; that is, they are formed between atoms in different molecules. They are somewhat ionic in character because they are formed by electrostatic attraction. Hydrogen bonds are much weaker than the ionic or covalent bonds that unite atoms to atoms to form compounds. Despite their weakness, hydrogen bonds are of great chemical importance.

Figure 13.4, part (a), shows two water molecules linked by a hydrogen bond, and part (b) shows six water molecules linked together by hydrogen bonds. A dash (—) is used for the covalent bond and a dotted line (----) for the hydrogen bond. In water, one molecule is linked to another through hydrogen bonds, forming a three-dimensional aggregate of water molecules. This molecular

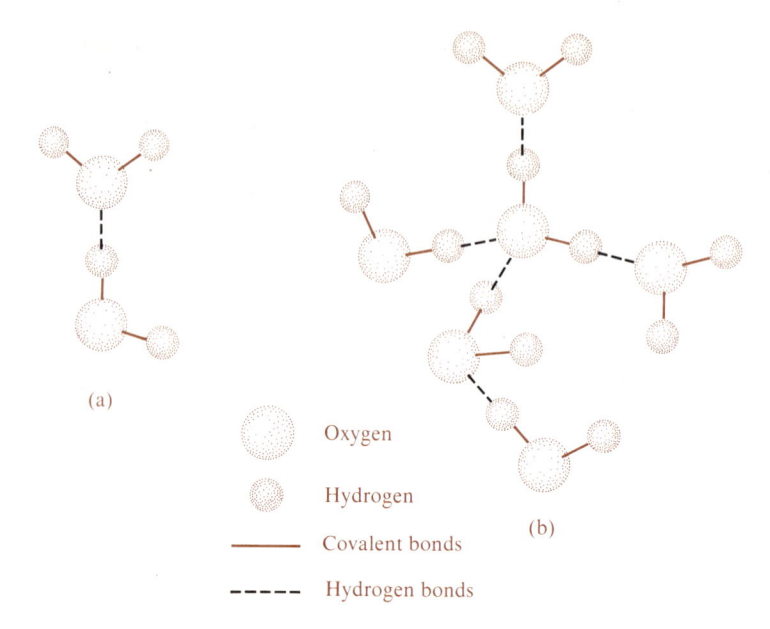

Figure 13.4. Hydrogen bonding: Water in the liquid and solid states exists as aggregates in which the water molecules are linked together by hydrogen bonds.

bonding effectively gives water the properties of a much larger, heavier molecule, explaining in part its relatively high melting point, boiling point, heat of fusion, and heat of vaporization. As water is heated and energy is absorbed, hydrogen bonds are continually being broken until at 100°C, with the absorption of an additional 540 cal/g, water separates into individual molecules, going into the gaseous state. Sulfur, selenium, and tellurium are not sufficiently electronegative for their hydrogen compounds to behave like water. As a result, H-bonding in H_2S, H_2Se, and H_2Te is only of small consequence (if any) to their physical properties.

Fluorine, the most electronegative element, forms the strongest hydrogen bonds. This bonding is strong enough to link hydrogen fluoride molecules together as *dimers*, H_2F_2. The dimer structure may be represented in this way:

H—F---H—F
↑
Hydrogen bond

The existence of salts, such as KHF_2 and NH_4HF_2, verifies the hydrogen fluoride (bifluoride) structure, HF_2^- (F—H----F)$^-$, where one H atom is bonded to two F atoms through one covalent bond and one hydrogen bond.

Hydrogen bonding can occur between two different atoms that are capable of forming H-bonds. Thus, we may have an O----H—N or O—H----N linkage in which the H-bond is between an oxygen and a nitrogen atom. This form of the H-bond exists in certain types of protein molecules and many biologically active substances.

13.5 Formation of Water and Chemical Properties of Water

Water is very stable to heat; it decomposes to the extent of only about 1% at temperatures up to 2000°C. Pure water is a nonconductor of electricity. But when a small amount of sulfuric acid or sodium hydroxide is added, the solution is readily decomposed into hydrogen and oxygen by an electric current. Two volumes of hydrogen are produced for each volume of oxygen:

$$2\,H_2O(l) \xrightarrow[H_2SO_4 \text{ or NaOH}]{\text{Electrical energy}} 2\,H_2(g) + O_2(g)$$

Formation

Water is formed when hydrogen burns in air. Pure hydrogen burns very smoothly in air, but mixtures of hydrogen and air (or oxygen) are dangerous and explode when ignited. The reaction is strongly exothermic:

$$2\,H_2(g) + O_2(g) \longrightarrow 2\,H_2O(g) + 115.6 \text{ kcal}$$

Water is produced by a variety of other reactions, especially by (1) acid–base neutralizations, (2) combustion of hydrogen-containing materials, and (3) metabolic oxidation in living cells.

1. $HCl(aq) + NaOH(aq) \longrightarrow NaCl(aq) + H_2O$

2. $2\,C_2H_2(g) + 5\,O_2(g) \longrightarrow 4\,CO_2(g) + 2\,H_2O(g) + 289.6 \text{ kcal}$
   Acetylene

3. $C_6H_{12}O_6 + 6\,O_2 \xrightarrow{\text{Enzymes}} 6\,CO_2 + 6\,H_2O + 673 \text{ kcal}$
   Glucose

The reaction represented by equation 2 is strongly exothermic and is capable of producing very high temperatures. It is used in oxygen–acetylene torches to cut and weld steel and other metals. The overall reaction of glucose with oxygen represented by equation 3 is the reverse of photosynthesis. It is the overall reaction by which living cells obtain needed energy by metabolizing glucose to carbon dioxide and water.

Reactions with Metals and Nonmetals

The reactions of metals with water at different temperatures show that these elements vary greatly in their reactivity. Metals such as sodium, potassium, and calcium react with cold water to produce hydrogen and a metal hydroxide. A small piece of sodium added to water melts from the heat produced by the reaction, forming a silvery metal ball, which rapidly flits back and forth on the surface of the water. One must use caution when experimenting with this reaction, since the hydrogen produced is frequently ignited by the sparking of the sodium, and it will explode, spattering sodium. Potassium reacts even more vigorously than sodium. Calcium sinks in water and only liberates a gentle stream of hydrogen. The equations for these reactions are

$$2\,Na(s) + 2\,H_2O(l) \longrightarrow H_2\uparrow + 2\,NaOH(aq)$$
$$2\,K(s) + 2\,H_2O(l) \longrightarrow H_2\uparrow + 2\,KOH(aq)$$
$$Ca(s) + 2\,H_2O(l) \longrightarrow H_2\uparrow + Ca(OH)_2(aq)$$

Zinc, aluminum, and iron do not react with cold water but will react with steam at high temperatures, forming hydrogen and a metallic oxide. The equations are

$$Zn(s) + H_2O(steam) \longrightarrow H_2\uparrow + ZnO(s)$$
$$2\,Al(s) + 3\,H_2O(steam) \longrightarrow 3\,H_2\uparrow + Al_2O_3(s)$$
$$3\,Fe(s) + 4\,H_2O(steam) \longrightarrow 4\,H_2\uparrow + Fe_3O_4(s)$$

Copper, silver, and mercury are examples of metals that do not react with cold water or steam to produce hydrogen. We conclude from these reactions that sodium, potassium, and calcium are chemically more reactive than zinc, aluminum, and iron, which are more reactive than copper, silver, and mercury.

Certain nonmetals react with water under various conditions. For example, fluorine reacts violently with cold water, producing hydrogen fluoride and free oxygen. The reactions of chlorine and bromine are much milder, producing what is commonly known as "chlorine water" and "bromine water," respectively. Chlorine water contains HCl, HOCl, and dissolved Cl_2; the free chlorine gives it a yellow-green color. Bromine water contains HBr, HOBr, and dissolved Br_2; the free bromine gives it a red-brown color. Steam passed over hot coke (carbon) produces a mixture of carbon monoxide and hydrogen that is known as "water gas." Since water gas is combustible, it is useful as a fuel. It is also the starting material for the commercial production of several alcohols. The equations for these reactions are

$$2\,F_2(g) + 2\,H_2O(l) \longrightarrow 4\,HF(aq) + O_2(g)$$
$$Cl_2(g) + H_2O(l) \longrightarrow HCl(aq) + HOCl(aq)$$
$$Br_2(l) + H_2O(l) \longrightarrow HBr(aq) + HOBr(aq)$$
$$C(s) + H_2O(g) \xrightarrow{1000°C} CO(g) + H_2(g)$$

Reactions with Metal and Nonmetal Oxides

basic anhydride

Metal oxides that react with water to form bases are known as **basic anhydrides**. If we heat the corresponding base, we can reverse the direction of the reaction and drive off water, forming the anhydride again.

$$CaO(s) + H_2O \longrightarrow Ca(OH)_2(aq)$$
$$Na_2O(s) + H_2O \longrightarrow 2\,NaOH(aq)$$
$$Ca(OH)_2(s) \xrightarrow{\Delta} CaO(s) + H_2O\uparrow$$

Certain metal oxides, such as CuO and Al$_2$O$_3$, do not form basic solutions because the oxides are insoluble in water.

Anhydrides do not contain any hydrogen. Thus, to determine the formula of an anhydride, the elements of water, H$_2$O, are removed from the designated acid or base formulas until all the hydrogen is removed. The formula of the anhydride then consists of the remaining metal or nonmetal and oxygen atoms. In calcium hydroxide, removal of water as indicated leaves CaO as the anhydride:

$$\mathrm{Ca}\begin{array}{c}\boxed{\mathrm{O\ H}}\\ \mathrm{OH}\end{array} \xrightarrow{\Delta} \mathrm{CaO} + \mathrm{H_2O}$$

In sodium hydroxide, H$_2$O cannot be removed from one formula unit, so two formula units of NaOH must be used, leaving Na$_2$O as the formula of the anhydride:

$$\begin{array}{c}\mathrm{NaO}\ \boxed{\mathrm{H}}\\ \mathrm{Na}\ \boxed{\mathrm{OH}}\end{array} \xrightarrow{\Delta} \mathrm{Na_2O} + \mathrm{H_2O}$$

acid anhydride

Nonmetal oxides that react with water to form acids are known as **acid anhydrides**. Examples are

$$\mathrm{SO_2}(g) + \mathrm{H_2O} \rightleftharpoons \mathrm{H_2SO_3}(aq)$$
$$\mathrm{N_2O_5}(s) + \mathrm{H_2O} \longrightarrow 2\,\mathrm{HNO_3}(aq)$$

The foregoing are examples of typical reactions of water but are by no means a complete list of the known reactions of water.

13.6 Hydrates

hydrate

Solids that contain water molecules as part of their crystalline structure are known as **hydrates**. Formulas for hydrates are expressed by first writing the usual anhydrous (without water) formula for the compound, then adding a dot, followed by the number of water molecules present. An example is BaCl$_2 \cdot$2H$_2$O. Sometimes the formula for water is enclosed in parentheses and the formula written as BaCl$_2$(H$_2$O)$_2$. These formulas tell us that each formula unit of this salt contains one barium ion, two chloride ions, and two water molecules. A crystal of the salt contains many of these units in its crystalline lattice.

In naming hydrates, we first name the compound exclusive of the water and then add the term *hydrate*, with the proper prefix representing the number of water molecules in the formula. For example, BaCl$_2 \cdot$2H$_2$O is called *barium chloride dihydrate*. Hydrates are true compounds and follow the Law of Definite Composition. The gram-formula weight of BaCl$_2 \cdot$2H$_2$O is 244.3 g; it contains 56.20% barium, 29.06% chlorine, and 14.74% water.

water of hydration

Water in a hydrate is known as **water of hydration** or **water of crystallization**. Water molecules are bonded by electrostatic forces between polar water molecules and the positive or negative ions of the compound. These

232 Chapter 13

water of crystallization

forces are not as strong as covalent or ionic chemical bonds. As a result, water of crystallization may be removed by moderate heating of the crystal. A partially dehydrated or completely anhydrous compound may result. When $BaCl_2 \cdot 2H_2O$ is heated, it loses its water at about 100°C:

$$BaCl_2 \cdot 2H_2O \xrightarrow{100°C} BaCl_2 + 2H_2O\uparrow$$

When a solution of copper(II) sulfate ($CuSO_4$) is allowed to evaporate, beautiful blue crystals containing 5 moles of water per mole of $CuSO_4$ are formed. The formula for this hydrate is $CuSO_4 \cdot 5H_2O$; it is called *cupric sulfate pentahydrate*, or *copper(II) sulfate pentahydrate*. When $CuSO_4 \cdot 5H_2O$ is heated, water is lost and a pale green-white powder, anhydrous $CuSO_4$, is formed:

$$CuSO_4 \cdot 5H_2O \xrightarrow{250°C} CuSO_4 + 5H_2O\uparrow$$

When water is added to anhydrous copper(II) sulfate, the above reaction is reversed and the salt turns blue again. The formation of the hydrate is noticeably exothermic. Because of this outstanding color change, anhydrous copper(II) sulfate has been used as an indicator to detect small amounts of water.

The formula for plaster of paris is $(CaSO_4)_2 \cdot H_2O$. When mixed with the proper quantity of water, plaster of paris sets to a hard mass; it is therefore useful for making patterns for the reproduction of art objects, molds, and surgical casts. The chemical reaction is

$$(CaSO_4)_2 \cdot H_2O(s) + 3H_2O(l) \longrightarrow 2CaSO_4 \cdot 2H_2O(s)$$

The occurrence of hydrates is very commonplace in salts. Table 13.2 lists a number of common hydrates.

Table 13.2. Selected hydrates.

Hydrate	Name
$CaCl_2 \cdot 2H_2O$	Calcium chloride dihydrate
$Ba(OH)_2 \cdot 8H_2O$	Barium hydroxide octahydrate
$MgSO_4 \cdot 7H_2O$	Magnesium sulfate heptahydrate
$SnCl_2 \cdot 2H_2O$	Tin(II) chloride dihydrate
$CoCl_2 \cdot 6H_2O$	Cobalt(II) chloride hexahydrate
$Na_2CO_3 \cdot 10H_2O$	Sodium carbonate decahydrate
$(NH_4)_2C_2O_4 \cdot H_2O$	Ammonium oxalate monohydrate
$NaC_2H_3O_2 \cdot 3H_2O$	Sodium acetate trihydrate
$Na_2B_4O_7 \cdot 10H_2O$	Sodium tetraborate decahydrate
$K_4Fe(CN)_6 \cdot 3H_2O$	Potassium ferrocyanide trihydrate

13.7 Hygroscopic Substances; Deliquescence; Efflorescence

hygroscopic substances

Many anhydrous salts and other substances readily absorb water from the atmosphere. Such substances are said to be **hygroscopic**. This property can be observed in the following simple experiment: Spread a weighed 10–20 g sample of anhydrous copper(II) sulfate on a watch glass and set it aside so that the salt is exposed to the air. Then weigh the sample periodically for 24 hours, noting the increase in weight, the change in color, and the formation of the blue pentahydrate.

Some compounds continue to absorb water beyond the hydrate stage to form solutions. A substance that absorbs water from the air until it forms a solution is said to be **deliquescent**. A few granules of anhydrous calcium chloride or pellets of sodium hydroxide exposed to the air will appear moist in a few minutes, and within an hour will absorb enough water to form a puddle of solution. Phosphorus pentoxide (P_2O_5) picks up water so rapidly that it cannot be weighed accurately unless it is weighed in an anhydrous atmosphere.

deliquescence

Compounds that absorb water are very useful as drying agents. Refrigeration systems must be kept dry with such agents or the moisture will freeze and clog the tiny orifices in the mechanism. Bags of drying agents are often enclosed in packages containing iron or steel parts to absorb moisture and prevent rusting. Anhydrous calcium chloride, magnesium sulfate, sodium sulfate, calcium sulfate, silica gel, and phosphorus pentoxide are some of the compounds commonly used for drying liquids and gases containing small amounts of moisture.

The process by which crystalline materials spontaneously lose water when exposed to the air is known as **efflorescence**. Glauber's salt ($Na_2SO_4 \cdot 10H_2O$), a transparent crystalline salt, loses water when exposed to the air. One can actually observe these well-defined, large crystals crumbling away as they lose water, forming a white, noncrystalline-appearing powder. From our discussion of the decomposition of hydrates, we can predict that heat will increase the rate of efflorescence. The rate also depends on the concentration of moisture in the air. A dry atmosphere will allow the process to take place more rapidly.

efflorescence

13.8 Hydrogen Peroxide

Although both water and hydrogen peroxide are compounds of hydrogen and oxygen, their properties are very different. A hydrogen peroxide molecule (H_2O_2) is composed of two H atoms and two O atoms. Its composition by weight is 94.1% oxygen and 5.9% hydrogen. Pure hydrogen peroxide is a pale blue liquid that freezes at $-1.7°C$, boils at $151°C$, and has a density of 1.44 g/ml at $20°C$. It is miscible with water in all proportions. Water solutions of hydrogen peroxide are slightly acid. The structure of hydrogen peroxide may be represented as

$$\begin{array}{c} H \\ \ddot{\underset{\ddot{\cdot}}{O}}\!:\!\ddot{\underset{\ddot{\cdot}}{O}} \\ H \end{array} \quad \text{or} \quad \begin{array}{c} H \\ :\ddot{O}\!-\!\ddot{O}: \\ H \end{array}$$

Hydrogen peroxide is a common, useful source of oxygen, since it decomposes easily to give oxygen and water:

$$2\,H_2O_2(l) \longrightarrow 2\,H_2O(l) + O_2\!\uparrow + 46.0 \text{ kcal}$$

This decomposition is accelerated by heat and light, but may be minimized by storing peroxide solutions in brown bottles, keeping them cold, and adding stabilizers. The decomposition may also be accelerated by catalysts such as manganese dioxide.

The peroxide group, like an oxygen molecule (O_2), contains two oxygen atoms linked by a covalent bond. It has a -2 oxidation number and is written O_2^{2-} or $:\overset{..}{O}:\overset{..}{O}:^{2-}$; each O atom is considered to have an oxidation number of -1. Metal dioxides also contain two O atoms, but each is bonded individually to the metal ion. Thus, a metal dioxide contains two oxide ($:\overset{..}{O}:^{2-}$) ions.

Some discretion must be used when working with peroxide formulas. From their formulas, BaO_2 and TiO_2 appear to be similar compounds, but BaO_2 is a peroxide consisting of a $+2$ barium ion and a -2 peroxide ion whereas titanium dioxide is an oxide consisting of a $+4$ titanium ion and two -2 oxide ions. Peroxides may be distinguished from dioxides chemically because they generally yield H_2O_2 or O_2 when treated with acids or water. Two examples are the reactions of barium peroxide and sodium peroxide:

$$BaO_2(s) + H_2SO_4(aq) \longrightarrow H_2O_2(aq) + BaSO_4\downarrow$$
$$Na_2O_2(s) + 2\,H_2O(l) \longrightarrow H_2O_2(aq) + 2\,NaOH(aq)$$

A 3% solution of H_2O_2, which is commonly available at drug stores, is used as an antiseptic to cleanse open wounds. Somewhat stronger H_2O_2 solutions are widely used as bleaching agents for cotton, wood, and hair. For certain oxidation processes, the chemical industry uses a 30% solution. Concentrations of 85% and higher are used for oxidizing fuels in rocket propulsion. These highly concentrated solutions are extremely sensitive to decomposition and represent a fire hazard if allowed to come into contact with organic material.

Ozone is another compound containing multiple oxygen linkages. One molecule, O_3, contains three atoms of oxygen:

$:\overset{..}{O}::\overset{..}{O}:$ $\cdot\overset{..}{O}:\overset{\overset{..}{O}}{}:\overset{..}{O}\cdot$

Oxygen   Ozone

Ozone can be prepared by passing air or oxygen through an electrical discharge:

$$3\,O_2 \xrightarrow[\text{discharge}]{\text{Electrical}} 2\,O_3 \quad -68.4 \text{ kcal}$$

The characteristic pungent odor of ozone is noticeable in the vicinity of electrical machines and power transmission lines. Ozone is formed in the atmosphere during electrical storms and by the photochemical action of ultraviolet radiation on a mixture of nitrogen dioxide and oxygen. Ozone is not a desirable low altitude constituent of the atmosphere, since it is known to cause plant damage, cracking of rubber, and the formation of eye-irritating substances. However, in the stratosphere ozone interacts with ultraviolet radiation to form molecular and atomic oxygen and thus prevents most of this harmful radiation from reaching the earth's surface.

$$O_3 \xrightarrow[\text{radiation}]{\text{Ultraviolet}} O_2 + O$$

allotropy

Many elements exist in two or more molecular or crystalline forms. This phenomenon is known as **allotropy** (from the Greek *allotropia*, meaning "variety"). The individual forms of the element are known as allotropic forms or allotropes. Oxygen (O_2) and ozone (O_3) are allotropic forms of the element oxygen. Two other common elements that exhibit allotropy are sulfur and carbon. (Diamond and graphite are allotropic forms of carbon.)

13.9 Natural Waters

Natural fresh waters are not pure, but contain dissolved minerals, suspended matter, and sometimes harmful bacteria. The water supplies of large cities are usually drawn from rivers or lakes. Such water is generally unsafe to drink without treatment. To make such water potable (that is, safe to drink), it is treated by some or all of the following processes:

1. *Screening.* Removal of relatively large objects, such as trash, fish, and so on.
2. *Flocculation and sedimentation.* Chemicals, usually lime and alum (aluminum sulfate), are added to form a flocculent jellylike precipitate of aluminum hydroxide. This precipitate enmeshes most of the suspended fine matter in the water and carries it to the bottom of the sedimentation basin.
3. *Sand filtration.* Water is drawn from the top of the sedimentation basin and passed downward through fine sand filters. Nearly all the remaining suspended matter and bacteria are removed by the sand filters.
4. *Aeration.* Water is drawn from the bottom of the sand filters and is aerated by spraying. The purpose of this process is to remove objectionable odors and tastes.
5. *Disinfection.* In the final stage, chlorine gas is injected into the water to kill harmful bacteria before it is distributed to the public. Ozone is also used in some countries to disinfect water. In emergencies, water may be disinfected by simply boiling for a few minutes.

If the drinking water of children contains an optimum amount of fluoride ion, their teeth will be more resistant to decay. Therefore, in many communities NaF or Na_2SiF_6 is added to the water supply to bring the fluoride ion concentration up to the optimum level of about 1.0 part per million (ppm). Excessively high concentrations of fluoride ion can cause mottling of the teeth.

Water that contains dissolved calcium and magnesium salts is called *hard water*. One drawback of hard water is that ordinary soap does not lather well in it; the soap reacts with the calcium and magnesium ions to form an insoluble greasy scum. However, synthetic soaps, known as detergents or syndets, are available; they have excellent cleaning qualities and do not form precipitates with hard water. Hard water is also undesirable because it causes "boiler scale" to form on the walls of water heaters and steam boilers, which greatly reduces their efficiency.

Three techniques used to "soften" hard water are distillation, chemical precipitation, and ion exchange. In distillation, the water is boiled and the steam thus formed is then condensed to a liquid again, leaving the minerals behind in the distilling vessel. Figure 13.5 illustrates a simple laboratory distillation apparatus. Commercial stills are available that are capable of producing hundreds of litres of distilled water per hour.

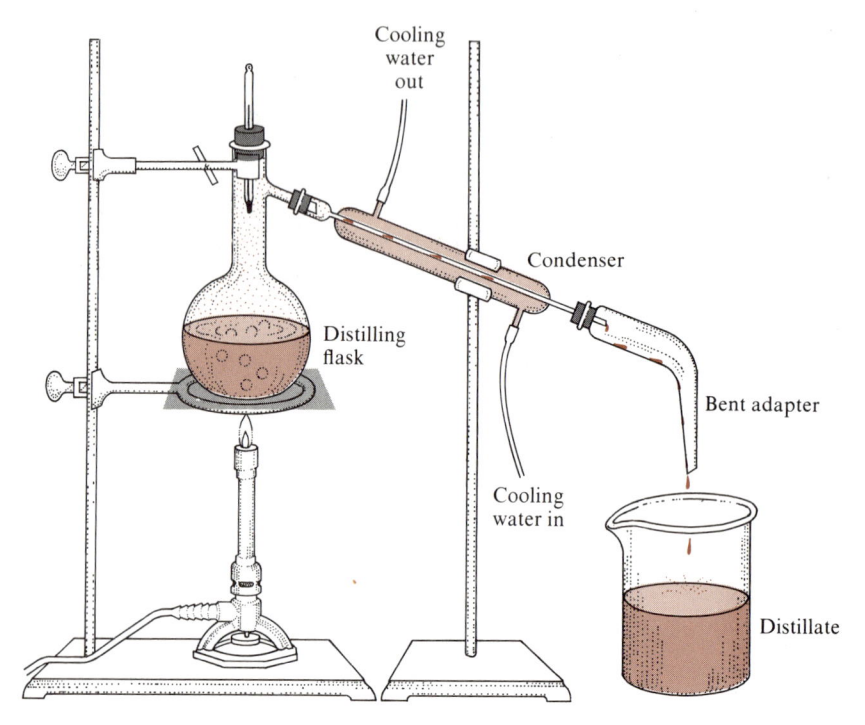

Figure 13.5. Simple laboratory setup for distillation of liquids.

Calcium and magnesium ions are precipitated from hard water by adding sodium carbonate and lime. Insoluble calcium carbonate and magnesium hydroxide are precipitated and are removed by filtration or sedimentation.

In the ion-exchange method, used in many households, hard water is effectively softened as it is passed through a bed or tank of zeolite. Zeolite is a complex sodium aluminum silicate. In this process, sodium ions replace objectionable calcium and magnesium ions, and the water is thereby softened:

$$Na_2Zeolite(s) + Ca^{2+}(aq) \longrightarrow CaZeolite(s) + 2\,Na^+(aq)$$

The zeolite is regenerated by back-flushing with concentrated sodium chloride solution, reversing the above reaction.

The sodium ions that are present in water softened either by chemical precipitation or by the zeolite process are not objectionable to most users of soft water.

In demineralization, both cations and anions are removed by a two-stage ion-exchange system. Special synthetic organic resins are used in the ion-exchange beds. In the first stage, metal cations are replaced by hydrogen ions. In the second stage, anions are replaced by hydroxide ions. The hydrogen and hydroxide ions react, and essentially pure, mineral-free water leaves the second stage (see Figure 13.6).

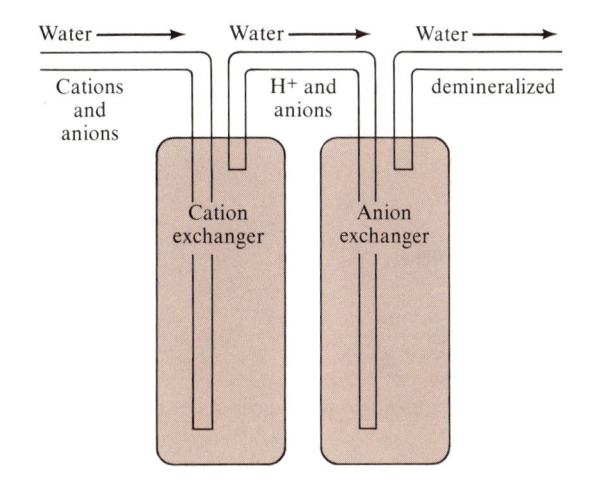

Figure 13.6. Demineralization of water: Water is passed through two beds of synthetic resin. In the cation exchanger, metal ions are exchanged for hydrogen ions. In the anion exchanger, anions are exchanged for hydroxide ions. The H^+ and OH^- ions react to form water, giving essentially pure, demineralized water.

The oceans represent an inexhaustible source of water; however, seawater contains about 3.5 lb of salts per 100 lb of water. This 35,000 parts per million (ppm) of dissolved salts makes seawater unfit for agricultural and domestic uses. Water that contains less than 1000 ppm of salts is considered reasonably good for drinking, and potable (fresh) water is already being obtained from the sea in many parts of the world. Continuous research is being done in an effort to make usable water from the oceans more abundant and economical.

13.10 Water Pollution

Polluted water was formerly thought of as water that was unclear, had a bad odor or taste, and contained disease-causing bacteria. However, such factors as increased population, industrial requirements for water, atmospheric pollution, and use of pesticides have greatly modified the problem of water pollution.

Many of the "newer" pollutants are not removed or destroyed by the usual water-treatment processes. For example, among the 66 organic compounds found in the drinking water of a major city on the Mississippi River, 3 are labelled slightly toxic, 17 moderately toxic, 15 very toxic, 1 extremely toxic, and 1 supertoxic. Two are known carcinogens (cancer-producing agents), 11 are suspect, and 3 are metabolized to carcinogens. The United States Public Health Service classifies water pollutants under eight broad categories. These are shown in Table 13.3.

Table 13.3. Classification of water pollutants.

Type of pollutant	Examples
Oxygen-demanding wastes	Decomposable organic wastes from domestic sewage and industrial wastes of plant and animal origin
Infectious agents	Bacteria, viruses, and other organisms from domestic sewage, animal wastes, and animal process wastes
Plant nutrients	Principally compounds of nitrogen and phosphorus
Organic chemicals	Large numbers of chemicals synthesized by industry; pesticides; chlorinated organic compounds
Other minerals and chemicals	Inorganic chemicals from industrial operations, mining, oil field operations, and agriculture
Radioactive substances	Waste products from mining and processing of radioactive materials, airborne radioactive fallout, increased use of radioactive materials in hospitals and research
Heat from industry	Large quantities of heated water returned to water bodies from power plants and manufacturing facilities after use for cooling purposes
Sediment from land erosion	Solid matter washed into streams and oceans by erosion, rain, and water runoff

Many outbreaks of disease or poisoning such as typhoid, dysentery, and cholera have been attributed directly to drinking water. Rivers and streams are a natural means for municipalities to dispose of their domestic and industrial waste products. Much of this water is used again by people downstream, and then discharged back into the water source. Then another community still farther downstream draws the same water and discharges its own wastes. Thus, along waterways such as the Mississippi and Delaware rivers, water is withdrawn and discharged many times. If this water is not properly treated, harmful pollutants will build up, causing epidemics of various diseases.

Mercury and its compounds have long been known to be highly toxic. Mercury gets into the body primarily through the foods we eat. Although it is not an essential mineral for the body, mercury accumulates in the blood, kidneys, liver, and brain tissues. Mercury in the brain causes serious damage to the central nervous system. The sequence of events that have led to incidents of mercury poisoning is as follows: Mercury and its compounds are used in many industries and in agriculture, primarily as a fungicide in the treatment of seeds. One of the largest uses is in the electrochemical conversion of sodium chloride brines to chlorine and sodium hydroxide, as represented by this equation:

$$2\,NaCl + 2\,H_2O \xrightarrow{Electrolysis} Cl_2 + 2\,NaOH + H_2$$

Although no mercury is shown in the chemical equation, it is used in the process for electrical contact, and small amounts are discharged along with spent brine solutions. Thus, considerable quantities of mercury, in low concentrations, have been discharged into lakes and other surface waters from the effluents of these manufacturing plants. The mercury compounds discharged into the water are converted by bacterial action and other organic compounds to methyl mercury, $(CH_3)_2Hg$, which then accumulates in the bodies of fish. Several major episodes of mercury poisoning that have occurred in the past years were the result of eating mercury-contaminated fish. The best way to control this contaminant is at the source, and much has been done since 1970 to eliminate the discharge of mercury in industrial wastes. In 1976, the Environmental Protection Agency banned the use of all mercury-containing insecticides and fungicides.

Many other major water pollutants have been recognized and steps have been taken to eliminate them. Three of these that pose serious problems are lead, detergents, and chlorine-containing organic compounds. Lead poisoning, for example, has been responsible for many deaths in past years. The toxic action of lead in the body is the inhibition of the enzyme necessary for the production of hemoglobin in the blood. The normal intake of lead into the body is through food. However, extraordinary amounts of lead can be ingested from water running through lead pipes and by using lead-containing ceramic containers for storage of food and beverages.

Keeping our lakes and rivers free from pollution is a very costly and complicated process. However, it has been clearly demonstrated that waterways rendered so polluted that the water is neither fit for human use nor able to sustain marine life can be successfully restored.

13.11 Evaporation

When beakers of water, ethyl ether, and ethyl alcohol are allowed to stand uncovered in an open room, the volumes of these liquids gradually decrease. The process by which this takes place is called *evaporation*.

Attractive forces exist between molecules in the liquid state. All these molecules, however, do not have the same kinetic energy. Molecules that have greater than average kinetic energy may overcome the attractive forces and break away from the surface of the liquid, flying off and becoming a gas. **Evaporation** is the escape of molecules from the liquid state to the gas or vapor state.

evaporation

In evaporation, molecules of higher than average kinetic energy escape from a liquid, leaving it cooler than it was before they escaped. For this reason, evaporation of perspiration is one way the human body cools itself and keeps its temperature constant. When volatile liquids such as ethyl chloride (C_2H_5Cl) are sprayed on the skin, they evaporate rapidly, cooling the area by removing heat. The numbing effect of the low temperature produced by evaporation of ethyl chloride allows it to be used as a local anesthetic for minor surgery.

240 Chapter 13

sublimation

Solids such as iodine, camphor, naphthalene (moth balls), and, to a small extent, even ice, will go directly from the solid to the gaseous state, bypassing the liquid state. This change is a form of evaporation and is called **sublimation**.

$$\text{Liquid} \xrightarrow{\text{Evaporation}} \text{Vapor}$$

$$\text{Solid} \xrightarrow{\text{Sublimation}} \text{Vapor}$$

13.12 Vapor Pressure

When a liquid evaporates in a closed system as shown in Figure 13.7, part (b), some of the molecules in the vapor or gaseous state strike the surface and return to the liquid state by the process of *condensation*. The rate of condensation increases until it is equal to the rate of evaporation. At this point, the space above the liquid is said to be saturated with vapor, and an equilibrium, or steady state, exists between the liquid and the vapor. The equilibrium equation is:

$$\text{Liquid} \xrightleftharpoons[\text{Condensation}]{\text{Evaporation}} \text{Vapor}$$

This equilibrium is dynamic; both processes—evaporation and condensation—are taking place, even though one cannot visually observe or measure a change. The number of molecules leaving the liquid in a given time interval is equal to the number of molecules returning to the liquid.

At the point of equilibrium, the molecules in the vapor exert a pressure like any other gas. The pressure exerted by a vapor in equilibrium with its

Figure 13.7. (a) Molecules in an open beaker may evaporate from the liquid and be dispersed into the atmosphere. Under this condition, evaporation will continue until all the liquid is gone. (b) Molecules leaving the liquid are confined to a limited space. With time, the concentration in the vapor phase will increase to a point at which an equilibrium between liquid and vapor is established.

vapor pressure

liquid is known as the **vapor pressure** of the liquid. The vapor pressure may be thought of as an internal pressure, a measure of the escaping tendency that molecules have to go from the liquid to the vapor state. The vapor pressure of a liquid is independent of the amount of liquid and vapor present, but it increases as the temperature rises (see Table 13.4).

Table 13.4. The vapor pressure of water, ethyl alcohol, and ethyl ether at various temperatures.

Temperature (°C)	Vapor pressure (mm Hg)		
	Water	Ethyl alcohol	Ethyl ether[a]
0	4.6	12.2	185.3
10	9.2	23.6	291.7
20	17.5	43.9	442.2
30	31.8	78.8	647.3
40	55.3	135.3	921.3
50	92.5	222.2	1276.8
60	152.9	352.7	1729.0
70	233.7	542.5	2296.0
80	355.1	812.6	2993.6
90	525.8	1187.1	3841.0
100	760.0	1693.3	4859.4
110	1074.6	2361.3	6070.1

[a] Note that the vapor pressure of ethyl ether at temperatures of 40°C and higher exceeds standard pressure, 760 mm Hg. This indicates that the substance has a low boiling point and therefore should be stored in a cool place in a tightly sealed container.

When equal volumes of water, ethyl ether, and ethyl alcohol are placed in separate beakers and allowed to evaporate at the same temperature, we observe that the ether evaporates faster than the alcohol, which in turn evaporates faster than the water. This order of evaporation is consistent with the fact that ether has a higher vapor pressure at any particular temperature than ethyl alcohol or water. One reason for this higher vapor pressure is that there is less attraction between ether molecules than there is between alcohol molecules or between water molecules. The vapor pressures of these three compounds at various temperatures are compared in Table 13.4.

volatile

Substances that evaporate readily are said to be **volatile**. A volatile liquid has a relatively high vapor pressure at room temperature. Ethyl ether is a very volatile liquid; water is not too volatile; mercury, which has a vapor pressure of 0.0012 mm Hg at 20°C, is essentially a nonvolatile liquid. Most substances that are normally solids are nonvolatile (solids that sublime are exceptions).

13.13 Boiling Point

The boiling temperature of a liquid is associated with its vapor pressure. We have seen that the vapor pressure increases as the temperature increases. When the internal or vapor pressure of a liquid becomes equal to the external pressure, the liquid boils. (By external pressure we mean the pressure of the atmosphere above the liquid.) The boiling temperature of a pure liquid remains constant as long as the external pressure does not vary.

The boiling point of water is 100°C. Table 13.4 shows that the vapor pressure of water at 100°C is 760 mm Hg, a figure we have seen many times before. The significant fact here is that the boiling point is the temperature at which the vapor pressure of the water or other liquid is equal to standard, or atmospheric, pressure at sea level. These relationships lead to the following definition: The **boiling point** is the temperature at which the vapor pressure of a liquid is equal to the external pressure above the liquid.

boiling point

We can readily see that a liquid has an infinite number of boiling points. When we give the boiling point of a liquid, we should also state the pressure. When we express the boiling point without stating the pressure, we mean it to be the **standard** or **normal boiling point** at standard pressure (760 mm Hg). Using Table 13.4 again, we see that the normal boiling point of ethyl ether is between 30°C and 40°C, and for ethyl alcohol it is between 70°C and 80°C, because, for each compound, 760 mm Hg pressure lies within these stated temperature ranges. At the normal boiling point, one gram of a liquid changing to a vapor (gas) absorbs an amount of energy equal to its heat of vaporization (see Table 13.5).

standard or normal boiling point

Table 13.5. Physical properties of ethyl chloride, ethyl ether, ethyl alcohol, and water.

	Boiling point (°C)	Melting point (°C)	Heat of vaporization (cal/g)	Heat of fusion (cal/g)
Ethyl chloride	13	−139	92.5	—
Ethyl ether	34.6	−116	83.9	—
Ethyl alcohol	78.4	−112	204.3	24.9
Water	100.0	0	540	80

The boiling point at various pressures may be evaluated by plotting the data of Table 13.4 on the graph in Figure 13.8, where temperature is plotted horizontally along the x axis and vapor pressure vertically along the y axis. The resulting curves are known as **vapor pressure curves**. Any point on these curves represents a vapor–liquid equilibrium at a particular temperature and pressure. We may find the boiling point at any pressure by tracing a horizontal line from the designated pressure to a point on the vapor pressure curve. From this point we draw a vertical line to obtain the boiling point on the temperature axis. Four such points are shown in Figure 13.8; they represent the normal boiling points of the four compounds at 760 mm Hg pressure.

vapor pressure curve

See if you can verify from the graph that the boiling points of ethyl chloride, ethyl ether, ethyl alcohol, and water at 600 mm Hg pressure are 8.5°C, 28°C, 73°C, and 93°C, respectively. By reversing this process, you can ascertain at what pressure a substance will boil at a specific temperature. The boiling point is one of the most commonly used physical properties for characterizing and identifying substances.

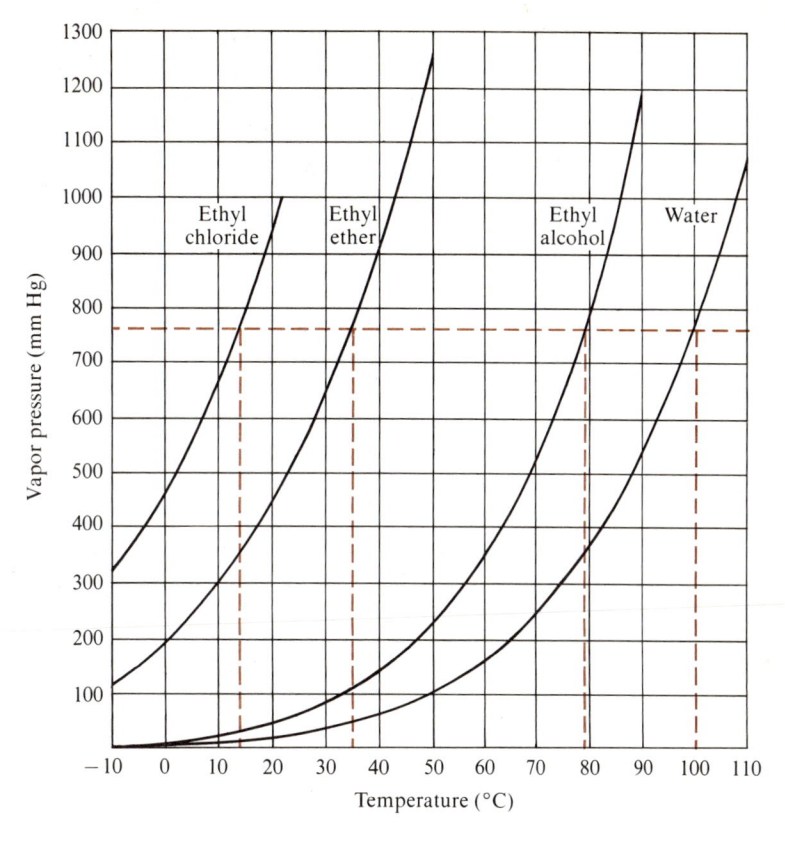

Figure 13.8. Vapor pressure–temperature curves for ethyl chloride, ethyl ether, ethyl alcohol, and water.

13.14 Freezing Point or Melting Point

freezing or melting point

As heat is removed from a liquid, the liquid becomes colder and colder, until a temperature is reached at which it begins to solidify. A liquid changing into a solid is said to be *freezing*, or *solidifying*. When a solid is heated continually, a temperature is reached at which the solid begins to liquefy. A solid changing into a liquid is said to be *melting*. The temperature at which the solid phase of a substance is in equilibrium with its liquid phase is known as the **freezing point** or **melting point** of that substance. The equilibrium equation is

$$\text{Solid} \underset{\text{Freezing}}{\overset{\text{Melting}}{\rightleftarrows}} \text{Liquid}$$

When a solid is slowly and carefully heated so that a solid–liquid equilibrium is maintained, the temperature will remain constant as long as both phases are present. One gram of a solid in changing into a liquid absorbs an amount of energy equal to its *heat of fusion* (see Table 13.5). The melting point is another physical property that is commonly used for characterizing substances.

The most common example of a solid–liquid equilibrium is ice and water (see Figure 13.1). In a well-stirred system of ice and water, the temperature remains at 0°C as long as both phases are present. The melting point is subject to changes in pressure, but is hardly affected unless the pressure change is very large.

It has been known for a long time that dissolved substances markedly decrease the freezing point of a liquid. For example, salt–water–ice equilibrium mixtures may be obtained at temperatures as low as −20°C, 20 degrees below the usual freezing point of water.

If, after all the solid has been melted, the liquid is heated continually, the temperature will rise until the liquid boils. The temperature will remain constant at the boiling point until all the liquid has boiled away. One gram of a liquid in changing into a gas at the normal boiling point absorbs an amount of energy equal to its *heat of vaporization* (see Table 13.5). The whole process of heating a substance (water) is illustrated graphically in Figure 13.9,

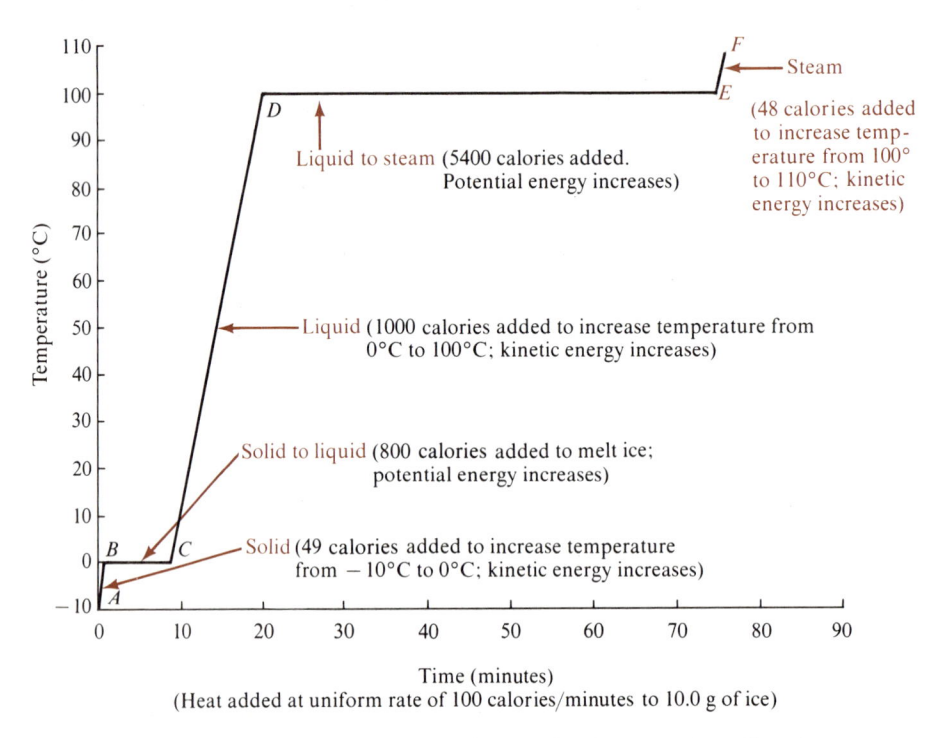

Time (minutes)
(Heat added at uniform rate of 100 calories/minutes to 10.0 g of ice)

Figure 13.9. Time–temperature phase diagram for the absorption of heat by a substance from the solid state to the gaseous state. Using water as an example, the interval *AB* represents the ice phase; *BC* interval, the melting of ice to water; *CD* interval, the elevation of the temperature of water from 0°C to 100°C; *DE* interval, the boiling of water to steam; and *EF* interval, the heating of steam.

where line *AB* represents the solid being heated and line *BC* represents the time during which the solid is melting and is in equilibrium with the liquid. Along line *CD* the liquid absorbs heat and finally, at point *D*, boils and continues to boil at a constant temperature (line *DE*). In the interval *EF*, all the water exists as steam and is being further heated or superheated.

Questions

A. Review the meanings of the new terms introduced in this chapter.
  1. Heat of fusion
  2. Heat of vaporization
  3. Hydrogen bond
  4. Basic anhydride
  5. Acid anhydride
  6. Hydrate
  7. Water of hydration
  8. Water of crystallization
  9. Hygroscopic substances
  10. Deliquescence
  11. Efflorescence
  12. Allotropy
  13. Evaporation
  14. Sublimation
  15. Vapor pressure
  16. Volatile
  17. Boiling point
  18. Standard or normal boiling point
  19. Vapor pressure curve
  20. Freezing or melting point

B. Answers to the following questions will be found in tables and figures.
  1. Compare the potential energy of the two states of water shown in Figure 13.1.
  2. In what state (solid, liquid, or gas) would H_2S, H_2Se, and H_2Te be at 0°C? (See Table 13.1.)
  3. The two thermometers in Figure 13.2 read 100°C. What is the pressure of the atmosphere?
  4. Draw a diagram of a water molecule and point out the areas that are the negative and positive ends of the dipole.
  5. Would the distillation setup in Figure 13.5 be satisfactory for separating salt and water? Ethyl alcohol and water? Explain.
  6. If the liquid in the flask in Figure 13.5 is ethyl alcohol and the atmospheric pressure is 543 mm Hg, what temperature would show on the thermometer?
  7. If water were placed in both containers in Figure 13.7, would they both have the same vapor pressure at the same temperature? Explain.
  8. In Figure 13.7, in which case, (a) or (b), will the atmosphere above the liquid reach a point of saturation?
  9. Suppose that a solution of ethyl ether and ethyl alcohol were placed in the closed bottle in Figure 13.7.
     (a) Would both substances be present in the vapor?
     (b) If the answer to part (a) is yes, which would have more molecules in the vapor?
  10. At approximately what temperature would each of the substances listed in Table 13.5 boil when the pressure is 30 mm Hg?
  11. Use the graph in Figure 13.9 to find the following:
     (a) The boiling point of water at 1000 mm Hg pressure
     (b) The normal boiling point of ethyl chloride
     (c) The boiling point of ethyl alcohol at 0.5 atm
  12. Consider Figure 13.9.
     (a) Why is line *BC* horizontal? What is happening in this interval?
     (b) What phases are present in the interval *BC*?

246 Chapter 13

(c) When heating is continued after point C, another horizontal line, DE, is reached at a higher temperature. What does this line represent?

C. Review questions.
1. List six physical properties of water.
2. What condition is necessary for water to have its maximum density? What is its maximum density?
3. Account for the fact that an ice–water mixture remains at 0°C until all the ice is melted, even though heat is applied to it.
4. Which contains less heat, ice at 0°C or water at 0°C? Explain.
5. Why does ice float in water? Would ice float in ethyl alcohol ($d = 0.79$ g/ml)? Explain.
6. If water molecules were linear instead of bent, would the heat of vaporization be higher or lower? Explain.
7. The heat of vaporization for ethyl ether is 83.9 cal/g and that for ethyl alcohol is 204.3 cal/g. Which of these compounds has hydrogen bonding? Explain.
8. Would there be more or less H-bonding if water molecules were linear instead of bent? Explain.
9. Which would show hydrogen bonding, ammonia (NH_3) or methane (CH_4)? Explain.
10. Hydrogen fluoride is believed to exist as H_6F_6 in the liquid state. Draw a possible structure for this molecule, illustrating the different bond formations. What would be its molecular weight?
11. Write equations to show how the following metals react with water: aluminum, calcium, iron, sodium, zinc. State the conditions for each reaction.
12. Is the formation of hydrogen and oxygen from water an exothermic or endothermic reaction? How do you know?
13. What is chlorine water and what chemical species are present in it?
14. (a) Write the formulas for the anhydrides of the following acids:
    H_2SO_3, H_2SO_4, HNO_3, $HClO_4$, H_2CO_3
    (b) Write the formulas for the anhydrides of the following bases:
    NaOH, KOH, $Ba(OH)_2$, $Ca(OH)_2$, $Mg(OH)_2$
15. Complete and balance the following equations:

    (a) $Mg(OH)_2 \xrightarrow{\Delta}$
    (b) $CH_3OH + O_2 \xrightarrow{\Delta}$
        Methyl alcohol
    (c) $Li + H_2O \longrightarrow$
    (d) $MgSO_4 \cdot 7H_2O \xrightarrow{\Delta}$
    (e) $HNO_3 + NaOH \longrightarrow$
    (f) $KOH \xrightarrow{\Delta}$
    (g) $Ba + H_2O \longrightarrow$
    (h) $Cl_2 + H_2O \longrightarrow$
    (i) $SO_3 + H_2O \longrightarrow$

16. Is the conversion of oxygen to ozone an exothermic or an endothermic reaction? How do you know?
17. Distinguish among an oxygen atom, an oxygen molecule, and an ozone molecule. How many electrons are in a peroxide ion?
18. How does ozone in the stratosphere protect the earth from excessive ultraviolet radiation?
19. Name each of the following hydrates:
    (a) $BaBr_2 \cdot 2 H_2O$ (c) $FePO_4 \cdot 4 H_2O$ (e) $FeSO_4 \cdot 7 H_2O$
    (b) $AlCl_3 \cdot 6 H_2O$ (d) $MgNH_4PO_4 \cdot 6 H_2O$ (f) $SnCl_4 \cdot 5 H_2O$
20. Explain how anhydrous copper(II) sulfate ($CuSO_4$) can act as an indicator for moisture.

Water and the Properties of Liquids 247

21. Compare the types of bonds in metal dioxides and metal peroxides.
22. Distinguish between deionized water and:
    (a) Hard water   (b) Soft water   (c) Distilled water
23. How can soap function to make soft water from hard water? What objections are there to using soap for this purpose?
24. What substance is commonly used to destroy bacteria in water?
25. What chemical, other than chlorine or chlorine compounds, can be used to disinfect water for domestic use?
26. Some organic pollutants in water can be oxidized by dissolved molecular oxygen. What harmful effect can result from this depletion of oxygen in the water?
27. Why should you not drink liquids that are stored in ceramic containers, especially unglazed ones?
28. Write the chemical equation showing how magnesium ions are removed by a zeolite water softener.
29. Write an equation to show how hard water containing calcium chloride ($CaCl_2$) is softened by using sodium carbonate (Na_2CO_3).
30. The vapor pressure at 20°C is given for the following compounds:

Methyl alcohol	96 mm Hg	Water	17.5 mm Hg
Acetic acid	11.7 mm Hg	Carbon tetrachloride	91 mm Hg
Benzene	74.7 mm Hg	Mercury	0.0012 mm Hg
Bromine	173 mm Hg	Toluene	23 mm Hg

    (a) Arrange these compounds in their order of increasing rate of evaporation.
    (b) Which substance listed would have the highest and which the lowest boiling point?
31. Explain why rubbing alcohol, warmed to body temperature, still feels cold when applied to your skin.
32. Suggest a method whereby water could be made to boil at 50°C.
33. If a dish of water initially at 20°C is placed in a living room maintained at 20°C, the water temperature will fall below 20°C. Explain.
34. Explain why a higher temperature is obtained in a pressure cooker than in an ordinary cooking pot.
35. What is the relationship between vapor pressure and boiling point?
36. From the point of view of the Kinetic-Molecular Theory, explain why vapor pressure increases with temperature.
37. Why does water have such a relatively high boiling point?
38. The boiling point of ammonia (NH_3) is −33.4°C and that of sulfur dioxide (SO_2) is −10.0°C. Which will have the higher vapor pressure at −40°C?
39. Explain what is occurring physically when a substance is boiling.
40. Explain why HF (bp 19.4°C) has a higher boiling point than HCl (bp −85°C), whereas F_2 (bp −188°C) has a lower boiling point than Cl_2 (bp −34°C).
41. Under which conditions are freezing point and melting point of a pure substance equal to each other?
42. Which of the following statements are correct?
    (a) The process of a substance changing directly from a solid to gas is called sublimation.
    (b) When water is decomposed, the volume ratio of H_2 to O_2 is 2:1, but the mass ratio of H_2 to O_2 is 1:8.
    (c) Hydrogen sulfide is a larger molecule than water.
    (d) The changing of ice into water is an exothermic process.
    (e) Water and hydrogen fluoride are both nonpolar molecules.
    (f) The main use of hydrogen peroxide is as an oxidizing agent.

248 Chapter 13

(g) $H_2O_2 \rightarrow 2\,H_2O + O_2$ represents a balanced equation for the decomposition of hydrogen peroxide.
(h) Steam at 100°C can cause more severe burns than liquid water at 100°C.
(i) The density of water is independent of temperature.
(j) Liquid A boils at a lower temperature than liquid B. This indicates that liquid A has a lower vapor pressure than liquid B at any particular temperature.
(k) Water boils at a higher temperature in the mountains than at sea level.
(l) No matter how much heat you put under an open pot of pure water on a stove, you cannot heat the water above its boiling point.
(m) The vapor pressure of a liquid at its boiling point is equal to the prevailing atmospheric pressure.
(n) The normal boiling temperature of water is 273°C.
(o) The amount of heat needed to change 1 mole of ice at 0°C to a liquid at 0°C is 1.44 kcal.

D. Review problems.
1. How many moles of water can be obtained from 100 g of each of these hydrates?
   (a) $CuSO_4 \cdot 5\,H_2O$ (b) $BaCl_2 \cdot 2\,H_2O$
2. When a person purchases washing soda ($Na_2CO_3 \cdot 10\,H_2O$) to use as a water softener, what percent of water is being bought?
3. How many calories are required to change 125 g of ice at 0°C to steam at 100°C?
4. How many calories of energy must be removed to change 75.0 g of water at 25°C to ice at 0°C?
5. The *molar heat of vaporization* is the number of calories required to change 1 mole of a substance at its boiling point to a vapor at the same temperature.
   (a) What is the molar heat of vaporization of water?
   (b) How many calories would be required to change 4.00 moles of water at 25°C to steam at 100°C?
6. Suppose 100 g of ice at 0°C is added to 250 g of water at 25°C. Is there sufficient ice to lower the temperature of the system to 0°C and still have ice remaining? Show evidence.
7. What weight of water must be decomposed to produce 50.0 litres of oxygen at STP?
8. (a) How many moles of oxygen can be obtained by decomposing 5.6 moles of hydrogen peroxide?
   (b) What volume will this oxygen occupy at 22°C and 650 mm Hg pressure?
9. How many litres of O_2 at STP can be obtained by decomposing 1.00 kg of 3.0% hydrogen peroxide?
10. Cadmium bromide ($CdBr_2$) forms a hydrate containing 20.9% water. Calculate the formula of this hydrate.
11. How many grams of $CuSO_4 \cdot 5\,H_2O$ need to be decomposed to obtain 22.0 ml of water ($d_{H_2O} = 1.00$ g/ml)?
12. What volume of hydrogen gas (at STP) can be obtained from 1000 g of aluminum reacting with steam? (See Section 13.5.)
13. How many grams of water will react with each of the following?
    (a) 1.00 mole of Na   (c) 1.00 g of Na
    (b) 1.00 mole of K   (d) 1.00 g of K
14. What is the pressure in a 1 litre vessel containing 0.50 g of water at 100°C?
15. Suppose one mole of water evaporates in one day. How many water molecules, on the average, leave the liquid each second?

16. A quantity of sulfuric acid is added to 100 ml of water. The final volume of the solution is 139 ml and it has a density of 1.38 g/ml. What weight of acid was added? Assume that the density of water is 1.00 g/ml.

E. Review exercises.
1. Can ice be colder than 0°C? Explain.
2. Why does a boiling liquid maintain a constant temperature when heat is continually being added?
3. At what temperature will copper have a vapor pressure of 760 mm Hg?
4. Why does a lake freeze from the top down?
5. What water temperature would you theoretically expect to find at the bottom of a very deep lake? Explain.
6. What reasons can you give for two compounds having approximately the same molecular weight but having very different boiling points?

14 Solutions

After studying Chapter 14 you should be able to:

1. Understand the terms listed under Question A at the end of the chapter.
2. Describe the different types of solutions that are possible based on the three states of matter.
3. List the general properties of solutions.
4. Outline the solubility rules for common mineral substances.
5. Describe and illustrate the process by which an ionic substance like sodium chloride dissolves in water.
6. Tell how temperature changes affect the solubilities of solids and gases in liquids.
7. Tell how changes of pressure affect the solubility of a gas in a liquid.
8. Identify and discuss the variables that affect the rate at which a solid dissolves in a liquid.
9. Determine by using a solubility graph or table whether a given solution is unsaturated, saturated, or supersaturated at a given temperature.
10. Calculate the weight percent or volume percent composition of a solution from appropriate data.
11. Calculate the amount of solute in a given quantity of a solution when given the weight percent or volume percent composition.
12. Calculate the molarity of a solution when given the volume of solution and moles of solute.
13. Calculate the weight of a substance needed to prepare a solution of specified volume and molarity.
14. Determine the resulting molarity when a given volume of a solution of known molarity is mixed with a specified volume of water or is mixed with a solution of different molarity.
15. From the equation for a reaction, relate the given weight, moles, solution volume, or gas volume of one substance to the corresponding quantities of any other substance appearing in the equation.
16. Understand the concepts of equivalent weight and normality and do calculations involving these concepts.
17. Relate the effect of a solute on the vapor pressure of a solvent to the freezing point and the boiling point of a solution.
18. Calculate the boiling point or freezing point of a solution from appropriate concentration data.

19. Calculate molality and molecular weight of a solute from boiling point or freezing point and weight concentration data.
20. Explain the phenomenon of osmosis.
21. Predict the direction of net solvent flow between solutions of known concentration separated by a semipermeable membrane.

14.1 Components of a Solution

solution

solute

solvent

The term **solution** is used in chemistry to describe a system in which one or more substances are homogeneously mixed or dissolved in another substance. A simple solution has two components, a solute and a solvent. The **solute** is the substance that is dissolved. The **solvent** is the dissolving agent and usually makes up the greater proportion of the solution. For example, when salt is dissolved in water to form a solution, salt is the solute and water is the solvent. Complex solutions containing more than one solute and/or more than one solvent are common.

14.2 Types of Solutions

From the three states of matter—solid, liquid, and gas—it is possible to have nine different types of solutions: solid dissolved in solid, solid dissolved in liquid, solid dissolved in gas, liquid dissolved in liquid, and so on. Of these, the most common solutions are solid dissolved in a liquid, liquid dissolved in liquid, gas dissolved in a liquid, and gas dissolved in a gas.

14.3 General Properties of Solutions

A true solution is one in which the dissolved solute is molecular or ionic in size, generally in the range of 1 to 10 Angstrom units (10^{-8} to 10^{-7} cm). The properties of a true solution are as follows:

1. It is a homogeneous mixture of two or more substances, solute and solvent.
2. It has a variable composition.
3. The dissolved solute is either molecular or ionic in size.
4. It may be either colored or colorless but is usually transparent.
5. The solute remains uniformly distributed throughout the solution and will not settle out with time.
6. The solute generally may be separated from the solvent by purely physical means (for example, by evaporation).

These properties are illustrated by water solutions of sugar and of potassium permanganate. Suppose that we prepare two sugar solutions, the first containing 10 g of sugar added to 100 ml of water and the second containing 20 g of sugar added to 100 ml of water. Each solution is stirred until all the solute dissolves, demonstrating that we may vary the composition of a solution. Every portion of the solution has the same sweet taste because the sugar molecules are uniformly distributed throughout. If confined so that no solvent is lost, the solution will taste and appear the same a week or a month later. The properties of the solution are unaltered after the solution is passed through filter paper. But by carefully evaporating the water, we may recover the sugar from the solution.

To observe the dissolving of potassium permanganate ($KMnO_4$) we affix a few crystals of $KMnO_4$ to paraffin wax or rubber cement at the end of a glass rod and submerge the entire rod, with the wax–permanganate end up, in a cylinder of water. Almost at once the beautiful purple color of dissolved permanganate ions (MnO_4^-) appears at the top of the rod and streams to the bottom of the cylinder as the crystals dissolve. The purple color at first is mostly at the bottom of the cylinder because potassium permanganate is denser than water. But after a while, the purple color disperses until it is evenly distributed throughout the solution. This demonstrates that molecules and ions move about freely and spontaneously (diffuse) in a liquid or solution. Once the solution is formed, it is permanent; the solute does not settle out.

The permanency of a solution is explained by the Kinetic-Molecular Theory (KMT). All matter, according to the KMT, is in some kind of motion at all temperatures above absolute zero, and the intensity of motion increases with increasing temperature. Once prepared, a solution remains homogeneous because of this constant random or thermal motion of the solute and solvent particles. This constant random motion is responsible for diffusion in liquids and gases.

14.4 Solubility

solubility

We use the term **solubility** to describe the amount of one substance that will dissolve in another. For example, 36.0 g of sodium chloride (NaCl) will dissolve in 100 g of water at 20°C. We say, then, that the solubility of NaCl in water is 36.0 g per 100 g of water at 20°C.

Solubility is often used in a relative way. We say that a substance is very soluble, moderately soluble, slightly soluble, or insoluble. Although these terms do not accurately indicate how much solute will dissolve, they are frequently used to describe the solubility of a substance qualitatively.

miscible

immiscible

Two other terms often used to describe solubility are miscible and immiscible. Liquids that are capable of mixing and forming a solution are **miscible**; those that do not form solutions or are generally insoluble in each other are **immiscible**. Methyl alcohol and water are miscible in each other in all proportions. Carbon tetrachloride and water are immiscible, forming two separate layers when they are mixed. Miscible and immiscible systems are illustrated in Figure 14.1.

The general rules for the solubility of common mineral substances are given in Table 14.1. The solubility of over 200 compounds is given in the

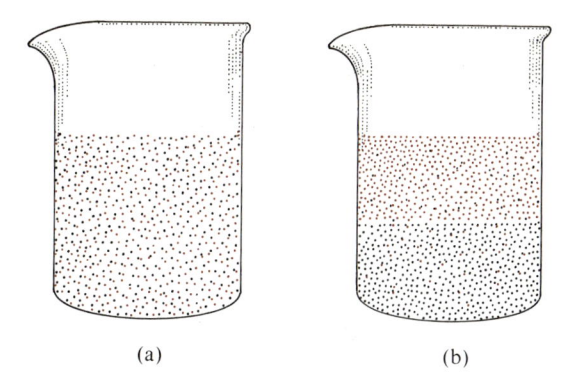

Figure 14.1. Miscible and immiscible systems: (a) miscible: H_2O and CH_3OH; (b) immiscible: H_2O and CCl_4. In a miscible system, a solution is formed, consisting of a single phase with the solute and solvent uniformly dispersed. An immiscible system is heterogeneous, and, in the case of two liquids, forms two liquid layers.

Table 14.1. General solubility rules for common mineral substances.[a]

Class	Solubility in cold water
Nitrates	All nitrates are soluble.
Acetates	All acetates are soluble.
Chlorides, Bromides, Iodides	All chlorides, bromides, and iodides are soluble except those of Ag, Hg(I), and Pb(II); $PbCl_2$ and $PbBr_2$ are slightly soluble in hot water.
Sulfates	All sulfates are soluble except those of Ba, Sr, and Pb; Ca and Ag sulfates are slightly soluble.
Carbonates, Phosphates	All carbonates and phosphates are insoluble except those of Na, K, and NH_4^+. Many bicarbonates and acid phosphates are soluble.
Hydroxides	All hydroxides are insoluble except those of the alkali metals and NH_4OH; $Ba(OH)_2$ and $Ca(OH)_2$ are slightly soluble.
Sodium salts, Potassium salts, Ammonium salts	All common salts of these ions are soluble.
Sulfides	All sulfides are insoluble except those of the alkali metals, ammonium, and the alkaline earth metals (Ca, Mg, Ba).

[a] When we say a substance is soluble, we mean that the substance is reasonably soluble. All substances have some solubility in water, although the amount of solubility may be very small; the solubility of silver iodide, for example, is about 1×10^{-8} mole AgI/litre H_2O.

254   Chapter 14

Solubility Table in Appendix IV. Solubility data for thousands of compounds may be found by consulting standard reference sources.[1]

concentration of a solution

The quantitative expression of the amount of dissolved solute in a particular quantity of solvent is known as the **concentration of a solution**. Several methods of expressing concentration will be described in Section 14.7.

14.5 Factors Related to Solubility

The entire concept of predicting solubilities is, at best, very complex and difficult. There are many variables, such as size of ions, charge on ions, interaction between ions, interaction between solute and solvent, and temperature, all of which bear upon the problem. Because of the factors involved, there are many exceptions to the general rules of solubility given in Table 14.1. However, the rules are very useful, because they do apply to a good many of the more common compounds that we encounter in the study of chemistry. Keep in mind that these are rules, not laws, and are therefore subject to exceptions. Fortunately, the solubility of a solute is relatively easy to determine experimentally. Four factors related to solubility are discussed below.

1. *The nature of the solute and solvent.*   The old adage that "like dissolves like" has merit, in a general way. Polar substances tend to be more miscible, or soluble, with other polar substances. Nonpolar substances tend to be miscible with other nonpolar substances and less miscible with polar substances. Thus, mineral acids, bases, and salts, which are polar, tend to be much more soluble in water, which is polar, than in solvents such as ether, carbon tetrachloride, or benzene, which are essentially nonpolar. Sodium chloride, a very polar substance, is soluble in water, slightly soluble in ethyl alcohol (less polar than water), and insoluble in ether and benzene. Pentane (C_5H_{12}), a nonpolar substance, is only slightly soluble in water but is very soluble in benzene and ether.

At the molecular level, the formation of a solution from two nonpolar substances, such as carbon tetrachloride and benzene, can be visualized as a process of simple mixing. The nonpolar molecules, having little tendency to either attract or repel one another, easily intermingle to form a homogeneous mixture.

Solution formation between polar substances is much more complex. For example, the process by which sodium chloride dissolves in water is illustrated in Figure 14.2. Water molecules are very polar and are attracted to other polar molecules or ions. When salt crystals (NaCl) are put into water, polar water molecules become attracted to the sodium and chloride ions on the crystal surfaces and weaken the attraction between Na^+ and Cl^- ions. The positive end of

---

[1]Two commonly used handbooks are *Lange's Handbook of Chemistry*, 12th ed. (New York: McGraw-Hill, 1979), and *Handbook of Chemistry and Physics*, 61st ed. (Cleveland; Chemical Rubber Co., 1980).

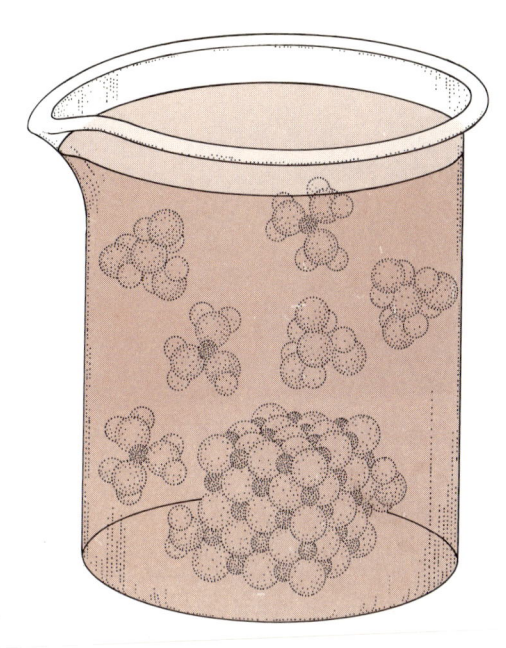

Figure 14.2. Dissolution of sodium chloride in water. Polar water molecules are attracted to Na$^+$ and Cl$^-$ ions in the salt crystal, weakening the attraction between the ions. As the attraction between the ions weakens, the ions move apart and become surrounded by water dipoles. The hydrated ions slowly diffuse away from the crystal to become dissolved in solution.

the water dipole is attracted to the Cl$^-$ ions, and the negative end of the water dipole to the Na$^+$ ions. The weaker attraction permits the ions to move apart, making room for more water dipoles. Thus, the surface ions are surrounded by water molecules, becoming hydrated ions, Na$^+$(aq) and Cl$^-$(aq), and slowly diffuse away from the crystals as dissolved ions in solution.

$$\text{NaCl (crystal)} \xrightarrow{\text{H}_2\text{O}} \text{Na}^+(aq) + \text{Cl}^-(aq)$$

Examination of the data in Table 14.2 reveals some of the complex questions relating to solubility. For example, some questions that may arise on examining the table are: Why are lithium halides, except for lithium fluoride (LiF), more soluble than sodium and potassium halides? Why, indeed, are the solubilities of LiF and sodium fluoride (NaF) so low in comparison to those of the other salts? Why does not the solubility of LiF, NaF, and NaCl increase proportionately with temperature, as the solubilities of the other salts do? Sodium chloride is appreciably soluble in water, but is insoluble in concentrated hydrochloric acid (HCl) solution. On the other hand, LiF and NaF are not very soluble in water but are quite soluble in hydrofluoric acid (HF) solution—why? These questions will not be answered directly here, but it is hoped that your curiosity will be aroused to the point that you will do some reading and research on the properties of solutions.

Table 14.2. Solubility of alkali metal halides in water.

Salt	Solubility (g salt/100 g H$_2$O) 0°C	100°C
LiF	0.12	0.14 (at 35°C)
LiCl	67	127.5
LiBr	143	266
LiI	151	481
NaF	4	5
NaCl	35.7	39.8
NaBr	79.5	121
NaI	158.7	302
KF	92.3 (at 18°C)	Very soluble
KCl	27.6	57.6
KBr	53.5	104
KI	127.5	208

2. **The effect of temperature on solubility.** Most solutes have a limited solubility in a specific amount of solvent at a fixed temperature. The temperature of the solvent has a marked effect on the amount of solute that will dissolve. For *most solids* dissolved in a liquid, an increase in temperature results in an increase in solubility (see Figure 14.3). However, the solubility of a *gas* in a liquid always

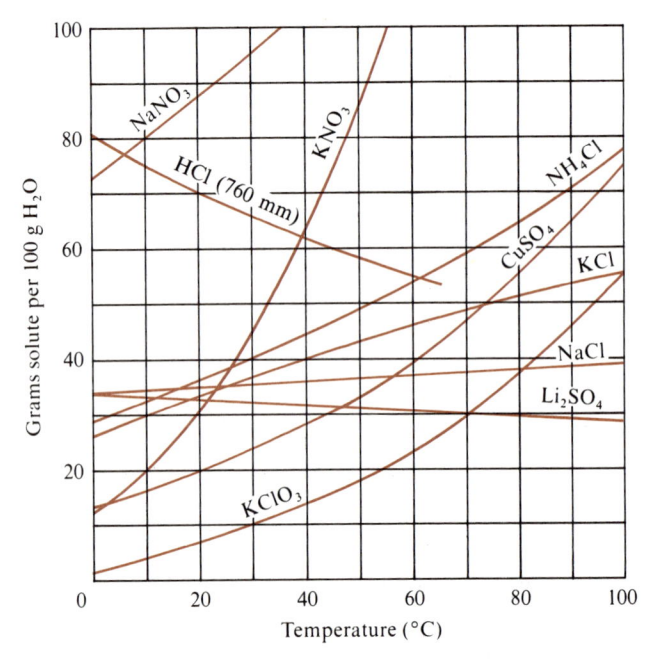

Figure 14.3. Solubility of various compounds in water.

decreases as the temperature *increases*. This is explained in terms of the KMT by assuming that in order to dissolve, the gas molecules must form "bonds" of some sort with the molecules of the liquid. An increase in temperature decreases the solubility of the gas because it increases the kinetic energy (speed) of the gas molecules and thereby decreases their ability to form "bonds" with the liquid molecules.

3. The effect of pressure on solubility. Small changes in pressure have little effect on the solubility of solids in liquids but have a marked effect on the solubility of gases in liquids. The solubility of a gas in a liquid is directly proportional to the pressure of that gas above the solution. Thus, the amount of a gas that is dissolved in solution will double if the pressure of that gas over the solution is doubled. For example, carbonated beverages contain dissolved carbon dioxide at pressures greater than atmospheric pressure. When a bottle of carbonated soda is opened, the pressure is immediately reduced to the atmospheric pressure, and the excess dissolved carbon dioxide bubbles out of the solution.

4. Rate of dissolving. The rate at which a solid solute dissolves is affected by (a) particle size of the solute, (b) temperature, (c) agitation or stirring, and (d) concentration of the solution.

   (a) *Particle size.* A solid can dissolve only at the surface that is in contact with the solvent. Because the surface to volume ratio increases as size decreases, smaller crystals dissolve faster than large ones. For example, if a salt crystal 1 cm on a side (6 cm^2 surface area) is divided into 1000 cubes, each 0.1 cm on a side, the total surface of the smaller cubes is 60 cm^2—a tenfold increase in surface area (see Figure 14.4).
   (b) *Temperature.* In most cases, the rate of dissolving of a solid increases with temperature. This is because of kinetic effects. The solvent molecules, moving more rapidly at higher temperatures, strike the solid surfaces more often and harder causing the rate of dissolving to increase.

0.1 cm cube
Area = 0.06 cm^2
1000 × 0.06 cm^2 = 60.0 cm^2

1 cm

1 cm

Surface area of
this cube is
6 × 1 cm^2 = 6 cm^2

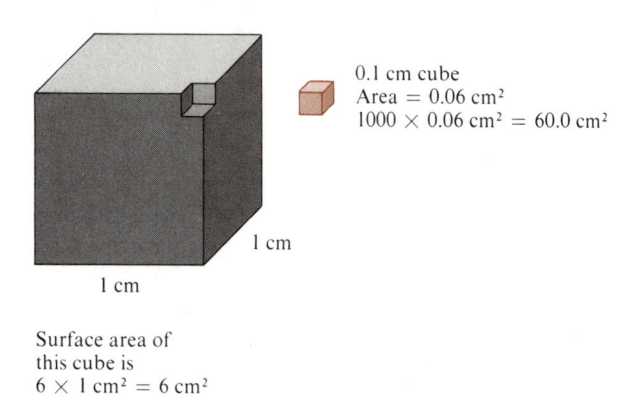

Figure 14.4. Surface area of crystals: A crystal 1 cm on a side has a surface area of 6 cm^2. Subdivided into 1000 smaller crystals, each with 0.1 cm on a side, the total surface area is increased to 60.0 cm^2.

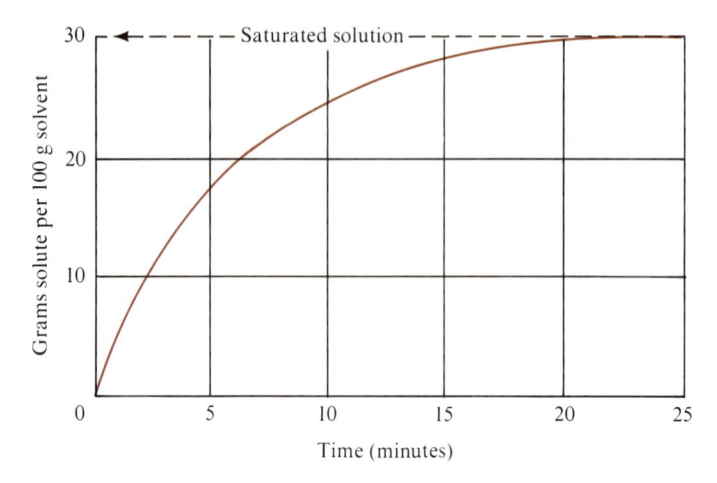

Figure 14.5. Rate of dissolution of a solid solute in a solvent. The rate is maximum at the beginning and decreases as the concentration approaches the saturation point.

(c) *Agitation or stirring.* The effect of agitation or stirring is kinetic. When a solid is first put into water, the only solvent with which it comes in contact is in the immediate vicinity. As the solid dissolves, the amount of dissolved solute around the solid becomes more and more concentrated and the rate of dissolving slows down. If the mixture is not stirred, the dissolved solute diffuses very slowly throughout the entire solution; weeks may pass before the solid is entirely dissolved. Through stirring, the dissolved solute is distributed rapidly throughout the solution and more solvent is brought into contact with the solid, causing it to dissolve more rapidly.

(d) *Concentration of the solution.* When the solute and solvent are first mixed, the rate of dissolving is at its maximum. As the concentration of the solution increases and the solvent becomes more nearly saturated with the solute, the rate of dissolving decreases greatly. The rate of dissolving is pictured graphically in Figure 14.5. Note that about 17 g dissolve in the first 5 minute interval but only about 1 g dissolves in the fourth 5 minute interval. Although different solutes show different rates, the rate of dissolving always becomes very slow as the concentration approaches the saturation point.

14.6 Solutions: A Reaction Zone

Many solids must be put in solution in order to undergo appreciable chemical reactions. We can easily write the equation for the double-replacement reaction between sodium chloride and silver nitrate:

$$NaCl + AgNO_3 \longrightarrow AgCl + NaNO_3$$

But suppose we mix solid NaCl and solid AgNO$_3$ and look for a chemical change. Some reaction may occur, but if it does, it is quite slow and essentially undetectable. In fact, the crystalline structures of NaCl and AgNO$_3$ are so different that we can separate them by tediously picking out the two different kinds of crystals from the mixture. But if we dissolve the sodium chloride and silver nitrate separately in water and mix the two solutions, we observe the immediate formation of a white, curdy precipitate of silver chloride.

Molecules or ions must come into intimate contact or collide with one another in order to react. In the foregoing example, the two solids did not react because the ions were securely locked within the crystal structures. But when the sodium chloride and silver nitrate are dissolved, their crystal lattices are broken down and the ions become mobile. When the two solutions are mixed, the mobile Ag^+ and Cl^- ions come into contact and react to form insoluble AgCl, which precipitates out of solution. The soluble Na^+ and NO_3^- ions remain in solution but form the crystalline salt $NaNO_3$ when the water is evaporated:

$$NaCl(aq) + AgNO_3(aq) \longrightarrow AgCl\downarrow + NaNO_3(aq)$$

$$(Na^+ + Cl^-) + (Ag^+ + NO_3^-) \xrightarrow{H_2O} AgCl\downarrow + Na^+ + NO_3^-$$

Sodium chloride solution · Silver nitrate solution · Silver chloride · Sodium nitrate in solution

The mixture of the two solutions provided a zone or space in which the Ag^+ and Cl^- ions could react. (See Chapter 15 for further discussion of ionic reactions.)

Solutions also function as diluents in reactions in which the undiluted reactants would combine with each other too violently. Moreover, a solution of known concentration provides a convenient method for delivering specific amounts of reagents.

14.7 Concentration of Solutions

The concentration of a solution gives us information concerning the amount of solute dissolved in a unit volume of solution. Because reactions are often conducted in solution, it is important to understand the methods of expressing concentration and to know how to prepare solutions of particular concentrations.

1. **Dilute and concentrated solutions.** When we say that a solution is *dilute* or *concentrated*, we are expressing, in a relative way, the amount of solute present. One gram of salt and 2 g of salt in solution are both dilute solutions when compared to the same volume of a solution containing 20 g of salt. Ordinary concentrated hydrochloric acid (HCl) contains 12 moles of HCl per litre of solution. In some laboratories, the dilute acid is made by mixing equal volumes of water and the concentrated acid. In other laboratories, the concentrated acid is diluted with two or three volumes of water, depending on its use. The term **dilute solution**, then, describes a solution that contains a relatively small amount of dissolved solute. Conversely, a **concentrated solution** contains a relatively large amount of dissolved solute.

dilute solution

concentrated solution

2. Saturated, unsaturated, and supersaturated solutions. At a specific temperature there is a limit to the amount of solute that will dissolve in a given amount of solvent. When this limit is reached, the resulting solution is said to be *saturated*. For example, when we put 40.0 g KCl into 100 g H_2O at 20°C, we find that 34.0 g KCl dissolve and 6.0 g KCl remain undissolved. The solution formed is a saturated solution of KCl.

Two processes are occurring simultaneously in a saturated solution. The solid is dissolving into solution and, at the same time, the dissolved solute is crystallizing out of solution. This may be expressed as

$$\text{Solute (undissolved)} \rightleftharpoons \text{Solute (dissolved)}$$

When these two opposing processes are occurring at the same rate, the amount of solute in solution is constant and a condition of equilibrium is established between dissolved and undissolved solute. A **saturated solution** contains dissolved solute in equilibrium with undissolved solute.

saturated solution

It is especially important to state the temperature of a saturated solution. A solution that is saturated at one temperature may not be saturated at another. If the temperature of a saturated solution is changed, the equilibrium is disturbed, and the amount of dissolved solute will change to reestablish the equilibrium.

A saturated solution may be either dilute or concentrated, depending on the solubility of the solute. A saturated solution can be conveniently prepared by dissolving the solute at a temperature somewhat higher than room temperature. The amount of solute in solution should be in excess of its solubility at room temperature. When the solution cools, the excess solute crystallizes, leaving the solution saturated. In this case, the solute must be more soluble at higher temperatures and must not form a supersaturated solution. Examples expressing the solubility of saturated solutions at two different temperatures are given in Table 14.3.

unsaturated solution

A solution containing less solute per unit of volume than does its corresponding saturated solution is said to be **unsaturated**. In other words, more solute can be dissolved into an unsaturated solution without altering other conditions. Consider a solution made from 40 g of KCl and 100 g of water

Table 14.3. Saturated solutions at 20°C and 50°C.

Solute	Solubility (g solute/100 g H_2O)	
	20°C	50°C
NaCl	36.0	37.0
KCl	34.0	42.6
$NaNO_3$	88.0	114.0
$KClO_3$	7.4	19.3
$AgNO_3$	222.0	455.0
$C_{12}H_{22}O_{11}$	203.9	260.4

at 20°C (see Table 14.3). The solution formed will certainly be saturated and will contain about 6 g of undissolved salt, because the maximum amount of KCl that can dissolve in 100 g of water at 20°C is 34 g. If the solution is now heated and maintained at 50°C, all the salt will dissolve and, in fact, more can be dissolved. Thus, the solution at 50°C is unsaturated.

In some instances, solutions can be prepared that contain more solute than that of the saturated solution at a particular temperature. Such solutions are said to be **supersaturated**. However, we must qualify this definition by noting that a supersaturated solution is unstable. Disturbances such as jarring, stirring, scratching the walls of the container, or dropping in a "seed" crystal cause the supersaturation to break. When a supersaturated solution is disturbed, the excess solute crystallizes out rapidly, returning the solution to a saturated state.

supersaturated solution

Supersaturated solutions, while not easy to prepare, may be formed from selected substances by dissolving, in warm solvent, an amount of solute greater than that needed for a saturated solution at room temperature. The warm solution is then allowed to cool very slowly. With the proper solute and careful work, a supersaturated solution will result. Two substances commonly used to demonstrate this property are sodium thiosulfate pentahydrate, $Na_2S_2O_3 \cdot 5H_2O$, and sodium sulfate, Na_2SO_4 (from a saturated solution at 30°C).

3. Weight percent solution. This expression of concentration gives the percentage of solute by weight in a solution. It says that for a given weight of solution, a certain percentage of that weight is solute. Suppose that we take a bottle from the reagent shelf that reads "Sodium hydroxide, NaOH, 10%" (see Figure 14.6). This means that for every 100 g of this solution we use, 10 g will be NaOH and 90 g will be water. (Note that this is 100 g and not 100 ml

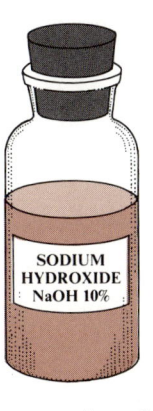

Figure 14.6. Weight percent concentration of solutions. The bottle contains 10 g of NaOH per 90 g of H_2O.

of solution.) We could also make this same concentration of solution by dissolving 2 g of NaOH in 18 grams of water. Weight percent concentrations are most generally used for solids dissolved in liquids.

$$\text{Weight percent} = \frac{\text{g solute}}{\text{g solute} + \text{g solvent}} \times 100\%$$

Illustrative problems follow.

Problem 14.1 What is the weight percent of sodium hydroxide in a solution that is made by dissolving 8.00 g of NaOH in 50.0 g of H_2O?

Grams of solute (NaOH) = 8.00 g

Grams of solvent (H_2O) = 50.0 g

$$\frac{8.00 \text{ g NaOH}}{8.00 \text{ g NaOH} + 50.0 \text{ g } H_2O} \times 100\% = 13.8\% \text{ NaOH solution}$$

Problem 14.2 What weights of potassium chloride (KCl) and water are needed to make 250 g of 5.00% solution?

The percentage expresses the weight of the solute.

250 g = Total weight of solution

5.00% of 250 g = 0.0500 × 250 g = 12.5 g KCl (solute)

250 g − 12.5 g = 237.5 g H_2O

Dissolving 12.5 g KCl in 237.5 g of H_2O gives a 5.00% KCl solution.

Problem 14.3 Suppose that 2.50 ml of 20.0% silver nitrate solution (d = 1.19 g/ml) was used to precipitate the chloride in a sample of salt water. What weight of $AgNO_3$ was used?

In this problem, a volume of solution is used. First, the weight of the solution is calculated from the volume and the density. Then, the weight of $AgNO_3$ can be determined from the weight percent.

$$d = \frac{\text{Mass}}{\text{Volume}} = \frac{g}{ml}$$

So,

$$g = ml \times d = 2.50 \text{ ml} \times \frac{1.19 \text{ g}}{ml} = 2.98 \text{ g}$$

Weight of solution = 2.98 g of 20.0% $AgNO_3$

Taking 20.0% of 2.98 g gives us the weight of $AgNO_3$ used:

2.98 g $AgNO_3$ × 0.200 = 0.596 g $AgNO_3$ (Answer)

When the concentration is given in percentage, it is assumed to be weight percent unless otherwise stated. The student should also be aware that the concentration expressed as weight percent is independent of the formula of the solute.

4. Volume percent. Solutions that are formulated from two liquids are often expressed as *volume percent* with respect to the solute. The label on a bottle of ordinary rubbing alcohol reads "Isopropyl alcohol, 70% by volume." Such a solution could be made by mixing 70 ml of alcohol and 30 ml of water. If we assume that these volumes are additive (which they are not, exactly), 1 litre of 70% isopropyl alcohol by volume will contain 700 ml of the alcohol.

$$\text{Volume percent} = \frac{\text{Volume of liquid in question}}{\text{Total volume of solution}} \times 100\%$$

5. Molarity. Weight percent solutions do not equate or express the number of formula or molecular weights of the solute in solution. For example, 1000 g of 10% NaOH solution contain 100 g of NaOH; 1000 g of 10% KOH solution contain 100 g of KOH. In terms of moles of NaOH and KOH, these solutions contain

$$\text{moles NaOH} = 100 \text{ g NaOH} \times \frac{1 \text{ mole NaOH}}{40.0 \text{ g NaOH}} = 2.50 \text{ moles NaOH}$$

$$\text{moles KOH} = 100 \text{ g KOH} \times \frac{1 \text{ mole KOH}}{56.1 \text{ g KOH}} = 1.78 \text{ moles KOH}$$

From the above figures, we see that the two 10% solutions do not contain the same number of moles of NaOH and KOH. Yet one mole of each of these two bases will neutralize the same amount of acid. As a result, we find that a 10% NaOH solution has more reactive alkali than a 10% KOH solution.

We need a method expressing concentration that will easily identify how many moles or formula weights of solute are present per unit of volume of solution. For this purpose, the molar method of expressing concentration is used.

1 molar solution

A **1 molar solution** contains 1 mole, or 1 gram-molecular weight, or 1 gram-formula weight of solute per litre of solution. For example, to make a 1 molar solution of sodium hydroxide (NaOH), we dissolve 40.0 g of NaOH (1 mole) in water and dilute the solution with more water to a volume of 1 litre. The solution contains 1 mole of the solute in 1 litre of solution and is said to be 1 molar (1 M) in concentration. Figure 14.7 illustrates the preparation of a 1 molar solution. Note that the volume of the solute and the solvent together is 1 litre.

molarity

The concentration of a solution may, of course, be varied by using more or less solute or solvent; but in any case, the **molarity** of a solution is the number of moles of solute per litre of solution. A capital M is the abbreviation for molarity. The units of molarity are moles per litre. The expression "2.0 M NaOH" means a 2.0 molar solution of NaOH (2.0 moles, or 80 g, of NaOH dissolved in 1 litre of solution).

$$\text{Molarity} = M = \frac{\text{number of moles of solute}}{\text{litre of solution}} = \frac{\text{moles}}{\text{litre}}$$

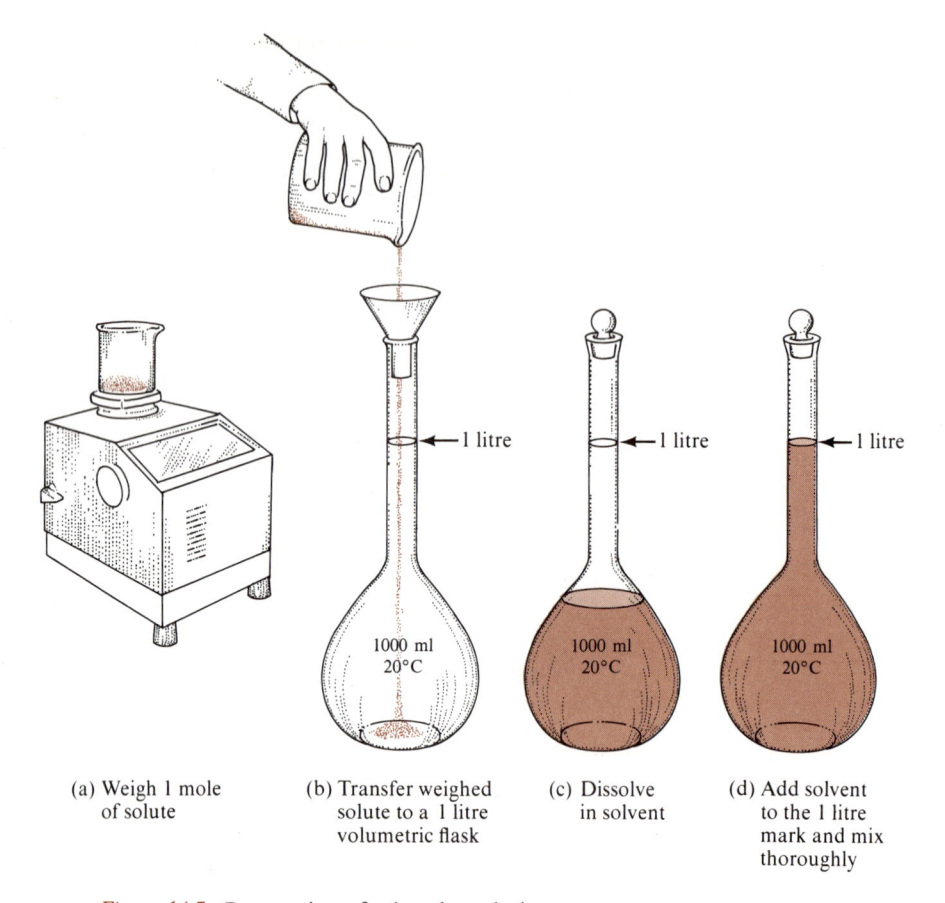

(a) Weigh 1 mole of solute (b) Transfer weighed solute to a 1 litre volumetric flask (c) Dissolve in solvent (d) Add solvent to the 1 litre mark and mix thoroughly

Figure 14.7. Preparation of a 1 molar solution.

Flasks that are calibrated to contain specific volumes at a particular temperature are used to prepare solutions of a desired concentration. These *volumetric flasks* have a calibration mark on the neck to indicate accurately the measured volume.

Suppose we want to make 500 ml of 1 M solution. This solution can be prepared by weighing 0.5 mole of the solute and diluting with water in a 500 ml volumetric flask. The molarity will be

$$M = \frac{0.5 \text{ mole solute}}{0.5 \text{ litre solution}} = 1 \text{ molar}$$

Thus, you can see that it is not necessary to have a litre of solution to express molarity. All we need to know is the number of moles of dissolved solute and the volume of solution. Thus, 0.001 mole NaOH in 10 ml of solution is 0.1 M:

$$\frac{0.001 \text{ mole}}{10 \text{ ml}} \times \frac{1000 \text{ ml}}{\text{litre}} = 0.1 \ M$$

When we stop to think that a balance is not calibrated in moles but in grams, we see that we really need to incorporate grams into the molarity formula. This is done by using the relationship

$$\text{Number of moles} = \frac{g}{\text{g-mol. wt}}$$

Substituting this relationship into our expression for molarity, we get

$$M = \frac{\text{moles}}{\text{litre}} = \frac{\text{g solute}}{\text{g-mol. wt solute} \times \text{litre solution}} = \frac{g}{\text{g-mol. wt} \times \text{litre}}$$

We can now weigh any amount of a solute that has a known formula, dilute it to any volume, and calculate the molarity of the solution using this formula.

The molarities of the concentrated acids commonly used in the laboratory are

HCl   12 M $HC_2H_3O_2$ 17 M
HNO_3 16 M H_2SO_4 18 M

Illustrative problems follow.

Problem 14.4   What weight of potassium hydroxide (KOH) is needed to prepare 1.00 litre of 1.00 M solution?

From the definition for molarity, 1 mole of KOH will be needed. This is 56.1 g, the gram-formula weight.

Data: Volume = 1.00 litre $M = \dfrac{1.00 \text{ mole}}{\text{litre}}$ g-mol. wt = $\dfrac{56.1 \text{ g}}{\text{mole}}$

litres ⟶ moles ⟶ grams

$$1.00 \text{ litres} \times \frac{1.00 \text{ mole}}{\text{litre}} \times \frac{56.1 \text{ g}}{\text{mole}} = 56.1 \text{ g KOH} \quad \text{(Answer)}$$

Problem 14.5   What is the molarity of a solution containing 1.4 moles of acetic acid ($HC_2H_3O_2$) in 250 ml of solution?

We may reason that if there are 1.4 moles in 250 ml, there will be four times that amount, or 5.6 moles, in 1 litre, because 250 ml is contained four times in 1 litre:

$$1 \text{ litre} = 1000 \text{ ml}; \quad \frac{1000 \text{ ml}}{250 \text{ ml}} = 4$$

The concentration of the solution would be 5.6 M (4 × 1.4).

By the unit conversion method, we note that the concentration given in the problem statement is 1.4 moles per 250 ml (moles/ml). Since molarity = moles/litre, the needed conversion is:

$$\frac{\text{moles}}{\text{ml}} \longrightarrow \frac{\text{moles}}{\text{litre}}$$

$$\frac{1.40 \text{ moles}}{250 \text{ ml}} \times \frac{1000 \text{ ml}}{\text{litre}} = \frac{5.6 \text{ moles}}{\text{litre}} = 5.6 \text{ M} \quad \text{(Answer)}$$

266 Chapter 14

Problem 14.6 What is the molarity of a solution made by dissolving 2.00 g of potassium chlorate ($KClO_3$) in enough water to make 150 ml of solution?

Use the formula

$$M = \frac{g}{\text{g-mol. wt} \times \text{litre}}$$

and substitute the given data:

$$g = 2.00$$
$$\text{g-mol. wt } KClO_3 = 122.6 \text{ g/mole}$$
$$150 \text{ ml} = 0.150 \text{ litre}$$

$$M = 2.00 \text{ g } KClO_3 \times \frac{1 \text{ mole } KClO_3}{122.6 \text{ g } KClO_3} \times \frac{1}{0.150 \text{ litre}}$$

$$= \frac{0.109 \text{ mole } KClO_3}{\text{litre}} = 0.109 \, M \quad \text{(Answer)}$$

This problem may also be solved using the unit conversion method. The steps in the conversions must lead to units of moles/litre.

$$\frac{\text{g } KClO_3}{\text{ml}} \longrightarrow \frac{\text{g } KClO_3}{\text{litre}} \longrightarrow \frac{\text{moles } KClO_3}{\text{litre}} = M$$

$$\frac{2.00 \text{ g } KClO_3}{150 \text{ ml}} \times \frac{1000 \text{ ml}}{\text{litre}} \times \frac{1 \text{ mole}}{122.6 \text{ g } KClO_3} = \frac{0.109 \text{ mole}}{\text{litre}} = 0.109 \, M$$

Problem 14.7 How many grams of potassium hydroxide are required to prepare 600 ml of 0.450 M KOH solution?

Data:

$$\text{Volume} = 0.600 \text{ litre} \qquad M = \frac{0.450 \text{ mole}}{\text{litre}} \qquad \text{g-mol. wt } KOH = \frac{56.1 \text{ g } KOH}{\text{mole}}$$

The calculation is

$$0.600 \text{ litre} \times \frac{0.450 \text{ mole}}{\text{litre}} \times \frac{56.1 \text{ g } KOH}{\text{mole}} = 15.1 \text{ g } KOH \quad \text{(Answer)}$$

Problem 14.8 How many millilitres of 2.0 M HCl will react with 28.0 g of NaOH?

Step 1. Write and balance the equation for the reaction:

$$HCl(aq) + NaOH(aq) \longrightarrow NaCl(aq) + H_2O(aq)$$

The equation states that 1 mole of HCl reacts with 1 mole of NaOH.

Step 2. Find the number of moles of NaOH in 28.0 g of NaOH.

$$\text{moles} = \frac{g}{\text{g-mol. wt}} = \frac{28.0 \text{ g NaOH}}{40.0 \text{ g/mole}} = 0.70 \text{ mole NaOH}$$

$$28.0 \text{ g NaOH} = 0.70 \text{ mole NaOH}$$

Step 3. Solve for moles and volume of HCl needed. From Steps 1 and 2 we see that 0.70 mole of HCl will react with 0.70 mole of NaOH, because the ratio of moles reacting is 1:1. We know that 2.0 M HCl contain 2.0 moles of HCl per litre; therefore, the volume that contains 0.70 mole of HCl will be less than 1 litre.

moles NaOH ⟶ moles HCl ⟶ litres HCl ⟶ ml HCl

$$0.70 \text{ mole NaOH} \times \frac{1 \text{ mole HCl}}{1 \text{ mole NaOH}} \times \frac{1 \text{ litre HCl}}{2 \text{ moles HCl}} = 0.350 \text{ litre HCl}$$

0.350 litre HCl × 1000 ml/litre = 350 ml HCl

Therefore, 350 ml of 2.0 M HCl contain 0.70 mole of HCl and will react with 0.70 mole, or 28.0 g, of NaOH.

Problem 14.9 What volume of 0.250 M solution can be prepared from 16.0 g of potassium carbonate (K_2CO_3)?

Step 1. Solving the equation

$$M = \frac{g}{\text{g-mol. wt} \times \text{litre}}$$

for litres, we obtain

$$\text{litres} = \frac{g}{\text{g-mol. wt} \times M}$$

Step 2. Substitute the data given into the equation and solve.

Data: g K_2CO_3 = 16.0   g-mol. wt K_2CO_3 = 138.2 g/mole

M = 0.250 mole/litre

$$\text{litres} = \frac{16.0 \text{ g}}{138.2 \text{ g/mole} \times 0.250 \text{ mole/litre}} = 0.463 \text{ litre, or 463 ml}$$

Or, by the unit conversion method,

$$16.0 \text{ g } K_2CO_3 \times \frac{1 \text{ mole } K_2CO_3}{138.1 \text{ g } K_2CO_3} \times \frac{1 \text{ litre}}{0.250 \text{ mole } K_2CO_3} = 0.463 \text{ litre}$$

Thus, a 0.250 M solution can be made by dissolving 16.0 g of K_2CO_3 in water and diluting to 463 ml.

Chemists often find it necessary to dilute solutions from one concentration to another by adding more solvent to the solution. If a solution is diluted by adding pure solvent, the volume of the solution increases but the number of moles of solute in the solution remains the same. Thus, the moles/litre (molarity) of the solution decreases. It is important to read a problem carefully to distinguish between (1) how much solvent must be added to dilute a solution to a particular concentration and (2) to what volume a solution must be diluted to prepare a solution of a particular concentration.

Problem 14.10 Calculate the molarity of a sodium hydroxide solution that is prepared by mixing 100 ml of 0.20 M NaOH with 150 ml of water.

268   Chapter 14

This is a dilution problem. If we double the volume of a solution by adding water, we cut the concentration in half. Therefore, the concentration of the above solution should be less than 0.10 M. In the dilution, the moles of NaOH remain constant; the molarity and volume change.

Step 1.   Calculate the moles of NaOH in the original solution.

$$\text{moles} = \text{litres} \times M = 0.100 \text{ litre} \times \frac{0.20 \text{ mole NaOH}}{\text{litre}} = 0.020 \text{ mole NaOH}$$

Step 2.   Solve for the new molarity, taking into account that the total volume of the solution is 250 ml (0.250 litre).

$$M = \frac{0.020 \text{ mole NaOH}}{0.250 \text{ litre}} = 0.080 \; M \text{ NaOH} \quad \text{(Answer)}$$

Alternate Solution: When the moles of solute in a solution before and after dilution are the same, then the moles before and after dilution may be set equal to each other:

$$\text{moles}_1 = \text{moles}_2$$

where moles_1 = moles before dilution and moles_2 = moles after dilution. Then

$$\text{moles}_1 = \text{litres}_1 \times M_1 \qquad \text{moles}_2 = \text{litres}_2 \times M_2$$
$$\text{litres}_1 \times M_1 = \text{litres}_2 \times M_2$$

or $V_1 \times M_1 = V_2 \times M_2$, where both volumes are in the same units. For this problem,

$V_1 = 0.100$ litre $M_1 = 0.20 \; M$
$V_2 = 0.250$ litre $M_2 = M_2$ (unknown)

Then:

$$0.100 \text{ litre} \times 0.20 \; M = 0.250 \text{ litre} \times M_2$$

Solving for M_2, we get

$$M_2 = \frac{0.100 \text{ litre} \times 0.20 \; M}{0.250 \text{ litre}} = 0.080 \; M \text{ NaOH}$$

Problem 14.11   What weight of silver chloride (AgCl) will be precipitated by adding sufficient silver nitrate (AgNO$_3$) to react with 1500 ml of 0.400 M BaCl$_2$ (barium chloride) solution?

$$2 \, \text{AgNO}_3(aq) + \underset{1 \text{ mole}}{\text{BaCl}_2(aq)} \longrightarrow \underset{2 \text{ moles}}{2 \, \text{AgCl}\downarrow} + \text{Ba(NO}_3)_2(aq)$$

The fact that BaCl$_2$ is in solution means that we need to consider the volume and concentration of the solution in order to know the number of moles of BaCl$_2$ reacting.

Step 1.   Determine the number of moles of BaCl$_2$ in 1500 ml of 0.400 M solution:

$$M = \frac{\text{moles}}{\text{litre}}$$

$$\text{moles} = \text{litre} \times M$$

$$1.500 \text{ litres} \times \frac{0.400 \text{ mole BaCl}_2}{\text{litre}} = 0.600 \text{ mole BaCl}_2$$

Step 2. Calculate the moles of AgCl formed by using the mole-ratio method:

$$0.600 \text{ mole BaCl}_2 \times \frac{2 \text{ moles AgCl}}{1 \text{ mole BaCl}_2} = 1.20 \text{ moles AgCl}$$

Step 3. Convert the moles of AgCl to grams:

$$1.20 \text{ moles AgCl} \times \frac{143.3 \text{ g AgCl}}{\text{mole AgCl}} = 172 \text{ g AgCl} \quad \text{(Answer)}$$

---

normality

1 normal solution

6. Normality. Normality is another way of expressing the concentration of a solution. It is based on an alternate chemical unit of mass called the *equivalent weight*. The **normality** of a solution is the concentration expressed as the number of equivalent weights (equivalents) of solute per litre of solution. A **1 normal (1 N) solution** contains 1 equivalent weight of solute per litre of solution. Normality is widely used in analytical chemistry because it simplifies many of the calculations involving solution concentration.

Every substance may be assigned an equivalent weight; it may be equal either to its molecular weight or to some small integral fractional part of its molecular weight (that is, the molecular weight divided by 2, 3, 4, and so on). To gain an understanding of the meaning of equivalent weight, let us start by considering these two reactions:

$$\text{HCl}(aq) + \text{NaOH}(aq) \longrightarrow \text{NaCl}(aq) + \text{H}_2\text{O}$$
  1 mole   1 mole
  (36.5 g)   (40.0 g)

$$\text{H}_2\text{SO}_4(aq) + 2\,\text{NaOH}(aq) \longrightarrow \text{Na}_2\text{SO}_4(aq) + 2\,\text{H}_2\text{O}$$
  1 mole   2 moles
  (98.1 g)   (80.0 g)

We note first that 1 mole of hydrochloric acid (HCl) reacts with 1 mole of sodium hydroxide (NaOH) and 1 mole of sulfuric acid (H_2SO_4) reacts with 2 moles of NaOH. If we make 1 molar solutions of these acids, 1 litre of 1 M HCl will react with 1 litre of 1 M NaOH. From this, we can see that H_2SO_4 has twice the chemical capacity of HCl when reacting with NaOH. We can, however, adjust these acid solutions to be equal in reactivity by dissolving only 0.5 mole of H_2SO_4 per litre of solution. By doing this, we find that we are required to use 49.0 g of H_2SO_4 per litre (instead of 98.1 g of H_2SO_4 per litre) to make a solution that is equivalent to one made from 36.5 g of HCl per litre. These weights, 49.0 g of H_2SO_4 and 36.5 g of HCl, are chemically equivalent and are known as the equivalent weights of these substances, because they react with the same amount of NaOH (40.0 g). The equivalent weight of HCl is equal to its molecular weight, but that of H_2SO_4 is one-half its molecular weight. Table 14.4 summarizes these relationships.

270 Chapter 14

Table 14.4. Comparison of molar and normal solutions of HCl and H_2SO_4 reacting with NaOH.

	Molecular weight	Concentration	Volumes that react	Equivalent weight	Concentration	Volumes that react
HCl	36.5	1 M	1 litre	36.5	1 N	1 litre
NaOH	40.0	1 M	1 litre	40.0	1 N	1 litre
H_2SO_4	98.1	1 M	1 litre	49.0	1 N	1 litre
NaOH	40.0	1 M	2 litres	40.0	1 N	1 litre

Expressions for normality follow. Notice the similarity to the molar solution definition.

$$\text{Normality} = N = \frac{\text{Number of equivalents of solute}}{1 \text{ litre of solution}} = \frac{\text{equivalents}}{\text{litre}}$$

where

$$\text{Number of equivalents of solute} = \frac{\text{grams of solute}}{\text{equivalent weight of solute}}$$

Then

$$N = \frac{\text{g solute}}{\text{eq wt solute} \times \text{litre solution}} = \frac{\text{g}}{\text{eq wt} \times \text{litre}}$$

Thus, 1 litre of solution containing 36.5 g of HCl would be 1 N, and 1 litre of solution containing 49.0 g of H_2SO_4 would also be 1 N. A solution containing 98.1 g of H_2SO_4 (1 mole) per litre would be 2 N when reacting with NaOH in the above equation.

Consider the following reactions, in which an excess of HCl is present. Hydrogen actually exists as H_2 molecules, but for convenience of considering the data, the hydrogen produced is shown as the number of atomic weights of hydrogen released per atomic weight of metal reacting. Table 14.5 summarizes the pertinent data for these reactions.

$$Na(s) + HCl(aq) \longrightarrow NaCl(aq) + H^0(g)$$
$$Ca(s) + 2\,HCl(aq) \longrightarrow CaCl_2(aq) + 2\,H^0(g)$$
$$Al(s) + 3\,HCl(aq) \longrightarrow AlCl_3(aq) + 3\,H^0(g)$$

In each of the above reactions, the equivalent weight of the reacting metals is the weight that reacts with 1 equivalent weight of the acid, liberates 1 atomic weight of H atoms, or involves the transfer of 1 mole of electrons in the reaction. One atomic weight of Na metal lost 1 electron per atom in going to NaCl; 1 at. wt of Ca metal lost 2 electrons per atom in going to $CaCl_2$; 1 at. wt of Al metal lost 3 electrons per atom in going to $AlCl_3$. In each reaction, 1 at. wt of H^+ gained 1 electron per atom in going to free hydrogen.

$$\text{eq wt} = \frac{\text{at. wt Na}}{1} = \frac{\text{at. wt Ca}}{2} = \frac{\text{at. wt Al}}{3} = \frac{\text{at. wt H}}{1}$$

Table 14.5. Equivalent weight of sodium, calcium, and aluminum in reaction with hydrochloric acid.

Metal	Atomic weight (amu)	Number of atomic weights of hydrogen liberated per atomic weight of metal	Equivalent weight of metal (amu)
Na	23.0	1	$\frac{23.0}{1} = 23.0$
Ca	40.1	2	$\frac{40.1}{2} = 20.0$
Al	27.0	3	$\frac{27.0}{3} = 9.0$

equivalent weight

Two definitions of **equivalent weight** can now be stated:

1. The equivalent weight is that weight of a substance which will react with, combine with, contain, replace, or in any other way be equivalent to 1 gram-atomic weight of hydrogen.
2. In oxidation–reduction reactions, the gram-equivalent weight is that weight of a substance which loses or gains Avogadro's number of electrons.

The equivalent weight of a substance may be variable; its value is dependent on the reaction that the substance is undergoing. Consider the reactions represented by these equations:

$$\text{NaOH} + \text{H}_2\text{SO}_4 \longrightarrow \text{NaHSO}_4 + \text{H}_2\text{O} \tag{1}$$

$$2\,\text{NaOH} + \text{H}_2\text{SO}_4 \longrightarrow \text{Na}_2\text{SO}_4 + 2\,\text{H}_2\text{O} \tag{2}$$

In reaction (1), 1 mole of sulfuric acid furnishes 1 g-at. wt of hydrogen. Therefore, the equivalent weight of sulfuric acid is the formula weight, namely 98.1 g. But in reaction (2), 1 mole of H_2SO_4 furnishes 2 g-at. wt of hydrogen. Therefore, the equivalent weight of the sulfuric acid is one-half the formula weight, or 49.0 g.

14.8 Colligative Properties of Solutions

When two solutions are prepared, one containing 1 mole (60.0 g) of urea (NH_2CONH_2) and the other containing one mole (342 g) of sucrose ($\text{C}_{12}\text{H}_{22}\text{O}_{11}$) in 1 kg of water, the freezing point of each solution is $-1.86°\text{C}$, not $0°\text{C}$ as in pure water. Urea and sucrose are very different substances, yet each lowers the freezing point of the water by the same amount. The only thing apparently common to these two solutions is that each contains 1 mole (6.02×10^{23} molecules) of solute and 1 kg of solvent. In fact, if we dissolved one mole of any other solute (provided that it is one that does not produce ions in solution) in 1 kg of water, the freezing point of the resulting solution would be $-1.86°\text{C}$.

This leads us to conclude that the freezing point depression for a solution containing 6.02×10^{23} solute molecules (particles) and 1 kg of water is a constant, namely 1.86°. Freezing point depression is a general property of solutions. When solutions are prepared from 1 kg of any solvent and 1 mole of any (nonionized) solute, the freezing point will be lower than that of the pure solvent. Furthermore the amount by which the freezing point is depressed is the same for all solutions made with a given solvent; that is, each solvent shows a characteristic *freezing point depression constant*. Freezing point depression constants for several solvents are given in Table 14.6.

Table 14.6. Freezing point depression and boiling point elevation constants of selected solvents.

Solvent	Freezing point of pure solvent (°C)	Freezing point depression constant $\left(\dfrac{°C,\ kg\ solvent}{mole\ solute}\right)$	Boiling point of pure solvent (°C)	Boiling point elevation constant $\left(\dfrac{°C,\ kg\ solvent}{mole\ solute}\right)$
Water	0.00	1.86	100.0	0.52
Acetic acid	16.6	3.90	118.5	3.07
Benzene	5.5	5.1	80.1	2.53
Camphor	178	40	208.2	5.95

colligative properties

The solution formed by the addition of a nonvolatile solute to a solvent has a lower freezing point, a higher boiling point, and a lower vapor pressure than that of the pure solvent. All these effects are related and are known as colligative properties. The **colligative properties** are properties that depend only upon the number of solute atoms or molecules in a solution and not on the nature of those atoms or molecules. Freezing point depression, boiling point elevation, vapor pressure lowering, and osmotic pressure are colligative properties of solutions.

The colligative properties of a solution can be considered in terms of vapor pressure. The vapor pressure of a pure liquid depends on the tendency of molecules to escape from its surface. Thus, if 10% of the molecules in a solution are nonvolatile solute molecules, the vapor pressure of the solution is 10% lower than that of the pure solvent. The vapor pressure is lower because the surface of the solution contains 10% nonvolatile molecules and 90% of the volatile solvent molecules. A liquid boils when its vapor pressure equals the pressure of the atmosphere. Thus, we can see that the solution just described as having a lower vapor pressure will have a higher boiling point than the pure solvent. The solution with a lowered vapor pressure does not boil until it has been heated above the boiling point of the solvent (see Figure 14.8). Each solvent has its own characteristic boiling point elevation constant (see Table 14.6). The boiling point elevation constant is based on a solution that contains 1 mole of solute particles per kilogram of solvent. For example, the boiling point elevation for a solution containing 1 mole of solute particles per kilogram of water is 0.52°C. This means that this water solution will boil at 100.52°C.

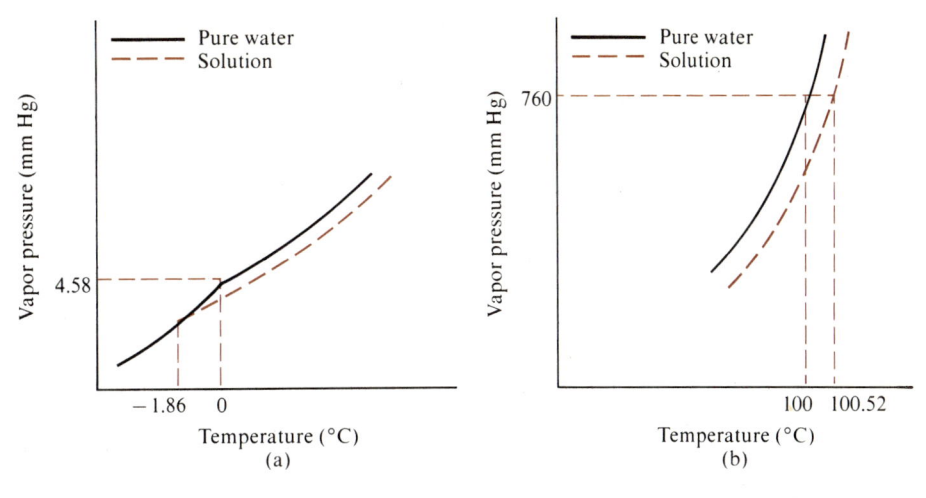

Figure 14.8. Vapor pressure curves of pure water and water solutions, showing (a) freezing point depression and (b) boiling point elevation effects. (Concentration: 1 mole of solute per kilogram of water.)

The freezing behavior of a solution can also be considered in terms of lowered vapor pressure. Figure 14.8 shows the vapor pressure relationships of ice, water, and a solution containing 1 mole of solute per kilogram of water. The freezing point of water is at the intersection of the water and ice vapor pressure curves; that is, at the point where water and ice have the same vapor pressure. Because the vapor pressure of water is lowered by the solute, the vapor pressure curve of the solution does not intersect the vapor pressure curve of ice until the solution has been cooled below the freezing point of pure water. Thus, it is necessary to cool the solution below 0°C in order to freeze out ice.

The foregoing discussion dealing with freezing point depressions is restricted to *un-ionized* substances. The discussion of boiling point elevations is restricted to *nonvolatile* and un-ionized substances. The colligative properties of ionized substances (Electrolytes, Chapter 15) are not explored at this point.

Some practical applications involving colligative properties are (1) use of salt–ice mixtures to provide low freezing temperatures for homemade ice cream, (2) use of salt or calcium chloride to melt ice from streets, (3) use of ethylene glycol and water mixtures as antifreeze in automobile radiators (ethylene glycol also raises the boiling point of radiator fluid and thus allows the engine to operate at a higher temperature).

Both the freezing point depression and the boiling point elevation are directly proportional to the number of moles of solute per kilogram of solvent. When we deal with the colligative properties of solutions, another concentration expression, *molality*, is used. The **molality** (m) of a solute is the number of moles of solute per kilogram of solvent:

molality

$$m = \frac{\text{moles solute}}{\text{kg solvent}}$$

Note that a lowercase *m* is used for molality concentrations, while a capital *M* is used for molarity. The difference between molality and molarity is that molality refers to moles of solute *per kilogram of solvent*, whereas molarity refers to moles of solute *per litre of solution*. For un-ionized substances, the colligative properties of a solution are directly proportional to its molality.

The following equations show the relationship of freezing point depression, molecular weight, and solution concentration. The symbol Δt_f indicates the change in the freezing point of the solution with respect to the freezing point of the pure solvent; K_f is the freezing point depression constant of the solvent.

$$\Delta t_f = K_f m \quad \text{or} \quad m = \frac{\Delta t_f}{K_f}$$

$$\Delta t_f = K_f \times \frac{\text{moles solute}}{\text{kg solvent}} = K_f \times \frac{\text{g solute}}{\text{Mol. wt solute}} \times \frac{1}{\text{kg solvent}}$$

This equation is commonly used to calculate the freezing points of solutions and the molecular weights of compounds. For boiling point elevation calculations substitute Δt_b for Δt_f and K_b for K_f, where Δt_b is the observed boiling point elevation and K_b is the boiling point elevation constant of the solvent.

Problem 14.12   A solution is made by dissolving 100 g of ethylene glycol ($C_2H_6O_2$) in 200 g of water. What is the freezing point of this solution?

Data: K_f(for water) = 1.86°C/mole/kg   Mol. wt $C_2H_6O_2$ = 62.0 g/mole

$$\Delta t_f = 1.86°C/\text{mole/kg} \times \frac{100 \text{ g}}{62.0 \text{ g/mole}} \times \frac{1}{0.200 \text{ kg}} = 15.0°C$$

Since 15.0°C is the freezing point depression, it must be subtracted from 0°C, the freezing point of the pure solvent. Therefore, the freezing point of the solution is −15°C.

Problem 14.13   A solution made by dissolving 3.25 g of a compound of unknown molecular weight in 100.0 g of water has a freezing point of −1.46°C. What is the molecular weight of the compound?

Data: $\Delta t_f = 1.46°C$ $K_f = 1.86°C/\text{mole/kg}$

Solving the general formula for the molecular weight of the solute, we obtain

$$\text{Mol. wt solute} = K_f \times \frac{\text{g solute}}{\Delta t_f} \times \frac{1}{\text{kg solvent}}$$

$$= 1.86°C/\text{mole/kg} \times \frac{3.25 \text{ g}}{1.46°C} \times \frac{1}{0.100 \text{ kg}} = 41.4 \text{ g/mole} \quad \text{(Answer)}$$

14.9   Osmosis and Osmotic Pressure

When red blood cells are put in distilled water, they gradually swell and, in time, may burst. If red blood cells are put in a 5% urea (or a 5% salt) solution, they gradually shrink and take on a wrinkled appearance. The cells behave in

semipermeable membrane

osmosis

this fashion because they are enclosed in semipermeable membranes. A **semipermeable membrane** is one that allows the passage of water (solvent) molecules through it in either direction but prevents the passage of solute molecules or ions. When two solutions of different concentrations (or water and a water solution) are separated by a semipermeable membrane, water diffuses through the membrane from the solution of lower concentration into the solution of higher concentration. The diffusion of water, either from a dilute solution or from pure water, through a semipermeable membrane into a solution of higher concentration is called **osmosis**.

All solutions exhibit *osmotic pressure*. Osmotic pressure is another colligative property; it is dependent only on the concentration of the solute particles and is independent of their nature. The osmotic pressure of a solution can be measured by determining the amount of counterpressure needed to prevent osmosis; this pressure can be very large. The osmotic pressure of a solution containing 1 mole of solute particles in 1 kg of water is about 22.4 atm, which is about the same as the pressure exerted by 1 mole of a gas confined in a volume of 1 litre at 0°C.

Osmosis has a role in many biological processes. Semipermeable membranes occur commonly in living organisms. But artificial or synthetic membranes can also be made. Ordinary cellophane that has been treated to remove the waterproof coating is a good semipermeable membrane. The roots of plants are covered with tiny structures called root hairs; soil water enters the plant by osmosis, passing through the semipermeable membranes covering the root hairs.

Osmosis can be demonstrated with the simple laboratory setup shown in Figure 14.9. As a result of osmotic pressure, water passes through the cellophane membrane into the thistle tube, causing the solution level to rise. In osmosis, water is always transferred from a less concentrated to a more concentrated solution; that is, the effect is toward equalization of the concentration on both sides of the membrane. It should also be noted that the effective movement of water in osmosis is always from the region of *higher water concentration* to the region of *lower water concentration*.

Osmosis can be explained by assuming that a semipermeable membrane has passages that permit water molecules—but no other molecules or ions—to pass in either direction. Both sides of the membrane are constantly being struck by water molecules in random motion. The number of water molecules entering the membrane is proportional to the number of water molecule–membrane impacts per unit of time. Since solute molecules or ions reduce the concentration of water, there are more water molecules, and more water molecule impacts, on the side with the lower solute concentration (more dilute solution). The greater number of water molecule–membrane impacts on the dilute side is thus responsible for the net transfer of water to the more concentrated solution. Again note that the overall process involves the net transfer, by diffusion through the membrane, of water molecules from a region of higher water concentration (dilute solution) to one of lower water concentration (more concentrated solution).

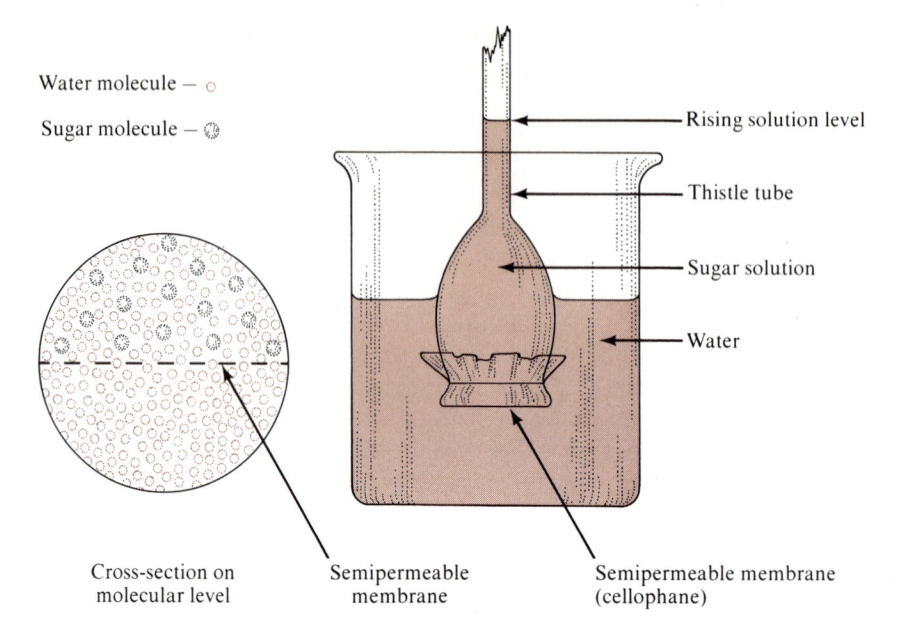

Figure 14.9. Laboratory demonstration of osmosis; as a result of osmosis, water passes through the membrane into the thistle tube, causing the solution level to rise.

This explanation is a simplified picture of osmosis. Remember that no one has ever seen the hypothetical passages that allow water molecules, and no other kinds of molecules or ions, to pass through them! Alternative explanations have been proposed. Our discussion has been confined to water solutions, but osmotic pressure is a general colligative property and osmosis is known to occur in nonaqueous systems.

physiological saline solution

A 0.90% (0.15 M) sodium chloride solution is known as a **physiological saline solution** because it is *isotonic* with blood plasma; that is, it has the same osmotic pressure as blood plasma. Since each mole of NaCl yields about 2 moles of ions when in solution, the solute particle concentration in physiological saline solution is nearly 0.30 molar. Five percent glucose solution (0.28 molar) is also approximately isotonic with blood plasma. Blood cells neither swell nor shrink in an isotonic solution. The cells described in the first paragraph of this section swelled in water because the water was *hypotonic* to cell plasma. The cells shrank in 5% urea solution because the urea solution was *hypertonic* to the cell plasma. In order to prevent possible injury to blood cells by osmosis, fluids for intravenous use are usually made up at approximately isotonic concentration.

Questions

A. Review the meanings of the new terms introduced in this chapter.
 1. Solution
 2. Solute
 3. Solvent
 4. Solubility
 5. Miscible
 6. Immiscible

Solutions 277

7. Concentration of a solution
8. Dilute solution
9. Concentrated solution
10. Saturated solution
11. Unsaturated solution
12. Supersaturated solution
13. 1 molar solution
14. Molarity
15. Normality
16. 1 normal solution
17. Equivalent weight
18. Colligative properties
19. Molality
20. Semipermeable membrane
21. Osmosis
22. Physiological saline solution

B. Answers to the following questions will be found in tables and figures.
1. Make a sketch indicating the orientation of water molecules (a) about a single sodium ion and (b) about a single chloride ion in solution.
2. Which of the substances listed below are reasonably soluble and which are insoluble?
   (a) KOH   (c) ZnS   (e) Na_2CrO_4 (g) $MgCO_3$ (i) $Fe(NO_3)_3$
   (b) $NiCl_2$ (d) $AgC_2H_3O_2$ (f) PbI_2 (h) $CaCl_2$ (j) $BaSO_4$
3. Estimate the number of grams of sodium fluoride that would dissolve in 100 g of water at 50°C.
4. What is the solubility of each of the substances listed below at 25°C (see Figure 14.3)?
   a) Potassium chloride   (b) Potassium chlorate   (c) Potassium nitrate
5. What would be the total surface area if the 1 cm cube in Figure 14.4 were cut into cubes 0.01 cm on a side?
6. At which temperatures—10°C, 20°C, 30°C, 40°C, and 50°C—would you expect a solution made from 63 g of ammonium chloride and 150 g of water to be unsaturated? (See Figure 14.3.)
7. Explain why the rate of dissolving decreases as shown in Figure 14.5.
8. Does the bottle in Figure 14.6 necessarily contain 90 g of water? Explain.
9. Would the volumetric flasks in Figure 14.7 be satisfactory for preparing normal solutions? Explain.
10. Assume the thistle tube in Figure 14.9 contains 1.0 M sugar solution and that the water in the beaker has just been replaced by a 2.0 M solution of urea. Would the solution level in the thistle tube (a) continue to rise, (b) remain constant, or (c) fall? Explain.

C. Review questions.
1. Name and distinguish between the two components of a solution.
2. Explain why the solute does not settle out of a solution.
3. Is it possible to have one solid dissolved in another solid? Explain.
4. Explain how a colored salt such as $KMnO_4$ can be used to demonstrate the spontaneous movement (diffusion) of ions in solution.
5. Why is air considered to be a solution?
6. Explain why carbon tetrachloride will dissolve benzene but will not dissolve sodium chloride.
7. Tea and Coca Cola are popular drinks over most of the globe. Tea is drunk either hot or iced, but Coca Cola is never served hot. Why not?
8. In which will a teaspoonful of sugar dissolve more rapidly, 200 ml of iced tea or 200 ml of hot coffee? Explain your answer in terms of the KMT.
9. Which will dissolve more rapidly in 200 ml of water, 25 g of rock salt (large crystals) or 25 g of table salt? Explain.
10. What is the effect of pressure on the solubility of the following:
    (a) Solids in liquids   (b) Gases in liquids
11. Is the rate of dissolving zero in a saturated solution? Explain.

278 Chapter 14

12. What is the effect of an increase in temperature on the solubility of a gas in a liquid?
13. Explain why there is no apparent reaction when crystals of $AgNO_3$ and NaCl are mixed, but a reaction is apparent immediately when solutions of $AgNO_3$ and NaCl are mixed.
14. What do we mean when we say that concentrated hydrochloric acid (HCl) is 12 M?
15. Will 1 litre of 1 M NaCl contain more chloride ions than 0.5 litre of 1 M $MgCl_2$? Explain.
16. Will 1 litre of 1 M HCl neutralize 1 litre of 1 M KOH?
17. Will 1 litre of 1 N H_2SO_4 neutralize 1 litre of 1 N NaOH? Explain.
18. Explain in terms of vapor pressure why the boiling point of a solution containing a nonvolatile solute is higher than that of the pure solvent.
19. Explain in terms of vapor pressure why the freezing point of a solution containing a solute is lower than that of the pure solvent.
20. Which would be colder, a glass of water and crushed ice or a glass of Seven-Up and crushed ice? Explain.
21. Which would be the most effective in lowering the freezing point of 400 ml of water?
    (a) 100 g of ethyl alcohol (C_2H_5OH) or 100 g of ethylene glycol [$C_2H_4(OH)_2$]
    (b) 75.0 g of ethyl alcohol or 60.0 g of methyl alcohol (CH_3OH)
22. Explain in terms of the KMT how a semipermeable membrane functions when placed between pure water and a 10% sugar solution.
23. Which has the higher osmotic pressure, a solution containing 100 g of urea (CH_4ON_2) in 1 kg of H_2O or one containing 150 g of glucose ($C_6H_{12}O_6$) in 1 kg of H_2O?
24. Explain why a lettuce leaf in contact with a salad dressing containing salt and vinegar soon becomes wilted and limp while another lettuce leaf in contact with plain water remains crisp.
25. Which of the following statements are correct?
    (a) A solution differs from a compound in that it can have a variable concentration.
    (b) Benzene and water do not mix; they are said to be miscible.
    (c) Most common nitrates and acetates are soluble in water.
    (d) Compared to 12 M HCl, 3 M HCl is a dilute solution.
    (e) A solution may be saturated at several different temperatures.
    (f) Hydrogen bromide, a gas, is more soluble in water at 40°C than at 20°C.
    (g) A 10% solution of NaCl contains 10 g of NaCl in 100 ml of solution.
    (h) 1 mole of solute in 1 litre of solution has the same concentration as 0.1 mole of solute in 100 ml of solution.
    (i) When 100 ml of 0.200 M HCl is diluted by adding 100 ml of water, the number of moles of HCl in the final solution is one-half the number of moles of HCl in the original solution.
    (j) 1 mole of HCl will react with the same amount of NaOH as 1 mole of HNO_3.
    (k) 1 mole of H_2SO_4 will react with twice as much NaOH as 1 mole of HNO_3.
    (l) A solution containing 0.2 mole CH_3OH in 100 g of water will freeze at $-1.86°C$.
    (m) A solution containing a nonvolatile solute has a higher boiling point than the pure solvent as a result of having a lower vapor pressure than the pure solvent.

D. Review problems.
1. What is the weight percent of a saturated solution of copper(II) sulfate ($CuSO_4$) at 20°C?
2. Calculate the weight percent of the following solutions:
   (a) 10 g KCl + 100 g H_2O (c) 30 g $MgCl_2$ + 150 g H_2O
   (b) 20 g KCl + 100 g H_2O (d) 0.50 g $KMnO_4$ + 7.5 g H_2O
3. How much solute is present in the following solutions?
   (a) 50 g of 6.0% NaCl   (b) 250 g of 15.0% KCl
4. Physiological saline solutions used in intravenous injections have a concentration of 0.90% NaCl. How many grams of NaCl are needed to prepare 400 g of this solution?
5. A solution made from 10.0 g of $NaNO_3$ and 100 g of H_2O is allowed to evaporate in an open beaker at 20°C. What weight of water must evaporate for the solution to become saturated?
6. Will 100 g of 6.0% KOH solution be a sufficient amount of base to neutralize 100 ml of 1.0 M HCl? Show proof.
7. A sugar syrup solution contains 22.0% sugar and has a density of 1.09 g/ml. How many grams of sugar are there in 1.00 litre of this syrup?
8. The density of 24.0% H_2SO_4 solution is 1.17 g/ml. What weight and what volume of this solution will contain 50.0 g of H_2SO_4?
9. Determine the molarity of the following solutions:
   (a) 140 g $CaBr_2$ in 1.00 litre of solution
   (b) 48.0 g NH_4Cl in 400 ml of solution
   (c) 12.0 g NaOH in 200 ml of solution
   (d) 25.0 g $BaCl_2 \cdot 2\,H_2O$ in 1500 ml of solution
10. Calculate the number of moles of solute in each of the following solutions:
    (a) 2.00 litres of 3.00 M $CaCl_2$ (d) 15.0 ml of 12.0 M HNO_3
    (b) 225 ml of 1.50 M $KC_2H_3O_2$ (e) 25.0 ml of 0.320 M NaOH
    (c) 2000 ml of 0.250 M $AgNO_3$ (f) 2.50 litres of 1.50 M KF
11. Determine the weight of solute in each of these solutions:
    (a) 2.5 litres of 0.10 M KCl   (c) 1.0 litre of 1.0 M H_3PO_4
    (b) 125 ml of 1.50 M Na_2SO_4 (d) 4.00 ml of 0.600 M $Zn(NO_3)_2$
12. What is the molarity of concentrated hydrobromic acid if the solution is 48.0% HBr and has a density of 1.50 g/ml?
13. Calculate the volume of concentrated reagent required to prepare the diluted solutions indicated:
    (a) 12 M HCl to prepare 100 ml of 4.0 M HCl
    (b) 16 M HNO_3 to prepare 200 ml of 6.0 M HNO_3
    (c) 18 M H_2SO_4 to prepare 1000 ml of 3.0 M H_2SO_4
    (d) 17 M $HC_2H_3O_2$ to prepare 25 ml of 3.0 M $HC_2H_3O_2$
14. To what volume must 26.0 g of zinc nitrate [$Zn(NO_3)_2$] be diluted to make a 1.50 M solution?
15. What will be the molarity of each of the solutions made by mixing 200 ml of 0.50 M H_2SO_4 with the following?
    (a) 200 ml of H_2O
    (b) 200 ml of 0.80 M H_2SO_4
    (c) 300 ml of 0.80 M H_2SO_4
16. In the reaction
    $6FeCl_2 + K_2Cr_2O_7 + 14HCl \longrightarrow 6FeCl_3 + 2CrCl_3 + 2KCl + 7H_2O$
    (a) How many moles of $FeCl_3$ will be produced from 1.0 mole of $FeCl_2$?
    (b) How many moles of $CrCl_3$ will be produced from 1.0 mole of $FeCl_2$?

(c) How many moles of $K_2Cr_2O_7$ will react with 0.040 mole of $FeCl_2$?
(d) How many millilitres of 0.080 M $K_2Cr_2O_7$ will react with 0.040 mole of $FeCl_2$?
(e) What volume of 6.0 M HCl is needed to react with 0.040 mole $FeCl_2$?

17. Using the equation below, calculate
$$2\,KMnO_4 + 16\,HCl \longrightarrow 2\,MnCl_2 + 5\,Cl_2 + 8\,H_2O + 2\,KCl$$
    (a) The moles of $MnCl_2$ produced from 1.0 mole of $KMnO_4$
    (b) The moles of Cl_2 produced from 1.0 mole of $KMnO_4$
    (c) The moles of HCl that will react with 1.0 mole of $KMnO_4$
    (d) The number of millilitres of 0.100 M HCl that will react with 50.0 ml of 0.250 M $KMnO_4$
    (e) The volume of Cl_2 (gas) at STP that will be produced when 100 ml of 3.00 M HCl react

18. $BaCl_2(aq) + K_2CrO_4(aq) \longrightarrow BaCrO_4\downarrow + 2\,KCl(aq)$
    (a) What weight of $BaCrO_4$ can be obtained by reacting 50.0 ml of 0.250 M $BaCl_2$ solution?
    (b) What volume of 1.0 M K_2CrO_4 solution is needed to react with the 50.0 ml of 0.250 M $BaCl_2$ solution?

19. $3\,Cu(s) + 8\,HNO_3(aq) \longrightarrow 3\,Cu(NO_3)_2(aq) + 2\,NO\uparrow + 4\,H_2O$
    (a) How many grams of Cu will react with 100 ml of 4.00 M HNO_3 solution?
    (b) What volume of NO gas (at STP) will be produced in the reaction in part (a)?

20. How many moles of hydrogen will be liberated from 100 ml of 3.00 M HCl reacting with an excess of magnesium?
$$Mg(s) + 2\,HCl(aq) \longrightarrow MgCl_2(aq) + H_2\uparrow$$

21. When 250 ml of hydrochloric acid were treated with an excess of magnesium, 6.00 litres (at STP) of hydrogen gas were liberated. Calculate the molarity of the HCl solution.

22. The compounds $Mg(OH)_2$ and $Al(OH)_3$ are used in antacid formulations to neutralize excess acid in the stomach. On an equal mass basis, show which base is more effective in neutralizing hydrochloric acid in the stomach.

23. Calculate the equivalent weight of the acid and the base in each of the reactions below:
    (a) $2\,HCl + Ca(OH)_2 \longrightarrow CaCl_2 + 2\,H_2O$
    (b) $3\,HCl + Al(OH)_3 \longrightarrow AlCl_3 + 3\,H_2O$
    (c) $H_2SO_4 + Mg(OH)_2 \longrightarrow MgSO_4 + 2\,H_2O$
    (d) $H_2SO_4 + LiOH \longrightarrow LiHSO_4 + H_2O$
    (e) $H_3PO_4 + KOH \longrightarrow KH_2PO_4 + H_2O$

24. Calculate the equivalent weight of the metal in each of the following reactions:
    (a) $2\,Rb + 2\,H_2O \longrightarrow 2\,RbOH + H_2\uparrow$
    (b) $Mg + H_2O \xrightarrow{\Delta} MgO + H_2\uparrow$
    (c) $Zn + 2\,HBr \longrightarrow ZnBr_2 + H_2\uparrow$
    (d) $Pb + CuSO_4 \longrightarrow PbSO_4 + Cu$
    (e) $Fe_2O_3 + 3\,H_2 \xrightarrow{\Delta} 2\,Fe + 3\,H_2O$
    (f) $2\,Ga + 3\,Cl_2 \longrightarrow 2\,GaCl_3$

25. (a) What is the freezing point of a solution that contains 1.200 g of urea (CH_4ON_2) in 10.0 g of H_2O?
    (b) Calculate the molality of this solution.

26. (a) What is the freezing point of a solution that contains 40.0 g of ethylene glycol ($C_2H_6O_2$) in 100.0 g of H_2O?
    (b) What is the boiling point of this solution?

27. What is (a) the freezing point and (b) the boiling point of a solution containing 3.07 g of naphthalene ($C_{10}H_8$) in 20.0 g of benzene?
28. The freezing point of a solution of 4.50 g of an unknown compound dissolved in 20.0 g of acetic acid is 13.2°C. Calculate the molecular weight of the compound.
29. When 5.40 g of a compound are dissolved in 60.0 g of H_2O, the solution has a freezing point of −0.930°C. The empirical formula of the compound is CH_2O. What is its molecular formula?
30. What (a) weight and (b) volume of ethylene glycol ($C_2H_6O_2$, $d = 1.11$ g/ml) should be added to 10.0 litres of H_2O in an automobile radiator to protect it from freezing at −20.0°C?

15 Ionization. Acids, Bases, Salts

After studying Chapter 15 you should be able to:

1. Understand the terms listed in Question A at the end of the chapter.
2. Define an acid and a base in terms of Arrhenius, Brønsted–Lowry, and Lewis theories.
3. When given the reactants, complete and balance equations for the reactions of acids with bases, metals, metal oxides, and carbonates.
4. When given the reactants, complete and balance equations for the reaction of an amphoteric hydroxide with either a strong acid or a strong base.
5. Write balanced equations for the reaction of sodium hydroxide or potassium hydroxide with zinc and with aluminum.
6. Classify common compounds as electrolytes or nonelectrolytes.
7. Write equations for the dissociation or ionization of acids, bases, and salts in water.
8. Describe and write equations for the ionization of water.
9. Given pH as an integer, give the H^+ molarity and vice versa.
10. Use the simplified log scale given in the chapter to estimate pH values from corresponding H^+ molarities.
11. Calculate the molarity or volume of an acid or base solution from appropriate titration data.
12. Write balanced molecular, total ionic, and net ionic equations for neutralization reactions.
13. List the rules for writing net ionic equations.
14. Discuss colloids and describe methods for their preparation.
15. Describe the characteristics that distinguish true solutions, colloidal dispersions, and mechanical suspensions.
16. Tell how colloidal dispersions can be (a) stabilized and (b) how they can be precipitated.

15.1 Acids and Bases

The word *acid* is derived from the Latin *acidus*, meaning "sour" or "tart," and is also related to the Latin word *acetum*, meaning "vinegar." Vinegar has been known since antiquity as the product of the fermentation of wine and apple cider. The sour constituent of vinegar is acetic acid ($HC_2H_3O_2$).

Some of the characteristic properties commonly associated with acids are the following: Water solutions of acids are sour to the taste and are capable of changing the color of litmus, a vegetable dye, from blue to red. Water solutions of nearly all acids are able to react with: (1) metals such as zinc and magnesium to produce hydrogen gas; (2) bases to produce water and a salt; and (3) carbonates to produce carbon dioxide. These properties are due to hydrogen ions, H^+, released by the acid in a water solution.

Classically, a *base* is a substance capable of liberating hydroxide ions, OH^-, in water solution. Hydroxides of the alkali metals (Group IA) and alkaline earth metals (Group IIA), such as LiOH, NaOH, KOH, $Ca(OH)_2$, and $Ba(OH)_2$, are the most common inorganic bases. Water solutions of bases are called *alkaline solutions* or *basic solutions*. They have the following properties: a bitter or caustic taste; a slippery, soapy feeling; the ability to change litmus from red to blue; and the ability to interact with acids to form a salt and water.

Several theories have been proposed to answer the question "What is an acid and a base?" One of the earliest, most significant of these theories was advanced in a doctoral thesis in 1884 by Svante Arrhenius (1859–1927), a Swedish scientist, who stated that an acid is a hydrogen-containing substance that dissociates to produce hydrogen ions, and that a base is a hydroxide-containing substance that dissociates to produce hydroxide ions in aqueous solutions. Arrhenius postulated that the hydrogen ions were produced by the dissociation of acids in water; and hydroxide ions were produced by the dissociation of bases in water.

$$HA \longrightarrow H^+ + A^-$$
Acid

$$MOH \longrightarrow M^+ + OH^-$$
Base

Thus, an acid solution contains an excess of hydrogen ions and a base an excess of hydroxide ions.

In 1923, the Brønsted–Lowry proton transfer theory was introduced by J. N. Brønsted, a Danish chemist (1897–1947), and T. M. Lowry, an English chemist (1874–1936). This theory states that an acid is a proton donor and a base is a proton acceptor.

Consider the reaction of hydrogen chloride gas with water to form hydrochloric acid.

$$HCl(g) + H_2O(l) \longrightarrow H_3O^+(aq) + Cl^-(aq) \tag{1}$$

In the course of the reaction, HCl donates, or gives up, a proton to form a Cl^- ion and H_2O accepts a proton to form the H_3O^+ ion. Thus, HCl is an acid and H_2O is a base, according to the Brønsted–Lowry theory.

A hydrogen ion, H^+, is nothing more than a bare proton and does not exist by itself in an aqueous solution. In water, a proton combines with a polar water molecule to form a hydrated hydrogen ion, H_3O^+ [$H(H_2O)^+$],

hydronium ion

commonly called a **hydronium ion**. The proton is attracted to a polar water molecule, forming a bond with one of the two pairs of unshared electrons:

$$H^+ + H\colon\!\overset{\displaystyle\cdot\cdot}{\underset{\displaystyle H}{\ddot{O}}}\colon \longrightarrow \left[H\colon\!\overset{\displaystyle\cdot\cdot}{\underset{\displaystyle H}{\ddot{O}}}\colon\!H\right]^+$$

Hydronium ion

Note the electron structure of the hydronium ion. For simplicity of expression in equations, we often use H^+ instead of H_3O^+, with the explicit understanding that H^+ is always hydrated in solution.

Whereas the Arrhenius theory is restricted to aqueous solutions, the Brønsted–Lowry approach has application in all media and has become the more important theory when the chemistry of substances in solutions other than water is studied. Ammonium chloride (NH_4Cl) is a salt, yet its water solution has an acidic reaction. From this test we must conclude that NH_4Cl has acidic properties. The Brønsted–Lowry explanation shows that the ammonium ion, NH_4^+, is a proton donor, and water is the proton acceptor:

$$NH_4^+ \rightleftharpoons NH_3 + H^+ \qquad (2)$$
Acid　　　　Base　　Acid

$$NH_4^+ + H_2O \longrightarrow H_3O^+ + NH_3 \qquad (3)$$
Acid　　Base　　　　Acid　　Base

The Brønsted–Lowry theory also applies to certain cases where no solution is involved. For example, in the reaction of hydrogen chloride and ammonia gases, HCl is the proton donor and NH_3 is the base. (Remember that "(g)" after a formula in equations stands for a gas.)

$$HCl(g) + NH_3(g) \longrightarrow NH_4^+ + Cl^- \qquad (4)$$
Acid　　　Base　　　　Acid　　Base

In equations (1), (3), and (4), a conjugate acid and base are produced as products. The formulas of a conjugate acid–base pair differ by one proton (H^+). In equation (1), the conjugate base of the acid HCl is Cl^-, and the conjugate acid of the base H_2O is H_3O^+. In equation (3), the conjugate acid of the base H_2O is H_3O^+, and the conjugate base of the acid NH_4^+ is NH_3. In equation (4), HCl–Cl^- and NH_4^+–NH_3 are the conjugate acid–base pairs.

$$HCl + H_2O \longrightarrow H_3O^+ + Cl^-$$
Acid　　Base　　Conjugate　Conjugate
　　　　　　　　acid　　　　base

A more general concept of acids and bases was introduced by Gilbert N. Lewis (1875–1946). The Lewis theory deals with the way in which a substance with an unshared pair of electrons reacts in an acid–base type of reaction. According to this theory a base is any substance that has an unshared pair of electrons (electron-pair donor) and an acid is any substance that will attach itself to or accept a pair of electrons. In the reaction

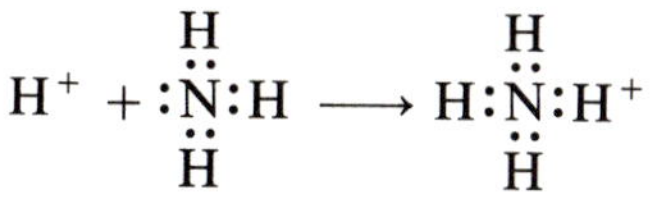

$$H^+ + :\underset{H}{\overset{H}{N}}:H \longrightarrow H:\underset{H}{\overset{H}{N}}:H^+$$

Acid   Base

H^+ is a Lewis acid and $:NH_3$ is a Lewis base. According to the Lewis theory, substances other than proton donors (for example, BF_3) behave as acids:

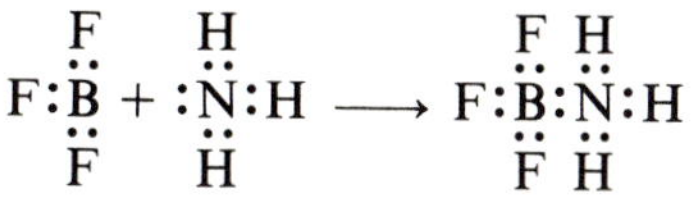

$$F:\underset{F}{\overset{F}{B}} + :\underset{H}{\overset{H}{N}}:H \longrightarrow F:\underset{F}{\overset{F}{B}}:\underset{H}{\overset{H}{N}}:H$$

Acid   Base

The Lewis and Brønsted–Lowry bases are identical, because to accept a proton, a base must have an unshared pair of electrons.

The three theories are summarized in Table 15.1. These theories explain how acid–base reactions occur. We will generally use the theory that best explains the reaction that is under consideration. Most of our examples will refer to aqueous solutions. It is important to realize that in an aqueous acidic solution, the H^+ ion concentration is always greater than the OH^- ion concentration. And, vice versa, in an aqueous basic solution, the OH^- ion concentration is always greater than the H^+ ion concentration.

Table 15.1. Summary of acid–base definitions according to Arrhenius, Brønsted–Lowry, and G. N. Lewis theories.

Theory	Acid	Base
Arrhenius	A hydrogen-containing substance that produces hydrogen ions in aqueous solution	A hydroxide-containing substance that produces hydroxide ions in aqueous solution
Brønsted–Lowry	A proton (H^+) donor	A proton (H^+) acceptor
Lewis	Any species that will bond to an unshared pair of electrons (electron-pair acceptor)	Any species that has an unshared pair of electrons (electron-pair donor)

15.2 Reactions of Acids

In aqueous solutions it is the H^+ or H_3O^+ ions that are responsible for the characteristic reactions of acids. All the following reactions are in an aqueous medium.

(a) **Reaction with metals.** Acids react with metals that lie above hydrogen in the activity series of elements to produce hydrogen and a salt (see Section 17.5).

Acid + Metal ⟶ Hydrogen + Salt

$2\ HCl(aq) + Ca(s) \longrightarrow H_2\uparrow + CaCl_2(aq)$

$H_2SO_4(aq) + Mg(s) \longrightarrow H_2\uparrow + MgSO_4(aq)$

$6\ HC_2H_3O_2(aq) + 2\ Al(s) \longrightarrow 3\ H_2\uparrow + 2\ Al(C_2H_3O_2)_3(aq)$

Acids such as nitric acid (HNO_3) are oxidizing substances (see Chapter 17) and react with metals to produce water instead of hydrogen. For example,

$3\ Zn(s) + 8\ HNO_3(\text{dilute}) \longrightarrow 3\ Zn(NO_3)_2(aq) + 2\ NO(g) + 4\ H_2O$

(b) **Reaction with bases.** The interaction of an acid and a base is called a *neutralization reaction*. In aqueous solutions, the products of this reaction are water and a salt.

Acid + Base ⟶ Salt + Water

$HBr(aq) + KOH(aq) \longrightarrow KBr(aq) + H_2O$

$2\ HNO_3(aq) + Ca(OH)_2(aq) \longrightarrow Ca(NO_3)_2(aq) + 2\ H_2O$

$2\ H_3PO_4(aq) + 3\ Ba(OH)_2(aq) \longrightarrow Ba_3(PO_4)_2\downarrow + 6\ H_2O$

(c) **Reaction with metal oxides.** This reaction is closely related to that of an acid with a base. With an aqueous acid, the products are water and a salt.

Acid + Metal oxide ⟶ Salt + Water

$2\ HCl(aq) + Na_2O(s) \longrightarrow 2\ NaCl(aq) + H_2O$

$H_2SO_4(aq) + MgO(s) \longrightarrow MgSO_4(aq) + H_2O$

$6\ HCl(aq) + Fe_2O_3(s) \longrightarrow 2\ FeCl_3(aq) + 3\ H_2O$

(d) **Reaction with carbonates.** Many acids react with carbonates to produce carbon dioxide, water, and a salt. Carbonic acid (H_2CO_3) is not the product because it is unstable and decomposes into water and carbon dioxide.

Acid + Carbonate ⟶ Salt + Water + Carbon dioxide

$2\ HCl(aq) + Na_2CO_3(aq) \longrightarrow 2\ NaCl(aq) + H_2O + CO_2\uparrow$

$H_2SO_4(aq) + MgCO_3(s) \longrightarrow MgSO_4(aq) + H_2O + CO_2\uparrow$

15.3 Reactions of Bases

The OH^- ions are responsible for the characteristic reactions of bases. All the following reactions are in an aqueous medium.

(a) **Reaction with acids.** Bases react with acids to produce a salt and water. See reaction of acids with bases in Section 15.2(b).

amphoteric

(b) **Amphoteric hydroxides.** Hydroxides of certain metals, such as zinc, aluminum, and chromium, are **amphoteric**; that is, they are capable of reacting as either an acid or a base. When treated with a strong acid, they behave like bases; when reacted with a strong base, they behave like acids.

$$Zn(OH)_2(s) + 2\,HCl(aq) \longrightarrow ZnCl_2(aq) + 2\,H_2O$$
$$Zn(OH)_2(s) + 2\,NaOH(aq) \longrightarrow Na_2ZnO_2(aq) + 2\,H_2O$$

(c) **Reaction of NaOH and KOH with certain metals.** Some amphoteric metals react directly with the strong bases sodium hydroxide and potassium hydroxide to produce hydrogen and a salt.

$$\text{Base} + \text{Metal} \longrightarrow \text{Salt} + \text{Hydrogen}$$
$$2\,NaOH(aq) + Zn(s) \longrightarrow Na_2ZnO_2(aq) + H_2\uparrow$$
$$6\,KOH(aq) + 2\,Al(s) \longrightarrow 2\,K_3AlO_3(aq) + 3\,H_2\uparrow$$

(d) **Reaction with salts.** Bases will react with many salts in solution due to the formation of insoluble metal hydroxides.

$$\text{Base} + \text{Salt} \longrightarrow \text{Metal hydroxide}\downarrow + \text{Salt}$$
$$2\,NaOH(aq) + MnCl_2(aq) \longrightarrow Mn(OH)_2\downarrow + 2\,NaCl(aq)$$
$$3\,NH_4OH(aq) + FeCl_3(aq) \longrightarrow Fe(OH)_3\downarrow + 3\,NH_4Cl(aq)$$
$$2\,KOH(aq) + CuSO_4(aq) \longrightarrow Cu(OH)_2\downarrow + K_2SO_4(aq)$$

15.4 Salts

Salts are very abundant in nature. Most of the rocks and minerals of the earth's mantle are salts of one kind or another. Huge quantities of dissolved salts also exist in the oceans. Salts may be considered to be compounds that have been derived from acids and bases. They consist of positive metal or ammonium ions (H^+ excluded) combined with negative nonmetal ions (OH^+ and O^{2-} excluded). The positive ion is the base counterpart and the nonmetal ion is the acid counterpart:

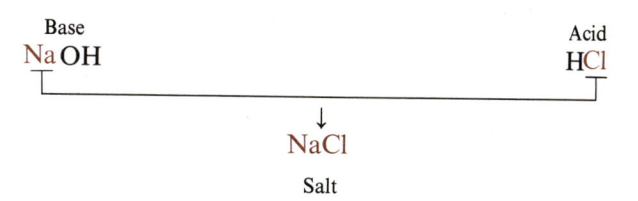

```
       Base                              Acid
       Na OH                              HCl
        └─────────────┬─────────────────┘
                      ↓
                     NaCl
                     Salt
```

Salts are generally crystalline and have high melting and boiling points.

From a single acid such as hydrochloric acid (HCl), we may produce many chloride salts by replacing the hydrogen with a metal ion (for example, NaCl, KCl, RbCl, $CaCl_2$, $NiCl_2$). The number of known salts greatly exceeds the number of known acids and bases. Salts are ionic compounds. If the hydrogen atoms of a binary acid are replaced by a nonmetal, the resulting compound has covalent bonding and is therefore not considered to be a salt (for example, PCl_3, S_2Cl_2, Cl_2O, NCl_3, ICl).

A review of Chapter 8 on the nomenclature of acids, bases, and salts may be beneficial at this point.

15.5 Electrolytes and Nonelectrolytes

Some of the most convincing evidence as to the nature of chemical bonding within a substance is the ability (or lack of ability) of a water solution of the substance to conduct electricity.

It can be readily shown that solutions of certain substances are conductors of electricity. A simple apparatus to demonstrate conductivity consists of a pair of electrodes connected to a voltage source through a light bulb and a switch (see Figure 15.1). When the switch is closed and the medium between the electrodes is a conductor of electricity, the light bulb glows. When chemically pure water is placed in the beaker and the switch is closed, the light does not glow, indicating that water is a nonconductor. When we dissolve a small amount of sugar in water and test the resulting solution, the light does not glow, showing that a sugar solution is also a nonconductor. But when a small amount of salt, NaCl, is dissolved in water and this solution is tested, the light glows brilliantly.

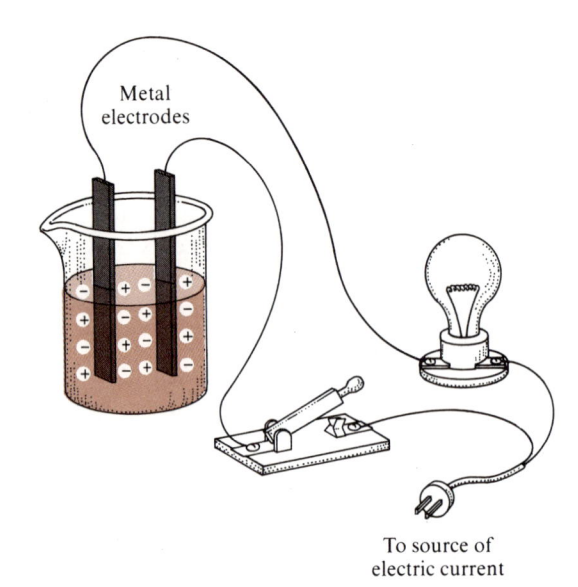

Figure 15.1. A simple conductivity apparatus for testing electrolytes and nonelectrolytes in solution. If the solution contains an electrolyte, the light will glow when the switch is closed.

Ionization, Acids, Bases, Salts 289

Thus, the salt solution conducts electricity. A fundamental difference exists between the chemical bonding of sugar and that of salt. Sugar is a covalently bonded (molecular) substance; salt is an electrovalently bonded (ionic) substance.

Substances whose aqueous solutions are conductors of electricity are called **electrolytes**. Substances whose solutions are nonconductors are known as **nonelectrolytes**. The classes of compounds that are electrolytes are acids, bases, and salts. Solutions of certain oxides are also conductors because they form an acid or a base when dissolved in water. One major difference between electrolytes and nonelectrolytes is that electrolytes exist as ions or are capable of producing ions in solution, whereas nonelectrolytes do not have this property. Solutions that contain a sufficient number of ions will conduct an electric current. Although pure water is a nonconductor, many city water supplies contain enough dissolved ionic matter to cause the light to glow dimly when tested in a conductivity apparatus. Table 15.2 lists some common electrolytes and nonelectrolytes.

Table 15.2. Representative electrolytes and nonelectrolytes.

Electrolytes	Nonelectrolytes
H_2SO_4	$C_{12}H_{22}O_{11}$ (sugar)
HCl	C_2H_5OH (ethyl alcohol)
HNO_3	$C_2H_4(OH)_2$ (ethylene glycol)
$NaOH$	$C_3H_5(OH)_3$ (glycerol)
$HC_2H_3O_2$	CH_3OH (methyl alcohol)
NH_4OH	$CO(NH_2)_2$ (urea)
K_2SO_4	O_2
$NaNO_3$	

⊙ *Acids, bases, and salts are electrolytes.*

15.6 Dissociation and Ionization of Electrolytes

Arrhenius received the 1903 Nobel Prize in chemistry for his work on electrolytes. He stated that a solution conducts electricity because the solute dissociates immediately upon dissolving into electrically charged particles called *ions*. The movement of these ions toward oppositely charged electrodes causes the solution to be a conductor. According to his theory, solutions that are relatively poor conductors contain electrolytes that are partly dissociated. Arrhenius also believed that ions exist in solution whether or not there is an electric current. In other words, the electric current does not cause the formation of ions. Positive ions, attracted to the cathode, are cations; and negative ions, attracted to the anode, are anions.

Positive ions are called cations.
Negative ions are called anions.

We have seen that sodium chloride crystals consist of sodium and chloride ions held together by ionic bonds. When placed in water, these ions are attracted by polar water molecules, which surround each ion as it dissolves. In water, the salt dissociates, forming hydrated sodium and chloride ions (see Figure 15.2). The sodium and chloride ions in solution are bonded to a specific number of water dipoles and have less attraction for each other than they had in the crystalline state. The equation representing this dissociation is

$$NaCl(s) + (x + y)H_2O \longrightarrow Na^+(H_2O)_x + Cl^-(H_2O)_y$$

A simplified dissociation equation in which the water is omitted but understood to be present is

$$NaCl \longrightarrow Na^+ + Cl^-$$

It is important to remember that sodium chloride exists in an aqueous solution as hydrated ions and not as NaCl units, although the formula as such is very often used in equations.

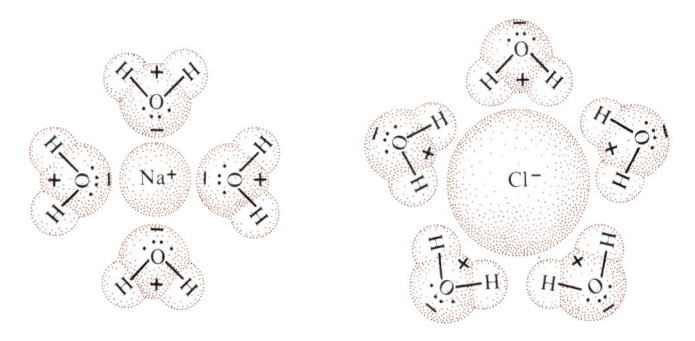

Figure 15.2. Hydrated sodium and chloride ions. When sodium chloride dissolves in water, each Na^+ and Cl^- ion becomes surrounded by water molecules. The negative end of the water dipole is attracted to the Na^+ ion, and the positive end is attracted to the Cl^- ion.

The chemical reactions of salts in solution are the reactions of their ions. For example, when sodium chloride and silver nitrate react and form a precipitate of silver chloride, only the Ag^+ and Cl^- ions participate in the reaction. The Na^+ and NO_3^- remain as ions in solution.

$$Ag^+ + Cl^- \longrightarrow AgCl\downarrow$$

In many cases, the number of molecules of water associated with a particular ion is known. For example, the blue color of the copper(II) ion is due to the hydrated ion $Cu(H_2O)_4^{2+}$. The hydration of ions can be demonstrated in a striking way with cobalt(II) chloride. When cobalt(II) chloride hexahydrate is dissolved in water, a pink solution forms due to the $Co(H_2O)_6^{2+}$ ions. If concentrated hydrochloric acid is added to this pink solution, the color gradually changes to blue. If water is now added to the blue solution, the color changes to pink again. These color changes are due to the exchange of water molecules and chloride ions on the cobalt ion. The complex ion $CoCl_4^{2-}$ is blue. Thus, the hydration of the cobalt ion is a reversible or equilibrium reaction (see Chapter 16). The equilibrium equation representing these changes is

$$Co(H_2O)_6^{2+} + 4\,Cl^- \rightleftharpoons CoCl_4^{2-} + 6\,H_2O$$
$$\text{Pink} \qquad\qquad\qquad \text{Blue}$$

Ionization is the formation of ions; it may occur as a result of a chemical reaction of certain substances with water. Glacial acetic acid (100% $HC_2H_3O_2$) is a liquid that behaves as a nonelectrolyte when tested by the method described in Section 15.5. But a water solution of acetic acid conducts an electric current, as indicated by the dull-glowing light of the conductivity apparatus. The equation for the reaction with water, forming hydronium and acetate ions, is

$$HC_2H_3O_2 + H_2O \rightleftharpoons H_3O^+ + C_2H_3O_2^-$$
$$\text{Acid} \qquad \text{Base} \qquad \text{Acid} \qquad \text{Base}$$

or, in the simplified equation,

$$HC_2H_3O_2 \rightleftharpoons H^+ + C_2H_3O_2^-$$

In the above ionization reaction, water serves not only as a solvent but also as a base according to the Brønsted–Lowry theory.

The bond in hydrogen chloride is predominantly covalent, but when dissolved in water, it reacts, forming hydronium and chloride ions:

$$HCl(g) + H_2O \longrightarrow H_3O^+ + Cl^-$$

When a hydrogen chloride solution is tested for conductivity, the light glows brilliantly, indicating many ions in the solution.

In each of the above two reactions with water, ionization occurs, producing ions in solution. The necessity for water in the ionization process may be demonstrated by dissolving hydrogen chloride in a nonpolar solvent such as benzene, and testing the solution for conductivity. The solution fails to conduct electricity, indicating that no ions are produced.

The terms *dissociation* and *ionization* are often used interchangeably to describe processes taking place in water. But, strictly speaking, the two are different. In the **dissociation** of a salt, the salt already exists as ions, but when dissolved in water, the ions separate or dissociate and increase in mobility. In the **ionization** process, ions are actually produced by the reaction of a compound with water.

dissociation

ionization

Electrolytes composed of two ions per formula unit dissociate to give two ions in solution; electrolytes composed of three ions per formula unit dissociate to give three ions in solution; and so on. The dissociation equations for several electrolytes are given below. In all cases, the ions are actually hydrated.

$NaOH \xrightarrow{H_2O} Na^+ + OH^-$ 2 ions in solution per formula unit

$Ca(OH)_2 \xrightarrow{H_2O} Ca^{2+} + 2\,OH^-$ 3 ions in solution per formula unit

$Na_2SO_4 \xrightarrow{H_2O} 2\,Na^+ + SO_4^{2-}$ 3 ions in solution per formula unit

$AlCl_3 \xrightarrow{H_2O} Al^{3+} + 3\,Cl^-$ 4 ions in solution per formula unit

$Fe_2(SO_4)_3 \xrightarrow{H_2O} 2\,Fe^{3+} + 3\,SO_4^{2-}$ 5 ions in solution per formula unit

One mole of NaCl will give 1 mole of Na^+ ions and 1 mole of Cl^- ions in solution, assuming complete dissociation of the salt. One mole of $CaCl_2$ will give 1 mole of Ca^{2+} ions and 2 moles of Cl^- ions in solution.

$NaCl \xrightarrow{H_2O} Na^+ + Cl^-$
1 mole   1 mole   1 mole

$CaCl_2 \xrightarrow{H_2O} Ca^{2+} + 2\,Cl^-$
1 mole   1 mole   2 moles

We have learned that when 1 mole of sucrose, a nonelectrolyte, is dissolved in 1000 g of water, the solution freezes at $-1.86°C$. When 1 mole of NaCl is dissolved in 1000 g of water, the freezing point of the solution is not $-1.86°C$, as might be expected, but is closer to $-3.72°C$ (-1.86×2). The reason for the lower freezing point is that 1 mole of NaCl in solution produces 2 moles of particles ($2 \times 6.02 \times 10^{23}$ ions) in solution. Thus, the freezing point lowering by 1 mole of NaCl is essentially equivalent to that produced by 2 moles of a nonelectrolyte. An electrolyte such as $CaCl_2$, which yields three ions in water, gives a freezing point depression about three times that of a nonelectrolyte. These freezing point data provide additional evidence that electrolytes dissociate when dissolved in water.

15.7 Strong and Weak Electrolytes

strong electrolyte

weak electrolyte

Electrolytes are classified as either strong or weak, depending on the degree or extent of dissociation or ionization. **Strong electrolytes** are essentially 100% ionized in solution; **weak electrolytes** are considerably less ionized (assuming 0.1 M solutions). Most electrolytes are either strong or weak, with a small number being classified as moderately strong or weak. Most salts are strong electrolytes. Acids and bases that are strong electrolytes (highly ionized) are called *strong acids* and *strong bases*. Acids and bases that are weak electrolytes (slightly ionized) are called *weak acids* and *weak bases*.

For equivalent concentrations, solutions of strong electrolytes contain many more ions than solutions of weak electrolytes. As a result, solutions of strong electrolytes are better conductors of electricity. Consider the two solutions, 1 M HCl and 1 M $HC_2H_3O_2$. Hydrochloric acid is almost 100% ionized;

acetic acid is about 1% ionized. Thus, HCl is a strong acid and $HC_2H_3O_2$ is a weak acid. Hydrochloric acid has about 100 times as many hydronium ions in solution as acetic acid, making the HCl solution much more acidic.

One can distinguish between strong and weak electrolytes experimentally by using the apparatus described in Section 15.5. A 1 M HCl solution causes the light to glow brilliantly, but a 1 M $HC_2H_3O_2$ solution causes only a dull glow. In a similar fashion, the strong base sodium hydroxide (NaOH) may be distinguished from the weak base ammonium hydroxide (NH_4OH). The ionization of a weak electrolyte in water is represented by an equilibrium equation showing that both the un-ionized and ionized forms are present in solution. In the equilibrium equation of $HC_2H_3O_2$ and its ions, the equilibrium is far to the left, since relatively few hydrogen and acetate ions are present in solution:

$$HC_2H_3O_2(aq) \rightleftharpoons H^+ + C_2H_3O_2^-$$

We have previously used a double arrow in an equation to represent reversible processes in the equilibrium between dissolved and undissolved solute in a saturated solution. A double arrow ($\rightleftharpoons$) is also used in the ionization equation of soluble weak electrolytes to indicate that the solution contains a considerable amount of the un-ionized compound in equilibrium with its ions in solution. (See Section 16.1 for a discussion of reversible reactions.) A single arrow is used to indicate that the electrolyte is essentially all in the ionic form in the solution. For example, nitric acid is a strong acid; nitrous acid is a weak acid. Their ionization equations in water may be indicated as

$$HNO_3(aq) \xrightarrow{H_2O} H^+ + NO_3^-$$

$$HNO_2(aq) \xrightleftharpoons{H_2O} H^+ + NO_2^-$$

Practically all soluble salts; acids such as sulfuric, nitric, and hydrochloric acids; and bases such as sodium, potassium, calcium, and barium hydroxides are strong electrolytes. Weak electrolytes include numerous other acids and bases such as acetic acid, nitrous acid, carbonic acid, and ammonium hydroxide. The terms *strong acid*, *strong base*, *weak acid*, and *weak base* refer to whether an acid or base is a strong or weak electrolyte. A list of strong and weak electrolytes is given in Table 15.3.

Table 15.3. Strong and weak electrolytes.

Strong electrolytes	Weak electrolytes
Most soluble salts	$HC_2H_3O_2$
H_2SO_4	H_2CO_3
HNO_3	HNO_2
HCl	H_2SO_3
HBr	H_2S
$HClO_4$	$H_2C_2O_4$
NaOH	H_3BO_3
KOH	HClO
$Ca(OH)_2$	NH_4OH
$Ba(OH)_2$	HF

15.8 Ionization of Water

The more we study chemistry, the more intriguing the little molecule of water becomes. Two equations commonly used to show how water ionizes are

$$H_2O + H_2O \rightleftharpoons H_3O^+ + OH^-$$
$$\text{Acid} \quad \text{Base} \quad \quad \text{Acid} \quad \text{Base}$$

and

$$H_2O \rightleftharpoons H^+ + OH^-$$

The first equation represents the Brønsted–Lowry concept, with water reacting as both an acid and a base, forming a hydronium ion and a hydroxide ion. The second equation is a simplified version, indicating that water ionizes to give a hydrogen and a hydroxide ion. Actually, the proton, H^+, is hydrated and exists as a hydronium ion. In either case, equal molar amounts of acid and base are produced so that water is neutral, having neither H^+ nor OH^- ions in excess. The ionization of water at 25°C produces an H^+ ion concentration of 1.0×10^{-7} mole per litre and an OH^- ion concentration of 1.0×10^{-7} mole per litre. These concentrations are usually expressed as

$$[H^+] \text{ or } [H_3O^+] = 1.0 \times 10^{-7} \text{ mole/litre}$$
$$[OH^-] = 1.0 \times 10^{-7} \text{ mole/litre}$$

These figures mean that about two out of every billion water molecules are ionized. This amount of ionization, small as it is, is a significant factor in the behavior of water in many chemical reactions.

15.9 Introduction to pH

The acidity of an aqueous solution depends on the concentration of hydrogen or hydronium ions. The acidity of solutions involved in a chemical reaction is often critically important, especially for biochemical reactions. The pH scale of acidity was devised to fill the need for a simple, convenient numerical way to state the acidity of a solution. Values on the pH scale are obtained by mathematical conversion of H^+ ion concentrations to pH by this expression:

$$pH = \log \frac{1}{[H^+]} \quad \text{or} \quad -\log[H^+]$$

pH

where $[H^+] = H^+$ or H_3O^+ ion concentration in moles per litre. The **pH** is defined as the logarithm (log) of the reciprocal of the H^+ or H_3O^+ ion concentration in moles per litre. The scale itself is based on the H^+ concentration in water at 25°C. At this temperature, water has an H^+ concentration of 1×10^{-7} mole/litre and is calculated to have a pH of 7.

$$\text{pH} = \log \frac{1}{[\text{H}^+]} = \log \frac{1}{[1 \times 10^{-7}]} = \log 1 \times 10^7 = 7$$

The pH of pure water at 25°C is 7 and is said to be neutral; that is, it is neither acidic nor basic, because the concentrations of H^+ and OH^- are equal. Solutions that contain more H^+ ions than OH^- ions have pH values less than 7, and solutions that contain less H^+ ions than OH^- ions have values greater than 7.

When $[H^+] = 1 \times 10^{-5}$ mole/litre, pH = 5
When $[H^+] = 1 \times 10^{-9}$ mole/litre, pH = 9

Instead of saying that the hydrogen ion concentration in the solution is 1×10^{-5} mole/litre, it is customary to say that the pH of the solution is 5. The smaller the pH value, the more acidic the solution.

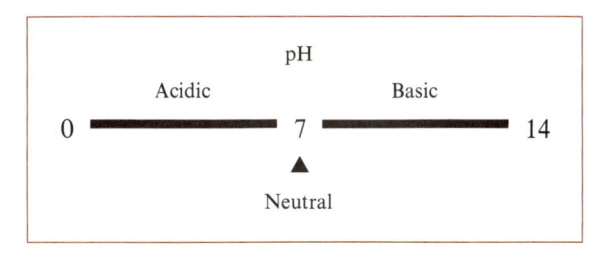

```
                      pH
          Acidic              Basic
    0 ━━━━━━━━━━ 7 ━━━━━━━━━━ 14
                      ▲
                   Neutral
```

A solution of a strong acid is more acidic (has more H^+) than a weak acid at the same molarity. The pH of 0.1 M HCl is 1.00 and that of 0.1 M $HC_2H_3O_2$ is 2.87, indicating that hydrochloric acid is a stronger acid than acetic acid. The $[H^+]$ and thus the pH varies with the degree of dilution of a solution. The following comparative data show that although acetic acid is a weak acid, its pH approaches that of hydrochloric acid (100% ionized) as the solution becomes more dilute; this indicates that a higher percentage of acetic acid molecules ionize as the solution becomes more dilute:

HCl solution	pH	$HC_2H_3O_2$ solution	pH
0.100 M	1.00	0.100 M	2.87
0.0100 M	2.00	0.0100 M	3.37
0.00100 M	3.00	0.00100 M	3.90

The pH scale, along with its interpretation, is given in Table 15.4. Note that a change of only 1 pH unit means a tenfold increase or decrease in H^+ ion concentration.

$$[H^+] = 1 \times 10^{-5}$$

When this number   pH = This number (5)
is exactly 1   pH = 5

$$[\text{H}^+] = 2 \times 10^{-5}$$

When this number is between 1 and 10

pH is between this number and next lower number (4 and 5)
pH = 4.7

Table 15.4. The pH scale for expressing acidity.

$[\text{H}^+]$ (mole/litre)	pH	
1×10^{-14}	14	↑
$0.0000000000001 = 1 \times 10^{-13}$	13	
1×10^{-12}	12	
$0.00000000001 = 1 \times 10^{-11}$	11	Basic
1×10^{-10}	10	
$0.000000001 = 1 \times 10^{-9}$	9	
1×10^{-8}	8	↓
$0.0000001 = 1 \times 10^{-7}$	7	Neutral
1×10^{-6}	6	↑
$0.00001 = 1 \times 10^{-5}$	5	
1×10^{-4}	4	Acid
$0.001 = 1 \times 10^{-3}$	3	
1×10^{-2}	2	
$0.1 = 1 \times 10^{-1}$	1	
$1.0 = 1 \times 10^{0}$	0	↓

Increasing acidity

Table 15.5 lists the pH of some common solutions.

Calculation of the pH value corresponding to any H^+ ion concentration requires the use of logarithms. However, if you are not familiar with the use of logarithms (logs), the following simplified log scale can be used to estimate the logarithms of various numbers:

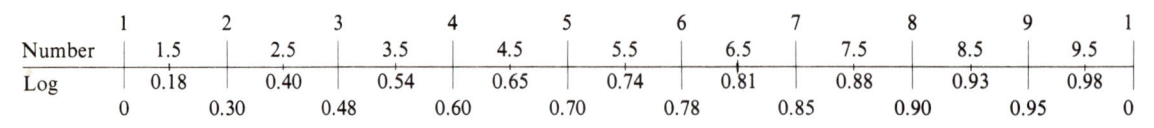

Number	1	1.5	2	2.5	3	3.5	4	4.5	5	5.5	6	6.5	7	7.5	8	8.5	9	9.5	1
Log	0	0.18	0.30	0.40	0.48	0.54	0.60	0.65	0.70	0.74	0.78	0.81	0.85	0.88	0.90	0.93	0.95	0.98	0

Table 15.5 The pH of some common solutions.

Solution	pH
0.1 M HCl	1.0
0.1 M $HC_2H_3O_2$	2.87
Blood	7.4
Urine	5.5–8.0
Lemon juice	2.3
Vinegar	2.8
Milk	6.6
Carbonated water	3.0
Tomatoes	4.0–4.5

Let us see how to use this log scale in calculating the pH of a solution with $[H^+] = 2 \times 10^{-5}$:

$$[H^+] = \underset{}{②} \times 10^{-⑤}$$

pH = This number minus the log of this number (which must be between 1 and 10)

pH = $5 - \log 2$

From the log scale, $\log 2 = 0.30$. Thus, pH = $5 - 0.30 = 4.7$ (Answer)

The measurement and control of pH is extremely important in many fields of science and technology. The proper soil pH is necessary to grow certain types of plants successfully. The pH of certain foods is too acid for some diets. Many biological processes are delicately controlled pH systems. The pH of human blood is regulated to very close tolerances by the uptake or release of H^+ by mineral ions such as HCO_3^- and CO_3^{2-}. Changes in the pH of the blood by as little as 0.4 pH unit result in death.

Compounds with colors that change at particular pH values are used as indicators in acid–base reactions. For example, phenolphthalein, an organic compound, is colorless in acid solution and changes to pink at a pH of 8.3. When a solution of sodium hydroxide is added to a hydrochloric acid solution containing phenolphthalein, the change in color (from colorless to pink) indicates that all the acid is neutralized. Commercially available pH test paper, such as shown in Figure 15.3, contains chemical indicators. The indicator in the paper takes on different colors when wetted with solutions of different pH. Thus, the pH of a solution can be estimated by placing a drop on the test paper and comparing the color of the test paper with a color chart calibrated at different pH values. Electronic pH meters of the type shown in Figure 15.4 are used for making rapid and highly precise pH determinations.

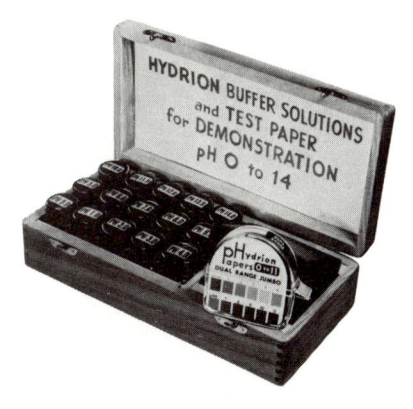

Figure 15.3. pH test paper for determining the approximate acidity of solutions. (Courtesy Micro Essential Laboratory, Inc.)

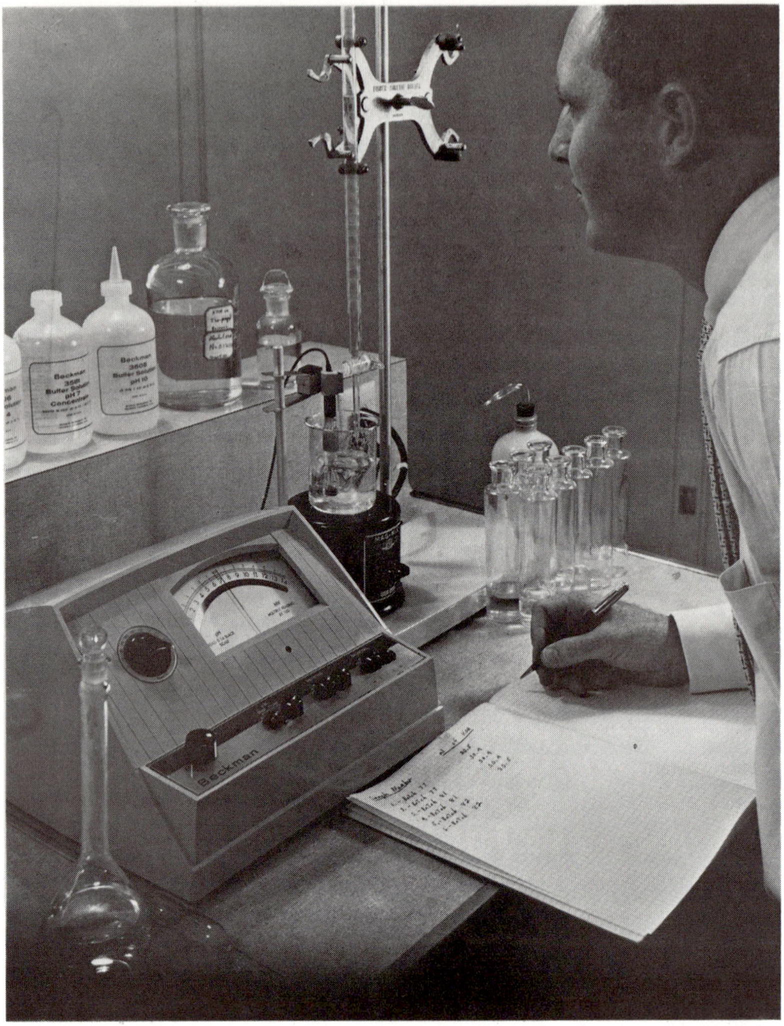

Figure 15.4. An electronic pH meter: Accurate measurements may be made by meters of this type. The scale is calibrated to read in both pH units and millivolts. (Courtesy Beckman Instruments, Inc. Zeromatic is a registered trademark.)

15.10 Neutralization

neutralization

The reaction of an acid and a base to form a salt and water is known as **neutralization**. We have seen this reaction before, but now, in the light of what we have learned about ions and ionization, let us reexamine what occurs during neutralization.

Consider the reaction that occurs when solutions of sodium hydroxide and hydrochloric acid are mixed. The ions present initially are Na^+ and OH^-

Ionization, Acids, Bases, Salts 299

from the base and H⁺ and Cl⁻ from the acid. The products, sodium chloride and water, exist as Na⁺ and Cl⁻ ions and H₂O molecules. A chemical equation representing this reaction is:

$$HCl(aq) + NaOH(aq) \longrightarrow NaCl(aq) + H_2O \tag{1}$$

This equation is a formula equation and does not show that ions are present. The following total ionic equation gives a much better representation of the reaction:

$$(H^+ + Cl^-) + (Na^+ + OH^-) \longrightarrow Na^+ + Cl^- + H_2O \tag{2}$$

spectator ions

Equation (2) shows that the Na⁺ and Cl⁻ ions did not react. These ions are **spectator ions** because they were present but did not take part in the reaction. The only reaction that occurred was that between the H⁺ and OH⁻ ions. Therefore, the equation for the neutralization can be written as this net ionic equation:

$$\underset{\text{Acid}}{H^+} + \underset{\text{Base}}{OH^-} \longrightarrow \underset{\text{Water}}{H_2O} \tag{3}$$

This simple net ionic equation (3) represents not only the reaction of sodium hydroxide and hydrochloric acid but also the reaction of any acid with any base in an aqueous solution. The driving force of a neutralization reaction is the ability of an H⁺ ion and an OH⁻ ion to react and form a molecule of un-ionized water.

titration

The amount of acid, base, or other species in a sample may be determined by titration. **Titration** is the process of measuring the volume of one reagent required to react with a measured weight or volume of another reagent. Let us consider the titration of an acid with a base. A measured volume of acid of unknown concentration is placed into a flask and a few drops of an indicator solution are added. Base solution of known concentration is slowly added from a buret to the acid until the indicator changes color (see Figure 15.5). The indicator selected is one that changes color when the stoichiometric quantity (according to the equation) of base has been added to the acid. At this point, known as the *end point of the titration*, the titration is complete and the volume of base used to neutralize the acid is read from the buret. The concentration or amount of acid in solution can be calculated from the titration data and the chemical equation for the reaction. Illustrative problems follow.

Problem 15.1 Suppose that 42.00 ml of 0.15 M NaOH solution is required to titrate 100 ml of hydrochloric acid solution. What is the molarity of the acid solution?

The equation for the reaction is

$$NaOH(aq) + HCl(aq) \longrightarrow NaCl(aq) + H_2O$$

In this neutralization, NaOH and HCl react in a 1:1 mole ratio. Therefore, the moles of HCl in solution are equal to the moles of NaOH required to react with it. First we calculate the moles of NaOH used, and from this value, the moles of HCl.

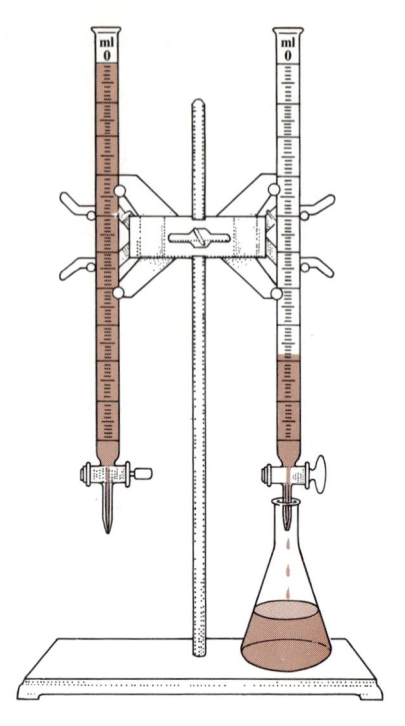

Figure 15.5. Graduated burets are used in titrations for neutralization of acids and bases as well as for many other volumetric determinations. Figure 15.4 illustrates a titration using a pH meter as indicator.

Data: 42.00 ml of 0.15 M NaOH   100 ml HCl   Molarity of acid = M (unknown)

Moles of NaOH:

$$M = \text{moles/litre} \quad 42.00 \text{ ml} = 0.04200 \text{ litre}$$

$$0.04200 \text{ litre} \times \frac{0.15 \text{ mole NaOH}}{\text{litre}} = 0.0063 \text{ mole NaOH}$$

Since NaOH and HCl react in a 1:1 mole ratio, 0.0063 mole of HCl was present in the 100 ml of HCl solution. Therefore, the molarity of the HCl is

$$M = \frac{\text{moles}}{\text{litre}} = \frac{0.0063 \text{ mole HCl}}{0.100 \text{ litre}} = 0.063 \, M \text{ HCl} \quad \text{(Answer)}$$

Problem 15.2 Suppose that 42.00 ml of 0.15 M NaOH solution is required to titrate 100 ml of sulfuric acid (H_2SO_4) solution. What is the molarity of the acid solution?

The equation for the reaction is

$$2 \, NaOH(aq) + H_2SO_4(aq) \longrightarrow Na_2SO_4(aq) + 2 \, H_2O$$

The same amount of base (0.0063 mole of NaOH) is used in this titration as in Problem 15.1. However, the mole ratio of acid to base in the reaction is 1:2. The moles of H_2SO_4 reacted can be calculated by using the mole-ratio method.

Ionization, Acids, Bases, Salts 301

Data: 42.00 ml of 0.15 M NaOH = 0.0063 mole NaOH

$$0.0063 \text{ mole NaOH} \times \frac{1 \text{ mole } H_2SO_4}{2 \text{ moles NaOH}} = 0.00315 \text{ mole } H_2SO_4$$

Therefore, 0.00315 mole of H_2SO_4 was present in 100 ml of H_2SO_4 solution. The molarity of the H_2SO_4 is

$$M = \frac{\text{moles}}{\text{litre}} = \frac{0.00315 \text{ mole } H_2SO_4}{0.100 \text{ litre}} = 0.0315 \; M \; H_2SO_4 \quad \text{(Answer)}$$

15.11 Writing Ionic Equations

molecular equation

total ionic equation

net ionic equation

In Section 15.10, we wrote the reaction of hydrochloric acid and sodium hydroxide in three different equations: (1) the molecular equation, (2) the total ionic equation, and (3) the net ionic equation. In the **molecular equation**, compounds are written in their molecular or normal formula expressions. In the **total ionic equation**, compounds are written in the form in which they are predominantly present: strong electrolytes as ions in solution; and nonelectrolytes, weak electrolytes, precipitates, and gases, in their molecular forms. In the **net ionic equation**, only those molecules or ions that have changed are included in the equation; ions or molecules that do not change are omitted. Up to now, we have been concerned only with balancing the individual elements when we balanced equations. Because we are using ions, which are electrically charged, net ionic equations are often not neutral in charge and end up with a net electrical charge. The net electrical charge of an ionic equation, as well as its atoms, should be in balance. Therefore, a balanced ionic equation will have the same net electrical charge on both sides of the equation, whether it is zero, positive, or negative.

Study the examples below. Note that all reactions are in solution.

(a) $HNO_3(aq) + KOH(aq) \longrightarrow KNO_3(aq) + H_2O$ Molecular equation
  $(H^+ + NO_3^-) + (K^+ + OH^-) \longrightarrow (K^+ + NO_3^-) + H_2O$ Total ionic equation
  $H^+ + OH^- \longrightarrow H_2O$ Net ionic equation

In example (a), HNO_3, KOH, and KNO_3 are soluble, strong electrolytes.

(b) $2\,AgNO_3(aq) + BaCl_2(aq) \longrightarrow 2\,AgCl\downarrow + Ba(NO_3)_2(aq)$
  $(2\,Ag^+ + 2\,NO_3^-) + (Ba^{2+} + 2\,Cl^-) \longrightarrow 2\,AgCl\downarrow + (Ba^{2+} + 2\,NO_3^-)$
  $Ag^+ + Cl^- \longrightarrow AgCl\downarrow$ Net ionic equation

Although silver chloride (AgCl) is an ionic salt, it is written in the molecular form because in example (b), most of the Ag^+ and Cl^- ions are no longer in solution, but have formed a precipitate of AgCl.

(c) $Na_2CO_3(aq) + H_2SO_4(aq) \longrightarrow Na_2SO_4(aq) + H_2O + CO_2\uparrow$
  $(2\,Na^+ + CO_3^{2-}) + (2\,H^+ + SO_4^{2-}) \longrightarrow (2\,Na^+ + SO_4^{2-}) + H_2O + CO_2\uparrow$
  $CO_3^{2-} + 2\,H^+ \longrightarrow H_2O + CO_2\uparrow$ Net ionic equation

In example (c), carbon dioxide (CO_2) is a gas and evolves from solution.

(d) $HC_2H_3O_2(aq) + NaOH(aq) \longrightarrow NaC_2H_3O_2(aq) + H_2O$
$HC_2H_3O_2 + (Na^+ + OH^-) \longrightarrow (Na^+ + C_2H_3O_2^-) + H_2O$
$HC_2H_3O_2 + OH^- \longrightarrow C_2H_3O_2^- + H_2O$ Net ionic equation

In example (d), acetic acid ($HC_2H_3O_2$), a weak acid, is written in the molecular form, but sodium acetate ($NaC_2H_3O_2$), a soluble salt, is written in the ionic form. The Na^+ ion is the only spectator ion in this reaction. Both sides of the net ionic equation have a -1 electrical charge.

(e) $Mg(s) + 2\,HCl(aq) \longrightarrow MgCl_2(aq) + H_2\uparrow$
$Mg + (2\,H^+ + 2\,Cl^-) \longrightarrow (Mg^{2+} + 2\,Cl^-) + H_2\uparrow$
$Mg + 2\,H^+ \longrightarrow Mg^{2+} + H_2\uparrow$ Net ionic equation

In example (e), the net electrical charge on both sides of the equation is $+2$.

(f) $H_2SO_4(aq) + Ba(OH)_2(aq) \longrightarrow BaSO_4\downarrow + 2\,H_2O$
$(2\,H^+ + SO_4^{2-}) + (Ba^{2+} + 2\,OH^-) \longrightarrow BaSO_4\downarrow + 2\,H_2O$
$2\,H^+ + SO_4^{2-} + Ba^{2+} + 2\,OH^- \longrightarrow BaSO_4\downarrow + 2\,H_2O$ Net ionic equation

In example (f), barium sulfate ($BaSO_4$) is a highly insoluble salt. If we conduct this reaction using the apparatus described in Section 15.5, the light, which glows brightly at first, will be extinguished when the reaction is complete, because essentially no ions are left in solution. The $BaSO_4$ precipitates out of solution, and water is a nonconductor.

Here is a list of rules to observe when writing ionic equations.

1. Strong electrolytes are written in their ionic form.
2. Weak electrolytes are written in their molecular form.
3. Nonelectrolytes are written in their molecular form.
4. Insoluble substances, precipitates, and gases are written in their molecular forms.
5. The net ionic equation should include only those substances that have undergone a chemical change.
6. Equations must be balanced, both in atoms and in electrical charge.

15.12 Colloids: Introduction

When we add sugar to a flask of water and shake, the sugar dissolves and forms a clear homogeneous *solution*. When we do the same experiment with very fine sand and water, the sand particles form a *suspension*, which settles when the shaking stops. For a third trial, let us use ordinary corn starch. Starch does not dissolve in cold water; but when the mixture is heated and stirred, the starch forms a cloudy, opalescent *dispersion*. This dispersion does not appear to be clear and homogeneous like the sugar solution. Yet it is not obviously heterogeneous and does not settle like the sand dispersion. In short, its properties are intermediate between those of the sugar solution and those of the sand suspension. The starch dispersion is actually a *colloid*. The name "colloid" is derived from the Greek *kolla*, meaning "glue," and was coined by the English scientist Thomas Graham in 1861. Graham classified solutes as crystalloids if they readily diffused through a parchment membrane and as colloids if they did not diffuse through the membrane.

colloid

As it is now used, the word **colloid** means a dispersion in which the dispersed particles are larger than the solute ions or molecules of a true solution and smaller than the particles of a mechanical suspension. The term does not imply a gluelike quality, although most glues actually are colloidal materials. The size of colloidal particles ranges from a lower limit of about 10 Å (1 nm, or 10^{-7} cm) to an upper limit of about 10,000 Å (1000 nm, or 10^{-4} cm).

The fundamental difference between a colloidal dispersion and a true solution is the size, not the nature, of the particles. The solute particles in a solution are ordinarily single ions or molecules, which may be hydrated to varying degrees. Colloidal particles are usually aggregations of ions or molecules. However, the molecules of some polymers such as proteins are large enough to be classified as colloidal particles when in solution. To appreciate fully the differences in relative size, the volumes, not just the linear dimensions, of colloidal particles and solute particles must be compared. The difference in volumes can be approximated by assuming that the particles are spheres. A large colloidal particle has a diameter of about 5000 Å, while a fair-sized ion or molecule has a diameter of about 5 Å. Thus, the diameter of the colloidal particle is 1000 times that of the solute "particle." Because the volumes of spheres are proportional to the cubes of their diameters, we can readily calculate that the volume of a colloidal particle can be up to a billion ($10^3 \times 10^3 \times 10^3 = 10^9$) times greater than that of a solution "particle"!

Colloids are mixtures in which one component, the *dispersed phase*, exists as discrete particles in the other component, the *dispersing phase*. The components of a colloidal dispersion are also sometimes called the *discontinuous phase* and the *continuous phase*.

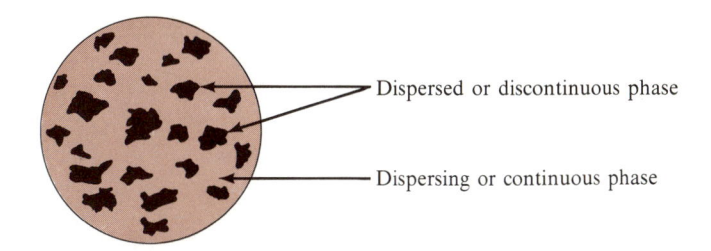

Dispersed or discontinuous phase

Dispersing or continuous phase

Each component, or phase, may exist as a solid, a liquid, or a gas. The components cannot be mutually soluble and both phases of the dispersion cannot be gases because such conditions would describe an ordinary solution. Hence, only eight types of colloidal dispersions, based on the physical states of the phases, are known. The eight types, together with specific examples, are listed in Table 15.6.

Table 15.6. Types of colloidal dispersions.

Type	Name	Examples
Gas in liquid	Foam	Whipped cream, soap suds
Gas in solid	Solid foam	Styrofoam, foam rubber, pumice
Liquid in gas	Liquid aerosol	Fog, clouds
Liquid in liquid	Emulsion	Milk, vinegar in oil salad dressing, mayonnaise
Liquid in solid	Solid emulsion	Cheese, opals, jellies
Solid in gas	Solid aerosol	Smoke, dust in air
Solid in liquid	Sol	India ink, gold sol
Solid in solid	Solid sol	Tire rubber, certain gems (for example, rubies)

15.13 Preparation of Colloids

Colloidal dispersions can be prepared by two general methods: (1) *dispersion*, the breaking down of larger particles to colloidal size; (2) *condensation*, the formation of colloidal particles from solutions.

Homogenized milk is a good example of a colloid prepared by dispersion. Milk, as drawn from the cow, is a poor emulsion of fat in water. The fat globules are so large that they rise and form a cream layer in a few hours. To avoid this separation of the cream, milk is homogenized. This is usually accomplished by pumping the milk through very small holes, or orifices, at high pressure. The violent shearing action of this treatment breaks the fat globules into particles well within the colloidal size range. The butterfat in homogenized milk remains dispersed indefinitely. Colloid mills, which reduce particles to colloidal size by a grinding or shearing action, are used in preparing many commercial products such as paints, cosmetics, and salad dressings.

The preparation of colloids by condensation frequently involves a precipitation reaction in a dilute solution. As an example, a good colloidal sulfur sol can be made by simply bubbling hydrogen sulfide into a solution of sulfur dioxide. Solid sulfur is formed and dispersed as a colloid.

$$SO_2 + 2\,H_2S \longrightarrow 3\,\underset{\substack{\text{Colloidal}\\\text{sulfur}}}{S} + 2\,H_2O$$

A colloidal dispersion is also easily made by adding iron(III) chloride solution to boiling water. The reddish-brown colloidal dispersion that is formed probably consists of iron(III) hydroxide and hydrated iron(III) oxide.

$$FeCl_3 + 3\,HOH \longrightarrow Fe(OH)_3 + 3\,HCl$$

$$Fe(OH)_3 \xrightarrow{H_2O} Fe_2O_3 \cdot xH_2O$$

A great many products for home use (insecticides, insect repellents, and deodorants, to name a few) are packaged as aerosols. The active ingredient, either a liquid or a solid, is dissolved in a liquified gas and sealed under pressure in a container fitted with a release valve. When this valve is opened, the pressurized solution is ejected. The liquified gas vaporizes, and the active ingredient is converted to a colloidal aerosol almost instantaneously.

15.14 Properties of Colloids

In 1827, Robert Brown (1773–1858), while observing strongly illuminated pollen under a high-powered microscope, noted that the pollen grains appeared to have a trembling, erratic motion. He determined later that this erratic motion was not confined to pollen but was characteristic of colloidal particles in general. The random motion of colloidal particles, first reported by Brown, is called **Brownian movement**. It is readily observed in cigarette smoke. The smoke is confined in a small transparent chamber and is illuminated with a strong beam of light at right angles to the optical axis of the microscope. The smoke particles appear as tiny randomly moving lights because the light is reflected from their surfaces. This motion is due to the continual bombardment of the smoke particles by air molecules. Since Brownian movement can be seen when colloidal particles are dispersed in either a gaseous or a liquid medium, it affords nearly direct visual proof that matter at the molecular level actually is moving randomly, as postulated by the Kinetic-Molecular Theory.

When an intense beam of light is passed through an ordinary solution and is viewed at an angle, the beam passing through the solution is hardly visible. A beam of light, however, is clearly visible and sharply outlined when it is passed through a colloidal dispersion (see Figure 15.6). This phenomenon is known as the **Tyndall effect**. It was first described by Michael Faraday in 1857 and later amplified by John Tyndall. The Tyndall effect, like Brownian movement, can be observed in nearly all colloidal dispersions. It occurs because the colloidal particles are large enough to scatter the rays of visible light. The ions or molecules of true solutions are too small to scatter light and, therefore, do not exhibit a noticeable Tyndall effect.

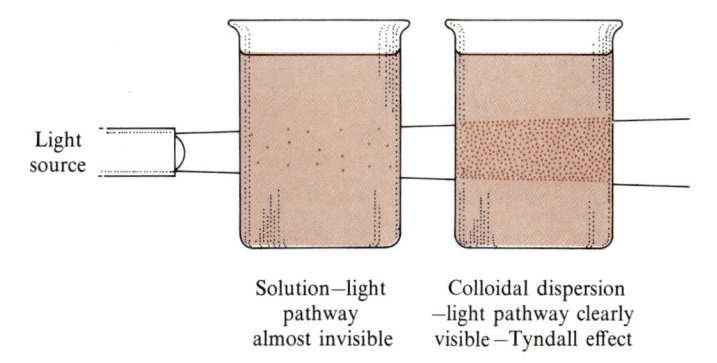

Figure 15.6. The Tyndall effect.

Another important characteristic of colloids is that the particles have relatively huge surface areas. We have seen in Section 14.5 that the surface area is increased tenfold when a 1 cm cube is divided into 1000 cubes with sides of 0.1 cm. When a 1 cm cube is divided into colloidal-size cubes measuring 10^{-6} cm, the surface area of the particles becomes a million times greater than that of the original cube.

Colloidal particles become electrically charged when they adsorb ions on their surfaces. This property is, of course, directly related to the large surface area of the particles. Adsorption occurs because the atoms or ions at the surface of a particle are not surrounded by other atoms or ions like those in the interior. Consequently, these surface ions attract and adsorb ions or polar molecules from the dispersion medium onto the surfaces of the colloidal particles. The particles in a given dispersion tend to adsorb ions having only one kind of charge. For example, cations are primarily adsorbed on ferric hydroxide sol, resulting in positively charged colloidal particles. On the other hand, the particles of an arsenic(III) sulfide (As_2S_3) sol primarily adsorb anions, resulting in negatively charged colloidal particles. The properties of true solutions, colloidal dispersions, and mechanical suspensions are summarized and compared in Table 15.7.

Adsorption should not be confused with *absorption*. Adsorption refers to the adhesion of molecules or ions to a surface, while absorption refers to the taking in of one material by another material.

Table 15.7. Comparison of the properties of true solutions, colloidal dispersions, and suspensions.

	Particle size (Å)	Ability to pass through filter paper	Ability to pass through parchment	Exhibits Tyndall effect	Exhibits Brownian movement	Settles out on standing	Appearance
True solution	<10	Yes	Yes	No	No	No	Transparent, homogeneous
Colloidal dispersion	10–10,000	Yes	No	Yes	Yes	Generally does not	Usually not transparent, but may appear to be homogeneous
Suspension	>10,000	No	No	—	No	Yes	Not transparent, heterogeneous

15.15 Stability of Colloids

The properties of the dispersed and dispersing phases are key factors in making a colloidal dispersion. The stability of different dispersions varies a great deal. We have noted that nonhomogenized milk is a colloid, yet it will separate on standing for a few hours. However, the particles of a good colloidal dispersion will remain in suspension indefinitely. As a case in point, a ruby-red gold sol has been kept in the British Museum for more than a century without noticeable settling. (This specimen is kept for historical interest because it was prepared by Michael Faraday.) The particles of a specific colloid remain dispersed because (1) they are bombarded by the molecules of the dispersing phase and (2) they are electrically charged. Molecular bombardment keeps the particles in motion (Brownian movement) so that gravity does not cause them to settle out. Since the colloidal particles have the same kind of electrical charge, they repel each other. This mutual repulsion prevents the dispersed particles from coalescing to larger particles, which would settle out of suspension.

In certain types of colloids, the presence of a material known as a protective colloid is necessary for stability. Egg yolk, for example, acts as a stabilizer or protective colloid in mayonnaise. The yolk adsorbs on the surfaces of the oil particles and prevents them from coalescing.

15.16 Applications of Colloidal Properties

Activated charcoal is frequently used as an adsorbent in gas masks. This material has an enormous surface area, approximately one million square centimetres per gram in some samples of charcoal. Hence, charcoal is very effective in selectively adsorbing the polar molecules of some poisonous gases. Charcoal can be used to adsorb impurities from liquids as well as from gases, and large amounts are used to remove substances that have objectionable tastes and odors from water supplies. In sugar refineries, activated charcoal is used to adsorb colored impurities from the raw sugar solutions.

The Cottrell process is widely used for dust and smoke control in many urban and industrial areas. This process, devised by an American, Frederick Cottrell (1877–1948), takes advantage of the fact that the particulate matter in dust and smoke is electrically charged. Air to be cleaned of dust or smoke is passed between electrode plates charged with high voltage. Positively charged particles are attracted to, neutralized, and thereby precipitated at the negative electrodes. Negatively charged particles are removed in the same fashion at the positive electrodes. Large Cottrell units are fitted with devices for automatic removal of precipitated material. In some installations, particularly at cement mills and smelters, the value of the dust collected may be sufficient to pay for the precipitation equipment. Small units, designed for removing dust and pollen from air in the home, are now on the market. Unfortunately, Cottrell units remove only particulate matter; they cannot remove gaseous pollutants such as carbon monoxide, sulfur dioxide, and nitrogen oxides.

dialysis

Thomas Graham found that a parchment membrane would allow the passage of true solutions but would prevent the passage of colloidal dispersions. Dissolved solutes can be removed from colloidal dispersions through the use of such a membrane by a process called **dialysis**. The membrane itself is called a *dialyzing membrane*. Many animal membranes can act as dialyzing membranes. Artificial membranes are made from such materials as parchment paper, collodion, or certain kinds of cellophane. Dialysis can be demonstrated by putting a colloidal starch dispersion and some copper(II) sulfate solution in a parchment paper bag and suspending it in running water. In a few hours, the blue color of the copper(II) sulfate has disappeared and only the starch dispersion remains in the bag.

An interesting application of dialysis has been the development of artificial kidneys. These devices are dialyzing units that are able to act as kidneys by removing soluble waste products from the blood. The blood of a patient suffering from partial kidney failure is bypassed through the artificial kidney machine for several hours. During passage through the machine, the soluble waste products are removed by dialysis.

Questions

A. Review the meanings of the new terms introduced in this chapter.
1. Hydronium ion
2. Amphoteric
3. Electrolyte
4. Nonelectrolyte
5. Dissociation
6. Ionization
7. Strong electrolyte
8. Weak electrolyte
9. pH
10. Neutralization
11. Spectator ions
12. Titration
13. Molecular equation
14. Total ionic equation
15. Net ionic equation
16. Colloid
17. Brownian movement
18. Tyndall effect
19. Dialysis

B. Answers to the following questions will be found in tables and figures.
1. In which situations will the light glow in the apparatus shown in Figure 15.1? The switch is closed and the aqueous solution contains:
   (a) HNO_3
   (b) C_2H_5OH
   (c) $C_2H_5OH + NH_4Cl$
   (d) $HC_2H_3O_2$
   (e) $KOH + K_2SO_4$
2. List the solutions in Table 15.5 in order of decreasing acidity.
3. Is it possible for the pH scale shown in Table 15.4 to be extended on both ends? Explain.
4. Between what whole numbers is the pH of a solution that has an H^+ ion concentration of 3.6×10^{-3} mole/litre?
5. Explain the orientation of the water molecules on the hydrated sodium and chloride ions shown in Figure 15.2.
6. Between what levels of H^+ ion concentration does the pH of blood lie?

C. Review questions.
1. Explain why hydrogen chloride is classified as an acid under the (a) Arrhenius, (b) Brønsted–Lowry, and (c) Lewis acid–base theories.
2. Why is a species that is a base by Brønsted–Lowry theory also a base by Lewis theory?
3. How does the Brønsted–Lowry theory define an acid? A base?
4. Can ammonia (NH_3) act as a base in the absence of water (a) by the Arrhenius definition, (b) by the Brønsted–Lowry definition, or (c) by the Lewis definition? Explain.
5. According to the Brønsted–Lowry theory, by what do the formulas of a conjugate acid–base pair differ?
6. Identify the conjugate acid–base pairs in the following equations:
   (a) $HNO_3 + H_2O \rightarrow H_3O^+ + NO_3^-$
   (b) $HC_2H_3O_2 + OH^- \rightarrow H_2O + C_2H_3O_2^-$

(c) $NH_3 + H_3O^+ \rightarrow NH_4^+ + H_2O$
(d) $NH_3 + HBr \rightarrow NH_4^+ + Br^-$
(e) $HC_2H_3O_2 + H_2O \rightarrow H_3O^+ + C_2H_3O_2^-$

7. Write the electron-dot structure for:
   (a) Chloride ion (b) Hydroxide ion (c) Sulfate ion
   Why are these ions considered to be bases according to the Brønsted–Lowry and Lewis acid–base theories?

8. Complete and balance the following equations:
   (a) $CaO(s) + H_2SO_4(aq) \rightarrow$
   (b) $Zn(s) + HCl(aq) \rightarrow$
   (c) $NH_3(g) + HC_2H_3O_2(g) \rightarrow$
   (d) $Ba(OH)_2(aq) + HClO_4(aq) \rightarrow$
   (e) $K_2CO_3(aq) + HI(aq) \rightarrow$
   (f) $Fe_2O_3(s) + HNO_3(aq) \rightarrow$

9. Complete and balance the following equations:
   (a) $NaOH(aq) + HCl(aq) \rightarrow$
   (b) $Ca(OH)_2(aq) + HBr(aq) \rightarrow$
   (c) $Ba(OH)_2(aq) + H_2SO_4(aq) \rightarrow$
   (d) $Ca(OH)_2(aq) + Na_2CO_3(aq) \rightarrow$
   (e) $NH_4OH(aq) + FeCl_3(aq) \rightarrow$

10. Which of the following are electrolytes and which are nonelectrolytes? Consider each substance to be mixed with water.
    (a) $Mg(NO_3)_2$ (g) HBr
    (b) CH_4 (insoluble) (h) K_2CrO_4
    (c) CO_2 (i) HCOOH (formic acid)
    (d) Na_2O (j) C_2H_5OH (ethyl alcohol)
    (e) C_6H_6 (insoluble) (k) H_3PO_4
    (f) H_2 (insoluble) (l) LiOH

11. Name all the compounds shown in Table 15.3.

12. Explain the following statement in terms of ionization and chemical bonding: When solutions of hydrogen chloride in water and in benzene are prepared, the water solution conducts an electric current but the benzene solution does not.

13. An aqueous solution of ethyl alcohol (C_2H_5OH) does not conduct an electric current but an aqueous solution of potassium hydroxide (KOH) does. What does this information tell us about the OH group in the alcohol?

14. Why does molten sodium chloride conduct an electric current?

15. How does a hydronium ion differ from a hydrogen ion?

16. How does hydrochloric acid differ from hydrogen chloride? Are both electrolytes?

17. Water solutions of HCl and of NaCl are strong electrolytes, but the processes by which HCl and NaCl dissolve in water are quite different. Explain.

18. Distinguish between the dissociation of ionic compounds and the ionization of molecular compounds.

19. Distinguish between weak and strong electrolytes.

20. Explain why ions are hydrated in aqueous solution.

21. Write simplified equations to show how the following compounds ionize or dissociate in water:
    (a) $NaNO_3$ (c) $HC_2H_3O_2$ (e) $HClO_4$ (g) $H_2C_2O_4$
    (b) NH_4Br (d) K_2SO_4 (f) $HClO$ (h) Na_3PO_4

22. Write the net ionic equation for the reaction of an acid with a base in an aqueous solution.

23. Rewrite the following unbalanced equations, converting them into balanced net ionic equations (all reactions are in water solution):
    (a) $BaCl_2(aq) + AgNO_3(aq) \rightarrow Ba(NO_3)_2(aq) + AgCl\downarrow$
    (b) $MgCO_3(s) + HCl(aq) \rightarrow MgCl_2(aq) + H_2O + CO_2\uparrow$
    (c) $Zn(s) + HC_2H_3O_2(aq) \rightarrow Zn(C_2H_3O_2)_2 + H_2\uparrow$
    (d) $H_2S(g) + Hg(C_2H_3O_2)_2(aq) \rightarrow HgS\downarrow + HC_2H_3O_2$
    (e) $CaCl_2(aq) + (NH_4)_2C_2O_4(aq) \rightarrow CaC_2O_4\downarrow + NH_4Cl(aq)$
24. Water contains both acid and base ions. Why is it neutral?
25. At 100°C water ionizes about ten times as much as at 25°C and contains about 1×10^{-6} mole of H^+ ions/litre. Is water acidic at this temperature? Explain.
26. Which is more acidic, $1\ M\ HNO_3$ or $1\ M\ H_2SO_4$?
27. Two drops (0.10 ml) of $1.0\ M$ HCl are added to pure water to make 1 litre of solution. What is the pH of this solution if the HCl is 100% ionized?
28. Samples of lemon juice and tomato juice have pH values of 2.3 and 4.3, respectively. How many times greater is the H^+ ion concentration in the lemon juice than in the tomato juice?
29. Explain why a solution containing 1 mole of acetic acid in 1 kg of water freezes at a lower temperature than one containing 1 mole of ethyl alcohol in 1 kg of water.
30. Three solutions are made by dissolving 1 mole of each of the following substances in 1 kg of water:
    (a) HCl   (b) $C_6H_{12}O_6$ (glucose)   (c) $CaCl_2$
    Arrange these solutions in decreasing order of their freezing points, placing the one with the highest freezing point first.
31. Based on the physical states of the dispersed phases, what type of colloidal dispersion cannot exist? Explain.
32. Give brief descriptions of the two general methods for preparing colloidal dispersions.
33. What are the principal reasons that colloidal particles, once dispersed, do not settle out?
34. How does homogenized milk differ physically from nonhomogenized milk?
35. Ozone is a serious air pollutant in some areas. Would a small Cottrell precipitation unit be of value in reducing the ozone concentration in a home in such an area? Explain.
36. Would activated charcoal be more efficient at 0°C or at 100°C in absorbing gaseous hydrogen sulfide? Explain.
37. Which of the following statements are correct?
    (a) Seawater will boil at a higher temperature than pure water.
    (b) The concentration of Cl^- in a $0.50\ M\ AlCl_3$ solution is $1.5\ M$.
    (c) $H^+(aq) + OH^-(aq) \rightarrow H_2O$ represents a neutralization reaction.
    (d) Water can act as both a Brønsted–Lowry acid and a Brønsted–Lowry base.
    (e) The conjugate base of NH_4^+ is NH_3.
    (f) The conjugate acid of Cl^- is HCl.
    (g) Sodium acetate ($NaC_2H_3O_2$) is a weak electrolyte.
    (h) Na^+, Ca^{2+}, Al^{3+}, and NH_3 are all considered to be cations.
    (i) One mole of $CaCl_2$ contains $3 \times 6.02 \times 10^{23}$ ions.
    (j) A solution with $[H^+] = 1 \times 10^{-8}$ mole/litre is acidic.
    (k) A $0.001\ M$ HCl solution has a pH of 3.
    (l) Equal volumes of $0.10\ M$ HCl and $0.10\ M\ HC_2H_3O_2$ will react with the same volume of $0.20\ M$ NaOH solution.

(m) A 1.0 molal solution of HCl will freeze at a lower temperature than a 1.0 molal solution of $HC_2H_3O_2$.
(n) The size of colloidal particles ranges from 10^{-4} to 10^{-1} cm.
(o) Adsorption refers to the adhesion of molecules or ions to a surface.
(p) The Tyndall effect can be observed because colloidal particles are large enough to scatter the rays of visible light.

D. Review problems.
1. Calculate the molarity of the ions in each of the salt solutions listed below. Consider each salt to be 100% dissociated. For example, calculate both the Na^+ and Cl^- ion molarity in a NaCl solution.
   (a) 0.10 M NaCl
   (b) 0.32 M KNO_3
   (c) 1.25 M Na_2SO_4
   (d) 0.68 M $CaCl_2$
   (e) 0.22 M $FeCl_3$
   (f) 0.75 M $MgSO_4$
   (g) 0.050 M $(NH_4)_3PO_4$
   (h) 0.050 M $Al_2(SO_4)_3$
2. What is the molar concentration of a magnesium bromide ($MgBr_2$) solution that has a bromide ion molarity of 0.526 M?
3. How many Al^{3+} and Cl^- ions are there in 10.0 ml of 0.001 M $AlCl_3$ solution? Assume 100% dissociation.
4. Given the data for the following five titrations, calculate the molarity of HCl in titrations (a), (b), and (c), and the molarity of NaOH in titrations (d) and (e):

Volume of HCl (ml)	Molarity of HCl	Volume of NaOH (ml)	Molarity of NaOH
(a) 25.00	—	35.00	0.150
(b) 32.40	—	18.50	0.425
(c) 9.50	—	48.00	0.235
(d) 24.00	0.260	24.30	—
(e) 12.20	0.260	49.80	—

5. What would be the concentration of each ion in a solution made by dissolving 40.0 g of magnesium chloride ($MgCl_2$) in sufficient water to make 250 ml of solution? How many moles of silver chloride (AgCl) could be precipitated from this solution?
6. The concentration of a barium hydroxide solution is determined by titration with hydrochloric acid:

$$Ba(OH)_2(aq) + 2HCl(aq) \longrightarrow BaCl_2(aq) + 2H_2O$$

It was found that 32.8 ml of 0.200 M HCl were required to titrate 25.0 ml of $Ba(OH)_2$ solution. What is the molarity of the $Ba(OH)_2$ solution?
7. How many moles of solute and solvent are used in preparing a solution containing 210 g of benzene (C_6H_6) and 1000 g of carbon tetrachloride (CCl_4)?
8. What volume of each component should be used to prepare the solution in Problem 7? (Densities at 20°C: $C_6H_6 = 0.879$ g/ml; $CCl_4 = 1.595$ g/ml.)
9. A sample of pure sodium bicarbonate weighing 0.420 g was dissolved in water and neutralized with hydrochloric acid; 42.5 ml of the acid were required. Compute the molarity of the acid.

$$NaHCO_3(aq) + HCl(aq) \longrightarrow NaCl(aq) + CO_2\uparrow + H_2O$$

10. What volume of hydrogen gas can be obtained at 27°C and 600 mm Hg pressure by reacting 3.00 g Zn metal with 100 ml of 0.300 M HCl?

11. Calculate the pH of the following solutions:
    (a) $[H^+] = 0.001\ M$
    (b) $[H^+] = 0.10\ M$
    (c) $[H^+] = 1 \times 10^{-7}\ M$
    (d) $[H^+] = 8.5 \times 10^{-10}\ M$
    (e) $[H^+] = 0.50\ M$
12. Calculate the pH of the following:
    (a) Orange juice, $[H^+] = 2 \times 10^{-4}\ M$
    (b) Vinegar, $[H^+] = 1.6 \times 10^{-3}\ M$
    (c) Black coffee, $[H^+] = 6.3 \times 10^{-6}\ M$
    (d) Limewater, $[H^+] = 3.2 \times 10^{-11}\ M$
13. Suppose that 30.0 g of acetic acid, $(HC_2H_3O_2)$ are dissolved in 100 g of water, forming a solution that has a density of 1.03 g/ml. What is the volume of the solution?
14. A solution is made by dissolving 12.0 g of barium chloride $(BaCl_2)$ in sufficient water to make 50.0 ml of solution. The density of the solution is 1.203 g/ml. Calculate the weight percent and the molarity of the solution.
15. What volume of concentrated (18.0 M) sulfuric acid must be used to prepare 20.0 litres of 3.00 M solution?
16. How many grams of silver iodide (AgI) will be precipitated when 10.0 ml of 1.00 M KI and 19.0 ml of 0.500 M Ag NO_3 are mixed together?

16 Chemical Equilibrium

After studying Chapter 16 you should be able to:

1. Understand the terms listed in Question A at the end of the chapter.
2. Describe a reversible reaction.
3. Explain why the rate of the forward reaction decreases and the rate of the reverse reaction increases as a chemical reaction approaches equilibrium.
4. State the principle of Le Chatelier.
5. Tell how the speed of a chemical reaction is affected by the following: (a) changes in concentration of reactants, (b) changes of pressure on gaseous reactants, (c) changes of temperature, and (d) presence of a catalyst.
6. Write the equilibrium constant expression for a chemical reaction from the balanced equation.
7. Calculate the concentration of one substance in an equilibrium when given the equilibrium constant and the concentrations of all the other substances.
8. Calculate the concentrations of all the chemical species in a solution of a weak acid when given the percent ionization or the ionization constant.
9. Compare the relative strengths of acids from their ionization constants.
10. Using the ion product constant for water, K_w, calculate $[H^+]$, $[OH^-]$, pH, or pOH when given any one of these quantities.
11. Calculate the solubility product constant, K_{sp}, of a slightly soluble salt when given its solubility, or vice versa.
12. Calculate the solubility of a salt when given the K_{sp} value.
13. Compare the relative solubilities of salts when given their solubility products.
14. Explain how a buffer solution is able to counteract the addition of small amounts of either H^+ or OH^- ions.
15. Explain the relative energy diagram of a reaction in terms of activation energy, exothermic or endothermic reaction, and the effect of a catalyst.

16.1 Reversible Reactions

In the preceding chapters we have treated chemical reactions mainly as reactants going to products. However, many reactions do not go to completion. One reason why reactions do not go to completion is that many of them are reversible; that is, when the products are formed, they react to produce the starting reactants.

We have encountered several reversible systems. One is the vaporization of a liquid by heating and subsequent condensation by cooling:

$$\text{Liquid} + \text{Heat} \longrightarrow \text{Vapor}$$
$$\text{Vapor} + \text{Cooling} \longrightarrow \text{Liquid}$$

Another is the crystallization of an aqueous salt solution, which may be considered the reverse of the dissolving and dissociation of the salt:

$$\text{NaCl}(s) \xrightarrow{H_2O} \text{Na}^+(aq) + \text{Cl}^-(aq) \quad \text{(Dissociation)}$$
$$\text{Na}^+(aq) + \text{Cl}^-(aq) \longrightarrow \text{NaCl}(s) \quad \text{(Crystallization)}$$

Weak electrolytes are ionized to a small degree because of the reversible reaction of their ions to form the un-ionized compound. A 1 M solution of acetic acid illustrates this behavior:

$$HC_2H_3O_2 + H_2O \longrightarrow H_3O^+ + C_2H_3O_2^- \quad \text{(Forward reaction 1\%)}$$
$$H_3O^+ + C_2H_3O_2^- \longrightarrow HC_2H_3O_2 + H_2O \quad \text{(Reverse reaction 99\%)}$$

These two reactions may be represented by a single equation with a double arrow, $\rightleftarrows$, to indicate that they are taking place at the same time:

$$HC_2H_3O_2 + H_2O \rightleftarrows H_3O^+ + C_2H_3O_2^- \quad \text{(This single equation represents both the forward and reverse reactions)}$$

The interconversion of nitrogen dioxide (NO_2) and dinitrogen tetroxide (N_2O_4) offers visible evidence of the reversibility of a reaction. NO_2 is a reddish-brown gas that changes, with cooling, to N_2O_4, a yellow liquid boiling at 21.2°C and to a colorless solid melting at −11.2°C. The reaction is reversible by heating the N_2O_4.

$$2\,NO_2(g) \xrightarrow{\text{Cooling}} N_2O_4(l)$$
$$N_2O_4(l) \xrightarrow{\text{Heating}} 2\,NO_2(g)$$

This reversible reaction can readily be demonstrated by sealing samples of NO_2 in two tubes and placing one tube in warm water and the other in ice water (see Figure 16.1). Heating promotes disorder or randomness in a system, so we would expect more NO_2, a gas, to be present at higher temperatures.

A **reversible reaction** is one in which the products formed in a chemical reaction also react to produce the original reactants. Both the forward reaction and the reverse reaction occur simultaneously. The forward reaction is also called *the reaction to the right*, and the reverse reaction is called *the reaction to the left*. A double arrow is used in the equation to indicate that the reaction is reversible.

reversible reaction

Chemical Equilibrium 315

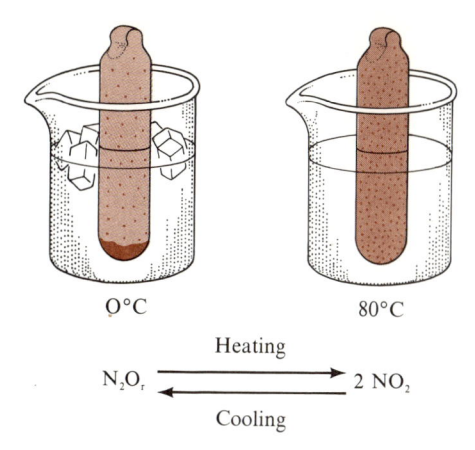

$$N_2O_4 \underset{\text{Cooling}}{\overset{\text{Heating}}{\rightleftharpoons}} 2\,NO_2$$

Figure 16.1. Reversible reaction of nitrogen dioxide (NO_2) and dinitrogen tetroxide (N_2O_4). More of the reddish-brown NO_2 molecules are visible in the tube that is heated than in the tube that is cooled.

16.2 Rates of Reaction

Every reaction has a rate, or speed, at which it proceeds. Some are fast and some are extremely slow. The study of reaction rates and reaction mechanisms is known as **chemical kinetics**.

chemical kinetics

The speed of a reaction is variable and depends on the concentration of the reacting species, the temperature, the presence of catalytic agents, and the nature of the reactants. Consider the hypothetical reaction

$A + B \longrightarrow C + D$ (Forward reaction)
$C + D \longrightarrow A + B$ (Reverse reaction)

where a collision between A and B is necessary for a reaction to occur. The rate at which A and B react depends on the concentration or the number of A and B molecules present; it will be fastest, for a fixed set of conditions, when they are first mixed. As the reaction proceeds, the number of A and B molecules available for reaction decreases, and the rate of reaction slows down. If the reaction is reversible, the speed of the reverse reaction is zero at first, and gradually increases as the concentrations of C and D increase. As the number of A and B molecules decreases, the forward rate slows down, because A and B cannot find one another as often in order to accomplish a reaction. To counteract this diminishing rate of reaction, an excess of one reagent is often used to keep the reaction from becoming unreasonably slow. Collisions between molecules may be likened to the scooters or "dodge'ems" found at amusement parks. When many cars are on the floor, collisions occur frequently, but if only a few cars are present, collisions can usually be avoided.

16.3 Chemical Equilibrium

equilibrium

chemical equilibrium

Any system at **equilibrium** represents a dynamic state in which two or more opposing processes are taking place at the same time and at the same rate. A chemical equilibrium is a dynamic system in which two or more chemical reactions are going on at the same time and at the same rate. When the rate of the forward reaction is exactly equal to the rate of the reverse reaction, a condition of **chemical equilibrium** exists (see Figure 16.2). The concentration of the products is not changing and the system appears to be at a standstill because the products are reacting at the same rate at which they are being formed.

Chemical Equilibrium:
Rate of forward reaction = Rate of reverse reaction

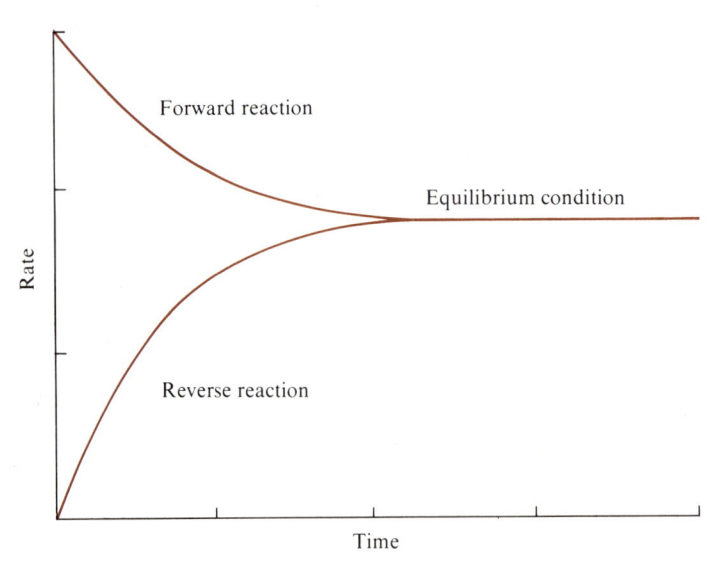

Figure 16.2. The graph illustrates that the rates of the forward and reverse reactions become equal at some point in time. The forward reaction rate decreases as a result of decreasing amounts of reactants. The reverse reaction rate starts at zero and increases as the amount of product increases. When the two rates become equal, a state of chemical equilibrium has been reached.

A saturated salt solution is in a condition of equilibrium:

$$NaCl(s) \rightleftharpoons Na^+(aq) + Cl^-(aq)$$

At equilibrium, salt crystals are continuously dissolving, and Na^+ and Cl^- ions are continuously crystallizing. Both processes are occurring at the same rate.

The ionization of weak electrolytes represents another common chemical equilibrium system:

$$HC_2H_3O_2 + H_2O \rightleftharpoons H_3O^+ + C_2H_3O_2^-$$

In this reaction, the equilibrium is established in a 1 M solution when the forward reaction has gone about 1%—that is, when only 1% of the acetic acid molecules in solution have ionized. Therefore, only a relatively few ions are present, and the acid behaves as a weak electrolyte. In any acid–base equilibrium system, the position of equilibrium favored is toward the weaker conjugate acid and base. In the ionization of acetic acid, the $HC_2H_3O_2$ is a weaker acid than the H_3O^+, and H_2O is a weaker base than $C_2H_3O_2^-$.

The reactions represented by

$$H_2 + I_2 \rightleftharpoons 2\,HI$$

provide another good example of a chemical equilibrium. Theoretically, 1.00 mole of hydrogen should react with 1.00 mole of iodine to yield 2.00 moles of hydrogen iodide. Actually, when 1.00 mole of H_2 and 1.00 mole of I_2 are reacted at 700 K, only 1.58 moles of HI are present when equilibrium is attained. Since 1.58 is 79% of the theoretical yield of 2.00 moles of HI, the forward reaction is only 79% complete at equilibrium. The equilibrium mixture will also contain 0.21 mole each of unreacted H_2 and I_2 because only 79% has reacted (1.00 mole − 0.79 mole = 0.21 mole).

$$H_2 + I_2 \xrightarrow{700\,K} 2\,HI$$
1.00 mole   1.00 mole   2.00 moles

(This would represent the condition if the reaction were 100% complete; 2.00 moles of HI would be formed and no H_2 and I_2 would be left unreacted.)

$$H_2 + I_2 \xrightleftharpoons{700\,K} 2\,HI$$
0.21 mole   0.21 mole   1.58 moles

(This represents the actual equilibrium attained starting with 1.00 mole each of H_2 and I_2. It shows that the forward reaction is only 79% complete.)

16.4 Principle of Le Chatelier

principle of Le Chatelier

In 1888, the French chemist Henri Le Chatelier (1850–1936) set forth a simple far-reaching generalization on the behavior of equilibrium systems. This generalization is known as the **principle of Le Chatelier** and states: If the conditions of a system in equilibrium are altered, then processes occur in the system that tend to counteract the change. In other words, if a stress is applied to a system in equilibrium, the system will behave in such a way as to relieve that stress and restore equilibrium, but under a new set of conditions.

The application of Le Chatelier's principle helps us to predict the effect of changing conditions in chemical reactions. We will examine the effect of changes in concentration, temperature, and pressure.

16.5 Effect of Concentration on Reaction Rate and Equilibrium

The way in which the rate of a chemical reaction depends on the concentration of the reactants must be determined experimentally. Many simple, one-step reactions occur as the result of a collision between two molecules or ions. The rate of such one-step reactions can be altered by changing the concentration of the reactants or products. An increase in concentration of the reactants provides more individual reacting species for collisions and results in an increase in the rate of reaction.

An equilibrium is disturbed when the concentration of one or more of its components is changed. As a result, the concentration of all the species will change and a new equilibrium mixture will be established. Consider the hypothetical equilibrium represented by the equation

$$A + B \rightleftharpoons C + D$$

where A and B react in one step to form C and D. When the concentration of B is increased, the following occurs:

1. The rate of the reaction to the right (forward) increases. This rate is proportional to the concentration of A times the concentration of B.
2. Therefore, the rate to the right is greater than the rate to the left.
3. Reactants A and B are used faster than they are produced; C and D are produced faster than they are used.
4. After a period of time, rates to the right and left become equal, and the system is again in equilibrium.
5. In the new equilibrium, the concentration of A is less and the concentrations of C and D are greater than in the original equilibrium.

Conclusion: The equilibrium has shifted to the right.

Applying this change in concentration to the equilibrium mixture of 1.00 mole of hydrogen and 1.00 mole of iodine from Section 16.3, we find that when an additional 0.20 mole of I_2 is added, the yield of HI is 85% (1.70 moles) instead of 79%. A comparison of the two systems after the new equilibrium mixture is reached follows.

Original equilibrium	New equilibrium
1.00 mole H_2 + 1.00 mole I_2	1.00 mole H_2 + 1.20 moles I_2
Yield: 79% HI	Yield: 85% HI (based on H_2)
Equilibrium mixture contains:	Equilibrium mixture contains:
1.58 moles HI	1.70 moles HI
0.21 mole H_2	0.15 mole H_2
0.21 mole I_2	0.35 mole I_2

Analyzing this new system, we see that when the 0.20 mole I_2 was added, the equilibrium shifted to the right in order to counteract the change of I_2

concentration. Some of the H_2 reacted with added I_2 and produced more HI, until an equilibrium mixture was established again. When I_2 was added, the concentration of I_2 increased, the concentration of H_2 decreased, and the concentration of HI increased. What do you think would be the effects of adding (a) more H_2 and (b) more HI?

The equation

$$Fe^{3+}(aq) + SCN^-(aq) \rightleftharpoons Fe(SCN)^{2+}(aq)$$
Pale yellow   Colorless   Red

represents an equilibrium that is used in certain analytical procedures as an indicator because of the readily visible, intense red color of the complex $Fe(SCN)^{2+}$ ion. A very dilute solution of iron(III), Fe^{3+}, and thiocyanate, SCN^-, is light red in color. When the concentration of Fe^{3+} or SCN^- is increased, the equilibrium shift to the right is observed by an increase in the intensity of the color, resulting from the formation of additional $Fe(SCN)^{2+}$.

If either Fe^{3+} or SCN^- is removed from solution, the equilibrium will shift to the left, and the solution will become lighter in color. When Ag^+ is added to the above solution, a white precipitate of silver thiocyanate (AgSCN) is formed, thus removing SCN^- ion from the equilibrium:

$$Ag^+(aq) + SCN^-(aq) \longrightarrow AgSCN\downarrow$$

The system accordingly responds to counteract the change in SCN^- concentration by shifting the equilibrium to the left. This shift is evident by a decrease in the intensity of the red color due to a decreased concentration of $Fe(SCN)^{2+}$.

Let us now consider the effect of changing the concentrations in the equilibrium mixture of chlorine water. The equilibrium equation is

$$Cl_2(aq) + 2 H_2O \rightleftharpoons HOCl(aq) + H_3O^+ + Cl^-(aq)$$

The variation in concentrations and the equilibrium shifts are tabulated below. An X in the second or third column indicates the reagent that is increased or decreased. The fourth column indicates the direction of the equilibrium shift.

| | Concentration | | |
Reagent	Increase	Decrease	Equilibrium shift
Cl_2	—	X	Left
H_2O	X	—	Right
HOCl	X	—	Left
H_3O^+	—	X	Right
Cl^-	X	—	Left

Consider the equilibrium in a 0.1 M acetic acid solution:

$$HC_2H_3O_2 + H_2O \rightleftharpoons H_3O^+ + C_2H_3O_2^-$$

In this solution, the concentration of the hydronium ion (H_3O^+), which is a measure of the acidity, is 1.34×10^{-3} mole/litre, corresponding to a pH of 2.87.

What will happen to the acidity when 0.1 mole of sodium acetate (NaC$_2$H$_3$O$_2$) is added to 1 litre of 0.1 M HC$_2$H$_3$O$_2$? When NaC$_2$H$_3$O$_2$ dissolves, it dissociates into sodium (Na$^+$) and acetate (C$_2$H$_3$O$_2^-$) ions. The acetate ion from the salt is a common ion to the acetic acid equilibrium system and increases the total acetate ion concentration in the solution. As a result, the equilibrium shifts to the left, decreasing the hydronium ion concentration and lowering the acidity of the solution. Evidence of this decrease in acidity is shown by the fact that the pH of a solution that is 0.1 M in HC$_2$H$_3$O$_2$ and 0.1 M in NaC$_2$H$_3$O$_2$ is 4.74. The pH of several different solutions of HC$_2$H$_3$O$_2$ and NaC$_2$H$_3$O$_2$ is shown below. Each time the acetate ion is increased, the pH increases, showing a further shift in the equilibrium toward un-ionized acetic acid.

Solution	pH
1 litre 0.1 M HC$_2$H$_3$O$_2$	2.87
1 litre 0.1 M HC$_2$H$_3$O$_2$ + 0.1 mole NaC$_2$H$_3$O$_2$	4.74
1 litre 0.1 M HC$_2$H$_3$O$_2$ + 0.2 mole NaC$_2$H$_3$O$_2$	5.05
1 litre 0.1 M HC$_2$H$_3$O$_2$ + 0.3 mole NaC$_2$H$_3$O$_2$	5.23

A secondary reaction of the acetate ion, which also aids in reducing the acidity, is its reaction with water, forming un-ionized HC$_2$H$_3$O$_2$ and a hydroxide (OH$^-$) ion:

$$C_2H_3O_2^- + H_2O \rightleftharpoons HC_2H_3O_2 + OH^-$$

The OH$^-$ ion produced reacts with H$_3$O$^+$ to decrease the acidity. Proof that sodium acetate produces OH$^-$ ions in solution is shown by the fact that a 0.1 M NaC$_2$H$_3$O$_2$ solution is alkaline, having a pH of 8.87.

In summary, we can say that when the concentration of a reagent on the left side of an equation is increased, the equilibrium shifts to the right. When the concentration of a reagent on the right side of an equation is increased, the equilibrium shifts to the left. In accordance with Le Chatelier's principle, the equilibrium always shifts in the direction that tends to reduce the concentration of the added reactant.

16.6 Effect of Pressure on Reaction Rate and Equilibrium

Changes in pressure significantly affect the reaction rate only when one or more of the reactants or products is a gas. In these cases the effect of increasing the pressure of the reacting gases is equivalent to increasing their concentrations. In the reaction

$$CaCO_3(s) \xrightleftharpoons{\Delta} CaO(s) + CO_2(g)$$

calcium carbonate decomposes into calcium oxide and carbon dioxide when heated above 825°C. Increasing the pressure of the equilibrium system by adding CO_2 or by decreasing the volume speeds up the reverse reaction and causes the equilibrium to shift to the left. The increased pressure gives the same effect as that caused by increasing the concentration of CO_2, the only gaseous substance in the reaction.

We have seen that when the pressure on a gas is increased, its volume is decreased. In a system composed entirely of gases, an increase of the pressure will cause the reaction and the equilibrium to shift to the side that contains the smaller volume or smaller number of molecules. This is because the increase in pressure is partially relieved by the system's shifting its equilibrium toward the side in which the substances occupy the smaller volume.

Prior to World War I, Fritz Haber in Germany invented the first major process for the fixation of nitrogen. In the Haber process nitrogen and hydrogen are reacted together in the presence of a catalyst at moderately high temperature and pressure to produce ammonia. The catalyst consists of iron and iron oxide with small amounts of potassium and aluminum oxides. For this process, Haber received the Nobel Prize in chemistry in 1918.

$$N_2(g) + 3H_2(g) \rightleftharpoons 2NH_3(g) + 22.1 \text{ kcal}$$

| 1 mole | 3 moles | 2 moles |
| 1 volume | 3 volumes | 2 volumes |

The left side of the equation in the Haber process represents four volumes of gas combining to give two volumes of gas on the right side of the equation. An increase in the total pressure on the system shifts the equilibrium to the right. This increase in pressure results in a higher concentration of both reactants and products. Since there are fewer moles of NH_3 than there are moles of N_2 and H_2, the equilibrium shifts to the right when the pressure is increased.

Ideal conditions for the Haber process are 200°C and 1000 atm pressure. However, at 200°C the rate of reaction is very slow, and at 1000 atm extraordinarily heavy equipment is required. As a compromise the reaction is run at 400–600°C and 200–350 atm pressure, which gives a reasonable yield at a reasonable rate. The effect of pressure on the yield of ammonia at one particular temperature is shown in Table 16.1.

When the total number of gaseous molecules on both sides of an equation is the same, a change in pressure does not cause an equilibrium shift. The

Table 16.1. The effect of pressure in the conversion of H_2 and N_2 to NH_3 at 450°C. The starting ratio of H_2 to N_2 is 3 moles to 1 mole.

Pressure (atm)	Yield of NH_3 (%)
10	2.04
30	5.80
50	9.17
100	16.4
300	35.5
600	53.4
1000	69.4

following reaction is an example:

$$N_2(g) + O_2(g) \rightleftharpoons 2\,NO(g)$$

1 mole	1 mole	2 moles
1 volume	1 volume	2 volumes
6.02×10^{23} molecules	6.02×10^{23} molecules	$2 \times 6.02 \times 10^{23}$ molecules

When the pressure on this system is increased, the rate of both the forward and the reverse reactions will increase because of the higher concentrations of N_2, O_2, and NO. But the equilibrium will not shift, because the increase in concentration of molecules is the same on both sides of the equation and the decrease in volume is the same on both sides of the equation.

16.7 Effect of Temperature on Reaction Rate and Equilibrium

An increase in temperature is generally accompanied by an increased rate of reaction. Molecules at elevated temperatures are more energetic and have more kinetic energy; thus, their collisions are more likely to result in a reaction. However, we cannot assume that the rate of a desired reaction will keep increasing indefinitely as the temperature is raised. High temperatures may cause the destruction or decomposition of the reactants and products or may initiate reactions other than the one desired. For example, when calcium oxalate (CaC_2O_4) is heated to 500°C, it decomposes into calcium carbonate and carbon monoxide:

$$CaC_2O_4(s) \xrightarrow{500°C} CaCO_3(s) + CO\uparrow$$

If calcium oxalate is heated to 850°C, the products are calcium oxide, carbon monoxide, and carbon dioxide:

$$CaC_2O_4(s) \xrightarrow{850°C} CaO(s) + CO\uparrow + CO_2\uparrow$$

When heat is applied to a system in equilibrium, the reaction that absorbs heat is favored. When the process, as written, is endothermic, the forward reaction is increased. When the reaction is exothermic, the reverse reaction is favored. In this sense heat may be treated as a reactant in endothermic reactions or as a product in exothermic reactions. Therefore, temperature is analogous to concentration when applying Le Chatelier's principle to heat effects on a chemical reaction.

Hot coke (C) is a very good reducing agent. In the reaction

$$C(s) + CO_2(g) + Heat \rightleftharpoons 2\,CO\uparrow$$

very little, if any, CO is formed at room temperature. At 1000°C, the equilibrium mixture contains about an equal number of moles of CO and CO_2. At higher temperatures, the equilibrium shifts to the right, increasing the yield of CO. The reaction is endothermic and, as can be seen, the equilibrium is shifted to the right at higher temperatures.

Phosphorus trichloride reacts with dry chlorine gas to form phosphorus pentachloride. The reaction is exothermic:

$$PCl_3(l) + Cl_2(g) \rightleftharpoons PCl_5(s) + \text{Heat}$$

Heat must continually be removed during the reaction to obtain a good yield of the product. According to the principle of Le Chatelier, heat will cause the product, PCl_5, to decompose, reforming PCl_3 and Cl_2. The equilibrium mixture at 200°C contains 52% PCl_5 and at 300°C the mixture contains 3% PCl_5, verifying that heat causes the equilibrium mixture to shift to the left.

When the temperature of a system is raised, the rate of reaction increases because of increased kinetic energy and more frequent collisions of the reacting species. In a reversible reaction, the rate of both the forward and reverse reactions is increased by an increase in temperature; however, the reaction that absorbs heat increases to a greater extent, and the equilibrium shifts to favor that reaction. The following examples illustrate these effects:

$$4\,HCl + O_2 \rightleftharpoons 2\,H_2O + 2\,Cl_2 + 28.4 \text{ kcal} \tag{1}$$

$$H_2 + Cl_2 \rightleftharpoons 2\,HCl + 44.2 \text{ kcal} \tag{2}$$

$$CH_4 + 2\,O_2 \rightleftharpoons CO_2 + 2\,H_2O + 212.8 \text{ kcal} \tag{3}$$

$$N_2O_4 + 14 \text{ kcal} \rightleftharpoons 2\,NO_2 \tag{4}$$

$$2\,CO_2 + 135.2 \text{ kcal} \rightleftharpoons 2\,CO + O_2 \tag{5}$$

$$H_2 + I_2 + 12.4 \text{ kcal} \rightleftharpoons 2\,HI \tag{6}$$

Reactions (1), (2), and (3) are exothermic; an increase in temperature will cause the equilibrium to shift to the left. Reactions (4), (5), and (6) are endothermic; an increase in temperature will cause the equilibrium to shift to the right.

16.8 Effect of Catalysts on Reaction Rate and Equilibrium

catalyst

A **catalyst** is a substance that influences the speed of a reaction and that may be recovered essentially unchanged at the end of the reaction. A catalyst does not shift the equilibrium of a reaction; it affects only the speed at which the equilibrium is reached. If a catalyst does not affect the equilibrium, then it follows that it must affect the speed of both the forward and the reverse reactions equally.

The reaction between phosphorus trichloride (PCl_3) and sulfur is highly exothermic, but it is so slow that very little product, thiophosphoryl chloride ($PSCl_3$), is obtained, even after prolonged heating. When a catalyst, such as aluminum chloride ($AlCl_3$), is added, the reaction is complete in a few seconds:

$$PCl_3(l) + S(s) \xrightarrow{AlCl_3} PSCl_3(l)$$

We have already demonstrated that manganese dioxide used as a catalyst increases the rates of decomposition of potassium chlorate and hydrogen peroxide.

Catalysts are extremely important to industrial chemistry. Hundreds of chemical reactions that are otherwise too slow to be of practical value have been put to commercial use once a suitable catalyst was found. But it is in the area of biochemistry that catalysts are of supreme importance. Nearly all the chemical reactions associated with all forms of life are completely dependent on biochemical catalysts known as *enzymes*.

16.9 Equilibrium Constants

Law of Mass Action

The **Law of Mass Action** states that the rate of a chemical reaction is proportional to the concentration of the reacting species. This simply means that the higher the concentration of reactants, the more frequently they collide and form products. For the equilibrium system in which A and B react in one step to give C and D,

$$A + B \rightleftharpoons C + D$$

the rates of the forward and reverse reactions can be expressed as

$$\text{Rate}_f = k_f \times \text{Concentration of A} \times \text{Concentration of B}$$
$$\text{Rate}_r = k_r \times \text{Concentration of C} \times \text{Concentration of D}$$

where Rate_f is the rate of the forward reaction, Rate_r is the rate of the reverse reaction, and k_f and k_r are the proportionality rate constants. For dilute solutions, the unit of concentration is moles per litre; for gases, it is either moles per litre or pressure. To simplify the rate expressions, we place the formula of each substance in brackets to indicate the concentration of each substance. The concentrations of A, B, C, and D are then expressed as [A], [B], [C], and [D], respectively.

At equilibrium,

$$\text{Rate}_f = \text{Rate}_r$$

Then

$$\text{Rate}_f = k_f[A][B]$$
$$\text{Rate}_r = k_r[C][D]$$

$$\frac{k_f}{k_r} = \frac{[C][D]}{[A][B]}$$

Since both k_f and k_r are constants, k_f/k_r is also a constant and is known as the **equilibrium constant**, abbreviated K_{eq}:

equilibrium constant

$$K_{eq} = \frac{[C][D]}{[A][B]}$$

This expression reads as follows: The equilibrium constant, K_{eq}, is equal to the product of the concentration of C and the concentration of D divided by the product of the concentration of A and the concentration of B. Consider the equilibrium

$$2A \rightleftharpoons C + D$$

The forward reaction is dependent on the collision of two A molecules. Therefore,

$$\text{Rate}_f = k_f[A][A] = k_f[A]^2$$
$$\text{Rate}_r = k_r[C][D]$$

$$K_{eq} = \frac{k_f}{k_r} = \frac{[C][D]}{[A]^2}$$

In this equilibrium, the equilibrium constant is equal to the product of the concentrations of C times D divided by the concentration of A squared.

For the general reaction

$$nA + mB \rightleftharpoons pC + qD$$

where n, m, p, and q are the small whole numbers in the balanced equation, the equilibrium constant expression is

$$K_{eq} = \frac{[C]^p[D]^q}{[A]^n[B]^m}$$

Observe that the concentration of each substance is raised to a power that is the same as its numerical coefficient in the balanced equation. It is conventional that the concentrations of the substances in the numerator are those of the products (the substances on the right side of the equation as written); the concentrations of the reactants are in the denominator.

16.10 Ionization Constants

ionization constant

As a first application of an equilibrium constant, let us consider the constant for acetic acid in solution. Because it is a weak acid, an equilibrium is established between molecular $HC_2H_3O_2$ and its ions in solution. The constant is called the **ionization constant**, K_i, a special type of equilibrium constant. The concentration of water in the solution does not change appreciably, so we may use the following simplified equation to set up the constant:

$$HC_2H_3O_2 \rightleftharpoons H^+ + C_2H_3O_2^-$$

The ionization constant expression is

$$K_i = \frac{[H^+][C_2H_3O_2^-]}{[HC_2H_3O_2]}$$

It states that the ionization constant, K_i, is equal to the product of the hydrogen ion (H^+) concentration times the acetate ion ($C_2H_3O_2^-$) concentration divided by the concentration of the un-ionized acetic acid ($HC_2H_3O_2$).

326 Chapter 16

At 25°C, a 0.1 M $HC_2H_3O_2$ solution is 1.34% ionized and has a hydrogen ion concentration of 1.34×10^{-3} mole/litre. From this information, we can calculate the ionization constant for acetic acid.

A 0.10 M solution contains 0.10 mole of acetic acid per litre. Of this 0.10 mole, only 1.34%, or 1.34×10^{-3} mole, is ionized. This gives an H^+ ion concentration of 1.34×10^{-3} mole/litre. Since each molecule of acid that ionizes yields one H^+ and one $C_2H_3O_2^-$, the concentration of $C_2H_3O_2^-$ ions is also 1.34×10^{-3} mole/litre. This ionization leaves $0.10 - 0.00134 = 0.09866$ mole/litre of un-ionized acetic acid.

	Initial concentration	Equilibrium concentration
$[HC_2H_3O_2]$	0.10 mole/litre	0.09866 mole/litre
$[H^+]$	0	0.00134 mole/litre
$[C_2H_3O_2^-]$	0	0.00134 mole/litre

Substituting these concentrations in the equilibrium expression, we obtain the value for K_i:

$$K_i = \frac{[H^+][C_2H_3O_2^-]}{[HC_2H_3O_2]}$$

$$= \frac{[1.34 \times 10^{-3}][1.34 \times 10^{-3}]}{[0.09866]} = 1.8 \times 10^{-5}$$

The low magnitude of this constant indicates that the position of the equilibrium is far toward the un-ionized acetic acid. In fact, a 0.1 M acetic acid solution is 98.66% un-ionized.

Once the value for this constant is established, it can be used to describe any system containing H^+, $C_2H_3O_2^-$, and $HC_2H_3O_2$ in equilibrium at 25°C. The ionization constants for several other weak acids are listed in Table 16.2.

Table 16.2. Ionization constants (K_i) of weak acids at 25°C

Acid	Formula	K_i
Acetic	$HC_2H_3O_2$	1.8×10^{-5}
Benzoic	$HC_7H_5O_2$	6.3×10^{-5}
Carbolic (phenol)	HC_6H_5O	1.3×10^{-10}
Cyanic	$HCNO$	2.0×10^{-4}
Formic	$HCHO_2$	1.8×10^{-4}
Hydrocyanic	HCN	4.0×10^{-10}
Hypochlorous	$HClO$	3.5×10^{-8}
Nitrous	HNO_2	4.5×10^{-4}

Problem 16.1   What is the ionization constant expression for nitrous acid?

First, write the simplified ionization equation:

$$HNO_2(aq) \rightleftharpoons H^+ + NO_2^- \quad \text{(Simplified)}$$

The format of the ionization constant expression, K_i, is the product of the concentrations of the substances on the right side divided by the product of the concentrations of the substances on the left side of the equation. Thus,

$$K_i = \frac{[H^+][NO_2^-]}{[HNO_2]} \quad \text{(Answer)}$$

Problem 16.2   What is the H^+ ion concentration in a 0.50 M $HC_2H_3O_2$ solution? The ionization constant, K_i, for $HC_2H_3O_2$ is 1.8×10^{-5}.

To solve this problem, first write the equilibrium equation and the K_i expression:

$$HC_2H_3O_2 \rightleftharpoons H^+ + C_2H_3O_2^-$$

$$K_i = \frac{[H^+][C_2H_3O_2^-]}{[HC_2H_3O_2]} = 1.8 \times 10^{-5}$$

Let $[H^+] = Y$. Then $[C_2H_3O_2^-]$ will also equal Y, because one acetate ion is produced for each hydrogen ion. The $[HC_2H_3O_2]$ will then be $0.50 - Y$, the starting concentration minus the amount that ionized.

$$[H^+] = [C_2H_3O_2^-] = Y \qquad [HC_2H_3O_2] = 0.50 - Y$$

Substituting these values into the K_i expression, we obtain

$$K_i = \frac{(Y)(Y)}{0.50 - Y} = \frac{Y^2}{0.50 - Y} = 1.8 \times 10^{-5}$$

An exact solution of this equation for Y requires the use of the quadratic formula. However, an approximate solution is readily obtained if we first assume that Y is small compared to 0.50. Then $0.50 - Y$ will be equal to approximately 0.50. The equation now becomes

$$\frac{Y^2}{0.50} = 1.8 \times 10^{-5}$$

$$Y^2 = 1.8 \times 10^{-5} \times 0.50 = 0.90 \times 10^{-5} = 9.0 \times 10^{-6}$$

$$Y = \sqrt{9.0 \times 10^{-6}} = 3.0 \times 10^{-3}$$

Thus, the $[H^+]$ is approximately 3.0×10^{-3} mole/litre in a 0.50 M $HC_2H_3O_2$ solution. The exact solution to this problem, using a quadratic equation, gives a value of 2.99×10^{-3} for $[H^+]$, showing that we were justified in neglecting Y compared to 0.50.

16.11   Ion Product Constant for Water

We have seen that water ionizes to a slight degree. This ionization is represented by these equilibrium equations:

$$H_2O + H_2O \rightleftharpoons H_3O^+ + OH^- \tag{1}$$
$$H_2O \rightleftharpoons H^+ + OH^- \tag{2}$$

Equation (1) is the more accurate representation of the equilibrium since free protons (H^+) do not exist in water. Equation (2) is a simplified and widely used

328 Chapter 16

representation of the water equilibrium. The actual concentration of H^+ produced in pure water is very minute and amounts to only 1×10^{-7} mole per litre at 25°C. In pure water,

$$[H^+] = [OH^-] = 1 \times 10^{-7} \text{ mole/litre}$$

since both ions are produced in equal molar amounts, as shown in equation (2).

K_w The $H_2O \rightleftarrows H^+ + OH^-$ equilibrium exists in water and in all water solutions. A special equilibrium constant called the *ion product constant for water*, K_w, applies to this equilibrium. The constant K_w is defined as the product of the H^+ ion concentration and the OH^- ion concentration, each in moles per litre:

$$K_w = [H^+][OH^-]$$

The numerical value of K_w is 1×10^{-14}, since for pure water at 25°C

$$K_w = [H^+][OH^-] = [1 \times 10^{-7}][1 \times 10^{-7}] = 1 \times 10^{-14}$$

The value of K_w for all water solutions at 25°C is the constant 1×10^{-14}. It is important to realize that as the concentration of one of these ions, H^+ or OH^-, increases, the other decreases. However, the product of the H^+ and OH^- concentrations always equals the constant 1×10^{-14}. This relationship can be seen in the examples shown in Table 16.3. If the concentration of one ion is known, the concentration of the other can be calculated by use of the K_w expression.

Table 16.3. Relationship of H^+ and OH^- concentrations in water solutions.

$[H^+]$	$[OH^-]$	K_w	pH	pOH
1×10^{-2}	1×10^{-12}	1×10^{-14}	2	12
1×10^{-4}	1×10^{-10}	1×10^{-14}	4	10
2×10^{-6}	5×10^{-9}	1×10^{-14}	5.7	8.3
1×10^{-7}	1×10^{-7}	1×10^{-14}	7	7
1×10^{-9}	1×10^{-5}	1×10^{-14}	9	5

Problem 16.3   What is the concentration of (a) H^+ and (b) OH^- in a 0.001 M HCl solution? Assume that the HCl is 100% ionized.

(a) Since all the HCl is ionized, $H^+ = 0.001$ mole/litre.

$$\text{HCl} \longrightarrow H^+ + Cl^-$$
    0.001 M   0.001 M   0.001 M

$$[H^+] = 1 \times 10^{-3} \quad \text{(Answer)}$$

(b) To calculate the $[OH^-]$ in this solution, solve the K_w expression for OH^- and substitute in the values for K_w and $[H^+]$.

$$K_w = [H^+][OH^-]$$

$$[OH^-] = \frac{K_w}{[H^+]} = \frac{1 \times 10^{-14}}{1 \times 10^{-3}} = 1 \times 10^{-11} \text{ mole/litre} \quad \text{(Answer)}$$

Problem 16.4 What is the pH of a 0.01 M NaOH solution? Assume that the NaOH is 100% ionized. Since all the NaOH is ionized, OH⁻ = 0.01 mole/litre.

$$NaOH \longrightarrow Na^+ + OH^-$$
$$0.01\ M \qquad 0.01\ M \quad 0.01\ M$$

In order to find the pH of the solution, we must first calculate the H^+ concentration. This is done by using the K_w expression. Solve for $[H^+]$ and substitute the values for K_w and $[OH^-]$.

$$K_w = [H^+][OH^-]$$

$$[H^+] = \frac{K_w}{[OH^-]} = \frac{1 \times 10^{-14}}{1 \times 10^{-2}} = 1 \times 10^{-12} \text{ mole/litre}$$

$$pH = \log \frac{1}{[H^+]} = \log \frac{1}{10^{-12}} = \log 10^{12} = 12 \quad \text{(Answer)}$$

The pH can also be calculated by the method shown in Section 15.9:

$$[H^+] = 1 \times 10^{-12}$$
$$pH = 12 - \log 1 = 12 - 0 = 12 \quad \text{(Answer)}$$

Just as pH is used to express the acidity of a solution, pOH is used to express the basicity of an aqueous solution. The pOH is related to the OH⁻ ion concentration in the same way that the pH is related to the H^+ ion concentration:

$$pOH = \log \frac{1}{[OH^-]} \quad \text{or} \quad -\log[OH^-]$$

Thus, a solution in which $[OH^-] = 1 \times 10^{-2}$, as in Problem 16.4, will have pOH = 2.

In pure water, where $[H^+] = 1 \times 10^{-7}$ and $[OH^-] = 1 \times 10^{-7}$, the pH is 7 and the pOH is 7. The sum of the pH and pOH is 14.

$$pH + pOH = 14$$

This relationship holds in all aqueous solutions and is illustrated in the examples in Table 16.3.

16.12 Solubility Product Constant

solubility product constant

The **solubility product constant**, abbreviated K_{sp}, is another application of the equilibrium constant. It is derived from the equilibrium between a slightly soluble substance and its ions in solution. The following example illustrates how this constant is evaluated.

The solubility of silver chloride (AgCl) in water is 1.3×10^{-5} mole per litre at 25°C. The equation for the equilibrium between AgCl and its ions in solution is

$$AgCl(s) \rightleftharpoons Ag^+ + Cl^-$$

The equilibrium constant expression is

$$K_{eq} = \frac{[Ag^+][Cl^-]}{[AgCl(s)]}$$

The amount of solid AgCl does not affect the equilibrium system provided that some is present. In other words, the concentration of solid silver chloride is constant, whether 1 mg or 10 g of the salt is present. Therefore, the product obtained by multiplying the two constants K_{eq} and $[AgCl(s)]$ is also a constant. This constant is called the solubility product constant, K_{sp}.

$$K_{eq} \times [AgCl(s)] = [Ag^+][Cl^-] = K_{sp}$$
$$K_{sp} = [Ag^+][Cl^-]$$

The K_{sp} is equal to the product of the Ag^+ ion times the Cl^- ion concentrations, each in moles/litre. When 1.3×10^{-5} mole/litre of AgCl dissolves, it produces 1.3×10^{-5} mole/litre each of Ag^+ and Cl^-. From these concentrations, the K_{sp} can be evaluated.

$$[Ag^+] = 1.3 \times 10^{-5} \text{ mole/litre} \qquad [Cl^-] = 1.3 \times 10^{-5} \text{ mole/litre}$$
$$K_{sp} = [Ag^+][Cl^-] = [1.3 \times 10^{-5}][1.3 \times 10^{-5}] = 1.7 \times 10^{-10}$$

Once this K_{sp} value is established, it can be used to describe other systems containing Ag^+ and Cl^-. For example, if silver nitrate, $AgNO_3$, is added to a saturated AgCl solution until the Ag^+ concentration is 0.1 M, what will be the Cl^- ion concentration remaining in solution? The addition of $AgNO_3$ puts Ag^+ ions into solution and causes the AgCl equilibrium to shift to the left, reducing the Cl^- ion concentration. This process of increasing the concentration of one of the ions in an equilibrium, thereby causing the other ion to decrease in concentration, is known as the **common ion effect**.

common ion effect

We use the K_{sp} to calculate the Cl^- ion concentration remaining in solution. The K_{sp} is constant at a particular temperature and remains the same no matter how we change the concentration of the species involved.

$$K_{sp} = [Ag^+][Cl^-] = 1.7 \times 10^{-10} \qquad [Ag^+] = 0.1 \text{ mole/litre}$$

We then substitute the concentration of Ag^+ ion into the K_{sp} expression and calculate:

$$[0.1][Cl^-] = 1.7 \times 10^{-10}$$
$$[Cl^-] = \frac{1.7 \times 10^{-10}}{0.1}$$
$$[Cl^-] = 1.7 \times 10^{-9} \text{ mole/litre}$$

This calculation shows a 10,000-fold reduction of Cl^- ions in solution. It illustrates that Cl^- ions may be quantitatively removed from solution with an excess of Ag^+ ions. The equilibrium equations and the K_{sp} expressions for several

Table 16.4. Solubility product constants (K_{sp}) at 25°C.

Compound	K_{sp}
AgCl	1.7×10^{-10}
AgBr	5×10^{-13}
AgI	8.5×10^{-17}
$AgC_2H_3O_2$	2×10^{-3}
Ag_2CrO_4	1.9×10^{-12}
$BaCrO_4$	8.5×10^{-11}
$BaSO_4$	1.5×10^{-9}
CaF_2	3.9×10^{-11}
CuS	9×10^{-45}
$Fe(OH)_3$	6×10^{-38}
PbS	7×10^{-29}
$PbSO_4$	1.3×10^{-8}
$Mn(OH)_2$	2.0×10^{-13}

other substances are given below. Table 16.4 lists K_{sp} values for these and several other substances.

$$AgBr(s) \rightleftharpoons Ag^+ + Br^- \qquad K_{sp} = [Ag^+][Br^-]$$
$$BaSO_4(s) \rightleftharpoons Ba^{2+} + SO_4^{2-} \qquad K_{sp} = [Ba^{2+}][SO_4^{2-}]$$
$$Ag_2CrO_4(s) \rightleftharpoons 2\,Ag^+ + CrO_4^{2-} \qquad K_{sp} = [Ag^+]^2[CrO_4^{2-}]$$
$$CuS(s) \rightleftharpoons Cu^{2+} + S^{2-} \qquad K_{sp} = [Cu^{2+}][S^{2-}]$$
$$Mn(OH)_2(s) \rightleftharpoons Mn^{2+} + 2\,OH^- \qquad K_{sp} = [Mn^{2+}][OH^-]^2$$
$$Fe(OH)_3(s) \rightleftharpoons Fe^{3+} + 3\,OH^- \qquad K_{sp} = [Fe^{3+}][OH^-]^3$$

Note that the concentration of each substance in the K_{sp} expressions is raised to a power that is the same number as its numerical coefficient in the balanced equilibrium equation, for example, $[Ag^+]^2$ in Ag_2CrO_4. When the product of the molar concentration of the ions in solution, each raised to its proper power, is greater than the K_{sp} for that substance, precipitation should occur. If the ion product is less than the K_{sp} value, no precipitation will occur.

Problem 16.5 Write K_{sp} expressions for AgI and PbI_2, both of which are slightly soluble salts.

First write the equilibrium equations:

$$AgI(s) \rightleftharpoons Ag^+ + I^-$$
$$PbI_2(s) \rightleftharpoons Pb^{2+} + 2\,I^-$$

Since the concentration of the solid crystals is constant, the K_{sp} equals the product of the molar concentrations of the ions in solution. In the case of PbI_2, the $[I^-]$ must be squared.

$$K_{sp} = [Ag^+][I^-]$$
$$K_{sp} = [Pb^{2+}][I^-]^2$$

Problem 16.6 The K_{sp} value for lead sulfate ($PbSO_4$, g-mol. wt = 305.9 g/mole) is 1.3×10^{-8}. Calculate the solubility of $PbSO_4$ in grams per litre.

First find the solubility in moles per litre from the K_{sp} value and then convert moles per litre to grams per litre. Let S equal moles per litre of $PbSO_4$ in a saturated solution. Since the salt is completely dissociated,

$$PbSO_4(s) \rightleftharpoons Pb^{2+} + SO_4^{2-}$$
$$[Pb^{2+}] = S \quad [SO_4^{2-}] = S$$
$$K_{sp} = [Pb^{2+}][SO_4^{2-}] = 1.3 \times 10^{-8}$$
$$[S][S] = 1.3 \times 10^{-8}$$
$$[S]^2 = 1.3 \times 10^{-8}$$
$$S = \sqrt{1.3 \times 10^{-8}} = 1.14 \times 10^{-4} \text{ mole/litre}$$

Therefore, the solubility of $PbSO_4$ is 1.14×10^{-4} mole/litre. The weight in grams that dissolves is

$$g = \text{moles} \times \text{g-mol. wt}$$

$$g = \frac{1.14 \times 10^{-4} \text{ mole}}{\text{litre}} \times \frac{305.9 \text{ g}}{\text{mole}} = 3.5 \times 10^{-2} \text{ g/litre}$$

The solubility of $PbSO_4$ is 0.035 g/litre.

16.13 Buffer Solutions

The control of pH within narrow limits is critically important in many chemical applications and vitally important in many biological systems. For example, human blood must be maintained between a pH of 7.35 and 7.45 for the efficient transport of oxygen from the lungs to the cells. This narrow pH range is maintained by buffer systems within the blood.

buffer solution

A solution that resists changes of pH when diluted or when an acid or base is added is called a **buffer solution**. Two common types of buffer solutions are (1) a weak acid together with a salt of that weak acid and (2) a weak base together with a salt of that weak base.

The action of a buffer system can be understood by considering a solution of acetic acid and sodium acetate. The weak acid, $HC_2H_3O_2$, is mostly un-ionized and is in equilibrium with its ions in solution. The salt, $NaC_2H_3O_2$, is completely ionized.

$$HC_2H_3O_2(aq) \rightleftharpoons H^+(aq) + C_2H_3O_2^-(aq)$$
$$NaC_2H_3O_2 \longrightarrow Na^+(aq) + C_2H_3O_2^-(aq)$$

Since the salt is completely ionized, the solution contains a much higher concentration of acetate ions than would be present if only acetic acid were in solution. The acetate ion represses the ionization of acetic acid and also reacts with water, causing the solution to have a higher pH (be more basic) than an acetic acid solution (see Section 16.5). Thus, a 0.1 M acetic acid solution has a pH of 2.87, but a solution that is 0.1 M in acetic acid and 0.1 M in sodium acetate has a pH of 4.74.

A buffer solution has a built-in mechanism that counteracts the effect of adding acid or base. Consider the effect of adding HCl or NaOH to an acetic

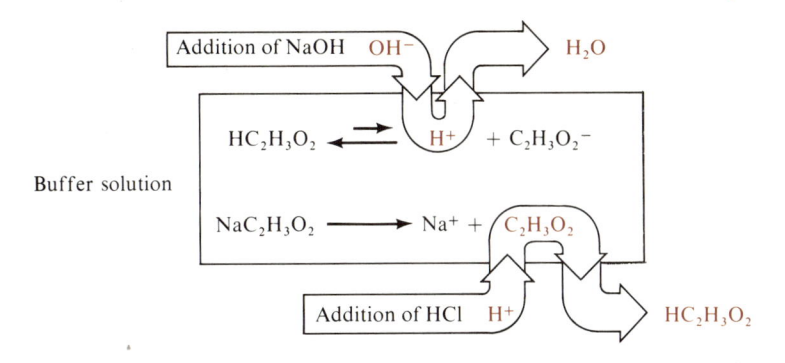

Figure 16.3. The effect of adding HCl and NaOH to an acetic acid–sodium acetate buffer solution. The added H^+ from HCl is removed from solution by forming un-ionized acetic acid. The added OH^- from NaOH is removed by reacting with H^+ to form water.

acid–sodium acetate buffer. When HCl is added, the acetate ions of the buffer combine with the H^+ ions of HCl to form un-ionized acetic acid, thus neutralizing the added acid and maintaining the approximate pH of the solution. When NaOH is added, the H^+ ions in the buffer combine with the OH^- ions to form water. Additional acetic acid then ionizes (Le Chatelier's principle) to restore the H^+ ions and maintain the approximate pH of the solution. The action of this buffer system in counteracting added acid or added base is illustrated in Figure 16.3.

Data comparing the changes in pH caused by adding HCl and NaOH to pure water and to an acetic acid–sodium acetate buffer solution are shown in Table 16.5.

Table 16.5. Changes in pH caused by the addition of HCl and NaOH to pure water and to an acetic acid–sodium acetate buffer solution.

Solution	pH	Change in pH
H_2O (1000 ml)	7	—
H_2O + 0.01 mole HCl	2	5
H_2O + 0.01 mole NaOH	12	5
Buffer solution (1000 ml), 0.10 M $HC_2H_3O_2$ + 0.10 M $NaC_2H_3O_2$	4.74	—
Buffer + 0.01 mole HCl	4.66	0.08
Buffer + 0.01 mole NaOH	4.83	0.09

16.14 Mechanism of Reactions

mechanism of a reaction

How a reaction occurs—that is, the manner or route by which it proceeds—is known as the **mechanism of the reaction**. The mechanism shows us the path, or course, the atoms and molecules take to arrive at the products.

334 Chapter 16

Our aim here is not to study the mechanisms themselves but to show that chemical reactions occur by specific routes.

When hydrogen and iodine are mixed at room temperature, we observe no appreciable reaction. In this case, the reaction takes place as a result of a collision between an H_2 and an I_2 molecule, but at room temperature the collisions do not result in reaction because the molecules do not have sufficient energy to react with each other. We might say that an energy barrier to reaction exists. As heat is added, the kinetic energy of the molecules increases. When molecules of H_2 and I_2 having sufficient energy collide, an intermediate, known as the **activated complex**, is formed. The amount of energy needed to form the activated complex is known as the **activation energy**. The complex, H_2I_2, is in a metastable form and has an energy level higher than that of the reactants or that of the product. It can decompose to form either the reactants or the product. Three steps constitute the mechanism of the reaction: (1) collision of an H_2 and an I_2 molecule; (2) formation of the activated complex, H_2I_2; and (3) decomposition to the product, HI. The various steps in the formation of HI are shown in Figure 16.4. Figure 16.5 illustrates the energy relationships in this reaction.

The reaction of hydrogen and chlorine proceeds by a different mechanism. When H_2 and Cl_2 are mixed and kept in the dark, essentially no product is formed. But if the mixture is exposed to sunlight or ultraviolet radiation, it reacts very rapidly. The overall reaction is

$$H_2(g) + Cl_2(g) \longrightarrow 2\ HCl(g)$$

This reaction proceeds by what is known as a *free radical mechanism*. A **free radical** is a neutral atom or group containing one or more unpaired electrons. Both atomic chlorine ($:\ddot{\underset{..}{Cl}}\cdot$) and atomic hydrogen (H·) have an unpaired electron and are free radicals. The reaction occurs in three steps.

Step 1. Initiation:

$$:\ddot{\underset{..}{Cl}}:\ddot{\underset{..}{Cl}}: + h\nu \longrightarrow\ :\ddot{\underset{..}{Cl}}\cdot\ +\ :\ddot{\underset{..}{Cl}}\cdot$$

Chlorine free radicals

In this step a chlorine molecule absorbs energy in the form of a photon, $h\nu$, of light or ultra violet radiation. The energized chlorine molecule then splits into two chlorine free radicals.

Step 2. Propagation:

$$:\ddot{\underset{..}{Cl}}\cdot + H:H \longrightarrow HCl\ +\ H\cdot$$

Hydrogen free radical

$$H\cdot +\ :\ddot{\underset{..}{Cl}}:\ddot{\underset{..}{Cl}}: \longrightarrow HCl + :\ddot{\underset{..}{Cl}}\cdot$$

This step begins when a chlorine free radical reacts with a hydrogen molecule to produce a molecule of hydrogen chloride and a hydrogen free radical. The hydrogen radical next reacts with another chlorine molecule to form hydrogen chloride and another chlorine free radical. This chlorine radical can repeat the process by reacting with another hydrogen molecule, and the reaction continues to propagate itself in this manner until one or both of the reactants is used up.

Margin terms:
- activated complex
- activation energy
- free radical

Chemical Equilibrium   335

$$H_2 + I_2 \longrightarrow [H_2I_2] \longrightarrow 2HI$$

[Activated complex]

[Activated complex]

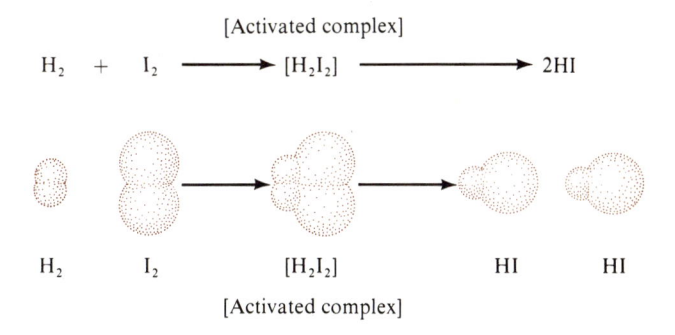

H₂   I₂   [H₂I₂]   HI   HI

Figure 16.4. Mechanism of the reaction between hydrogen and iodine: H_2 and I_2 molecules of sufficient energy unite, forming the intermediate activated complex, which decomposes to the product, hydrogen iodide.

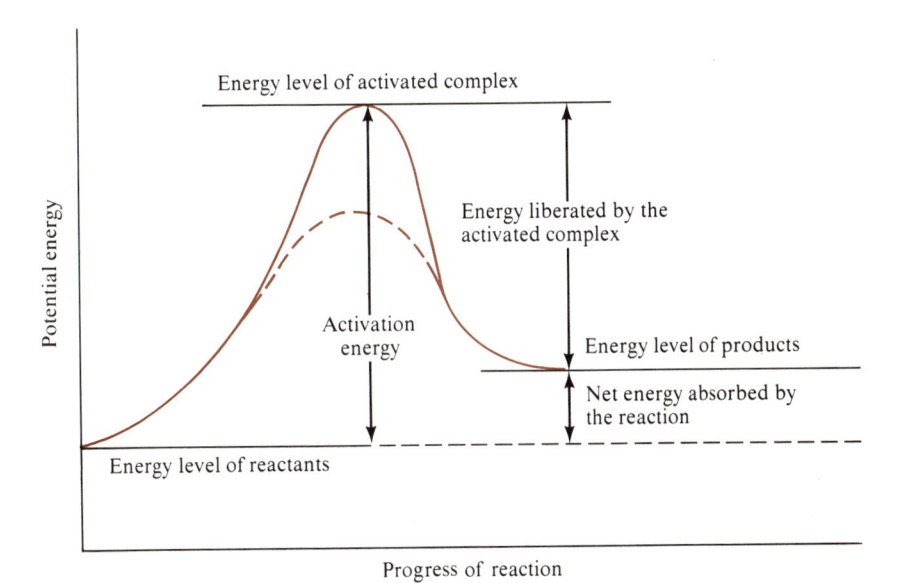

Figure 16.5. Relative energy diagram for the reaction between hydrogen and iodine.

$$H_2 + I_2 \longrightarrow [H_2I_2] \longrightarrow 2HI$$

Energy equal to the activation energy is put into the system in forming the activated complex, H_2I_2. When this complex decomposes, it liberates energy, forming the product. In this case, the product is at a higher energy level than the reactants, indicating that the reaction is endothermic and that energy is absorbed during the reaction. The dotted line represents the effect that a catalyst would have on the reaction. The catalyst lowers the activation energy, thereby increasing the rate of the reaction.

Step 3. Termination:

$$:\!\ddot{\underset{..}{Cl}}\!\cdot\ +\ :\!\ddot{\underset{..}{Cl}}\!\cdot\ \longrightarrow\ Cl_2$$

$$H\cdot\ +\ H\cdot\ \longrightarrow\ H_2$$

$$H\cdot\ +\ :\!\ddot{\underset{..}{Cl}}\!\cdot\ \longrightarrow\ HCl$$

Hydrogen and chlorine free radicals can react in any of the three ways shown. Unless further activation occurs, the formation of hydrogen chloride will terminate when the radicals form molecules. In an exothermic reaction such as that between hydrogen and chlorine, there is usually enough heat and light energy available to maintain the supply of free radicals, and the reaction will continue until at least one reactant is exhausted.

Questions

A. Review the meanings of the new terms introduced in this chapter.
   1. Reversible reaction
   2. Chemical kinetics
   3. Equilibrium
   4. Chemical equilibrium
   5. Principle of Le Chatelier
   6. Catalyst
   7. Law of Mass Action
   8. Equilibrium constant, K_{eq}
   9. Ionization constant, K_i
   10. K_w
   11. Solubility product constant, K_{sp}
   12. Common ion effect
   13. Buffer solution
   14. Mechanism of a reaction
   15. Activated complex
   16. Activation energy
   17. Free radical

B. Answers to the following questions will be found in tables and figures.
   1. Explain how each tube in Figure 16.1 illustrates the principle of Le Chatelier.
   2. How would the contents of the two tubes of Figure 16.1 differ in appearance if the one at the left was at $-25°C$ and the one at the right was at $25°C$?
   3. Using the same rate and time scale, resketch Figure 16.2 as it would appear if a catalyst were used to double the forward reaction rate.
   4. Which is the strongest and which is the weakest acid listed in Table 16.2? (The strength of an acid is indicated by its degree of ionization.)
   5. Classify each of the solutions in Table 16.3 as being acid, basic, or neutral.
   6. How many times greater is the H^+ ion concentration in the first solution of Table 16.3 than in pure water?
   7. Using Table 16.4, tabulate the relative order of molar solubilities of AgCl, AgBr, AgI, $AgC_2H_3O_2$, $PbSO_4$, $BaSO_4$, $BaCrO_4$, CuS, and PbS. List the most soluble first.
   8. Using Figure 16.3 and Table 16.5, explain how the acetic acid–sodium acetate buffer system maintains a nearly constant pH when 0.01 mole of HCl is added to 1 litre of the buffer solution.
   9. How would Figure 16.4 be modified if the reaction were exothermic?

C. Review questions.
   1. Express the following reversible systems in equation form:
      (a) Water at $0°C$
      (b) A saturated solution of potassium bromide (KBr)
      (c) A closed system containing liquid and gaseous methanol (CH_3OH)
   2. What constitutes equilibrium in a chemical system?

3. Sodium chloride precipitates when gaseous hydrogen chloride is passed into a saturated NaCl solution. Would a precipitate form if gaseous HCl were passed into a saturated solution of potassium chloride (KCl)? Explain.
4. Will a precipitate form when gaseous hydrogen chloride is passed into a saturated ammonium chloride (NH$_4$Cl) solution? Explain.
5. Explain why the rate of a reaction increases when the concentration of one of the reactants is increased.
6. Explain why a decrease in temperature causes the rate of a chemical reaction to decrease.
7. If pure hydrogen iodide (HI) is placed in a sealed vessel at 700 K, will it decompose? Explain.
8. What changes take place in the concentration of each of the substances when some ammonia is removed from the equilibrium system shown below?

$$4\,NH_3(g) + 5\,O_2(g) \rightleftarrows 4\,NO(g) + 6\,H_2O(g)$$

9. Consider the equilibrium represented by this equation:

$$N_2(g) + 3\,H_2(g) \rightleftarrows 2\,NH_3(g) + \text{Heat}$$

What will be the effect on the equilibrium when each of the following changes is made?
   (a) More N$_2$ is added.
   (b) Some NH$_3$ is removed.
   (c) The pressure is increased.
   (d) The temperature is decreased.

10. In the equations below, in which direction (left or right) will the equilibrium shift when the following changes are made? The temperature is increased; the pressure is increased; a catalyst is added.
    (a) $3\,O_2(g) + 64.8\text{ kcal} \rightleftarrows 2\,O_3(g)$
    (b) $CH_4(g) + Cl_2(g) \rightleftarrows CH_3Cl(g) + HCl(g) + 26.4\text{ kcal}$
    (c) $2\,NO(g) + 2\,H_2(g) \rightleftarrows N_2(g) + 2\,H_2O(g) + 159\text{ kcal}$
    (d) $2\,SO_3(g) + 47\text{ kcal} \rightleftarrows 2\,SO_2(g) + O_2(g)$
    (e) $4\,NH_3(g) + 3\,O_2(g) \rightleftarrows 2\,N_2(g) + 6\,H_2O(g) + 366\text{ kcal}$

11. Explain what occurs when the pure substances A and B are mixed and react to establish the equilibrium

$$A + B \rightleftarrows C + D$$

12. A 1.0 M HC$_2$H$_3$O$_2$ solution is 0.42% ionized. A 0.10 M HC$_2$H$_3$O$_2$ solution is 1.34% ionized. Explain these figures, using the ionization equation and equilibrium principles.
13. A more concentrated acetic acid (HC$_2$H$_3$O$_2$) solution ionizes to a smaller degree than a less concentrated HC$_2$H$_3$O$_2$ solution. Does this fact contradict Le Chatelier's principle? Explain. (See Question 12.)
14. Write the equilibrium constant expression for each of the following reactions:
    (a) $H_2(g) + I_2(g) \rightleftarrows 2\,HI(g)$
    (b) $N_2(g) + 3\,H_2(g) \rightleftarrows 2\,NH_3(g)$
    (c) $PCl_5(g) \rightleftarrows PCl_3(g) + Cl_2(g)$
    (d) $HClO_2(aq) \rightleftarrows H^+ + ClO_2^-$
    (e) $NH_4OH(aq) \rightleftarrows NH_4^+ + OH^-$
    (f) $4\,NH_3(g) + 5\,O_2(g) \rightleftarrows 4\,NO(g) + 6\,H_2O(g)$
15. What effect, if any, will increasing the OH$^-$ ion concentration have upon the following?
    (a) pH   (b) pOH   (c) [H$^+$]   (d) K_w

16. Write the solubility product expression (K_{sp}) for each of the following substances:
    (a) AgBr (c) $PbCl_2$ (e) $Al(OH)_3$ (g) $FePO_4$
    (b) $CaSO_4$ (d) Ag_2CO_3 (f) As_2S_3 (h) $Ca_3(PO_4)_2$
17. Explain why silver acetate ($AgC_2H_3O_2$) is more soluble in nitric acid than in water. (*Hint:* Write the equilibrium equation first and then consider the effect of the acid on the acetate ion.)
18. Explain why the solution becomes basic when the salt, sodium acetate, is dissolved in pure water. (*Hint:* A small amount of $HC_2H_3O_2$ is formed.)
19. One of the important pH-regulating systems in the blood consists of a carbonic acid–sodium bicarbonate buffer:

$$H_2CO_3(aq) \rightleftarrows H^+(aq) + HCO_3^-(aq)$$
$$NaHCO_3(aq) \rightarrow Na^+(aq) + HCO_3^-(aq)$$

Explain how this buffer resists changes in pH when (a) excess acid and (b) excess base get into the blood stream.

20. The action of a catalyst does not shift the position of an equilibrium (change the value of the equilibrium constant). Why not?
21. Explain and illustrate the formation and decomposition of an activated complex in the reaction of H_2 and I_2 molecules to form HI molecules.
22. Describe and illustrate the role of free radicals in the reaction of hydrogen and chlorine to form hydrogen chloride.
23. Which of the following statements are correct?
    (a) In an equilibrium reaction, equilibrium is established when the concentrations of the reactants and products are equal.
    (b) The effect of a catalyst in an equilibrium system is to lower the activation energy of the reaction.
    (c) Increasing the pressure on a reaction in the gaseous state increases the rate of reaction.
    (d) A large value for an equilibrium constant means a high concentration of the products and a low concentration of the reactants at equilibrium.

    Statements (e)–(k) pertain to the equilibrium system

    $$2\,NO(g) + O_2(g) \rightleftarrows 2\,NO_2(g) + 27\text{ kcal}$$

    (e) As the temperature on the system is increased, the amount of O_2 increases.
    (f) As the pressure on the system is decreased, the amount of NO_2 decreases.
    (g) As the pressure on the system is increased, the amount of NO decreases.
    (h) If a catalyst is added to the system, the heat of reaction decreases.
    (i) When some NO_2 is removed from the system, the concentration of all three substances will be less when equilibrium is reestablished.
    (j) The equilibrium constant expression for the reaction is
    $$K_{eq} = \frac{[NO_2]^2}{[NO]^2[O_2]^2}$$
    (k) The reaction as shown is endothermic.
    (l) A solution with an H^+ ion concentration of 1×10^{-4} mole/litre has a pOH of 10.
    (m) An aqueous solution that has an OH^- ion concentration of 1×10^{-3} mole/litre has an H^+ ion concentration of 1×10^{-11} mole/litre.
    (n) $K_w = [H^+][OH^-] = 1 \times 10^{-14}$
    (o) As solid $BaSO_4$ is added to a saturated solution of $BaSO_4$, the magnitude of its K_{sp} increases.

Chemical Equilibrium

D. Review problems.
1. What is the maximum amount of hydrogen iodide (HI) that can be produced when 1.40 moles of H_2 and 1.20 moles of I_2 are reacted?
2. (a) How many moles of hydrogen iodide (HI) will be produced when 1.50 moles of H_2 and 1.50 moles of I_2 are reacted at 700 K? (The reaction is 79% complete.)
   (b) If 0.20 mole of I_2 is added to this system, the yield of HI is 85%. What will be the new concentration (in moles) of H_2, I_2, and HI?
3. Three (3.00) g of hydrogen and 100 g of iodine are reacted at 500 K. After equilibrium is reached, analysis shows that there are 32.0 g of HI in the flask. How many moles of H_2, I_2, and HI are present in this equilibrium mixture? (See Section 16.3.)
4. Five (5.00) moles of pure hydrogen iodide are placed in a reaction vessel at 700 K. What concentration of H_2, I_2, and HI will be present when equilibrium is reached? (See Section 16.3.)
5. If the velocity of a reaction doubles for every 10 degrees that the temperature rises, how much faster will it proceed at 100°C than at 20°C?
6. One hundred grams of phosphorus pentachloride (PCl_5) are sealed in a flask and heated. At equilibrium, 12.0 g of PCl_5 remain undecomposed.

   $PCl_5(s) \rightleftarrows PCl_3(l) + Cl_2(g)$

   (a) How many moles of chlorine (Cl_2) are present in the equilibrium mixture?
   (b) What volume will this amount of Cl_2 occupy at 200°C and 1 atm pressure?
7. When 1.00 mole each of H_2 and I_2 are reacted in a 1.00 litre flask at 700 K to produce HI, the reaction is 79% complete at equilibrium. Calculate the K_{eq} for this reaction.
8. Calculate the ionization constant for the acids listed in the table. Each is a monoprotic acid and ionizes as follows: $HA \rightleftarrows H^+ + A^-$.

Acid	Acid concentration	$[H^+]$
Propanoic, $HC_3H_5O_2$	0.10 M	1.16×10^{-3} mole/litre
Hydrofluoric, HF	0.10 M	8.5×10^{-3} mole/litre
Hydrocyanic, HCN	0.20 M	8.94×10^{-6} mole/litre

9. Calculate the hydrogen ion concentration and the pH in 0.20 M $HC_2H_3O_2$ solution. The ionization constant, K_i, for $HC_2H_3O_2$ is 1.8×10^{-5}.
10. Calculate the nitrite ion concentration in a 0.25 M solution of nitrous acid (HNO_2). The ionization constant, K_i, for HNO_2 is 4.5×10^{-4}.
11. What is the hydrogen ion concentration in 100 ml of a solution that contains 0.50 g of acetic acid? (The ionization constant, K_i, for $HC_2H_3O_2$ is 1.8×10^{-5}.)
12. What is the percent ionization of acetic acid in a 0.010 M $HC_2H_3O_2$ solution?
13. What is the pH of a 0.010 M $HC_2H_3O_2$ solution?
14. Given the following solubility data, calculate the solubility product constant for each substance:
    (a) $CuCO_3$, 1.58×10^{-5} mole/litre
    (b) ZnS, 3.5×10^{-12} mole/litre
    (c) MgF_2, 2.7×10^{-9} mole/litre
    (d) Ag_2CO_3, 1.27×10^{-4} mole/litre
    (e) $MgCO_3$, 2.7×10^{-6} g/litre
    (f) $CaSO_4$, 0.67 g/litre
    (g) $Zn(OH)_2$, 2.33×10^{-4} g/litre
    (h) Ag_3PO_4, 6.73×10^{-3} g/litre
15. Calculate the solubility of $AgC_2H_3O_2$ in moles/litre and grams/100 ml. (See Table 16.4 for K_{sp}.)

16. Given the following solubility products, calculate the molar solubility for each substance:
    (a) $PbCO_3$, $K_{sp} = 1.5 \times 10^{-15}$
    (b) $SrSO_4$, $K_{sp} = 7.6 \times 10^{-7}$
    (c) $PbCl_2$, $K_{sp} = 1.6 \times 10^{-5}$
    (d) $Mg(OH)_2$, $K_{sp} = 8.9 \times 10^{-12}$
17. Which has the greater molar solubility?
    (a) $CaCO_3$ or $MgCO_3$
    (b) $BaCO_3$ or Ag_2CO_3
    K_{sp} values: $CaCO_3$, 4.7×10^{-9}; $MgCO_3$, 1.0×10^{-15}; $BaCO_3$, 1.6×10^{-9}; Ag_2CO_3, 8.2×10^{-12}
18. Calculate the H^+ ion concentration and the pH of buffer solutions which are 0.10 M in $HC_2H_3O_2$ and contain sufficient sodium acetate to make the $C_2H_3O_2^-$ ion concentration equal to:
    (a) 0.050 M
    (b) 0.10 M
    Use the K_i expression of $HC_2H_3O_2$ in your calculations.

17 Oxidation–Reduction

After studying Chapter 17 you should be able to:

1. Understand the terms listed in Question A at the end of the chapter.
2. Assign oxidation numbers to all the atoms in a given compound or ion.
3. Determine what is being oxidized and what is being reduced in an oxidation–reduction reaction.
4. Identify the oxidizing agent and the reducing agent in an oxidation–reduction reaction.
5. Balance oxidation–reduction equations in molecular and in ionic form.
6. List the general principles concerning the activity series of the metals.
7. Use the activity series to determine whether a proposed single-replacement reaction will occur.
8. Distinguish between an electrolytic and a voltaic cell.
9. Draw a voltaic cell that will produce electric current from an oxidation–reduction reaction involving two metals and their salts.
10. Identify the anode reaction and the cathode reaction in a given electrolytic or voltaic cell.
11. Write equations for the overall chemical reaction and for the oxidation and reduction reactions involved in the discharging or charging of a lead storage battery.
12. Explain how the charge condition of a lead storage battery can be estimated with the aid of a hydrometer.

17.1 Oxidation Number

The oxidation number of an atom (sometimes called its oxidation state) can be considered to represent the number of electrons lost, gained, or unequally shared by the atom. Oxidation numbers can be zero, positive, or negative. When the oxidation number of an atom is zero, the atom has the same number of electrons assigned to it as there are in the free neutral atom. When the oxidation number is positive, the atom has fewer electrons assigned to it than there are in the neutral atom. When the oxidation number is negative, the atom has more electrons assigned to it than there are in the neutral atom.

The oxidation number of an atom that has lost or gained electrons to form an ion is the same as the plus or minus charge of the ion. In covalent compounds, where electrons are shared between two atoms, the atoms are assigned oxidation numbers by a somewhat arbitrary system based on the relative electronegativities of the atoms. When two atoms share a pair of electrons, the atom with the higher electronegativity has a greater attraction for the electrons. Therefore, when a pair of electrons is shared unequally between two atoms, both electrons are assigned to the more electronegative element. Each element is then assigned an oxidation number based on the number of electrons gained or lost compared to the neutral atom.

In the ionic compound sodium chloride (NaCl), where one electron has completely transferred from a Na atom to a Cl atom, the oxidation number of Na is clearly established to be $+1$, and for Cl it is -1. In magnesium chloride ($MgCl_2$), two electrons have completely transferred from the Mg atom to the Cl atoms; thus, the oxidation number of Mg is $+2$.

In symmetrical covalent molecules such as H_2 and Cl_2,

$$H:H \qquad :\ddot{C}l:\ddot{C}l:$$

electrons are shared equally between the two atoms. Neither atom is more positive or negative than the other, therefore, each is assigned an oxidation number of zero.

In compounds with covalent bonds, such as NH_3 and H_2O,

$$\begin{array}{c} H \\ :\ddot{N}:H \\ \ddot{H} \end{array} \quad \text{Shared pairs of electrons} \quad \begin{array}{c} H:\ddot{O}: \\ \ddot{H} \end{array}$$

the pairs of electrons are unequally shared between the atoms and are attracted toward the more electronegative elements, N and O. This unequal sharing causes the N and O atoms to be relatively negative with respect to the H atoms. At the same time, it causes the H atoms to be relatively positive with respect

Table 17.1. Arbitrary rules for assigning oxidation numbers.

1. All elements in their free state (uncombined with other elements) have an oxidation number of zero (for example, Na, Cu, Mg, H_2, O_2, Cl_2, N_2).
2. H is $+1$, except in metal hydrides, where it is -1 (for example, NaH, CaH_2).
3. O is -2, except in peroxides, where it is -1, and in OF_2, where it is $+2$.
4. The metallic element in an ionic compound has a positive oxidation number.
5. In covalent compounds, the negative oxidation number is assigned to the most electronegative atom.
6. The algebraic sum of the oxidation numbers of the elements in a compound is zero.
7. The algebraic sum of the oxidation numbers of the elements in a polyatomic ion is equal to the charge of the ion.

Table 17.2. Oxidation numbers of atoms in selected compounds.

Ion or compound	Oxidation number
H_2O	H, +1; O, −2
SO_2	S, +4; O, −2
CH_4	C, −4; H, +1
CO_2	C, +4; O, −2
$KMnO_4$	K, +1; Mn, +7; O, −2
Na_3PO_4	Na, +1; P, +5; O, −2
$Al_2(SO_4)_3$	Al, +3; S, +6; O, −2
NO	N, +2; O, −2
BCl_3	B, +3; Cl, −1
SO_4^{2-}	S, +6; O, −2
NO_3^-	N, +5; O, −2
CO_3^{2-}	C, +4; O, −2

to the N and O atoms. In H_2O, both pairs of shared electrons are assigned to the O atom, giving it two electrons more than the neutral O atom. At the same time, each H atom is assigned one electron less than the neutral H atom. Therefore, the O atom is assigned an oxidation number of −2, and each H atom is assigned an oxidation number of +1. In NH_3, the three pairs of shared electrons are assigned to the N atom, giving it three electrons more than the neutral N atoms. At the same time, each H atom has one electron less than the neutral atom. Therefore, the N atom is assigned an oxidation number of −3, and each H atom is assigned an oxidation number of +1.

The assignment of correct oxidation numbers to elements is essential for balancing oxidation–reduction equations. Restudy Sections 7.11–7.14, regarding oxidation numbers, oxidation number tables, and the determination of oxidation numbers from formulas. Table 7.4 (page 119) lists relative electronegativities of the elements. Rules for assigning oxidation numbers are given in Section 7.11 (page 127) and are summarized in Table 17.1, given here. Examples showing oxidation numbers in compounds and ions are given in Table 17.2.

17.2 Oxidation–Reduction

oxidation–reduction

redox

oxidation

reduction

Oxidation–reduction, also known as **redox**, is a chemical process in which the oxidation number of an element is changed. The process may involve the complete transfer of electrons to form ionic bonds or only a partial transfer or shift of electrons to form covalent bonds.

Oxidation occurs whenever the oxidation number of an element increases as a result of losing electrons. Conversely, **reduction** occurs whenever the oxidation number of an element decreases as a result of gaining electrons. For example, a change in oxidation number from +2 to +3 or from −1 to 0 is oxidation; a change from +5 to +2 or from −2 to −4 is reduction (see Figure 17.1). Oxidation and reduction occur simultaneously in a chemical reaction; one cannot take place without the other.

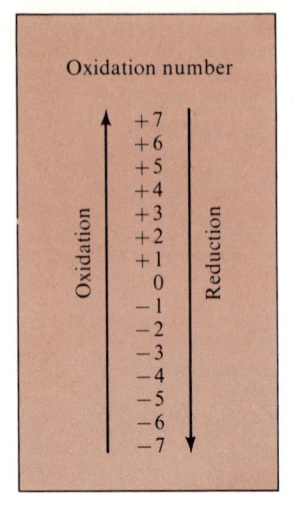

Figure 17.1. Oxidation and reduction: Oxidation results in an increase in the oxidation number, and reduction results in a decrease in the oxidation number.

Many combination, decomposition, and single-replacement reactions involve oxidation–reduction. Let us examine the combustion of hydrogen and oxygen from this point of view:

$$2\,H_2 + O_2 \longrightarrow 2\,H_2O$$

Both reactants, hydrogen and oxygen, are elements in the free state and have an oxidation number of zero. In the product water, hydrogen has been oxidized to $+1$ and oxygen reduced to -2. The substance that does the oxidizing is known as the **oxidizing agent**. The substance that does the reducing is the **reducing agent**. In this reaction, the oxidizing agent is free oxygen and the reducing agent is free hydrogen. In the reaction

$$Zn(s) + H_2SO_4(aq) \longrightarrow ZnSO_4(aq) + H_2\uparrow$$

metallic zinc is oxidized and hydrogen ions are reduced. Zinc is the reducing agent and hydrogen ions the oxidizing agent. Electrons are transferred from the zinc metal to the hydrogen ions. The reaction is better expressed as

$$Zn^0 + 2\,H^+ + SO_4^{2-} \longrightarrow Zn^{2+} + SO_4^{2-} + H_2^0\uparrow$$

Oxidation: *Increase in oxidation number*
              Loss of electrons

Reduction: *Decrease in oxidation number*
              Gain of electrons

The oxidizing agent is reduced and gains electrons. The reducing agent is oxidized and loses electrons. The loss and gain of electrons is a characteristic feature of all redox reactions.

17.3 Balancing Oxidation–Reduction Equations

Many simple redox equations may be easily balanced by inspection, or trial and error.

$$Na + Cl_2 \longrightarrow NaCl \quad \text{(Unbalanced)}$$
$$2\,Na + Cl_2 \longrightarrow 2\,NaCl \quad \text{(Balanced)}$$

Balancing this equation is certainly not complicated. But as we study more complex reactions and equations, such as

$$P + HNO_3 + H_2O \longrightarrow NO + H_3PO_4 \quad \text{(Unbalanced)}$$
$$3\,P + 5\,HNO_3 + 2\,H_2O \longrightarrow 5\,NO + 3\,H_3PO_4 \quad \text{(Balanced)}$$

the trial-and-error method of finding the proper numbers to balance the equation would take an unnecessarily long time.

The systematic method of balancing oxidation–reduction equations is based on the transfer of electrons between the oxidizing and reducing agents. Consider the first equation again.

$$Na^0 + Cl_2^0 \longrightarrow Na^+Cl^- \quad \text{(Unbalanced)}$$

In this reaction, sodium metal loses one electron per atom when it changes to a sodium ion. At the same time, chlorine gains one electron per atom. Because chlorine is diatomic, two electrons per molecule are needed to form a chloride ion from each atom. These electrons are furnished by two sodium atoms. Stepwise, the reaction may be written as two half-reactions, the oxidation half-reaction and the reduction half-reaction:

Oxidation half-reaction $2\,Na^0 \longrightarrow 2\,Na^+ + 2\,e^-$

Reduction half-reaction $\underline{Cl_2^0 + 2\,e^- \longrightarrow 2\,Cl^-}$

$2\,Na^0 + Cl_2^0 \longrightarrow 2\,Na^+Cl^-$

When the two half-reactions, each containing the same number of electrons, are added together algebraically, the electrons cancel out. In this reaction there are no excess electrons; the two electrons lost by the two sodium atoms are utilized by chlorine. In all redox reactions, the loss of electrons by the reducing agent must equal the gain of electrons by the oxidizing agent. Sodium is oxidized; chlorine is reduced. Chlorine is the oxidizing agent; sodium is the reducing agent.

The following examples illustrate a systematic method of balancing redox equations.

Example 1   Balance the equation.

$$Sn + HNO_3 \longrightarrow SnO_2 + NO_2 + H_2O \quad \text{(Unbalanced)}$$

(1) The first step is to assign oxidation numbers to each element in order to identify the elements that are being oxidized and those that are being reduced. Write the oxidation

numbers below each element in order to avoid confusing them with the charge on an ion or radical.

$$\underset{0\quad +1\,+5\,-2}{Sn\,+\,H\,N\,O_3} \longrightarrow \underset{+4\,-2}{Sn\,O_2} + \underset{+4\,-2}{N\,O_2} + \underset{+1\,-2}{H_2O}$$

Note that the oxidation numbers of Sn and N have changed.

(2) Now write two new equations, using only the elements that change in oxidation number. Then add electrons to bring the equations into electrical balance. One equation represents the oxidation step; the other represents the reduction step. The oxidation step produces electrons; the reduction step uses electrons.

Oxidation $\quad Sn^0 \longrightarrow Sn^{4+} + 4e^-\quad$ (Sn^0 loses 4 electrons)
Reduction $\quad N^{5+} + 1e^- \longrightarrow N^{4+}\quad$ (N^{5+} gains 1 electron)

(3) Now multiply the two equations by the smallest integral numbers that will make the loss of electrons by the oxidation step equal to the number of electrons gained in the reduction step. In this reaction, the oxidation step is multiplied by 1 and the reduction step by 4. The equations become

Oxidation $\quad Sn^0 \longrightarrow Sn^{4+} + 4e^-\quad$ (Sn^0 loses 4 electrons)
Reduction $\quad 4\,N^{5+} + 4e^- \longrightarrow 4\,N^{4+}\quad$ ($4\,N^{5+}$ gain 4 electrons)

We have now established the ratio of the oxidizing to the reducing agent as being four atoms of N to one atom of Sn.

(4) Now transfer the coefficient that appears in front of each substance in the balanced oxidation–reduction equations to the corresponding substance in the original equation. We need to use 1 Sn, 1 SnO_2, 4 HNO_3, and 4 NO_2:

$$Sn + 4\,HNO_3 \longrightarrow SnO_2 + 4\,NO_2 + H_2O \quad \text{(Unbalanced)}$$

(5) In the usual manner, balance the remaining elements that are not oxidized or reduced to give the final balanced equation:

$$Sn + 4\,HNO_3 \longrightarrow SnO_2 + 4\,NO_2 + 2\,H_2O \quad \text{(Balanced)}$$

In balancing the final elements, we must not change the ratio of the elements that were oxidized and reduced. We should make a final check to ensure that both sides of the equation have the same number of atoms of each element. The final balanced equation contains 1 atom of Sn, 4 atoms of N, 4 atoms of H, and 12 atoms of O on each side.

Since each new equation may present a slightly different problem and since proficiency in balancing equations requires practice, we will work through a few more examples.

Example 2 Balance the equation.

$$I_2 + Cl_2 + H_2O \longrightarrow HIO_3 + HCl \quad \text{(Unbalanced)}$$

(1) Assign oxidation numbers:

$$\underset{0\quad\ \ 0\quad\ +1\,-2}{I_2 + Cl_2 + H_2O} \longrightarrow \underset{+1\,+5\,-2}{H\,I\,O_3} + \underset{+1\,-1}{H\,Cl}$$

The oxidation numbers of I_2 and Cl_2 have changed.

(2) Write oxidation and reduction steps, balancing with electrons:

Oxidation $\quad I_2 \longrightarrow 2\,I^{5+} + 10\,e^-$ $\quad$ (I_2 loses 10 electrons)
Reduction $\quad Cl_2 + 2\,e^- \longrightarrow 2\,Cl^-$ $\quad$ (Cl_2 gains 2 electrons)

(3) Adjust loss and gain of electrons so that they are equal. Multiply oxidation step by 1 and reduction step by 5.

Oxidation $\quad I_2 \longrightarrow 2\,I^{5+} + 10\,e^-$ $\quad$ (I_2 loses 10 electrons)
Reduction $\quad 5\,Cl_2 + 10\,e^- \longrightarrow 10\,Cl^-$ $\quad$ (5 Cl_2 gain 10 electrons)

(4) Transfer the coefficients from the balanced redox equations into the original equation. We need to use 1 I_2, 2 HIO_3, 5 Cl_2, and 10 HCl.

$$I_2 + 5\,Cl_2 + H_2O \longrightarrow 2\,HIO_3 + 10\,HCl \quad \text{(Unbalanced)}$$

(5) Balance the remaining elements, H and O:

$$I_2 + 5\,Cl_2 + 6\,H_2O \longrightarrow 2\,HIO_3 + 10\,HCl \quad \text{(Balanced)}$$

Check: The final balanced equation contains 2 atoms of I, 10 atoms of Cl, 12 atoms of H, and 6 atoms of O on each side.

Example 3 Balance the equation.

$$K_2Cr_2O_7 + FeCl_2 + HCl \longrightarrow CrCl_3 + KCl + FeCl_3 + H_2O \quad \text{(Unbalanced)}$$

(1) Assign oxidation numbers (Cr and Fe have changed):

$$K_2Cr_2O_7 + FeCl_2 + H\,Cl \longrightarrow CrCl_3 + K\,Cl + FeCl_3 + H_2O$$
$$+1\ +6\ -2 \quad +2\ -1 \quad +1\ -1 \quad\quad +3\ -1 \quad +1\,-1 \quad +3\,-1 \quad +1\,-2$$

(2) Write the oxidation–reduction steps:

Oxidation $\quad Fe^{2+} \longrightarrow Fe^{3+} + 1\,e^-$ $\quad$ (Fe^{2+} loses 1 electron)
Reduction $\quad Cr^{6+} + 3\,e^- \longrightarrow Cr^{3+}$
$\quad$ or $\quad 2\,Cr^{6+} + 6\,e^- \longrightarrow 2\,Cr^{3+}$ $\quad$ (2 Cr^{6+} gain 6 electrons)

(3) Balance the loss and gain of electrons. Multiply the oxidation step by 6 and the reduction step by 1 to equalize the transfer of electrons.

Oxidation $\quad 6\,Fe^{2+} \longrightarrow 6\,Fe^{3+} + 6\,e^-$ $\quad$ (6 Fe^{2+} lose 6 electrons)
Reduction $\quad 2\,Cr^{6+} + 6\,e^- \longrightarrow 2\,Cr^{3+}$ $\quad$ (2 Cr^{6+} gain 6 electrons)

(4) Transfer the coefficients from the balanced redox equations into the original equation. (Note that one formula unit of $K_2Cr_2O_7$ contains two Cr atoms.) We need to use 1 $K_2Cr_2O_7$, 2 $CrCl_3$, 6 $FeCl_2$, and 6 $FeCl_3$.

$$K_2Cr_2O_7 + 6\,FeCl_2 + HCl \longrightarrow 2\,CrCl_3 + KCl + 6\,FeCl_3 + H_2O \quad \text{(Unbalanced)}$$

(5) Balance the remaining elements in this order: K, Cl, H, O.

$$K_2Cr_2O_7 + 6\,FeCl_2 + 14\,HCl \longrightarrow 2\,CrCl_3 + 2\,KCl + 6\,FeCl_3 + 7\,H_2O \quad \text{(Balanced)}$$

Check: The final balanced equation contains 2 K atoms, 2 Cr atoms, 7 O atoms, 6 Fe atoms, 26 Cl atoms, and 14 H atoms on each side.

Example 4   Try the following equation, which has a little different twist to it.

$$Cu + HNO_3 \longrightarrow Cu(NO_3)_2 + NO + H_2O \quad \text{(Unbalanced)}$$

(1) Assign oxidation numbers [Cu and N (in NO) have changed]:

$$Cu + HNO_3 \longrightarrow Cu(NO_3)_2 + NO + H_2O$$
$$ 0 +5 +2\ +5 +2$$

(2) Write the oxidation–reduction steps:

Oxidation $Cu^0 \longrightarrow Cu^{2+} + 2\,e^-$ (Cu^0 loses 2 electrons)
Reduction $N^{5+} + 3\,e^- \longrightarrow N^{2+}$ (N^{5+} gains 3 electrons)

(3) Balance the loss and gain of electrons. Multiply the oxidation step by 3 and the reduction step by 2 to equalize the loss and gain of electrons.

Oxidation $3\,Cu^0 \longrightarrow 3\,Cu^{2+} + 6\,e^-$ (3 Cu^0 lose 6 electrons)
Reduction $2\,N^{5+} + 6\,e^- \longrightarrow 2\,N^{2+}$ (2 N^{5+} gain 6 electrons)

(4) Transfer the coefficients from the balanced redox equations into the original equation.

$$3\,Cu + 2\,HNO_3 \longrightarrow 3\,Cu(NO_3)_2 + 2\,NO + H_2O \quad \text{(Unbalanced)}$$

(5) Balance the remaining elements. In doing this, we notice that there are 8 N atoms on the right side of the equation and 2 N atoms on the left. This imbalance indicates that 6 more HNO_3 molecules are needed on the left and also that 6 NO_3^- ions did not enter into the redox reaction. The use of 8 HNO_3 in the balanced equation does not destroy the ratio of 3 Cu/2 HNO_3 needed for oxidation–reduction. The balanced equation is

$$3\,Cu + 8\,HNO_3 \longrightarrow 3\,Cu(NO_3)_2 + 2\,NO + 4\,H_2O \quad \text{(Balanced)}$$

Check: The final balanced equation contains 3 Cu atoms, 8 H atoms, 8 N atoms, and 24 O atoms on each side.

17.4   Balancing Ionic Redox Equations

The main difference in balancing ionic versus molecular redox equations is the handling of ions. In addition to having the same number of each kind of element, the net charges on both sides of the final equation must be equal to each other. In assigning oxidation numbers, we must be careful to consider the charge on the ions. In many respects, balancing ionic equations is much simpler than balancing molecular equations.

Example 5   Balance the equation.

$$Fe^{2+} + Br_2 \longrightarrow Fe^{3+} + Br^- \quad \text{(Unbalanced)}$$

You might try to balance this equation simply by placing a 2 in front of the Br^-:

$$Fe^{2+} + Br_2 \longrightarrow Fe^{3+} + 2\,Br^- \quad \text{(Unbalanced)}$$

However, the equation is not balanced, because the electrical charges on the left and the right sides of the equation are not equal. The left side has a charge of $+2$ and the right side has a charge of $(+3) + (-2) = +1$. The net charge on each side is determined by adding the charges of all the ions. The equation is correctly balanced by the use of electrons, as follows:

Oxidation $Fe^{2+} \longrightarrow Fe^{3+} + 1\,e^-$ (Fe^{2+} loses 1 electron)
Reduction $Br_2 + 2\,e^- \longrightarrow 2\,Br^-$ (2 Br gain 2 electrons)

Equalize the loss and gain of electrons:

Oxidation $2\,Fe^{2+} \longrightarrow 2\,Fe^{3+} + 2\,e^-$ (2 Fe lose 2 electrons)
Reduction $Br_2 + 2\,e^- \longrightarrow 2\,Br^-$ (2 Br gain 2 electrons)

Finally,

$$2\,Fe^{2+} + Br_2 \longrightarrow 2\,Fe^{3+} + 2\,Br^- \quad \text{(Balanced)}$$

Net charge: $(+4) + (0) = +4 \quad (+6) + (-2) = +4$

The balanced equation contains the same number of each kind of atom and the same electrical charge on each side of the equation. The charge on each side is $+4$.

Example 6 Try the more complex equation

$$MnO_4^- + S^{2-} + H^+ \longrightarrow Mn^{2+} + S^0 + H_2O \quad \text{(Unbalanced)}$$

First, assign oxidation numbers:

$$MnO_4^- + S^{2-} + H^+ \longrightarrow Mn^{2+} + S^0 + H_2O$$
$\,+7 \quad\; -2 \quad\quad\quad\quad\; +2 \quad\;\; 0$

Oxidation $S^{2-} \longrightarrow S^0 + 2\,e^-$ (S^{2-} loses 2 electrons)
Reduction $Mn^{7+} + 5\,e^- \longrightarrow Mn^{2+}$ (Mn^{7+} gains 5 electrons)

Multiply the oxidation step by 5 and the reduction step by 2 to balance the loss and gain of electrons:

Oxidation $5\,S^{2-} \longrightarrow 5\,S^0 + 10\,e^-$ ($5\,S^{2-}$ lose 10 electrons)
Reduction $2\,Mn^{7+} + 10\,e^- \longrightarrow 2\,Mn^{2+}$ ($2\,Mn^{7+}$ gain 10 electrons)

Transfer balanced redox coefficients to the original equation:

$$2\,MnO_4^- + 5\,S^{2-} + H^+ \longrightarrow 2\,Mn^{2+} + 5\,S^0 + H_2O \quad \text{(Unbalanced)}$$

At this point there remain to be balanced the electrical charge, the H atoms, and the O atoms. First the electrical charges are balanced by the use of additional H^+ ions. The H and O atoms are then balanced by the use of H_2O molecules as needed. We find that $16\,H^+$ ions are needed and $8\,H_2O$ molecules are needed to bring the equation into balance.

$$2\,MnO_4^- + 5\,S^{2-} + 16\,H^+ \longrightarrow 2\,Mn^{2+} + 5\,S^0 + 8\,H_2O \quad \text{(Balanced)}$$

$(-2) + (-10) + (+16) = +4 \quad (+4) + (0) + (0) = +4$

Check: Both sides have a net charge of $+4$ and contain the same number of atoms of each element. The equation is balanced.

The H^+ ions in the equation show that the reaction of Example 6 is in an acid solution. Therefore, H^+ ions and water molecules are used in balancing the ionic equation. For a reaction in an alkaline solution, OH^- ions and H_2O molecules are used as needed to balance the ionic equation.

17.5 Activity Series of Metals

Knowledge of the relative chemical reactivity of the elements is useful for predicting the course of many chemical reactions.

Calcium reacts with cold water and magnesium reacts with steam to produce hydrogen in each case. Calcium, therefore, is considered to be a more reactive metal than magnesium.

$$Ca(s) + 2\,H_2O(l) \longrightarrow Ca(OH)_2(aq) + H_2\uparrow$$

$$Mg(s) + \underset{\text{Steam}}{H_2O(g)} \longrightarrow MgO(s) + H_2\uparrow$$

The difference in their activity is attributed to the relative ease with which each loses its two valence electrons. It is apparent that calcium loses these electrons more easily than magnesium and is therefore more reactive and/or more readily oxidized than magnesium.

When a strip of copper is placed in a solution of silver nitrate ($AgNO_3$), free silver begins to plate out on the copper. After the reaction has continued for some time, we can observe a blue color in the solution, indicating the presence of copper(II) ions. If a strip of silver is placed in a solution of copper(II) nitrate [$Cu(NO_3)_2$], no reaction is visible. The equations are

$$Cu^0 + 2\,AgNO_3(aq) \longrightarrow 2\,Ag^0 + Cu(NO_3)_2(aq)$$

$Cu^0 + 2\,Ag^+ \longrightarrow 2\,Ag^0 + Cu^{2+}$ Net ionic equation

$Cu^0 \longrightarrow Cu^{2+} + 2\,e^-$ Oxidation of Cu^0

$Ag^+ + e^- \longrightarrow Ag^0$ Reduction of Ag^+

$Ag^0 + Cu(NO_3)_2(aq) \longrightarrow$ No reaction

In the reaction between Cu and $AgNO_3$, electrons are transferred from Cu^0 atoms to Ag^+ ions in solution. Since copper has a greater tendency than silver to lose electrons, an electrochemical force is exerted upon silver ions to accept electrons from copper atoms. When an Ag^+ ion adds an electron, it is reduced to a Ag^0 atom and is no longer soluble in solution. At the same time, Cu^0 is oxidized and goes into solution as Cu^{2+} ions. From this reaction we can conclude that copper is more reactive than silver.

Metals such as sodium, magnesium, zinc, and iron, which react with solutions of acids to liberate hydrogen, are more reactive than hydrogen. Metals such as copper, silver, and mercury, which do not react with solutions of acids to liberate hydrogen, are less reactive than hydrogen. By studying a series of reactions such as those given above, we may list metals according to their chemical activity, placing the most active at the top and the least active at the bottom. This list is called the **Activity Series of Metals**. Table 17.3 shows some of the common metals in the series. The arrangement corresponds to the ease with which the elements listed are oxidized or lose electrons. The most easily oxidizable element is listed first. More extensive tables are available in chemistry reference books.

activity series of metals

Table 17.3. Activity Series of Metals.

↑ Ease of oxidation

K	→	K$^+$	+	e^-
Ba	→	Ba^{2+}	+	$2e^-$
Ca	→	Ca^{2+}	+	$2e^-$
Na	→	Na$^+$	+	e^-
Mg	→	Mg^{2+}	+	$2e^-$
Al	→	Al^{3+}	+	$3e^-$
Zn	→	Zn^{2+}	+	$2e^-$
Cr	→	Cr^{3+}	+	$3e^-$
Fe	→	Fe^{2+}	+	$2e^-$
Ni	→	Ni^{2+}	+	$2e^-$
Sn	→	Sn^{2+}	+	$2e^-$
Pb	→	Pb^{2+}	+	$2e^-$
H$_2$	→	2 H$^+$	+	$2e^-$
Cu	→	Cu^{2+}	+	$2e^-$
As	→	As^{3+}	+	$3e^+$
Ag	→	Ag$^+$	+	e^-
Hg	→	Hg^{2+}	+	$2e^-$
Au	→	Au^{3+}	+	$3e^-$

The general principles governing the arrangement and use of the activity series are as follows:

1. The reactivity of the metals listed decreases from top to bottom.
2. A free metal can displace the ion of a second metal from solution provided that the free metal is above the second metal in the activity series.
3. Free metals above hydrogen react with nonoxidizing acids in solution to liberate hydrogen gas.
4. Free metals below hydrogen do not liberate hydrogen from acids.
5. Conditions such as temperature and concentration may affect the relative position of some of these elements.

Two examples of the application of the Activity Series follow.

Example 1 Will zinc metal react with dilute sulfuric acid?

From Table 17.3, we see that zinc is above hydrogen; therefore, zinc atoms will lose electrons more readily than hydrogen atoms. Hence, zinc atoms will reduce hydrogen ions from the acid to form hydrogen gas and zinc ions. In fact, these reagents are commonly used for the laboratory preparation of hydrogen. The equation is

$$Zn(s) + H_2SO_4(aq) \longrightarrow ZnSO_4(aq) + H_2\uparrow$$
$$Zn + 2\,H^+ \longrightarrow Zn^{2+} + H_2\uparrow \qquad \text{Net ionic equation}$$

Example 2 Will a reaction occur when copper metal is placed in an iron(II) sulfate solution?

No, copper lies below iron in the series, loses electrons less easily than iron, and therefore will not replace iron(II) ions from solution. In fact, the reverse is true. When an iron nail is dipped into a copper(II) sulfate solution, it becomes coated with free copper. The equations are

$$Cu(s) + FeSO_4(aq) \longrightarrow \text{No reaction}$$
$$Fe(s) + CuSO_4(aq) \longrightarrow FeSO_4(aq) + Cu\downarrow$$

From Table 17.3 we may abstract the following pair in their relative position to each other.

$$Fe \longrightarrow Fe^{2+} + 2\,e^-$$
$$Cu \longrightarrow Cu^{2+} + 2\,e^-$$

According to the second principle listed above on the use of the activity series, we can predict that free iron will react with copper(II) ions in solution to form free copper metal and iron(II) ions in solution.

$$Fe + Cu^{2+} \longrightarrow Fe^{2+} + Cu$$

17.6 Electrolytic and Voltaic Cells

electrolysis
electrolytic cell

The process in which electrical energy is used to bring about a chemical change is known as **electrolysis**. An **electrolytic cell** uses electrical energy to produce a nonspontaneous chemical reaction. There are many applications of electrical energy in the chemical industry—for example, in the production of sodium, sodium hydroxide, chlorine, fluorine, magnesium, aluminum, and pure hydrogen and oxygen, and in the purification and electroplating of metals.

What happens when an electric current is passed through a solution? Let us consider a hydrochloric acid solution in a simple electrolysis cell, as shown in Figure 17.2. The cell consists of a battery connected to two electrodes that are immersed in a solution of hydrochloric acid. The cathode is attached to the negative pole of the battery and becomes the negative electrode. The anode is attached to the positive pole and becomes the positive electrode. The battery supplies electrons to the cathode.

When the switch is closed, the electric circuit is completed; positive hydronium ions (H_3O^+) migrate toward the cathode, where they pick up electrons and evolve hydrogen gas. At the same time, the negative chloride ions (Cl^-) migrate toward the anode, where they lose electrons, completing the cycle, and evolve chlorine gas.

Reaction at the cathode
$$H_3O^+ + 1\,e^- \longrightarrow H^0 + H_2O$$
$$H^0 + H^0 \longrightarrow H_2\uparrow$$

Reaction at the anode
$$Cl^- \longrightarrow Cl^0 + 1\,e^-$$
$$Cl^0 + Cl^0 \longrightarrow Cl_2\uparrow$$

Net reaction
$$2\,HCl \xrightarrow{\text{Electrolysis}} H_2\uparrow + Cl_2\uparrow$$

Note that oxidation–reduction has taken place. Chloride ions lose electrons (are oxidized) at the anode, and hydronium ions gain electrons (are reduced) at the cathode. In electrolysis, oxidation always occurs at the anode and reduction at the cathode.

When sodium chloride brines are electrolyzed, the products are sodium hydroxide, hydrogen, and chlorine. The overall reaction is

$$2\,Na^+ + 2\,Cl^- + 2\,H_2O \xrightarrow{\text{Electrolysis}} 2\,Na^+ + 2\,OH^- + H_2\uparrow + Cl_2\uparrow$$

The net ionic equation is

$$2\ Cl^- + 2\ H_2O \longrightarrow 2\ OH^- + H_2\uparrow + Cl_2\uparrow$$

During the electrolysis, Na^+ ions move toward the cathode and Cl^- ions move toward the anode. The anode reaction is similar to that of hydrochloric acid; chlorine is liberated.

$$2\ Cl^- \longrightarrow Cl_2\uparrow + 2\ e^-$$

Even though Na^+ ions are attracted by the cathode, the facts show that hydrogen is liberated there. No evidence of metallic sodium is found, but the area around the cathode tests alkaline from accumulated OH^- ions. The reaction at the cathode is believed to be

$$2\ H_2O + 2e^- \longrightarrow H_2\uparrow + 2\ OH^-$$

If the electrolysis is allowed to continue until all the chloride is reacted, the solution remaining will contain only sodium hydroxide. Large tonnages of sodium hydroxide and chlorine are made by this process.

When molten sodium chloride (without water) is subjected to electrolysis, metallic sodium and chlorine gas are formed:

$$2\ Na^+ + 2\ Cl^- \xrightarrow{Electrolysis} 2\ Na + Cl_2\uparrow$$

An important electrochemical application is electroplating of metals. Electroplating is the art of covering a surface or an object with a thin adherent electrodeposited metal coating. Electroplating is done for protection of the surface of the base metal or for a purely decorative effect. The layer deposited is surprisingly thin, varying from as little as 5×10^{-5} cm to 2×10^{-3} cm, depending on the metal and the intended use. The object to be plated is set up as the cathode and is immersed in a solution containing ions of the metal to be plated. When an electric current passes through the solution, metal ions migrating to the cathode are reduced, depositing on the object as the free metal. In most cases the metal deposited on the object is replaced in the solution by using an anode of the same metal. The following equations show the chemical changes in the electroplating of nickel:

Reaction at the cathode	$Ni^{2+} + 2e^- \longrightarrow Ni(s)$	Ni plated out on an object
Reaction at the anode	$Ni(s) \longrightarrow Ni^{2+} + 2\ e^-$	Ni replenished to solution

Metals most commonly used in commercial electroplating are copper, nickel, zinc, lead, cadmium, chromium, tin, gold, and silver.

In the electrolytic cell shown in Figure 17.2, an electric current flows through the circuit when the switch is closed. The driving force responsible for the current is supplied by the battery (or other source of direct current). Electrons are moving through the wires and electrodes, and ions (H_3O^+ and Cl^-) are moving in the solution. As a result of the transfer of electrons and ions, hydrochloric acid is converted to hydrogen and chlorine. In electrolytic

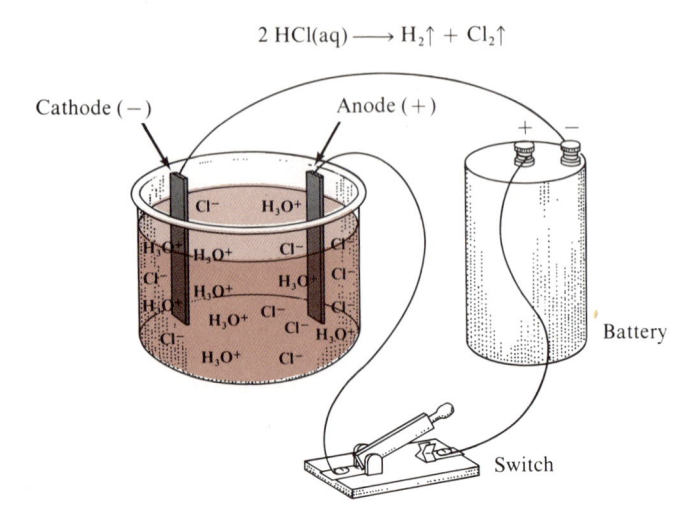

Figure 17.2. Electrolysis: During the electrolysis of a hydrochloric acid solution, positive hydronium ions are attracted to the cathode, where they gain electrons and form hydrogen gas. Chloride ions migrate to the anode, where they lose electrons and form chlorine gas. The equation for this process is

$$2\,HCl \longrightarrow H_2\uparrow + Cl_2\uparrow$$

processes of this kind, electrical energy is used to bring about nonspontaneous redox reactions. The hydrogen and chlorine produced have more potential energy than was present in the hydrochloric acid before electrolysis.

Conversely, some spontaneous redox reactions can be made to supply useful amounts of electrical energy. When a piece of zinc is put in a copper(II) sulfate solution, the zinc quickly becomes coated with metallic copper. We expect this to happen because zinc is above copper in the activity series; copper(II) ions are therefore reduced by zinc atoms:

$$Zn^0 + Cu^{2+} \longrightarrow Zn^{2+} + Cu^0$$

This reaction is clearly a spontaneous redox reaction. But simply dipping a zinc rod into a copper(II) sulfate solution will not produce useful electric current! However, when we carry out this reaction in the cell shown in Figure 17.3, an electric current is produced. The cell consists of a piece of zinc immersed in a zinc sulfate solution and connected by a wire through a voltmeter to a piece of copper immersed in copper(II) sulfate solution. The two solutions are connected by a salt bridge. Such a cell produces an electric current and a potential of about 1.1 volts when both solutions are 1.0 M in concentration. A cell that produces electric current from a spontaneous chemical reaction is called a **voltaic cell**.

voltaic cell

Although the zinc–copper voltaic cell is no longer used commercially, it was used to energize the first transcontinental telegraph lines. Such cells are the direct ancestors of the many different kinds of "dry" cells that operate portable radio and television sets, automatic cameras, tape recorders, and so on.

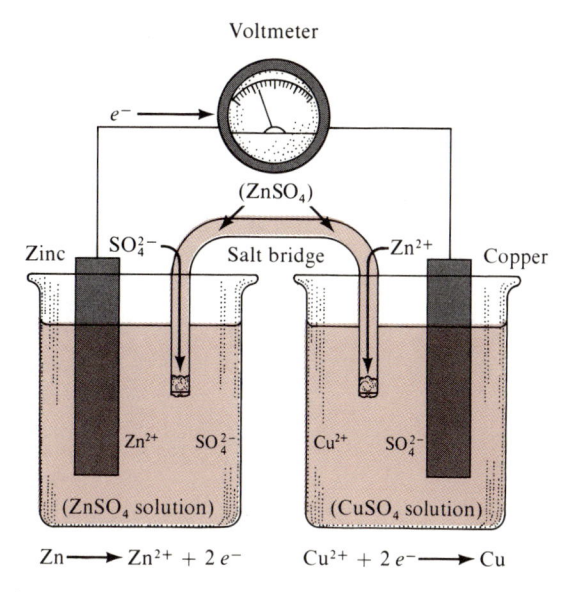

Figure 17.3. Zinc–copper voltaic cell. The cell has a potential of 1.1 volts when ZnSO₄ and CuSO₄ solutions are 1.0 M. The salt bridge provides electrical contact between the two half-cells.

The driving force responsible for the electric current in the zinc–copper cell originates in the great tendency of zinc atoms to lose electrons relative to the tendency of copper(II) ions to gain electrons. In the cell shown in Figure 17.3, zinc atoms lose electrons and are converted to zinc ions at the zinc electrode surface; the electrons flow through the wire to the copper electrode. Here, copper(II) ions pick up electrons and are reduced to copper atoms, which plate out on the copper electrode. Sulfate ions flow from the CuSO₄ solution via the salt bridge into the ZnSO₄ solution to complete the circuit. The equations for the reactions of this cell are

Anode	$Zn^0 \longrightarrow Zn^{2+} + 2\,e^-$	(Oxidation)
Cathode	$Cu^{2+} + 2\,e^- \longrightarrow Cu^0$	(Reduction)
Net ionic	$Zn^0 + Cu^{2+} \longrightarrow Zn^{2+} + Cu^0$	
Overall	$Zn + CuSO_4 \longrightarrow Cu + ZnSO_4$	

The redox reaction, the movement of electrons in the metallic or external part of the circuit, and the movement of ions in the solution or internal part of the circuit of the copper–zinc cell are very similar to the actions that occur in the electrolytic cell of Figure 17.2. The only important difference is that the reactions of the zinc–copper cell are spontaneous. This is the crucial difference between all voltaic and electrolytic cells. Voltaic cells use chemical reactions to produce electrical energy, and electrolytic cells use electrical energy to produce chemical reactions.

An ordinary automobile storage battery is an energy reservoir. The charged battery acts as a voltaic cell and through chemical reactions furnishes electrical energy to operate the starter, lights, radio, and so on. When the engine

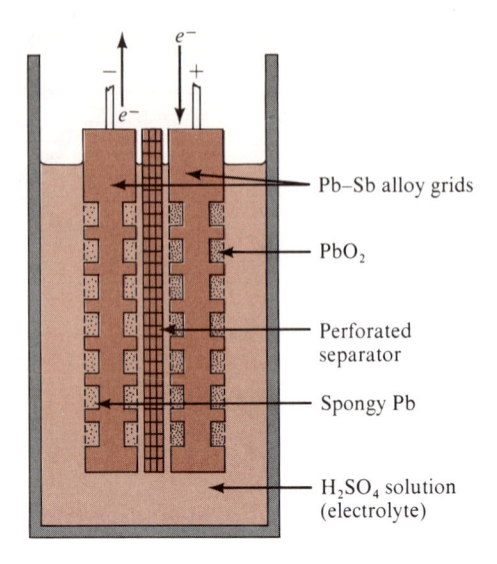

Figure 17.4. Cross-sectional diagram of a lead storage battery cell.

is running, a generator, or alternator, produces and forces an electric current through the battery and, by electrolytic chemical action, restores the battery to the charged condition.

The cell unit consists of a lead plate filled with spongy lead and a lead dioxide plate, both immersed in dilute sulfuric acid solution, which serves as the electrolyte (see Figure 17.4). When the cell is discharging, or acting as a voltaic cell, these reactions occur.

Pb plate (anode) $\quad Pb^0 \longrightarrow Pb^{2+} + 2e^-$ (Oxidation)

PbO$_2$ plate (cathode) $\quad PbO_2 + 4H^+ + 2e^- \longrightarrow Pb^{2+} + 2H_2O$ (Reduction)

Net ionic redox reaction $\quad Pb^0 + PbO_2 + 4H^+ \longrightarrow 2Pb^{2+} + 2H_2O$

Precipitation reaction on plates $\quad Pb^{2+} + SO_4^{2-} \longrightarrow PbSO_4$

Since lead(II) sulfate is insoluble, the Pb^{2+} ions combine with SO_4^{2-} ions to form a coating of $PbSO_4$ on each plate. The overall chemical reaction of the cell is

$$Pb + PbO_2 + 2H_2SO_4 \xrightarrow{\text{Discharge cycle}} 2PbSO_4\downarrow + 2H_2O$$

If the active material on both plates is converted to $PbSO_4$, the cell is discharged and no more electrical energy is to be had. The cell can be recharged by reversing the chemical reaction. This is accomplished by forcing an electric current through the cell in the opposite direction. Lead sulfate and water are reconverted to lead, lead dioxide, and sulfuric acid:

$$2\,PbSO_4 + 2\,H_2O \xrightarrow[\text{cycle}]{\text{Charge}} Pb + PbO_2 + 2\,H_2SO_4$$

The electrolyte in a lead storage battery is a 38% by weight sulfuric acid solution having a density of 1.29 g/ml. As the battery is discharged, sulfuric acid is removed, thereby decreasing the density of the electrolyte solution. The state of charge or discharge of the battery can be estimated by measuring the density of the electrolyte solution with a hydrometer. When the density has dropped to about 1.05 g/ml, the battery needs recharging.

In an actual battery, each cell consists of a series of cell units of alternating lead–lead dioxide plates separated and supported by wood, asbestos, or glass wool. The energy storage capacity of a single cell is limited and its electrical potential is only about 2 volts. Therefore, a bank of six cells is connected in series to provide the 12 volt output of the usual automobile battery.

Questions

A. Review the meanings of the new terms introduced in this chapter.
   1. Oxidation–reduction
   2. Redox
   3. Oxidation
   4. Reduction
   5. Oxidizing agent
   6. Reducing agent
   7. Activity Series of Metals
   8. Electrolysis
   9. Electrolytic cell
   10. Voltaic cell

B. Answers to the following questions will be found in tables and figures.
   1. Would you expect nickel to replace hydrogen in either cold water or steam? Explain. (See Table 17.3.)
   2. The reaction between powdered aluminum and iron(III) oxide (in the Thermite process), producing molten iron, is very exothermic.
      (a) Write the equation for the chemical reaction that occurs.
      (b) Explain in terms of Table 17.3 why a reaction occurs.
      (c) Would you expect a reaction between aluminum and chromium(III) oxide (Cr_2O_3)?
   3. Write equations for the chemical reaction of each of the following metals with dilute solutions of hydrochloric and sulfuric acids: aluminum, chromium, gold, iron, magnesium, mercury, and zinc. If a reaction will not occur, write "no reaction" as the product. (See Table 17.3.)
   4. A $NiCl_2$ solution (instead of HCl) is placed in the apparatus shown in Figure 17.2. Write equations for:
      (a) The anode reaction
      (b) The cathode reaction
      (c) The net electrochemical reaction

C. Review questions.
   1. Which of the following half-reactions are oxidation and which are reduction? Supply the proper number of electrons to balance each equation.
      (a) $SO_3^{2-} + H_2O \rightarrow SO_4^{2-} + 2\,H^+$
      (b) $Al^{3+} \rightarrow Al$
      (c) $2\,H_2O \rightarrow O_2 + 4\,H^+$
      (d) $NO_3^- + 10\,H^+ \rightarrow NH_4^+ + 3\,H_2O$
      (e) $Mn^{2+} + 4\,H_2O \rightarrow MnO_4^- + 8\,H^+$
      (f) $Cr_2O_7^{2-} + 14\,H^+ \rightarrow 2\,Cr^{3+} + 7\,H_2O$

2. For the following equations:
   (a) Identify the oxidizing agent, the reducing agent, and the substances oxidized and reduced.
   (b) Assign oxidation numbers to each element in each oxidation and reduction step.
   (c) Write equations for the half-reactions of each oxidation and reduction.

   (1) $Cu + S \rightarrow CuS$

   (2) $Zn + HgO \rightarrow ZnO + Hg$

   (3) $Fe + Br_2 \rightarrow FeBr_3$

   (4) $H_2O + Cl_2 \rightarrow HCl + HOCl$

   (5) $(NH_4)_2Cr_2O_7 \xrightarrow{\Delta} Cr_2O_3 + H_2O + N_2$

   (6) $KBr + H_2SO_4 \rightarrow Br_2 + K_2SO_4 + SO_2 + H_2O$

   (7) $H^+ + NO_3^- + CuS \rightarrow Cu^{2+} + H_2O + NO + S$

   (8) $H^+ + CH_3OH + MnO_4^- \rightarrow MnO_2 + CO_2 + H_2O$

   (9) $H_2O + CrI_3 + Cl_2 \rightarrow CrO_4^{2-} + IO_3^- + Cl^-$

   (10) $H_2O + Si + OH^- \rightarrow SiO_3^{2-} + H_2$

3. Balance the following redox equations:
   (a) $Fe_2O_3 + CO \rightarrow Fe + CO_2$
   (b) $Ag + HNO_3 \rightarrow AgNO_3 + NO + H_2O$
   (c) $MnO_2 + HBr \rightarrow MnBr_2 + Br_2 + H_2O$
   (d) $CuO + NH_3 \rightarrow N_2 + Cu + H_2O$
   (e) $NiO + NH_3 \rightarrow Ni + N_2 + H_2O$
   (f) $H_2C_2O_4 + KBrO_3 \rightarrow CO_2 + KBr + H_2O$
   (g) $Cl_2 + KOH \rightarrow KClO_3 + KCl + H_2O$
   (h) $H_2O_2 + KMnO_4 + H_2SO_4 \rightarrow O_2 + MnSO_4 + K_2SO_4 + H_2O$
   (i) $HNO_3 + H_2S \rightarrow S + NO + H_2O$
   (j) $Cu_2S + HNO_3 \rightarrow Cu(NO_3)_2 + NO_2 + H_2O + S$
   [*Note:* Three elements are changing oxidation numbers.]

4. Balance the following ionic redox equations: [*Note:* Supply H^+ or OH^- ions and H_2O molecules if needed to balance.]
   (a) $H^+ + Cu + SO_4^{2-} \rightarrow Cu^{2+} + SO_2 + H_2O$
   (b) $Zn + Cr_2O_7^{2-} \rightarrow Zn^{2+} + Cr^{3+} + H_2O$
   (c) $H^+ + Cu + NO_3^- \rightarrow Cu^{2+} + NO$
   (d) $H^+ + Zn + AsO_4^{3-} \rightarrow AsH_3 + Zn^{2+}$
   (e) $H_2O + Al + OH^- \rightarrow AlO_2^- + H_2$
   (f) $H_2O + CrO_4^{2-} + Fe(OH)_2 \rightarrow Cr(OH)_3 + Fe(OH)_3 + OH^-$
   (g) $H^+ + ClO_3^- + Cl^- \rightarrow Cl_2$
   (h) $NO_3^- + I^- \rightarrow NO + I_2$

5. Why are oxidation and reduction said to be complementary processes?

6. When molten $CaCl_2$ is electrolyzed, calcium metal and chlorine gas are produced. Write equations for the two half-reactions that occur at the electrodes. Label the anode half-reaction and the cathode half-reaction.

7. Why is direct current instead of alternating current used in the electroplating of metals?

8. What property of lead dioxide and lead(II) sulfate makes it unnecessary to have salt bridges in the cells of a lead storage battery?

9. In one type of alkaline cell used to power devices such as portable radios, Hg^{2+} ions are reduced to metallic mercury when the cell is being discharged. Does this occur at the anode or at the cathode? Explain.
10. Explain why the density of the electrolyte in a lead storage battery decreases during the discharge cycle.
11. Which of these statements are correct?
    (a) An atom of an element in the uncombined state has an oxidation number of zero.
    (b) The oxidation number of molybdenum in Na_2MoO_4 is $+4$.
    (c) The oxidation number of an ion is the same as the electrical charge on the ion.
    (d) The process in which an atom or an ion loses electrons is called reduction.
    (e) The reaction $Fe^{3+} + e^- \rightarrow Fe^{2+}$ is a reduction reaction.
    (f) In the reaction $2\,Al + 3\,CuCl_2 \rightarrow 2\,AlCl_3 + 3\,Cu$, aluminum is the oxidizing agent.
    (g) In a redox reaction the oxidizing agent is reduced and the reducing agent is oxidized.
    (h) $Cu^0 \rightarrow Cu^{2+}$ is a balanced oxidation equation.
    (i) In the electrolysis of sodium chloride brine (solution), Cl_2 gas is formed at the cathode and hydroxide ions at the anode.
    (j) In any cell, electrolytic or voltaic, reduction takes place at the cathode and oxidation occurs at the anode.
    (k) In the Zn–Cu voltaic cell, the reaction at the anode is $Zn \rightarrow Zn^{2+} + 2e^-$.

    The statements in (l) to (o) pertain to the Activity Series:

    Ba   Mg   Zn   Fe   H   Cu   Ag

    (l) The reaction $Zn + MgCl_2 \rightarrow Mg + ZnCl_2$ is a spontaneous reaction.
    (m) Barium is a more reactive metal than copper.
    (n) Silver metal will react with acids to liberate hydrogen gas.
    (o) Iron is a better reducing agent than zinc.

D. Review problems.
1. What weight and volume of Cl_2 will react with 40.0 g of KOH in the equation given in Question C.3, part (g)?
2. How many moles of SO_2 are formed by reacting 3.0 moles of Cu with sulfuric acid? [See the equation given in Question C.4, part (a).]
3. How many millilitres of 0.100 M $KMnO_4$ solution are needed to oxidize 8.50 g of H_2O_2? What volume of O_2 gas will be formed? [See the equation given in Question C.3, part (h).] Assume STP conditions.
4. What weight of copper is formed when 16.0 litres of ammonia gas are reacted with copper(II) oxide? [See the equation given in Question C.3, part (d).] Assume STP.
5. What weight of bromine (at STP) can be obtained from 50.0 g of HBr according to the equation in Question C.3, part (c)?
6. A sample of crude potassium iodide was analyzed and found to contain 72.0% KI. What maximum weight of iodine can be obtained by oxidizing 500 g of this iodide?

$$KI + H_2SO_4 \rightarrow I_2 + H_2S + K_2SO_4 + H_2O$$

7. How many moles of Fe^{2+} can be oxidized to Fe^{3+} by 0.75 mole of Cl_2 according to the following equation?

$$Fe^{2+} + Cl_2 \rightarrow Fe^{3+} + Cl^-$$

18 Radioactivity and Nuclear Chemistry

After studying Chapter 18 you should be able to:

1. Understand the terms listed in Question A at the end of the chapter.
2. Outline the historical development of nuclear chemistry, giving the major contributions of Henri Becquerel, Marie Curie, Ernest Rutherford, Paul Villard, E. O. Lawrence, Irene Joliot-Curie, Otto Hahn, F. Strassman, and Edwin McMillan.
3. Write balanced nuclear chemical equations using isotopic notation.
4. List the characteristics that distinguish alpha, beta, and gamma rays from the standpoint of mass, charge, relative velocities, and penetrating power.
5. Describe the effect of a magnetic field on alpha particles, beta particles, and gamma rays.
6. Describe a radioactive disintegration series and predict what isotope would be formed by the loss of specified numbers of alpha and beta particles from a given radioactive element.
7. Distinguish between radioactive disintegration and nuclear fission reactions.
8. Tell how the fission of $^{235}_{92}U$ can lead to a chain reaction and why a critical mass is needed for the chain reaction.
9. Explain how the energy from nuclear fission is converted to electrical energy.
10. Tell what the essential difference is between the fission reactions of a nuclear reactor and those of an atomic bomb.
11. Explain what is meant by nuclear fusion and why large-scale efforts to develop controlled nuclear fusion are underway.
12. Determine the amount of a radioactive isotope remaining after a given period of time when the starting amount and half-life are given.
13. Calculate the mass defect and binding energy of a given isotope from appropriate data.
14. Explain how nuclear radiation can cause acute, long-term, and genetic damage to living organisms.
15. Describe the formation of carbon-14 in the atmosphere and tell how this isotope is used to determine the age of carbon-containing material.
16. Explain the use of radioactive tracers.

Up to this point, we have considered only atomic changes that involve the electron structures of atoms and molecules. In this chapter we introduce an entirely different aspect of the atom—reactions of the nucleus. Since the dramatic ending of World War II, nuclear chemistry has become a tremendously important branch of science. The immense amounts of energy available from nuclear reactions can be used for constructive or destructive purposes. The controlled release of nuclear energy has become an important source of power. Nuclear power plants are in operation or are being built in virtually every industrialized nation in the world. The use of radioactive tracers is a routine technique in industry and in biological and medical research and applications.

18.1 Discovery of Radioactivity

One of the most important steps leading to the discovery of radioactivity was made in 1895 by Wilhelm Konrad Roentgen (1845–1923). Roentgen discovered X rays when he observed that a vacuum discharge tube, enclosed in a thin, black cardboard box, caused a nearby piece of paper coated with barium platinocyanide to glow with a brilliant luminescence. From this and other experiments he concluded that certain rays, which he called X rays, were emitted from the discharge tube, penetrated the box and caused the salt to glow. Roentgen also showed that X rays could penetrate other bodies and affect photographic plates. This observation led to the development of X-ray photography.

Shortly after this discovery, Antoine Henri Becquerel (1852–1908) attempted to show a relationship between X rays and the luminescence of uranium salts. In one of his experiments, he wrapped a photographic plate in black paper, placed a sample of uranium salt upon it, and exposed it to sunlight. The developed photographic plate showed that rays emitted from the salt had penetrated the paper. Later, Becquerel prepared to repeat the experiment, but because the sunlight was intermittent, he placed the entire setup in a drawer. Several days later he decided to develop the photographic plate, expecting to find it only slightly affected. He was amazed to observe an intense image on the plate. He repeated the experiment in total darkness and obtained the same results, proving that the uranium salt emitted rays that affected the photographic plate without its being exposed to sunlight. In this way, Becquerel discovered radioactivity. The name radioactivity was given to this phenomenon 2 years later (in 1898) by Marie Curie. Becquerel later showed that the rays coming from uranium were able to ionize air and were also capable of penetrating thin sheets of metal.

In 1898, Marie Sklodowska Curie (1867–1934) and her husband, Pierre Curie (1859–1906), turned their research interests to radioactivity. In a short time, the Curies discovered two new elements, polonium and radium, both of which are radioactive. To confirm their work on radium, they processed 1 ton of pitchblende residue ore to obtain 0.1 g of pure radium chloride, which they used for atomic weight determinations.

In 1899, Ernest Rutherford began to investigate the nature of the rays emitted from uranium. He found two rays, which he called *alpha* and *beta rays*. Soon after, he realized that uranium, while emitting these rays, was changing into another element. By 1912, over 30 radioactive isotopes (radioisotopes) were known, and many more are known today. The *gamma ray*, a third ray emitted from radioactive materials and similar to an X ray, was discovered by Paul Villard in 1900. After the description of the nuclear atom by Rutherford, the phenomenon of radioactivity was attributed to reactions taking place in the nuclei of atoms.

The symbolism and notation described for isotopes in Chapter 5 is very useful in nuclear chemistry and is briefly reviewed here.

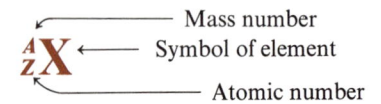

$$_Z^A X \quad \begin{array}{l} \longleftarrow \text{Mass number} \\ \longleftarrow \text{Symbol of element} \\ \longleftarrow \text{Atomic number} \end{array}$$

For example, $^{238}_{92}U$ represents a uranium isotope with an atomic number of 92 and a mass number of 238. This isotope is also designated as U-238 or uranium-238. Table 18.1 shows the isotopic notations for several particles that are often associated with nuclear chemistry.

Table 18.1. Symbols in isotopic notation for several particles (and small isotopes) associated with nuclear chemistry.

Particle	Symbol	Atomic number, Z	Atomic mass, A
Neutron	$_0^1 n$	0	1
Proton	$_1^1 H$	1	1
Beta particle (electron)	$_{-1}^0 e$	−1	0
Positron (positive electron)	$_{+1}^0 e$	1	0
Alpha particle (helium nucleus)	$_2^4 He$	2	4
Deuteron (heavy hydrogen nucleus)	$_1^2 H$	1	2

18.2 Natural Radioactivity

radioactivity

radioactive decay

Radioactivity is the spontaneous emission of radiations from the nucleus of an atom. Elements having this property are said to be radioactive. Radioactive elements continually undergo **radioactive decay**, or disintegration, to form different elements. The chemical properties of an element are associated with its electronic structure, but radioactivity is a property of the nucleus. Therefore, neither ordinary changes of temperature and pressure nor the chemical or physical state of an element have any effect on its radioactivity.

The principal emissions from the nuclei of radioisotopes are known as alpha rays (or particles), beta rays (or particles), and gamma rays. Upon losing an alpha or beta particle, the radioactive element changes into a different element. This process will be explained in detail later.

Each radioactive isotope disintegrates at a specific and constant rate, which is expressed in units of half-life. The **half-life** ($t_{1/2}$) is the time required for one-half of a specific amount of a radioactive element to disintegrate. The half-life of the elements may vary from a fraction of a second to billions of years. For example, $^{238}_{92}U$ has a half-life of 4.5×10^9 years and $^{226}_{88}Ra$ has a half-life of 1620 years. To illustrate, if we start today with 1.0 g of $^{226}_{88}Ra$, we will have 0.50 g of $^{226}_{88}Ra$ remaining at the end of 1620 years; at the end of another 1620 years, 0.25 g will remain, and so on.

$$1.0 \text{ g } ^{226}_{88}Ra \xrightarrow[1620 \text{ years}]{t_{1/2}} 0.50 \text{ g } ^{226}_{88}Ra \xrightarrow[1620 \text{ years}]{t_{1/2}} 0.25 \text{ g } ^{226}_{88}Ra$$

The half-lives of the various radioisotopes of the same element are different from each other. Those for radium are shown below.

Radium isotope	Half-life
Ra-223	11.2 days
Ra-224	3.64 days
Ra-226	1620 years
Ra-228	6.7 years

All isotopes of elements that have atomic numbers greater than 83 are naturally radioactive, although some of these isotopes have extremely long half-lives. Some of the naturally occurring isotopes of elements 81, 82, and 83 are radioactive and some are stable. Only a few naturally occurring isotopes of elements that have atomic numbers less than 81 are radioactive. However, no stable isotopes of element 43 (technetium) or of element 61 (promethium) are known.

Radioactivity is believed to be a result of an unstable ratio of neutrons to protons in the nucleus. Stable isotopes of elements up to about atomic number 20 generally have about a 1:1 neutron to proton ratio. In elements above number 20 the neutron to proton ratio in the stable isotopes gradually increases to about 1.5:1 in element number 83 (bismuth).

18.3 Properties of Alpha, Beta, and Gamma Rays

(a) Alpha particle. An **alpha particle** consists of two protons and two neutrons, has a mass of about 4 amu, a charge of +2, and is considered to be a doubly charged helium atom. It is usually given one of the following symbols: α, He^{2+}, or ^4_2He. When an alpha particle is emitted from the nucleus, a different element is formed. The atomic number of the new element is 2 less and the mass is 4 amu less than the starting element. For example, when $^{238}_{92}U$ loses an alpha particle, $^{234}_{90}Th$ is formed, because two neutrons and two protons are lost from the uranium nucleus. This may be written as an equation:

$$^{238}_{92}U \longrightarrow {}^{234}_{90}Th + \alpha \quad \text{or} \quad {}^{238}_{92}U \longrightarrow {}^{234}_{90}Th + {}^{4}_{2}He$$

For the loss of an alpha particle from $^{226}_{88}Ra$, the equation is

$$^{226}_{88}Ra \longrightarrow {}^{222}_{86}Rn + {}^{4}_{2}He \quad \text{or} \quad {}^{226}_{88}Ra \longrightarrow {}^{222}_{86}Rn + \alpha$$

A nuclear equation, like a chemical equation, consists of reactants and products and must be balanced. To have a balanced nuclear equation, the sum of the mass numbers (superscripts) on both sides of the equation must be equal, and the sum of the atomic numbers (subscripts) on both sides of the equation must be equal.

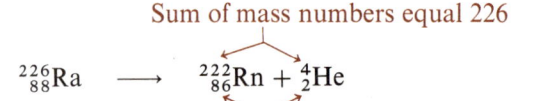

Sum of mass numbers equal 226

$$^{226}_{88}Ra \longrightarrow {}^{222}_{86}Rn + {}^{4}_{2}He$$

Sum of atomic numbers equals 88

What new element will be formed when $^{230}_{90}Th$ loses an alpha particle? This loss is equivalent to two protons and two neutrons. The new element will have a mass of 226 amu (230 − 4) and will contain 88 protons (90 − 2) so that its atomic number is 88. Locate the corresponding element on the periodic chart. It is $^{226}_{88}Ra$ or radium-226.

beta particle

(b) Beta particle. The **beta particle** is identical in mass and charge to an electron; its charge is −1. Both a beta particle and a proton are produced by the decomposition of a neutron. The beta particle leaves and the proton remains in the nucleus. When an atom loses a beta particle from its nucleus, a new element is formed, having essentially the same mass but an atomic number that is 1 greater than the starting element. The beta particle is written as β or $_{-1}^{0}e$. Examples of equations in which a beta particle is lost are given below:

$$^{234}_{90}Th \longrightarrow {}^{234}_{91}Pa + \beta$$
$$^{234}_{91}Pa \longrightarrow {}^{234}_{92}U + {}^{0}_{-1}e$$
$$^{210}_{82}Pb \quad {}^{210}_{83}Bi + \beta$$

See if you can determine what element is formed when $^{237}_{93}Np$ loses a beta particle.

gamma ray

(c) Gamma ray. **Gamma rays** are photons of energy with properties similar to those of X rays. They have no electrical charge and no measurable mass. Gamma rays emanate from the nucleus in many radioactive changes along with either alpha or beta particles. The designation for a gamma ray is γ. Gamma radiation does not result in a change of atomic number or the mass of an element.

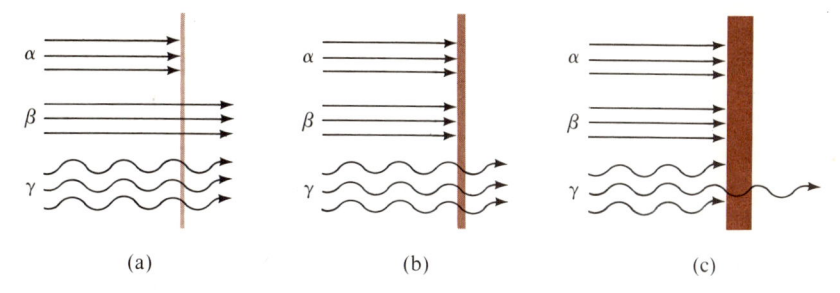

Figure 18.1. Relative penetrating ability of alpha, beta, and gamma radiation. (a) Thin sheet of paper; (b) thin sheet of aluminum; (c) 5 cm lead block.

The ability of radioactive rays to pass through various objects is in proportion to the speed at which they leave the nucleus. Gamma rays travel at the velocity of light (186,000 miles per second) and are capable of penetrating several inches of lead. The velocities of beta particles are variable, the fastest being about nine-tenths the velocity of light. Alpha particles have velocities less than one-tenth the velocity of light. Figure 18.1 illustrates the relative penetrating power of these rays. A few sheets of paper will stop alpha particles; a thin sheet of aluminum will stop both alpha and beta particles; and a 5 centimetre block of lead will reduce, but not completely stop, gamma radiation. In fact, it is difficult to stop all gamma radiation. Table 18.2 summarizes the properties of alpha, beta, and gamma radiation.

Table 18.2. Characteristics of nuclear radiation.

Radiation	Symbol	Mass (amu)	Electrical charge	Velocity	Composition
Alpha	α $_2^4He$	4	+2	Variable—less than 10% the speed of light	Identical to He^{2+}
Beta	β $_{-1}^0e$	$\frac{1}{1837}$	−1	Variable—up to 90% the speed of light	Identical to an electron
Gamma	γ	None	0	Speed of light	Photons or electromagnetic waves of energy

The classical experiment proving that alpha and beta particles are oppositely charged was conducted by Marie Curie. A radioactive source was placed in a hole in a lead block, and the rays given off were allowed to pass between the poles of a strong electromagnet. The paths of the charged particles were deflected as they passed through the field of the electromagnet and were finally detected by striking a photographic plate (see Figure 18.2). The lighter beta

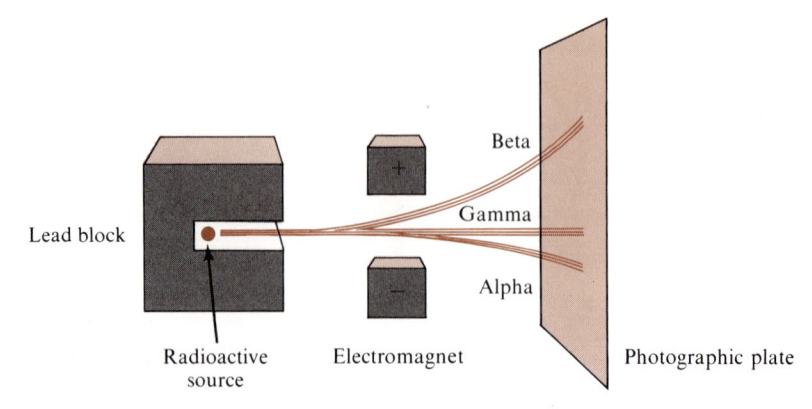

Figure 18.2. The effect of an electromagnetic field on the three radioactive rays. Lighter beta particles are deflected considerably more than alpha particles. Alpha and beta particles are deflected in opposite directions. Gamma radiation is not affected by the electromagnetic field.

Table 18.3. The uranium disintegration series. The elements are listed in the order in which they are formed from each other by radioactive decay. For example, $^{238}_{92}U$ loses alpha particle, forming $^{234}_{90}Th$, which in turn loses a beta particle, forming $^{234}_{91}Pa$, and so on.

$$^{238}_{92}U \xrightarrow{\alpha} {}^{234}_{90}Th \xrightarrow{\beta} {}^{234}_{91}Pa \xrightarrow{\beta} \text{and so on}$$

Element	Symbol	Radiation	Half-life
Uranium	$^{238}_{92}U$	α	4.5×10^9 years
Thorium	$^{234}_{90}Th$	β	24.1 days
Protactinium	$^{234}_{91}Pa$	β	1.18 min
Uranium	$^{234}_{92}U$	α	2.48×10^5 years
Thorium	$^{230}_{90}Th$	α	8.0×10^4 years
Radium	$^{226}_{88}Ra$	α	1.62×10^3 years
Radon	$^{222}_{86}Rn$	α	3.82 days
Polonium	$^{218}_{84}Po$	α	3.05 min
Lead	$^{214}_{82}Pb$	β	26.8 min
Bismuth	$^{214}_{83}Bi$	β	19.7 min
Polonium	$^{214}_{84}Po$	α	1.6×10^{-4} sec
Lead	$^{210}_{82}Pb$	β	22 years
Bismuth	$^{210}_{83}Bi$	β	5.0 days
Polonium	$^{210}_{84}Po$	α	138 days
Lead	$^{206}_{82}Pb$	Stable	Stable

particles were deflected considerably more than the alpha particles, and in the opposite direction; but the gamma rays were not affected by the electromagnet and struck the photographic plate in paths straight out of the lead block.

18.4 Radioactive Disintegration Series

The naturally occurring radioactive elements above lead (Pb) fall into three orderly disintegration series. Each series proceeds from one element to the next by a loss of either an alpha or a beta particle, finally ending in a nonradioactive form of lead. The uranium series starts with $^{238}_{92}U$ and ends with $^{206}_{82}Pb$. The thorium series starts with $^{232}_{90}Th$ and ends with $^{208}_{82}Pb$. The actinium series starts with $^{235}_{92}U$ and ends with $^{207}_{82}Pb$. A fourth series, the neptunium series, starts with the synthetic element $^{241}_{94}Pu$ and ends with the stable bismuth isotope $^{209}_{83}Bi$. The uranium series is shown in Table 18.3; gamma radiation, which accompanies alpha and beta radiation, is not shown in the table.

By using such a series and the half-life of its members, scientists have been able to approximate the age of certain geologic deposits by comparing the amount of $^{238}_{92}U$ with the amount of $^{208}_{82}Pb$ and other isotopes in the series that are present in a particular geologic formation. Rocks found in Canada and Finland have been calculated to be about 3.0×10^9 (3 billion) years old. Some meteorites have been determined to be 4.5×10^9 years old.

18.5 Transmutation of Elements

transmutation

Transmutation is the conversion of one element into another by either natural or artificial means. Transmutation occurs spontaneously in natural radioactive disintegrations. Alchemists tried for centuries to convert lead and mercury into gold by artificial means. But transmutation by artificial means was not achieved until 1919, when Ernest Rutherford succeeded in bombarding the nuclei of nitrogen atoms with alpha particles and produced oxygen isotopes and protons. The nuclear equation for this transmutation can be written as

$$^{14}_{7}N + \alpha \longrightarrow {}^{17}_{8}O + {}^{1}_{1}H \quad \text{or} \quad {}^{14}_{7}N + {}^{4}_{2}He \longrightarrow {}^{17}_{8}O + {}^{1}_{1}H$$

It is believed that the alpha particle enters the nitrogen nucleus, forming the compound nucleus $^{18}_{9}F$, which then decomposes into the products.

Rutherford's experiments opened the door to nuclear transmutations of all kinds. Atoms were bombarded by alpha particles, neutrons, protons, deuterons ($^{2}_{1}H$), electrons, and so on. Massive instruments were developed for accelerating these particles to very high speeds and energies to aid their penetration of the nucleus. Some of these instruments are the famous cyclotron, developed by E. O. Lawrence at the University of California; the Van de Graaf electrostatic generator; the betatron; and the electron and proton synchrotron.

With these instruments, many nuclear transmutations became possible. A few are listed in equation form below.

$$^{7}_{3}Li + ^{1}_{1}H \longrightarrow 2\,^{4}_{2}He$$
$$^{40}_{18}Ar + ^{1}_{1}H \longrightarrow ^{40}_{19}K + ^{1}_{0}n$$
$$^{23}_{11}Na + ^{1}_{1}H \longrightarrow ^{23}_{12}Mg + ^{1}_{0}n$$
$$^{114}_{48}Cd + ^{2}_{1}H \longrightarrow ^{115}_{48}Cd + ^{1}_{1}H$$
$$^{2}_{1}H + ^{2}_{1}H \longrightarrow ^{3}_{1}H + ^{1}_{1}H$$
$$^{209}_{83}Bi + ^{2}_{1}H \longrightarrow ^{210}_{84}Po + ^{1}_{0}n$$
$$^{16}_{8}O + ^{1}_{0}n \longrightarrow ^{13}_{6}C + ^{4}_{2}He$$
$$^{27}_{13}Al + ^{4}_{2}He \longrightarrow ^{30}_{15}P + ^{1}_{0}n$$

18.6 Artificial Radioactivity

Irene Joliot-Curie, a daughter of Pierre and Marie Curie, and her husband, Frederic Joliot, discovered that aluminum reacting with alpha particles produced a radioactive phosphorus isotope, $^{30}_{15}P$:

$$^{27}_{13}Al + ^{4}_{2}He \longrightarrow ^{30}_{15}P + ^{1}_{0}n$$

Phosphorus-30 has a half-life of 2.5 min and decays to silicon-30, $^{30}_{14}Si$, with the emission of a positron (positive electron):

$$^{30}_{15}P \longrightarrow ^{30}_{14}Si + ^{0}_{+1}e$$

artificial radioactivity

induced radioactivity

The radioactivity of isotopes produced in this manner is known as **artificial radioactivity** or **induced radioactivity**. Artificial radioisotopes behave like natural radioactive elements in that they disintegrate in a definite fashion and have a specific half-life for each isotope. The Joliot-Curies received the Nobel Prize in chemistry in 1935 for the discovery of artificial or induced radioactivity.

A list of some commonly used radioisotopes and their mode of decay is given in Table 18.4.

Table 18.4. Some commonly used radioisotopes.

Radioisotope	Radiation emitted	Half-life
$^{14}_{6}C$	β	5668 years
$^{32}_{15}P$	β	14.3 days
$^{35}_{16}S$	β	87 days
$^{60}_{27}Co$	β	5.3 years
$^{90}_{38}Sr$	β	28 years
$^{131}_{53}I$	β	8.1 days
$^{137}_{55}Cs$	β	30 years

18.7 Measurement of Radioactivity

Radiation from radioactive sources is so energetic that it is called *ionizing radiation*. When it strikes an atom or a molecule, one or more electrons are knocked off and an ion is created. One of the common instruments used to detect and measure radioactivity, the Geiger counter, depends on this fact. The instrument consists of a Geiger–Müller detecting tube and a counting device. The detector tube is a pair of oppositely charged electrodes in an argon gas-filled chamber fitted with a thin window. When radiation, such as a beta particle, passes through the window into the tube, some argon is ionized and a momentary pulse of current (discharge) flows between the electrodes. These current pulses are electronically amplified in the counter and appear as signals in the form of audible clicks, flashing lights, meter deflections, or numerical readouts. Figure 18.3 illustrates a simple Geiger counter.

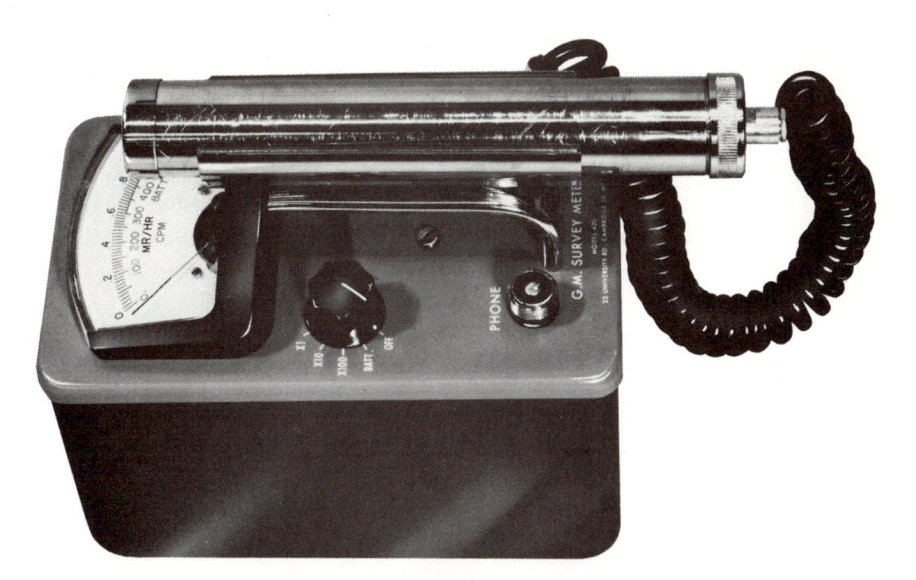

Figure 18.3. A Geiger counter, or battery-operated portable Geiger–Müller survey meter. Radioactivity is indicated in a headphone by clicks and by a graduated meter indicating the number of counts per minute or hour. The Geiger tube located on top is the detector; the rest of the instrument is the counting device. (Courtesy Atomic Accessories, Inc.)

one curie

The curie is the unit used to express the amount of radioactivity produced by an element. **One curie** is defined as the quantity of radioactive material giving 3.7×10^{10} disintegrations per second. The basis for this figure is pure radium, which has an activity of 1 curie per gram. Because the curie is such a large quantity, the millicurie and microcurie, representing one-thousandth and one-millionth of a curie, respectively, are more practical and more commonly used.

18.8 Nuclear Fission

nuclear fission

In **nuclear fission**, a heavy atomic nucleus, when struck in the right way by a neutron, splits into two or more large fragments. The fragments are called *fission products*. As the atom splits, it releases energy and two or three neutrons, each of which can cause another nuclear fission. The first instance of nuclear fission was reported in January 1939 by the German scientists Otto Hahn and F. Strassmann. Detecting isotopes of barium, krypton, cerium, and lanthanum after bombarding uranium with neutrons, the scientists were led to believe that the uranium nucleus had been split.

Characteristics of nuclear fission are

1. Upon absorption of a neutron, a heavy nucleus splits into two or more smaller nuclei (fission products).
2. The mass of the nuclei formed range from about 70 to 160 amu.
3. Two or more neutrons are produced from the fission of each atom.
4. Large quantities of energy are produced as a result of the conversion of a small amount of mass into energy.
5. All nuclei formed are radioactive and give off beta and gamma radiations.

One suggested process by which this fission takes place is illustrated in Figure 18.4. When a heavy nucleus captures a neutron, the energy increase may be sufficient to cause deformation of the nucleus until the mass finally splits into two fragments, releasing radiant energy and usually two or more neutrons.

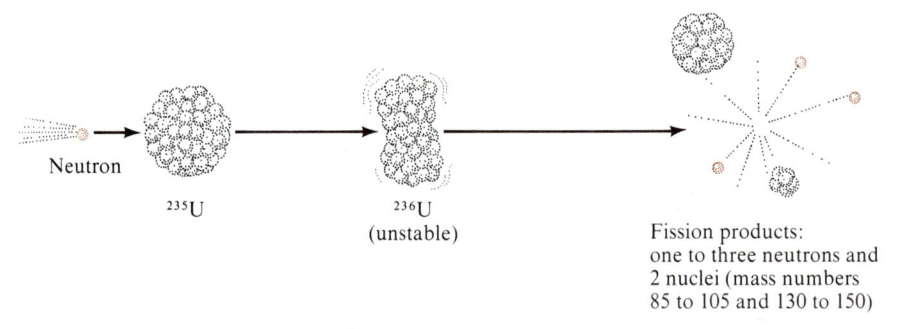

Figure 18.4. The fission process: When a neutron is captured by a heavy nucleus, the nucleus becomes more unstable. The more energetic nucleus begins to deform, resulting in fission. Two nuclear fragments and three neutrons are produced by this fission process.

In a typical fission reaction, a $^{235}_{92}U$ nucleus captures a neutron and forms unstable $^{236}_{92}U$. This $^{236}_{92}U$ nucleus undergoes fission, quickly disintegrating into two fragments such as $^{139}_{56}Ba$ and $^{94}_{36}Kr$ and three neutrons. The three neutrons in turn may be captured by three other $^{235}_{92}U$ atoms, each of which undergoes fission, producing nine neutrons, and so on. A reaction of this kind, where the products cause the reaction to continue or magnify, is known as a **chain reaction**. For a chain reaction to continue, there must be enough fissionable material present so that each atomic fission causes, on the average, at least one additional

chain reaction

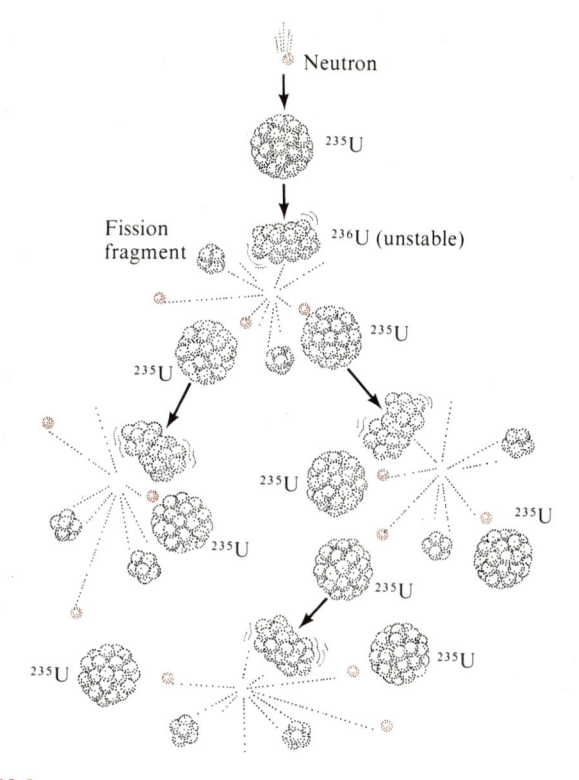

Figure 18.5. Fission and chain reaction of $^{235}_{92}U$. Each fission produces two major fission fragments and three neutrons, which may be captured by other $^{235}_{92}U$ nuclei, continuing the chain reaction.

critical mass

fission. The minimum quantity of an element needed to support a self-sustaining chain reaction is called the **critical mass**. Since energy is released in each atomic fission, chain reactions constitute a possible source of a steady supply of energy. A chain reaction is illustrated in Figure 18.5. Two of the many possible ways U-235 may fission are shown by the equations below.

$$^{235}_{92}U + ^{1}_{0}n \longrightarrow ^{139}_{56}Ba + ^{94}_{36}Kr + 3\,^{1}_{0}n$$

$$^{235}_{92}U + ^{1}_{0}n \longrightarrow ^{144}_{54}Xe + ^{90}_{38}Sr + 2\,^{1}_{0}n$$

18.9 Nuclear Power

Nearly all electricity is produced by machines consisting of a turbine linked by a drive shaft to an electrical generator. The energy required to run the turbine may be supplied from falling water, used in hydroelectric power plants, or by steam generated by heat from fuel, used in thermal power plants. Thermal power plants burn fossil fuel—coal, oil, or natural gas.

The world's demand for energy, largely from fossil fuels, has continued to grow at an ever-increasing rate for about 250 years. Even at present rates of consumption, the estimated world supply of fossil fuels is sufficient for only a few centuries. Although the United States has large coal deposits, it is currently importing about 40% of its oil supply. We clearly need to develop alternate energy sources. At present, uranium is by far the most productive alternate energy source; and about 12% of the electrical energy used in the United States is generated in power plants using uranium fuel.

The major disadvantage of nuclear-fueled power plants is that they produce highly radioactive waste in the form of isotopes some of which have half-lives of thousands of years. As yet, no general agreement has been reached on how to safely dispose of these dangerous wastes.

A nuclear power plant is a thermal power plant in which heat is produced by a nuclear reactor instead of by combustion of fossil fuel. The major components of a nuclear reactor are (1) an arrangement of nuclear fuel, called the reactor core; (2) a control system, which serves to regulate the rate of fission and thereby the rate of heat generation; and (3) a cooling system, which serves to remove the heat from the reactor and also keep the core at the proper temperature. One type of reactor uses metal slugs containing uranium enriched from the normal 0.7% U-235 to about 3% U-235. These slugs are placed into a graphite block to form a nuclear pile. The self-sustaining fission reaction is moderated by the graphite and by adjustable control rods containing substances that slow down and capture some of the neutrons produced. Ordinary water, heavy water, and sodium are typical coolants used. Energy obtained from nuclear reactions in the form of heat is used in the production of steam to drive turbines for generating electricity.

Reactors used for commercial power production in the United States use uranium that is enriched with the relatively scarce fissionable U-235 isotope. Because the supply of U-235 is limited, the breeder reactor is being developed. Breeder reactors convert nonfissionable uranium-238 into fissionable plutonium-239 (see Section 18.10). Several breeder reactors are already in operation in Europe and Great Britain.

18.10 The Atomic Bomb

The atomic bomb is a fission bomb; it operates on the principle of a very fast chain reaction that releases a tremendous amount of energy. An atomic bomb and a nuclear reactor both depend on self-sustaining nuclear fission chain reactions. The essential difference is that in a bomb, the fission is "wild," or uncontrolled, whereas in a nuclear reactor the fission is moderated and carefully controlled. A minimum critical mass of fissionable material is needed for a bomb or a major explosion will not occur. When a quantity smaller than the critical mass is used, too many neutrons formed in the fission step escape without combining with another nucleus, and a chain reaction does not occur. Therefore, the fissionable material of an atomic bomb must be stored as two or more subcritical masses and brought together to form the critical mass at the desired time of explosion. The temperature developed in an atomic bomb is believed to be about 10 million °C. Figure 18.6 shows an atomic bomb of the "Little Boy" type, the World War II nuclear weapon detonated over Hiroshima, Japan.

Figure 18.6. Atomic bomb of the "Little Boy" type detonated over Hiroshima, Japan, in World War II. The bomb is 28 in. in diameter and 120 in. long. The first nuclear weapon ever detonated, it weighed about 9000 lb and had a yield equivalent to approximately 20,000 tons of high explosives. (Courtesy U.S. Department of Energy, Los Alamos Scientific Laboratory.)

The isotopes used in atomic bombs are U-235 and Pu-239. Uranium deposits contain about 0.7% of the U-235 isotope, the remainder being U-238. Uranium-238 does not undergo fission, except with very high-energy neutrons. It was discovered, however, that U-238 captures a low-energy neutron without undergoing fission and that the product, U-239, changes to $^{239}_{94}$Pu (plutonium) by a beta decay process. Plutonium-239 readily undergoes fission upon capture of a neutron and is therefore useful for nuclear weapons. The equations for the nuclear transformations are

$$^{238}_{92}\text{U} + ^{1}_{0}n \longrightarrow \,^{239}_{92}\text{U} \xrightarrow{-\beta} \,^{239}_{93}\text{Np} \xrightarrow{-\beta} \,^{239}_{94}\text{Pu}$$

The hazards of an atomic bomb explosion include not only shock waves from the explosive pressure and tremendous heat, but also intense radiation in the form of alpha, beta, gamma, and ultraviolet rays. Gamma rays and X rays can penetrate deeply into the body, causing burns, sterilization, and mutation of the genes, which may have an adverse effect on future generations. Both radioactive fission products and unfissioned material are present after the explosion. If the bomb explodes near the ground, many tons of dust are lifted into the air. Radioactive material adhering to this dust, known as *fallout*, is spread by air currents over wide areas of the land and constitutes a lingering source of radiation hazard.

374   Chapter 18

The possibility of a nuclear war is certainly one of the most awesome threats facing civilization today. Only two rather primitive atomic bombs were used to destroy the Japanese cities of Hiroshima and Nagasaki and to bring World War II to an early end. Many thousands of nuclear weapons, some at least several hundred times more powerful than the Hiroshima and Nagasaki bombs, are now in existence. The threat of nuclear war is further complicated by the fact that the number of nations possessing nuclear weapons is steadily increasing.

18.11   Nuclear Fusion

nuclear fusion

The process of uniting the nuclei of two light elements to form one heavier nucleus is known as **nuclear fusion**. Such reactions are used for producing energy because the mass of the two individual nuclei that are fused into a single nucleus is greater than the mass of the nucleus formed by their fusion. The mass differential is liberated in the form of energy. Fusion reactions are responsible for the tremendous energy output of the sun. Thus, aside from relatively small amounts from nuclear fission and radioactivity, fusion reactions are the ultimate source of our energy, even that obtained from fossil fuels. They are also responsible for the devastating power of the thermonuclear, or hydrogen, bombs.

Fusion reactions require temperatures on the order of tens of millions of degrees for initiation. Such temperatures are present in the sun but have been produced only momentarily on earth. For example, the hydrogen, or fusion, bomb is triggered by the temperature of an exploding fission bomb.

Two typical fusion reactions are

$$^{3}_{1}H + ^{2}_{1}D \longrightarrow ^{4}_{2}He + ^{1}_{0}n + \text{Energy}$$

Tritium   Deuterium

$$^{3}_{1}H + ^{1}_{1}H \longrightarrow ^{4}_{2}He + \text{Energy}$$

3.01495   1.00782   4.00260
 amu   amu   amu

The total mass of the reactants in the second equation is 4.02277 amu, which is 0.02017 amu greater than the mass of the product. This difference in mass is manifested in the great amount of energy liberated.

During the past 30 years, a large amount of research on controlled nuclear fusion reactions has been done in the United States and in other countries, especially the Soviet Union. So far, the goal of controlled nuclear fusion has not been attained, although the required ignition temperature has been reached in several devices. Evidence to date leads us to believe that it is possible to develop a practical fusion power reactor. Fusion power, if it can be developed, will be far superior to fission power because: (1) Virtually infinite amounts of energy are to be had from fusion. Uranium supplies for fission power are limited, but heavy hydrogen, or deuterium (the most likely fusion fuel), is relatively abundant. It is estimated that the deuterium in a cubic mile of seawater is an energy resource (as fusion fuel) greater than the petroleum reserves of the entire world. (2) From an environmental viewpoint, fusion power is expected to be much "cleaner" than fission power. This is because fusion reactions, in contrast to uranium and plutonium fission reactions, do not produce large amounts of long-lived and dangerously radioactive isotopes.

18.12 Mass–Energy Relationship in Nuclear Reactions

Since large amounts of energy are released in nuclear reactions, significant amounts of mass are converted to energy. We stated earlier that the amount of mass converted to energy in chemical changes is considered insignificant. In fission reactions, about 0.1% of the mass is converted into energy. In fusion reactions, as much as 0.5% of the mass may be changed into energy. The Einstein equation, $E = mc^2$, may be used to calculate the energy liberated, or available, when the mass loss is known. For example, in the reaction

$$^{7}_{3}Li \; + \; ^{1}_{1}H \longrightarrow \; ^{4}_{2}He \; + \; ^{4}_{2}He \; + \text{Energy}$$
$$7.0160 \text{ g} \quad 1.0078 \text{ g} \quad 4.0026 \text{ g} \quad 4.0026 \text{ g}$$

the mass difference between the reactants and the products $(8.0238 - 8.0052)$ is 0.0186 g. The energy equivalent to this amount of mass is 4.0×10^{11} cal. By comparison, this is more than 4 million times greater than the 9.4×10^4 cal of energy obtained from the complete combustion of 12.0 g (1 mole) of carbon.

The mass of a nucleus is actually less than the sum of the masses of the protons and neutrons that make up the nucleus. The energy equivalent to this difference in mass (mass defect) is known as the *binding energy* of the nucleus. This is the energy that would be required to break a nucleus up into its individual protons and neutrons. The higher the binding energy, the more stable is the nucleus. Elements of intermediate atomic masses have high binding energies. For example, iron (element number 26) has a very high binding energy and therefore has a very stable nucleus. Just as electrons attain less energetic and more stable arrangements through ordinary chemical reactions, neutrons and protons attain less energetic and more stable arrangements through nuclear fission or fusion reactions. Thus, when uranium undergoes fission, the products have less mass (and greater binding energy) than the original uranium. In like manner, when hydrogen and lithium fuse to form helium, the helium has less mass (and greater binding energy) than the hydrogen and lithium. It is this conversion of mass to energy that accounts for the very large amounts of energy associated with both nuclear fission and fusion reactions.

18.13 Transuranium Elements

transuranium elements

The elements following uranium on the periodic chart and having atomic numbers greater than 92 are known as the **transuranium elements**. All of them are synthetic radioactive elements; none of them occur naturally.

The first transuranium element, number 93, was discovered in 1939 by Edwin M. McMillan (1907–   ) at the University of California while he was investigating the fission of uranium. He named it neptunium for the planet Neptune. In 1941, element 94, plutonium, was identified as a beta decay product of neptunium:

$$^{238}_{93}\text{Np} \longrightarrow ^{238}_{94}\text{Pu} + ^{0}_{-1}e$$

$$^{239}_{93}\text{Np} \longrightarrow ^{239}_{94}\text{Pu} + ^{0}_{-1}e$$

Plutonium is one of the most important fissionable elements known today.

Element 104 was first reported in 1964 by a Russian research group, but their work has not yet been independently confirmed. They have suggested the name kurchatovium (Ku) after the late Russian physicist Igor Kurchatov (1903–1960). In 1969, a research group at the Lawrence Radiation Laboratory of the University of California announced the synthesis and positive identification of two isotopes of element 104. They suggested the name rutherfordium (Rf) for this element. Nuclear reactions for the syntheses of element 104 are given below:

$$^{242}_{94}\text{Pu} + ^{22}_{10}\text{Ne} \longrightarrow ^{260}_{104}\text{Ku} + 4\,^{1}_{0}n$$

$$^{249}_{98}\text{Cf} + ^{12}_{6}\text{C} \longrightarrow ^{257}_{104}\text{Rf} + 4\,^{1}_{0}n$$

$$^{249}_{98}\text{Cf} + ^{13}_{6}\text{C} \longrightarrow ^{259}_{104}\text{Rf} + 3\,^{1}_{0}n$$

In April 1970, the Lawrence Radiation group, headed by Albert Ghiorso, reported the discovery of element 105. They suggested that it be named Hahnium (Ha) after the German chemist Otto Hahn, who won the Nobel Prize for discovering nuclear fission. Element 105 was synthesized by bombarding Cf-249

Table 18.5. Transuranium elements.

Element	Symbol	Atomic number	Discovery date
Neptunium	Np	93	1939
Plutonium	Pu	94	1941
Americium	Am	95	1944
Curium	Cm	96	1944
Berkelium	Bk	97	1949
Californium	Cf	98	1950
Einsteinium	Es	99	1953
Fermium	Fm	100	1953
Mendelevium	Md	101	1955
Nobelium	No	102	1957
Lawrencium	Lr	103	1961
Kurchatovium[a]	Ku[a]	104	1964
Rutherfordium[a]	Rf[a]	104	1969
Hahnium[a]	Ha[a]	105	1970
—	—	106	1974

[a] Unofficial name and symbol.

with a stream of N-15 nuclei; it has a half-life of 1.6 seconds.

$$^{249}_{98}Cf + ^{15}_{7}N \longrightarrow ^{260}_{105}Ha + 4\,^{1}_{0}n$$

Both the Russian and American nuclear scientists announced the synthesis of element 106 in 1974.

Table 18.5 lists all the presently known transuranium elements.

18.14 Biological Effects of Radiation

Radiation that has enough energy to dislocate bonding electrons and create ions when passing through matter is classified as *ionizing radiation*. Alpha, beta, and gamma rays, along with X rays, fall into this classification. Ionizing radiation can damage or kill living cells. This damage is particularly devastating when it occurs in the cell nuclei and affects molecules involved in cell reproduction. The overall effects of radiation on living organisms fall into these general categories: (1) acute, or short-term, effects; (2) long-term effects, and (3) genetic effects.

Acute radiation damage. High levels of radiation, especially of gamma or X rays, produce nausea, vomiting, and diarrhea. The effect has been likened to a sunburn throughout the body. If the dosage is high enough, death will occur in a few days. The damaging effects of radiation appear to be centered in the nuclei of the cells, and cells that are undergoing rapid cell division are most susceptible to damage. It is for this reason that cancers are often treated with gamma radiation from a cobalt-60 source. Cancerous cells are multiplying rapidly and are destroyed by a level of radiation that does not seriously damage normal cells.

Long-term radiation damage. Protracted exposure to low levels of any form of ionizing radiation can weaken the organism and lead to the onset of malignant tumors, even after fairly long time delays. There is evidence that a number of early workers in radioactivity and X-ray technology may have had their lives shortened by long-term radiation damage.

A number of women who, in the early 1920s, had been employed to paint luminous numbers on watch dials, died some years later from the effects of radiation. These women had ingested radium by using their lips to point the brushes used on the job. Radium was retained in their bodies and, as an alpha emitter with a half-life of about 1620 years, continued to inflict radiation damage.

Strontium-90 isotopes are present in the fallout from atmospheric testing of nuclear weapons. Strontium is in the same periodic table group as calcium, and its chemical behavior is similar to that of calcium. Hence, when foods contaminated with Sr-90 are eaten, Sr-90 ions are laid down in the bone tissue along with ordinary calcium ions. Strontium-90 is a beta emitter with a half-life of 28 years. Blood cells that are manufactured in bone marrow are affected by by the radiation from Sr-90. Hence, there is concern that Sr-90 accumulation in the environment may cause an increase in the incidence of leukemia and bone cancers. Fortunately, the United States and the Soviet Union agreed to stop atmospheric testing of nuclear weapons several years ago; however, some countries are still doing testing in the atmosphere.

Genetic effects. All the information needed to create an individual of a particular species—be it a bacterial cell or a human being—is contained within the nucleus of a cell. This genetic information is encoded in the structure of DNA (deoxyribonucleic acid) molecules, which make up genes. The DNA molecules form precise duplicates when cells divide, thus passing genetic information from one generation to the next (see Section 31.7). Radiation can damage DNA molecules. If the damage is not severe enough to prevent reproduction, a mutation (a sudden heritable variation in the offspring) may result. Most mutation-induced traits are undesirable. Unfortunately, if the bearer of the altered genes survives to reproduce, these traits are passed along to succeeding generations. In other words, the genetic effects of increased radiation exposure are found in future generations, not in the present generation.

Because radioactive rays are hazardous to health and living tissue, special precautions must be taken in designing laboratories and nuclear reactors, in disposing of waste materials, and in monitoring the radiation exposure of people working in this field. For example, personnel working in areas of hazardous radiation wear film badges or pocket dosimeters to provide them with an accurate indication of cumulative radiation exposure.

18.15 Applications of Radioisotopes

To date, the largest uses of radioactive materials have been for making weapons and for the generation of electricity in nuclear power plants. Aside from these major uses, radioisotopes have innumerable applications. They are used extensively in chemical, physical, biological, and medical research. Radioisotopes now serve in a wide variety of almost routine technological applications in medicine and various branches of industry including chemical, petroleum, and metallurgical processing. A few of these applications are briefly described here.

For many years, radium has been used in the treatment of cancer; cobalt-60 and cesium-137 are now extensively used for therapy in this field. The effectiveness of radiation therapy for cancer is dependent on the fact that rapidly growing or dividing malignant cells are more susceptible to radiation damage than are normal cells. Therefore, radiation therapy is effective when it can be applied at levels that will destroy cancer cells but not normal cells.

Compounds containing a radioactive isotope are described as being *labeled* or *tagged*. These compounds undergo their normal chemical reactions. In addition, their location can be detected because of their radioactivity. When such compounds are given to a plant or to an animal, the movement of the radioisotope can be traced through the organism by the use of a Geiger counter or other detecting device. The following are examples of biological research in which such tracer techniques have been employed: the rate of phosphate intake by plants, using radiophosphorus; the utilization of carbon dioxide in photosynthesis, using radioactive carbon; the accumulation of iodine in the thyroid gland, using radioactive iodine; and the absorption of iron by the hemoglobin of the blood, using radioactive iron. In chemistry, the applications are unlimited: the study of reaction mechanisms, the measurement of the rates of chemical reactions, and the determination of physical constants are just a few of the areas of application.

Agricultural research scientists use gamma radiation from cobalt-60 or other sources to develop disease-resistant and highly productive grains and other crops. The seeds are exposed to gamma radiation to induce mutations.

(a)

(b)

Figure 18.7. Small doses of radiation have successfully inhibited the sprouting of onions and potatoes. Sprouting in onions can be inhibited by a dose of 2000–4000 rads of radiation; potatoes require about 7500 rads. There is no detectable change in the product at these dosage levels. The onions (a) and potatoes (b) illustrated have been stored over 5 months. In each case, sprouting was greatest in the control, which was not irradiated. The control basket of potatoes is the upper left one; those in each basket (left to right) have been treated with a larger dose of radiation. (Courtesy Agriculture Canada.)

The most healthy and vigorous plants grown from the irradiated seed are then selected and propagated to obtain new and improved varieties for commercial use.

Radioactivity has been used to control and, in some areas, to eliminate the screw-worm fly. The larvae of this obnoxious insect pest burrow into wounds in livestock. The female fly, like a queen bee, mates only once. When large numbers of gamma ray-sterilized male flies are released at the proper time in an area infested with screw-worm flies, the majority of the females mate with sterile males. As a consequence, the flies fail to reproduce sufficiently to maintain their numbers. This technique has also been used recently to eradicate the Mediterranean fruit fly, which had appeared in an area near the Los Angeles International Airport.

An interesting outgrowth of the use of radioisotope techniques is *radiocarbon dating*. The method is based on the decay rate of carbon-14 and was devised by the American chemist W. F. Libby, who received the Nobel Prize in chemistry in 1960 for this work. The principle of radiocarbon dating is as follows: Carbon dioxide in the atmosphere contains a fixed ratio of radioactive carbon-14 to ordinary carbon-12. This is because carbon-14 is produced at a steady rate in the atmosphere by bombardment of nitrogen-14 by neutrons from cosmic ray sources.

$$^{14}_{7}N + ^{1}_{0}n \longrightarrow ^{14}_{6}C + ^{1}_{1}H$$

Plants that consume carbon dioxide during photosynthesis and animals that eat the plants contain the same proportion of radiocarbon as long as they are alive. When an organism dies, the amount of carbon-12 remains fixed, but the carbon-14 content diminishes according to its half-life (5668 years). Thus, the ratio of C-14 to C-12 gives data relative to the age of the object being dated. In 5668 years, one-half the radiocarbon initially present will have undergone decomposition. In 11,336 years, one-fourth of the original C-14 will be left. The age of fossil material, archaeological specimens, and old wood can be determined by this method. The age of specimens from ancient Egyptian tombs calculated by radiocarbon dating correlates closely with the chronological age established by Egyptologists. Charcoal samples obtained at Darrington Walls, a wood-henge in Great Britain, were determined to be about 4000 years old. Radiocarbon dating instruments currently in use enable researchers to date specimens back as far as 50,000 years.

Questions

A. Review the meanings of the new terms introduced in this chapter.
1. Radioactivity
2. Radioactive decay
3. Half-life
4. Alpha particle
5. Beta particle
6. Gamma ray
7. Transmutation
8. Artificial radioactivity
9. Induced radioactivity
10. One curie
11. Nuclear fission
12. Chain reaction
13. Critical mass
14. Nuclear fusion
15. Transuranium element

Radioactivity and Nuclear Chemistry   381

B. *Answers to the following will be found in tables and figures.*
   1. To afford protection from radiation injury, which kind of radiation requires:
      (a) The most shielding?   (b) The least shielding?
   2. Why is an alpha particle deflected less than a beta particle in passing through a magnetic field?
   3. List the first five isotopes in Table 18.4 in order of increasing half-life.
   4. What is the half-life of strontium-90?
   5. Name three pairs of isotopes that might be obtained by fissioning three U-235 atoms.
   6. List the transuranium elements that are named for:
      (a) People   (b) Geographical place names   (c) Planets

C. *Review questions.*
   1. Identify the following people and their associations with the early history of radioactivity:
      (a) Henri Becquerel   (c) Wilhelm Roentgen   (e) Paul Villard
      (b) Marie and Pierre Curie   (d) Ernest Rutherford
   2. Why is the radioactivity of an element unaffected by the usual factors that affect the rates of chemical reactions, such as ordinary changes of temperature and concentration?
   3. Which of the following elements do not have stable isotopes?
      (a) Californium   (d) Palladium   (g) Thorium
      (b) Francium   (e) Platinum   (h) Vanadium
      (c) Lead   (f) Technetium   (i) Xenon
   4. Explain the term *half-life*.
   5. Tell how alpha, beta, and gamma radiation are distinguished from the standpoint of:
      (a) Charge   (c) Nature of particle or ray
      (b) Relative mass   (d) Relative penetrating power
   6. How are the mass and the atomic number of a nucleus affected by the loss of the following?
      (a) An alpha particle   (b) A beta particle
   7. Distinguish between natural and artificial radioactivity.
   8. Write nuclear equations for the alpha decay of:
      (a) $^{218}_{85}At$ (b) $^{221}_{87}Fr$ (c) $^{192}_{78}Pt$ (d) $^{210}_{84}Po$
   9. Write nuclear equations for the beta decay of:
      (a) $^{14}_{6}C$ (b) $^{137}_{55}Cs$ (c) $^{239}_{93}Np$ (d) $^{90}_{38}Sr$
   10. Stable $^{208}_{82}Pb$ is formed from $^{232}_{90}Th$ in the thorium disintegration series by successive $\alpha, \beta, \beta, \alpha, \alpha, \alpha, \alpha, \beta, \beta, \alpha$ particle emissions. Write the symbol (including mass and atomic number) for each isotope formed in this series.
   11. The isotope $^{237}_{93}Np$ loses a total of seven alpha particles and four beta particles. What isotope remains after these losses?
   12. Was there more or less lead on earth 1 billion years ago than there is today? Explain.
   13. Write nuclear equations for the following:
      (a) Conversion of $^{13}_{6}C$ to $^{14}_{6}C$ (c) Conversion of $^{30}_{15}P$ to $^{30}_{14}Si$
      (b) Conversion of $^{3}_{1}H$ to $^{4}_{2}He$
   14. Complete and balance the following nuclear equations:
      (a) $^{27}_{13}Al + ^{4}_{2}He \longrightarrow ^{30}_{15}P +$
      (b) $^{27}_{14}Si \longrightarrow _{+1}^{0}e +$

(c) $ + {}^{2}_{1}H \longrightarrow {}^{13}_{7}N + {}^{1}_{0}n$

(d) $ \longrightarrow {}^{82}_{36}Kr + {}^{0}_{-1}e$

(e) ${}^{66}_{29}Cu \longrightarrow {}^{66}_{30}Zn +$

(f) ${}^{0}_{-1}e + \longrightarrow {}^{7}_{3}Li$

15. What is the essential difference between the nuclear reactions in a nuclear reactor and those in an atomic bomb?
16. Why must a certain minimum amount of fissionable material be present before a self-supporting chain reaction can occur?
17. Explain why radioactive rays are classified as ionizing radiation.
18. Give a brief description of the biological hazards associated with radioactivity.
19. Strontium-90 has been found to occur in radioactive fallout. Why is there so much concern about this radioisotope being found in cow's milk (see Table 18.5)?
20. What is a radioactive tracer? How is it used?
21. Describe the radiocarbon method for dating archaeological artifacts.
22. Which of the following statements are correct?
    (a) Radioactivity was discovered by Marie Curie.
    (b) There are 59 neutrons in an atom of ${}^{59}_{28}Ni$.
    (c) The loss of a beta particle by an atom of ${}^{75}_{33}As$ forms an isotope of increased atomic number.
    (d) The emission of an alpha particle from the nucleus of an atom lowers its atomic number by 4 and lowers its atomic mass by 2.
    (e) Emission of gamma radiation from the nucleus of an atom leaves both the atomic number and the atomic mass unchanged.
    (f) There are relatively few naturally occurring radioactive isotopes with atomic numbers below 81.
    (g) The longer the half-life of a radioisotope, the more slowly it decays.
    (h) The beta ray has the greatest penetrating power of all the rays emitted from the nucleus of an atom.
    (i) The symbol ${}^{0}_{+1}e$ is used to indicate a positron, which is a positively charged particle with the mass of an electron.
    (j) The disintegration of ${}^{226}_{88}Ra$ into ${}^{214}_{83}Po$ involves the loss of three alpha particles and two beta particles.
    (k) If 1.0 g of a radioisotope has a half-life of 7.2 days, the half-life of 0.50 g of that isotope is 3.6 days.
    (l) If the mass of a radioisotope is reduced by radioactive decay from 12 g to 0.75 g in 22 hours, the half-life of the isotope is 5.5 hours.
    (m) A very high temperature is required to initiate nuclear fusion reactions.
    (n) Radiocarbon dating of archaeological artifacts is based on an increase in the ratio of carbon-12 to carbon-14.

D. Review problems.
   1. One gram of Ra-226 was sealed in a vault in 1980. In what year will a chemist of the future find one-eighth of the radium remaining?
   2. If radium costs $50,000 a gram, how much will 5.00 g of ${}^{226}RaCl_2$ cost if the price is based on the radium content?
   3. The half-life of Sn-119 is 3 hr. If you prepared 100 g of this isotope, how much would be left after a period of 24 hr?
   4. An archaeological specimen was analyzed and found to contain 6.25% of its original ${}^{14}_{6}C$. How old is it?

5. What is the half-life of an isotope if 16 g of the isotope decays to 0.50 g in 2 hr?
6. Calculate (a) the mass defect and (b) the binding energy of $_2^4He$. Mass data: $_2^4He = 4.0026$ g; $n = 1.0087$ g; $p = 1.0073$ g; $e^- = 0.00055$ g; 1.0 g $= 2.2 \times 10^{13}$ cal (from $E = mc^2$).

E. Review exercises.
1. The half-life of plutonium-244 is 76 million years. If the age of the earth is about 5 billion years, discuss the feasibility of this isotope's being found as a naturally occurring element.
2. Suggest why nuclear reactions using alpha particles are easier to carry out with lighter elements than with heavier ones.
3. Agricultural researchers have produced improved crop species by cross-breeding. Explain how the same results might be accomplished by exposing seeds to radioactive rays.
4. How might radioactivity be used to locate a leak in an underground pipe?

19 Chemistry of Some Selected Metals

After studying Chapter 19 you should be able to:

1. Understand the terms listed in Question A at the end of the chapter.
2. List the alkali metals (Group IA) and compare their atomic radii, densities, melting points, and first ionization energies.
3. List the alkaline earth metals (Group IIA) and compare their atomic radii, densities, melting points, and first ionization energies.
4. Write balanced equations for the cathode reaction, anode reaction, and total reaction for the electrolysis of (a) molten sodium chloride, (b) molten potassium hydroxide, and (c) aqueous sodium chloride.
5. Write balanced chemical equations for the reaction of any alkali metal with (a) water, (b) any halogen, and (c) hydrogen.
6. Write formulas for (a) baking soda, (b) borax, (c) caustic soda, (d) chile saltpeter, (e) halite, (f) hypo, (g) lye, (h) potash, (i) saltpeter, (j) soda ash, and (k) trona.
7. Describe the preparation of magnesium metal from seawater and write equations for the reactions involved.
8. Explain, using chemical equations, why it is difficult to extinguish a magnesium fire with water or carbon dioxide.
9. Write formulas for (a) barite, (b) calcite, (c) calcium carbide, (d) Epsom salts, (e) fluorspar, (f) gypsum, (g) limestone, (h) magnesia, (i) magnesite, (j) plaster of paris, (k) quicklime, and (l) slaked lime.
10. List the alkali metals and alkaline earth metals that are biologically essential to the human body.
11. Explain why colors appear when certain elements are heated strongly in a nonluminous flame and list the specific colors obtained from Li, Na, K, Rb, Ca, Sr, and Ba.
12. List essential raw materials needed for pig iron production and write equations for the pertinent blast furnace reactions.
13. State the essential chemical difference between pig iron and steel.
14. Describe the principal steel-making processes and tell how the basic open-hearth process or the basic oxygen process (BOP) affects the amounts of carbon, sulfur, and phosphorus in the steel that is produced.
15. Explain cathodic protection, and list two metals commonly used to protect iron and write the appropriate equations for the anode reaction of each.
16. Describe the production of aluminum metal from bauxite ore by the Hall–Heroult process and write equations for the reactions involved.

17. Explain why aluminum is more expensive than iron or steel even though it is more abundant than iron in the earth's crust.
18. Explain why aluminum and magnesium can be used as structural materials even though they are very reactive elements.
19. Write balanced equations pertaining to aluminum chemistry—for example, (a) aluminum chloride with potassium metal, (b) aluminum metal and iron oxide (thermite welding), (c) aluminum hydroxide and hydrochloric acid, and (d) aluminum hydroxide and sodium hydroxide.
20. Apply the principles learned in studying stoichiometry, gases, and solutions to problems involving the chemistry of metals.

19.1 The Alkali Metals

Physical Properties

alkali metal

The elements of Group IA in the periodic table—lithium, sodium, potassium, rubidium, cesium, and francium—are known as the **alkali metals** or the *sodium family*. They are named *alkali metals* because all of them form strong alkalies. An alkali is a substance that has the properties of a base. All Group IA metals are silvery-white in color when freshly cut, are soft and ductile, and are good conductors of electricity. Lithium is the hardest, being only a little softer than lead. Francium is radioactive; it is so rare that comparatively little is known about its properties.

None of the alkali metals are found free in nature. They are among the most reactive of the elements. Their chemical activity increases with size and atomic number. Because these metals react readily with oxygen and moisture, they are stored in a vacuum or under an inert liquid such as kerosene, benzene, or toluene. The similarity in chemical properties of this family is due to a similar electronic structure. An atom of an alkali metal has only one valence electron. Therefore, these elements form ions having only a +1 oxidation number. Variations in their properties and reactivities are due to differences in mass, size, and ease with which they lose their outer-shell electron. Some of the physical properties of the alkali metals are shown in Table 19.1.

Metallic sodium has excellent thermal and electrical conductivity. These properties are put to use in special industrial applications. Molten sodium, for

Table 19.1. Physical properties of the alkali metals.

Element	Atomic number	Atomic weight	Electron configuration	Metallic atomic radius (Å)	Radius of M$^+$ ion (Å)	Density (0°C, g/ml)	Melting point (°C)	Boiling point (°C)
Li	3	6.941	2, 1	1.58	0.60	0.53	186.0	1,336
Na	11	22.990	2, 8, 1	1.86	0.95	0.97	97.7	880
K	19	39.098	2, 8, 8, 1	2.38	1.33	0.86 (at 20°C)	63.6	760
Rb	37	85.468	2, 8, 18, 8, 1	2.53	1.48	1.53	39.0	700
Cs	55	132.905	2, 8, 18, 18, 8, 1	2.72	1.69	1.90	28.4	670

example, is used as a heat transfer agent in nuclear power stations. Heat from the nuclear reactor core is absorbed by the sodium and transferred to steam-generating equipment for production of electrical energy.

Sodium is not quite as good a conductor of electricity as silver, copper, aluminum, or gold. However, for some applications in the electrical power industry, sodium is more economical to use than the usual electrical conductors. Sodium metal encased in steel has replaced large copper bus bars in some power stations. Polyethylene tubes filled with sodium metal are in experimental use as replacements for conventional copper and aluminum cables.

Alkali Metal Mineral Deposits

Salts or minerals of the alkalies, especially sodium, are widespread throughout the earth. The chief minerals of lithium are spodumene [$LiAl(SiO_3)_2$], amblygonite [$LiAl(F,OH)PO_4$], and lepidolite [$KLi_2AlSi_4 \cdot O_{10}(F,OH)_2$]. They are found principally in South Africa; Manitoba and Quebec, Canada; and South Dakota and North Carolina.

The most important compound of sodium is sodium chloride. It occurs in seawater to the extent of about 3% and as rock salt deposits all over the globe. Most sodium chloride is obtained from subterranean mineral deposits, although the sea remains the largest potential source. The mineral name for sodium chloride is *halite*. Trona ($Na_2CO_3 \cdot NaHCO_3 \cdot 2H_2O$) is found in large deposits at Searles Lake, California. Large deposits of sodium nitrate (Chile saltpeter, $NaNO_3$) are located in Chile and Peru. Other sodium salts found in dry lake beds include sodium sulfate (Na_2SO_4) and sodium tetraborate (borax, $Na_2B_4O_7 \cdot 10H_2O$).

The Dead Sea has the highest concentration (30%) of mineral salts of any natural sea or ocean. Large amounts of KCl and NaCl, along with other valuable chemicals, are obtained from this concentrated pool of mineral salts at a chemical plant located at Sodom, Israel.

Major deposits of potassium salts are located at Stassfurt, Germany; Searles Lake, California; and in the Carlsbad, New Mexico, area. The principal minerals are sylvite (KCl), carnalite ($KCl \cdot MgCl_2 \cdot 6H_2O$), polyhalite ($K_2SO_4 \cdot MgSO_4 \cdot 2CaSO_4 \cdot 2H_2O$), and longbeinite ($K_2SO_4 \cdot 2MgSO_4$).

Small amounts of rubidium and cesium are found in potassium salt deposits. Sodium and potassium are the sixth and seventh most abundant elements in the earth's crust.

Preparation

The alkali metals are usually prepared by electrolysis of the molten chlorides or hydroxides. Graphite and iron electrodes are used; the alkali metal deposits at the cathode and floats to the surface of the more dense salt. The cathode and anode reactions of molten Na^+Cl^- are shown below.

Cathode reaction	$2 Na^+ + 2 e^- \longrightarrow 2 Na^0$
Anode reaction	$2 Cl^- \longrightarrow Cl_2^0 \uparrow + 2 e^-$
Total reaction	$2 Na^+ + 2 Cl^- \longrightarrow 2 Na^0 + Cl_2^0 \uparrow$

The electrolysis of aqueous sodium chloride gives sodium hydroxide, hydrogen, and chlorine as products.

Potassium metal may be prepared by electrolysis of potassium hydroxide. Potassium and hydrogen are liberated at the cathode and oxygen at the anode. Potassium is also prepared by the reaction of sodium with potassium chloride.

$$2\,KOH \xrightarrow{\text{Electrolysis}} 2\,K + H_2\uparrow + O_2\uparrow$$

$$Na + KCl \longrightarrow K + NaCl$$

More than 100,000 tons of sodium are produced annually in the United States. Some lithium and potassium are produced, but they are much more expensive to produce than sodium, the price being 60–70 times that of sodium. Rubidium and cesium are not produced on a commercial scale.

Chemical Properties

The chemical properties of an alkali metal depend on the ease with which it loses its outer-shell electron. The ion formed when an electron is lost from an alkali metal has a noble gas structure and, therefore, has great stability. We have seen (Chapter 7) that the amount of energy required to remove an electron from an atom is known as the ionization energy. A study of ionization energies shows that the values for the alkali metals are lower than those for all the other metals (see Table 7.2), indicating their relatively greater reactivity. Table 19.2 shows the ionization energies for the alkali metals and some other elements for comparison. A value of 90 kcal/mole for cesium compared with 118 kcal/mole for sodium means that cesium loses an electron more easily than does sodium, which verifies the fact that cesium is more reactive than sodium. The alkali metals are excellent reducing agents.

Reaction with water. The reaction of the alkali metals with water is vigorous and strongly exothermic, with lithium being the least reactive of the

Table 19.2. Ionization energies of the alkali metals. These values represent the energies required to remove the first and the second electron from one gram-atomic weight of the element causing the formation of positive ions. Values for hydrogen, helium, and magnesium are included for comparison.

Element	Ionization energy (kcal/mole) First electron	Second electron
Li	124	1,744
Na	118	1,091
K	100	733
Rb	96	634
Cs	90	548
H	314	—
He	567	1,254
Mg	176	346

group. Potassium, rubidium, and cesium are so highly reactive that they will react with ice at $-100°C$. In the equation below, M represents an alkali metal:

$$2\,M(s) + 2\,H_2O \longrightarrow 2\,MOH(aq) + H_2\uparrow$$

Alkali metals react violently with dilute acids. The hydrogen gas that is produced may ignite and explode.

All the alkali metal hydroxides, MOH, are readily soluble in water. Lithium hydroxide is the least soluble and cesium hydroxide is the most soluble. All are strong bases. Cesium hydroxide is the strongest base known—so strong that it readily attacks glass and therefore is generally stored in silver or platinum containers. Cesium is the most electropositive element; it stands first in the Activity Series of Metals.

Reaction with hydrogen. Rubidium and cesium react violently with hydrogen at room temperature. The other three metals require elevated temperatures in order to form the hydride. The general equation is

$$2\,M(s) + H_2(g) \longrightarrow 2\,MH(s)$$

The compounds that are formed are called *hydrides*. Lithium hydride and sodium hydride are useful sources of hydrogen in organic chemistry reactions. When the hydrides are reacted with water, free hydrogen and metal hydroxides are formed.

$$LiH(s) + H_2O \longrightarrow LiOH(aq) + H_2\uparrow$$

Reaction with halogens. Lithium and sodium react slowly with chlorine at room temperature. Molten sodium burns with a brilliant yellow flame in a chlorine atmosphere to form sodium chloride:

$$2\,Na + Cl_2 \longrightarrow 2\,NaCl$$

Potassium, rubidium, and cesium react vigorously with all the halogens, forming metal halides.

Table 19.3. Important compounds of lithium.

Compound	Uses
Lithium chloride, LiCl	Starting material for the manufacture of Li metal; a 35–40% aqueous solution is used in air-conditioning units for humidity control
Lithium carbonate, Li_2CO_3	Ceramic glazes
Lithium hydride, LiH	Source of H_2; a reducing agent
Lithium hydroxide, LiOH	Excellent absorbent for CO_2: $LiOH + CO_2 \rightarrow Li_2CO_3$
Lithium stearate, $LiC_{18}H_{35}O_2$	Added to lubricating oils and greases to maintain their viscosity at low temperatures

Reaction with oxygen. When heated, all five alkali metals react vigorously with oxygen. Lithium forms only the monoxide, Li_2O; sodium and potassium form peroxides, Na_2O_2 and K_2O_2. Cesium explodes spontaneously when it is in contact with air or oxygen. By way of comparison, the combustion of the alkali metals is a more violent reaction that the combusion of magnesium in air.

Compounds

This section describes some of the more important and interesting compounds of the alkali metals and their uses. In these compounds, the metal has an oxidation number of $+1$. Lithium compounds are listed in Table 19.3.

Sodium compounds. The principal sodium compounds of commerce are sodium chloride, sodium carbonate, and sodium hydroxide. Sodium chloride, because of its abundance and low cost, is the starting material for manufacturing metallic sodium and other sodium compounds. Several sodium compounds are listed in Table 19.4.

Table 19.4. Important compounds of sodium.

Compound	Uses
Sodium chloride, NaCl	Starting material for preparation of metallic sodium and other sodium compounds; table salt; source of chlorine gas; dietary need for the human body
Sodium carbonate, Na_2CO_3	Also known as *soda ash*; used in manufacture of soap, cleansers, glass; water softening; pulp and paper industry
Sodium hydroxide, NaOH	Also known as *caustic soda* and *lye*; used in manufacture of soap, rayon, and cellulose film; most common base used in the chemical laboratory
Sodium bicarbonate, $NaHCO_3$	Also known as *baking soda*; used in baking as a leavening agent; source of CO_2 in some fire extinguishers
Sodium nitrate, $NaNO_3$	Known as *Chile saltpeter*; formerly used to make HNO_3; now used mainly as a fertilizer
Sodium bromide, NaBr	Photographic emulsions
Sodium iodide, NaI	Photographic emulsions and medicinals
Sodium tetraborate, $Na_2B_4O_7 \cdot 10H_2O$	Known as *borax*; used as a water softener and cleansing agent
Sodium thiosulfate, $Na_2S_2O_3$	Known as *hypo*; used as a fixing agent in photography
Tetrasodium lead, Na_4Pb	Synthesis of tetraethyl lead for "ethyl" gasoline
Sodium sulfate, Na_2SO_4	Diluent for synthetic detergents, many of which contain more than 50% Na_2SO_4
Sodium peroxide, Na_2O_2	Oxidizing agent; preparation of hydrogen peroxide

Potassium compounds. Potassium compounds were known in early times; the name *potash* was given to potassium carbonate, which was obtained by leaching the ashes of burned wood and plants in iron pots. Mineral salts of potassium are selectively absorbed from the soil by plants. Animals require both sodium and potassium in their diets; plants require only potassium for proper growth. Table 19.5 lists some potassium compounds.

Table 19.5. Important compounds of potassium.

Compound	Uses
Potassium chloride, KCl	Starting material for preparation of most potassium compounds; fertilizer
Potassium hydroxide, KOH	Preparation of soft soaps; textile industry
Potassium carbonate, K_2CO_3	Manufacture of glass and soap
Potassium iodide, KI	Photographic emulsions; medicinals
Potassium nitrate, KNO_3	Known as *saltpeter*; formerly used as the oxidant in black gunpowder; now used for glass, ceramic glazes, fertilizers, and in the curing and pickling of meat
Potassium hydrogen tartrate, $KHC_4H_4O_6$	Known as *cream of tartar*; present in grape juice; used in baking powder
Potassium sulfate, K_2SO_4	Fertilizer

19.2 Alloys

alloy

metallurgy

The majority of metals and metal products used in the home and industry are not pure. Pure metals do not have the tensile strength, wearing qualities, or corrosion resistance of mixtures of metals. An **alloy** is a metallic material made by blending a metal with other elements. Alloys are usually formed by combining the materials in a molten state so that a uniform mixture can be obtained. Alloys are of great practical importance to industry. **Metallurgy** is the science of obtaining and refining metals from their ores, and the study of the properties and applications of metals.

Because of their reactivity, alkali metals are not ordinarily used to make alloys for structural or mechanical use. However, for chemical use the alkali metals are often alloyed with some other metal in order to reduce their reactivity. Sodium–lead (Na–Pb) alloys containing as high as 30% Na are used. The reaction of water with a 30% Na, 70% Pb alloy is not violent at all. Potassium–lead alloys containing up to about 12% potassium also react calmly with water.

amalgam

Metal alloys of mercury are called **amalgams**; they may be either solid or liquid, depending upon the amount of mercury used. Many metal amalgams are known. Sodium amalgam is used in numerous reactions in place of sodium. The amalgam is a good reducing agent, and reactions with it are easy to control. In the days of panning for gold, mercury was used to dissolve (amalgamate) the gold and free it from the last traces of sand.

Two lead alloys containing alkali metals are used for axle bearings in railroad engines and cars. One is composed of 95.8% Pb, 1.3% Na, 0.12% Sb, and 0.08% Sn; the other contains 98.76% Pb, 0.6% Na, 0.04% Li, and 0.6% Cu.

It is possible to make an almost limitless number of alloys. Table 19.6 gives some examples of the thousands of alloys that are available on today's market.

Table 19.6. Composition of selected alloys.

Alloy	Percent composition
Stainless steel	74 Fe, 18 Cr, 8 Ni, 0.18 C (many others are known)
Storage battery plates	94 Pb, 6 Sb
Coinage silver	90 Ag, 10 Cu
Plumber's solder	67 Pb, 33 Sn
Babbitt metal	90 Sn, 7 Sb, 3 Cu
Yellow brass	67 Cu, 33 Zn
10 carat gold	42 Au, 38–46 Cu, 12–20 Ag
18 carat gold	75 Au, 10–20 Ag, 5–15 Cu
Coinage (U.S.)	25 Ni, 75 Cu
Nichrome	60 Ni, 40 Cr
Stellite	55 Co, 15 W, 25 Cr, 5 Mo
Spring steel	98.6 Fe, 1 Cr, 0.4 C

19.3 The Alkaline Earth Metals

alkaline earth metal

The elements of Group IIA of the periodic table—beryllium, magnesium, calcium, strontium, barium, and radium—are called the **alkaline earth metals**. They were so named because of the alkaline properties of their oxides and because early chemists called all insoluble nonmetallic substances that were unaffected by fire "earths."

Group IIA elements have typical metallic properties; for example, they are malleable, ductile, have a silvery color, are shiny when cut, and are good conductors of heat and electricity. As a group, they are generally harder and less reactive than Group IA metals. However, all the alkaline earth metals are too reactive to be found free in nature. Only beryllium and magnesium are stable in air at ordinary temperatures. Barium and strontium are so reactive that they are usually stored under kerosene or some other inert solvent. Radium is highly radioactive. Some of the physical properties of the alkaline earth metals are shown in Table 19.7.

Occurrence

Calcium and magnesium are the most common Group IIA elements, and in order of abundance in the earth's crust, rank fifth and eighth, respectively.

Table 19.7. Physical properties of the alkaline earth metals.

Element	Atomic number	Atomic weight	Electron configuration	Metallic atomic radius (Å)	Radius of M^{2+} ion (Å)	Density (20°C, g/ml)	Melting point (°C)	Boiling point (°C)
Be	4	9.012	2,2	1.12	0.31	1.85	1,278	2,970
Mg	12	24.305	2, 8,2	1.60	0.65	1.74	651	1,107
Ca	20	40.08	2, 8, 8,2	1.97	0.99	1.55	843	1,487
Sr	38	87.62	2, 8, 18, 8,2	2.15	1.13	2.6	769	1,384
Ba	56	137.34	2, 8, 18, 18, 8,2	2.22	1.35	3.6	725	1,140
Ra	88	226	2, 8, 18, 32, 18, 8,2	2.46[a]	—	4.4[a]	700	<1,737

[a] Estimated.

The other alkaline earth metals are much less common. Alkaline earth metals occur as compounds in both igneous and sedimentary rocks. The sea is a virtually inexhaustible source of magnesium, since this element is present in seawater to the extent of about 0.13% (6 million tons of magnesium per cubic mile of seawater). A list of the most commonly occurring minerals of the alkaline earth metals is given in Table 19.8.

Table 19.8. Principal naturally occurring minerals of the alkaline earth metals.

Mineral	Chemical composition
Beryllium	
Beryl	$Be_3Al_2(SiO_3)_6$
Chrysoberyl	$BeO \cdot Al_2O_3$
Magnesium	
Magnesite	$MgCO_3$
Dolomite	$CaCO_3 \cdot MgCO_3$
Epsom salt	$MgSO_4 \cdot 7H_2O$
Asbestos	$H_4Mg_3Si_2O_9$
Talc	$Mg_3(Si_4O_{10})(OH)_2$
Calcium	
Calcite (limestone)	$CaCO_3$
Gypsum	$CaSO_4 \cdot 2H_2O$
Fluorspar	CaF_2
Apatite	$Ca_5F(PO_4)_3$
Strontium	
Celestite	$SrSO_4$
Strontianite	$SrCO_3$
Barium	
Barite	$BaSO_4$
Witherite	$BaCO_3$

Preparation

All the alkaline earth metals can be prepared in the laboratory by electrolysis of their fused chlorides. Electrolysis is also the method usually used for commercial production of these elements. Only magnesium, calcium, and beryllium are produced in significant quantities. On a commercial scale, calcium and magnesium are obtained by electrolysis of their fused chlorides. Beryllium is made commercially by electrolyzing a melted mixture of beryllium chloride and sodium chloride. Sodium chloride is added to lower the melting point of the beryllium chloride. Since sodium ions are more difficult to reduce than beryllium ions, no free sodium is produced.

$$BeCl_2 \xrightarrow[NaCl]{Electrolysis} Be + Cl_2$$

Only very small quantities of strontium and barium are needed for industrial use. These two metals are generally prepared not by electrolysis, but instead by reducing the oxides (SrO or BaO) with aluminum metal in an electrically heated vacuum furnace. Relatively pure strontium and barium are obtained in this process, since at the furnace temperature, either strontium or barium metal will distill away from the solid aluminum oxide that is also produced.

$$3\,SrO + 2\,Al \xrightarrow[Vacuum]{>1000°C} Al_2O_3 + 3\,Sr$$

The commercial production of magnesium is of special interest because it is the only metal that is obtained from seawater. Originally, magnesium was produced from brines pumped from underground deposits. Later, a relatively simple process for precipitating magnesium ions from seawater was developed and put into operation on a large scale by the Dow Chemical Company near Freeport, Texas. The principal steps and chemical reactions involved are as follows:

Step 1. The process starts by dredging the ocean floor to obtain oyster shells, which are essentially pure calcium carbonate. The oyster shells are cleaned and heated to produce calcium oxide.

$$CaCO_3(s) \xrightarrow{\Delta} CaO(s) + CO_2\uparrow$$

Step 2. The calcium oxide is added to seawater (in large vats) and forms calcium ions, which precipitate the highly insoluble magnesium hydroxide from the seawater.

$$CaO(s) + HOH \longrightarrow Ca^{2+}(aq) + 2\,OH^-(aq)$$
$$Mg^{2+}(aq) + 2\,OH^-(aq) \longrightarrow Mg(OH)_2\downarrow$$

Step 3. The magnesium hydroxide is separated by filtration and reacted with hydrochloric acid to form magnesium chloride.

$$Mg(OH)_2(s) + 2\,HCl \longrightarrow MgCl_2 + 2\,H_2O$$

Step 4. The resulting magnesium chloride solution is evaporated to yield the hexahydrate MgCl$_2\cdot$6H$_2$O, which is further dehydrated to reduce the water content to about 1.5 moles H$_2$O/mole MgCl$_2$.

Step 5. The almost anhydrous salt is placed in electrolytic cells, melted, and electrolyzed to form magnesium and chlorine. The magnesium floats on the molten mix and is drawn off periodically. Chlorine is collected, converted to hydrochloric acid, and recycled back to the process.

$$MgCl_2 \xrightarrow{Electrolysis} Mg + Cl_2$$

A flow diagram for this process is shown in Figure 19.1.

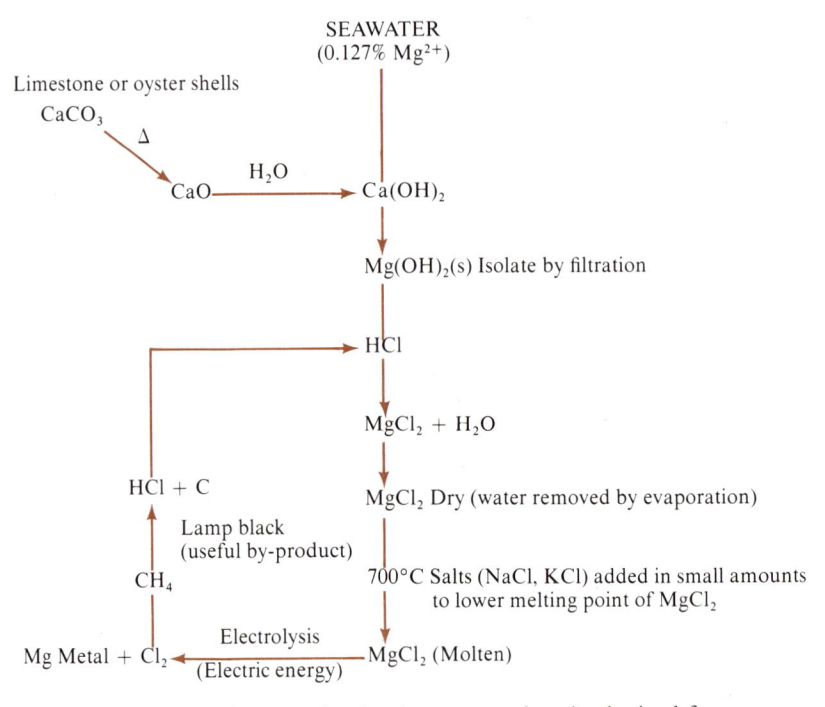

Figure 19.1. Flow diagram showing how magnesium is obtained from seawater.

Chemical and Physical Properties

All the Group IIA elements react by losing the outermost pair of s electrons to form +2 ions. Hence, the formulas of their compounds are similar (BeSO$_4$, MgSO$_4$, CaSO$_4$, and so on). However, these elements differ widely in reactivity.

Beryllium is the least reactive of the alkaline earth metals. It is resistant to oxidation and stable in air at ordinary temperatures, but oxidizes rapidly at about 800°C.

$$2\,Be + O_2 \xrightarrow{800°C} 2\,BeO$$

Beryllium has many special properties that make it useful in modern technology. Because it is light, strong, resistant to corrosion, and has a high melting point, beryllium is used in aircraft and space vehicles. Springs made of copper–beryllium alloys (~2% Be) are virtually fatigue-free, and for this reason are used in precision instruments. Because of its low atomic number, beryllium is very permeable to X-rays and is therefore used for windows in X-ray tubes. Beryllium is also useful in nuclear technology. It is capable of slowing neutrons and is used as a moderator in nuclear reactors. When bombarded with alpha particles, beryllium is a good source of neutrons.

$$^{9}_{4}Be + {}^{4}_{2}He \longrightarrow {}^{12}_{6}C + {}^{1}_{0}n$$

Magnesium metal is being used on an increasing scale because of its high strength-to-weight ratio and corrosion resistance. Magnesium is resistant to atmospheric corrosion, not because it is unreactive, but because it does react with oxygen. When magnesium is exposed to air, it becomes coated almost immediately with a thin, transparent layer of MgO. Once it is coated with oxide, the magnesium metal is protected, at ordinary temperatures, from further corrosion.

Care must be exercised in the use of magnesium metal because, despite its apparent inertness to atmospheric corrosion, magnesium is a very reactive metal. It will, for example, react slowly with boiling water to form magnesium hydroxide and hydrogen:

$$Mg + 2H_2O(l) \xrightarrow{\sim 100°C} Mg(OH)_2 + H_2$$

Magnesium and steam react to form magnesium oxide and hydrogen:

$$Mg + H_2O(g) \longrightarrow MgO + H_2$$
(Steam)

Magnesium burns brilliantly in air or in oxygen. This property is easily demonstrated by taking a pair of tongs and holding a short piece of magnesium ribbon in a Bunsen flame. (Proceed with caution!) Fortunately, any piece of magnesium large enough to be used as structural material is difficult to ignite. This is because magnesium is a good heat conductor, and to be ignited it must be heated above its melting point of 651°C. When a large piece of the metal is heated at one point, the heat is conducted away so rapidly that the temperature does not readily reach the ignition point.

A magnesium fire, once started, is very hard to extinguish. Hot magnesium reacts exothermically with common fire extinguishing agents such as water or carbon dioxide. It will even react with sand (SiO_2). In effect, the hot magnesium takes the oxygen needed for combustion away from the "fire extinguishing" material—water, carbon dioxide, or sand—and thus continues to burn.

$$Mg + H_2O(g) \longrightarrow MgO + H_2 + 86.0 \text{ kcal}$$
$$2Mg + CO_2 \longrightarrow 2MgO + C + 193.5 \text{ kcal}$$
$$2Mg + SiO_2 \longrightarrow 2MgO + Si + 82.2 \text{ kcal}$$

The alkaline earth metals that are heavier than magnesium are too reactive to be used as structural materials. They react with cold water to form free hydrogen and the corresponding metal hydroxide. Reactivity increases from calcium to radium. Calcium is the only one of the heavier Group IIA elements that is produced in commercially significant amounts. It is used as a minor constituent of some alloys and as a reducing agent. For example, the plates in some storage batteries are made from a lead–calcium alloy containing about 0.1% calcium. Vanadium metal is obtained by reduction of vanadium pentoxide with calcium.

$$5\,Ca + V_2O_5 \xrightarrow{\Delta} 2\,V + 5\,CaO$$

Other metals, including titanium and uranium, may also be prepared by reduction of their oxides with calcium.

Compounds

Calcium carbonate is of special importance and occurs in various natural forms such as marble, calcite, travertine, and ordinary limestone. It also occurs extensively in the mineral *dolomite*, which has the formula $CaCO_3 \cdot MgCO_3$. Large areas of the earth's surface are covered with layers of limestone up to several thousand feet thick.

Calcium carbonate is probably the most extensively used industrial raw material. For centuries, *lime burning* has been practiced on an industrial scale. From a chemical standpoint, lime burning amounts to simple thermal decomposition of calcium carbonate. The calcium carbonate, usually limestone, is heated to about 1000°C in a large, inclined, tubular furnace known as a *lime kiln*. Calcium oxide and carbon dioxide are produced.

$$CaCO_3 \xrightarrow{\Delta} CaO + CO_2$$

Calcium oxide is known as *quicklime*. Huge quantities of this substance are used for such purposes as making mortar and plaster, softening hard water, making paper, refining sugar, and in many chemical manufacturing operations. Large amounts of calcium oxide are also used in making glass and portland cement. In the manufacture of these substances, limestone is usually heated directly with sand and other materials. Reactions of the following kind occur:

$$CaCO_3 \longrightarrow CaO + CO_2$$
$$CaO + SiO_2 \longrightarrow CaSiO_3$$

(Calcium silicate)

Calcium hydroxide, also known as *slaked lime*, is the cheapest base or source of hydroxide ions. It is made by reacting calcium oxide and water.

$$CaO + H_2O \longrightarrow Ca(OH)_2$$

Calcium sulfate, which occurs mainly as the mineral *gypsum*, is another common naturally occurring calcium compound. Gypsum, $CaSO_4 \cdot 2H_2O$, is much less abundant than calcium carbonate. However, the several hundred

Table 19.9. Important compounds of Group IIA elements.

Compound	Uses
Beryllium oxide, BeO	Refractory material used in crucibles; neutron reflector in nuclear reactors
Beryllium aluminum silicate, $3BeO \cdot Al_2O_3 \cdot 6SiO_2$	The precious and semiprecious gems emerald, aquamarine, and beryl have this approximate composition
Magnesium chloride, $MgCl_2$	Principal source of magnesium metal; occurs in underground deposits and dissolved in seawater
Magnesium oxide, MgO	Also known as *magnesia*; refractory material used in furnace linings and other high-temperature applications; milk of magnesia, which is used as an antacid, is a mixture of MgO and $Mg(OH)_2$
Magnesium carbonate, $MgCO_3$	Also known as *magnesite*; used as an insulating material; source of magnesium for other magnesium compounds
Magnesium sulfate, $MgSO_4$	Occurs as the heptahydrate ($MgSO_4 \cdot 7H_2O$) and is known as *epsom salts*; this salt is used as a purgative; also used in the tanning and dyeing industries and as a fertilizer for magnesium-deficient soils
Calcium carbonate, $CaCO_3$	Largest volume raw material in chemical industry; flux in steel-making; manufacture of cement, mortar, and glass; building material
Calcium oxide, CaO	Also known as *quicklime*; used for mortar and plaster, paper-making, sugar refining, water treating, and others
Calcium hydroxide, $Ca(OH)_2$	Known as *slaked lime*; cheapest source of OH^- ions
Calcium phosphate, $Ca_3(PO_4)_2$	Used as fertilizer; source of phosphoric acid and phosphorus compounds
Calcium sulfate, $CaSO_4$	Occurs as gypsum ($CaSO_4 \cdot 2H_2O$); used to make plaster of paris and building materials, especially plasterboard
Calcium fluoride, CaF_2	Known as *fluorspar* or *fluorite*; used as a flux in metallurgy; principal source of fluorine for other fluorine compounds
Calcium chloride, $CaCl_2$	Very hygroscopic; used to reduce dust on unpaved roads and to reduce danger of dust explosions in mines; also used to melt ice on streets
Calcium carbide, CaC_2	Source of acetylene
Strontium nitrate, $Sr(NO_3)_2$	Produces an intense red color in flares and fireworks
Barium sulfate, $BaSO_4$	Also known as *barite*; high-density material used in oil-well drilling muds; paint pigment; filler for paper-making; also used in X-ray photography of the gastrointestinal tract
Barium nitrate, $Ba(NO_3)_2$	Used to produce a brilliant green color in flares and fireworks

square miles of "sand" in the Great White Sands Desert in New Mexico is composed of finely divided gypsum. Gypsum when heated forms plaster of paris.

$$2\,CaSO_4 \cdot 2\,H_2O \longrightarrow (CaSO_4)_2 \cdot H_2O + 3\,H_2O$$

Gypsum   Plaster of paris
(Calcium sulfate   (Calcium sulfate
dihydrate)   hemihydrate)

To make the familiar surgical casts used in setting broken bones or to make plaster molds and impressions, the foregoing reaction is reversed by wetting powdered plaster of paris to form a solid cast of calcium sulfate dihydrate.

$$(CaSO_4)_2 \cdot H_2O + 3\,H_2O \longrightarrow 2\,CaSO_4 \cdot 2\,H_2O$$

Large amounts of calcium sulfate are also used in the manufacture of cement and plasterboard, which are used in the construction and building trades.

Table 19.9 lists the uses of some of the more important and interesting compounds of the Group IIA elements.

Biological Properties

Calcium and magnesium are required for many essential biological processes in both animal and plant life. Magnesium is present in the chlorophyll of green plants, and it is required for the activity of some enzymes involved in the metabolism of animals. Calcium is needed for nitrogen utilization by plants. Calcium serves the human body in a variety of ways; for example, it is present in bones and teeth as calcium phosphate and it has an essential role in the clotting of blood.

Strontium is believed to be a biologically nonessential element, but is not toxic to humans. However, the isotope Sr-90, which is present in nuclear fallout, is extremely hazardous due to its radioactivity (Section 18.14).

Soluble beryllium and barium salts are poisonous to humans. Barium salts are so toxic that a quantity of less than 1 g of barium chloride may be fatal to an adult. An exception to this general toxicity is barium sulfate, which is given orally prior to X-ray or fluoroscopic examination of the gastrointestinal tract. Barium sulfate is used for this purpose because (1) barium is a heavy metal and therefore is opaque to X rays, and (2) it is so highly insoluble that no barium ions are absorbed into the body from the gastrointestinal tract.

19.4 Flame Tests and Spectroscopy

All Group IA elements and calcium, strontium, and barium from Group IIA can be detected by simple flame tests. A Bunsen burner and a piece of clean platinum wire are the only essential apparatus needed to make flame tests. When salts of the elements are volatilized in the burner flame, the characteristic colors are visible.

Element	Color
Li	Red
Na	Bright yellow (persistent)
K	Violet (short duration)
Rb	Violet
Ca	Brick red (short duration)
Sr	Crimson
Ba	Light green

Salts of some of these elements are used in fireworks and flares. For example, the crimson color of railroad signal flares and highway emergency flares is due to strontium salts.

The color of flames is related to the electronic structure of atoms and is the basis of spectroscopy. At ordinary temperatures, electrons are at the lowest possible, or ground-state, energy levels. Electrons can exist only at certain definite quantized energy levels within atoms (Section 5.6). For Group IA elements, most Group IIA elements, and a few other elements, the temperature of a Bunsen flame is high enough to promote some ground-state electrons to higher, or excited, energy levels. In falling back to the ground-state energy level, each electron releases exactly the same amount of energy (in the form of radiation) that was absorbed when it was promoted to the excited energy level. If the wavelength of this radiation falls within the visible range of the spectrum (4000 to 7000 Å), it is seen as light of a definite color corresponding to a particular wavelength.

Colored flames are a qualitative manifestation of a property of matter that has been exploited by spectroscopy both to identify elements and to give detailed information concerning the structure of atoms. When light produced by atoms heated to high temperatures (an electric arc is needed for some elements) is passed through a prism (or a ruled grating), it is found to consist of narrow bands called spectral lines. Each band (or line) corresponds to the energy emitted when an electron falls from a specific excited state to a specific lower, or ground-state, energy level. Since the number of electrons in each element is different, each element has a characteristic set of spectral lines that is different from that of every other element. This set of spectral lines identifies the element and provides a means of determining the energy levels that are occupied by electrons within the atoms of that element.

19.5 Iron

Physical Properties, Occurrence, and Preparation

Iron is the fourth most abundant element, following oxygen, silicon, and aluminum. It has been the key element to the development of modern civilization. Iron was known in prehistoric times and its known use dates back almost 8000 years. The first iron to be used was probably obtained from meteorites.

Iron was produced from ores as early as 1300 B.C. The first ironworks in America was built in Jamestown, Virginia, in 1619. By 1850, the American production of pig iron was about 560,000 tons annually (from 377 furnaces). Furnace production was 1–6 tons per day. Today, some furnaces can produce more than 4000 tons in a single day.

Pure iron is a silvery white, relatively soft, very ductile metal. It has a melting point of 1532°C and it boils at 3000°C. One of its most distinguishing characteristics is its magnetism, which it loses when it is heated to 770°C. In the presence of a magnetic field or electric current, iron shows greater magnetic properties than any other element. (Cobalt and nickel are also strongly magnetic.)

Very little iron is used in its pure form; most of it is used in the form of steel, an iron–carbon alloy. Steel may contain small amounts of other elements such as Mn, Si, Cr, Ni, V, W, Mo, Co, and Ti. These alloying elements enhance certain qualities of the steel such as toughness, hardness, corrosion resistance, and so on. The chief iron-bearing minerals are listed in Table 19.10. Large iron ore deposits are found on all continents. The most important iron ore-producing area in the United States is the Mesabi range in northern Minnesota.

Table 19.10. Iron-bearing minerals.

Name	Composition
Magnetite	Fe_3O_4
Hematite	Fe_2O_3
Limonite	$Fe_2O_3 \cdot xH_2O$
Siderite	$FeCO_3$
Ilmenite	$FeTiO_3$
Pyrites	FeS_2

The first step in the production of steel is the reduction of iron ores to iron, commonly called *pig iron*. The ore is reduced in a blast furnace (see Figure 19.2.) A typical blast furnace is a firebrick-lined structure more than 100 ft high and about 25 ft in diameter. Temperatures in the furnace range from about 200°C at the top to about 1900°C near the bottom. The furnace is charged at the top with iron ore, coke, and limestone. As the ore descends through the furnace, it is reduced to free iron by reaction with hot, reducing gas (CO) in countercurrent flow. The reducing gas is produced by burning coke with preheated air blasted in through *tuyeres* or nozzles near the bottom of the furnace. The free iron melts and forms a pool of liquid at the bottom of the furnace. This molten iron is covered by a protective layer of liquid slag. (Limestone is added to the charge for the primary purpose of forming this slag.) Most of the siliceous impurities (SiO_2) from the ore are taken up in forming the slag. Once formed, the slag protects the molten iron from being reoxidized by the incoming air. The molten pig iron and the slag are tapped (drawn off) from the bottom of the furnace periodically. The more important chemical reactions that occur in the blast furnace are summarized on page 401.

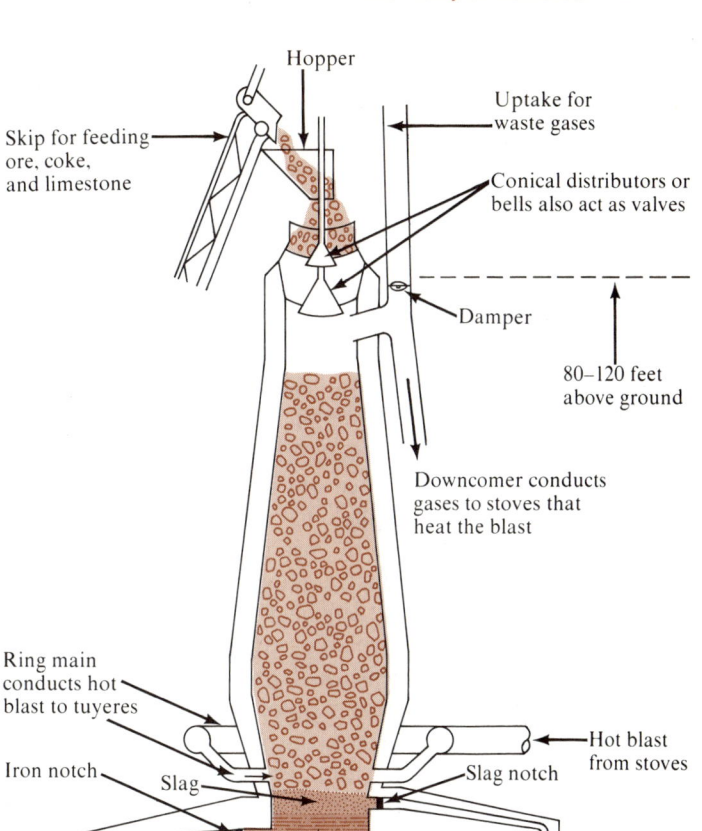

Figure 19.2. Blast furnace for production of pig iron.

$$2\,C + O_2 \longrightarrow 2\,CO$$
Coke

$$3\,Fe_2O_3 + CO \longrightarrow 2\,Fe_3O_4 + CO_2$$
Iron ore

$$Fe_3O_4 + CO \longrightarrow 3\,FeO + CO_2$$
$$FeO + CO \longrightarrow Fe + CO_2$$
$$CaCO_3 \longrightarrow CaO + CO_2$$
Limestone

$$CaO + SiO_2 \longrightarrow \underset{\text{Slag}}{CaSiO_3}$$

Pig iron contains free Mn, Si, and P as impurities. These elements are produced by the reduction of their oxides with hot coke.

Steel

Steel is a malleable alloy of iron that contains up to 1.7% carbon as the essential alloying element. In addition to carbon, steel may contain other alloying elements. Pig iron contains on the average about 1% Si, 0.03% S, 0.27% P, 2.4% Mn, and 4.6% C, the balance being Fe. These impurities must be removed or lowered to carefully controlled levels to convert pig iron into steel.

Three principal processes have been used in the past 125 years to make steel: (1) the Bessemer process, (2) the open-hearth process, and (3) the basic oxygen process (BOP). The Bessemer process was developed by Henry Bessemer (1813–1898) in England in the mid-1800s. The Bessemer converter is a large pear-shaped vessel lined with firebrick. In the process, air is blasted through the molten pig iron in the converter through intake ports at the bottom. Carbon and silicon are burned out in the blast. Sulfur and phophorus are not removed.

Historically, the Bessemer process was of great importance because it made possible the production of the large quantities of steel needed for the growth of industry. However, Bessemer steel was often of low quality because (1) carbon and silicon levels were not properly controlled, and (2) sulfur and phosphorus, which produce brittleness, were not removed.

The open-hearth process was developed in the late 1800s. Its advantages over the Bessemer process are that it produces about 20 times as much steel per batch and gives a product with a more precisely controlled composition and with less impurities. The process is conducted in a large furnace resembling a huge oven. The furnace is charged with molten pig iron, scrap iron, limestone, and iron oxides. A mixture of gas and air is burned and recirculated over the surface of the molten mass to provide the high temperature needed in the process. The iron oxide converts carbon to CO, which escapes and burns. Sulfur, phosphorus, and silicon are also converted to oxides. Limestone ($CaCO_3$) is converted to lime and takes up these oxides to form a slag on the surface of the molten steel. The entire process takes about 10 to 12 hours. This allows ample time for testing and the necessary quality control. Near the end of the process, a scavenger such as manganese is added to remove excess oxygen, which if allowed to remain, would cause the steel to be brittle. An ecological advantage of this process is that it uses some scrap iron or steel and thereby conserves our natural iron ores.

Modern steel-making research, especially in Linz, Austria, led to the development of the L-D or basic oxygen process (BOP). The process, first operated commercially in 1952, uses a pear-shaped furnace. The furnace is charged with molten pig and scrap iron and a flux consisting of limestone and fluorspar (CaF_2) to form a slag. A long, water-cooled lance is lowered into the furnace until its tip is 4 to 8 feet from the surface of the white-hot metal. Nearly pure oxygen under a pressure of 150 pounds per square inch is blown into the melt, converting the impurities to oxides. Carbon oxidizes to carbon monoxide, which in turn burns to CO_2. The oxides of sulfur, phosphorus, and silicon are absorbed in the slag. The process is highly exothermic and no outside source of energy is needed. This part of the refining process takes 20 minutes, and a batch of high-quality steel can be made in 50 minutes. The BOP plant put

Table 19.11. The effects of selected alloying elements on the properties of steel.

Alloying element	Effect on steel
Carbon	Makes tempering possible; increases tensile strength (up to 0.83% C); increases hardness
Cobalt	Imparts high-temperature strength
Chromium	Improves hardness, abrasion resistance, corrosion resistance, and high-temperature properties
Manganese	Imparts hardness, toughness, and resistance to wear and abrasion
Molybdenum	Imparts hardness, shock resistance, and high-temperature strength
Nickel	Imparts corrosion resistance; increases toughness and high-temperature strength; reduces brittleness at very low temperatures
Silicon	Increases strength without affecting ductility; modifies magnetic and electrical properties; high Si concentrations impart resistance to acid corrosion
Tungsten	Imparts hardness, which is retained at high temperatures
Vanadium	Increases strength and toughness; improves heat-treating characteristics

into operation by the Ford Motor Company in 1964 is capable of producing steel at the rate of 300 tons an hour.

No Bessemer steel is made in the United States today. In August 1969, the United States production of BOP steel surpassed that produced by the open-hearth process. World production of raw steel by all processes is over 650 million tons per year. United States steel production in 1978 was 137 million tons.

Elements such as manganese, aluminum, vanadium, and silicon are added to molten steel to remove traces of dissolved oxygen. However, certain elements may be added in larger portions to produce alloy steels with specific properties. Some important alloying elements and their effects on steel are given in Table 19.11.

Chemical Properties

Iron is a fairly reactive metal and forms two principal series of compounds: iron(II), or ferrous compounds, and iron(III), or ferric compounds. The two oxidation states Fe^{2+} and Fe^{3+} are the most common, but compounds are known in which iron has a +4 and a +6 oxidation number.

Iron dissolves in both dilute hydrochloric and dilute sulfuric acids to give Fe^{2+} compounds and hydrogen gas. When cold dilute nitric acid (an oxidizing acid) is reacted with iron, iron(II) nitrate and ammonium nitrate are formed; with concentrated nitric acid, iron(III) nitrate and nitrogen oxides are formed.

$$Fe + 2\,HCl \longrightarrow FeCl_2 + H_2\uparrow$$
$$Fe + H_2SO_4 \longrightarrow FeSO_4 + H_2\uparrow$$
$$4\,Fe + 10\,HNO_3 \longrightarrow 4\,Fe(NO_3)_2 + NH_4NO_3 + 3\,H_2O$$
$$Fe + 6\,HNO_3 \longrightarrow Fe(NO_3)_3 + 3\,NO_2 + 3\,H_2O$$

Iron reacts with many nonmetals, such as sulfur and the halogens, to form binary compounds:

$$Fe + S \longrightarrow FeS$$
$$2\,Fe + 3\,Cl_2 \longrightarrow 2\,FeCl_3$$

It reacts with steam to form magnetic iron oxide and hydrogen:

$$\underset{\text{Hot}}{3\,Fe} + \underset{\text{Steam}}{4\,H_2O} \longrightarrow Fe_3O_4 + 4\,H_2\uparrow$$

Magnetic iron oxide is considered to contain both $+2$ and $+3$ iron, $Fe_2O_3 \cdot FeO$. Solutions of ferrous salts are difficult to maintain in the $+2$ state due to the ease of oxidation of Fe^{2+} to Fe^{3+} by oxygen of the air. The equation for the net ionic reaction may be represented as follows:

$$4\,Fe^{2+} + O_2 \longrightarrow 4\,Fe^{3+} + 2\,O^{2-}$$

Corrosion of iron, commonly called *rusting*, is a serious economic problem that causes losses of billions of dollars annually in the United States. Rusting occurs on the surface of the iron, where it is exposed to oxygen and moisture in the atmosphere. The iron is transformed into reddish-brown rust, a hydrated iron(III) oxide ($Fe_2O_3 \cdot xH_2O$). As the rust is formed, it flakes off the surface, allowing the corrosion to penetrate deeper into the iron. Both oxygen and water are needed for rusting. Iron at first is oxidized to iron(II) hydroxide. This, in turn, is rapidly oxidized to iron(III) hydroxide. The latter loses water to form iron(III) oxide, containing variable amounts of water of hydration.

$$2\,Fe + O_2 + 2\,H_2O \longrightarrow 2\,Fe(OH)_2$$
$$4\,Fe(OH)_2 + O_2 + 2\,H_2O \longrightarrow 4\,Fe(OH)_3$$
$$Fe(OH)_3 \longrightarrow Fe_2O_3 \cdot xH_2O$$

Table 19.12. Selected reactions for differentiating iron(II) ions (ferrous ions) from iron(III) ions (ferric ions).

Reagent	Reaction or product formed with Fe^{2+}	Reaction or product formed with Fe^{3+}
Sodium hydroxide, NaOH	$Fe(OH)_2$, a white precipitate that darkens rapidly	$Fe(OH)_3$, a reddish-brown precipitate
Potassium thiocyanate, KCNS[a]	No apparent reaction	Blood-red solution formed, $Fe(SCN)^{2+}$
Potassium ferrocyanide, $K_4Fe(CN)_6$	$Fe_2[Fe(CN)_6]$, a white precipitate is formed in the absence of Fe^{3+} and O_2	Dark blue precipitate[b]
Potassium ferricyanide, $K_3Fe(CN)_6$	Dark blue precipitate[b]	Brown solution

[a] The test with KCNS is extremely sensitive and can be used to detect concentrations of Fe^{3+} of 10^{-5} M.

[b] The precipitates formed by Fe^{3+} with $K_4Fe(CN)_6$ and by Fe^{2+} with $K_3Fe(CN)_6$ are believed to have the same formula, $KFe[Fe(CN)_6]$.

Selected reactions used to differentiate between ferrous and ferric ions are given in Table 19.12.

Since iron is so widely used and since damaging corrosion occurs so readily, a great deal of research is continually being done to improve methods for protecting iron from corrosion. The most common general methods are (1) protective coatings, (2) alloying, and (3) cathodic protection. Protective coatings of paint, enamel, tar, grease, or metal are commonly used to prevent contact with the atmosphere. Some coatings impart a decorative value as well as corrosion resistance—for example, the chromium plating on automobile bumpers. The term *galvanized* is applied to iron or steel that is protected by a coating of zinc metal.

Highly corrosion-resistant iron alloys are produced industrially. Eighteen-eight (18% Cr and 8% Ni) stainless steel provides almost complete protection from atmospheric corrosion. The high cost of this alloy prevents its general use, although it is used in many special applications. An iron–silicon alloy containing about 12% Si (known as *duriron*) is very resistant to the action of acids.

The ordinary rusting or corrosion of iron can be considered an electrochemical process in which iron acting as an anode is oxidized in this fashion:

$$\text{Anode reaction} \quad \text{Fe}^0 \longrightarrow \text{Fe}^{2+} + 2e^-$$

To afford *cathodic protection* to a steel pipeline or tank, magnesium (or zinc) stakes or rods are driven into the earth and are connected by wires to the pipeline or tank. This in effect sets up an electrochemical cell, with the more easily oxidized magnesium metal acting as the anode. Electrons flow from the magnesium to the iron, causing the iron to become negatively charged or cathodic with respect to the magnesium. Under these conditions, the magnesium, instead of iron, acts as the anode and is oxidized:

$$\text{Mg}^0 \longrightarrow \text{Mg}^{2+} + 2e^-$$

Oxidation does not occur at the iron cathode, and thus no corrosion occurs on the iron pipe or tank. The following reduction reaction is believed to occur at the iron cathode:

$$2\,\text{H}_2\text{O} + \text{O}_2 + 4e^- \longrightarrow 4\,\text{OH}^-$$

The magnesium or zinc rods or stakes must be replaced periodically, since they are consumed in affording protection to the iron or steel. The iron hulls of ships can be protected by fastening blocks of magnesium metal to them. The zinc coating on galvanized iron actually provides cathodic protection for the iron.

19.6 Aluminum

Occurrence and Physical Properties

Aluminum (generally spelled *aluminium* in countries other than the United States) is the most abundant metal and the third most abundant element in the earth's crust, being exceeded only by oxygen and silicon. Aluminum is not found as the free metal but occurs mainly in clays, feldspars, and granite as complex aluminum silicates and in bauxite ores as hydrated aluminum oxide. Pure aluminum oxide (Al_2O_3) is known as *alumina*.

Alumina occurs naturally as the mineral corundum. Corundum is a very hard material, standing next to diamond on the hardness scale, and it is used as an abrasive. Emery, also widely used as an abrasive, is an impure corundum consisting of Al_2O_3 and Fe_3O_4. Rubies and sapphires are much sought-after precious gemstones. They consist of corundum colored by small amounts of other metallic oxides. Artificial rubies and sapphires are produced that are practically indistinguishable from the natural stones. In addition to their use as gemstones and bearings for watches and other instruments, artificial rubies and sapphires are used in lasers for the production of intense beams of coherent light.

Although aluminum is the most abundant metal in the earth's crust, free aluminum metal was not known until 1825. In that year the Danish scientist Hans Christian Oersted succeeded in making some impure samples of aluminum metal by heating aluminum chloride with potassium–mercury amalgam. However, this was not a practical method for commercial aluminum production. Research continued, and the Oersted process was improved, first by eliminating the mercury, and then by substituting cheaper sodium for potassium. However, the reduction of aluminum by an alkali metal was an inherently costly process, and aluminum remained nearly as expensive as gold and platinum during most of the 19th century. Finally in 1886, a relatively cheap method for the electrolytic production of aluminum was devised. Since then, aluminum has been used in ever-increasing amounts. Production increased tremendously during World War II, when millions of tons of aluminum were consumed in the manufacture of aircraft. World production of aluminum was only 7700 tons in 1900, but it now amounts to more than 15 million tons per year. United States production of aluminum exceeds 4 million tons per year.

Aluminum is a Group IIIA element. It is silvery in color, ductile, malleable, and has a melting point of 660°C and a boiling point of 2467°C. Aluminum has a density of only 2.7 g/ml, a value about one-third that of other common metals such as iron, copper, and zinc. Because it is a light-weight metal and has the ability to form a variety of useful alloys, aluminum has many important commercial applications.

Aluminum is an excellent conductor of heat. As an electrical conductor, it is surpassed only by silver and copper. However, when a comparison is made with wires having the same mass per unit of length, aluminum wire, being larger, surpasses both copper and silver as an electrical conductor. For this reason—and because of the increasing scarcity of copper—high-voltage transmission cables are now usually made of aluminum wires reinforced with a steel core for additional strength.

Production

The principal aluminum ore is bauxite, which contains 40 to 60% alumina together with impurities of silica, iron oxide, quartz, and so on. Bauxite is found in many parts of the world. In addition, nepheline, a sodium or potassium aluminum silicate, is available in almost unlimited quantities in Siberia. However, aluminum is considerably more difficult to obtain from nepheline than from bauxite.

Aluminum is obtained by the electrolysis of alumina dissolved in an electrolyte consisting mainly of fused cryolite (Na_3AlF_6). The process was developed simultaneously by Charles M. Hall (1863–1914) in the United States and by Paul L. Heroult (1863–1914) in France in 1886. Both men were only 22 years old at that time. Their basic electrolytic process is still in use today, and economically feasible aluminum is dependent on inexpensive electrical energy.

Bauxite ores containing hydrated aluminum oxide ($Al_2O_3 \cdot 2H_2O$) must first be purified, since the Hall–Heroult process requires essentially pure alumina. In the Bayer process for bauxite purification, the ore is first treated with concentrated caustic soda (NaOH). Because of its amphoteric nature, the hydrated alumina dissolves, forming soluble sodium aluminate:

$$Al_2O_3 \cdot 2H_2O + 2\,NaOH \longrightarrow \underset{\text{Sodium aluminate}}{2\,NaAlO_2(aq)} + 3\,H_2O$$

The insoluble impurities are separated from the sodium aluminate solution. The aluminum is then precipitated as $Al(OH)_3$ by the addition of carbon dioxide, separated, and calcined at 1200°C to give anhydrous alumina (Al_2O_3):

$$2\,NaAlO_2 + 3\,H_2O + CO_2 \longrightarrow 2\,Al(OH)_3\downarrow + Na_2CO_3$$

$$2\,Al(OH)_3 \xrightarrow{1200°C} Al_2O_3 + 3\,H_2O\uparrow$$

The purified alumina is dissolved in molten cryolite (about 950°C) and electrolyzed between carbon electrodes. Molten aluminum is formed at the cathode, sinks to the bottom of the cell, and is periodically drawn off and poured into molds. Oxygen is formed at the anode and combines with the carbon to form CO and CO_2 (see Figure 19.3). The aluminum obtained by the Hall–Heroult process is 99.5% pure.

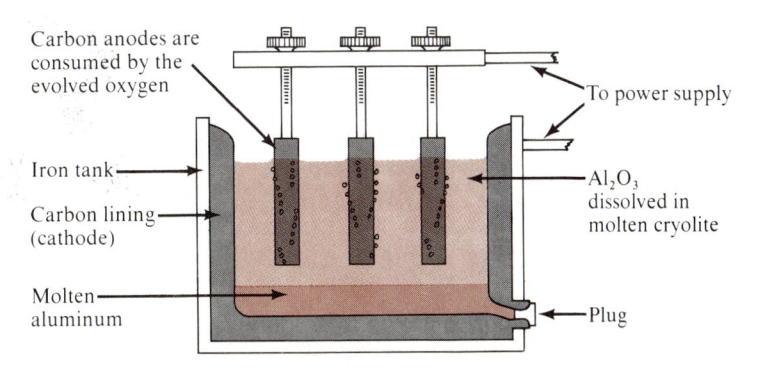

Figure 19.3. Electrolytic production of aluminum.

Chemical Properties

Since aluminum is a Group IIIA metal, its common oxidation state is +3. It is a very reactive element, but when in contact with air, aluminum forms a protective oxide film that prevents further reaction. This film is so thin (0.001 to 0.1 micrometre) and transparent that the aluminum retains its bright metallic luster.

At temperatures around 800°C, aluminum reacts vigorously in oxygen to form Al_2O_3 and in air to form both Al_2O_3 and AlN.

$$4\ Al + 3\ O_2 \longrightarrow 2\ Al_2O_3 + 798\ kcal$$

Aluminum does not react with water at room temperature, but at 180°C it oxidizes rapidly.

$$2\ Al + 3\ H_2O \longrightarrow Al_2O_3 + 3\ H_2$$

The reaction of aluminum with dry chlorine or hydrogen chloride produces aluminum chloride ($AlCl_3$):

$$2\ Al + 3\ Cl_2 \longrightarrow 2\ AlCl_3$$
$$2\ Al + 6\ HCl \longrightarrow 2\ AlCl_3 + 3\ H_2$$

Anhydrous $AlCl_3$ is used as a catalytic reagent in many organic reactions.

At high temperatures aluminum is a good reducing agent and is capable of reducing the oxides of many other metals (Goldschmidt reaction). For example, Cr_2O_3, WO_3, and MoO_3 are readily reduced by aluminum:

$$2\ Al + Cr_2O_3 \longrightarrow Al_2O_3 + 2\ Cr$$

In the thermite welding process, a mixture of powdered aluminum and iron oxide is reacted, producing a temperature of about 3000°C. The molten iron produced by the reaction is used to join or repair (weld) iron or steel articles such as railroad rails. Burning magnesium or an ignition powder is used to start the thermite reaction.

$$2\ Al + Fe_2O_3 \longrightarrow Al_2O_3 + 2\ Fe + 202\ kcal$$

Aluminum reacts readily with hydrochloric acid, slowly with dilute sulfuric acid, but is rendered passive and does not react with even concentrated nitric acid. In fact, nitric acid is shipped in aluminum-lined containers.

When aluminum salts react with bases, aluminum hydroxide forms as a gelatinous white precipitate:

$$Al^{3+} + 3\ OH^- \longrightarrow Al(OH)_3\downarrow$$

This hydroxide is amphoteric and will dissolve in either a strong base or a strong acid:

$$Al(OH)_3 + NaOH \longrightarrow NaAlO_2 + 2\ H_2O$$
$$2\ Al(OH)_3 + 6\ HCl \longrightarrow 2\ AlCl_3 + 3\ H_2O$$

Applications of Aluminum and Its Compounds

The properties of lightness, strength when alloyed, corrosion resistance, and good heat and electrical conductivity make aluminum and its alloys useful in almost every facet of modern society. Aluminum is used in the building trades as a structural and decorative material; in the construction of aircraft, automobiles, ships, and other kinds of vehicles; and in cooking utensils, kitchen equipment, and household appliances. Aluminum foil is a common article in most households. More than half a million tons of aluminum are used each year in making containers for foods and beverages. Photographic flashbulbs are filled with thin strips of foil or fine wire made of aluminum. Aluminum is used in paints and in coating reflective surfaces. The mirror in the giant 200 inch telescope on Mt. Palomar is surfaced with a very thin coating of aluminum.

Aluminum compounds also have many applications. The double salt of potassium and aluminum sulfates $[K_2SO_4 \cdot Al_2(SO_4)_3 \cdot 24H_2O]$ is commonly called *potassium alum* or *alum*. It is generally represented by the simpler formula $KAl(SO_4)_2 \cdot 12H_2O$. Beautiful, colorless octahedral crystals of potassium alum are easily prepared by dissolving equal molar quantities of K_2SO_4 and $Al_2(SO_4)_3$ crystals in water and recrystallizing by evaporating the water.

When we speak of alum, we are usually referring to potassium alum, but a variety of other alums are known. Sodium or ammonium ions can be substituted for the potassium ions in alum to obtain sodium or ammonium alum. Other tripositive ions can be substituted for the aluminum ion; for example, the violet-colored chrome alum has the formula $KCr(SO_4)_2 \cdot 12H_2O$.

Alums have a wide variety of useful applications. They are, for example, used as a source of acid in baking powders, as mordants in dyeing textiles, in weighting and sizing paper, in fireproofing fabrics, in water treatment, as an astringent, and as hardeners in photography.

Concern is sometimes expressed over the possibility of ingesting dangerous quantities of aluminum from acid foods that have been cooked or stored in aluminum utensils. There appears to be no danger from this source, since aluminum is essentially inert in the body. People have taken aluminum hydroxide preparations as antacids for many years without apparent ill effects; these preparations have also been used in the treatment of peptic ulcers.

Certain aluminum salts exert an astringent or constricting action on the skin and mucous membranes. This property is exploited in many commercial antiperspirant preparations. Substances called *aluminum chlorohydrates* are the principal active agents in many of these formulations. Aluminum chlorohydrates are actually aluminum hydroxy chlorides having formulas of the type $Al_2(OH)_5Cl \cdot 2H_2O$ or $Al(OH)_2Cl$. These substances prevent or retard perspiration by constricting the openings of the sweat glands.

Questions

A. *Review the meanings of the new terms introduced in this chapter.*

1. Alkali metal
2. Alloy
3. Metallurgy
4. Amalgam
5. Alkaline earth metal
6. Steel

B. Review questions.
  1. Explain, in terms of electronic structure, why potassium reacts more violently with water than does lithium.
  2. Why is sodium metal intended for laboratory use usually kept under kerosene?
  3. How would the ionization energy of francium be expected to compare with that of cesium?
  4. Write equations for the oxidation and reduction half-reactions that occur when lithium reacts with hydrogen.
  5. What is oxidized and what is reduced during the electrolysis of potassium hydroxide?
  6. Write the equations for the anode reaction and the cathode reaction for the electrolysis of:
     (a) Molten sodium chloride   (b) An aqueous sodium chloride solution
  7. What is the starting raw material used in the preparation of most sodium chemicals? Why?
  8. Write the formulas of an alkali metal oxide, hydroxide, acetate, bicarbonate, and carbonate using a different metal for each compound.
  9. Which would a farmer be more likely to buy in tonnage quantities—sodium sulfate or potassium sulfate? Justify your answer.
  10. Give several reasons for making alloys.
  11. Give the formula and at least one use for each of the following:
      (a) Caustic soda   (d) Magnesia   (g) Slaked lime
      (b) Chile saltpeter   (e) Plaster of paris (h) Soda ash
      (c) Epsom salts   (f) Quicklime
  12. On the basis of electronic structure explain why a Group IIA metal is less reactive than the Group IA metal that precedes it in the periodic table.
  13. Describe the process and write equations for the principal chemical reactions involved in the preparation of magnesium metal from seawater.
  14. Why can a magnesium fire not be fought successfully with conventional fire extinguishing agents such as water or carbon dioxide? Supplement your answer with appropriate equations.
  15. Explain why it is possible to use barium sulfate, but not barium chloride, as an aid to X-ray photography of the gastrointestinal tract.
  16. Write equations for the following reactions related to the preparation of magnesium:
      (a) The cathode reaction   (c) The overall chemical reaction
      (b) The anode reaction
  17. Briefly discuss the nature and origin of the colored flames that are used for the quantitative detection of certain Group IA and Group IIA elements.
  18. List the three raw materials needed for pig iron production in a blast furnace and explain the role of each.
  19. What is the essential chemical difference between pig iron and steel?
  20. Bare aluminum and magnesium metal structures are not damaged by prolonged exposure to the atmosphere. Yet bare iron, generally less reactive than either aluminum or magnesium, is severely corroded by long exposure to the atmosphere. Explain.
  21. What three general methods are employed to protect iron from corrosion?
  22. Write the equation for the anode reaction that occurs when steel is afforded cathodic protection by magnesium.
  23. Which of the following elements will afford cathodic protection to iron? Explain.
      (a) Aluminum   (b) Copper   (c) Tin   (d) Zinc

24. Outline the Hall–Heroult process for producing aluminum from bauxite ores. Write equations for the pertinent chemical reactions.
25. Why is aluminum more expensive than iron even though aluminum is a more abundant element than iron?
26. In the Hall–Heroult process of aluminum production, oxide ions are oxidized to free oxygen at the anode, but free oxygen is not given off. Explain.
27. When a solution of $AlCl_3$ is reacted with NH_4OH, a white precipitate is formed that does not dissolve in excess NH_4OH. But when NaOH is used, the white precipitate formed dissolves in excess NaOH.
    (a) What is the white precipitate?
    (b) Explain why the two bases behave differently when excess amounts are added to the precipitate.
    (c) Write an equation for the reaction that occurred when excess NaOH was added.
28. At high temperatures, aluminum reacts with nitrogen.
    (a) Write the equation for the reaction.
    (b) Draw the Lewis electron-dot diagram for the compound that is formed.
29. Write formulas for the following substances:
    (a) Magnesium nitride (d) Sodium chrome alum
    (b) Hematite   (e) Aluminum dihydroxy chloride
    (c) Ammonium alum

C. *Review problems.*
1. (a) How many kilograms of sodium chloride must be electrolyzed to produce 5.00 kg of sodium metal?
   (b) How many moles of chlorine are produced along with the 5.00 kg of sodium from part (a)?
2. Sodium hydroxide is produced by electrolyzing an aqueous solution of sodium chloride.
   (a) How many grams of sodium chloride are needed to produce 1.00 lb of NaOH?
   (b) How many litres of Cl_2 (at STP) are produced along with the 1.00 lb of NaOH from part (a)?
3. Seawater contains about 1.3 milligrams of magnesium per litre. How many gallons of seawater must be processed to obtain 1.00 metric ton (1000 kg, or 2200 lb) of magnesium if 85% of the metal can be recovered?
4. Analysis of a pure mineral indicated that it contained 72.4% Fe. Was the mineral hematite, magnetite, or siderite?
5. How many tons of bauxite ore are needed to prepare 100 tons of aluminum if the ore contains 57.5% Al_2O_3?

20 Chemistry of Some Selected Nonmetals

After studying Chapter 20 you should be able to:

1. Understand the terms listed in Question A at the end of the chapter.
2. List the halogens (Group VIIA) and compare the trends in properties from top to bottom within the group with respect to physical state and color, atomic and ionic radii, boiling point, and electronegativity.
3. Identify the halogen that has the highest electronegativity (and first ionization energy) and explain why it has an especially high electronegativity.
4. Explain why the boiling point of hydrogen fluoride is abnormally high when compared to that of other hydrogen halides.
5. Explain why hydrofluoric acid is packed in wax or plastic-lined bottles.
6. Write balanced chemical equations for typical reactions of halogens with (a) metals, (b) hydrogen, and (c) water.
7. Write balanced equations for the preparation of halogens by (a) electrolysis of the appropriate salt and (b) reaction of the appropriate salt (fluorides excepted) with sulfuric acid and manganese dioxide.
8. Explain and illustrate with appropriate equations how chlorine in water acts as a bleaching (and disinfecting) agent.
9. Show with appropriate equations how bromine is obtained from seawater.
10. Describe the role of bromine compounds in gasoline.
11. Explain why pure HBr or HI cannot be prepared by heating NaBr or NaI with concentrated sulfuric acid.
12. Give examples of and explain the formation of interhalogen compounds.
13. Arrange the halogens in order of increasing strength as oxidizing agents and the halide ions in order of increasing strength as reducing agents.
14. Discuss the biological effects or role of free halogens and of halide ions.
15. List the principal source of each halogen.
16. Write balanced chemical equations for the preparation of oxygen from (a) silver oxide, (b) lead(IV) oxide, (c) sodium nitrate, (d) hydrogen peroxide, (e) barium peroxide, and (f) sodium peroxide and water.
17. Describe the Frasch process of mining sulfur.
18. Describe and explain a safe procedure for diluting concentrated sulfuric acid.
19. Describe the contact process for manufacturing sulfuric acid.

20. Write equations illustrating the reactions of sulfuric acid as an (a) acid, (b) dehydrating agent, (c) oxidizing agent, and (d) reagent in the preparation of other acids.
21. Tell which element of Group VA is the most metallic and which is the least metallic.
22. Describe the Haber process for making ammonia and the Ostwald process for making nitric acid.
23. Diagram and discuss the nitrogen cycle.
24. Balance redox equations involving the chemistry of Groups VA, VIA, and VIIA elements.

20.1 The Halogen Family

General Properties

halogen family

The **halogen family**, Group VIIA of the periodic table, consists of five elements: fluorine (F), chlorine (Cl), bromine (Br), iodine (I), and astatine (At). None of these are found as free elements in nature. The term *halogen* is derived from the Greek words *halos* and *gen*, which mean "salt producer." Chlorine, bromine, and iodine are easily prepared in the laboratory; fluorine is much more difficult to prepare. Astatine is a radioactive element found only in very small amounts as a radioactive decay product of other elements.

All the halogen atoms have seven electrons in their outermost shell and are followed, in each case, in the periodic table by a noble gas. Consequently, their most stable oxidation number is -1. All the halogens except fluorine are known to exist in several oxidation states. A halogen atom has a great tendency to fill its outer orbital electron vacancy either by gaining an electron, as in the formation of a sodium halide, Na^+X^- (X = F, Cl, Br, I), or by sharing electrons and forming a covalent bond. In the free state, the halogens exist as diatomic molecules, F_2, Cl_2, Br_2, and I_2.

$$:\!\ddot{F}\!:\!\ddot{F}\!: \quad :\!\ddot{Cl}\!:\!\ddot{Cl}\!: \quad :\!\ddot{Br}\!:\!\ddot{Br}\!: \quad :\!\ddot{I}\!:\!\ddot{I}\!:$$

The differences in reactivity of the halogens are due to differences in their electronegativities. Fluorine, as the most electronegative, has the greatest attraction for electrons and is the most reactive, followed by chlorine, bromine, and iodine, respectively (see Table 20.1).

Since fluorine has the highest electronegativity, it is the most characteristically nonmetallic of all the elements. Although iodine is not a metal, the general trend of the elements to become less electronegative (from top to bottom within a periodic group), and hence more metallic in characteristics, is clearly evident among the halogens. Fluorine and chlorine are gases, bromine is a liquid, and iodine is a solid with a characteristic metallic luster.

Some of the properties of the halogens are tabulated in Table 20.1.

Table 20.1. Physical characteristics of the halogens.

Property	Fluorine	Chlorine	Bromine	Iodine
Molecular formula	F_2	Cl_2	Br_2	I_2
Atomic number	9	17	35	53
Atomic weight	18.998	35.453	79.904	126.904
Electron configuration	2, 7	2, 8, 7	2, 8, 18, 7	2, 8, 18, 18, 7
Atomic radius (Å)	0.72	0.99	1.14	1.33
Ionic (X^-) radius (Å)	1.36	1.81	1.95	2.16
Appearance	Pale yellow gas	Yellow-green gas	Reddish-brown liquid	Greyish-black solid
Density (g/ml)	1.11 (liquid at bp)	1.57 (liquid at bp)	3.14	4.94
Melting point (°C)	−219.6	−102.4	−7.2	113.6
Boiling point (°C)	−187.9	−34.0	58.8	184.5
Relative electronegativity	4.0	3.0	2.8	2.5

Discovery of the Halogens

Because of its extreme reactivity, the discovery of fluorine eluded the efforts of many skilled chemists until 1886, when it was finally prepared by the French chemist Henri Moissan (1852–1907). He prepared fluorine by electrolysis of a solution of potassium hydrogen fluoride (KHF_2) in anhydrous hydrogen fluoride.

Chlorine was first prepared in 1774 by Karl W. Scheele (1742–1786), a Swedish chemist, who reacted hydrochloric acid with pyrolusite (MnO_2). The substance that Scheele prepared and named *dephlogistinated marine acid* was believed to be a compound until 1810, when the English scientist Sir Humphry Davy (1778–1829) gave positive proof that it was an element.

$$MnO_2(s) + 4\ HCl(aq) \longrightarrow Cl_2\uparrow + MnCl_2(aq) + 2\ H_2O$$

Antoine Jérôme Balard (1802–1876) was the first to announce, in 1826, that he had prepared the element bromine. He obtained it by passing chlorine into naturally occurring brines and distilling the bromine that was formed. Other workers had previously prepared bromine, but had either failed to recognize it as a new element or did not make an official report of their experiments.

Iodine was discovered in 1811 by Bernard Courtois (1777–1838). While treating the saline extract from the ashes of brown sea algae with sulfuric acid, Courtois first noticed the beautiful violet vapors of iodine rising from the reaction.

The names of the halogens were derived from the following sources: fluorine, from the Latin *fluere*, meaning "to flow"; chlorine, from the Greek *chloros*, meaning "light green"; bromine, from the Greek *bromos*, meaning "bad smell" or "stench"; iodine, from the Greek *iodes*, meaning "violet-like."

Fluorine

The principal minerals in which fluorine is found are fluorspar (fluorite), CaF_2, and cryolite, Na_3AlF_6. The bones and teeth of mammals contain fluorine, and the element is also present as a trace quantity in some natural waters. According to some authorities, fluorine is the thirteenth most abundant element in the earth's crust, occurring there to the extent of 0.078%.

Elemental fluorine was not manufactured in the United States until World War II, when it became needed in order to separate U-235 and U-238 isotopes in the development of the atomic bomb. Today fluorine is made by the electrolysis of a mixture of liquid anhydrous hydrogen fluoride and potassium fluoride, the latter acting as the electrolyte. One of the biggest problems encountered in preparing and handling fluorine is its corrosive characteristic; elemental fluorine attacks all metals. Fortunately, the metal vessels used in the electrochemical cells form a metal fluoride coating that protects the cell from further corrosion.

Fluorine is the most reactive of all the halogens and the most reactive nonmetal, having an electronegativity greater than oxygen. Nearly all the elements, including some of the noble gases, form compounds with fluorine. The most reactive metals react with fluorine at room temperature and the less reactive ones require higher temperatures. There are two known oxides: oxygen difluoride (OF_2) and dioxygen difluoride (O_2F_2). Three silver compounds—silver subfluoride (Ag_2F), silver fluoride (AgF), and silver difluoride (AgF_2)—and two copper salts—copper(I) fluoride (CuF) and copper(II) fluoride (CuF_2)—have been synthesized. Sulfur hexafluoride (SF_6) and nitrogen trifluoride (NF_3) have exceptional stability and are difficult to decompose. Antimony trifluoride (SbF_3), cobalt trifluoride (CoF_3), and silver difluoride (AgF_2) are important compounds used in the fluorination of other substances. Boron trifluoride (BF_3) is a catalyst commonly used in many organic chemistry reactions. Illustrating its extreme reactivity, solid fluorine reacts explosively with liquid hydrogen even at $-252°C$, forming hydrogen fluoride:

$$H_2(g) + F_2(g) \longrightarrow 2\,HF(g)$$

When fluorine reacts with water, hydrogen fluoride, oxygen difluoride, hydrogen peroxide, and oxygen are formed:

$$5\,F_2 + 5\,H_2O \longrightarrow 8\,HF + OF_2 + H_2O_2 + O_2$$

This reaction shows that fluorine is a very strong oxidizing agent.

Hydrogen fluoride is the most important commercial product of fluorine. It is a colorless gas, having a boiling point of 19.4°C and a melting point of $-83°C$. Hydrogen fluoride is dangerous to inhale, since it causes edema of the lungs; the eyes are also quite sensitive to this compound. The gas is very soluble in water, forming a solution of hydrofluoric acid, a weak acid. The industrial source of hydrogen fluoride is fluorspar and sulfuric acid:

$$CaF_2(s) + H_2SO_4(conc.) \longrightarrow 2\,HF\uparrow + CaSO_4(s)$$

Although fluorine has properties typical of the halogens, many of its properties are unique. In fact, the first member of each family of elements has some properties that are markedly atypical of that family.

Earlier we described hydrogen bonding, the ability of highly electronegative atoms (nitrogen, oxygen, and fluorine) to attract and form a weak bond with hydrogen atoms already bonded to another element. Fluorine is the only

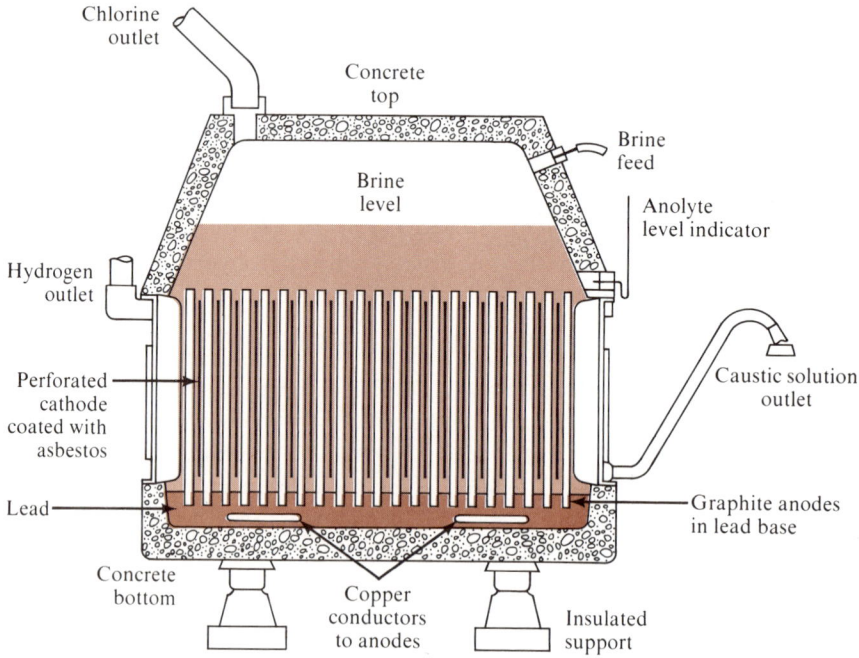

Figure 20.1. Hooker electrochemical cell for manufacturing chlorine from sodium chloride brine. The cell operates continuously and can produce over 2000 lb of chlorine daily. The chlorine, which is about 97.5% pure when it leaves the cell, is washed, dried, liquefied by refrigeration, and stored or delivered by tank cars. Sodium hydroxide and hydrogen are equally important products of this process. (Courtesy Hooker Chemical Corporation.)

halogen sufficiently electronegative to form hydrogen bonds. In the gaseous state, hydrogen fluoride molecules are bonded together by hydrogen bonds and exist in polymeric forms, $(HF)_x$, where x has values ranging up to 6.

$$6\ HF \rightleftharpoons H_6F_6$$

Like water, the abnormally high boiling point of hydrogen fluoride (bp 19.4°C), in comparison with the other hydrogen halides (for example, bp of HI = −50.6°C), is indicative of a much heavier substance than single hydrogen fluoride molecules.

Another unique property of hydrogen fluoride is its ability to etch glass. The art of etching glass with hydrogen fluoride was practiced as early as the 16th century. The chemical reaction that takes place is the dissolving of silicon dioxide in glass, forming silicon tetrafluoride:

$$SiO_2(s) + 4\ HF(g) \longrightarrow SiF_4\uparrow + 2\ H_2O$$

The enamel and dentine of one's teeth contain small amounts of fluorine. The presence of fluoride ions is necessary for the proper development of growing children's teeth. However, too much fluoride can cause mottled teeth. Sodium fluoride is used to supplement water containing little or no fluoride.

Other important uses of fluorine compounds are as refrigerants and plastics. Dichlorodifluoromethane (CCl_2F_2), known as Freon-12, is the commonly used refrigerant in household refrigerators and air conditioners. Polytetrafluoroethylene, $(C_2F_4)_n$, known as Teflon, is a plastic material with high resistance to chemical attack, good temperature stability, and self-lubricating qualities.

Chlorine

Chlorine is prepared commercially by electrolysis of either aqueous sodium chloride solution or molten sodium chloride. Figure 20.1 illustrates the Hooker cell, which is used to manufacture chlorine from brines.

To prepare chlorine in the laboratory, hydrochloric acid or a sodium chloride–sulfuric acid mixture is reacted with manganese dioxide:

$$4\ HCl(aq) + MnO_2(s) \longrightarrow MnCl_2(aq) + 2\ H_2O + Cl_2\uparrow$$
$$2\ NaCl(s) + 2\ H_2SO_4(aq) + MnO_2(s) \longrightarrow MnSO_4(aq) + Na_2SO_4(aq) + 2\ H_2O + Cl_2$$

A dilute potassium chlorate solution and hydrochloric acid also liberate chlorine, when heated gently:

$$KClO_3(aq) + 6\ HCl(aq) \xrightarrow{\Delta} KCl(aq) + 3\ H_2O + 3\ Cl_2\uparrow$$

Figure 20.2 shows a laboratory setup for preparing and collecting dry chlorine. Chlorine is about 2.5 times as heavy as air and is collected by the upward displacement of air.

Chlorine is quite reactive, combining directly or indirectly with many metals to form metal chlorides. All the common metal chlorides, except silver, lead, and mercury(I), are soluble in water. This property is used to separate $AgCl$, $PbCl_2$, and $HgCl_2$ from other metals for purposes of identification

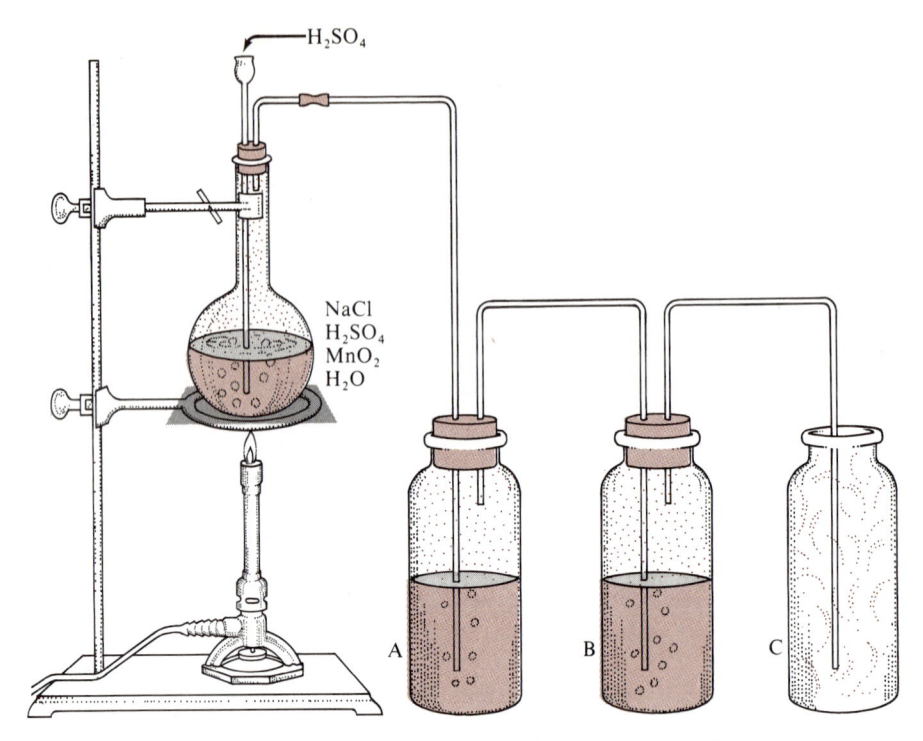

Figure 20.2. Laboratory preparation of chlorine: Reactants in the generator are sodium chloride, sulfuric acid, manganese dioxide, and water. Bottle A contains water; it is used to remove any hydrogen chloride carried over with the product. Bottle B contains concentrated sulfuric acid, which functions as a drying agent, removing moisture from the product. Dry chlorine gas is collected in bottle C.

With nonmetals, chlorine forms compounds that have covalent bonds. A few of these are listed below:

S_2Cl_2	Sulfur monochloride	BCl_3	Boron trichloride
Cl_2O	Dichlorine monoxide	CCl_4	Carbon tetrachloride
NCl_3	Nitrogen trichloride	$SiCl_4$	Silicon tetrachloride
PCl_3	Phosphorus trichloride	ICl	Iodine monochloride
PCl_5	Phosphorus pentachloride	C_2Cl_4	Tetrachloroethylene

Table 20.2 lists the oxides of chlorine.

Table 20.2. Oxides of chlorine.

Oxide	Formula	Oxidation number of Cl	Melting point (°C)	Boiling point (°C)
Dichlorine monoxide	Cl_2O	+1	−116	2
Dichlorine trioxide	Cl_2O_3	+3	—	—
Chlorine dioxide	ClO_2	+4	−59	11
Dichlorine hexoxide	Cl_2O_6	+6	3.5	203
Dichlorine heptoxide	Cl_2O_7	+7	−91.5	80

One of the most important compounds of chlorine is hydrogen chloride (HCl), which is prepared by combining hydrogen and chlorine or by reacting sodium chloride, or other metal chlorides, with sulfuric acid. Hydrogen burns smoothly in an atmosphere of pure chlorine, but mixtures of hydrogen and chlorine ranging from 14 to 92% chlorine are considered to be explosive.

$$H_2(g) + Cl_2(g) \longrightarrow 2\,HCl(g) + 44.2\,kcal$$

At low temperatures, sodium chloride and sulfuric acid react to form sodium hydrogen sulfate and hydrogen chloride. In commercial processes, the reaction is run at high temperature in order to use the second hydrogen of sulfuric acid. The salt thus formed is sodium sulfate.

$$NaCl + H_2SO_4 \xrightarrow{\text{Low temp.}} NaHSO_4 + HCl\uparrow$$

$$\underline{NaCl + NaHSO_4 \xrightarrow{\text{High temp.}} Na_2SO_4 + HCl\uparrow}$$

$$2\,NaCl + H_2SO_4 \longrightarrow Na_2SO_4 + 2\,HCl\uparrow \quad \text{(Overall reaction)}$$

Hydrogen chloride is very soluble in water, forming hydrochloric acid. The concentrated acid contains 36 to 38% HCl and has a density of about 1.19 g/ml. A demonstration of the solubility of hydrogen chloride in water is illustrated in Figure 20.3, the hydrogen chloride fountain. The top flask contains dry hydrogen chloride; the bottom flask, water. When a small amount of water is squirted into the top flask by the attached medicine dropper, hydrogen chloride dissolves in this water, creating a partial vacuum. The water from the

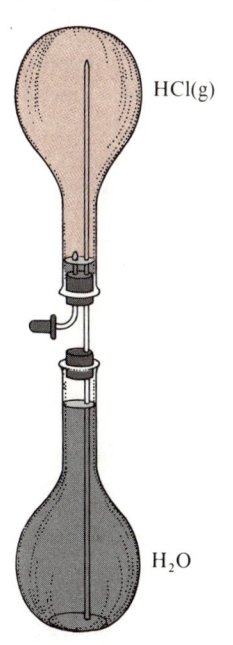

Figure 20.3. Hydrogen chloride fountain: Water from the lower flask rises into the upper flask because of the vacuum created when hydrogen chloride gas dissolves in water. The reaction is initiated by priming with a small amount of water from the attached medicine dropper.

bottom flask is then continuously drawn up into the top flask in a fountainlike effect as the hydrogen chloride dissolves.

Technical-grade hydrochloric acid is called *muriatic acid* and is usually yellow in color because of a small amount of dissolved ferric ion. Hydrochloric acid is not only used as an acid, but also for preparing other chlorides and for "pickling" (cleaning) metal surfaces. The acidity of human gastric juice is due to hydrochloric acid, which is a necessary aid in the digestion of foods.

Chlorine is a powerful oxidizing agent; because of its germicidal action, it is used in the purification of water. Sodium hypochlorite (NaOCl), an important household and commercial bleach and disinfectant, is prepared by reacting chlorine and sodium hydroxide below room temperature. At higher temperatures, sodium chlorate (NaClO$_3$) is formed.

$$Cl_2(g) + 2\,NaOH(aq) \xrightarrow{\text{Low temp.}} NaOCl(aq) + NaCl(aq) + H_2O$$

$$3\,Cl_2(g) + 6\,NaOH(aq) \xrightarrow{\text{High temp.}} NaClO_3(aq) + 5\,NaCl(aq) + 3\,H_2O$$

Potassium hydroxide and chlorine behave in the same way. Sodium chlorate is used as an oxidizing agent in the generation of chlorine dioxide and in the manufacture of explosives, dyes, and matches.

Chlorine dissolved in water is commonly called *chlorine water*. Its solubility is about 1% by weight at 10°C. Part of the dissolved chlorine reacts with water, producing hypochlorous and hydrochloric acids:

$$Cl_2(g) + H_2O \rightleftharpoons HOCl(aq) + HCl(aq)$$

The behavior of chlorine water as a bleaching agent is due to the decomposition of unstable hypochlorous acid, which liberates *nascent* (atomic) oxygen (O), a powerful oxidant. The oxygen atom either attacks the substance being bleached or combines with another atom to form an oxygen molecule. In the reaction, the chlorine atom is reduced from an oxidation number of $+1$ to -1.

$$HOCl(aq) \longrightarrow HCl(aq) + (O)$$

An acidified sodium hypochlorite solution reacts similarly to chlorine water and is used as a source of chlorine for purifying swimming pools. The equation for the reaction is

$$NaOCl(aq) + HCl(aq) \longrightarrow NaCl(aq) + HOCl(aq)$$

Chlorine is less reactive than fluorine but more reactive than bromine and iodine. It oxidizes bromides and iodides to the free elements:

$$Cl_2(g) + 2\,NaBr(aq) \longrightarrow Br_2(l) + 2\,NaCl(aq)$$
$$Cl_2(g) + 2\,NaI(aq) \longrightarrow I_2(s) + 2\,NaCl(aq)$$

Thousands of organic compounds contain chlorine. The total amount of chlorine used annually in the United States amounts to more than 10 million tons, and the demand is still increasing. Organic chlorine compounds are used for refrigeration, gasoline additives, insecticides, pesticides, solvents, paints, and plastics.

Bromine

Bromine is the only nonmetallic element that is liquid at ordinary temperatures. Its deposits are widely scattered throughout the earth, with the highest concentrations located at Stassfurt, Germany; at Searles Lake, California; and in the Dead Sea brines. The major source of bromine is the ocean, where the concentration is about 67 mg/litre. Less than 0.1 g/litre seems like a small amount, until one calculates that the bromine in 1 cubic mile of seawater amounts to 308,000 tons. Currently, more than 200,000 tons of bromine are produced annually.

The bromine industry has grown greatly since the discovery of tetraethyl lead, used in gasoline. Ethylene dibromide ($C_2H_4Br_2$) is used in conjunction with tetraethyl lead in gasoline to keep lead from depositing on automobile cylinders and spark plugs. Over 90% of the bromine produced goes into the production of ethylene dibromide.

Bromine may be prepared in the laboratory by oxidizing a bromide–sulfuric acid mixture with manganese dioxide (see Figure 20.4):

$$2\,KBr + 2\,H_2SO_4 + MnO_2 \longrightarrow Br_2 + K_2SO_4 + MnSO_4 + 2\,H_2O$$

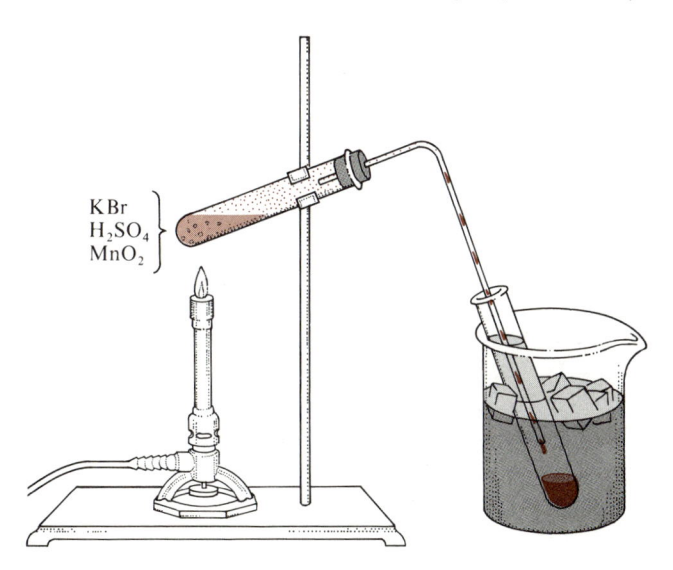

Figure 20.4. Laboratory preparation of bromine. Potassium bromide, sulfuric acid, and manganese dioxide are heated together. The bromine distilling out of the generator is collected in cold water as a liquid.

$$2\,KBr + 2\,H_2SO_4 + MnO_2 \longrightarrow Br_2 + K_2SO_4 + MnSO_4 + 2\,H_2O$$

Bromine is manufactured commercially by passing chlorine into large volumes of seawater. The bromine formed is removed by bubbling a stream of air through the chlorinated seawater. The air–bromine mixture is then passed into a sodium carbonate solution, where the bromine reacts to form sodium bromide and sodium bromate. Bromine is again formed when this mixture is acidified and heated with sulfuric acid. The equations are

$$Cl_2 + 2\,Br^- \longrightarrow Br_2 + 2\,Cl^-$$
$$3\,Br_2 + 6\,Na_2CO_3 + 3\,H_2O \longrightarrow 5\,NaBr + NaBrO_3 + 6\,NaHCO_3$$
$$5\,NaBr + NaBrO_3 + 3\,H_2SO_4 \longrightarrow 3\,Br_2 + 3\,Na_2SO_4 + 3\,H_2O$$

The chemical reactions of bromine are similar to those of chlorine, except that bromine is not as strong an oxidizing agent. With water, it forms hypobromous and hydrobromic acids:

$$Br_2(l) + H_2O \rightleftharpoons \underset{\text{Hypobromous acid}}{HOBr(aq)} + \underset{\text{Hydrobromic acid}}{HBr(aq)}$$

These two acids are colorless, but *bromine water* appears reddish-brown because of dissolved Br_2 in solution. The equations for the reactions with sodium hydroxide are:

$$Br_2 + 2\,NaOH \xrightarrow{\text{Low temp.}} NaOBr + NaBr + H_2O$$
$$3\,Br_2 + 6\,NaOH \xrightarrow{\text{High temp.}} 5\,NaBr + NaBrO_3 + 3\,H_2O$$

Bromine is more reactive than iodine and oxidizes iodides in solution to free iodine:

$$Br_2 + 2\,I^- \longrightarrow I_2 + 2\,Br^-$$

Pure hydrogen bromide cannot be made from bromide salts and sulfuric acid, because hydrogen bromide is a fairly good reducing agent and reacts further with the acid to form sulfur dioxide and bromine:

$$2\,NaBr + H_2SO_4 \longrightarrow Na_2SO_4 + 2\,HBr$$
$$2\,HBr + H_2SO_4 \longrightarrow Br_2 + SO_2 + 2\,H_2O$$

Evidence of bromine formation is observed from its reddish-brown color. Hydrogen bromide can be prepared from (1) hydrogen and bromine, (2) sodium bromide and phosphoric acid, or (3) phosphorus tribromide and water. The last two are convenient laboratory methods.

$$H_2(g) + Br_2(g) \longrightarrow 2\,HBr(g) + 17.4\text{ kcal} \tag{1}$$
$$3\,NaBr + H_3PO_4 \longrightarrow Na_3PO_4 + 3\,HBr \tag{2}$$
$$PBr_3 + 3\,H_2O \longrightarrow \underset{\substack{\text{Phosphorous}\\\text{acid}}}{H_3PO_3} + 3\,HBr \tag{3}$$

The hydrogen bromide dissolves in water to form hydrobromic acid, which is a strong acid.

Although it cannot compete with chlorine in price, bromine is a good disinfectant and bleaching agent. Some of its compounds are used as pharmaceuticals (for example, mercurochrome), fumigants, tear gas, refrigerants, dyes, and photographic emulsions.

Precautions must be taken when working with bromine, because its fumes are poisonous and suffocating and it produces painful sores when spilled on the skin.

Iodine

Iodine is a bluish-black solid that forms violet vapors when heated. Compounds of iodine are found in seawater, where they are selectively absorbed and concentrated by certain types of seaweed (kelp). The first commercial source was from ashes of seaweed. The main commercial source of iodine in the United States is from brine wells in Michigan. The world's major supply comes from Chile, where it occurs in sodium nitrate deposits as sodium iodate ($NaIO_3$). To obtain iodine, the salt deposit is first leached with hot water to form a concentrated brine solution. Sodium nitrate is then crystallized from this brine. Iodine is obtained by treating the remaining solution, which contains the more soluble sodium iodate, with sodium hydrogen sulfite:

$$2\,NaIO_3 + 5\,NaHSO_3 \longrightarrow I_2 + 3\,NaHSO_4 + 2\,Na_2SO_4 + H_2O$$

Oil well brines are treated with chlorine to liberate iodine:

$$Cl_2 + 2\,I^- \longrightarrow I_2 + 2\,Cl^-$$

In the laboratory, iodine is easily prepared and collected by heating a mixture containing sodium iodide, sulfuric acid, and manganese dioxide. Figure 20.5 shows this mixture being heated in a beaker. The iodine vapors are condensed and collected on the chilled porcelain dish covering the beaker.

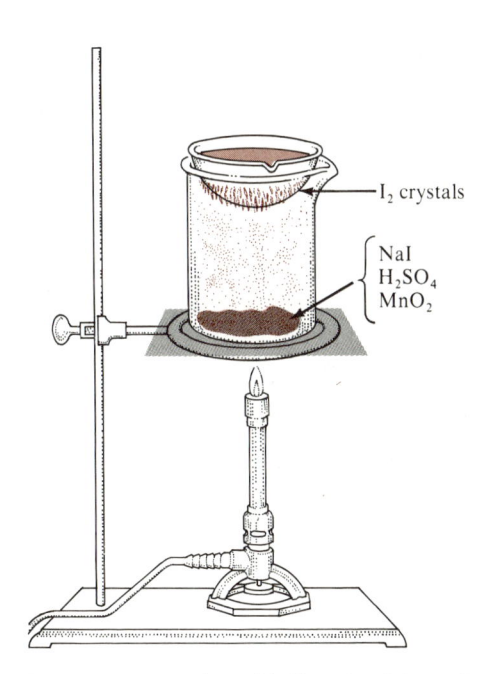

$$2\,NaI + 2\,H_2SO_4 + MnO_2 \longrightarrow I_2 + Na_2SO_4 + MnSO_4 + 2\,H_2O$$

Figure 20.5. Laboratory preparation of iodine. A mixture of sodium iodide, sulfuric acid, and manganese dioxide is gently heated in a beaker. Iodine is formed and vaporizes. Beautiful violet vapors of iodine leave the mixture and condense as crystals on the cooled porcelain dish.

Chemically, iodine is less reactive than the other three halogens, but it still combines with most of the elements, both metals and nonmetals, to form iodides:

$$2\,Na(s) + I_2(s) \longrightarrow 2\,NaI(s)$$
$$2\,Sb(s) + 3\,I_2(g) \longrightarrow 2\,SbI_3(s)$$
$$2\,P(s) + 3\,I_2(s) \longrightarrow 2\,PI_3(s)$$

Pure hydrogen iodide cannot be prepared from sodium iodide and sulfuric acid because of a secondary reaction between hydrogen iodide and the acid, forming iodine and hydrogen sulfide.

$$NaI + H_2SO_4 \longrightarrow HI + NaHSO_4$$
$$\underline{8\,HI + H_2SO_4 \longrightarrow 4\,I_2 + H_2S + 4\,H_2O}$$
$$8\,NaI + 9\,H_2SO_4 \longrightarrow 4\,I_2 + H_2S + 4\,H_2O + 8\,NaHSO_4 \quad \text{(Overall reaction)}$$

Hydrogen iodide is usually made from phosphorus triiodide and water:

$$PI_3 + 3\,H_2O \longrightarrow H_3PO_3 + 3\,HI\uparrow$$

The reaction of hydrogen and iodine is endothermic.

$$H_2 + I_2 + 12.4\text{ kcal} \longrightarrow 2\,HI\uparrow$$

Hydrogen iodide is very soluble in water and forms a strong acid solution known as hydriodic acid. The iodide ion in an acid medium is a strong reducing agent. For example, hydrogen iodide reduces sulfuric acid to hydrogen sulfide; the sulfur atom being reduced from $+6$ to -2. Although the copper(II) ion, Cu^{2+}, is not normally considered to be a strong oxidizing agent, iodide ions will reduce it to the copper(I) state, forming copper(I) iodide and iodine, as shown in the equation below. This type of reaction does not occur with the other halide ions.

$$2\,Cu^{2+} + 4\,I^- \longrightarrow Cu_2I_2\downarrow + I_2$$

Iodine is only slightly soluble in water, but it is very soluble in organic solvents such as alcohol, carbon tetrachloride, and carbon disulfide. The brown alcohol solution is known as *tincture* of iodine, and is used as an antiseptic and disinfectant for cuts and scratches. The carbon tetrachloride and carbon disulfide solutions are violet. Iodine is soluble in aqueous potassium iodide because of the formation of the soluble complex triiodide ion, I_3^-. The solution is labeled "I_2 in KI" and is brown in color.

$$I_2(s) + KI(aq) \longrightarrow KI_3(aq)$$
$$I_2(s) + I^- \longrightarrow I_3^-(aq)$$

An intense blue complex is formed when iodine is added to a starch suspension. This iodine–starch color is used as a very sensitive test for iodine; it is a useful indicator that allows iodine to be used in the analysis of other substances.

The thyroid gland is richer in iodine than any other part of the human body. A deficiency of iodine causes this gland to enlarge, creating the condition known as goiter. Iodine and iodides have been used for many years as an effective treatment for goiter. Iodine in the regular diet may be supplemented by iodized salt, which contains a small amount of a soluble iodide.

Silver iodide (AgI) is very sensitive to light; it is used in high-speed photographic films. In general, iodine compounds are very reactive, but their widespread use has been limited by their high cost.

Interhalogen Compounds

The ability of halogen atoms to share electrons raises the possibility of two different kinds of halogen atoms combining to form a molecule. All the possible interhalogen combinations containing one atom of each halogen are known, except for iodine monofluoride (IF). They are chlorine monofluoride (ClF), bromine monofluoride (BrF), bromine monochloride (BrCl), iodine monochloride (ICl), and iodine monobromide (IBr). Because of different electronegativities, the bonding electrons are not equally shared between the atoms in a molecule. For example, in ICl, the chlorine atom is slightly more negative than the iodine atom; hence, chlorine is assigned an oxidation number of -1. Some of these compounds are quite stable and easy to prepare. Iodine monochloride may be prepared by passing chlorine gas into a weighed quantity of iodine until the calculated (theoretical) weight increase has been attained:

$$I_2(s) + Cl_2(g) \longrightarrow 2\ ICl(l)$$

With continued addition of chlorine, iodine trichloride is formed:

$$ICl(l) + Cl_2(g) \longrightarrow ICl_3(s)$$

Other known interhalogen compounds are chlorine trifluoride (ClF_3), bromine trifluoride (BrF_3), bromine pentafluoride (BrF_5), iodine pentafluoride (IF_5), and iodine heptafluoride (IF_7). The chemistry of the interhalogens containing only two atoms is very similar to that of the free halogens. For example:

$$BrCl(l) + H_2O \rightleftharpoons HOBr(aq) + HCl(aq)$$

Comparison of Halogen Reactivities

This section summarizes some of the properties of the halogens and their compounds by comparing their relative reactivities.

Oxidizing strength. Fluorine is capable of oxidizing chlorides, bromides, and iodides to the free halogen. Chlorine will oxidize both bromides and iodides, and bromine can oxidize iodides. The order of oxidizing strength, therefore, is $F_2 > Cl_2 > Br_2 > I_2$. This order is in agreement with their respective electronegativities.

$$F_2 + 2\,Cl^- \longrightarrow Cl_2 + 2\,F^-$$
$$Cl_2 + 2\,Br^- \longrightarrow Br_2 + 2\,Cl^-$$
$$Br_2 + 2\,I^- \longrightarrow I_2 + 2\,Br^-$$

The oxidizing strengths of chlorate, bromate, and iodate ions also agree with this order ($ClO_3^- > BrO_3^- > IO_3^-$). The equations are

$$2\,ClO_3^- + Br_2 \longrightarrow 2\,BrO_3^- + Cl_2$$
$$2\,ClO_3^- + I_2 \longrightarrow 2\,IO_3^- + Cl_2$$
$$2\,BrO_3^- + I_2 \longrightarrow 2\,IO_3^- + Br_2$$

Reducing strength. The relative ability of an ion or molecule to lose electrons and become oxidized is a measure of its reducing strength. From the fact that oxidation and reduction are opposite phenomena, with respect to electrons, we can conclude that a strong oxidizing agent would not likely be a strong reducing agent. We can deduce from this that the order of reducing strength of the halogens should be the reverse of their oxidizing strength. As reducing agents, the order is $I^-\ \ Br^- > Cl^- > F^-$. This order may be proven by a few selective reactions.

Nitrous acid is a very mild oxidant that will be reduced by hydriodic acid and none of the other hydrogen halides.

$$2\,HNO_2(aq) + 2\,HI(aq) \longrightarrow I_2(s) + 2\,NO\uparrow + 2\,H_2O$$

$$HF,\ HCl,\ \text{or}\ HBr + HNO_2 \longrightarrow \text{No reaction}$$

Concentrated sulfuric acid is reduced by hydrobromic acid (HBr) and hydriodic acid (HI), but not by hydrochloric acid (HCl) and hydrofluoric acid (HF). Nitric acid will oxidize HCl, HBr, and HI, but not HF.

$$6\,HCl(aq) + 2\,HNO_3(aq) \longrightarrow 3\,Cl_2\uparrow + 2\,NO\uparrow + 4\,H_2O$$

Acidity of hydrohalic acids. Hydrochloric, hydrobromic, and hydriodic acids are all strong acids. Hydrofluoric acid is a weak acid. A 1 M solution of hydrofluoric acid is between 2 and 3% ionized.

Reaction of halogens with water. Fluorine reacts violently with water, forming as products H_2O_2, OF_2, HF, and O_2. Chlorine and bromine react with water to form HOCl + HCl and HOBr + HBr, respectively. Iodine is only slightly soluble in water but readily dissolves in potassium iodide solution because of the formation of the soluble triiodide ion I_3^-.

Oxy-acids of the halogens. The oxy-acids of the halogens are listed in Table 20.3. No oxy-acids of fluorine have been isolated. In general, the salts are much more stable than the acids.

Table 20.3. Oxy-acid of the halogens.

Fluorine	Bromine	Iodine	Chlorine	Name of chlorine acid	Name and formula of sodium salt	
—	HBrO	HIO	HClO	Hypochlorous acid	NaClO	Sodium hypochlorite
—	—	—	HClO$_2$	Chlorous acid	NaClO$_2$	Sodium chlorite
—	HBrO$_3$	HIO$_3$	HClO$_3$	Chloric acid	NaClO$_3$	Sodium chlorate
—	—	H$_5$IO$_6$	HClO$_4$	Perchloric acid	NaClO$_4$	Sodium perchlorate

Chemistry of Some Selected Nonmetals 427

Solubility of silver halides. These data point out, again, the uniqueness of fluorine in contrast to the other halogens.

Halide	Solubility (g/100 ml H_2O)
AgF	182 (15°C)
AgCl	8.9×10^{-5} (20°C)
AgBr	2×10^{-5} (20°C)
AgI	3×10^{-7} (20°C)

Biological effects. As free elements, all of the halogens are toxic; fluorine being the most dangerous and iodine the least dangerous of the group.

Many halogen compounds are known to have specific physiological or biological effects in nutrition and medicine. Our diets must contain certain amounts of chloride and iodide ions for normal health. Fluoride ions are needed for sound tooth structure, but are definitely harmful when excessive amounts are consumed. Bromide is known to be present in animal tissue, but its function is not definitely established.

20.2 The Sulfur Family

sulfur family

The elements of Group VIA in the periodic table are known as the **sulfur family**. This family includes oxygen, sulfur, selenium, tellurium, and polonium.

Oxygen

Oxygen is the most abundant and widespread of all the elements on earth. It occurs both as free oxygen gas and in combined forms with other elements. There are three naturally occurring isotopes of oxygen—$^{16}_{8}O$, $^{17}_{8}O$, and $^{18}_{8}O$—of which $^{16}_{8}O$ is by far the most abundant. The oxygen molecule is diatomic. The following electron-dot structures are frequently used to represent this molecule:

$$:\ddot{O}:\ddot{O}: \quad \text{and} \quad :\ddot{O}::\ddot{O}:$$

Since these structures do not adequately account for the bonding between oxygen atoms, additional bonding theories have been advanced. These theories, however, are quite complex and are beyond the scope of this discussion.

Oxygen is found combined with more elements than any other single element, and it will form compounds with all the elements except some of the noble gases. Water is 88.9% oxygen; the human body is 60% oxygen by weight. The atmosphere contains about 21% oxygen by volume and about 23% oxygen by weight. Table 20.4 shows the composition of normal air.

Table 20.4. Gaseous components that are always present in normal air.

Constituent	Concentration
N_2	78.08 volume percent
O_2	20.95 volume percent
Ar	0.93 volume percent
CO_2	0.03 volume percent
Ne	18.18 parts per million by volume
He	5.24 parts per million by volume
Kr	1.14 parts per million by volume
Xe	0.09 parts per million by volume
H_2	0.5 parts per million by volume
CH_4	2 parts per million by volume
N_2O	0.5 parts per million by volume
H_2O (vapor)	Variable parts per million by volume

Oxygen is a colorless, odorless, tasteless gas, with a boiling point of −183°C and a melting point of −218.4°C. As a liquid or solid, it has a pale blue color. Liquid oxygen is *paramagnetic*; that is, it is attracted by a magnet. Oxygen gas is slightly heavier than air. Its solubility in water is 0.03 ml of gas per gram of water at 20°C, a factor very important in sustaining marine life.

Preparation of oxygen. We have seen that oxygen can be prepared by heating mercuric oxide or potassium chlorate and by the electrolysis of water. Oxygen can also be prepared from certain other oxygen-containing compounds, several examples of which are given below. Some oxygen-containing compounds such as magnesium oxide, copper(II) oxide, and zinc oxide are very stable and do not liberate oxygen when heated.

$$2\,Ag_2O(s) \xrightarrow{160°C} 4\,Ag(s) + O_2\uparrow$$

$$3\,MnO_2(s) \xrightarrow{950°C} Mn_3O_4(s) + O_2\uparrow$$

$$2\,PbO_2(s) \xrightarrow{300°C} 2\,PbO(s) + O_2\uparrow$$

$$2\,NaNO_3(s) \xrightarrow{380°C} 2\,NaNO_2(s) + O_2\uparrow$$

Oxygen is also readily obtained by the decomposition of peroxides.

$$2\,Na_2O_2(s) + 2\,H_2O \longrightarrow 4\,NaOH(s) + O_2\uparrow + Heat$$

$$2\,H_2O_2(aq) \xrightarrow{\Delta} 2\,H_2O + O_2\uparrow$$

$$2\,BaO_2(s) \xrightarrow{\Delta} 2\,BaO(s) + O_2\uparrow$$

Oxygen is obtained commercially from air. Air is first freed of carbon dioxide and moisture and then liquefied to give a mixture that is essentially liquid oxygen and nitrogen. Oxygen is separated from nitrogen by fractional distillation; nitrogen (bp −195.8°C), which has a lower boiling temperature than oxygen, distills off first. Oxygen prepared by this process is about 99.5% pure. The U.S. production of oxygen in 1978 was over 17 million tons.

One of the outstanding properties of oxygen is its ability to support combustion. By **combustion** we generally mean the act of burning or, commonly, the union of oxygen with other substances accompanied by the evolution of light and heat. Oxygen is consumed during combustion but does not burn. We are highly dependent on heat, electrical, and mechanical energy in our daily lives. Most of this energy is obtained from combustion of fossil fuels.

Animal life as we know it is dependent on oxygen. Absorbed in the blood, oxygen is carried by way of the arteries to all tissues of the body, where it is used in metabolism. Metabolism includes the oxidation of carbohydrates, fats, and proteins and provides the energy needed to carry on the normal life processes.

Certain additional aspects of oxygen, including peroxides and allotropic forms, have been discussed in Chapter 13.

Sulfur

Occurrence and physical properties. Sulfur has been known for thousands of years; evidence of its use dates back 16 to 20 centuries B.C. It is mentioned in the Bible, where it is called *brimstone*, a name also sometimes used today. Much mysticism surrounded the use of sulfur in the early days when priests burned sulfur in their religious ceremonies. Burning sulfur was also used in medieval times to purge and to purify rooms contaminated by disease and sick people. The Chinese used sulfur in their early gunpowder mixtures.

Sulfur occurs both in the free state and in combination with other elements. Some of its naturally occurring compounds include iron pyrites and other metal sulfides, sulfate minerals, and hydrogen sulfide gas. It is also found in petroleum deposits. An essential element in life processes, sulfur is present in living matter; in the blood; in some amino acids of proteins; and in eggs, milk, and other products of living organisms.

Sulfur is found principally in Sicily, Louisiana, Texas, Canada, and volcanic regions of Japan and Mexico. Sulfur is mined in the United States by the Frasch process illustrated in Figure 20.6. The world production of sulfur in 1900 was about 2–2.5 million tons; at present, more than 12 million tons of sulfur are produced in the United States annually.

Sulfur is widely used in many phases of agriculture and industry. More than four-fifths of all sulfur produced is made into sulfuric acid. Sulfur is used in other phases of the chemical industry, in agriculture as a fertilizer, in pesticides, in the steel industry, in petroleum refining, in rubber vulcanizing, in paper and pulp manufacturing, and in producing life-saving sulfa drugs, to mention only a few of its widespread applications.

Sulfur is a bright yellow, odorless solid. It occurs as four stable isotopes—$^{32}_{16}S$, $^{33}_{16}S$, $^{34}_{16}S$, and $^{36}_{16}S$—and several artificial radioisotopes. The S-32 isotope accounts for about 95% of natural sulfur. The atomic number of sulfur is 16; its electronic structure is $1s^2 2s^2 2p^6 3s^2 3p^4$, with six electrons in the outer shell, like oxygen. In the solid form, sulfur exists in a staggered ring structure containing eight atoms connected by covalent bonds. The molecular aggregate is

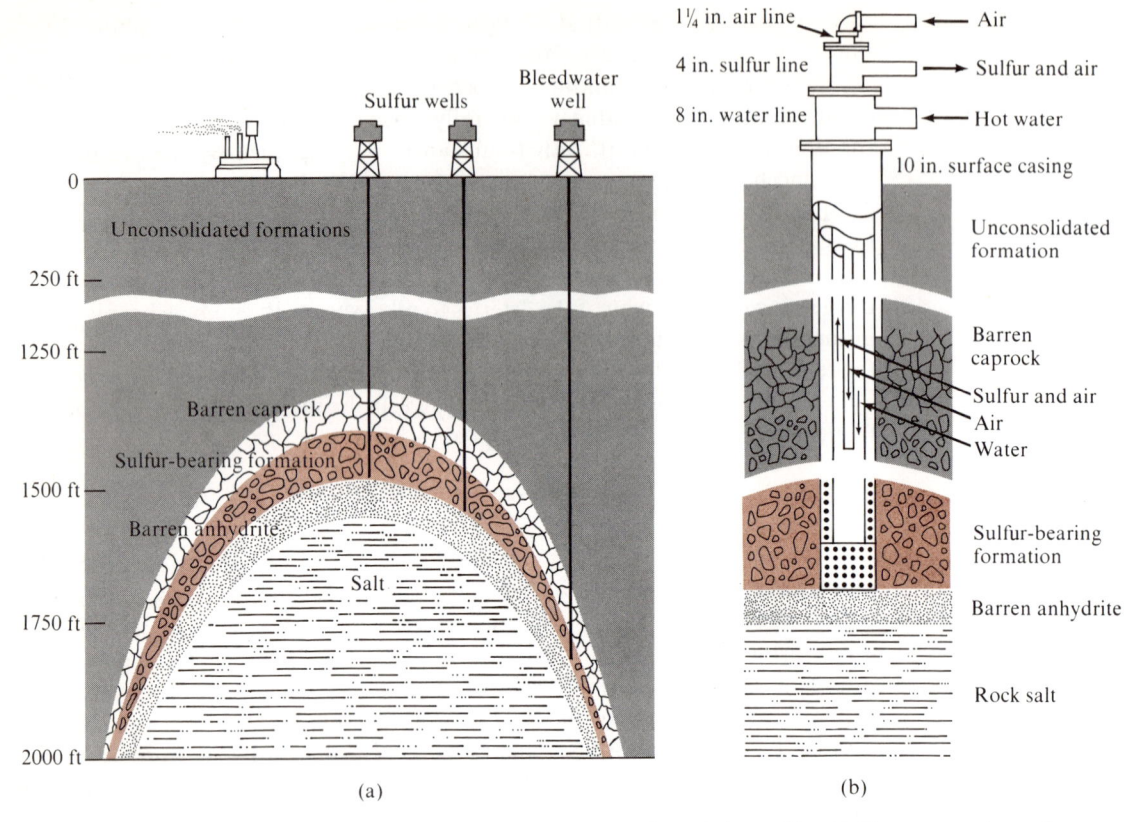

Figure 20.6. The Frasch process for mining sulfur. Sulfur occurs along the coast of the Gulf of Mexico in subterranean salt-dome deposits. It is mined by sinking a 10-inch shaft to the deposit, melting the sulfur with superheated water, and forcing the molten sulfur to the surface by hot compressed air. (Courtesy Freeport Sulphur Company. A division of Freeport Minerals Company.)

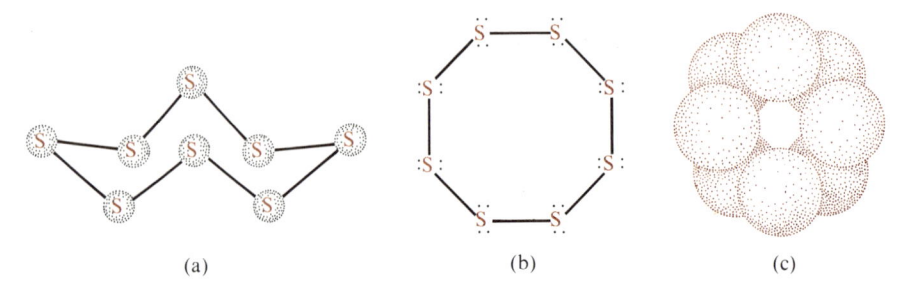

Figure 20.7. Diagrams of the S_8 molecule: (a) Puckered-ring formation with bond angles of 105°; (b) electron-dot structure; (c) compact molecular-orbital structure. The latter most nearly represents the S_8 molecule.

actually S_8, but in many of its reactions sulfur is represented by a single S. Figure 20.7 shows several ways of representing the S_8 molecule. Sulfur is insoluble in water but soluble in organic solvents such as carbon disulfide and benzene. Several allotropic forms of sulfur are known. The most common are rhombic, monoclinic, and amorphous sulfur. Rhombic sulfur melts at 112.8°C; the monoclinic form melts at 119°C. Sulfur boils at 444.6°C. Figure 20.8 illustrates crystals of rhombic and monoclinic sulfur.

Rhombic
(a)

Monoclinic
(b)

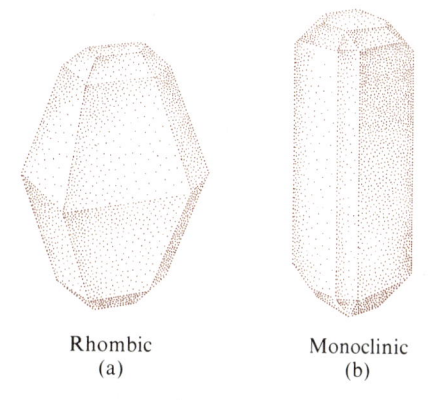

Figure 20.8. Crystal structure of rhombic and monoclinic sulfur.

Sulfur undergoes unique and interesting changes when it is heated. When a test tube of sulfur is carefully heated at 115°C, it melts and forms a straw-colored, mobile liquid that can be poured easily. As the temperature of the liquid is increased, the color of the sulfur darkens, and at about 160°C the viscosity (resistance to flow) begins to increase rapidly. At 187°C the viscosity reaches a maximum, and the dark reddish-brown sulfur is so viscous that it will not pour when the test tube is inverted. As sulfur is heated above 187°C, the color remains dark, the viscosity decreases rapidly, and near its boiling point, sulfur again becomes a mobile liquid. When sulfur near its boiling point is rapidly cooled by pouring into cold water, an amber-colored, rubbery solid called *plastic* or *amorphous sulfur* is formed.

The changes in viscosity are explained on the basis of molecular composition. Rhombic and monoclinic sulfur consist of S_8 rings. Liquid sulfur, near its melting point, also is composed principally of S_8 rings and has low viscosity. At higher temperatures, the rings begin to break up and form long chains of sulfur atoms (polymers) represented by S_x. Since viscosity increases with increasing molecular size, the chains apparently reach a maximum length at 187°C. Above 187°C the chains begin to break into shorter lengths, resulting in a lower viscosity. Gaseous sulfur is known to contain S_2 molecules.

Chemical properties of sulfur. Since sulfur is less electronegative than oxygen, its compounds are less ionic and more covalent than those of oxygen. Ionic sulfur compounds are formed with only the most electropositive metals, the alkali metals and the alkaline earth metals (Group IIA). The oxidation state of sulfur ranges from -2 to $+6$. It combines directly with many elements, both metals and nonmetals; for example,

$$Zn(s) + S(s) \xrightarrow{\Delta} ZnS(s)$$

$$Fe(s) + S(s) \xrightarrow{\Delta} FeS(s)$$

$$H_2(g) + S(l) \xrightarrow{200°C} H_2S(g)$$

$$C(s) + 2\,S(g) \xrightarrow{3000°C} CS_2(g)$$

$$2\,S(s) + Cl_2(g) \longrightarrow S_2Cl_2(l)$$

Disulfur dichloride (also known as sulfur monochloride), the product in the last equation above, is a very toxic, golden-yellow liquid. It is used with carbon disulfide in one of the processes for the manufacture of carbon tetrachloride:

$$CS_2(l) + 2\,S_2Cl_2(l) \longrightarrow CCl_4(l) + 6\,S(s)$$

The sulfur formed is recovered and recycled through the process. Sulfur also combines directly with fluorine and bromine but not with iodine. There are at least ten known oxides of sulfur, two of which are well known—sulfur dioxide (SO_2) and sulfur trioxide (SO_3).

The prefix *thio* is used to indicate the presence of sulfur in a compound. More specifically, it means that sulfur is used in place of oxygen in an analogous compound. Two examples of thio compounds are sodium thiosulfate and thiourea. Their formulas and the formulas of their oxygen analogues are shown below.

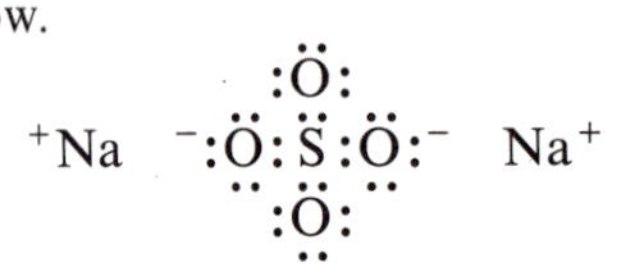

Sodium sulfate, Na_2SO_4    Sodium thiosulfate, $Na_2S_2O_3$

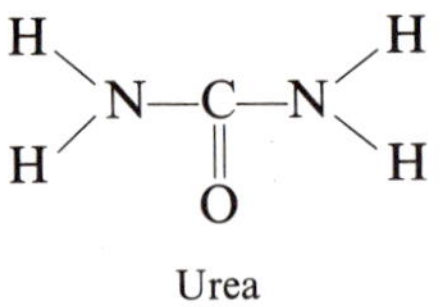

Urea    Thiourea

The central sulfur atom in sodium thiosulfate has a $+6$ oxidation number, whereas the thio sulfur has a -2 oxidation number, the same as the oxygen it replaced. Many other compounds with multiple sulfur linkages are known.

Sulfur burns in air or oxygen with a small, bluish flame to form sulfur dioxide:

$$S(l) + O_2(g) \longrightarrow SO_2(g)$$

Sulfur dioxide is also formed by reacting iron pyrites (FeS_2) or other metal sulfides with oxygen (air) at elevated temperatures:

$$4\,FeS_2 + 11\,O_2 \xrightarrow{\Delta} 2\,Fe_2O_3 + 8\,SO_2$$

Sulfur dioxide is a colorless gas that is irritating to the throat, causing choking and coughing. It also causes eye irritation and is known to be damaging to plant life. Sulfur dioxide as an air pollutant is discussed in Chapter 21. The gas liquefies at $-10°C$ and is shipped as a liquid in tank cars, drums, and cylinders.

Sulfur dioxide is the anhydride of sulfurous acid (H_2SO_3) and forms the acid when dissolved in water. It is a weak acid, decomposing easily into sulfur dioxide and water.

$$SO_2(g) + H_2O \rightleftharpoons H_2SO_3(aq)$$

Salts of sulfurous acid, called *sulfites* or *bisulfites*, liberate sulfur dioxide when treated with strong acids.

$$\underset{\text{Sodium sulfite}}{Na_2SO_3} + 2\,HCl(aq) \longrightarrow 2\,NaCl(aq) + H_2O + SO_2\uparrow$$

$$\underset{\text{Sodium bisulfite}}{NaHSO_3} + HCl(aq) \longrightarrow NaCl(aq) + H_2O + SO_2\uparrow$$

Much industrial sulfur dioxide goes into the production of sulfuric acid. It is also used as a bleaching agent in the pulp and paper industry, in refining sugar, and in processing dried fruit. Stored grain and cereals are treated with sulfur dioxide to fumigate them against rodents and insects. The boiling temperature and ease of liquefaction make sulfur dioxide useful as a refrigerant, especially in large refrigeration units.

Hydrogen sulfide (H_2S) is the sulfur analog of water, but that is as far as the resemblance goes. It is normally a gas; the liquid boils at $-60°C$. It has a disagreeable odor, like rotten eggs, and is very toxic. The gas is slightly soluble in water, forming an extremely weak acid solution called hydrosulfuric acid. Hydrogen sulfide is an excellent reducing agent; it is easily oxidized to free sulfur or other higher oxidation states of sulfur. The electronic structure of hydrogen sulfide is similar to that of water, except that the angle between the two hydrogen atoms and the central sulfur atom is 92° instead of 105°.

In the laboratory, hydrogen sulfide is prepared from iron(II) sulfide and dilute hydrochloric or sulfuric acid:

$$FeS + 2\,HCl \longrightarrow FeCl_2 + H_2S\uparrow$$

In nature, hydrogen sulfide is found in natural gas wells, petroleum deposits, and volcanic gases.

Hydrogen sulfide forms insoluble precipitates with many metals and is used extensively in the laboratory as an analytical reagent for the separation and identification of certain metals. Some of these metal sulfides have distinctive colors. A few are listed in Table 20.5.

Hydrogen sulfide burns in air with a small, bluish flame, forming sulfur dioxide and water:

$$2\,H_2S(g) + 3\,O_2(g) \longrightarrow 2\,SO_2\uparrow + 2\,H_2O$$

Table 20.5. Colors of selected metal sulfides.

Metal sulfide	Formula	Color
Cadmium sulfide	CdS	Bright yellow
Silver sulfide	Ag$_2$S	Dark brown
Lead(II) sulfide	PbS	Black
Copper(II) sulfide	CuS	Black
Arsenic(III) sulfide	As$_2$S$_3$	Yellow
Antimony(III) sulfide	Sb$_2$S$_3$	Orange
Zinc sulfide	ZnS	White
Nickel(II) sulfide	NiS	Black

The tarnishing of silver is due to small amounts of hydrogen sulfide in the atmosphere or sulfur in foodstuffs. This brown tarnish is silver sulfide.

$$2\,Ag(s) + H_2S(g) \longrightarrow Ag_2S(s) + H_2(g)$$

Hydrogen sulfide is usually detected by its rotten-egg odor, but it may be identified chemically by passing the gas over a strip of filter paper that has been dipped into a lead acetate solution. The formation of black lead(II) sulfide is a positive indication of hydrogen sulfide.

$$H_2S(g) + Pb(C_2H_3O_2)_2(aq) \longrightarrow PbS\downarrow + 2\,HC_2H_3O_2(aq)$$

Sulfuric acid. It has been said that the economic stature of a nation may be measured by its consumption of sulfuric acid. In the United States, sulfuric acid is one of the major industrial chemicals produced, exceeding 39 million tons annually.

The uses of sulfuric acid (H$_2$SO$_4$) are so numerous and varied that one can hardly begin to enumerate them. Large amounts are used to convert insoluble phosphates into soluble phosphates for agricultural fertilizers, and for pickling of metal surfaces. Sulfuric acid is used in the production of hydrochloric acid, ammonium sulfate, and aluminum sulfate. It is also used in petroleum refining; the manufacture of rayon; the making of explosives (nitroglycerin and TNT), alcohols, detergents, metal sulfates; textile printing; dyeing; and in scores of other industries. Dilute sulfuric acid is the electrolyte used in the millions of ordinary automobile storage batteries.

Sulfuric acid is a colorless, oily liquid with a density of 1.84 g/ml. Its freezing point is 10.4°C; its boiling point is 338°C. Sulfuric acid was formerly called *oil of vitriol*, because it was first obtained from distilling green vitriol (FeSO$_4 \cdot$ 7H$_2$O). The acid decomposes as it boils, liberating its anhydride, sulfur trioxide (SO$_3$):

$$H_2SO_4(l) \xrightarrow{\Delta} H_2O(g) + SO_3\uparrow$$

Ordinary concentrated sulfuric acid contains about 96 to 98% H$_2$SO$_4$; the remainder is water.

Sulfuric acid is generally manufactured by the *contact process*. The essential steps are as follows:

1. Sulfur is burned in air to produce sulfur dioxide:

$$S + O_2 \longrightarrow SO_2$$

2. The sulfur dioxide is further oxidized to sulfur trioxide in a contact chamber in the presence of a vanadium oxide (V_2O_5) catalyst.

$$2\,SO_2 + O_2 \xrightarrow{V_2O_5} 2\,SO_3$$

3. The sulfur trioxide is dissolved in concentrated sulfuric acid to form pyrosulfuric acid ($H_2S_2O_7$). Solutions of SO_3 in H_2SO_4 are known as *fuming sulfuric acid* or *oleum*.

$$SO_3 + H_2SO_4 \longrightarrow H_2S_2O_7$$

4. Calculated amounts of water are added to the pyrosulfuric acid to produce the desired concentration of sulfuric acid.

$$H_2S_2O_7 + H_2O \longrightarrow 2\,H_2SO_4$$

Concentrated sulfuric acid reacts vigorously and rapidly with water, forming the hydrates $H_2SO_4 \cdot H_2O$ and $H_2SO_4 \cdot 2H_2O$. Because the reaction is extremely exothermic, care must be exercised in preparing dilute solutions. The acid should always be added, slowly and with stirring, to the water. When water is added to the acid, it remains on top of the acid because of their relative densities, and localized boiling occurs, accompanied by violent spattering.

Sulfuric acid ionizes in water in two steps, giving a strongly acidic solution:

Step 1. $H_2SO_4 + H_2O \longrightarrow H_3O^+ + HSO_4^-$

Step 2. $HSO_4^- + H_2O \rightleftharpoons H_3O^+ + SO_4^{2-}$

The reaction in Step 1 goes 100%. The reaction in Step 2 is reversible and goes only to about 10% in a 0.1 M solution. This mode of ionization explains the existence of the bisulfate ion (HSO_4^-) and the formation of bisulfate salts such as $NaHSO_4$.

The chemical properties of sulfuric acid may be grouped into four categories: its reactions (1) as an acid, (2) as an oxidizing agent, (3) as a dehydrating agent, and (4) in the preparation of other acids.

1. *Reactions as an acid.* A water solution of sulfuric acid contains hydronium ions and therefore shows the usual acid reactions by reacting with metals, carbonates, bases, and so on. Several examples are

$Zn(s) + H_2SO_4(aq) \longrightarrow ZnSO_4(aq) + H_2\uparrow$

$Cu(s) + H_2SO_4\ (dilute)(aq) \longrightarrow$ No reaction

$CaCO_3(s) + H_2SO_4(aq) \longrightarrow CaSO_4(aq) + CO_2\uparrow + H_2O$

$NaOH(aq) + H_2SO_4(aq) \longrightarrow NaHSO_4(aq) + H_2O$

$2\,NaOH(aq) + H_2SO_4(aq) \longrightarrow Na_2SO_4(aq) + 2H_2O$

2. *Oxidizing agent.* Concentrated sulfuric acid, especially when hot, is a good oxidizing agent; metals reacting with it do not liberate hydrogen. Because of its oxidizing strength, metals below hydrogen in the activity series, which ordinarily do not react with the dilute acid, will react with the hot concentrated acid; for example,

$$Cu + 2\,H_2SO_4(conc.) \xrightarrow{\Delta} CuSO_4 + 2\,H_2O + SO_2\uparrow$$

Reacting with a stronger reducing agent such as zinc, the sulfur is reduced from $+6$ to -2:

$$4\,Zn + 5\,H_2SO_4(conc.) \xrightarrow{\Delta} 4\,ZnSO_4 + H_2S\uparrow + 4\,H_2O$$

Even carbon is oxidized by hot sulfuric acid:

$$C + 2\,H_2SO_4(conc.) \xrightarrow{\Delta} CO_2\uparrow + 2\,SO_2\uparrow + 2\,H_2O$$

3. *Dehydrating agent.* Sulfuric acid has a great affinity for water. This property makes concentrated sulfuric acid an excellent drying agent. Gases, which do not react with sulfuric acid and which contain water vapor, may be dried by bubbling through concentrated sulfuric acid.

Sulfuric acid is such a good dehydrating agent that it will remove water from sugar (or other carbohydrates), leaving carbon as a residue:

$$C_{12}H_{22}O_{11} + 11\,H_2SO_4 \longrightarrow 12\,C + 11\,H_2SO_4\cdot H_2O$$

This interesting reaction is easily demonstrated by adding 25 ml of concentrated sulfuric acid to a 50 ml beaker half filled with sugar. After several seconds have elapsed, a black cylinder of carbon rises to a height greater than that of the beaker (see Figure 20.9). This dehydration reaction also occurs when cellulose products such as wood, cotton, and paper turn black when they come into contact with sulfuric acid.

The following is another striking demonstration of the ability of sulfuric acid to absorb water: (1) Add concentrated sulfuric acid to a small flask until it is nearly filled. (2) Put the flask in a beaker and set the beaker in a sealed desiccator

Figure 20.9. When concentrated sulfuric acid is poured onto sugar, the sugar is dehydrated, leaving a column of black carbon.

containing some water. After a few days, the acid will have absorbed sufficient water to fill the flask and cause the liquid to overflow into the beaker.

4. *Preparation of other acids.* The high boiling point of sulfuric acid makes it ideal for preparing acids with lower boiling points. Salts of chlorides or nitrates are mixed with sulfuric acid and heated. The more volatile products, hydrogen chloride (HCl, bp $-84.9°C$) and hydrogen nitrate (HNO_3, bp $86°C$), distill off long before the boiling temperature of sulfuric acid is reached.

$$NaCl + H_2SO_4 \xrightarrow{\Delta} NaHSO_4 + HCl\uparrow$$
$$KNO_3 + H_2SO_4 \xrightarrow{\Delta} KHSO_4 + HNO_3\uparrow$$

Other Elements of the Sulfur Family

Selenium and tellurium resemble sulfur much more than they resemble oxygen. Both elements are less electronegative and tend to be more metallic than sulfur. Their hydrogen compounds, hydrogen selenide (H_2Se) and hydrogen telluride (H_2Te), have vile odors and are more acidic than hydrosulfuric acid.

Several allotropic forms of selenium are known. One form, gray selenium, shows the interesting property of having low electric conductivity in darkness and a much higher conductivity when exposed to light. This is the basis for the operation of the selenium photoelectric cell, which is used in controlling automatic equipment of all kinds. Selenium is also used to impart a red color to glass, glazes, and enamels.

Tellurium has a definite metallic luster and is a better conductor of electricity than selenium or sulfur. Its major uses are in metal alloys, glass, and dyestuffs.

The heaviest element of the family, polonium, is very rare and claims fame only as the first radioactive element discovered by Pierre and Marie Curie.

20.3 The Nitrogen Family

nitrogen family

The **nitrogen family**, periodic Group VA, includes the elements nitrogen, phosphorus, arsenic, antimony, and bismuth. These elements, especially nitrogen and phosphorus, are fairly well known. Nitrogen is the most abundant element in the atmosphere and phosphorus is the 11th most abundant element in the earth's crust. Compounds of nitrogen and phosphorus are essential for all forms of life.

The electron structures of these elements show that they each have five electrons in their outer shell, two s and three p electrons. This structure suggests that their oxidation numbers will range from -3 to $+5$. In fact, nitrogen and phosphorus are known to exist in every possible oxidation state from -3 to $+5$.

The elements of Group VA become increasingly metallic from top to bottom in the group. Nitrogen is a colorless gas with virtually no suggestion of metallic properties, but bismuth is distinctly metallic in character.

Nitrogen

Nitrogen occurs as two stable isotopes, $^{14}_{7}N$ and $^{15}_{7}N$; the lighter isotope accounts for more than 99% of the nitrogen atoms. Nitrogen is found in the atmosphere, living organic matter, decayed matter, ammonia, ammonium salts, and nitrate deposits. It was discovered to be an element in 1772 by Daniel Rutherford (1749–1819), a Scottish physician and botanist. Henry Cavendish (1731–1810), the discoverer of hydrogen, Priestly, and Scheele also independently discovered nitrogen at about the same time.

Elemental nitrogen is a diatomic molecule in the gaseous state. Its electron structure contains three shared pairs of electrons, forming a triple bond between the nitrogen atoms:

$$\cdot \ddot{N} \cdot \qquad :N:::N: \qquad :N{\equiv}N:$$

Nitrogen atom   Nitrogen molecule

The common physical properties of nitrogen are listed in Table 20.6.

Table 20.6. Physical properties of the nitrogen family.

Element	Melting point (°C)	Boiling point (°C)	Density (g/ml)
Nitrogen	−210.1	−195.8	0.81[a]
Phosphorus	44.2	280	1.82[b]
Arsenic	817	615[c]	5.7
Antimony	630.5	1,380	6.6
Bismuth	271.3	1,450	9.8

[a] Liquid at boiling point.
[b] White phosphorus.
[c] Sublimes.

Nitrogen can be prepared in the laboratory by decomposing ammonium nitrite; the gas is collected by the displacement of water:

$$NH_4NO_2(s) \xrightarrow{\Delta} N_2\uparrow + 2\,H_2O$$

Nitrogen is also formed when organic amino compounds, such as urea, are reacted with nitrous acid:

$$\underset{\text{Urea}}{CO(NH_2)_2(aq)} + \underset{\text{Nitrous acid}}{2\,HNO_2(aq)} \longrightarrow 2\,N_2\uparrow + CO_2\uparrow + 3\,H_2O$$

The atmosphere is composed of about four-fifths nitrogen and is the commercial source of the element. Nitrogen is obtained as a coproduct in the preparation of oxygen from liquid air. The nitrogen obtained in this manner contains very small amounts of noble gases.

Liquid nitrogen, because of its low boiling temperature, is stored and shipped in Dewar flasks. The liquid is used primarily as a coolant. Purified nitrogen gas is stored and shipped in high-pressure steel cylinders. The production of nitrogen in the United States exceeds 14 million tons annually.

The approximate volume of nitrogen in the air can be demonstrated by burning phosphorus in air in a tube inverted in water. Phosphorus combines with the oxygen in the tube, forming P_4O_{10}, which dissolves in water to give phosphoric acid. Water rises in the tube to occupy the volume of the consumed oxygen. The gas remaining is essentially nitrogen, occupying about four-fifths of the volume of the tube.

$$4\ P(s) + 5\ O_2(g) \longrightarrow P_4O_{10}(s)$$
$$P_4O_{10}(s) + 6\ H_2O \longrightarrow 4\ H_3PO_4(aq)$$

The high energy of dissociation (225 kcal/mole) is evidence of the stability of the nitrogen molecule. As a result, free nitrogen is chemically unreactive except at high temperatures or in the presence of certain catalysts.

Nitric oxide is formed when air is passed through an electric arc or when nitrogen and oxygen are reacted at about 3000°C:

$$N_2(g) + O_2(g) \rightleftharpoons 2\ NO(g)$$

This reaction accounts for the formation of oxides of nitrogen during electrical storms. In the Haber process, synthetic ammonia is manufactured by passing nitrogen and hydrogen gases over a catalyst at elevated temperatures and pressures:

$$N_2(g) + 3\ H_2(g) \rightleftharpoons 2\ NH_3(g) + 22\ \text{kcal}$$

Nitrogen combines directly with many metals to form metal nitrides when the gas is passed over the heated metal:

$$6\ Li + N_2 \xrightarrow{\Delta} 2\ Li_3N$$
$$3\ Mg + N_2 \xrightarrow{\Delta} Mg_3N_2$$
$$3\ Ca + N_2 \xrightarrow{\Delta} Ca_3N_2$$
$$2\ Al + N_2 \xrightarrow{\Delta} 2\ AlN$$

Compounds of Nitrogen

Hundreds of useful nitrogen compounds are known. Some of the more important ones, produced and used in large tonnages, are ammonia (NH_3), nitric acid (HNO_3), ammonium nitrate (NH_4NO_3), ammonium sulfate [$(NH_4)_2SO_4$], ammonium phosphate [$(NH_4)_3PO_4$], urea [$CO(NH_2)_2$], sodium nitrate ($NaNO_3$), and calcium cyanamide ($CaCN_2$).

The current production of ammonia in the United States is over 17 million tons annually. By far the greatest consumption of ammonia is in the manufacture of fertilizers or in its direct use as a fertilizer. Other industrial uses of ammonia or its derivatives include the manufacture of nitric acid, plastics, refrigerants, explosives, pulp and paper, and textiles.

The electron-dot structures of ammonia and ammonium ion are

$$H:\ddot{N}:H \atop \ddot{H}$$ $$\left[{H:\ddot{N}:H \atop \ddot{H}} \right]^+$$

Ammonia   Ammonium ion

Contrasting these two groups, we see that ammonia, having an open pair of electrons, is a Lewis base and reacts with a proton as shown below.

$$NH_3 + H^+ \longrightarrow NH_4^+$$
$$NH_3 + HCl(g) \longrightarrow NH_4^+ + Cl^-$$
$$NH_3 + H_2O \rightleftharpoons NH_4OH(aq) \rightleftharpoons NH_4^+ + OH^-$$

A water solution of ammonia, known as *ammonium hydroxide*, is alkaline. A few drops of phenolphthalein indicator produces a red color in an ammonium hydroxide solution. When this solution is boiled, it gradually becomes colorless, because of the decomposition of ammonium hydroxide and the subsequent loss of ammonia (boiling point, −33.4°C) from the solution:

$$NH_4OH(aq) \xrightarrow{\Delta} NH_3\uparrow + H_2O$$

The ammonium ion, on the other hand, is an acid, furnishing a proton in solution. A water solution of ammonium chloride is acidic.

$$NH_4^+ \rightleftharpoons NH_3 + H^+$$
$$NH_4^+ + Cl^- + H_2O \rightleftharpoons NH_3 + H_3O^+ + Cl^-$$

The known oxides of nitrogen are as follows:

N_2O	Dinitrogen oxide	Colorless gas
NO	Dinitrogen monoxide (nitric oxide)	Colorless gas
N_2O_3	Dinitrogen trioxide	Red-brown gas
NO_2	Nitrogen dioxide	Red-brown gas
N_2O_4	Dinitrogen tetroxide	Yellow liquid, colorless solid
N_2O_5	Dinitrogen pentoxide	White solid

Dinitrogen pentoxide is the anhydride of nitric acid; it reacts vigorously with water to form the acid:

$$N_2O_5 + H_2O \longrightarrow 2\,HNO_3$$

Nitric acid is produced in the United States in excess of 8 million tons annually. The pure material is a colorless liquid; it has a melting point of −41.6°C, a boiling point of 86°C, and a density (at 20°C) of 1.503 g/ml. Nitric acid has a high vapor pressure at room temperature, fumes strongly in moist air, and is miscible with water in all proportions. The concentrated acid contains 68–70% HNO_3 and has a specific gravity of 1.42. The concentrated acid often appears yellow because it contains dissolved nitrogen dioxide. This NO_2 is formed by slow decomposition of the acid when it is exposed to light:

$$4\,HNO_3 \xrightarrow{Light} 4\,NO_2 + 2\,H_2O + O_2$$

The structure for nitric acid may be represented as follows:

$$H:\ddot{\underset{\cdot\cdot}{O}}:N::O: \quad \text{or} \quad H:\ddot{\underset{\cdot\cdot}{O}}:N=O:$$
$$\phantom{H:\ddot{O}:N::}:\ddot{\underset{\cdot\cdot}{O}}: \downarrow$$
$$\phantom{H:\ddot{O}:N::O:aaaaaaaaa}:\ddot{\underset{\cdot\cdot}{O}}:$$

In an older process for making nitric acid, nitrates (chile saltpeter) and sulfuric acid were used:

$$NaNO_3 + H_2SO_4 \xrightarrow{\Delta} HNO_3\uparrow + NaHSO_4$$

The reactants were heated in cast-iron retorts, and the nitric acid fumes were condensed and collected in stoneware vessels. Now most nitric acid is made by the Ostwald process. In this process, an ammonia–air mixture is passed over a platinum catalyst at about 800°C. Ammonia is first oxidized to nitric oxide, which in turn is further oxidized to nitrogen dioxide. The latter is then absorbed in water to form the acid. The equations are

$$4\,NH_3(g) + 5\,O_2(g) \longrightarrow 4\,NO(g) + 6\,H_2O(g)$$
$$2\,NO(g) + O_2(g) \longrightarrow 2\,NO_2(g)$$
$$3\,NO_2(g) + H_2O \longrightarrow 2\,HNO_3(aq) + NO(g)$$

The acid produced is about 60% HNO_3. The nitric oxide formed in the last step is recycled back through the process.

Nitric acid is both a strong acid and a powerful oxidizing agent. It will attack all metals except gold, platinum, iridium, rhodium, and tantalum. As an acid, it reacts with bases to form nitrate salts:

$$NaOH(aq) + HNO_3(aq) \longrightarrow NaNO_3(aq) + H_2O$$
$$NH_3(aq) + HNO_3(aq) \longrightarrow NH_4NO_3(aq)$$
$$Na_2CO_3(aq) + 2\,HNO_3(aq) \longrightarrow 2\,NaNO_3(aq) + H_2O + CO_2\uparrow$$

Nitric acid reacts with metals to form a variety of nitrate reduction products, depending on the concentration of the acid and the reducing ability of the metal. Hydrogen is not formed by the reaction of nitric acid with metals. With concentrated acid and metals, nitrogen dioxide gas is always formed. With dilute acid and metals such as copper and mercury, nitric oxide gas is formed. With very dilute acid and stronger reducing metals such as zinc or aluminum, nitric acid is reduced to the ammonium ion. In this last case, the oxidation number of nitrogen is reduced from $+5$ to -3. The following equations illustrate these three different reduction reactions:

$$Hg + 4\,H^+ + 4\,NO_3^- \longrightarrow Hg^{2+} + 2\,NO_3^- + 2\,NO_2\uparrow + 2\,H_2O$$
$$\text{Conc. } HNO_3$$

$$3\,Cu + 8\,H^+ + 8\,NO_3^- \longrightarrow 3\,Cu^{2+} + 6\,NO_3^- + 2\,NO\uparrow + 4\,H_2O$$
$$\text{Dilute } HNO_3$$

$$4\,Zn + 10\,H^+ + 10\,NO_3^- \longrightarrow 4\,Zn^{2+} + NH_4^+ + 9\,NO_3^- + 3\,H_2O$$
$$\text{Dilute } HNO_3$$

A mixture of 3 parts of concentrated hydrochloric acid to 1 part of concentrated nitric acid is known as *aqua regia* (royal water); it will attack and dissolve such metals as gold and platinum.

When nitric acid comes into contact with the skin, it reacts with the proteins to cause destruction of the tissues, producing a yellow discoloration of the skin.

The major uses for nitric acid are for manufacturing ammonium nitrate fertilizers, explosives, plastics, pharmaceuticals, dye intermediates, and organic nitro compounds, and for pickling stainless steel in steel refining.

The Nitrogen Cycle

The process by which nitrogen is circulated and recirculated from the atmosphere through living organisms and back to the atmosphere is known as the **nitrogen cycle**.

Nitrogen compounds are required by all living organisms. Despite the fact that the atmosphere is about four-fifths nitrogen, man and all other animals as well as the higher plants are unable to utilize free nitrogen. Higher plants require inorganic nitrogen compounds and animals must have nitrogen in the form of organic compounds.

Atmospheric nitrogen is *fixed*—that is, converted into chemical compounds that are useful to higher forms of life—by three general routes.

1. *Bacterial action.* Certain soil bacteria are capable of converting N_2 into nitrates. Most of these bacteria live in the soil in association with legumes (for example, peas, beans, clover). Some free-living soil bacteria and the blue-green algae, which live in water, are also capable of fixing nitrogen.
2. *Lightning.* The high temperature of lightning flashes causes the formation of substantial amounts of nitric oxide in the atmosphere. This NO is dissolved in rain water and is eventually converted to nitrate ions in the soil. The combustion of fuels also provides temperatures high enough to form NO in the atmosphere. The total amount of nitrogen fixed by combustion is relatively insignificant on a worldwide basis. However, NO produced by combustion, especially in automobile engines, is a serious air pollution problem in some areas.
3. *Chemical fixation.* Chemical processes have been devised for making nitrogen compounds directly from atmospheric nitrogen. By far the most important of these is the Haber process for making ammonia from nitrogen and hydrogen. This process is the major means of production for the millions of tons of nitrogen fertilizers that are produced synthetically each year.

A schematic diagram of the nitrogen cycle is shown in Figure 20.10. Starting with the atmosphere, the cycle begins with the fixation of nitrogen by any one of the three routes. In the soil, nitrogen or nitrogen compounds are converted to nitrates, taken up by higher plants, and converted to organic compounds. The plants eventually die or are eaten by animals. During the life of the animal, part of the nitrogen from plants in its diet is returned to the soil in the form of fecal and urinary excreta. Eventually, after death, both plants and animals are decomposed by bacterial action. Part of the nitrogen from plant and animal tissues is returned to the atmosphere as free nitrogen and part is retained in the soil. The cycle thus continues.

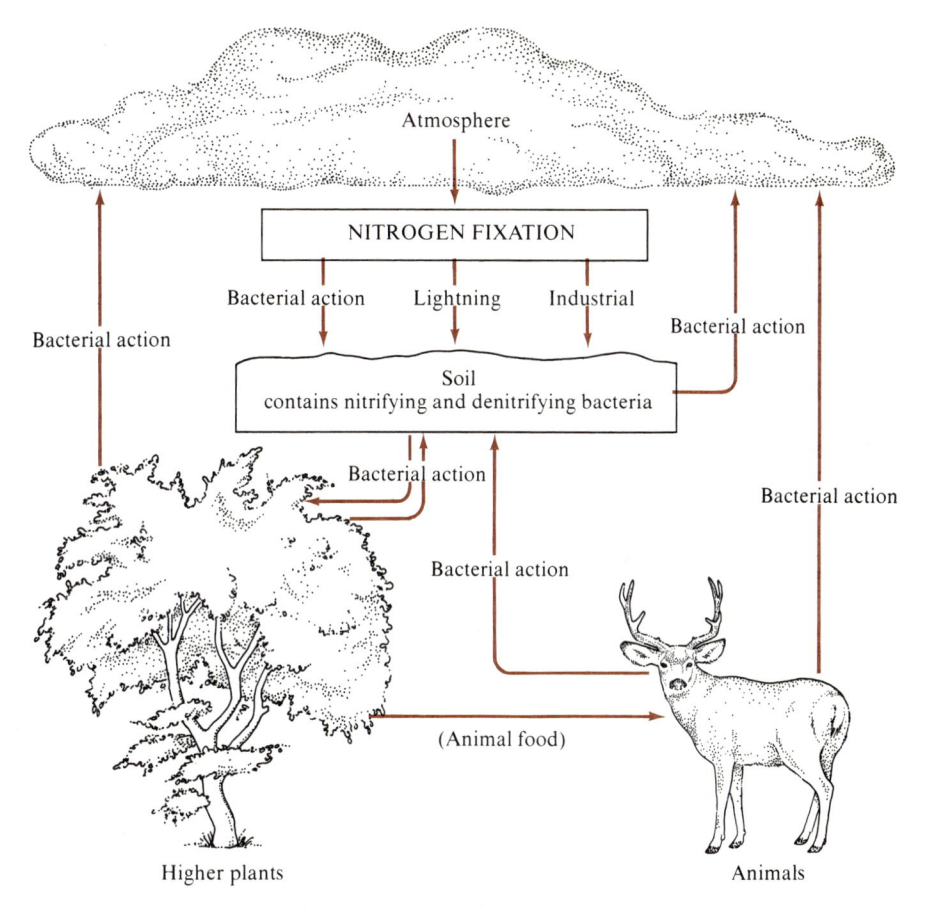

Figure 20.10. The nitrogen cycle.

Other Elements of the Nitrogen Family

The physical properties of the other four elements of Group VA—phosphorus, arsenic, antimony, and bismuth—differ markedly from those of nitrogen (see Table 20.6).

Phosphorus. Phosphorus is the 11th most abundant element in the earth's crust. Free phosphorus is never found in nature. The element occurs mainly in phosphate minerals. Calcium phosphate is present in the bones and teeth of animals, and small amounts of phosphorus compounds are present in the cells of all plants and animals. The free element exists in several allotropic modifications—white, red, and black being the most common. The white form is tetratomic, P_4; the red and black forms exist as polymeric chains of phosphorus atoms.

White phosphorus is a soft, waxy solid, insoluble in water but soluble in carbon disulfide. It is generally stored under water to keep it out of contact with air, in which it fumes and ignites spontaneously, burning to the pentoxide.

White phosphorus often appears yellow because of its gradual changing to the more stable red form. Red phosphorus is formed by heating white phosphorus to 250°C. The red modification is much less reactive and more stable than the white and need not be stored under water. Although white phosphorus is extremely poisonous, the red form is not. Red phosphorus is not soluble in carbon disulfide.

Another crystalline form, which is less reactive than the red, is black phosphorus. It is produced by subjecting white phosphorus to very high pressures. It resembles graphite in appearance, is insoluble in carbon disulfide, and is a fair conductor of heat and electricity.

Two principal oxides of phosphorus are P_4O_6 and P_4O_{10}. These oxides are commonly called phosphorus trioxide and phosphorus pentoxide, respectively. These names have been carried over from an earlier time when their molecular formulas were believed to be P_2O_3 and P_2O_5. Many different oxyacids of phosphorus are known; several are listed below.

H_3PO_2	Hypophosphorous acid	$H_4P_2O_6$	Hypophosphoric acid
HPO_2	Metaphosphorous acid	HPO_3	Metaphosphoric acid
H_3PO_3	Orthophosphorous acid	H_3PO_4	Orthophosphoric acid
$H_4P_2O_5$	Pyrophosphorous acid	$H_4P_2O_7$	Pyrophosphoric acid

Orthophosphoric and orthophosphorous acids are commonly called phosphoric and phosphorous acids, respectively. Their electron structures are given below. Phosphoric acid is commercially available as a syrupy solution containing 85% H_3PO_4.

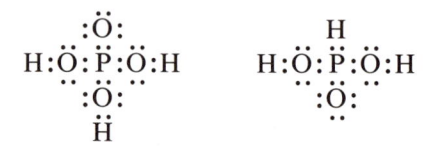

$$\begin{array}{cc} \ddot{\text{:O:}} & \text{H} \\ \text{H:}\ddot{\text{O}}\text{:}\ddot{\text{P}}\text{:}\ddot{\text{O}}\text{:H} & \text{H:}\ddot{\text{O}}\text{:}\ddot{\text{P}}\text{:}\ddot{\text{O}}\text{:H} \\ \ddot{\text{:O:}} & \ddot{\text{:O:}} \\ \text{H} & \end{array}$$

Phosphoric acid   Phosphorous acid

Phosphorus compounds are indispensable for both plants and animals. Although phosphorus is fairly abundant in the earth's crust, many soils do not contain enough soluble phosphates for optimum plant growth. The worldwide need to produce more food has brought about an increased demand for phosphate fertilizers. Consequently, about two-thirds of the phosphates produced are used in fertilizers. More than 5 million tons of phosphates (as P_2O_5) are applied to United States soils annually. Other uses of phosphorus are in matches, water softening, prevention of boiler scale, fireproofing, petroleum additives, and insecticides.

Arsenic, antimony, and bismuth. Although nitrogen and phosphorus are nonmetals, arsenic, antimony, and bismuth become progressively more metallic, with bismuth having properties approaching those of a true metal. All three were known in ancient times, and all are found in nature, both in the free and the combined states. Their principal ores are the sulfides. Their most common oxidation states are $+3$ and $+5$, and their oxides resemble nitrogen and phosphorus in composition: As_4O_6, As_2O_5, Sb_4O_6, Sb_2O_5, Bi_2O_5.

Arsenic, antimony, and bismuth are mainly used in alloys to increase the hardness of metals. Lead shot is hardened by adding up to 0.5% arsenic; type metal contains 15–18% antimony; pewter contains 6% bismuth and 1.7% antimony; babbitt metal contains 7% antimony. Low-melting alloys containing over 50% bismuth are used in safety sprinkler heads for fire protection. Arsenic and antimony compounds, although very toxic, have been successfully used as medicinals in controlled dosages. Certain arsenicals, such as paris green and calcium and lead arsenates, are used as weed killers and insecticides.

Questions

A. Review the meanings of the new terms introduced in this chapter.
  1. Halogen family
  2. Sulfur family
  3. Combustion
  4. Nitrogen family
  5. Nitrogen cycle

B. Review questions.
  1. Explain in terms of electronic structure why the reactivity of Group VIIA elements decreases as we go from top to bottom in the group.
  2. Explain in terms of electronic structure why fluorine is the most electronegative element.
  3. Use Lewis dot diagrams to illustrate how the halogen elements, designated X, can form both covalent bonds (X_2, HX) and electrovalent bonds (NaX).
  4. Give the physical state and describe the appearance of each of the first four halogens at 25°C.
  5. Why is the boiling point of HF abnormally high when compared to that of the other hydrogen halides?
  6. Explain why hydrofluoric acid cannot be stored in ordinary glass bottles; illustrate your explanation with an appropriate equation. [*Hint*: Glass contains some silicon dioxide.]
  7. Write balanced chemical equations showing how chlorine can be prepared by the following reactions:
     (a) Electrolysis of molten salt (NaCl)
     (b) Electrolysis of an aqueous salt (NaCl) solution
  8. What is chlorine water? Explain how chlorine water acts as a bleaching agent.
  9. (a) Explain how gaseous HCl can be prepared from sulfuric acid and sodium chloride.
     (b) Gaseous HBr cannot be prepared by reacting sulfuric acid and sodium bromide. Why not?
  10. Tell how bromine is obtained from seawater; give balanced equations for the reactions involved.
  11. (a) What is the role of bromine compounds in motor fuels?
      (b) Why is the annual demand for bromine in the United States decreasing?
  12. Arrange the halogens in order of increasing strength as oxidizing agents. Justify the order in terms of electronic structure.
  13. Why is iodine more soluble in potassium iodide solution than it is in pure water?
  14. (a) Assign oxidation numbers to the following: ClF, ICl, IF_7, F_2, BrF_3.
      (b) Why are substances such as IF called interhalogen compounds?

15. Complete and balance the following equations:
    (a) $MnO_2 + HBr \rightarrow$
    (b) $HNO_2 + HI \rightarrow$
    (c) $Cl_2 + KBr \rightarrow$
    (d) $KClO_3 + HCl \rightarrow$
    (e) $NaI + H_2SO_4 \rightarrow$
16. Write formulas for the following compounds:
    (a) Iodine pentafluoride
    (b) Potassium bromate
    (c) Hypochlorous acid
    (d) Chloric acid
    (e) Sodium perchlorate
17. What ill effects, if any, would occur from a lack of the following ions in the diet?
    (a) Fluoride ions
    (b) Chloride ions
    (c) Bromide ions
    (d) Iodide ions
18. List the four gases to be found in "dry" air in order of abundance. Which of these gases, if any, are necessary for our existence? Explain.
19. What is the source of most of the oxygen produced for commercial use?
20. Why is a stream of air usually kept bubbling through an aquarium containing tropical fish?
21. Describe the Frasch process for mining sulfur. What physical properties of sulfur make this process possible?
22. Explain why the viscosity of liquid sulfur increases markedly as it is heated in the range of 160–187°C.
23. How does *amorphous* sulfur differ from ordinary sulfur; how is amorphous sulfur prepared?
24. Write balanced chemical equations for the preparation of sulfur dioxide by three different types of chemical reactions.
25. Complete and balance:
    (a) $FeS + HCl \rightarrow$
    (b) $Ag + H_2S \rightarrow$
    (c) $Cu + H_2SO_4(conc.) \xrightarrow{\Delta}$
    (d) $NaCl + H_2SO_4(conc.) \xrightarrow{\Delta}$
    (e) $H_2 + S \xrightarrow{\Delta}$
    (f) $SO_3 + H_2O \rightarrow$
    (g) $SO_3 + H_2SO_4(conc.) \xrightarrow{\Delta}$
    (h) $Cu_2S + O_2 \xrightarrow{\Delta}$
    (i) $FeS + O_2 \xrightarrow{\Delta}$
    (j) $NaNO_3 + H_2SO_4 \xrightarrow{\Delta}$
26. Explain why free sulfur can act either as an oxidizing agent or as a reducing agent.
27. Explain why the lowest oxidation state of sulfur is -2 and the highest oxidation state is $+6$.
28. In the preparation of H_2S from FeS, HNO_3 cannot be used instead of HCl. Why not?
29. Complete and balance the following equations:
    (a) $SO_2 + O_2 \rightarrow$
    (b) $SbCl_3 + H_2S \rightarrow$
    (c) $H_2S + SO_2 \rightarrow$

(d) $H_2S + H_2SO_4(conc.) \rightarrow S + H_2O$
(e) $H_2S + O_2 \rightarrow S + H_2O$

30. Describe the contact process for manufacturing sulfuric acid and give the appropriate chemical equations.
31. Write equations to illustrate the reaction of sulfuric acid acting as:
    (a) An acid (b) An oxidizing agent (c) A dehydrating agent
32. Can wet NH_3 gas be dried by bubbling it through concentrated sulfuric acid? Explain.
33. Describe a safe procedure for diluting concentrated sulfuric acid in the laboratory.
34. Why do the oxidation numbers of Group VA elements range from -3 to $+5$?
35. Does the tendency to become more metallic increase or decrease from top to bottom in Group VA? Explain.
36. Draw Lewis-dot diagrams for:
    (a) A nitrogen molecule
    (b) An ammonia molecule
    (c) A nitride ion
    (d) An ammonium ion
    (e) A nitrate ion
37. In the fractional distillation of liquid air, does oxygen or nitrogen boil off first? Explain.
38. Describe the Haber process for making ammonia.
39. Describe the Ostwald process for making nitric acid.
40. What is aqua regia and why is it so named?
41. Complete and balance equations for the following reactions:
    (a) $NH_3 + HCl \rightarrow$
    (b) $Ag + HNO_3 \rightarrow$
    (c) $P_4O_{10} + H_2O \rightarrow$
    (d) $Mg + HNO_3 \rightarrow$
    (e) $NH_4NO_2 \xrightarrow{\Delta}$
42. Hydrochloric acid cannot be used to prepare HNO_3 from KNO_3. Why not?
43. Outline and discuss the nitrogen cycle in nature.
44. Which Group VA elements are absolutely essential to our health and well-being?

C. Review problems.
   1. What volume of SO_2 (at STP) can be obtained by combustion of 150 g of sulfur?
   2. What volume of chlorine (at STP) can be obtained by the electrolysis of 155 g of sodium chloride?
   3. The water coming from a large reservoir is being treated with 0.20 ppm (parts per million) of chlorine. Calculate the number of grams of chlorine required each hour when water is being pumped from the reservoir at a rate of 300,000 litres per hour.
   4. How many kilograms of nitric acid can be produced by the Ostwald process from 10,000 litres of ammonia (at STP)?
   5. How many moles of bromine are liberated when 15.0 litres of chlorine gas (at STP) are dissolved in seawater? Assume that 80.0% of the chlorine gas reacts with bromide ions present in the seawater.
   6. Calculate the molarity of each of the following concentrated acids:
      (a) Sulfuric acid: $d = 1.84$ g/ml, containing 96% H_2SO_4 by weight
      (b) Nitric acid: $d = 1.42$ g/ml, containing 70% HNO_3 by weight
      (c) Hydrochloric acid: $d = 1.19$ g/ml, containing 37% HCl by weight

21 Air Pollution

After studying Chapter 21 you should be able to:

1. Understand the terms listed in Question A at the end of the chapter.
2. Explain the characteristics of a London-type and a Los Angeles-type smog.
3. List the major source or sources of the following air pollutants: (a) CO, (b) SO_2, (c) NO_x, (d) hydrocarbons, and (e) particulate matter.
4. Explain the primary effect of aerosols in the atmosphere.
5. Give the physiological effects of each of the following air pollutants: (a) SO_2, (b) CO, (c) NO_x, and (d) O_3.
6. Show by equations how sulfur dioxide in the atmosphere is converted to a sulfuric acid mist.
7. Show equations and explain why marble statues deteriorate rapidly in large urban areas.
8. Cite evidence to show that the automobile is the principal source of carbon monoxide in urban atmospheres.
9. Explain why it is dangerous to operate an automobile engine in a closed garage.
10. Explain the physiological effects of cigarette smoking that can lead to inefficiency in driving an automobile.
11. Explain the mechanism by which carbon monoxide interferes with the normal functioning of the blood.
12. Explain the presence of nitric oxide in automobile exhaust.
13. List the names and formulas of the two most important oxides of nitrogen that are involved in air pollution; give their physical characteristics and equations for the reactions by which they are formed.
14. Describe the function of ozone in the stratosphere.
15. Discuss the effect of fluorocarbons on the ozone in the stratosphere.
16. Identify the major photochemical oxidants in the formation of smog, and explain their role in air pollution.
17. Describe the phenomenon of atmospheric temperature inversion.
18. Explain how atmospheric temperature inversion contributes to air pollution.
19. Explain the role of each of the following in air pollution: (a) a large urban community, (b) low wind velocity, (c) bright sunlight, (d) an industrial city surrounded by high hills, and (e) a large difference between day and night temperatures.

20. Explain why early air pollution control devices on automobiles did not decrease the brown haze of photochemical smog despite the reduction of hydrocarbons and carbon monoxide emissions.

21.1 Introduction

Our planet earth is endowed with a vast capability for absorbing waste products. In the preindustrial era waste products that found their way into the atmosphere, rivers and oceans, and the land were broken down, modified, or recycled so that the environment was not greatly affected. However, with the advent of industrialization and with huge increases in human population, the volume and nature of waste products changed greatly. We have recently (and none too soon) come to the realization that the natural depositories (atmosphere, oceans, and land) cannot absorb unlimited amounts of pollutants without threatening the very existence of human beings.

pollutant

Pollutants are substances that get into an environment in which they are not naturally present. Primitive people had very little impact on the natural environment. Perhaps the first human-generated pollution of the atmosphere was caused by grass or forest fires either accidentally or intentionally set. Today, the combustion of fuel in automobiles is the major source of atmospheric pollution in urban areas.

The addition of small amounts of contaminants to an environment can be tolerated. However, when the level of contamination becomes dangerous to health and well-being, action must be taken to alleviate the situation. The people of the United States, and of the world, are now confronted with hazardous levels of several major types of pollution.

Environmental problems and deterioration are evident in (1) atmospheric pollution, (2) water pollution, (3) chemical contamination of soil and food supplies, and (4) excessive noise levels. All these environmental problems are related to industrialization and a human population that is too large for specific geographic areas. The following discussion is confined to only one aspect of environmental contamination—that of air pollution. A great many of the facts and figures refer to the Los Angeles, California, region. This is because air pollution data there have been systematically collected for a longer period of time than in most other areas. In addition, the Los Angeles Basin has been the site of much pioneering research on air pollution and its control.

The air in the "normal" atmosphere contains about 21% oxygen and 78% nitrogen by volume; the remaining 1% is made up of CO_2, Ar, He, Ne, Kr, Xe, H_2, CH_4, and N_2O. However, practically all air is contaminated to some extent.

The composition of normal dry air is shown in Table 21.1. Note the two columns expressing concentration. For high concentrations, volume percent is used. For small concentrations, part per million (ppm) is used for convenience of expression. Data in either parts per million or percentage can be interchanged on the basis of $1\% = 10^4$ ppm.

Table 21.1. Composition of normal dry air.

Component	Content (volume percent)	Content (ppm)
N_2	78.08	—
O_2	20.95	—
Ar	0.934	—
CO_2	0.033	—
He	—	18.18
Ne	—	5.24
Kr	—	1.14
Xe	—	0.087
H_2	—	0.5
CH_4	—	2
N_2O	—	0.5

$$\text{Volume percent} \times \frac{10^4 \text{ ppm}}{1\%} = \text{ppm}$$

$$20.95\% \text{ } O_2 \times \frac{10^4 \text{ ppm}}{1\%} = 209{,}500 \text{ ppm } O_2$$

Thus, air contains 209,500 ppm O_2. Parts per million is commonly used in air pollution work, since concentration of contaminants in the air is generally low, falling into the range below 100 ppm.

Common atmospheric contaminants are classified as dust, fumes, mist, odor, smoke, and vapor. These contaminants consist of hundreds of different substances and come from the end products and by-products produced by people living in a highly technological society. We began by indiscriminantly dumping all types of waste products into the atmosphere. As urban populations grew and cities attracted more industry, the amounts of waste products entering the atmosphere increased to concentrations that were hazardous to people, other animals, and plant life. Such conditions as eye irritation, reduced visibility, and damage to rubber products and vegetation were some of the first noticeable effects of air pollution.

21.2 Major Air Pollution Episodes

Known episodes of bad air date back several hundreds of years. In the pamphlet *Fumifugium: On the Inconvenience of Aer, and Smoake of London Dissipated. Together with Some Remedies Humbly Proposed*, published over 300 years ago, John Evelyn wrote on the nature of bad air and some of its effects, and proposed methods of control. More recent episodes described below have intensified the urgency for control of air pollution.

London, England.   The city of London is widely known for its dense, "pea soup" fog. London has experienced several air pollution disasters dating back more than 100 years.

For centuries, sulfur-containing soft coal was burned in industries and homes in London. This coal released considerable amounts of smoke as well as other pollutants into the atmosphere. During the first week in December 1952, an unusually cold fog hung over the city of London. To keep warm, Londoners burned much more coal than usual. The increased concentration of smoke, sulfur dioxide, and other pollutants together with the heavy fog caused a major air pollution disaster. Over 4000 deaths were attributed to this episode. Most of the fatalities were elderly people, the very young, and those with respiratory diseases. Many others became violently ill. Ten years later, in 1962, another such episode occurred in which 750 died. The mortality figures were lower in 1962 because of the Clean Air Act, limiting production of smoke. In addition, the public was now aware of the problem and refrained from going out of doors during the siege.

smog

The word **smog** (smoke + fog) was coined to describe the atmospheric condition brought about by smoke and fog. This term is now used in a broad sense to mean any adverse air pollution condition.

Meuse River Valley, Belgium. The Meuse Valley in Belgium is a narrow river valley about 15 miles long, lined with hills about 300 feet high. This valley is highly industrialized and contains such installations as coke ovens, steel mills, a zinc smelter, and a sulfuric acid plant. From December 1 to December 5, 1950, an unusually calm weather condition existed in the valley. This allowed effluents from the factories to accumulate in the atmosphere. Thousands of people became ill and 60 died. Again, the dead were mostly older people and those with heart and respiratory diseases. The pollutants causing the disaster were believed to be chiefly the oxides of sulfur.

Donora, Pennsylvania. Donora is a small, industrialized community located in a valley on the Monongahela River about 30 miles south of Pittsburgh.

During an atmospheric temperature inversion (see Section 21.7) from October 27 to October 31, 1948, pollutants from factories were trapped and accumulated in the atmosphere of Donora, causing a hazardous air pollution problem. Most people coughed and suffered from sore throats, chest constrictions, headaches, burning eyes, vomiting, and nausea. Out of a population of 14,000 about 6000 became ill and many of them were hospitalized. Twenty persons died (about 8–10 times the normal rate), and several hundred domestic animals died. The people affected the most severely were the elderly and those with heart and lung problems. The pollutants causing death and illness were never conclusively determined, but the sulfur compounds SO_2, SO_3, and H_2SO_4 were present in the air in abnormally high concentrations.

21.3 Atmospheric Pollutants

particulate matter

Atmospheric pollutants occur in the form of dust, fumes, mists, odors, smoke, or vapor. These may be broadly classified into two types—gases and **particulate matter** (liquid droplets and solid particles).

452   Chapter 21

Particle size is generally expressed in micrometres (μm), where

$$1 \text{ micrometre} = 1 \text{ } \mu m = 1 \times 10^{-6} \text{ m} = 1 \times 10^{-4} \text{ cm}$$

Gases are molecular in size (less than 0.01 μm in diameter), while particulate matter may range in size from 0.01 to 100 μm in diameter.

aerosol

Particulate matter that remains suspended in the air for long periods of time is known as an **aerosol**. Aerosols are primarily dusts, fumes, and mists; they are responsible for reduced light and lowered visibility in the atmosphere. Particles in the 0.3–0.6 μm range cause reduced visibility as a result of light scattering, while larger particles reduce visibility by absorption of light. The United States Public Health Service has been collecting data on the nature and extent of air pollution for over 20 years. Their data generally show that the larger the city, the greater the amount of particulate matter in the atmosphere. For example, particulate emissions from all sources in Los Angeles County, California, were about 450 tons per day in 1971 and about 130 tons per day in 1973. Major sources of atmospheric pollution in the United States are given in Table 21.2.

Table 21.2.   Sources of atmospheric pollution in the United States in millions of tons per year (1975).

	Carbon monoxide	Particulates	Sulfur dioxide	Hydrocarbons	Nitrogen oxides
Motor vehicles	67.8	0.9	0.4	10.0	8.2
Transportation (miscellaneous)	9.6	0.4	0.4	1.7	2.5
Fuel combustion in stationary sources	1.2	6.6	26.3	1.4	12.4
Industrial processes	9.4	8.7	5.7	3.5	0.7
Solid waste disposal	3.3	0.6	<0.1	0.9	0.2
Miscellaneous	4.9	0.8	0.1	13.4	0.2
Totals	96.2	18.0	32.9	30.9	24.2

Data from United States Environmental Protection Agency, *National Air Quality and Emissions Trends Report*, Nov. 1976.

21.4   Gaseous Pollutants

Many different substances occur as gaseous pollutants. Some of the major contaminants resulting from industry and motor vehicles are sulfur dioxide, volatile hydrocarbons, carbon monoxide, nitrogen oxides, and ozone. Any gas, such as chlorine, hydrogen chloride, ammonia, and so on, will escape into the atmosphere if it is not contained, but these constitute only occasional hazards when they escape accidentally.

Sulfur Dioxide

Sulfur dioxide is a colorless gas with a suffocating odor. It is easily detected by most people by taste and odor in concentrations between 0.3 and 1 ppm.

Many chemistry students are familiar with SO_2 as the gaseous product of burning sulfur in air. Sulfur dioxide was one of the earliest recognized and controlled air pollutants. The deaths and illnesses in the air pollution episodes at Donora, Meuse Valley, and London have been attributed to SO_2 and its oxidation products.

Most unrefined fuels contain varying amounts of sulfur. Many ores (for example, zinc, copper, and iron) are sulfides. Consequently, when sulfur-containing fuels are burned in the air and when the metal sulfides are smelted, SO_2 is released. For example,

$$2\,ZnS + 3\,O_2 \longrightarrow 2\,ZnO + 2\,SO_2$$

Before emission regulations were imposed, the effluent gases from steam electric power plants that burned sulfur-containing fuels were heavily laden with SO_2. The problem was recognized and steps were taken to control these emissions. Emission standards limiting SO_2 concentration in stack gases were established in Los Angeles County, California, in 1947. Since 1959, power plants in Los Angeles County have been required by law to burn fuels with low sulfur content.

Sulfur dioxide in the air is slowly oxidized to sulfur trioxide (SO_3). This reaction is greatly accelerated by sunlight (ultraviolet radiation):

$$2\,SO_2 + O_2 \xrightarrow{\text{uv light}} 2\,SO_3$$

Sulfur trioxide combines rapidly with moisture in the air to form a mist of tiny droplets of sulfuric acid. These acid droplets cause reduced visibility. The acid may also combine with other contaminants in the air to form particulate sulfates, such as ammonium sulfate.

$$H_2O + SO_3 \longrightarrow H_2SO_4$$
$$2\,NH_3 + H_2SO_4 \longrightarrow (NH_4)_2SO_4$$

Sulfur dioxide is definitely known to be injurious to plants and to be a health hazard to humans and other animals. Its main effect on humans is to cause irritation of the upper respiratory tract and bronchial constriction, making it difficult to breathe. Persons suffering from bronchial disorders such as asthma and emphysema are seriously affected by concentrations of 0.1 to 0.2 ppm SO_2. Most healthy individuals suffer bronchial spasms when exposed to 5 to 10 ppm SO_2 for 1 hr. Sulfur dioxide is also intensely irritating to the eyes.

Sulfuric acid in aerosol form is absorbed deep in the lungs, causing damage to the sensitive tissues. Sulfuric acid is a strong acid and as a mist in the atmosphere causes considerable corrosion damage to steel, nickel, zinc, marble, and limestone surfaces.

Both SO_2 and H_2SO_4 are injurious to vegetation, causing yellowing and dropping of leaves. Injury to plants may be acute or chronic, depending on the concentration and length of exposure to the pollutants. Total destruction of vegetation over areas of many square miles has been experienced in regions surrounding certain metal sulfide smelters with high SO_2 emission.

Sulfur dioxide is not a major problem in automobile exhaust because most refined gasolines contain very little sulfur.

Hydrocarbons

Gasoline, kerosene, and diesel fuel used to power motor vehicles are essentially all hydrocarbons. Hydrocarbons are organic compounds consisting of carbon and hydrogen. They enter the atmosphere from automobile exhaust, evaporation from gasoline tanks and carburetors, petroleum refinery operations, storage tanks, and petroleum solvents. Most of the hydrocarbons in the atmosphere come from motor vehicles. A substantial amount also comes from users of petroleum solvents in paints and from other industries.

Currently, there is no evidence to show that the paraffins (saturated open-chain hydrocarbons) in concentrations known in the atmosphere are toxic to humans. In fact, methane (marsh gas) is a normal constituent in the atmosphere (2 ppm). The olefins (unsaturated hydrocarbons of the ethylene series) and the aromatic hydrocarbons such as benzene and toluene do contribute to air pollution problems. The olefins and aromatic hydrocarbons react with oxides of nitrogen and ozone in the presence of sunlight to produce photochemical smog. Among the reaction products are eye irritants and particulate matter, causing reduced visibility on smoggy days. More details of the process that produces photochemical smog are considered in Section 21.6.

Carbon Monoxide

Carbon monoxide (CO) is a colorless, odorless, highly poisonous gas. A small amount of CO enters the atmosphere from natural sources. High concentrations in the atmosphere result mainly from incomplete combustion of carbon and carbon compounds.

$$C + O_2 \longrightarrow 2\,CO \quad \text{(Incomplete combustion)}$$
$$2\,C_8H_{18} + 17\,O_2 \longrightarrow 16\,CO + 18\,H_2O \quad \text{(Incomplete combustion)}$$
$$2\,C_8H_{18} + 25\,O_2 \longrightarrow 16\,CO_2 + 18\,H_2O \quad \text{(Complete combustion)}$$

The primary source of CO in the atmosphere in urban communities is automobile exhaust. Los Angeles County Air Pollution Control District (now part of the South Coast Air Quality Management District) data show that the CO emitted into the area's atmosphere is almost exclusively from motor vehicles, since stationary sources are essentially completely controlled.

Many deaths have resulted from CO released in enclosed areas, and a number of these were caused by improperly adjusted heating equipment. If a person breathes a concentration of 1000 ppm (0.1%) for 1 hour, a state of unconsciousness results, with death following after about 4 hours. A concentration of about 1% (10,000 ppm) can cause death within a few minutes. Concentrations of a few parts per million are dangerous if breathed for a long time. The air quality standard adopted by the United States Environmental Protection Agency (EPA) states that the CO concentration should not exceed 9 ppm for a period of 8 hours or 35 ppm for 1 hour.

The body obtains vitally needed oxygen from a hemoglobin–oxygen complex circulating in the blood. This complex is called *oxyhemoglobin* (HbO_2). Carbon monoxide is toxic because it also combines with the hemoglobin in the blood, forming a carboxyhemoglobin complex (COHb), and thereby limits the blood's capacity to transport oxygen. This causes oxygen starvation throughout the body. The carboxyhemoglobin complex is about 200 times more stable than the HbO_2 complex. Thus, CO preferentially combines with hemoglobin when both O_2 and CO are present:

$$O_2 + Hb \rightleftharpoons HbO_2 \quad \text{(Oxyhemoglobin)}$$
$$CO + Hb \rightleftharpoons COHb \quad \text{(Carboxyhemoglobin)}$$

The concentration of COHb complex in the blood is directly proportional to the CO concentration in the inhaled air and the length of time this air is breathed. Fortunately, the body absorbs CO slowly and several hours are required to establish an equilibrium between the COHb complex and the CO inhaled in the air (see Figure 21.1).

A concentration of 2% COHb in blood shows some evidence of causing abnormal behavioral performance in humans. Greater concentrations (2–5%)

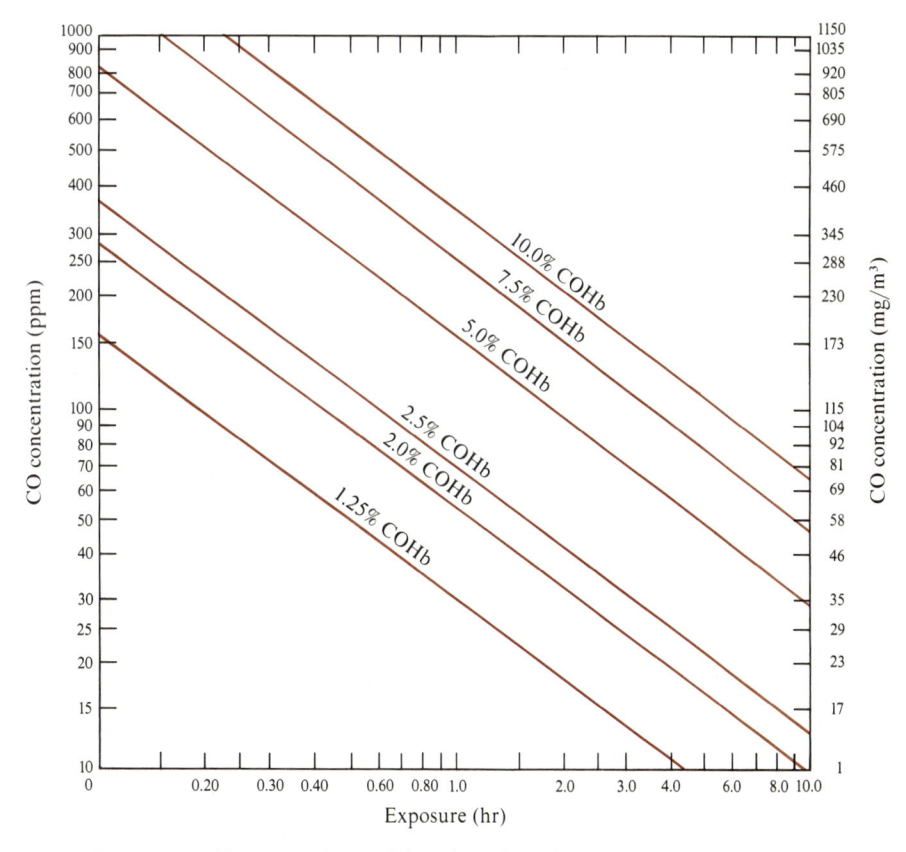

Figure 21.1. Concentration and duration of continuous CO exposure required to produce blood COHb concentrations of 1.25, 2.0, 2.5, 5.0, 7.5, and 10.0% in healthy male subjects engaging in sedentary activity. (Data from United States Department of Health, Education, and Welfare, *Air Quality Criteria for Carbon Monoxide.*)

lead to central nervous system effects—impaired vision and misjudgment of reaction time and distance. Still higher concentrations can cause cardiac and pulmonary problems, fatigue, drowsiness, loss of consciousness, and finally death by asphyxiation.

Since CO is one of the principal pollutants from motor vehicles, it constitutes a particular hazard on crowded freeways and on congested city streets. Numerous studies have shown that daily concentrations of CO follow the pattern of vehicular traffic density in highly populated urban areas. The concentration of CO has exceeded 50 ppm on many days in several cities.

Cigarette smoke contains a CO concentration of up to 20,000 ppm. Cigarette smokers inhale from 200 to 400 ppm CO every time they take a puff. It is estimated that a heavy smoker (two packs or more per day) has more than 5% of the hemoglobin in his or her body bound with CO.

Over 100 million tons of CO are emitted into the air in the United States every year. If this CO accumulated in the atmosphere, the air would soon be hazardous to breathe. Since worldwide concentrations of CO appear to be constant, there must be a mechanism in the environment whereby the CO is removed from the atmosphere. It was thought at one time that CO might be dissolved by the oceans. But the oceans are now recognized as being a source of CO. (Certain marine organisms are known to give off carbon monoxide.) Oxidation of CO to CO_2 by oxygen in the atmosphere is a very slow process. Green plants utilize CO_2, not CO. So what is causing the CO to disappear? Experiments have shown that microorganisms found in ordinary soil may be responsible for removing CO from the atmosphere. Carbon monoxide was rapidly removed from samples of air in contact with ordinary soil, but when the soil was sterilized, CO was not removed from similar air samples. Unfortunately, large metropolitan areas have greatly reduced this natural method of eliminating CO by their vast expanses of cement and asphalt-paved areas.

Nitrogen Oxides

Several oxides of nitrogen are known, but two of them—nitric oxide (NO) and nitrogen dioxide (NO_2)—are serious air pollution hazards. Collectively, NO and NO_2 are referred to as NO_x.

Nitric oxide is a colorless, odorless gas and is only slightly toxic. It is formed from the oxygen and nitrogen in the air during combustion processes. Nitric oxide is produced during the burning of fuels from all sources—automobiles, aircraft engines, home and commercial furnaces, and large power plants. The rate of formation of NO increases with the combustion temperature:

$$N_2 + O_2 + \text{Heat} \longrightarrow 2\,NO$$

The equilibrium concentration of NO in this equation is very dependent on the combustion temperature. Concentration values at various temperatures are shown in the following table. Temperatures around 2200°C are attained in high-compression internal-combustion engines in automobiles. Bunsen burners can create temperatures above 1500°C.

Temperature (°C)	Concentration of NO (ppm)
20	<0.001
427	0.3
527	2.0
1,538	3,700
2,200	25,000

Nitrogen dioxide is a reddish-brown gas with a pungent, sweet odor. It is far more toxic than NO. Its effect on humans ranges from mild lung irritation, to pulmonary edema (accumulation of fluid in the tissues), to death, depending on the concentration and the duration of exposure. Even short exposure is dangerous. Because of its color and its absorption of light, NO_2 can significantly reduce visibility and can be detected in smog as a brown haze. Nitrogen dioxide is also injurious to plants. Exposure to 0.3 to 0.5 ppm NO_2 for periods of 10 to 22 days suppresses plant growth. Higher concentrations cause visible leaf injury.

The concentration of NO_x in urban atmospheres is considerably greater than in rural atmospheres, indicating that NO_x is primarily produced by people. Motor vehicles are the largest source of NO_x. In Los Angeles County, for example, motor vehicles account for more than 50% of the total emission of NO_x. This amount is expected to be reduced sharply in the next 10 years as motor vehicle exhaust controls on new cars go into effect.

Nitric oxide will oxidize slowly in the air to form NO_2, but in the presence of certain reactive hydrocarbons and adequate sunlight, this oxidation is accelerated. The NO_2 formed absorbs more light energy and triggers the formation of photochemical smog (see Section 21.6).

$$2\,NO + O_2 \rightleftharpoons 2\,NO_2$$

Nitrogen dioxide can react with moisture in the air to produce nitrous and nitric acids:

$$2\,NO_2 + H_2O \longrightarrow HNO_2 + HNO_3$$

Nitrogen dioxide can also be oxidized to N_2O_5, which in turn will react with water to form nitric acid. The nitric acid then becomes particulate matter in the atmosphere as ammonium or metal nitrates.

$$4\,NO_2 + O_2 \longrightarrow 2\,N_2O_5 \xrightarrow{2\,H_2O} 4\,HNO_3$$

Ozone

Ozone is a colorless, pungent, extremely toxic gas. It is composed of three atoms of oxygen per molecule (O_3) and thus is an allotropic form of oxygen. Ozone is a vigorous oxidizing agent. The term *oxidant*, commonly used in air pollution discussions, refers primarily to ozone, although other oxygen-containing substances enter into photochemical smog reactions.

Ozone is normally present in the atmosphere at about 0.05 ppm at sea level. It can be produced by an electric discharge in air or oxygen (such as lightning) and can be detected by its odor in the vicinity of electric motor and high-voltage transmission lines.

Normal humans show few ill effects from ozone concentrations of less than 0.1 ppm. But in higher concentrations ozone can cause coughing, choking, headache, fatigue, and reduced resistance to respiratory infection. Concentrations between 10 and 20 ppm are fatal to humans.

Danger to vegetation and materials due to oxidants in smog is a widespread and serious problem. The National Air Pollution Control Administration (NAPCA) estimates the annual crop damage resulting from smog at more than $100 million. Ozone is particularly destructive to rubber. The damage is more noticeable in rubber products that are under stress. In fact, an early method of measuring atmospheric ozone was based on the depth of cracks formed on stretched rubber bands. In the Los Angeles area particularly, automobile tires developed cracks that greatly reduced their life span. This problem was overcome by adding antioxidants to the rubber. However, other rubber products such as hoses, gloves, and wire insulation are still affected.

21.5 Ozone in the Stratosphere

The stratosphere is the second lower layer of the earth's atmosphere. It is a layer around the earth that extends from about 11 to 50 km (6.8 to 31 miles) above the equator. The stratosphere contains a relatively high concentration of ozone (O_3), which is produced by the reaction of oxygen with high-energy radiation from the sun. Oxygen molecules are split into oxygen atoms, which then react with other oxygen molecules to form ozone:

$$O_2 \xrightarrow{\text{Sunlight}} \underset{\text{Oxygen atoms}}{O + O}$$

$$O_2 + O \longrightarrow O_3$$

Ozone itself is not very stable and is decomposed back to oxygen molecules and oxygen atoms. This occurs when ozone molecules absorb ultraviolet radiation from the sun:

$$O_3 \xrightarrow[\text{radiation}]{\text{uv}} O_2 + O + \text{Heat}$$

Ultraviolet radiation from the sun is highly damaging to living tissues of plants and animals. However, the ozone layer in the stratosphere shields the earth from solar ultraviolet radiation and thus prevents most of this lethal radiation from reaching the earth's surface.

Scientists have become concerned about the growing hazard to the ozone layer caused by fluorocarbons released into the atmosphere. Fluorocarbons such as the Freons CCl_2F_2 and CCl_3F (used in aerosol spray cans and in refrigeration and air conditioning units) are stable compounds in the lower atmosphere. However, when these compounds diffuse into the stratosphere, they absorb ultraviolet radiation and produce chlorine atoms. Chlorine atoms are capable of rapid reaction with ozone to form unstable chlorine monoxide and oxygen. Chlorine monoxide can react with atomic oxygen to reform chlorine atoms, which can then react with more ozone. Thus, some of the protective ozone layer could be, or has already been, destroyed. The equations for these reactions are:

$$CCl_2F_2 \xrightarrow{\text{uv radiation}} CClF_2 + \underset{\text{Atomic chlorine}}{Cl}$$

$$CCl_3F \xrightarrow{\text{uv radiation}} CCl_2F + Cl$$

$$Cl + O_3 \longrightarrow \underset{\text{Chlorine monoxide}}{ClO} + O_2 + \text{Heat}$$

$$ClO + O \longrightarrow Cl + O_2$$

The amount of ozone damage in the stratosphere is a highly controversial issue. Some scientists claim that 80% or more of the released fluorocarbons are destroyed before they reach the stratosphere. Other scientists claim that 1% of the ozone layer has already been destroyed and this figure could go up to 15% by the year 2000. This might result in disastrous effects on life and major climatic changes due to more heat passing into the lower atmosphere. A number of states have already banned the use of fluorocarbons as aerosol propellants.

21.6 Photochemical Smog

Ozone, in addition to being produced by lightning and electrical discharges, is also produced in the lower atmosphere by the interaction of nitrogen dioxide, sunlight, and oxygen. In this photochemical process, ultraviolet energy is first absorbed by nitrogen dioxide to form an energized molecule (NO_2^*). This energized molecule then decomposes to nitric oxide and atomic oxygen [O]. Atomic oxygen is an extremely reactive species, which combines with atmospheric oxygen to form ozone. One of the reactions of ozone in the atmosphere is with NO to again form NO_2.

$$NO_2 \xrightarrow{\text{Sunlight}} NO_2^* \quad \text{(Energized)}$$
$$NO_2^* \longrightarrow NO + \underset{\text{Atomic oxygen}}{O}$$

$$O + O_2 \longrightarrow O_3$$
$$O_3 + NO \longrightarrow NO_2 + O_2$$

The photochemical cycle of NO_2 and O_3 is shown in Figure 21.2. Ozone is continually being formed and used up during this photochemical cycle and thus cannot begin to accumulate in the atmosphere until all the NO is oxidized to NO_2.

$$NO_2 \xrightarrow{\text{uv radiation}} NO + O$$
$$NO + O_3 \leftarrow O_2$$

Figure 21.2. Photochemical formation of ozone.

The photochemical cycle shown in Figure 21.2 is disrupted by the presence of reactive hydrocarbons in the atmosphere. These hydrocarbons react rapidly with atomic oxygen and ozone to produce a variety of oxidation products such as aldehydes (for example, formaldehyde and acrolein), peroxyacylnitrates (PAN), and aerosols, which cause reduced visibility on smoggy days. Aldehydes and PAN are some of the compounds responsible for the eye irritation resulting from photochemical smog.

Formaldehyde: $H_2C=O$
Acrolein: $CH_2=CH-CHO$
Peroxyacetylnitrate (PAN): $CH_3C(O)O-O-NO_2$

All the chemical reactions involved in photochemical smog are not completely understood, but the fact that hydrocarbons are an essential ingredient is well documented. Figure 21.3 illustrates the flow of materials in the production of photochemical smog.

To summarize, the main effects of photochemical smog are eye irritation, damage to vegetation and materials, reduced visibility, and possible injury to the respiratory systems of humans and other animals. Photochemical smog has been principally associated with air pollution in Los Angeles, California.

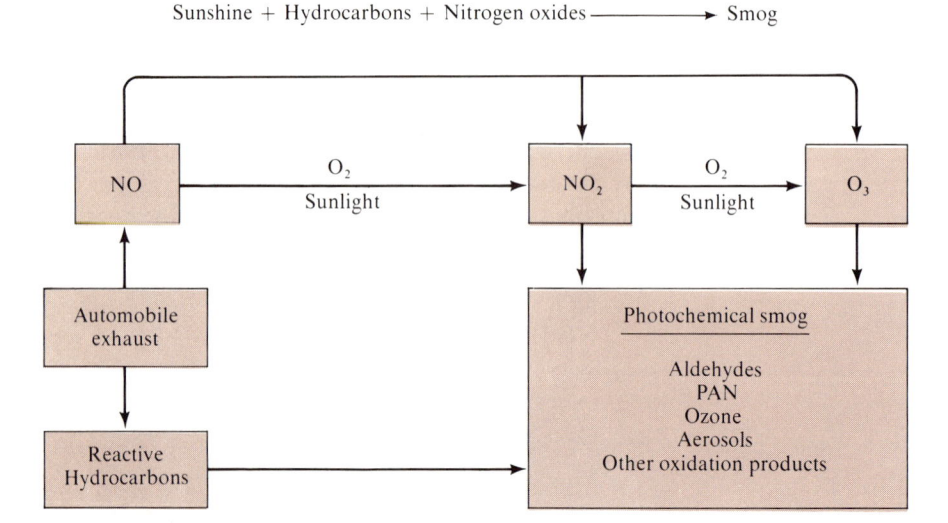

Sunshine + Hydrocarbons + Nitrogen oxides ⟶ Smog

NO $\xrightarrow[\text{Sunlight}]{O_2}$ NO_2 $\xrightarrow[\text{Sunlight}]{O_2}$ O_3

Automobile exhaust → NO
Reactive Hydrocarbons →
Photochemical smog: Aldehydes, PAN, Ozone, Aerosols, Other oxidation products

Figure 21.3. Formation of photochemical smog.

Figure 21.4. The effect of air pollution on visibility. The two pictures illustrate downtown Los Angeles, California, on a clear day and on a day when the temperature inversion layer is below 1000 ft, holding pollution down. (Courtesy South Coast Air Quality Management District.)

However, it has now been shown to be present in many other cities in the world. Figure 21.4 shows the effect of air pollution on visibility in Los Angeles.

21.7 Atmospheric Temperature Inversion

In addition to pollution sources, meteorology, climate, and topography of an area are important factors contributing to the accumulation of pollutants in the atmosphere. In areas with sufficient wind velocity, the contaminants discharged into the atmosphere are swept away and diluted with fresh air before they can cause a problem. Areas that lie within a basin or a valley are especially prone to air pollution problems.

The temperature and the density of the air normally decrease with altitude. Thus, air heated by radiation from the earth's surface expands and rises,

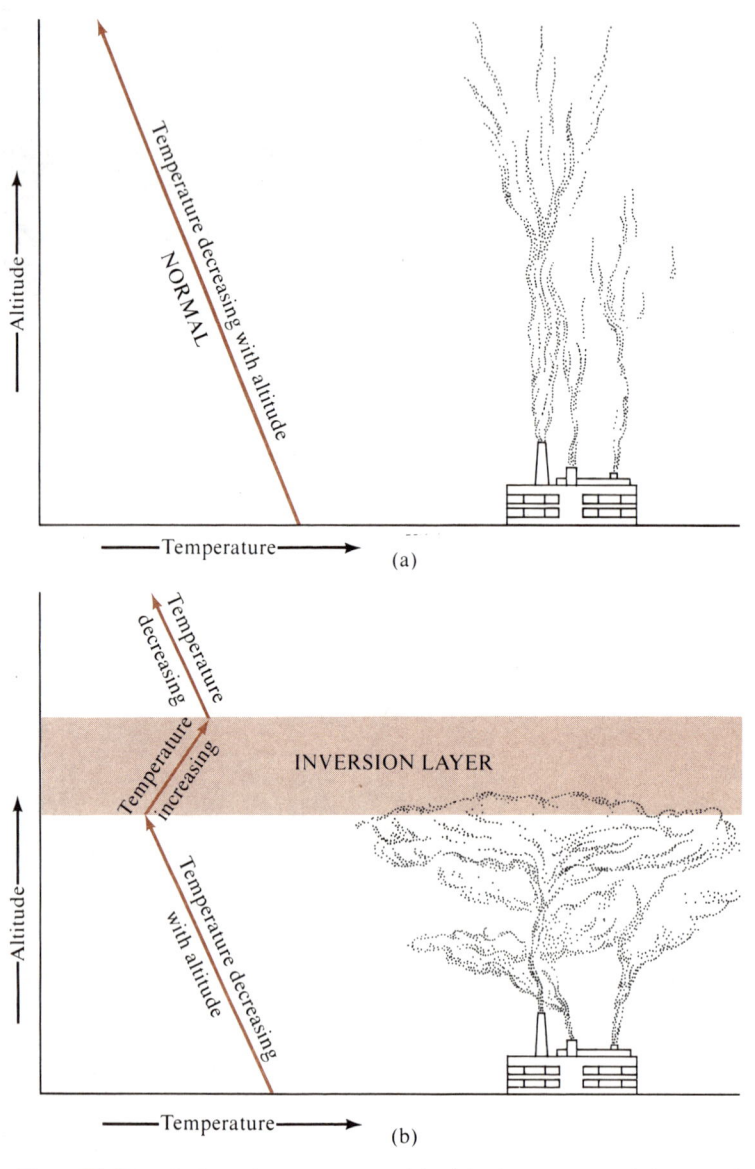

Figure 21.5. Variation of temperature with altitude. (a) Normal decrease in temperature of the air with altitude. (b) A temperature inversion layer of warm air over cooler air. When this condition occurs, circulation of the air is confined below the base of the inversion layer, and air pollutants can accumulate in the atmosphere near ground level.

generating air currents that disperse pollutants into the upper atmosphere. For example, under normal temperature conditions, a smoke plume will rise into the atmosphere and soon be dispersed by vertical mixing with the air.

temperature inversion

The atmospheric condition in which a layer of warm air is present over colder air is known as a **temperature inversion**. In some geographic areas, temperature inversions occur frequently. A temperature inversion effectively puts a lid over an area by confining the circulation of the air to the space between the ground and the base of the inversion layer.

Inversions may occur because of the rapid cooling of the ground at night. The air at the surface of the ground will cool to a temperature below that of the air at higher altitudes, creating a temperature inversion. Areas along the coast also have inversions because of the relatively cool surface air from ocean waters. Figure 21.5 shows the variations of temperature with altitude under both normal and temperature inversion conditions. On days of low-altitude inversion and low wind velocity, contaminants emitted into the atmosphere are trapped under the inversion layer and increase in concentration, causing air pollution problems.

The average altitude of the inversion layer in the Los Angeles Basin is under 0.46 km (1500 ft) on about 50% of the days each year and under 0.76 km (2500 ft) on about 70% of the days each year. Air pollution control authorities use data on the inversion layer altitude to predict days of potentially unhealthful air due to pollutants. On such days, the severity of air pollution can be limited by curtailment of the sources that produce pollutants.

21.8 Air Pollution Control

That there is a need to control the sources of air pollution is beyond question. Sources that must be controlled fall into two categories: (1) stationary sources and (2) moving sources. In the category of stationary sources we have thousands of industrial processing plants such as steel and related industries; chemical, pulp and paper, and agricultural industries; residential and commercial heating and power-generating plants; and the disposal of combustible refuse by open burning. Moving sources include motor vehicles of all kinds, trains, and airplanes.

Controls now in effect did not come about easily. Not many industrial corporations were willing to spend large amounts of capital for the research and control equipment necessary to reduce atmospheric pollution. The establishment of controls on industry was hampered by economic, political, technological, and legal pressures. Finally, controls had to be made mandatory by legislation. This generally occurred first at the local level, then at the state level, and finally at the federal level. In general, legislative controls were enacted when pollution problems became so acute that a hazard to the health and welfare of the general public was apparent. It is true that not all areas of the nation need pollution controls. But since the majority of the populace live in large metropolitan areas where pollution is a serious problem, and since we are a very mobile people, national regulations governing the emission of contaminants into the atmosphere are necessary.

The United States Environmental Protection Agency (EPA) published a list of *Air Quality Standards* on April 30, 1971. Two types of national air standards were established—primary (necessary to protect the public health) and secondary (necessary to protect the public welfare and the ecology from known or anticipated adverse effects of a pollutant). These standards for common air pollutants are given in Table 21.3.

Table 21.3. National primary and secondary air quality standards established by the United States Environmental Protection Agency.

Air pollutant	National standard	Maximum allowable short-period concentrations and averaging time			Significant harm to health	
		$\mu g/m^3$	ppm	Averaging time (hr)	ppm	Time (hr)
Photochemical oxidant, O_3	Primary[a]	160	0.12	1	0.60	1
	Secondary[b]	160	0.12	1		
Carbon monoxide	Primary	10,000	9	8	50	8
		40,000	35	1		
	Secondary	10,000	9	8	75	4
		40,000	35	1	125	1
Nitrogen dioxide	Primary	100	0.05	Annual mean	0.5	24
	Secondary	100	0.05	Annual mean	2.0	1
Sulfur dioxide	Primary	365	0.14	24	1.0	24
	Secondary	1,300	0.50	3		
Hydrocarbons (nonmethane)	Primary	160	0.24	3(6–9 A.M.)		
	Secondary	160	0.24	3(6–9 A.M.)	—	
Particulate matter	Primary	260	—	24	1000 $\mu g/m^3$	24
	Secondary	150	—	24		

[a] Primary standard—necessary to protect the public health.
[b] Secondary standard—necessary to protect the public welfare and the ecology from known or anticipated adverse effects of a pollutant.

One of the first areas in which serious attempts were made to control air pollution was Los Angeles County, California. The Los Angeles County Air Pollution Control District was legally established in 1947. The district adopted control laws and regulations and began enforcement activities in February 1948. Stationary sources are now controlled to the extent that they emit only about 10% of the air pollutants in the Los Angeles Basin. The other 90% comes from motor vehicles. Control measures are currently preventing more than 10,000 tons of pollutants from entering the Los Angeles County atmosphere every day.

Air-monitoring networks are being set up all over the country. Los Angeles has an air-monitoring network that records contaminants 24 hours a day. This monitoring system is the heart of a three-stage "Episode Criteria" system established in 1955 for the purpose of preventing excessive build-up of atmospheric pollutants. Whenever an episode level is predicted, the public is notified through the press, radio, television, and other mass media and is asked to participate in abatement actions. Depending on the severity of the episode and the pollutant causing the episode, the air pollution control officer is authorized to curtail public and private activities in order to reduce the concentration of the pollutant to safe levels. The concentration levels at which alerts are called are given in Table 21.4.

Table 21.4. Episode criteria levels for the South Coast Air Quality Management District (Los Angeles, Orange, Riverside, and San Bernardino Counties, California). Adopted May 6, 1977. Action to curtail use of equipment producing pollution is declared when the level of any of the pollutants listed is anticipated or reached.

Contaminant	Averaging time	Stage 1	Stage 2	Stage 3
Photochemical oxidant (including ozone)	1 hour	0.20 ppm	0.35 ppm	0.50 ppm
Oxidant in combination with sulfur dioxide	1 hour	0.20 ppm	0.35 ppm	0.50 ppm
Carbon monoxide	1 hour	40 ppm	75 ppm	100 ppm for 1 hour and predicted to persist for one additional hour
	12 hours	20 ppm	35 ppm	50 ppm
Sulfur dioxide	1 hour	0.5 ppm	1.0 ppm	2.0 ppm
	24 hours	0.2 ppm	0.7 ppm	0.9 ppm
Oxidant in combination with sulfate	24 hours (sulfate)		25 $\mu g/m^3$	
	1 hour (oxidant)		0.20 ppm	

Motor vehicle pollution control methods are presently in use. Crankcase emission controls became mandatory beginning with 1963 automobiles. Emission of pollution from crankcase blowby was reduced essentially to zero by use of a closed crankcase ventilation system. Beginning with 1971 models, motor vehicles were required to have controls for evaporative losses from fuel tanks and carburetors. By use of these controls, hundreds of tons of hydrocarbons are prevented from entering the atmosphere each day by direct evaporation from gasoline.

Control of automobile exhaust is a much more difficult problem. Several systems have been tried to bring about complete combustion of hydrocarbons and carbon monoxide, ranging from carburetor modifications to the use of auxiliary air injection pumps. Recirculation of exhaust gas has been proven to reduce oxides of nitrogen. However, to meet federal exhaust emission requirements (see Table 21.5) automobile exhaust gases are now also being circulated through catalytic converters. The catalyst has a dual function—it must be able to oxidize hydrocarbons and CO to CO_2 and to reduce NO_x to free N_2 gas.

Table 21.5. Federal standards for exhaust emission requirements for passenger cars. (All figures are in grams per mile.)

Year	Hydrocarbons	Carbon monoxide	Nitrogen oxides
Pre-1966 (no controls)	11.0	80.0	4.0
1972	3.4	39	—
1973	3.4	39	3
1974	3.4	39	3
1975	1.5	15	3
1976	1.5	15	3
1977	1.5	15	2.0
1978[a]	0.41	3.4	0.4

[a] Effective for 1978 and beyond unless changed by Congressional action.

The catalysts used in the catalytic converters are easily "poisoned" (rendered noneffective) by heavy metals such as lead. Since tetraethyl lead [$Pb(C_2H_5)_4$] is (or was) used in most gasolines, lead-free fuels were developed to avoid poisoning the catalyst. This was accomplished, automobile engines were modified to burn the unleaded fuel, and owners of 1975 and later model automobiles are instructed to use only unleaded fuel.

A secondary, but important, benefit of using unleaded fuels is the parallel reduction in atmospheric lead concentrations. This downtrend in atmospheric lead is projected to continue to decrease as older automobiles requiring leaded fuels are replaced and as more stringent state and federal regulations limiting fuel lead content take effect. Lead is a cumulative poison since the body retains some of the lead that we breathe and swallow. Over a long period of time concentrations of lead leading to serious health effects may be attained. Lead poisoning can result in kidney damage, damage to the nervous system, and an increased likelihood of death from brain damage.

The effectiveness of crankcase and evaporative controls has been generally excellent almost from their initial use. However, the control of hydrocarbons and carbon monoxide emissions from new automobiles has been poor, with many vehicles exceeding emission standards while new, and others not being able to maintain standards for more than 10,000–12,000 miles. One unfortunate effect of reducing hydrocarbon emissions was a large increase in the emission of oxides of nitrogen. This, of course, enhances photochemical oxidant formation. Beginning with 1971 models, the required control of NO_x emissions began to reduce photochemical smog. The use of the catalytic converter is helping to solve this problem.

Questions

A. Review the meanings of the new terms introduced in this chapter.
   1. Pollutant
   2. Smog
   3. Particulate matter
   4. Aerosol
   5. Temperature inversion

B. Review questions.
   1. What are the major differences between London smog and Los Angeles smog?
   2. What is the major source (or sources) of each of the following air pollutants?
      (a) CO   (b) SO_2 (c) NO_x (d) Hydrocarbons   (e) Particulate matter
   3. What is the most noticeable effect of aerosols in the atmosphere?
   4. Set up a table showing the physiological effects of each of the following air pollutants:
      (a) SO_2 (b) CO   (c) NO_x (d) O_3
   5. Explain how sulfur dioxide in the atmosphere is converted to a sulfuric acid mist (aerosol). Write chemical equations for the reactions.
   6. Explain why early Greek and Roman marble statues deteriorate rather rapidly when moved to large urban areas.
   7. What evidence shows that the automobile is the principal source of carbon monoxide in urban atmospheres?
   8. Why is it extremely hazardous to operate an automobile engine in a closed garage?
   9. In what way can cigarette smoking contribute to inefficiency in driving an automobile?
   10. In what specific way does carboxyhemoglobin interfere with normal functioning of the blood?
   11. Internal combustion engines burn hydrocarbons. Why is nitric oxide present in automobile exhaust?
   12. Explain how ozone is formed in the stratosphere.
   13. What would be the effect of a significant reduction of the ozone concentration in the stratosphere?
   14. What is photochemical smog?
   15. What substance is the major photochemical oxidant in the formation of smog? How does it enter into the air pollution problem?
   16. What is an atmospheric temperature inversion?
   17. Explain how atmospheric temperature inversion contributes to air pollution.
   18. Briefly explain how each of the following factors contributes to air pollution:
      (a) A large urban community
      (b) Low wind velocity
      (c) A bright sunny day
      (d) An industrial city located in a valley surrounded by high hills
      (e) A large difference between day and night temperatures
   19. Early air pollution control devices on automobiles decreased hydrocarbon and carbon monoxide emissions. Explain why these devices failed to decrease the brown haze in photochemical smog.
   20. Briefly describe the methods used to control air pollution in a large urban community such as Los Angeles, California.
   21. What devices are in use today to control pollution from motor vehicles? What pollutants are controlled by each of these devices?
   22. Explain how the catalytic converter is contributing to the reduction of photochemical smog.

468 Chapter 21

23. Which of the following statements are correct?
   (a) Pollutants are substances that get into an environment in which they are not normally present.
   (b) The primary effect of aerosols in the atmosphere is reduced light and lowered visibility.
   (c) Oxides of nitrogen cause effects on humans varying from mild lung irritation to pulmonary edema to death.
   (d) Ozone and sulfur dioxide are examples of particulate matter in air pollution.
   (e) Ammonium sulfate is a form of particulate matter originating from sulfur dioxide in the air.
   (f) In concentrations of 1 to 10 ppm, ozone can cause coughing, choking, headache, fatigue, and reduced resistance to respiratory infection.
   (g) Sulfur dioxide is converted to a sulfuric acid mist in the atmosphere by the following reaction:

   $$SO_2 + H_2 + O_2 \longrightarrow H_2SO_4$$

   (h) The primary source of sulfur dioxide in the air is from automobile exhaust.
   (i) In highly populated urban areas, the concentration of carbon monoxide in the air varies directly as the density of vehicular traffic.
   (j) It is dangerous to operate an automobile engine in a closed garage because of the build-up of ozone concentration.
   (k) Cigarette smoking exposes the lungs to high concentrations of carbon monoxide, which can lead to impaired vision and misjudgment of reaction time and distance.
   (l) Carbon monoxide reduces the oxygen-carrying capacity of the blood because it uses up the oxygen in being oxidized to carbon dioxide.
   (m) Nitric oxide is present in automobile exhaust because nitrogen in the air is oxidized at the high temperatures in the engine.
   (n) Nitrogen dioxide, a reddish-brown gas, is much more toxic than nitric oxide, a colorless gas.
   (o) The concentration of nitrogen oxides in exhaust can be reduced by higher operating temperatures in the automobile engine.
   (p) Nitric oxide is the major photochemical oxidant in the formation of smog.
   (q) The atmospheric condition where a layer of cold air is present over warmer air is known as a temperature inversion.
   (r) On days of low-altitude inversion and low wind velocity, pollutants are trapped under the inversion layer and increase in concentration.
   (s) A large difference between day and night ground temperatures helps prevent a temperature inversion.
   (t) In the catalytic converter, carbon monoxide is oxidized to carbon dioxide, and nitric oxide is oxidized to nitrogen dioxide.

22 Organic Chemistry; Saturated Hydrocarbons

After studying Chapter 22 you should be able to:

1. Define and understand the terms in Question A at the end of the chapter.
2. Understand the tetrahedral nature of the carbon atom.
3. Explain why the concept of hybridization is used to describe the bonding of carbon in simple compounds such as methane.
4. Explain the bonding in alkanes.
5. Write the electron-dot structures of alkanes and halogenated alkanes.
6. Write the names and formulas of the first ten normal alkanes.
7. Write structural formulas and IUPAC names for the isomers of an alkane or a halogenated alkane.
8. Give the IUPAC name of a hydrocarbon or a halogenated hydrocarbon when given the structural formula or vice versa.
9. Write equations to show how an alkane is chlorinated to form a monochloroalkane by a free radical reaction mechanism.
10. Write equations for the halogenation of an alkane, giving all the possible monohalosubstitution products.
11. Write structural formulas and names for simple cycloalkanes.

22.1 Organic Chemistry: History and Scope

Chemists during the late 18th and 19th centuries were baffled by the fact that compounds obtained from animal and vegetable sources defied the established rules for inorganic compounds (namely, that compound formation was due to the attraction between positive and negative charged elements). They observed that a group of four elements—carbon, hydrogen, oxygen, and nitrogen—gave rise to a large number of different compounds that often were remarkably stable. This was in contrast to only one, or at most a few, known compounds composed of any other group of only two or three elements.

Since no organic compounds had been synthesized from inorganic substances and since there was no other explanation for the complex nature of organic compounds, chemists were led to believe that organic compounds were formed by some "vital force." The **vital force theory** held that organic substances could originate only from some form of living material. In 1828, Friedrich Wöhler (1800–1882), a German chemist, did a simple experiment that eventually proved to be the death blow to this theory. In attempting to

vital force theory

prepare ammonium cyanate (NH$_4$CNO) by heating cyanic acid (HCNO) and ammonia, Wöhler obtained a crystalline white substance which he identified as urea (H$_2$N–CO–NH$_2$). Urea is an authentic organic substance because it is a product of metabolism, and Wöhler knew that it had been isolated from urine. Although Wöhler's discovery was not immediately and generally recognized, the vital force theory was overthrown by this simple experiment since *one* organic compound had been made from nonliving materials.

After the work of Wöhler, it was apparent that no vital force, other than skill and knowledge, was needed to make organic chemicals in the laboratory and that inorganic as well as other organic substances could be used as raw materials. Today, **organic chemistry** simply designates the branch of chemistry that deals with carbon compounds but does not imply that these compounds must originate from some form of life. A few special kinds of carbon compounds—for example, carbon oxides, metal carbides, and metal carbonates—are excluded from the organic classification because their chemistry is more conveniently related to that of inorganic substances.

organic chemistry

The field of organic chemistry is vast, for it includes not only the composition of all living organisms but also of a great many other materials that we use daily. Examples of organic materials are foodstuffs (fats, proteins, carbohydrates); fuels of all kinds; fabrics (cotton, wool, rayon, nylon); wood and paper products; paints and varnishes; plastics; dyes; soaps and detergents; cosmetics; medicinals; rubber products; and explosives.

The sources of organic compounds are carbon-containing raw materials—petroleum and natural gas, coal, carbohydrates, fats, and oils. In the United States we produce about 250 billion pounds of organic chemicals per year from these sources. This amounts to more than 1100 pounds per year for every man, woman, and child in the United States. About 90% of this 250 billion pounds comes from petroleum and natural gas. Since there is a finite amount of petroleum and natural gas in the world we will, sometime in the future, have to rely on the other sources to make the voluminous amount of organic chemicals that we use. Fortunately, we know how to do this, but at much greater expense than from petroleum and natural gas.

An immense and ever-growing number of organic chemical compounds have been prepared and are described in the chemical literature. The number is estimated to be on the order of 3.4 million. The elements that make up most of the organic compounds are carbon, hydrogen, oxygen, nitrogen, the halogens (Cl, Br, I), and sulfur. There is no theoretical limit on the number of organic compounds that can exist. Several thousand different organic chemicals are available commercially.

22.2 The Need for Classification of Organic Compounds

It is physically impossible for anyone to study the properties of each of the hundreds of thousands of known organic compounds. Hence, organic compounds with similar structural features are grouped into series or classes.

Even within a single class there is such a large number of possible compounds that it is virtually impossible to study each individual compound. The properties of a few members of a series are studied, and the information obtained is used to make predictions about the behavior of other members of the series. For example, ethyl alcohol, a two-carbon alcohol, can be oxidized to a two-carbon acid. On the basis of this information, it can be predicted and found to be true that hexyl alcohol, a six-carbon alcohol, can be oxidized to a six-carbon acid, because the oxidizable structures of the two alcohols are identical. Some of the classes of organic compounds we will study are hydrocarbons, alcohols, phenols, aldehydes, ketones, ethers, carboxylic acids, esters, carbohydrates, and proteins. Each of these classes of compounds is identified by certain characteristic structural features.

22.3 The Carbon Atom: Tetrahedral Structure

The carbon atom is central to all organic compounds. The atomic number of carbon is 6 and its electron structure is $1s^2 2s^2 2p^2$. Two stable isotopes of carbon exist, C-12 and C-13. In addition, there are several radioactive isotopes; C-14 is the most widely known of these because of its use in radiocarbon dating. Having four electrons in its outer shell, carbon has oxidation numbers ranging from $+4$ to -4 and forms predominantly covalent bonds. Carbon occurs as the free element in diamond, graphite, coal, coke, carbon black, charcoal, and lampblack.

A carbon atom generally forms four covalent bonds. The most common geometric arrangement of these bonds is tetrahedral (see Figure 22.1). In this tetrahedral structure, the four covalent bonds are not planar about the carbon atom but are directed toward the corners of a regular tetrahedron. (A tetrahedron is a solid figure with four sides.) Figure 22.1 illustrates (a) a regular tetrahedron, (b) a carbon atom with its bonds in tetrahedral arrangement, (c) a carbon atom placed inside a regular tetrahedron, and (d) a model of a methane molecule with the carbon–hydrogen bonds tetrahedral. The angle between these bonds is 109.5°.

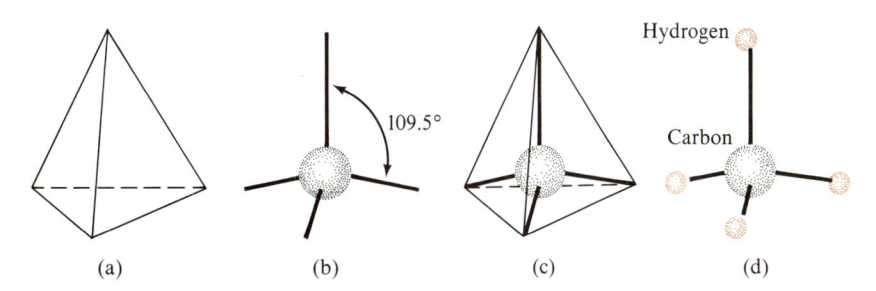

Figure 22.1. Tetrahedral structure of carbon: (a) A regular tetrahedron; (b) a carbon atom with tetrahedral bonds; (c) a carbon atom within a regular tetrahedron; (d) a methane molecule, CH_4.

22.4 Carbon–Carbon Bonds

With four outer-shell electrons, the carbon atom (·Ċ·), following the octet rule, forms four single covalent bonds by sharing electrons with other atoms. The structures of methane and carbon tetrachloride illustrate this point:

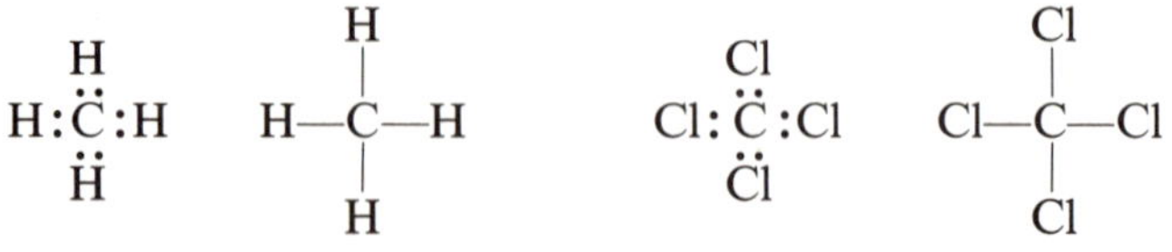

```
        H                         Cl
  H     |                         |
H:C:H  H—C—H    Cl:C:Cl    Cl—C—Cl
  H     |         Cl              |
        H                         Cl

    Methane         Carbon tetrachloride
```

Actually, the bond angles (109.5°) in these compounds are tetrahedral (see Figure 22.1), but for convenience of writing, the bonds are drawn at right angles. In methane, each bond is formed by the sharing of electrons between a carbon and a hydrogen atom. The ability of carbon to bond to itself is based on the sharing of electrons between carbon atoms. One, two, or three pairs of electrons can be shared between two carbon atoms, forming a single, double, or triple bond, respectively:

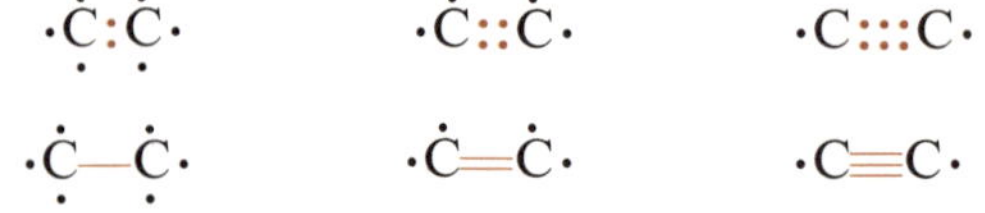

```
  ·C:C·         ·C::C·        ·C:::C·

  ·C—C·         ·C=C·         ·C≡C·

Single bond    Double bond    Triple bond
```

Each dash above represents a covalent bond. Carbon, more than any other element, has the ability to form short or very long chains of atoms covalently bonded together. This bonding ability is the main reason for the large number of organic compounds. Three examples are shown below. It is easy to see how, because of this bonding ability, a chain of carbon atoms can be formed, linking one carbon to another through covalent bonds.

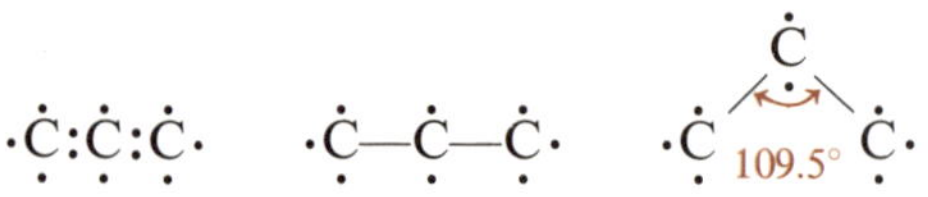

Three carbon atoms bonded by single bonds

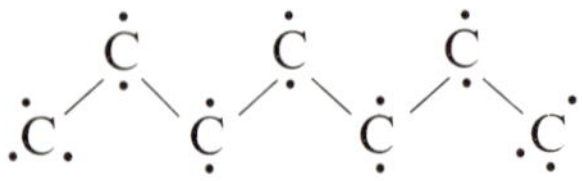

Seven-carbon chain

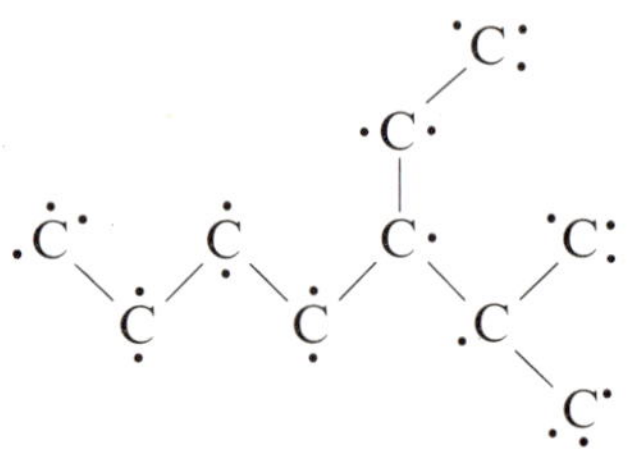

Ten carbon atoms bonded together

22.5 Hydrocarbons

hydrocarbons

Hydrocarbons are compounds that are composed entirely of carbon and hydrogen atoms bonded to each other by covalent bonds. Several classes of hydrocarbons are known. These include the alkanes, alkenes, alkynes, and aromatic hydrocarbons (see Figure 22.2).

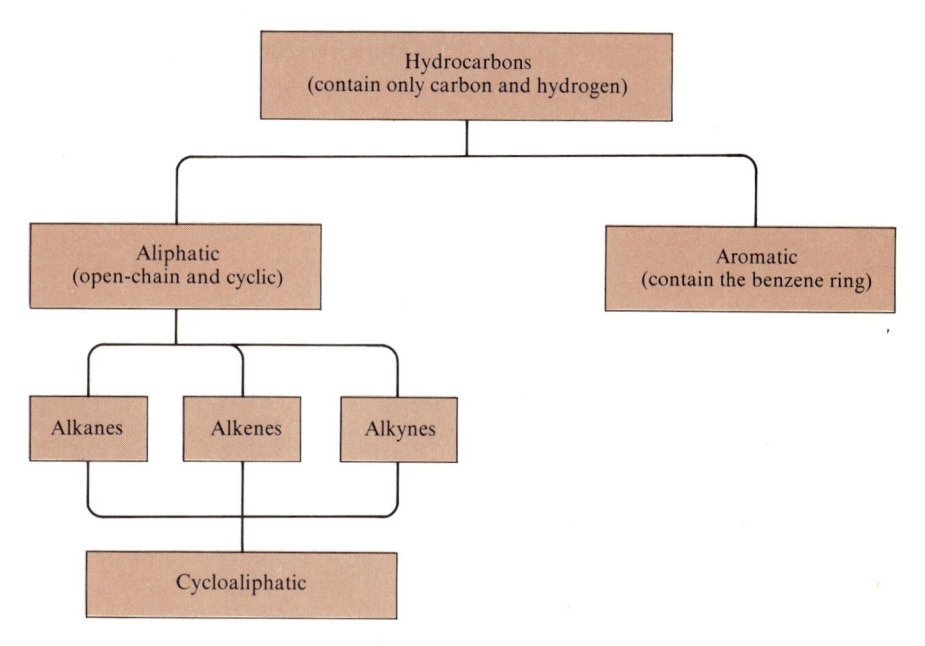

Figure 22.2. Classes of hydrocarbons.

Fossil fuels—natural gas, petroleum, and coal—are the principal sources of hydrocarbons. Natural gas is primarily methane with small amounts of ethane, propane, and butane. Petroleum is a complex mixture of hydrocarbons. Gasoline, kerosene, fuel oil, lubricating oil, paraffin wax, and petrolatum—all of which are simply mixtures of hydrocarbons—are separated from petroleum. Coal tar, a volatile product driven off in the process of making coke from coal for use in the steel industry, is the source of many valuable chemicals including the aromatic hydrocarbons, benzene, toluene, and naphthalene.

22.6 Saturated Hydrocarbons: Alkanes

alkanes

The **alkanes**, also known as *paraffins* or *saturated hydrocarbons*, are straight- or branched-chain hydrocarbons with only single covalent bonds between the carbon atoms. We shall study the alkanes in some detail because many other classes of organic compounds may be considered as derivatives of these substances. For example, it is necessary to learn the names of the first ten members of the alkane series because these names are used (with slight modifications) for corresponding compounds belonging to other classes.

Methane (CH$_4$) is the first member of the alkane series. Succeeding members having two, three, and four carbon atoms are ethane, propane, and butane, respectively. The names of the first four alkanes are of common or trivial origin and must simply be memorized; but the names beginning with the fifth member, pentane, are derived from Greek numbers and are relatively easy to recall. The names and formulas of the first ten members of the series are given in Table 22.1.

Table 22.1. Names, formulas, and physical properties of straight-chain alkanes.

Name	Formula, C_nH_{2n+2}	Boiling point (°C)	Melting point (°C)
Methane	CH$_4$	−161	−183
Ethane	C$_2$H$_6$	−88	−172
Propane	C$_3$H$_8$	−45	−187
Butane	C$_4$H$_{10}$	0	−138
Pentane	C$_5$H$_{12}$	36	−130
Hexane	C$_6$H$_{14}$	69	−95
Heptane	C$_7$H$_{16}$	98	−90
Octane	C$_8$H$_{18}$	125	−57
Nonane	C$_9$H$_{20}$	151	−54
Decane	C$_{10}$H$_{22}$	174	−30

Successive compounds in the alkane series differ in composition by one carbon and two hydrogen atoms. This difference between any two successive alkanes can be observed in Table 22.1. When each member of a series differs from the next higher member by a CH$_2$ group, as do the alkanes, the series is called a **homologous series**. The members of a homologous series are similar in structure but have a regular difference in formula. All common classes of organic compounds exist in homologous series. A homologous series can be represented by a general formula. For all open-chain alkanes the general formula is C_nH_{2n+2}, where n corresponds to the number of carbon atoms in the molecule. The formula of any specific alkane is easily determined from this general formula. Thus, for pentane, $n = 5$ and $2n + 2 = 12$, so the formula is C$_5$H$_{12}$. For hexadecane, a 16-carbon alkane, the formula is C$_{16}$H$_{34}$.

homologous series

22.7 Carbon Bonding in Alkanes

As was pointed out in Section 22.4, a carbon atom is capable of forming single covalent bonds with one, two, three, or four other carbon atoms. To understand this remarkable bonding ability, we must look at the electronic structure of carbon in some detail. The outer-shell electron orbital structure of an uncombined carbon atom is given as $2s^2 2p_x^1 2p_y^1$. But when a carbon atom is bonded to other atoms by single bonds (for example, to four hydrogen atoms), it would appear at first that there should be two different types of bonds—bonds involving the $2s$ electrons and bonds involving the $2p$ electrons of the carbon atom. However, this is not the case. All four carbon–hydrogen bonds are found to be identical.

$2p$ ↿ ↿ ___   A $2s$ electron $2p$ ↿ ↿ ↿   Hybridization $2sp^3$ ↿ ↿ ↿ ↿
$2s$ ↿⇂   is promoted to $2s$ ↿
                a $2p$ electron

Four carbon electrons in their ground-state orbitals

Four equivalent sp^3 orbitals—each contains one electron

Figure 22.3. Schematic hybridization of $2s^2 2p_x^1 2p_y^1$ orbitals of carbon to form four sp^3 electron orbitals.

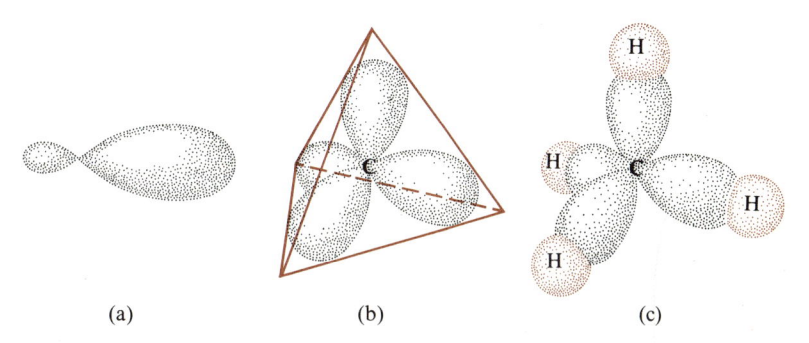

(a) (b) (c)

Figure 22.4. Tetrahedral nature of sp^3 orbitals: (a) A single sp^3 hybridized orbital; (b) four sp^3 hybridized orbitals in tetrahedral arrangement; (c) sp^3–s orbitals overlapping to form C–H bonds in methane.

If the carbon atom is to form four equivalent bonds, its electrons in the $2s$ and $2p$ orbitals must rearrange to four equivalent orbitals. To form the four equivalent orbitals, a $2s$ electron is promoted to a $2p$ orbital, giving carbon the electronic structure $2s^1 2p_x^1 2p_y^1 2p_z^1$. The $2s$ orbital and the three $2p$ orbitals then hybridize to form four equivalent hybrid orbitals, which are designated sp^3 orbitals. In this case, the superscript means that one s and three p orbitals are involved in forming the hybrid orbitals. This process is illustrated in Figure 22.3. It is these sp^3 orbitals that are directed toward the corners of a regular tetrahedron (see Figure 22.4).

A single bond is formed when one of the sp^3 orbitals overlaps an orbital of another atom. Thus, each C–H bond in methane is the result of the overlapping of a carbon sp^3 orbital and a hydrogen s orbital [Figure 22.4, part (c)]. Once the bond is formed, the pair of bonding electrons constituting it are said to be in a molecular orbital. In a similar way, a C–C single bond results from the overlap of sp^3 orbitals between two carbon atoms. This type of bond is called a sigma (σ) bond. A **sigma bond** exists if the cloud formed by the pair of bonding electrons is symmetrical about a straight line drawn between the nuclei of the bonded atoms.

sigma bond

22.8 Structural Formulas and Isomerism

The properties of an organic substance are dependent on its molecular structure. By structure we mean the way in which the atoms are bonded together within the molecule. The majority of organic compounds are made from relatively few elements—namely, carbon, hydrogen, oxygen, nitrogen, and the halogens. In these compounds, carbon is tetravalent, hydrogen is monovalent, oxygen is divalent, nitrogen is trivalent, and the halogens are monovalent. The valence bonds or points of attachment may be represented in structural formulas by a corresponding number of dashes attached to each atom:

—C—   H—   —O—   —N—   Cl—   Br—   I—   F—

Thus, carbon will have four bonds to each atom, nitrogen three bonds, oxygen two bonds, and hydrogen and the halogens one bond to each atom.

In an alkane, each carbon atom is joined to four other atoms by single covalent bonds. These bonds are separated by angles of 109.5° (the angles correspond to those formed by lines drawn from the center of a regular tetrahedron to its corners). Alkane molecules contain only carbon–carbon and carbon–hydrogen bonds and are essentially nonpolar. Alkane molecules are nonpolar because (1) carbon–carbon bonds are nonpolar since they are between like atoms; (2) carbon–hydrogen bonds are only slightly polar since there is only a small difference in electronegativity between carbon and hydrogen atoms; and (3) the bonds in an alkane are symmetrically directed toward the corners of a tetrahedron. Because of this nonpolarity, alkane molecules have very little intermolecular attraction and therefore relatively low boiling points when compared with other organic compounds of similar molecular weight.

Without the use of models or perspective drawings, the three-dimensional character of atoms and molecules is difficult to portray accurately. However, concepts of structure can be conveyed to some extent by electron-dot (or Lewis-dot) diagrams of structural formulas. Methane and ethane are represented by

electron-dot diagrams as

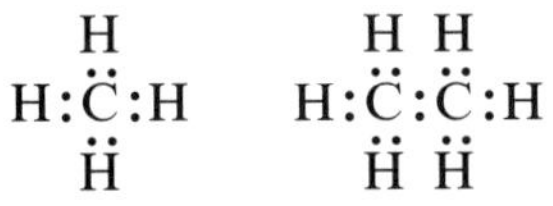

$$\begin{array}{cc} \text{H} & \text{H H} \\ \text{H:}\ddot{\text{C}}\text{:H} & \text{H:}\ddot{\text{C}}\text{:}\ddot{\text{C}}\text{:H} \\ \ddot{\text{H}} & \ddot{\text{H}}\;\ddot{\text{H}} \\ \text{Methane} & \text{Ethane} \end{array}$$

But it is more convenient to use conventional structural formulas representing electron pairs by single short lines:

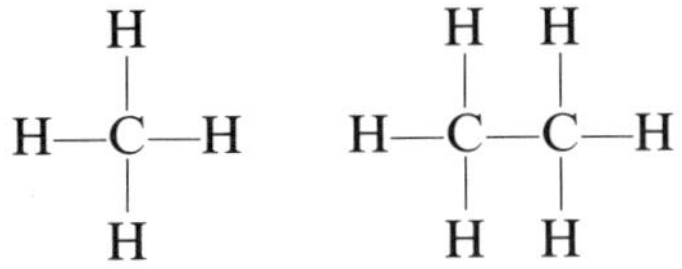

$$\begin{array}{cc} \text{H} & \text{H H} \\ | & | \;\; | \\ \text{H—C—H} & \text{H—C—C—H} \\ | & | \;\; | \\ \text{H} & \text{H H} \end{array}$$

To write the correct structural formula for propane (C_3H_8), the next member of the alkane series, we must determine how to place each atom in the molecule. An alkane contains only single bonds, and carbon is tetravalent. Therefore, each carbon atom must be bonded to four other atoms by either C–C or C–H bonds. Hydrogen is univalent and therefore must be bonded to only one carbon atom by a C–H bond, since C–H–C bonds do not occur, and an H–H bond would simply represent a hydrogen molecule. Applying this information, we find that the only possible structure for propane is

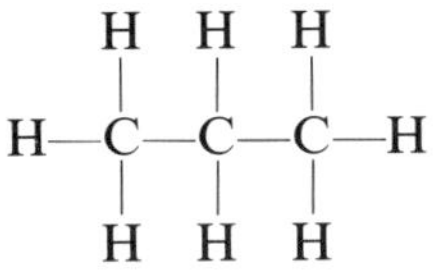

$$\begin{array}{c} \text{H H H} \\ |\;\;|\;\;| \\ \text{H—C—C—C—H} \\ |\;\;|\;\;| \\ \text{H H H} \end{array}$$

Propane

However, it is possible to write two structural formulas corresponding to the molecular formula C_4H_{10}:

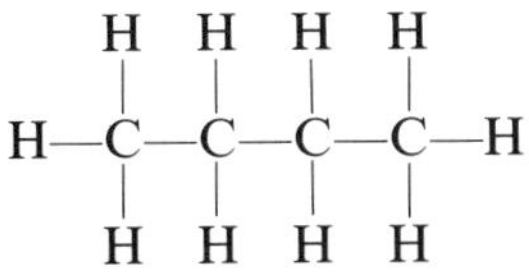

$$\begin{array}{ccc} \text{H H H H} & & \text{H}\;\;\text{H}\;\;\text{H} \\ |\;|\;|\;| & & \diagdown|\diagup \\ \text{H—C—C—C—C—H} & \text{and} & \text{C} \\ |\;|\;|\;| & & \text{H}\;\;|\;\;\text{H} \\ \text{H H H H} & & \text{H—C—C—C—H} \\ & & |\;\;|\;\;| \\ & & \text{H H H} \end{array}$$

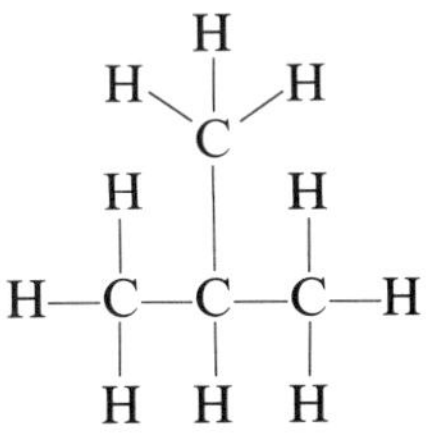

Normal butane   Isobutane

Two C_4H_{10} compounds with the structural formulas shown above actually exist. The butane with the unbranched carbon chain is called *normal butane* (abbreviated *n*-butane); it boils at 0.5°C and melts at −138.3°C. The branched-chain butane is called *isobutane*; it boils at −11.7°C and melts at −159.5°C. These differences in physical properties are sufficient to establish that the two compounds, both with the same molecular formula, are actually different substances. Models illustrating the structural arrangement of the atoms in methane, ethane, propane, butane, and isobutane are shown in Figure 22.5.

478   Chapter 22

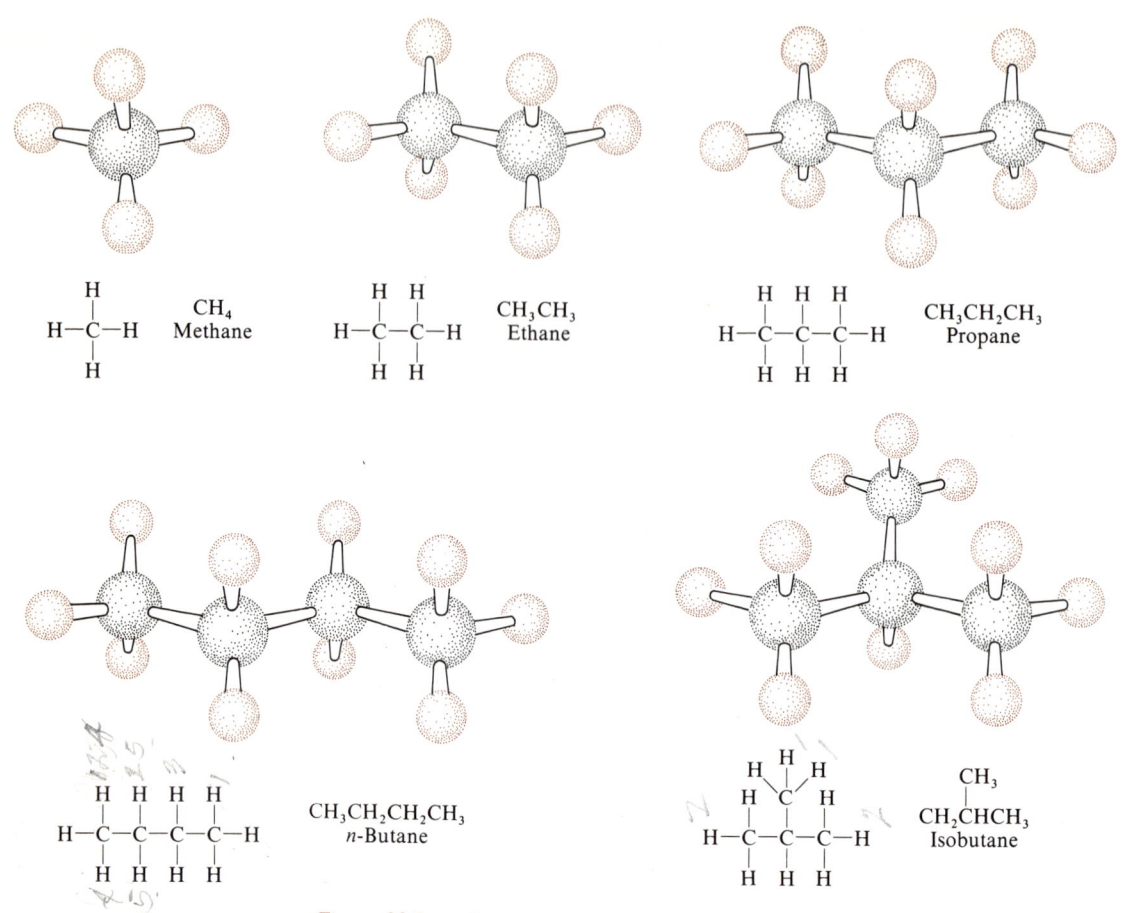

| H—C—H with H above and below   CH₄   Methane | H—C—C—H with H's   CH₃CH₃   Ethane | H—C—C—C—H with H's   CH₃CH₂CH₃   Propane |

| H—C—C—C—C—H with H's   CH₃CH₂CH₂CH₃   *n*-Butane | H—C—C—C—H with branch   CH₂CHCH₃ / CH₃   Isobutane |

Figure 22.5. Ball-and-stick models illustrating structural formulas of methane, ethane, propane, butane, and isobutane.

isomerism

isomers

This phenomenon of two or more compounds having the same molecular formula but different structures is called **isomerism**. The various individual compounds are called **isomers**. For example, there are 2 isomers of butane, C₄H₁₀. Isomerism is very common among organic compounds and is one reason for the large number of known compounds. There are 3 isomers of pentane, 5 isomers of hexane, 9 isomers of heptane, 18 isomers of octane, 35 isomers of nonane, and 75 isomers of decane. The phenomenon of isomerism is a very compelling reason for the use of structural formulas.

Isomers are compounds that have the same molecular formula but different structural formulas.

To save time and space in writing, condensed structural formulas are often used. In the condensed structural formulas, atoms and groups attached to a carbon atom are generally written to the right of that carbon atom. Examine these examples.

Molecular formula	Structural formula	Condensed structural formula
CH_4 Methane	H—C—H with H above and H below	CH_4
C_2H_6 Ethane	H—C—C—H with H's above and below each C	$CH_3—CH_3$ or CH_3CH_3
C_3H_8 Propane	H—C—C—C—H with H's above and below each C	$CH_3—CH_2—CH_3$ or $CH_3CH_2CH_3$
C_4H_{10} Butane	H—C—C—C—C—H with H's above and below each C	$CH_3—CH_2—CH_2—CH_3$, $CH_3CH_2CH_2CH_3$, or $CH_3(CH_2)_2CH_3$
C_4H_{10} 2-Methyl propane (Isobutane)	H—C—C—C—H with CH on top (H,H) and H's below	$CH_3—\overset{CH_3}{\overset{\|}{CH}}—CH_3$, $CH_3\overset{CH_3}{\overset{\|}{CH}}CH_3$, $CH_3CH(CH_3)CH_3$, $CH_3CH(CH_3)_2$, or $(CH_3)_3CH$

Let us interpret the condensed structural formula for propane:

$$\overset{1}{CH_3}—\overset{2}{CH_2}—\overset{3}{CH_3}$$

Carbon number 1 has three hydrogen atoms attached to it and is bonded to carbon number 2, which has two hydrogen atoms on it and is bonded to carbon number 3, which has three hydrogen atoms bonded to it.

Problem 22.1   There are three isomers of pentane, C_5H_{12}. Write structural formulas and condensed structural formulas for these isomers.

In a problem of this kind it is best to start by writing the carbon skeleton of the compound containing the longest continuous carbon chain. In this case it is five carbon atoms:

C—C—C—C—C

Now complete the structure by attaching hydrogen atoms around each carbon atom so that each carbon atom has four bonds attached to it. The carbon atoms at each end of the chain need three hydrogen atoms. The three inner carbon atoms each need two hydrogen atoms to give them four bonds.

480 Chapter 22

$$H-\underset{\underset{H}{|}}{\overset{\overset{H}{|}}{C}}-\underset{\underset{H}{|}}{\overset{\overset{H}{|}}{C}}-\underset{\underset{H}{|}}{\overset{\overset{H}{|}}{C}}-\underset{\underset{H}{|}}{\overset{\overset{H}{|}}{C}}-H$$

For the next isomer, start by writing a four-carbon chain and attach the fifth carbon atom to either of the middle carbon atoms—do not use the end ones.

$$\underset{}{\overset{\overset{C}{|}}{C}-C-C-C} \qquad C-\overset{\overset{C}{|}}{C}-C-C \qquad \text{Both of these structures represent the same compound.}$$

Now add the 12 hydrogen atoms to complete the structure:

(structural formula of 2-methylbutane with all H atoms shown)

For the third isomer, write a three-carbon chain, attach the other two carbon atoms to the central carbon atom, and complete the structure by adding the twelve hydrogen atoms:

(structural formula of 2,2-dimethylpropane with all H atoms shown)

The condensed structural formulas are derived from the structural formulas by placing the hydrogen atoms after the carbon atom to which they are attached:

$$CH_3CH_2CH_2CH_2CH_3 \qquad CH_3CH_2\overset{\overset{CH_3}{|}}{C}HCH_3 \qquad CH_3\overset{\overset{CH_3}{|}}{\underset{\underset{CH_3}{|}}{C}}CH_3 \quad \text{or} \quad C(CH_3)_4$$

or $CH_3(CH_2)_3CH_3$ or $CH_3CH_2CH(CH_3)_2$

22.9 Naming Organic Compounds

In the early stages of the development of organic chemistry, each new compound that was recognized was simply given a name, usually by the person who had isolated or synthesized it. Names were not systematic but often did carry some information—usually about the origin of the substance. Wood alcohol (methanol), for example, was so named because it was obtained by destructive distillation or pyrolysis of wood. Methane is formed during the decomposition of vegetable matter underwater in marshes, and was originally called *marsh gas*. Often, a single compound came to be known by several names.

For example, the active ingredient in alcoholic beverages has been called *alcohol, ethyl alcohol, methyl carbinol, grain alcohol, spirit,* and *ethanol.*

Beginning with a meeting in Geneva in 1892, an international system for naming compounds was developed. In its present form, the system recommended by the International Union of Pure and Applied Chemistry is systematic, generally unambiguous, and internationally accepted. It is called the IUPAC System. Despite the existence of the official IUPAC System, a great many well-established common, or trivial, names and abbreviations (such as TNT and DDT) are used because of their brevity and/or convenience. So it is necessary to have a knowledge of both the IUPAC System and many common names.

Table 22.2. Names and formulas of selected alkyl groups.

Formula	Name[a]	Formula	Name
CH_3-	Methyl	CH_3CH- with CH_3 branch	Isopropyl
CH_3CH_2-	Ethyl	CH_3CHCH_2- with CH_3 branch	Isobutyl
$CH_3CH_2CH_2-$	n-Propyl		
$CH_3CH_2CH_2CH_2-$	n-Butyl	CH_3CH_2CH- with CH_3 branch	sec-Butyl (secondary butyl)
$CH_3(CH_2)_3CH_2-$	n-Pentyl		
$CH_3(CH_2)_4CH_2-$	n-Hexyl	CH_3C- with three CH_3 branches	tert-Butyl (tertiary butyl)
$CH_3(CH_2)_5CH_2-$	n-Heptyl		
$CH_3(CH_2)_6CH_2-$	n-Octyl		
$CH_3(CH_2)_7CH_2-$	n-Nonyl		
$CH_3(CH_2)_8CH_2-$	n-Decyl		

[a] The lowercase *n* in *n*-propyl, *n*-butyl, and so on, means that the hydrocarbon chains are *normal;* that is, the alkyl group represents an unbranched hydrocarbon with a hydrogen atom missing from carbon number 1.

alkyl group

In order to name organic compounds systematically, you must be able to recognize certain common alkyl groups. **Alkyl groups** have the general formula C_nH_{2n+1} (one less hydrogen atom than the corresponding alkane). The name of the group is formed from the name of the corresponding alkane by simply dropping *-ane* and substituting a *-yl* ending. The names and formulas of selected alkyl groups up to and including four carbon atoms are given in Table 22.2. The letter "R" is often used in formulas to mean any of the many possible alkyl groups.

The following relatively simple rules are all that are needed to name a great many alkanes according to the IUPAC System. In later sections these rules will be extended to cover other classes of compounds, but advanced texts or references must be consulted for the complete system.

IUPAC rules for naming alkanes:

1. Select the longest continuous chain of carbon atoms as the parent compound, and consider all alkyl groups attached to it as branch chains that have replaced hydrogen atoms of the parent hydrocarbon. The name of the alkane consists of the name of the parent compound prefixed by the names of the branch-chain alkyl groups attached to it.
2. Number the carbon atoms in the parent carbon chain starting from the end closest to the first carbon atom that has an alkyl group substituted for a hydrogen atom.
3. Name each branch-chain alkyl group and designate its position on the parent carbon chain by a number (for example, 2-methyl means a methyl group attached to carbon number 2).
4. When the same alkyl group branch chain occurs more than once, indicate this by a prefix, di-, tri-, tetra-, and so forth, written in front of the alkyl group name (for example, *dimethyl* indicates two methyl groups). The numbers indicating the positions of these alkyl groups are separated by a comma, followed by a hyphen, and are placed in front of the name (for example, 2,3-dimethyl).
5. When several different alkyl groups are attached to the parent compound, list them in alphabetical order; for example, ethyl before methyl in 3-ethyl-4-methyloctane.

The compound shown below is commonly called isopentane. Consider naming it by the IUPAC System:

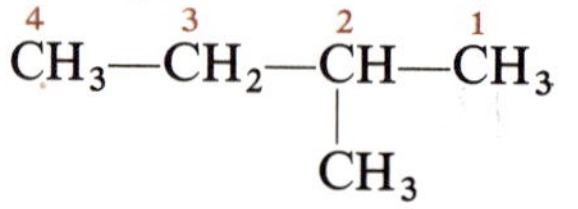

$$\overset{4}{C}H_3-\overset{3}{C}H_2-\overset{2}{C}H-\overset{1}{C}H_3$$
$$\underset{}{}|$$
$$CH_3$$

2-Methylbutane
(Isopentane)

The longest continuous chain contains four carbon atoms. Therefore, the parent compound is butane and the compound is named as a butane. The methyl group (CH₃—) attached to carbon number 2 is named as a prefix to butane, the "2-" indicating the point of attachment of the methyl group on the butane chain.

How would we write the structural formula for 2-methylpentane? An analysis of its name gives us this information.

1. The parent compound, pentane, contains five carbons. Write and number the five-carbon skeleton of pentane:

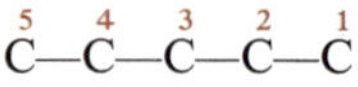

$$\overset{5}{C}-\overset{4}{C}-\overset{3}{C}-\overset{2}{C}-\overset{1}{C}$$

2. Put a methyl group on carbon number 2 (2-methyl in the name gives this information):

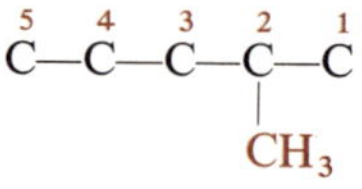

$$\overset{5}{C}-\overset{4}{C}-\overset{3}{C}-\overset{2}{C}-\overset{1}{C}$$
$$|$$
$$CH_3$$

3. Add hydrogens to give each carbon four bonds. The structural formula is

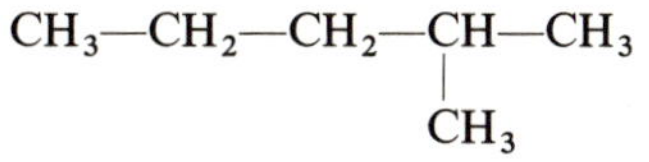

$$CH_3—CH_2—CH_2—CH—CH_3$$
$$|$$
$$CH_3$$

2-Methylpentane

Should this compound be called 4-methylpentane? No, since the IUPAC System specifically states that the parent carbon chain shall be numbered starting from the end nearest to the branch chain.

It is very important to understand that it is the sequence of atoms and groups that determines the name of a compound, and not the way the sequence is written. Each of the following formulas represents 2-methylpentane:

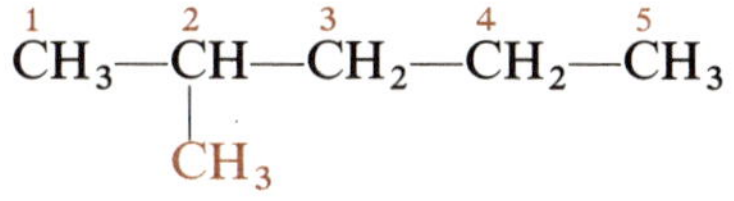

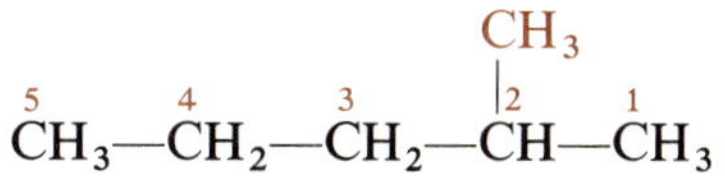

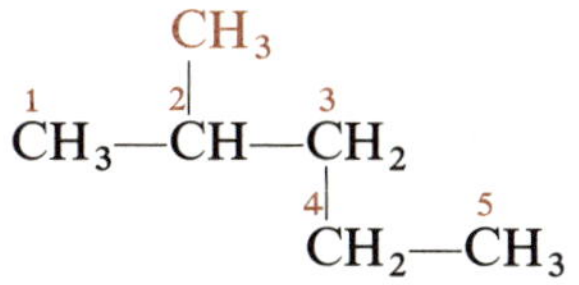

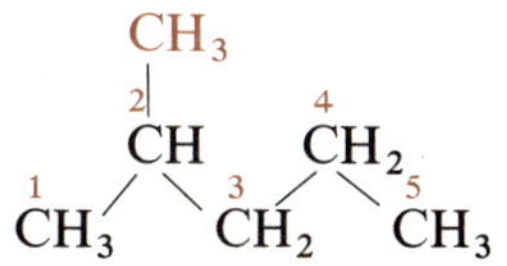

$$\overset{1}{C}H_3—\overset{2}{C}H—\overset{3}{C}H_2—\overset{4}{C}H_2—\overset{5}{C}H_3 \qquad \overset{5}{C}H_3—\overset{4}{C}H_2—\overset{3}{C}H_2—\overset{2}{C}H—\overset{1}{C}H_3$$
$$\underset{CH_3}{|} \qquad\qquad\qquad\qquad \overset{CH_3}{|}$$

$$\overset{1}{C}H_3—\overset{2}{C}H—\overset{3}{C}H_2$$
$$\underset{CH_3}{|}\qquad\underset{\overset{4}{C}H_2—\overset{5}{C}H_3}{|}$$

$$\overset{CH_3}{\underset{2}{|}}$$
$$\overset{CH}{}\quad \overset{CH_2}{}$$
$$\underset{1}{CH_3}\quad \underset{3}{CH_2}\quad \underset{5}{CH_3}$$

The following formulas and names demonstrate other aspects of the official nomenclature system:

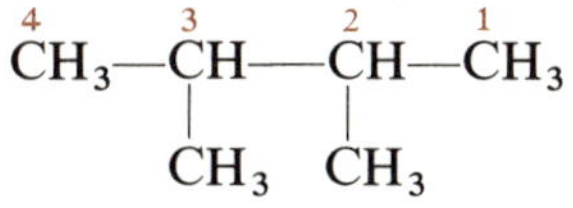

$$\overset{4}{C}H_3—\overset{3}{C}H—\overset{2}{C}H—\overset{1}{C}H_3$$
$$\underset{CH_3}{|}\ \underset{CH_3}{|}$$

2,3-Dimethylbutane

The name of this compound is 2,3-dimethylbutane. The longest carbon atom chain is four, indicating butane; dimethyl indicates two methyl groups; "2,3-" means that one CH_3 is on carbon 2 and one is on carbon 3.

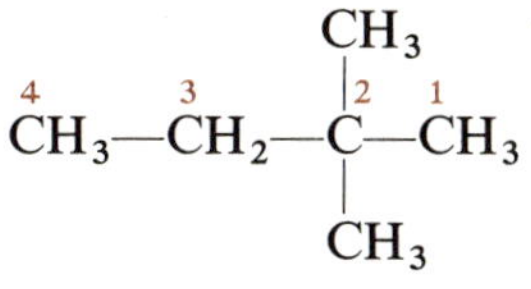

$$\overset{4}{C}H_3—\overset{3}{C}H_2—\overset{2}{C}—\overset{1}{C}H_3$$
$$\overset{CH_3}{|}$$
$$\underset{CH_3}{|}$$

2,2-Dimethylbutane

(Both methyl groups are on the same carbon atom; both numbers are required.)

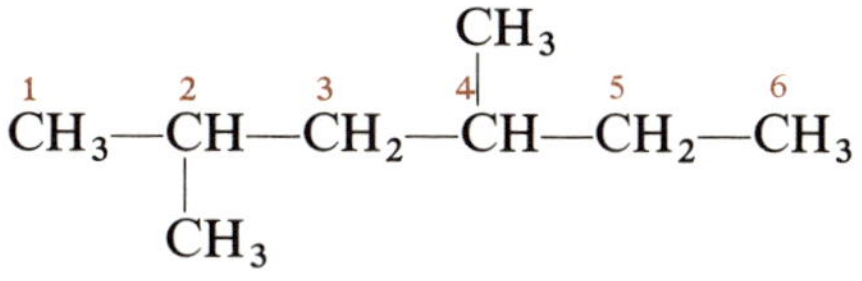

$$\overset{1}{C}H_3—\overset{2}{C}H—\overset{3}{C}H_2—\overset{4}{C}H—\overset{5}{C}H_2—\overset{6}{C}H_3$$
$$\underset{CH_3}{|}\qquad\qquad\overset{CH_3}{|}$$

2,4-Dimethylhexane

(The molecule is numbered from left to right.)

$$CH_3-\overset{3}{C}H-\overset{4}{C}H_2-\overset{5}{C}H_2-\overset{6}{C}H_3$$
$$\overset{2}{|}$$
$$CH_2$$
$$\overset{1}{|}$$
$$CH_3$$

3-Methylhexane

(There are six carbons in the longest continuous chain.)

$$\overset{8}{C}H_3-\overset{7}{C}H_2-\overset{6}{C}H_2-\overset{5}{C}H_2-\overset{4}{\underset{\underset{CH_3}{|}}{C}}\overset{\overset{CH_2-CH_3}{|}}{\underset{}{-}}\overset{3}{\underset{\underset{Cl}{|}}{C}H}-\overset{2}{\underset{\underset{CH_3}{|}}{C}H}-\overset{1}{C}H_3$$

3-Chloro-4-ethyl-2,4-dimethyloctane

Problem 22.2   Write the formulas for:
(a) 3-Ethylpentane   (b) 2,2,4-Trimethylpentane

(a) The name *pentane* indicates a five-carbon chain. Write five connecting carbon atoms and number them:

$$\overset{1}{C}-\overset{2}{C}-\overset{3}{C}-\overset{4}{C}-\overset{5}{C}$$

An ethyl group is written as CH_3CH_2-. Attach this group to carbon number 3:

$$\overset{1}{C}-\overset{2}{C}-\overset{3}{\underset{\underset{CH_2CH_3}{|}}{C}}-\overset{4}{C}-\overset{5}{C}$$

Now add hydrogen atoms to give each carbon atom four bonds. Carbons 1 and 5 each need three H atoms; carbons 2 and 4 each need two H atoms; and carbon 3 needs one H atom. The formula is complete.

$$CH_3CH_2\underset{\underset{CH_2CH_3}{|}}{C}HCH_2CH_3$$

(b) Pentane indicates a five-carbon chain. Write five connecting carbon atoms and number them:

$$\overset{1}{C}-\overset{2}{C}-\overset{3}{C}-\overset{4}{C}-\overset{5}{C}$$

There are three methyl groups (CH_3-) in the compound (trimethyl), two attached to carbon 2 and one attached to carbon 4. Attach these three methyl groups to their respective carbon atoms:

$$\overset{1}{C}-\overset{2}{\underset{\underset{CH_3}{|}}{\overset{\overset{CH_3}{|}}{C}}}-\overset{3}{C}-\overset{4}{\underset{}{\overset{\overset{CH_3}{|}}{C}}}-\overset{5}{C}$$

Now add H atoms to give each carbon atom four bonds. Carbons 1 and 5 each need three H atoms; carbon 2 does not need any H atoms; carbon 3 needs two H atoms; and carbon 4 needs one H atom. The formula is complete.

$$\begin{array}{c} CH_3CH_3 \\ || \\ CH_3CCH_2CHCH_3 \\ | \\ CH_3 \end{array}$$

Problem 22.3   Name the following compounds:

(a)
$$\begin{array}{c} CH_3 \\ | \\ CH_3CH_2CH_2CH_2CHCH_3 \end{array}$$

(b)
$$\begin{array}{c} CH_2CH_3 \\ | \\ CH_3CH_2CH_2CHCH_2CHCH_3 \\ | \\ CH_2CH_3 \end{array}$$

(a) The longest continuous carbon chain contains six carbon atoms. Thus, the parent name of the compound is hexane. Number the carbon chain from right to left so that the methyl group attached to carbon 2 is given the lowest possible number. With a methyl group on carbon 2 the name of the compound is 2-methylhexane.

(b) The longest continuous carbon chain contains eight carbon atoms.

$$\begin{array}{c} C-C \\ | \\ \overset{8}{C}-\overset{7}{C}-\overset{6}{C}-\overset{5}{C}-\overset{4}{C}-\overset{3}{C}-C \\ \overset{2}{|}\overset{1}{} \\ C-C \end{array}$$

Thus, the parent name is octane. As the chain is numbered, there is a methyl group on carbon 3 and an ethyl group on carbon 5. Thus, the name of the compound is 5-ethyl-3-methyloctane. Note that ethyl is named before methyl (alphabetical order).

22.10 Reactions of Alkanes

One single type of reaction of alkanes has inspired people to explore equatorial jungles, endure the heat and sandstorms of the deserts of Africa and the Middle East, mush across the frozen arctic, and drill holes in the earth more than 30,000 feet deep! These strenuous and expensive activities have been undertaken because alkanes, as well as other hydrocarbons, undergo combustion with oxygen with the evolution of large amounts of heat energy. Methane, for example, reacts with oxygen:

$$CH_4(g) + 2\,O_2(g) \longrightarrow CO_2(g) + 2\,H_2O(g) + 191.8 \text{ kcal}$$

The thermal energy can be converted to mechanical and electrical energy. In order of economic importance, combustion reactions overshadow all other reactions of alkanes. But combustion reactions are not usually of great interest to organic chemists, since carbon dioxide and water are the only chemical products of complete combustion.

Aside from their combustibility, alkanes are sluggish and limited in their reactivity. But with proper activation, such as high temperature and/or catalysts, alkanes can be made to react in a variety of ways. Some industrially important noncombustion reactions of alkanes are the following:

1. *Halogenation* (substitution of halogens for hydrogen): Halogenation is a general term. When a specific halogen, such as chlorine, is used, the reaction is called chlorination.

   $$RH + X_2 \longrightarrow RX + HX \quad (X = Cl \text{ or } Br)$$

   $$CH_3CH_3 + Cl_2 \longrightarrow \underset{\text{Chloroethane}}{CH_3CH_2Cl} + HCl$$

   This reaction yields alkyl halides (RX), which are useful as intermediates for the manufacture of other substances.

2. *Dehydrogenation* (removal of hydrogen):

   $$C_nH_{2n+2} \xrightarrow{700-900°C} C_nH_{2n} + H_2$$

   $$CH_3CH_2CH_3 \xrightarrow{\Delta} \underset{\text{Propene}}{CH_3CH=CH_2} + H_2$$

   This reaction yields alkenes, which, like alkyl halides, are useful chemical intermediates. Hydrogen is a valuable by-product.

3. *Cracking* (breaking up large molecules to form smaller ones):

   Example: $$C_{16}H_{34} \xrightarrow{\Delta} \underset{\text{Alkane}}{C_8H_{18}} + \underset{\text{Alkene}}{C_8H_{16}}$$
   $$\phantom{C_{16}H_{34} \xrightarrow{\Delta}}\text{Alkane}$$

4. *Isomerization* (rearrangement of molecular structures):

   Example:

   $$CH_3-CH_2-CH_2-CH_2-CH_3 \longrightarrow CH_3-CH_2-CH(CH_3)-CH_3$$

Halogenation is used extensively in the manufacture of petrochemicals (chemicals derived from petroleum and used for purposes other than fuels). The other three reactions—cracking, dehydrogenation, and isomerization—singly or in combination, are of great importance in the production of both motor fuels and petrochemicals.

A well-known reaction of methane and chlorine is shown by the equation

$$CH_4 + Cl_2 \longrightarrow \underset{\substack{\text{Chloromethane}\\\text{(Methyl chloride)}}}{CH_3Cl} + HCl$$

The equation tells us what the reactants and products are; but it gives no hint concerning the process involved in the reaction. Methane and chlorine react when exposed to ultraviolet radiation or when heated; the reaction therefore requires a large activation energy. But once begun, the chlorination goes very fast. Partly because of these facts, the reaction is believed to proceed by a mechanism in which highly reactive species, called free radicals, are formed as intermediate products. A **free radical** is an atom or a group of atoms that has an odd,

free radical

or unpaired, valence electron. Examples are

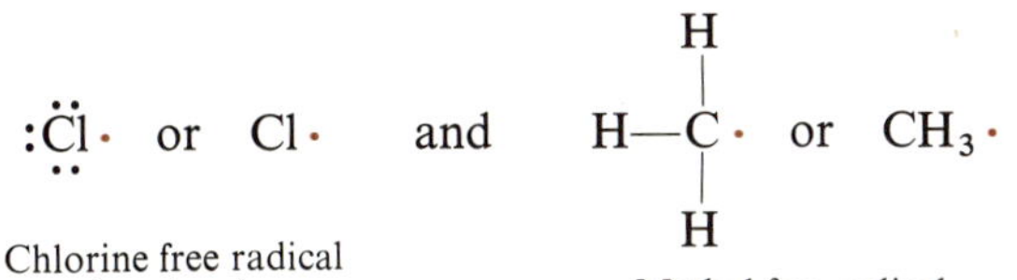

$:\ddot{\underset{..}{Cl}}\cdot$ or $Cl\cdot$ and $H-\underset{\underset{H}{|}}{\overset{\overset{H}{|}}{C}}\cdot$ or $CH_3\cdot$

Chlorine free radical   Methyl free radical
(or atom)

The chlorination of methane proceeds in three steps which are typical of free radical reactions.

1. *Initiation:* A chlorine molecule absorbs a quantum of light energy, and the energized molecule then splits into two chlorine free radicals:

$:\ddot{\underset{..}{Cl}}:\ddot{\underset{..}{Cl}}: + h\nu \longrightarrow 2:\ddot{\underset{..}{Cl}}\cdot$

Chlorine   Chlorine
molecule   free radicals

The symbol $h\nu$ stands for a quantity of energy (a photon) absorbed from light.

2. *Propagation:* A chlorine free radical attacks a methane molecule, forming a methyl free radical and a molecule of hydrogen chloride. The methyl free radical attacks a chlorine molecule to form a molecule of chloromethane and another chlorine free radical, which can then repeat this step:

$CH_4 + Cl\cdot \longrightarrow CH_3\cdot + HCl$
                            Methyl
                        free radical

$CH_3\cdot + Cl_2 \longrightarrow CH_3Cl + Cl\cdot$
                       Chloromethane

This propagation step may repeat hundreds or even thousands of times until one or both of the reactants are used up.

3. *Termination:* The chain reaction occurring in the propagation step is stopped when two free radicals combine to form a molecule with liberation of energy, perhaps in the form of heat.

$2\,Cl\cdot \longrightarrow Cl_2$
$CH_3\cdot + Cl\cdot \longrightarrow CH_3Cl$
$2\,CH_3\cdot \longrightarrow CH_3CH_3$
               Ethane

Up to this point, we have considered only the substitution of a single chlorine atom for a hydrogen atom on a methane molecule. By using larger amounts of chlorine and suitable reaction conditions, methane can be chlorinated to yield either all tetrachloromethane (carbon tetrachloride) or mixtures of mono-, di-, and trichloromethanes (Table 22.3).

$CH_4 \xrightarrow{Cl_2} CH_3Cl \xrightarrow{Cl_2} CH_2Cl_2 \xrightarrow{Cl_2} CHCl_3 \xrightarrow{Cl_2} CCl_4 + 4\,HCl$

Chapter 22

Table 22.3. Chlorination products of methane.

Formula	IUPAC name	Common name
CH_3Cl	Chloromethane	Methyl chloride
CH_2Cl_2	Dichloromethane	Methylene chloride
$CHCl_3$	Trichloromethane	Chloroform
CCl_4	Tetrachloromethane	Carbon tetrachloride

The term *monosubstitution product* refers to a new product in which one hydrogen atom from an organic molecule is substituted by one other atom or group of atoms. In hydrocarbons, for example, when we substitute one chlorine atom for a hydrogen atom, the new compound is a monosubstitution (monochloro substitution) product. In a like manner, we can have di-, tri-, tetra-, and so on, substitution products.

This kind of chlorination (or bromination) is general with alkanes. There are nine different chlorination products of ethane. See if you can write the structural formulas for all of them.

When propane is chlorinated, two isomeric monosubstitution products are obtained because a hydrogen atom may be replaced on either the first or second carbon.

$$CH_3CH_2CH_3 + Cl_2 \xrightarrow[25°]{\text{Light}} CH_3CH_2CH_2Cl + CH_3CHClCH_3 + HCl$$

1-Chloropropane 2-Chloropropane
(*n*-Propyl chloride) (Isopropyl chloride)
45% 55%

alkyl halide

The letter X is commonly used to indicate a halogen atom. The formula RX indicates a halogen atom attached to an alkyl group and represents the class of compounds known as the **alkyl halides**. When R = CH_3, CH_3X can be CH_3F, CH_3Cl, CH_3Br, or CH_3I.

Alkyl halides are named systematically in the same general way as alkanes. Halogen atoms are identified as *fluoro-*, *chloro-*, *bromo-*, or *iodo-*, and are named as substituents like side-chain alkyl groups. Study these examples:

CH_3—$CHCl$—CH_2—CH_3 CH_2Cl—$CHBr$—CH_3

2-Chlorobutane   2-Bromo-1-chloropropane

CH_3—CH_2—CH—$CHCl$—CH_3
                  |
                 CH_3

2-Chloro-3-methylpentane

Problem 22.4 How many monochlorosubstitution products can be obtained from pentane?

First write the formula for pentane:

$\overset{5}{C}H_3\overset{4}{C}H_2\overset{3}{C}H_2\overset{2}{C}H_2\overset{1}{C}H_3$

Now rewrite the formula five times substituting a Cl atom for an H atom on each C atom:

I   CH₃CH₂CH₂CH₂CH₂Cl   Cl on carbon 1
                    ¹

II CH₃CH₂CH₂CHClCH₃   Cl on carbon 2
               ²

III CH₃CH₂CHClCH₂CH₃   Cl on carbon 3
          ³

IV CH₃CHClCH₂CH₂CH₃   Cl on carbon 4
       ⁴

V   CH₂ClCH₂CH₂CH₂CH₃   Cl on carbon 5
     ⁵

Compound IV is identical to II and compound V is identical to I, because if we number the carbon chain from left to right, we find the Cl atoms substituted on carbons 1 and 2. By naming the compounds we find that both I and V are 1-chloropentane and that II and IV are both 2-chloropentane. Thus, there are three monochloro substitution products of pentane—compounds I, II, and III, 1-chloropentane, 2-chloropentane, and 3-chloropentane.

Other typical reactions of the alkanes, including combustion, are of the free radical type and require high temperatures (or moderately high temperatures and a catalyst) for initiation. Thus, mixtures of alkanes and oxygen (air) do not react until they are ignited. Once sufficient thermal energy has been supplied to break up some of the molecules and create free radicals, the combustion proceeds very rapidly. In fact, an explosion can occur if an appreciable volume of alkane–air mixture is at hand. Explosive combustion occurs because some of the intermediate reactions are highly exothermic. The heat raises the temperature and further increases an already rapid rate of combustion. Carbon dioxide and water are the only products of complete combustion of all hydrocarbons. But other products, including carbon monoxide, free carbon, alkenes, and various oxygenated compounds, occur at intermediate stages in combustion. Some of these substances are recognized as serious air pollutants; and if, for any reason, combustion is not complete, they are produced in varying amounts. Internal combustion engines have been notorious air pollution offenders because of (1) low oxygen to hydrocarbon ratios under some operating conditions and (2) insufficient reaction time in cylinders and exhaust manifold. Large-scale efforts are being made to solve this problem, and there are good reasons for believing that it will be solved satisfactorily (see Chapter 21).

22.11 Cycloalkanes

cycloalkane

Cyclic, or closed-chain, alkanes also exist. These substances, called **cycloalkanes**, *cycloparaffins*, or *naphthenes*, have the general formula C_nH_{2n}. Note that this series of compounds has two less hydrogen atoms than the open-chain alkanes. The bonds for the two missing hydrogen atoms are accounted for by an additional carbon–carbon bond in forming the cyclic ring of carbon atoms. Structures for the four smallest cycloalkanes are shown in Figure 22.6.

490   Chapter 22

$$CH_2 \atop CH_2—CH_2$$ $$CH_2—CH_2 \atop CH_2—CH_2$$ $$CH_2 \atop CH_2\quad CH_2 \atop CH_2—CH_2$$ $$CH_2 \atop CH_2\quad CH_2 \atop CH_2\quad CH_2 \atop CH_2$$

Cyclopropane
C_3H_6

Cyclobutane
C_4H_8

Cyclopentane
C_5H_{10}

Cyclohexane
C_6H_{12}

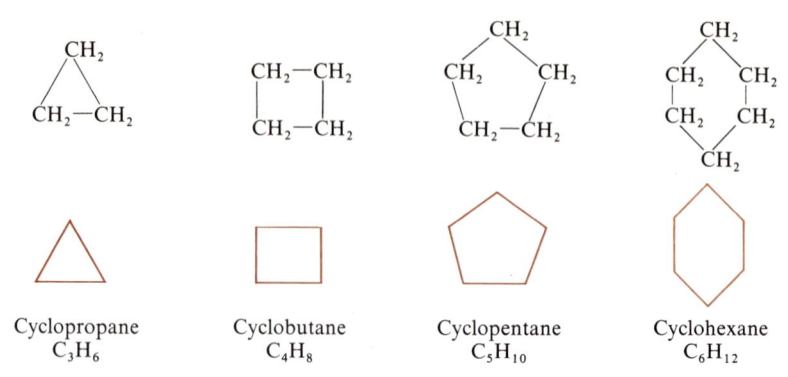

Figure 22.6. Cycloalkanes: In the line representations, each corner of the diagram represents a CH_2 group.

Cyclopropane

n-Hexane $CH_3CH_2CH_2CH_2CH_2CH_3$

Cyclohexane

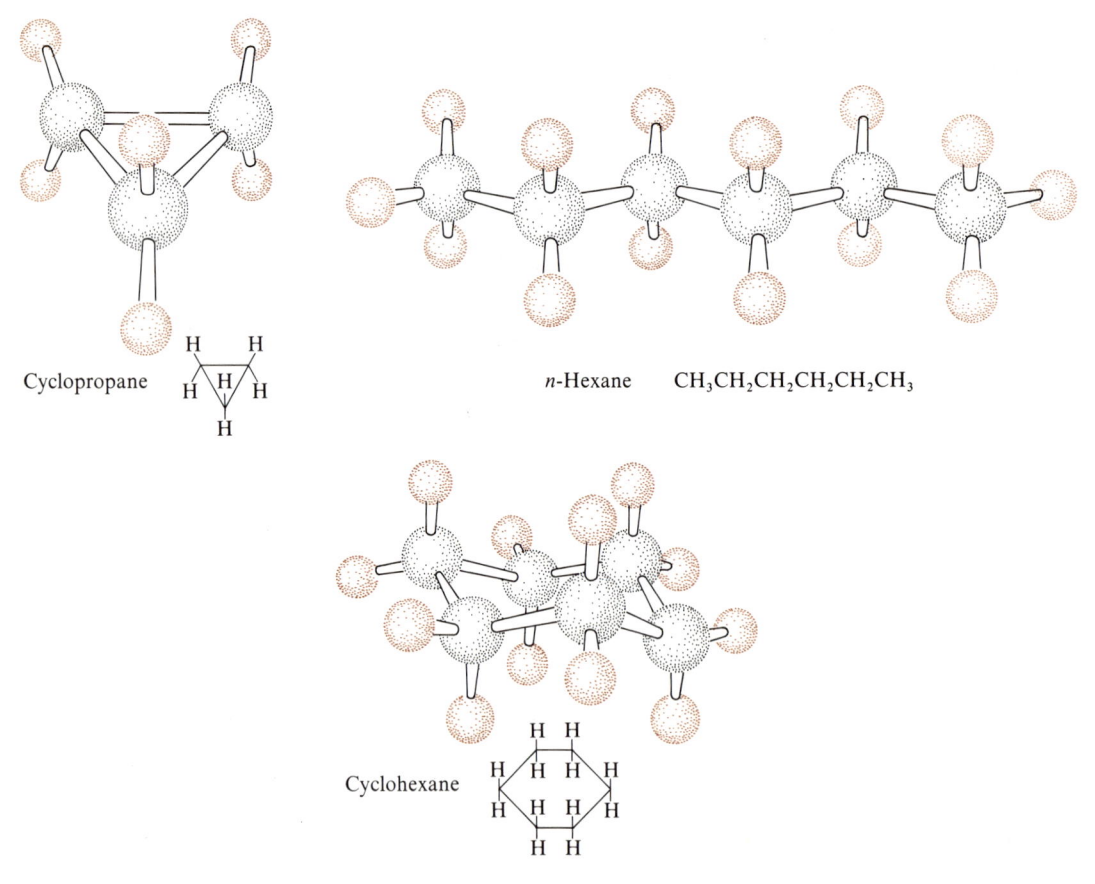

Figure 22.7. Ball-and-stick models illustrating cyclopropane, *n*-hexane, and cyclohexane. In cyclopropane all the carbon atoms are in one plane. The angle between carbon atoms is 60°, not the usual 109.5°; therefore, the cyclopropane ring is strained. In cyclohexane, the carbon–carbon bonds are not strained. This is because the molecule is puckered, with carbon–carbon bond angles of 109.5°, as found in normal hexane.

With the exception noted below for cyclopropane and cyclobutane, cycloalkanes are generally similar to open-chain alkanes in both physical properties and chemical reactivity. Cycloalkanes are saturated hydrocarbons.

The reactivity of cyclopropane, and to a lesser degree that of cyclobutane, is greater than that of other alkanes. This greater reactivity exists because the carbon–carbon bond angles in these substances deviate from the normal tetrahedral angle. The carbon atoms form a triangle in cyclopropane, and in cyclobutane they approximate a square. Cyclopropane molecules therefore have carbon–carbon bond angles of 60°, and in cyclobutane the bond angles are about 90°. In the open-chain alkanes and in larger cycloalkanes the carbon atoms are in a three-dimensional zigzag pattern in space and have normal (tetrahedral) bond angles of about 109.5°.

Bromine will add to cyclopropane readily and to cyclobutane to some extent. In this reaction the ring breaks and an open-chain dibromopropane is formed:

$$\underset{\text{Cyclopropane}}{\begin{array}{c} CH_2 \\ \diagup \diagdown \\ CH_2 - CH_2 \end{array}} + Br_2 \longrightarrow \underset{\text{1,3-Dibromopropane}}{BrCH_2CH_2CH_2Br}$$

Cyclopropane and cyclobutane react in this way because their carbon–carbon bonds are strained and therefore weakened. Cycloalkanes with rings having more than four carbon atoms do not react in this way since their molecules take the shape of nonplanar puckered rings. These rings can be considered to be formed by simply joining the end carbon atoms of the corresponding normal alkanes. The resulting cyclic molecules are nearly strain-free, with carbon atoms arranged in space so that the bond angles are close to 109.5° (Figure 22.7).

Cyclopropane is a useful general anesthetic and, along with certain other cycloalkanes, is used as an intermediate in some chemical syntheses. The high reactivity of cyclopropane requires great care in its use as an anesthetic in the operating room; it is an extreme fire and explosion hazard.

Questions

A. Study and understand the new terms introduced in this chapter.
   1. Vital force theory
   2. Organic chemistry
   3. Hydrocarbons
   4. Alkanes
   5. Homologous series
   6. Sigma bond
   7. Isomerism
   8. Isomers
   9. Alkyl group
   10. Free radical
   11. Alkyl halide
   12. Cycloalkane

B. Review questions.
   1. What is the major reason for the large number of organic compounds?
   2. Why is it believed that a carbon atom must form hybrid electron orbitals when it bonds to hydrogen atoms to form methane?
   3. Write electron-dot structures for:
      (a) CCl_4 (b) C_2Cl_6 (c) $CH_3CH_2CH_3$
   4. Write the names and formulas for the first ten normal alkanes.

5. The name of the compound of formula $C_{11}H_{24}$ is undecane. What is the formula for dodecane, the next higher homologue in the alkane series?
6. How many sigma bonds are in a molecule of
   (a) Ethane (b) Butane (c) Isobutane
7. Which of these formulas represent isomers?

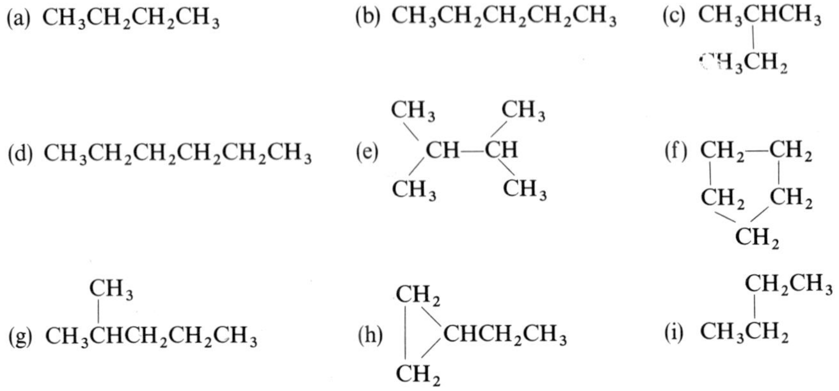

8. (a) How many methyl groups are in each formula in Question 7?
   (b) How many ethyl groups?
9. Which of these formulas represent the same compound?

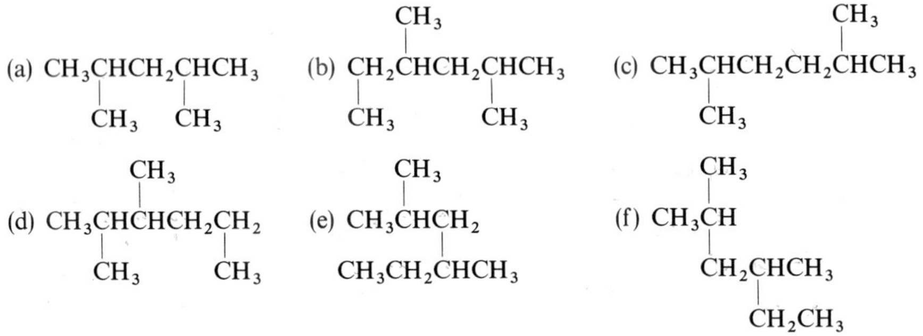

10. How many isomers are there having the formula:
    (a) CH_4 (c) CH_2Cl_2 (e) C_4H_9I
    (b) CH_3Br (d) C_2H_5Cl (f) $C_4H_8Cl_2$
    Draw structures for all the isomers.
11. What is the molecular weight of an alkane that contains 30 carbon atoms?
12. Write the structural formulas and IUPAC names of the five hexane isomers.
13. Give common and IUPAC names for the following:
    (a) CH_3CH_2Cl (c) $(CH_3)_2CHCH_2Cl$ (e) $(CH_3)_3CCl$
    (b) $CH_3CHClCH_3$ (d) $CH_3CH_2CH_2Cl$ (f) $CH_3CHClCH_2CH_3$
14. Draw structural formulas of the following compounds:
    (a) 2,4-Dimethylpentane (d) 4-Ethyl-2-methylhexane
    (b) 2,2-Dimethylpentane (e) 4-*t*-Butylheptane
    (c) 3-Isopropyloctane (f) 4-Ethyl-7-isopropyl-2,4,8-trimethyldecane
15. Write the balanced chemical equation for the complete combustion of butane.
16. Write equations to show how ethane is chlorinated to form chloroethane by a free radical reaction mechanism.

17. Give the IUPAC name for each of the following compounds:

(a) CH$_3$CH$_2$CHCH$_3$
           |
           CH$_3$

(b) (CH$_3$)$_2$CHCH$_2$CH(CH$_3$)$_2$

(c) $\underset{CH_2}{\overset{CH_2}{\triangleright}}$CHCH$_3$

(d) CH$_3$CH$_2$CH$_2$CHCH$_3$
                |
                CH$_3$CHCH$_3$

18. Complete the equations for (a) the monochlorination and (b) the complete combustion of butane.

(a) CH$_3$CH$_2$CH$_2$CH$_3$ + Cl$_2$ $\xrightarrow{h\nu}$

(b) CH$_3$CH$_2$CH$_2$CH$_3$ + O$_2$ $\xrightarrow{\Delta}$

19. The structure for hexane is CH$_3$CH$_2$CH$_2$CH$_2$CH$_2$CH$_3$. Draw structural formulas for all the monochlorohexanes, C$_6$H$_{13}$Cl, that have the same carbon structure as hexane.
20. Draw structures for the ten dichloro isomers of 2-methylbutane (C$_5$H$_{10}$Cl$_2$).
21. A hydrocarbon sample of formula C$_4$H$_{10}$ was brominated and four different monobromo compounds of formula C$_4$H$_9$Br were isolated. Was the sample a pure compound or a mixture of compounds? Explain your answer.
22. There are two cycloalkanes that have the formula C$_4$H$_8$. Draw their structures and name them.
23. Which of these statements are correct?
    (a) Alkane hydrocarbons are essentially nonpolar.
    (b) The C–H sigma bond in methane is made from an overlap of an *s* electron orbital and an *sp^3* electron orbital.
    (c) The valence electrons of every carbon atom in an alkane are in *sp^3* hybridized orbitals.
    (d) The four carbon–hydrogen bonds in methane are equivalent.
    (e) Hydrocarbons are composed of carbon and water.
    (f) In the alkane homologous series, the formula of each member differs from its preceding member by CH$_3$.
    (g) Carbon atoms can form single, double, and triple bonds with other carbon atoms.
    (h) The name for the alkane C$_5$H$_{12}$ is propane.
    (i) There are eight carbon atoms in a molecule of 3,3,4-trimethylpentane.
    (j) The general formula for an alkyl halide is RX.
    (k) The IUPAC name for CH$_3$CH$_2$CH$_2$CHClCH$_3$ is 4-chloropentane.
    (l) Isopropyl chloride is also called 2-chloropropane.
    (m) An uncombined chlorine atom is a free radical.
    (n) A methyl group that has an unpaired electron (CH$_3$·) is called a methyl free radical.
    (o) The products of complete combustion of a hydrocarbon are carbon monoxide and water.

23 Unsaturated Hydrocarbons: Alkenes, Alkynes, and Aromatic Hydrocarbons

After studying Chapter 23 you should be able to:

1. Understand the terms listed in Question A at the end of the chapter.
2. Explain the sp^2 and sp hybridization of carbon atoms.
3. Explain the formation of a pi bond.
4. Explain the formation of double and triple bonds.
5. Distinguish, by formulas, the difference between saturated and unsaturated hydrocarbons.
6. Name and write structural formulas of alkenes and alkynes.
7. Determine from structural formulas whether a compound can exist as geometric isomers.
8. Name geometric isomers by the *cis–trans* method.
9. Write equations for addition reactions of alkenes and alkynes.
10. Explain the formation of carbonium ions and the role they play in chemical reactions.
11. Apply Markovnikov's rule to the addition of HCl, HBr, HI, and H_2O/H^+ to alkenes and alkynes.
12. Explain Baeyer's test for unsaturation.
13. Distinguish, using simple chemical tests, among alkanes, alkenes, and alkynes.
14. Understand the octane number rating of gasolines.
15. Describe the nature of benzene and how its properties differ from open-chain unsaturated compounds.
16. Name monosubstituted, disubstituted, and polysubstituted benzene compounds.
17. Draw structural formulas of substituted benzene compounds.
18. Recognize the most common fused aromatic ring compounds.
19. Write equations for the following reactions of benzene and substituted benzenes: halogenation (chlorination or bromination), nitration, alkylation (Friedel–Craft reaction), and side-chain oxidation.
20. Describe and write equations for the mechanism by which benzene compounds are brominated in the presence of $FeBr_3$ or chlorinated in the presence of $FeCl_3$.

The bulk of the organic compounds manufactured in the United States are derived from seven starting materials: ethylene, propylene, benzene, butylene, toluene, xylene, and methane. Of the seven substances listed, all are unsaturated hydrocarbons except methane. Ethylene alone is the base for almost half the petrochemicals made in the country. About 28 billion pounds of ethylene are produced currently, making it the fifth ranking chemical in production volume in the United States.

23.1 Alkenes and Alkynes

alkene

alkyne

Both **alkenes** and **alkynes** are classified as unsaturated hydrocarbons. They are said to be unsaturated because, unlike alkanes, their molecules do not contain the maximum possible number of hydrogen atoms. Alkenes (also known as *olefins*) contain at least one double bond between adjacent carbon atoms. Alkynes (also known as *acetylenes*) contain at least one triple bond between adjacent carbon atoms. The simplest alkene is ethylene (or ethene, $CH_2=CH_2$) and the simplest alkyne is acetylene (or ethyne, $CH≡CH$). Both

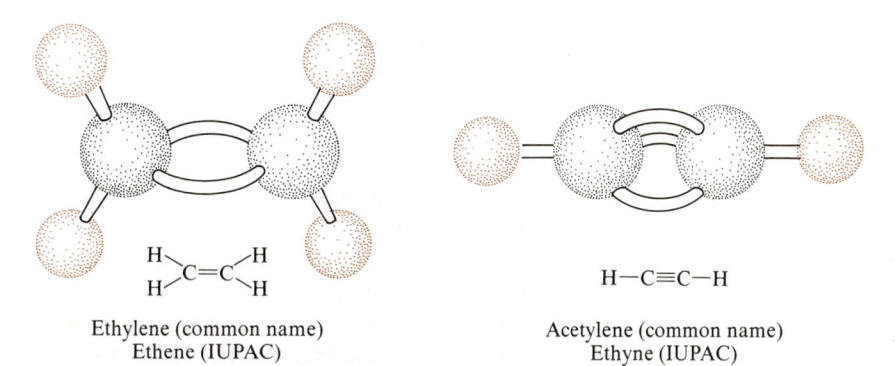

Ethylene (common name)
Ethene (IUPAC)

Acetylene (common name)
Ethyne (IUPAC)

ethylene and acetylene are the first members of homologous series in which the formulas of successive members differ by increments of $—CH_2—$ (for example, $CH_2=CH_2$, $CH_3CH=CH_2$, $CH_3CH_2CH=CH_2$, and so on). Huge quantities of alkenes are made by cracking and dehydrogenating alkanes in processing crude oils. These alkenes are used to manufacture motor fuels, polymers, and petrochemicals. Alkene molecules, like those of alkanes, have very little polarity. Hence, the physical properties of alkenes are very similar to those of the corresponding saturated hydrocarbons.

General formula for alkenes: C_nH_{2n}

General formula for alkynes: C_nH_{2n-2}

Note that the general formula for alkenes is the same as that for cycloalkanes. Alkenes and cycloalkanes have two hydrogen atoms less than alkanes. Table 23.1 gives the names and formulas for several alkenes and alkynes.

Table 23.1. Names and formulas for several alkenes and alkynes.

Common name	IUPAC name	Formula
Ethylene	Ethene	$CH_2{=}CH_2$
Propylene	Propene	$CH_3CH{=}CH_2$
Butylene	1-Butene	$CH_3CH_2CH{=}CH_2$
	2-Butene	$CH_3CH{=}CHCH_3$
Isobutylene	2-Methylpropene	$CH_3\underset{\underset{CH_3}{\mid}}{C}{=}CH_2$
Acetylene	Ethyne	$CH{\equiv}CH$
Methyl acetylene	Propyne	$CH_3C{\equiv}CH$
Ethyl acetylene	1-Butyne	$CH_3CH_2C{\equiv}CH$
Dimethyl acetylene	2-Butyne	$CH_3C{\equiv}CCH_3$

23.2 Bond Formation in Alkenes and Alkynes

In alkanes we saw that each carbon atom is bonded to four other atoms utilizing electrons in sp^3 hybrid orbitals. In ethylene each carbon atom is bonded to three other atoms—two hydrogen atoms and one carbon atom. The fourth electron of each carbon atom in ethylene is shared between the two carbon atoms forming the second bond of the carbon–carbon double bond.

The formation of double bonds in alkenes is explained by a different type of hybridization of electrons. The hybridization may be visualized in the following way. One of the $2s$ electrons of carbon is promoted to a $2p$ orbital forming the four half-filled orbitals, $2s^1 2p_x^1 2p_y^1 2p_z^1$. Three of these orbitals $(2s^1 2p_x^1 2p_y^1)$ hybridize forming three equivalent orbitals designated as sp^2. Thus, the four orbitals available for bonding are $2sp^2 2sp^2 2sp^2 2p$. This process is illustrated in Figure 23.1.

The three sp^2 hybrid orbitals form angles of 120° with each other and lie in a single plane. The remaining $2p$ orbital is oriented perpendicular to this plane, with one lobe above and one lobe below the plane (see Figure 23.2).

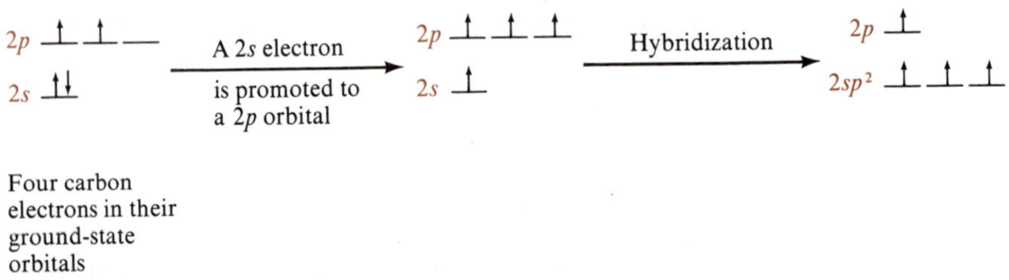

Figure 23.1. Schematic hybridization of $2s^2 2p_x^1 2p_y^1$ orbitals of carbon to form three sp^2 electron orbitals and one p electron orbital.

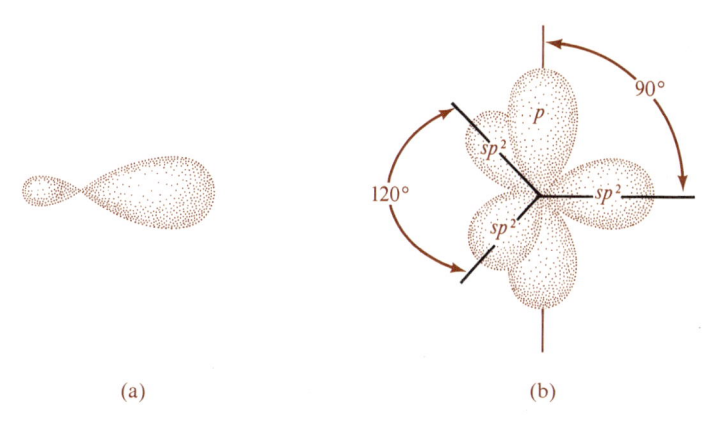

Figure 23.2. (a) A single sp^2 electron orbital and (b) a side view of three sp^2 orbitals all lying in the same plane with a p orbital perpendicular to the three sp^2 orbitals.

In the formation of a double bond, an sp^2 orbital of one carbon atom overlaps an identical sp^2 orbital of another carbon to form a sigma bond. At the same time, the two perpendicular p orbitals (one on each carbon atom) overlap to form a **pi (π) bond** between the two carbon atoms. This pi bond consists of two electron clouds, one above and one below the sigma bond (see Figure 23.3).

pi bond

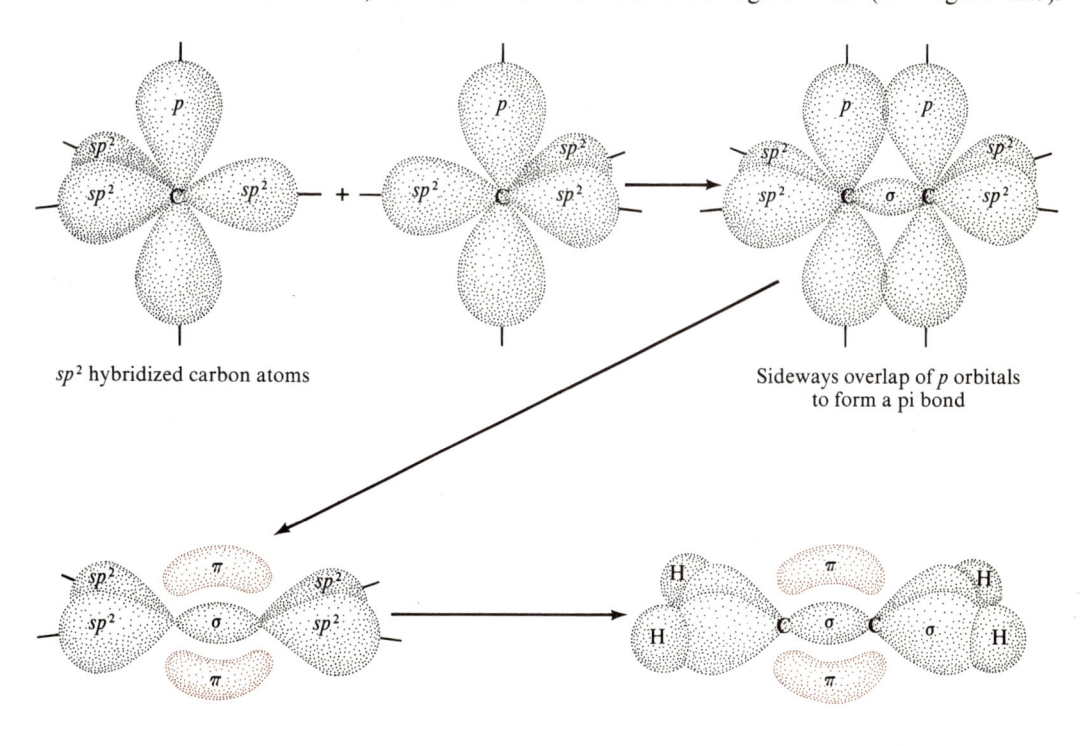

sp^2 hybridized carbon atoms

Sideways overlap of p orbitals to form a pi bond

Pi bond above and below C—C sigma bond

Ethylene

Figure 23.3. Pi (π) and sigma (σ) bonding in ethylene.

The H$_2$C=CH$_2$ molecule is completed as the remaining sp^2 orbitals (two on each carbon atom) overlap hydrogen s orbitals to form sigma bonds between the carbon and hydrogen atoms. Thus, there are five sigma bonds and one pi bond in an ethylene molecule.

In the formula commonly used to represent ethylene (CH$_2$=CH$_2$), no distinction is made between the sigma bond and the pi bond in the carbon–carbon double bond. Each bond is represented by a dash. However, these bonds are actually very different from each other. The sigma bond is formed by the overlap of sp^2 orbitals; the pi bond is formed by the overlap of p orbitals. The sigma bond electron cloud is distributed about a line joining the carbon nuclei, but the pi bond electron cloud is distributed above and below the sigma bond region (see Figure 23.3). The carbon–carbon pi bond is much weaker, and, as a consequence, much more reactive than the carbon–carbon sigma bond.

The formation of a triple bond between carbon atoms, as in acetylene, HC≡CH, may be visualized as follows:

1. A carbon atom 2s electron is promoted to a 2p orbital.
2. The remaining 2s orbital hybridizes with one of the 2p orbitals to form two equivalent orbitals known as sp orbitals. These two hybrid orbitals lie in a straight line passing through the center of the carbon atom. The remaining two unhybridized 2p orbitals are oriented at right angles to these sp orbitals and to each other.

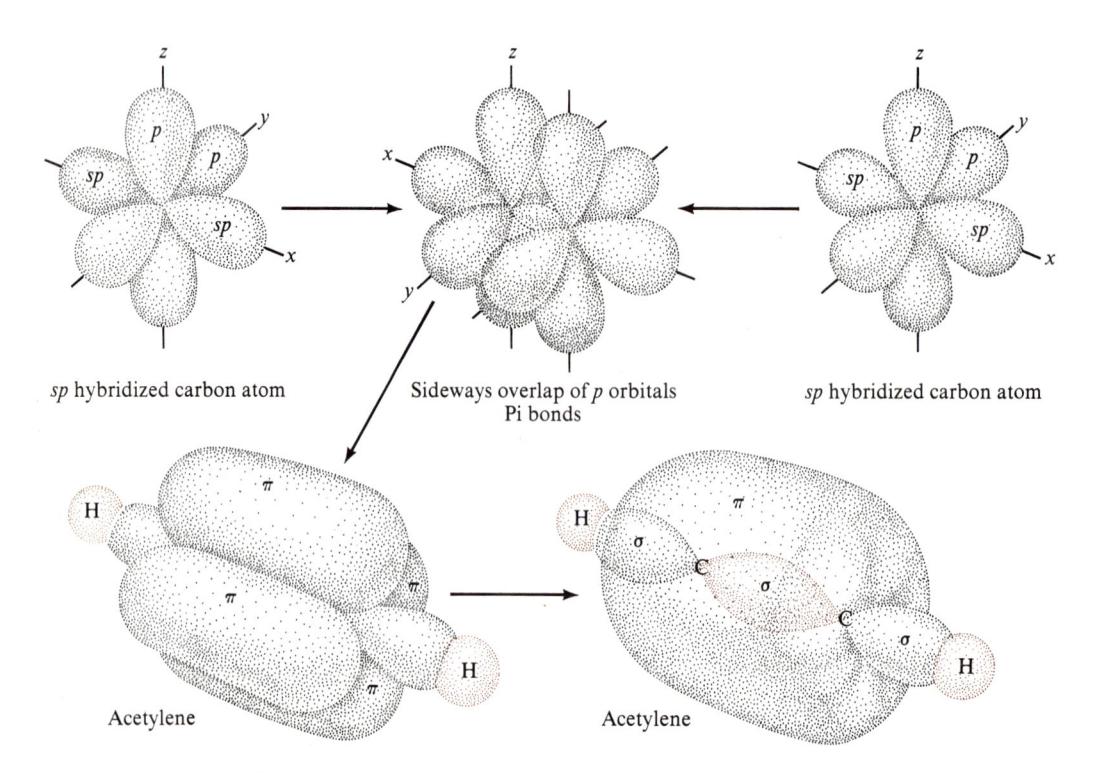

Figure 23.4. Pi (π) and sigma (σ) bonding in acetylene.

3. In forming carbon–carbon bonds, one carbon *sp* orbital overlaps an identical *sp* orbital on another carbon atom to establish the sigma bond between the two carbon atoms.
4. The remaining *sp* orbitals (one on each carbon atom) overlap *s* orbitals on hydrogens to form sigma bonds and establish the H—C—C—H bond sequence. Since all the atoms forming this sequence lie in a straight line, the acetylene molecule is linear.
5. Simultaneously, the two 2*p* orbitals on each carbon overlap to form two pi bonds. These two pi bond orbitals occupy sufficient space that they overlap each other to form a continuous tubelike electron cloud surrounding the sigma bond between the carbon atoms (Figure 23.4). These pi bond electrons (as in ethylene) are not as tightly held by the carbon nuclei as are the sigma bond electrons. Acetylene, consequently, is a very reactive substance.

23.3 Naming Alkenes and Alkynes

The names of alkenes and alkynes are derived from the names of the corresponding alkanes. To name an alkene (or alkyne) by the IUPAC System:

1. Select the longest carbon–carbon chain that contains the double or triple bond.
2. Name this parent compound as you would an alkane but change the *-ane* ending to *-ene* for an alkene or to *-yne* for an alkyne; for example, propane is changed to propene or propyne.
3. Number the carbon chain of the parent compound so that the double or triple bond carries the lowest possible numbers. Use the smaller of the two numbers on the double- or triple-bonded carbon atoms to indicate the position of the double or triple bond. Place this number in front of the alkene or alkyne name; for example, 2-butene means that the carbon–carbon double bond is between carbon numbers 2 and 3.
4. Side chains and other groups are treated as in naming alkanes, by numbering and assigning them to the carbon atom to which they are attached.

Study the following examples of named alkenes and alkynes:

$\overset{4}{C}H_3\overset{3}{C}H_2\overset{2}{C}H=\overset{1}{C}H_2$ $\overset{1}{C}H_3\overset{2}{C}H=\overset{3}{C}H\overset{4}{C}H_3$

   1-Butene   2-Butene

$\overset{4}{C}H_3\overset{\overset{\displaystyle CH_3}{|}}{\overset{3}{C}H}\overset{2}{C}H=\overset{1}{C}H_2$

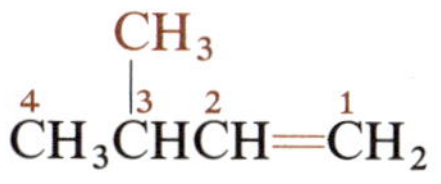

3-Methyl-1-butene

$\overset{4}{C}H_3\overset{3}{C}H_2\overset{2}{C}\equiv\overset{1}{C}H$ $\overset{1}{C}H_3-\overset{2}{C}\equiv\overset{3}{C}-\overset{\overset{\displaystyle CH_3}{|}}{\overset{4}{C}H}-\overset{\overset{\displaystyle |}{\underset{\displaystyle CH_3}{}}}{\overset{5}{C}H}-\overset{6}{C}H_3$

   1-Butyne   4,5-Dimethyl-2-hexyne

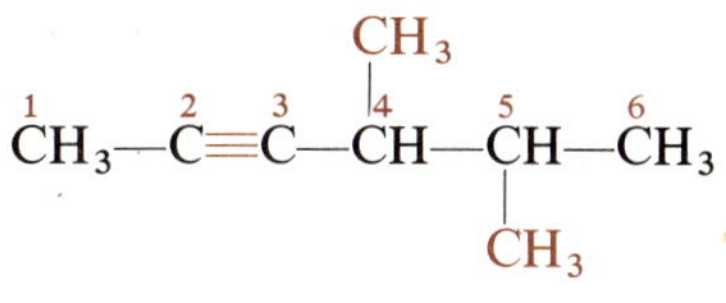

To write a structural formula from a systematic name, the naming process is, in effect, reversed. For example, how would we write the structural formula for 4-methyl-2-pentene? The name indicates:

1. Five carbons in the longest chain
2. A double bond between carbons 2 and 3
3. A methyl group on carbon 4

Write five carbon atoms in a row. Place a double bond between carbons 2 and 3, and place a methyl group on carbon 4:

$$\overset{1}{C}-\overset{2}{C}=\overset{3}{C}-\overset{4}{C}-\overset{5}{C}$$
$$\qquad\qquad\quad |$$
$$\qquad\qquad\; CH_3$$

Carbon skeleton

Now add hydrogen atoms to give each carbon atom four bonds. Carbons 1 and 5 each need three H atoms; carbons 2, 3, and 4 each need one H atom. The complete formula is

$$CH_3CH=CHCHCH_3$$
$$\qquad\qquad\quad\;\; |$$
$$\qquad\qquad\;\; CH_3$$

4-Methyl-2-pentene

Problem 23.1 Write structural formulas for (a) 2-pentene, (b) 7-methyl-2-octene, and (c) 3-hexyne.

(a) The prefix *pent-* indicates a five-carbon chain; the suffix *-ene* indicates a carbon–carbon double bond; the number 2 locates the double bond between carbons 2 and 3. Write five carbon atoms in a row and place a double bond between carbons 2 and 3:

$$\overset{1}{C}-\overset{2}{C}=\overset{3}{C}-\overset{4}{C}-\overset{5}{C}$$

Add hydrogen atoms to give each carbon atom four bonds. Carbons 1 and 5 each need three H atoms; carbons 2 and 3 each need one H atom; carbon 4 needs two H atoms. The complete formula is

$$CH_3CH=CHCH_2CH_3$$

2-Pentene

(b) Octene, like octane, indicates an eight-carbon chain. The chain contains a double bond between carbons 2 and 3 and a methyl group on carbon 7. Write eight carbon atoms in a row, place a double bond between carbons 2 and 3, and place a methyl group on carbon 7:

$$\overset{1}{C}-\overset{2}{C}=\overset{3}{C}-\overset{4}{C}-\overset{5}{C}-\overset{6}{C}-\overset{7}{C}-\overset{8}{C}$$
$$\qquad\qquad\qquad\qquad\qquad\quad |$$
$$\qquad\qquad\qquad\qquad\qquad CH_3$$

Now add hydrogen atoms to give each carbon atom four bonds. The complete formula is

$$CH_3CH=CHCH_2CH_2CH_2CHCH_3$$
$$\qquad\qquad\qquad\qquad\qquad\qquad |$$
$$\qquad\qquad\qquad\qquad\qquad\; CH_3$$

7-Methyl-2-octene

(c) The stem *hex-* indicates a six-carbon chain; the suffix *-yne* indicates a carbon–carbon triple bond; the number 3 locates the triple bond between carbons 3 and 4. Write six carbon atoms in a row and place a triple bond between carbons 3 and 4:

$$\overset{1}{C}-\overset{2}{C}-\overset{3}{C}\equiv\overset{4}{C}-\overset{5}{C}-\overset{6}{C}$$

Now add hydrogen atoms to give each carbon atom four bonds. Carbons 3 and 4 do not need any H atoms. The complete formula is

$$CH_3CH_2C\equiv CCH_2CH_3$$
3-Hexyne

23.4 Geometric Isomerism

There are only two dichloroethanes: 1,1-dichloroethane ($CHCl_2CH_3$) and 1,2-dichloroethane (CH_2ClCH_2Cl). But, surprisingly, there are three dichloroethenes; namely, 1,1-dichloroethene ($CCl_2\!=\!CH_2$) and *two* isomers of 1,2-dichloroethene ($CHCl\!=\!CHCl$). There is only one 1,2-dichloroethane because carbon atoms can rotate freely about a single bond. Thus, the structural formulas I and II given below represent the same compound. The chlorine atoms are simply shown in different relative positions in the two formulas due to rotation of the CH_2Cl group about the carbon–carbon single bond.

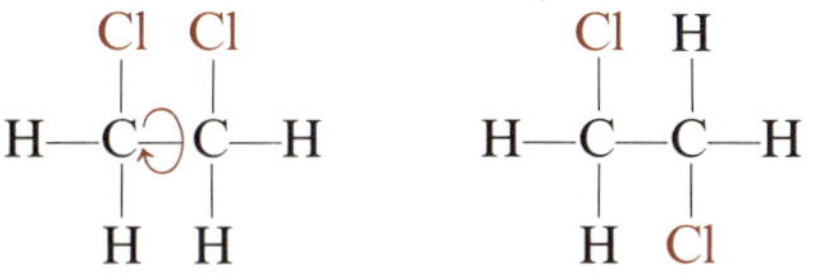

I   II
1,2-Dichloroethane

geometric isomers

Restricted rotation about a carbon–carbon double bond in a molecule gives rise to a type of isomerism known as *geometric isomerism*. Isomers that differ from each other only in the geometry of the molecules and not in the order of their atoms are known as **geometric isomers**. They are also called *cis–trans* isomers. Two isomers of 1,2-dichloroethene exist because of geometric isomerism.

For a further explanation let us look at the geometry of an ethylene molecule. This molecule is planar, or flat, with all six atoms lying in a single plane as in a rectangle:

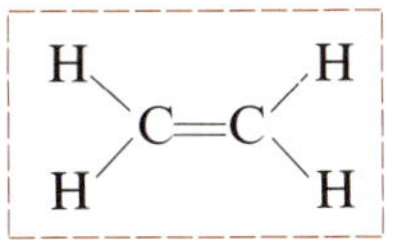

Since the hydrogen atoms are identical, only one structural arrangement is possible for ethylene. But if one hydrogen atom on each carbon atom is replaced by chlorine, for example, two different structural arrangements or geometric isomers are possible:

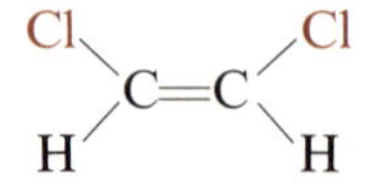

$$\underset{\substack{\text{cis-1,2-Dichloroethene} \\ \text{bp} = 60.1°\text{C}}}{\overset{\text{Cl}\text{Cl}}{\underset{\text{H}\text{H}}{\text{C}=\text{C}}}} \quad \text{and} \quad \underset{\substack{\text{trans-1,2-Dichloroethene} \\ \text{bp} = 48.4°\text{C}}}{\overset{\text{H}\text{Cl}}{\underset{\text{Cl}\text{H}}{\text{C}=\text{C}}}}$$

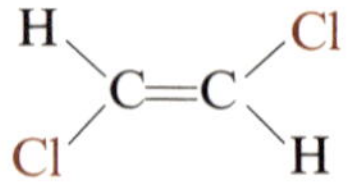

Both of these isomers are known and have been isolated. The fact that they have different boiling points as well as other different physical properties is proof that they are not the same compound. Note that in naming these geometric isomers, the prefix *cis* is used to designate the isomer having the substituent groups (chlorine atoms) on the same side of the double bond and the prefix *trans* is used to designate the isomer having the substituent groups (chlorine atoms) on the opposite sides of the double bond. Molecules of *cis*- and *trans*-1,2 dichloroethene are not superimposable. That is, we cannot pick up one molecule and place it over the other in such a way that all the atoms in each molecule occupy the same relative positions in space. Nonsuperimposability is a general test for isomerism which all kinds of isomers must meet.

Geometric isomers are possible with an alkene whenever the two groups attached to each carbon atom in the carbon–carbon double bond are different.

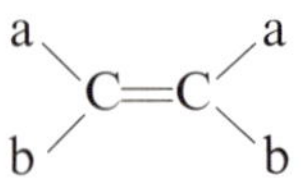

$$\underset{\text{cis isomer}}{\overset{\text{a}\text{a}}{\underset{\text{b}\text{b}}{\text{C}=\text{C}}}} \qquad \underset{\text{trans isomer}}{\overset{\text{a}\text{b}}{\underset{\text{b}\text{a}}{\text{C}=\text{C}}}}$$

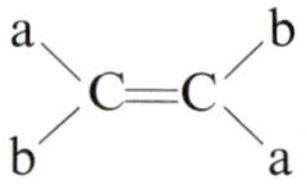

When two groups on one carbon atom are the same, geometric isomerism does not occur. Thus, there are no geometric isomers of ethene or propene.

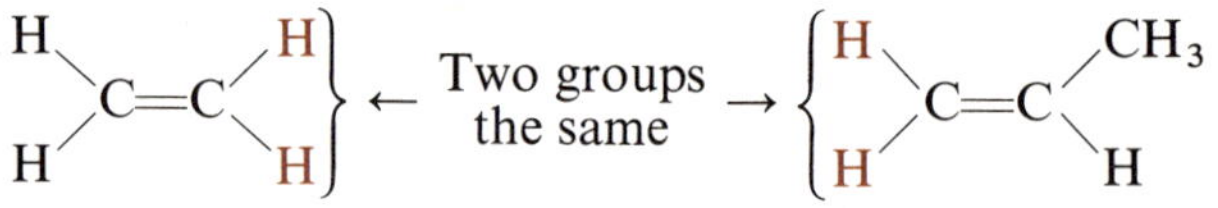

$$\left\{ \overset{\text{H}\text{H}}{\underset{\text{H}\text{H}}{\text{C}=\text{C}}} \right\} \leftarrow \begin{array}{c}\text{Two groups} \\ \text{the same}\end{array} \rightarrow \left\{ \overset{\text{H}\text{CH}_3}{\underset{\text{H}\text{H}}{\text{C}=\text{C}}} \right\}$$

Four structural isomers of butene (C_4H_8) are known. Two of these, 1-butene and 2-methylpropene, do not show geometric isomerism.

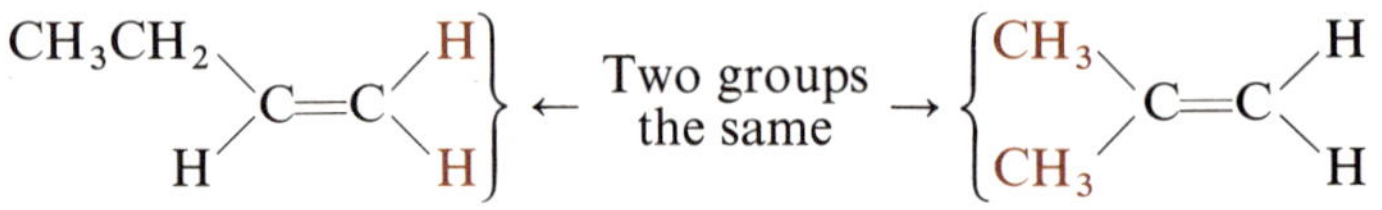

$$\underset{\text{1-Butene}}{\left\{ \overset{\text{CH}_3\text{CH}_2\text{H}}{\underset{\text{H}\text{H}}{\text{C}=\text{C}}} \right\}} \leftarrow \begin{array}{c}\text{Two groups} \\ \text{the same}\end{array} \rightarrow \underset{\text{2-Methylpropene}}{\left\{ \overset{\text{CH}_3\text{H}}{\underset{\text{CH}_3\text{H}}{\text{C}=\text{C}}} \right\}}$$

The other two butenes are the *cis–trans* isomers shown below.

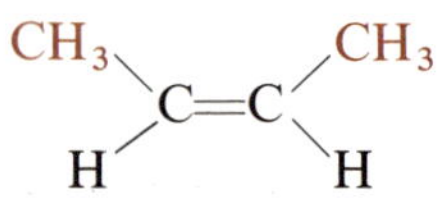

$$\underset{\text{cis-2-Butene}}{\overset{\text{CH}_3\text{CH}_3}{\underset{\text{H}\text{H}}{\text{C}=\text{C}}}} \quad \text{and} \quad \underset{\text{trans-2-Butene}}{\overset{\text{CH}_3\text{H}}{\underset{\text{H}\text{CH}_3}{\text{C}=\text{C}}}}$$

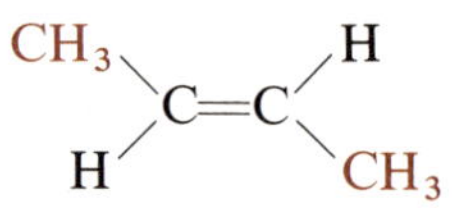

Geometric isomerism is one of the kinds of isomerism that is especially significant in biochemistry, because many substances in living organisms occur as only one of the possible geometric isomers.

Problem 23.2 Which of the following pentenes will show *cis–trans* isomerism? Draw and label the *cis–trans* isomers.

(a) $CH_2=CHCH_2CH_2CH_3$ (b) $CH_3CH_2\underset{\underset{CH_3}{|}}{C}=CH_2$ (c) $CH_3\underset{\underset{CH_3}{|}}{CH}CH=CH_2$

(d) $CH_3CH=CHCH_2CH_3$ (e) $CH_3CH=\underset{\underset{CH_3}{|}}{C}CH_3$

The formulas in parts (a), (b), and (c) each have two hydrogen atoms on one of the double-bond carbon atoms and therefore do not show geometric isomerism. Compound (e) has two methyl groups on one of the double-bond carbon atoms and therefore will not show geometric isomerism. Compound (d) has two different groups on each carbon atom of the double bond and therefore will exist in two geometric isomers. To show the two isomers, first draw two C=C groups in a planar arrangement:

$$\!>\!C\!=\!C\!<\qquad\!>\!C\!=\!C\!<$$

Now attach the H, CH$_3$, and CH$_3$CH$_2$ groups to make the two isomers:

$$\underset{\text{cis-2-Pentene}}{\overset{CH_3\diagdown\quad\diagup CH_2CH_3}{\underset{H\diagup\quad\diagdown H}{C=C}}} \qquad \underset{\text{trans-2-Pentene}}{\overset{CH_3\diagdown\quad\diagup H}{\underset{H\diagup\quad\diagdown CH_2CH_3}{C=C}}}$$

Problem 23.3 Is the compound below the *cis* or the *trans* isomer?

$$\overset{CH_3\diagdown\quad\diagup CH_3}{\underset{H\diagup\quad\diagdown CH_2CH_3}{C=C}}$$

In branched-chain alkenes, the *cis–trans* relationship ordinarily refers to the structure containing the longest carbon chain that includes the carbon–carbon double bond. In this case the longest chain is the 2-pentene in which the methyl group on carbon 2 and the ethyl group on carbon 3 are *trans* to each other. Thus, the name of the compound shown is *trans*-3-methyl-2-pentene. The *cis* isomer is

$$\overset{CH_3\diagdown\quad\diagup CH_2CH_3}{\underset{H\diagup\quad\diagdown CH_3}{C=C}}$$

23.5 Reactions of Alkenes

The alkenes are much more reactive than the corresponding alkanes. This greater reactivity is due to the carbon–carbon double bonds.

Addition

addition reaction

A reaction in which two substances join together to produce one compound is called an **addition reaction**. Addition at the carbon–carbon double bond is the most common reaction of alkenes. Hydrogen, halogens (Cl_2 or Br_2), hydrogen halides, sulfuric acid, and water are some of the reagents that can be added to unsaturated hydrocarbons. Ethylene (ethene), for example, reacts in this fashion.

$$CH_2{=}CH_2 + H_2 \xrightarrow[\text{1 atm}]{\text{Pt, 25°C}} CH_3{-}CH_3$$
Ethylene → Ethane

$$CH_2{=}CH_2 + Br{-}Br \longrightarrow CH_2Br{-}CH_2Br$$
1,2-Dibromoethane

Visible evidence of this reaction is the disappearance of the reddish-brown bromine as it reacts.

$$CH_2{=}CH_2 + HCl \longrightarrow CH_3CH_2Cl$$
Chloroethane (Ethyl chloride)

$$CH_2{=}CH_2 + HOSO_3H(conc.) \longrightarrow CH_3CH_2OSO_3H$$
Sulfuric acid → Ethyl hydrogen sulfate

$$CH_2{=}CH_2 + HOH \xrightarrow{H^+} CH_3CH_2OH$$
Ethanol (Ethyl alcohol)

(The H^+ indicates that the reaction is carried out under acid conditions.)

Note that the double bond is broken and the unsaturated alkene molecules become saturated by an addition reaction.

The preceding examples dealt with ethylene, but reactions of this kind can be made to occur on almost any molecule that contains a carbon–carbon double bond. The addition of hydrogen or of chlorine to a double bond site is represented as follows:

$$R{-}\underset{\underset{H}{|}}{\overset{\overset{H}{|}}{C}}{=}\underset{\underset{H}{|}}{\overset{\overset{H}{|}}{C}}{-}R + H_2 \xrightarrow[\text{1 atm}]{\text{Pt, 25°}} R{-}\underset{\underset{H}{|}}{\overset{\overset{H}{|}}{C}}{-}\underset{\underset{H}{|}}{\overset{\overset{H}{|}}{C}}{-}R$$

Hydrogenated product

$$R{-}\overset{\overset{H}{|}}{C}{=}\overset{\overset{H}{|}}{C}{-}R + Cl_2 \longrightarrow R{-}\underset{\underset{Cl}{|}}{\overset{\overset{H}{|}}{C}}{-}\underset{\underset{Cl}{|}}{\overset{\overset{H}{|}}{C}}{-}R$$

Chlorine addition product

If a symmetrical molecule such as Cl_2 is added to propene, only one product, 1,2-dichloropropane, is formed:

$$CH_2{=}CH{-}CH_3 + Cl_2 \longrightarrow CH_2Cl{-}CHCl{-}CH_3$$

But if an unsymmetrical molecule such as HCl is added to propene, two products are theoretically possible, depending upon which carbon atom adds the hydrogen. The two possible products are 1-chloropropane and 2-chloropropane. Experimentally, we find that 2-chloropropane is formed almost exclusively:

$$CH_3-CH=CH_2 + HCl \longrightarrow \begin{array}{l} CH_3CHClCH_3 \text{ (About 100\%)} \\ \\ CH_3CH_2CH_2Cl \text{ (Trace)} \end{array}$$

A single product is obtained because the reaction proceeds stepwise according to the following mechanism:

1. A proton (H$^+$) from HCl bonds to the number 1 carbon of propene utilizing the pi bond electrons. The intermediate formed is a positively charged alkyl group or carbonium ion. The positive charge is localized on the number 2 carbon atom of this carbonium ion.

$$CH_2=CH-CH_3 + HCl \longrightarrow CH_3-\overset{\oplus}{C}H-CH_3 + Cl^-$$
Isopropyl carbonium ion

2. The chloride ion then adds to the positively charged carbon atom to form a molecule of 2-chloropropane.

$$CH_3-\overset{\oplus}{C}H-CH_3 + Cl^- \longrightarrow CH_3-CHCl-CH_3$$
2-Chloropropane

carbonium ion

carbon cation

An ion in which a carbon atom has a positive charge is known as a **carbonium ion** or a **carbon cation**. There are four types of carbonium ions: methyl, primary (1°), secondary (2°), and tertiary (3°). Examples of these are given below.

$$\begin{array}{ccccc}
H & H & H & H & CH_3 \\
| & | & | & | & | \\
H-\overset{\oplus}{C} & CH_3-\overset{\oplus}{C} & CH_3CH_2-\overset{\oplus}{C} & CH_3-\overset{\oplus}{C}-CH_3 & CH_3-\overset{\oplus}{C} \\
| & | & | & | & | \\
H & H & H & H & CH_3
\end{array}$$

Methyl carbonium ion   Ethyl carbonium ion (Primary)   n-Propyl carbonium ion (Primary)   Isopropyl carbonium ion (Secondary)   t-Butyl carbonium ion (Tertiary)

A carbon atom is designated as primary if it is bonded to one carbon atom, secondary if it is bonded to two carbon atoms, and tertiary if it is bonded to three carbon atoms. In a primary carbonium ion the positive carbon atom is bonded to only one carbon atom. In a secondary carbonium ion the positive carbon atom is bonded to two carbon atoms. In a tertiary carbonium ion the positive carbon atom is bonded to three carbon atoms.

The order of stability of carbonium ions and hence the ease in which they are formed is tertiary > secondary > primary. Thus, in the reaction of propene and HCl, isopropyl carbonium ion (secondary) is formed as an intermediate in preference to n-propyl carbonium ion (primary).

Markovnikov's rule

In the middle of the 19th century, a Russian chemist, V. Markovnikov, observed reactions of this kind, and in 1869 he formulated a useful generalization now known as **Markovnikov's rule**. This rule in essence states:

When an unsymmetrical molecule such as HX(HCl) adds to a carbon–carbon double bond, the hydrogen from HX goes to the carbon atom that has the greater number of hydrogen atoms.

As you can see, the addition of HCl to propene discussed above follows Markovnikov's rule. The addition of HI to 2-methylpropene (isobutylene) is another example illustrating this rule:

$$CH_3-\underset{\underset{CH_3}{|}}{C}=CH_2 + HI \longrightarrow CH_3-\underset{\underset{I}{|}}{\overset{\overset{CH_3}{|}}{C}}-CH_3$$

2-Iodo-2-methylpropane
(*tert*-Butyl iodide)

General rules of this kind are useful in predicting the products of reactions. However, exceptions are known for most such rules.

Problem 23.4 Write equations for the addition of HCl to (a) 1-pentene and (b) 2-pentene.

(a) In the case of 1-pentene, $CH_3CH_2CH_2CH=CH_2$, the proton from HCl adds to carbon 1 to give the more stable secondary carbonium ion followed by the addition of Cl^- to give the product 2-chloropentane. The addition is directly in accordance with Markovnikov's rule.

$$CH_3CH_2CH_2CH=CH_2 + HCl \longrightarrow CH_3CH_2CH_2CHClCH_3$$
2-Chloropentane

(b) In 2-pentene, $CH_3CH_2CH=CHCH_3$, each carbon of the double bond has one hydrogen atom and the addition of a proton to either one forms a secondary carbonium ion. As a result, two isomeric products are formed in almost equal quantities:

$$CH_3CH_2CH=CHCH_3 + HCl \longrightarrow CH_3CH_2CH_2CHClCH_3 + CH_3CH_2CHClCH_2CH_3$$
2-Chloropentane   3-Chloropentane

Oxidation

Another typical reaction of alkenes is oxidation at the double bond. For example, when shaken with a cold, dilute solution of potassium permanganate ($KMnO_4$), an alkene is converted to a glycol (glycols are dihydroxy alcohols). Ethylene reacts in this manner.

$$\underset{\text{Ethylene}}{CH_2\!=\!CH_2} + \underset{\text{(Purple)}}{KMnO_4(aq)} + H_2O \longrightarrow \underset{\underset{\text{Ethylene glycol}}{OH \quad OH}}{CH_2\!-\!CH_2} + \underset{\text{(Brown)}}{MnO_2} + KOH$$

Baeyer's test makes use of this reaction to detect or confirm the presence of double (or triple) bonds in hydrocarbons. Evidence of reaction (positive Baeyer's test) is the disappearance of the purple color of permanganate ions. Baeyer's test is not specific for detecting unsaturation in hydrocarbons, since other classes of compounds may also give a positive Baeyer's test.

Carbon–carbon double bonds (ethylenic linkages) are found in many different kinds of molecules. Most of these substances will react with potassium permanganate and will undergo somewhat similar reactions with other oxidizing agents including oxygen in the air and, especially, with ozone. Such reactions are frequently troublesome. For example, premature aging and cracking of tires in smoggy atmosphere occurs because ozone attacks the ethylenic linkages in rubber molecules. Cooking oils and fats sometimes develop disagreeable odors and flavors because the oxygen of the air reacts with the double bonds present in these materials. Potato chips are especially subject to flavor damage caused by oxidation of the unsaturated cooking oils that they contain.

Polymerization

Addition polymerization is another important reaction of alkenes. In this reaction, simple alkene molecules add together to form giant macromolecules. Synthetic rubber, polyethylene, and many other polymers are made by addition polymerization. This reaction is discussed in Chapter 27.

23.6 Octane Rating of Gasoline

Gasoline does not contain appreciable amounts of alkenes. However, large quantities of alkenes are used to make the branched-chain alkanes that are used in high-quality gasoline. "Knocking," which is due to a too-rapid combustion or detonation of the air–gasoline mixture, is a severe problem in high-compression engines. The knock-resistance of gasolines, a quality that varies widely, is usually expressed in terms of *octane number* or *octane rating*.

Isooctane (2,2,4-trimethylpentane), because of its highly branched chain structure, is a motor fuel that is very resistant to knocking. Mixtures of isooctane and *n*-heptane, a straight-chain alkane that knocks badly, have been widely used as standards to establish octane ratings of gasolines. Isooctane is arbitrarily assigned an octane number of 100 and *n*-heptane an octane number of 0. To determine the octane rating, a gasoline is compared with mixtures of isooctane and *n*-heptane in a test engine. The octane number of the gasoline corresponds to the percentage of isooctane present in the isooctane and *n*-heptane mixture that matches the knocking characteristics of the gasoline.

Thus, a 90 octane gasoline would have knocking characteristics matching those of a mixture of 90% isooctane and 10% *n*-heptane.

When first used to establish octane numbers, isooctane was the most knock-resistant substance available. Substances even more resistant to knocking than isooctane have since been manufactured, and it is now possible to buy gasolines with octane numbers greater than 100.

Two general methods are available for increasing the octane number of gasoline. One method is to process the crude oil so as to increase the percentages of branched-chain alkanes and/or aromatics in the product. Both branched-chain alkanes and aromatic hydrocarbons have excellent antiknock properties, but these materials are present in limited quantities in most crude oils. So this method involves breaking up molecules and rearranging the fragments to form new molecules. Numerous processes, usually involving high pressure, high temperature, and special catalysts, have been developed for the production of high-octane gasoline by this general method.

The second method is the use of gasoline additives such as tetraethyl lead $[(C_2H_5)_4Pb]$ in "ethyl fluid." This approach is relatively inexpensive, very effective, and can increase the rating of some motor fuels by as much as 15 octane numbers. The highest-octane gasolines now commercially available are produced by combining the two methods; for example, "ethyl fluid" is added to gasoline stocks containing high percentages of branched-chain alkanes and aromatics.

Although lead compounds, specifically tetraethyl lead, are without question effective antiknock agents, their use in gasoline is currently open to the following serious objections:

1. In modern automobiles catalytic devices are used in exhaust systems to reduce the level of exhaust air pollutants (carbon monoxide, hydrocarbons, and nitrogen oxides). The use of leaded gasoline makes smog control more difficult because the lead in the exhaust deactivates most metal catalysts.
2. Lead is a poisonous substance. The continued use of lead in gasoline means that the environmental level of finely divided lead compounds is being increased. This environmental accumulation is mostly in the soil since the lead compounds in exhausts are not volatile enough to remain as atmospheric pollutants for long periods of time. Legal restrictions have been established to govern the maximum amounts of lead in gasoline.
3. Tetraethyl lead is a volatile poison. Its presence in gasoline constitutes an additional hazard to those normally present in handling this volatile and highly combustible material. For this reason leaded gasoline should never be used in gasoline stoves, lanterns, catalytic heaters, or similar devices.

23.7 Acetylenes: Preparation and Properties

Acetylene (ethyne), HC≡CH, is the simplest alkyne and is an industrial chemical of prime importance. It can be prepared from relatively inexpensive raw materials—lime and coke. These are heated to form calcium carbide, which is then reacted with water to form acetylene:

$$CaO + 3C \xrightarrow[\text{Electric furnace}]{2500°C} CaC_2 + CO$$

Calcium oxide (Lime)   Carbon (Coke)   Calcium carbide

$$CaC_2 + 2\,H_2O \longrightarrow CH{\equiv}CH + Ca(OH)_2$$
                                  Acetylene

Until about 1940, nearly all acetylene was made from calcium carbide. Cheaper methods of production, using methane as the raw material, were devised in Germany during World War II. When methane is subjected to high temperatures for a very short time (0.01–0.1 second) and the products are cooled rapidly, a reasonably high yield of acetylene is obtained. The equation for the reaction, which is endothermic, is

$$2\,CH_4 \xrightarrow{1500°C} CH{\equiv}CH + 3\,H_2 \quad -95.5\text{ kcal}$$

Methane   Acetylene

Special precautions must be taken when handling acetylene. Like all hydrocarbon gases, it forms explosive mixtures with air or oxygen. In addition to this hazard, acetylene is unstable at room temperature. When highly compressed or liquefied, acetylene may decompose violently (explode) either spontaneously or from a slight shock:

$$CH{\equiv}CH \longrightarrow H_2 + 2\,C + 54.3\text{ kcal}$$

To eliminate the danger of explosions, acetylene is dissolved under pressure in acetone and is packed in cylinders that contain a porous inert material.

Acetylene is used mainly (1) as fuel for oxyacetylene cutting and welding torches, and (2) as an intermediate in the manufacture of other substances. Both uses are dependent upon the great reactivity of acetylene. Acetylene and oxygen mixtures produce flame temperatures of about 2800°C. Acetylene readily undergoes addition reactions rather similar to those of ethylene. It will react with chlorine and bromine and will decolorize a permanganate solution (Baeyer's test). Either one or two molecules of bromine or chlorine may be added:

$$CH{\equiv}CH + Br_2 \longrightarrow CHBr{=}CHBr$$
                                  1,2-Dibromoethene

or

$$CH{\equiv}CH + 2\,Br_2 \longrightarrow CHBr_2{-}CHBr_2$$
                                  1,1,2,2-Tetrabromoethane

It is apparent that either unsaturated or saturated compounds may be obtained as addition products of acetylene. Often, unsaturated compounds capable of undergoing further reactions are made from acetylene. The monomer vinyl chloride, used to make the plastic polyvinyl chloride (PVC), can be made by simple addition of HCl to acetylene:

$$CH\equiv CH + HCl \longrightarrow CH_2=CHCl$$
Vinyl chloride
(Chloroethene)

Note: The common name for the $CH_2=CH-$ radical is *vinyl*.

Hydrogen chloride reacts with alkynes in a similar fashion to form substituted alkenes. The addition follows Markovnikov's rule. Alkynes can react with 1 or 2 moles of HCl. Consider the reaction of propyne with HCl:

$$CH_3C\equiv CH + HCl \longrightarrow CH_3CCl=CH_2$$
2-Chloropropene

$$CH_3CCl=CH_2 + HCl \longrightarrow CH_3CCl_2CH_3$$
2,2-Dichloropropane

Acetylenic hydrogen atoms—that is, hydrogen atoms attached to a carbon bearing a triple bond—are somewhat acidic and can be replaced, forming metal salts. Sodium, silver, and copper(I) are some of the metal salts that can be easily prepared. Acetylene reacts with molten sodium and with ammoniacal silver nitrate or copper(I) chloride forming insoluble precipitates according to the following equations:

$$2\,HC\equiv CH + 2\,Na(molten) \longrightarrow 2\,HC\equiv C^-Na^+(s) + H_2(g)$$
Sodium acetylide

$$HC\equiv CH + 2\,Ag(NH_3)_2NO_3 \longrightarrow AgC\equiv CAg(s) + 2\,NH_4NO_3 + 2\,NH_3$$
Silver carbide

$$HC\equiv CH + 2\,Cu(NH_3)_2Cl \longrightarrow CuC\equiv CCu(s) + 2\,NH_4Cl + 2\,NH_3$$
Copper(I) carbide

Silver and copper(I) carbides are unstable when dry and may be exploded by heat or shock. Compounds of the type $R-C\equiv CH$ yield metal acetylides, whereas compounds of the type $R-C\equiv C-R$ do not react.

The reactions of other alkynes are similar to those of acetylene. From an industrial viewpoint, although many other alkynes are known, acetylene is by far the most important.

23.8 Aromatic Hydrocarbons: Structure

aromatic compounds

Benzene and all substances that have structures and chemical properties resembling benzene are classified as **aromatic compounds**. The word *aromatic* originally referred to the rather pleasant odor possessed by many of these substances, but this meaning has been dropped. Benzene, the parent substance of the aromatic hydrocarbons, was first isolated by Michael Faraday in 1825; its correct molecular formula, C_6H_6, was established a few years later. The establishment of a reasonable structural formula that would account for the properties of benzene was a very difficult problem for chemists in the mid-19th century.

Finally, in 1865, August Kekulé proposed that the carbon atoms in a benzene molecule are arranged in a six-membered ring with one hydrogen atom bonded to each carbon atom and with three carbon–carbon double bonds, as shown below:

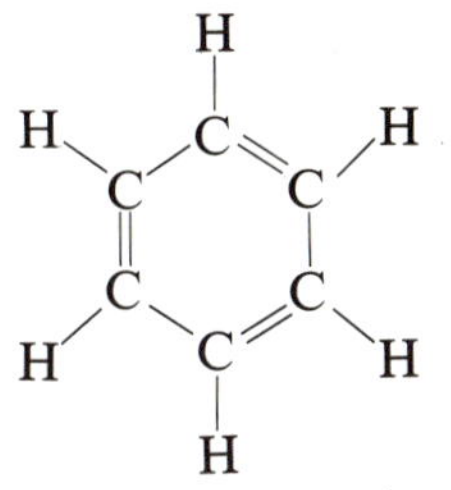

Kekulé soon realized that there should be two dibromobenzenes based on double bond positions relative to the two bromine atoms.

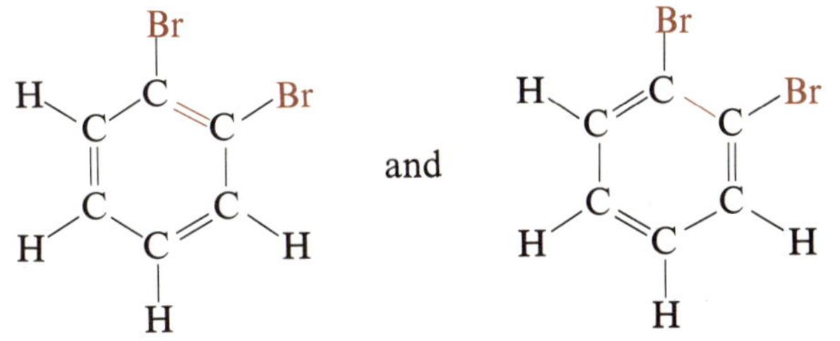

Since only one dibromobenzene (with bromine atoms on adjacent carbons) could be produced, Kekulé suggested that the double bonds are in rapid oscillation within the molecule. He also proposed that the structure of benzene could be represented in this fashion.

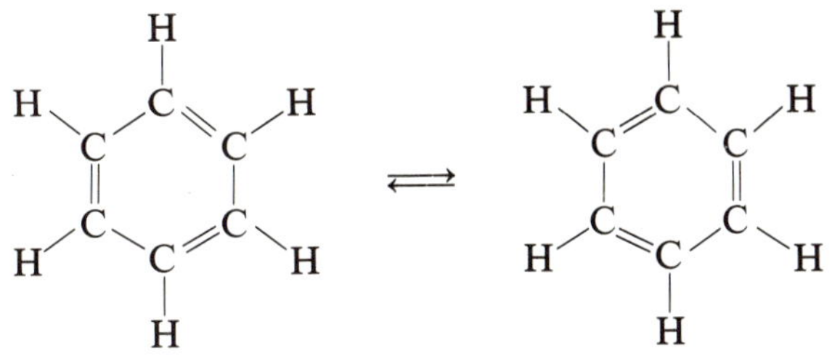

Kekulé's concepts are a landmark in the history of chemistry. They are the basis of the best representation of the benzene molecule devised in the 19th century; and they mark the beginning of our understanding of structure in aromatic compounds.

Kekulé's formulas do have one serious shortcoming in that they represent benzene and related substances as highly unsaturated compounds. Yet benzene does not react like a typical alkene (olefin); it does not decolorize bromine solutions rapidly, nor does it destroy the purple color of permanganate ions (Baeyer's test). Instead, the chemical behavior of benzene resembles that of an alkane. Its typical reactions are the substitution type, wherein a hydrogen atom is replaced by some other group; for example,

$$C_6H_6 + Cl_2 \xrightarrow{Fe} C_6H_5Cl + HCl$$

This problem was not fully resolved until the technique of X-ray diffraction, developed in the years following 1912, permitted us to determine the actual distances between the nuclei of carbon atoms in molecules. The center-to-center carbon atom distances in different kinds of hydrocarbon molecules are

   Ethane (single bond)   1.54 Å
   Ethylene (double bond)   1.34 Å
   Benzene   1.40 Å

Since only one carbon–carbon distance (bond length) is found in benzene, it is apparent that alternating single and double bonds do not exist in the benzene molecule.

Modern theory accounts for the structure of the benzene molecule in this way: The orbital hybridization of the carbon atoms is sp^2 (see structure of ethylene molecule). A planar hexagonal ring is formed by overlapping of two sp^2 orbitals on each of six carbon atoms. The other sp^2 orbital on each carbon atom overlaps an s orbital of a hydrogen atom, bonding the carbon to the hydrogen by a sigma bond. The remaining six p orbitals, one on each carbon atom, overlap each other and form doughnut-shaped pi electron clouds above and below the plane of the ring (see Figure 23.5). The electrons comprising these clouds are not attached to particular carbon atoms but are delocalized and associated with the entire molecule. This electronic structure imparts unusual stability to benzene and is responsible for many of the characteristic properties of aromatic compounds.

For convenience, the present-day chemist usually writes the structure of benzene as one or the other of these abbreviated forms.

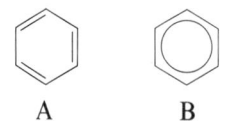

    A   B

In both representations, it is understood that there is a carbon atom and a hydrogen atom at each corner of the hexagon. The classical Kekulé structure is represented by A; and the modern molecular orbital structure is represented by B. Hexagons are used in representing the structural formulas of benzene derivatives—that is, substances in which one or more hydrogen atoms in the ring have been replaced by other atoms or groups. Chlorobenzene (C_6H_5Cl), for example, is written in this fashion:

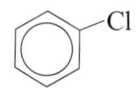

Chlorobenzene, C_6H_5Cl

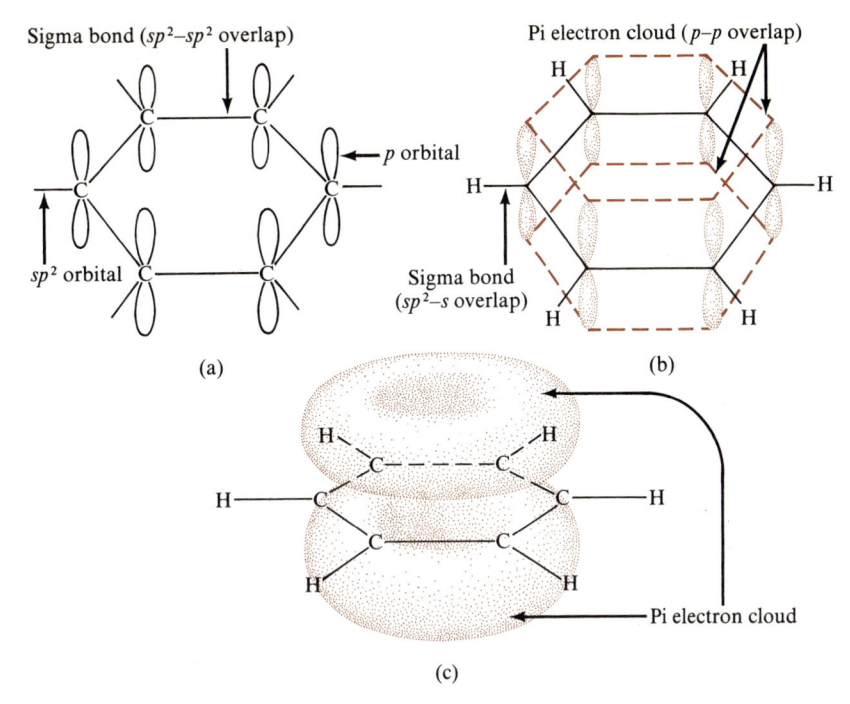

Figure 23.5. Bonding in a benzene molecule: (a) sp^2–sp^2 orbital overlap to form the carbon ring structure; (b) carbon–hydrogen bonds formed by sp^2–s orbital overlap and overlapping of p orbitals; (c) pi electron clouds above and below the plane of the carbon ring.

This notation indicates that the chlorine atom has replaced a hydrogen atom and is bonded directly to a carbon atom in the ring. Thus, the correct formula for chlorobenzene is C_6H_5Cl, not C_6H_6Cl.

23.9 Naming Aromatic Compounds

A substituted benzene is derived by replacing one or more hydrogen atoms of benzene by another atom or group of atoms. Thus, a monosubstituted benzene has the formula C_6H_5G, where G is the group replacing a hydrogen atom.

Monosubstituted Benzenes

Some monosubstituted benzenes are named by adding the name of the substituent group as a prefix to the word *benzene*. The name is written as one word. Several examples follow:

Nitrobenzene   Ethylbenzene   Chlorobenzene   Bromobenzene

Certain monosubstituted benzenes have special names. These are used as parent names for further substituted compounds, so they should be learned.

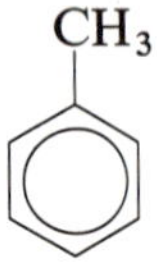

Toluene
(Methylbenzene)

Phenol
(Hydroxybenzene)

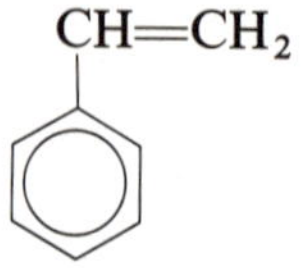

Styrene
(Vinylbenzene)

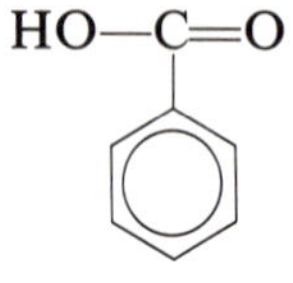

Benzoic acid
(Benzene carboxylic acid)

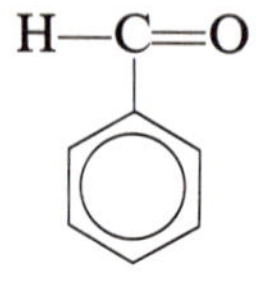

Benzaldehyde
(Benzene carboxaldehyde)

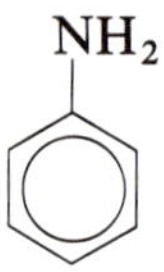

Aniline

All the hydrogen atoms in benzene are equivalent; therefore, it does not matter at which corner of the ring the substituted group is located. Each one of the formulas shown below represents chlorobenzene:

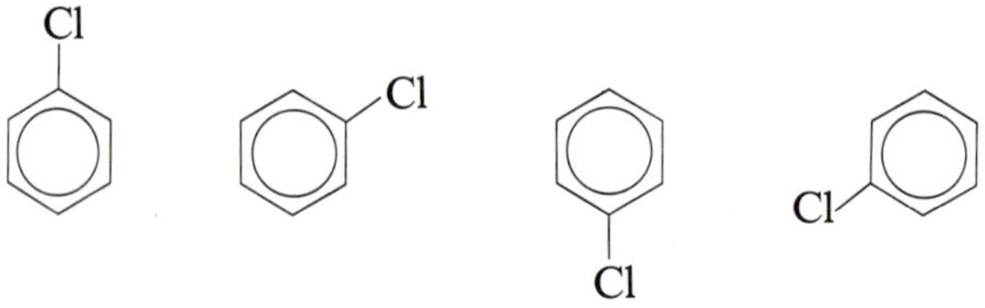

The C_6H_5— group is known as the phenyl group, and the name phenyl is used to name compounds that cannot easily be named as benzene derivatives. For example, the following compounds are named as derivatives of alkanes:

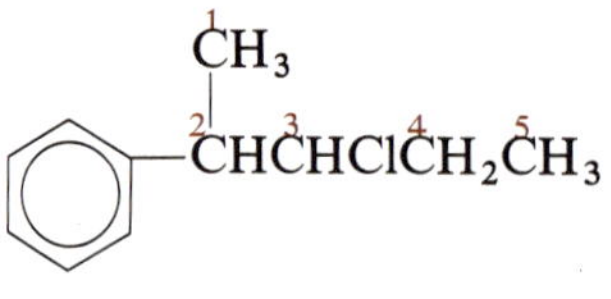

3-Chloro-2-phenylpentane

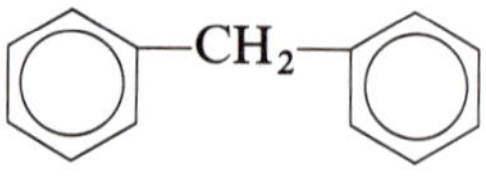

Diphenylmethane

Disubstituted Benzenes

When two substituent groups replace two hydrogen atoms in a benzene molecule, three different isomeric compounds are possible. The prefixes *ortho-*, *meta-*, and *para-* (abbreviated *o-*, *m-*, and *p-*) are used to name these disubstituted benzenes. In the *ortho* compound the substituents are located on adjacent carbon atoms. In the *meta* compound they are one carbon apart. And in the *para* compound the substituents are located on opposite sides of the ring:

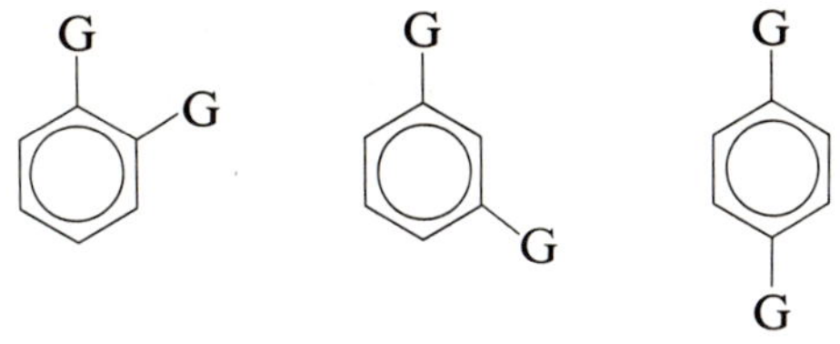

Unsaturated Hydrocarbons: Alkenes, Alkynes, and Aromatic Hydrocarbons 515

ortho   meta   para

Let us examine the dichlorobenzenes, $C_6H_4Cl_2$. Note that the three isomers have different physical properties, indicating that they are truly different substances.

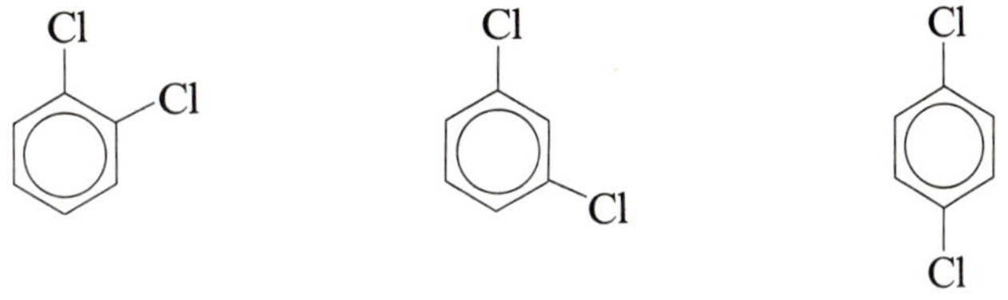

ortho-Dichlorobenzene
(1,2-Dichlorobenzene)
mp $-17.2°C$, bp $180.4°C$

meta-Dichlorobenzene
(1,3-Dichlorobenzene)
mp $-24.8°C$, bp $172°C$

para-Dichlorobenzene
(1,4-Dichlorobenzene)
mp $53.1°C$, bp $174.4°C$

The dimethylbenzenes have the special name *xylene*.

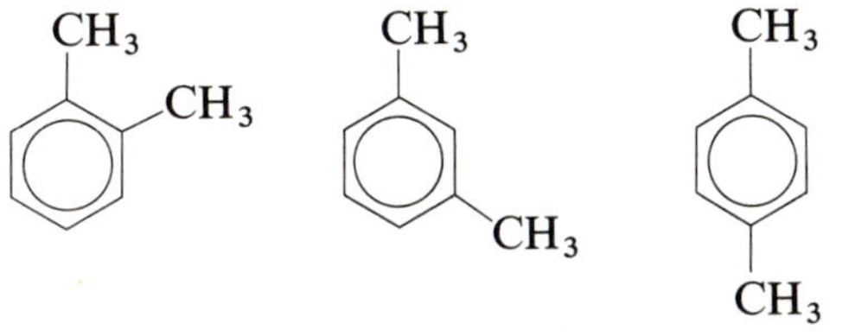

ortho-Xylene   *meta*-Xylene   *para*-Xylene

When one of the substituents corresponds to a monosubstituted benzene that has a special name, the disubstituted compound is named as a derivative of that parent compound. In the following examples the parent compounds are phenol, aniline, and toluene:

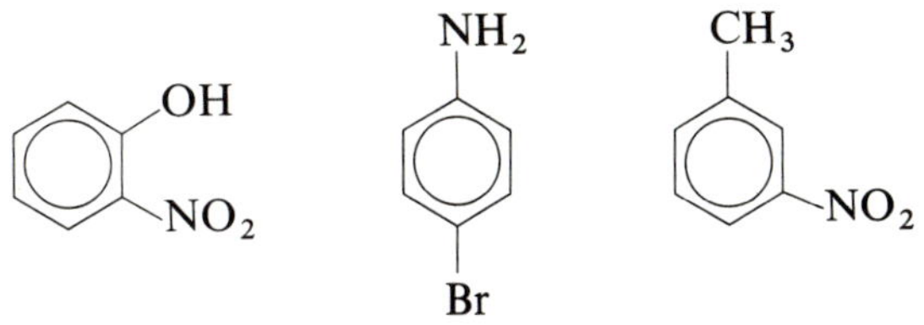

o-Nitrophenol   *p*-Bromoaniline   *m*-Nitrotoluene

Polysubstituted Benzenes

When there are more than two substituents on a benzene ring the carbon atoms in the ring are numbered, clockwise or counterclockwise, to give the lowest possible numbers to the substituent groups. When the compound is named as a derivative of one of the special parent compounds the substituent of the parent compound is considered to be on carbon 1 of the ring (the CH_3 group is on C-1 in 2,4,6-tribromotoluene). The following examples illustrate this system:

1,3,5-Trinitrobenzene

1,2,4-Tribromobenzene
(not 1,4,6-)

2,4,6-Trinitrotoluene
(TNT)

5-Bromo-2-chlorophenol

Problem 23.5 Write formulas for all the possible isomers of (a) chloronitrobenzene [$C_6H_4Cl(NO_2)$] and (b) tribromobenzene ($C_6H_3Br_3$).

(a) The name and formula indicate a chloro group (Cl) and a nitro group (NO_2) attached to a benzene ring. There are six positions in which to place these two groups. They can be *ortho*, *meta*, or *para* to each other.

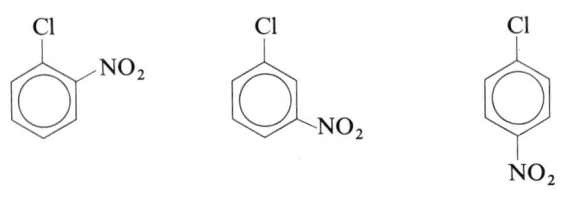

o-Chloronitrobenzene   *m*-Chloronitrobenzene   *p*-Chloronitrobenzene

(b) For tribromobenzene start by placing the three bromo groups in the 1, 2, and 3 positions; then the 1, 2, and 4 positions; and so on until all the possible isomers are formed. Name each isomer to check that no duplication of formulas have been written.

1,2,3-Tribromobenzene   1,2,4-Tribromobenzene   1,3,5-Tribromobenzene

There are only three isomers of tribromobenzene. If one erroneously writes the 1,2,5-compound, a further check will show that by numbering the ring as indicated it is in reality the 1,2,4-isomer.

1,2,5-Tribromobenzene
(Erroneous name)

1,2,4-Tribromobenzene
(Correct name)

23.10 Fused Aromatic Rings

fused aromatic ring systems

There are many other aromatic ring systems besides benzene. Their structures consist of two or more rings in which two carbon atoms are common to two rings. These compounds are known as **fused aromatic ring systems**. Three of the most common hydrocarbons in this category are naphthalene, anthracene, and phenanthrene. As you can see in the structures below, one hydrogen is attached to each carbon atom except at the carbons that are common to two rings.

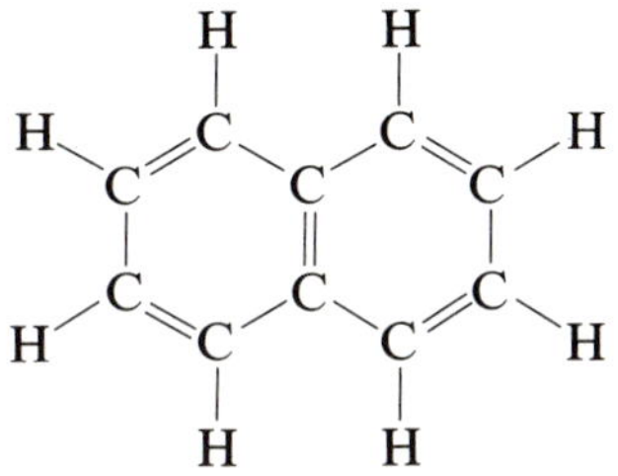

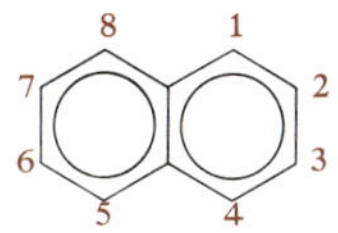

Napthalene, $C_{10}H_8$

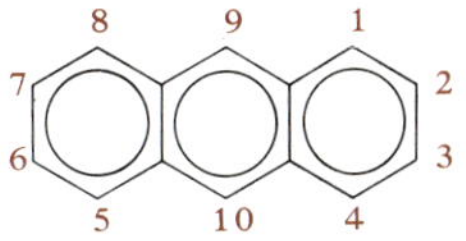

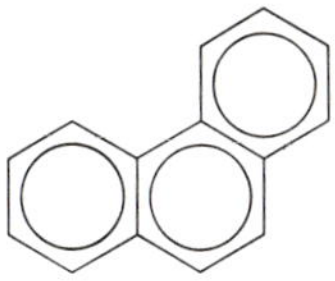

Anthracene, $C_{14}H_{10}$ Phenanthrene, $C_{14}H_{10}$

All three of these substances may be obtained from coal tar. Naphthalene is known as moth balls and has been used as a moth repellant for many years. A number of the polycyclic aromatic hydrocarbons (and benzene) have been shown to be carcinogenic (cancer-producing). Some of the more notable ones, found in coal tar, tar from cigarette smoke, and soot in urban environments, are shown below.

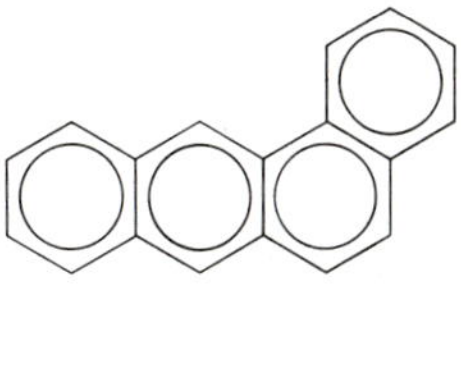

1,2-Benzanthracene 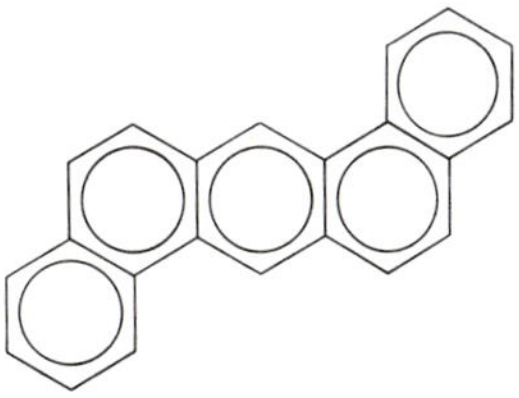   1,2,5,6-Dibenzanthracene

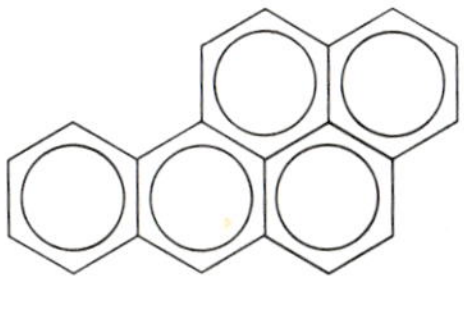

3,4-Benzpyrene 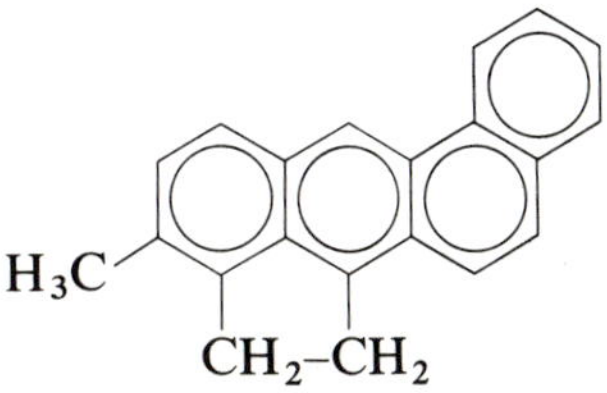   Methylcholanthrene

23.11 Sources and Properties of Aromatic Hydrocarbons

When coal is heated to high temperatures (450 to 1200°C) in the absence of air, coal gas and a complex mixture of condensable substances called *coal tar* are driven off. The aromatic hydrocarbons, such as benzene, toluene, xylene, naphthalene, and anthracene, were first obtained in quantity from coal tar. Since coal tar itself is a by-product of the manufacture of coke, the total amount of aromatics that can be obtained from this source is limited. The demand for aromatic hydrocarbons used in the production of a vast number of materials such as drugs, dyes, detergents, explosives, insecticides, plastics, and synthetic rubber became too great to satisfy from coal tar alone. Processes were devised to make aromatic hydrocarbons from the relatively inexpensive alkanes found in petroleum. Currently, about one-third of our benzene supply, and the greater portion of our toluene and xylene supplies, are obtained from petroleum. For example, toluene is made by cyclizing and dehydrogenating *n*-heptane:

$$CH_3CH_2CH_2CH_2CH_2CH_2CH_3 \xrightarrow[500°C]{Cr_2O_3\text{-}SiO_2\text{-}Al_2O_3} C_6H_5\text{-}CH_3 + 4H_2$$

n-Heptane → Toluene

Aromatic hydrocarbons are essentially nonpolar substances, insoluble in water, but soluble in many organic solvents. They are liquids or solids and usually have densities less than that of water. Aromatic hydrocarbons burn readily, usually with smoky yellow flames as a result of incomplete carbon combustion. Some are good motor fuels with excellent antiknock properties.

23.12 Reactions of Aromatic Hydrocarbons

The most characteristic reactions of aromatic hydrocarbons involve the substitution of some group for a hydrogen on one of the ring carbons. The following are examples of typical aromatic substitution reactions. In each of these reactions a functional group is substituted for a hydrogen atom.

1. *Halogenation* (chlorination or bromination):

$$C_6H_6 + X_2 \xrightarrow{FeX_3} C_6H_5\text{-}X + HX$$

Benzene + Chlorine or bromine → Chlorobenzene or bromobenzene

2. *Nitration*:

$$C_6H_6 + HO\text{-}NO_2 \xrightarrow{H_2SO_4} C_6H_5\text{-}NO_2 + H_2O$$

Benzene + Nitric acid → Nitrobenzene

3. *Alkylation* (Friedel–Craft reaction):

Benzene + CH$_3$CH$_2$Cl $\xrightarrow{\text{AlCl}_3}$ Ethylbenzene (C$_6$H$_5$—CH$_2$CH$_3$) + HCl

Chloroethane

From about 1860 onward, especially in Germany, a great variety of useful substances such as dyes, explosives, and drugs were synthesized from aromatic hydrocarbons by reactions of the type just described. These early syntheses were developed by trial-and-error methods. A good picture of the reaction mechanism, or step-by-step sequence of intermediate stages in the overall reaction, was not obtained until about 1940.

It is now recognized that aromatic substitution reactions generally proceed by a mechanism called *electrophilic substitution*. Three steps are involved: (1) the formation of an electrophile (electron-seeking group), (2) the attachment of the electrophile to the benzene ring forming a positively charged carbonium ion intermediate, and finally (3) the loss of a hydrogen ion from the carbonium ion to form the product. This reaction mechanism is illustrated in the chlorination of benzene catalyzed by iron(III) chloride.

$$\text{FeCl}_3 + \text{Cl}_2 \longrightarrow \text{FeCl}_4^- + \text{Cl}^+ \tag{1}$$

Iron(III) chloride   Chloronium ion

$$\text{C}_6\text{H}_6 + \text{Cl}^+ \longrightarrow [\text{C}_6\text{H}_6\text{Cl}]^+ \tag{2}$$

A carbonium ion

$$[\text{C}_6\text{H}_6\text{Cl}]^+ + \text{FeCl}_4^- \longrightarrow \text{C}_6\text{H}_5\text{Cl} + \text{FeCl}_3 + \text{HCl} \tag{3}$$

Chlorobenzene

The + sign on the carbonium ion does not mean that a positive charge is localized on a particular carbon atom in the ring; it means only that the $[\text{C}_6\text{H}_6\text{Cl}]^+$ entity is positively charged. The catalyst, FeCl$_3$, enters the reaction sequence in the first step by reacting with chlorine to form the electrophile, Cl$^+$, but is regenerated in the last step when chlorobenzene and hydrogen chloride are formed.

4. *Oxidation of side chains*: Carbon chains attached to an aromatic ring are fairly easy to oxidize. Reagents most commonly used to accomplish this in the laboratory are KMnO$_4$ or K$_2$Cr$_2$O$_7$ + H$_2$SO$_4$. No matter how long the side chain is, the carbon atom attached to the aromatic ring is oxidized to a carboxylic acid group, —COOH. For example, toluene, ethylbenzene, and propylbenzene

are all oxidized to benzoic acid:

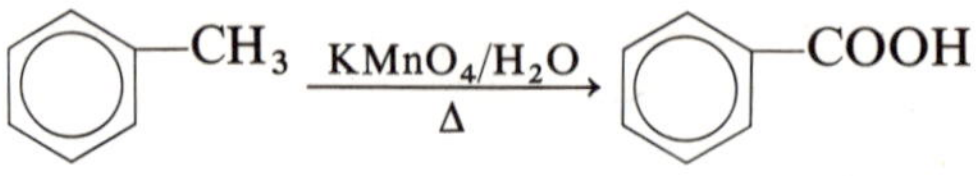

C₆H₅—CH₃ $\xrightarrow{\text{KMnO}_4/\text{H}_2\text{O}, \Delta}$ C₆H₅—COOH

Toluene → Benzoic acid

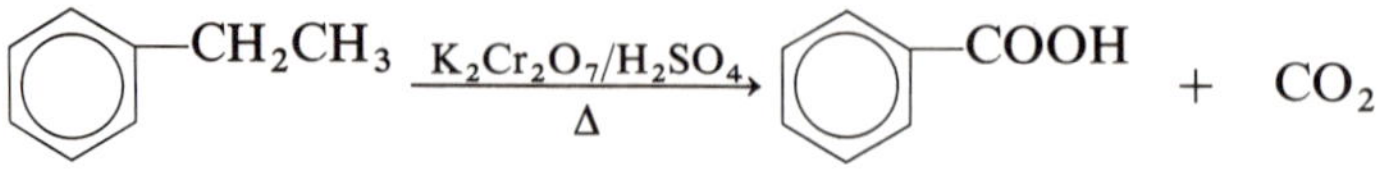

C₆H₅—CH₂CH₃ $\xrightarrow{\text{K}_2\text{Cr}_2\text{O}_7/\text{H}_2\text{SO}_4, \Delta}$ C₆H₅—COOH + CO₂

Ethylbenzene

C₆H₅—CH₂CH₂CH₃ $\xrightarrow{\text{K}_2\text{Cr}_2\text{O}_7/\text{H}_2\text{SO}_4, \Delta}$ C₆H₅—COOH + CH₃COOH

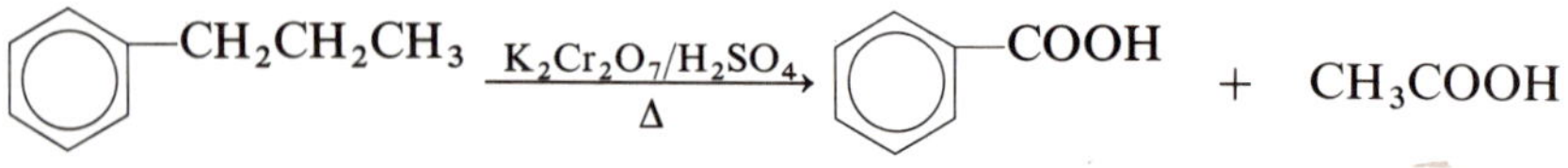

Propylbenzene

Aromatic substitution products of hydrocarbons are valuable in themselves and also serve as intermediates in the manufacture of other products. Thus, benzene can be nitrated to introduce a nitro group into the molecule. The product, nitrobenzene, is capable of a variety of reactions. Numerous drugs and dyes can be made by syntheses starting with nitrobenzene. These syntheses usually involve several steps. As an illustration, acetanilide, which is used in some pain- and fever-reducing drugs, can be made by the following sequence of reactions:

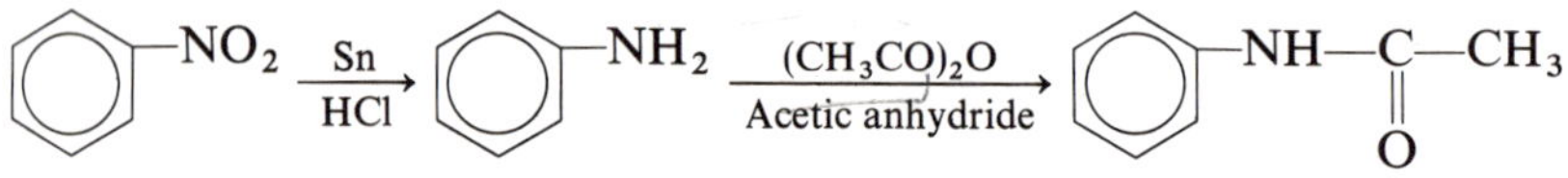

C₆H₅—NO₂ $\xrightarrow{\text{Sn}/\text{HCl}}$ C₆H₅—NH₂ $\xrightarrow{(\text{CH}_3\text{CO})_2\text{O}\;\text{Acetic anhydride}}$ C₆H₅—NH—C(=O)—CH₃

Nitrobenzene → Aniline → Acetanilide

Organic compounds containing nitro groups are toxic and may also have explosive properties. Therefore, special safety precautions must be taken when handling nitro compounds. A number of nitro compounds are synthesized specifically for use as explosives. For example, the high explosive trinitrotoluene, or TNT, is made by nitrating toluene:

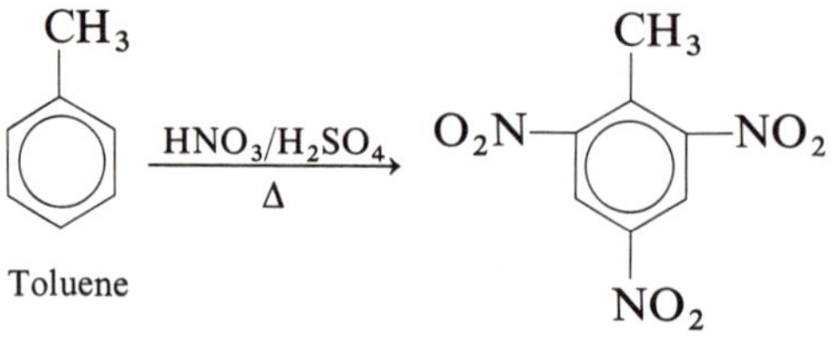

Toluene $\xrightarrow{\text{HNO}_3/\text{H}_2\text{SO}_4, \Delta}$ 2,4,6-Trinitrotoluene (TNT)

Questions

A. Study and understand the new terms introduced in this chapter.
1. Alkene
2. Alkyne
3. Pi bond
4. Geometric isomers
5. Addition reaction
6. Carbonium ion
7. Carbon cation
8. Markovnikov's rule
9. Aromatic compounds
10. Fused aromatic ring systems

B. Review questions.

1. The double bond in ethylene (C_2H_4) is made up of a sigma bond and a pi bond. Explain how the pi bond differs from the sigma bond.
2. Draw electron-dot (Lewis) diagrams to represent the following molecules:
   (a) Ethane   (b) Ethene   (c) Ethyne
3. Write the structural formulas and IUPAC names for all the pentenes. Be sure to check for all possible geometric isomers.
4. There are 17 possible isomeric hexenes including geometric isomers.
   (a) Write the structural formula for each isomer.
   (b) Name each isomer and include the prefix *cis* or *trans* where appropriate.
5. Draw structural formulas for the following:
   (a) 2,5-Dimethyl-3-hexene   (d) *cis*-1,2-Diphenylethene
   (b) 2-Ethyl-3-methyl-1-pentene (e) 3-Penten-1-yne
   (c) 4-Methyl-*cis*-2-pentene
6. Name the following compounds:

   (a)
   $$\begin{array}{c} H \\ \diagdown \\ CH_3 \end{array} C=C \begin{array}{c} H \\ \diagup \\ CHCH_2CH_3 \\ | \\ CH_3 \end{array}$$

   (b)
   $$\begin{array}{c} CH_3 \\ \diagdown \\ CH_3 \end{array} C=C \begin{array}{c} CH_3 \\ \diagup \\ CH_3 \end{array}$$

   (c) $CH_3CH_2CHCH=CH_2$
   $\quad\quad\quad\;\; |$
   $\quad\quad\quad\;\; CH$
   $\quad\quad\;\; CH_3 \;\; CH_3$

   (d)
   $$\begin{array}{c} CH_3CH_2 \\ \diagdown \\ H \end{array} C=C \begin{array}{c} CH_3 \\ \diagup \\ CH_2CH_3 \end{array}$$

   (e) $\text{C}_6\text{H}_5\text{—CH}_2\text{C}\equiv\text{CH}$

   (f) $CH_3CHBrCHBrC\equiv CCH_3$

7. Write the structural formulas and IUPAC names for all the (a) pentynes and (b) hexynes.
8. Why is it possible to obtain *cis* and *trans* isomers of 1,2-dichloroethene but not of 1,2-dichloroethane?
9. Which of the following molecules have structural formulas that permit *cis–trans* isomers to exist?
   (a) $(CH_3)_2C=CHCH_3$ (d) $CH_3CH=CHCl$
   (b) $CH_2=CHCl$ (e) $CCl_2=CBr_2$
   (c) $CH_3CH_2C\equiv CCH_3$ (f) $CH_2ClCH=CHCH_2Cl$
10. Complete the following equations:
    (a) $CH_3CH_2CH_2CH=CH_2 + Br_2 \rightarrow$
    (b) $CH_3CH_2CH_2CH=CH_2 + HCl \rightarrow$
    (c) $CH_3CH_2C=CHCH_3 + HI \rightarrow$
    $\quad\quad\quad\;\; |$
    $\quad\quad\quad\;\; CH_3$
    (d) $CH_3CH_2CH=CHCH_3 + HBr \rightarrow$
    (e) $CH_3CH_2CH=CH_2 + H_2O \xrightarrow{H^+}$
    (f) $\text{C}_6\text{H}_5\text{—CH}=CH_2 + HCl \rightarrow$

(g) $C_6H_5-CH=CH_2 + H_2 \xrightarrow[\text{1 atm}]{\text{Pt, 25°C}}$

(h) $CH_2=CHCl + HBr \rightarrow$

(i) $CH_3CH=CHCH_3 + KMnO_4 \xrightarrow{H_2O}$

11. Complete the following equations:

    (a) $CH_3C\equiv CH + 2H_2 \xrightarrow[\text{1 atm}]{\text{Pt, 25°C}}$

    (b) $CH_3C\equiv CCH_3 + Br_2 \text{ (1 mole)} \rightarrow$

    (c) $CH_3C\equiv CCH_3 + Br_2 \text{ (2 moles)} \rightarrow$

    (d) $CH\equiv CH + HCl \rightarrow \xrightarrow{HCl}$ (2 step reaction)

    (e) $CH_3C\equiv CH + HCl \rightarrow \xrightarrow{HCl}$ (2 step reaction)

    (f) $CH_3CH_2C\equiv CH + Ag(NH_3)_2NO_3 \rightarrow$

12. Cyclohexene (C_6H_{10}) has the structure

    and behaves in a manner similar to other alkenes. Write the formula and name for the product when cyclohexene reacts with:
    (a) Br_2 (b) HI   (c) H_2O, H^+ (d) $KMnO_4(aq)$

13. Write equations to show how 2-butyne can be converted to:
    (a) 2,3-Dibromobutane   (c) 2,2,3,3-Tetrabromobutane
    (b) 2,2-Dibromobutane

14. Two alkyl bromides are possible when 2-methyl-1-pentene is reacted with HBr. Which one will predominate? Why?

15. Your are given three unlabeled samples with the information that they are pentane, 1-pentene, and 1-pentyne. Describe a series of simple chemical tests that will identify these hydrocarbons.

16. Cyclohexane and 2-hexene both have the formula C_6H_{12}. How could you readily distinguish one from the other by chemical tests?

17. Why do many rubber products deteriorate rapidly in smog-ridden areas?

18. Explain the two different kinds of explosion hazards present when acetylene is being handled.

19. Write structural formulas for:
    (a) Benzene   (c) p-Xylene   (e) Aniline
    (b) Toluene   (d) Styrene   (f) Phenol

20. Write structural formulas for:
    (a) Ethylbenzene   (d) Naphthalene   (f) tert-Butylbenzene
    (b) Benzoic acid   (e) Anthracene   (g) 1,1-Diphenylethane
    (c) 1,3,5-Tribromobenzene

21. Write structural formulas and names for all the isomers of:
    (a) Trichlorobenzene ($C_6H_3Cl_3$)   (b) Dichlorobromobenzene ($C_6H_3Cl_2Br$)
    (c) The benzene derivatives of formula C_8H_{10}

22. (a) Write the structures for all the isomers that can be written by substituting a third chlorine atom in o-dichlorobenzene.

(b) Write the structure for all the isomers that can be written by substituting an additional chlorine atom in o-chlorobrombenzene.

23. Name the following compounds:

(a) 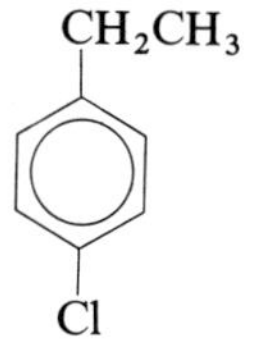 4-chloroethylbenzene structure with CH₂CH₃ and Cl

(b) 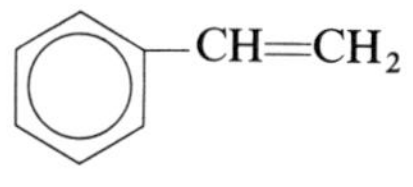 phenyl-CH=CH₂

(c) 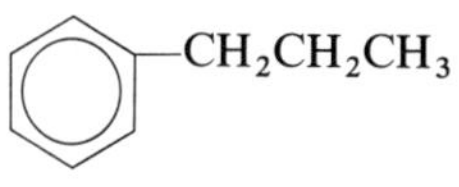 phenyl-CH₂CH₂CH₃

(d) 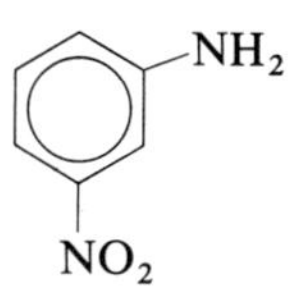 3-nitroaniline with NH₂ and NO₂

(e) 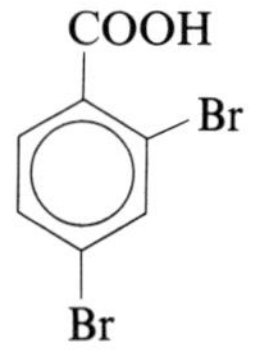 benzene with COOH, Br, Br

(f) 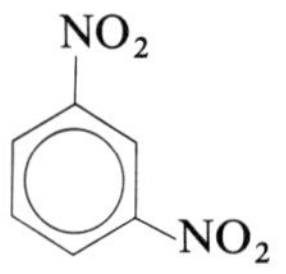 1,3-dinitrobenzene with two NO₂

(g) 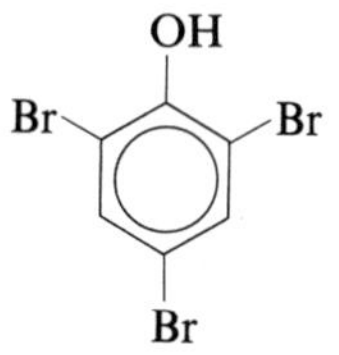 2,4,6-tribromophenol with OH and three Br

(h) 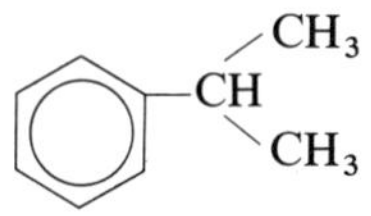 isopropylbenzene with CH(CH₃)₂

(i) 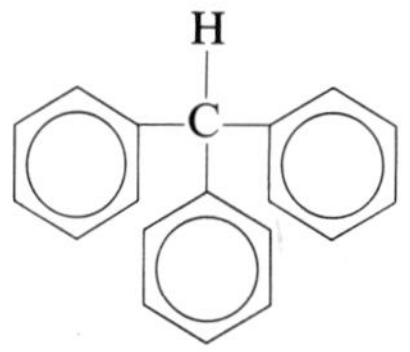 triphenylmethane structure with H and C

24. Explain how the reactions of benzene provide evidence that its structure does not include double bonds like those found in alkenes.

25. Complete the following equations and name the organic products:

(a) 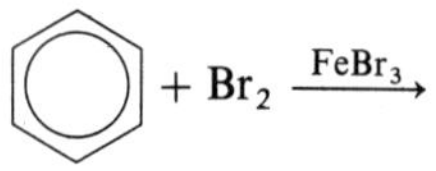 benzene + Br₂ $\xrightarrow{FeBr_3}$

(b) 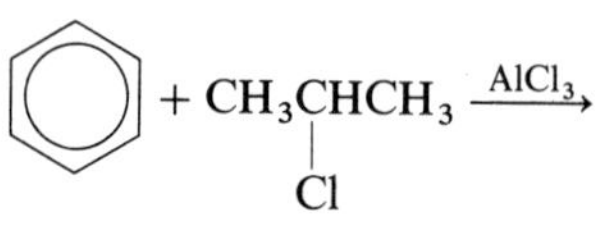 benzene + CH₃CHCH₃ with Cl $\xrightarrow{AlCl_3}$

(c) 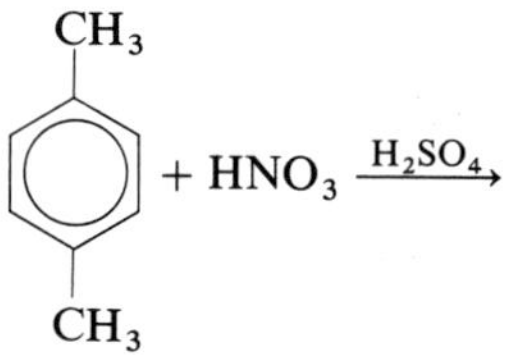 p-xylene (CH₃, CH₃) + HNO₃ $\xrightarrow{H_2SO_4}$

(d) 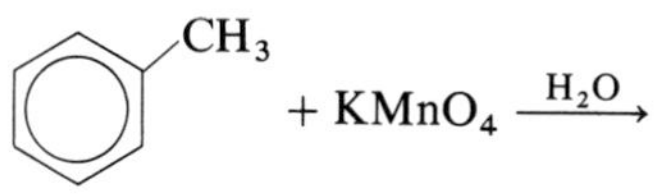 toluene with CH₃ + KMnO₄ $\xrightarrow{H_2O}$

26. Describe the reaction mechanism by which benzene is brominated in the presence of FeBr₃ [Question 25, part (a)]. Show equations.

27. In terms of historical events, why did the major source of aromatic hydrocarbons shift from coal tar to petroleum during the 10 year period 1935–1945?

28. Which of the following statements are correct?
    (a) The compound with the formula C_5H_{10} can be either an alkene or a cycloalkane.
    (b) If C_8H_{10} is an open-chain compound with multiple double bonds, it needs an additional ten hydrogen atoms to become a saturated hydrocarbon.
    (c) The pi bond is formed from two sp^2 electron orbitals.
    (d) A double bond consists of two equivalent bonds called pi bonds.
    (e) A triple bond consists of one sigma bond and two pi bonds.
    (f) The hybridized electron structure of a carbon atom in alkynes is *sp, sp, p, p*.
    (g) The acetylene molecule is linear.
    (h) A molecule of 2,3-dimethyl-1-pentene contains seven carbon atoms.
    (i) When an alkene is reacted with cold $KMnO_4$, a glycol is formed.
    (j) The disappearance of the purple color when $KMnO_4$ reacts with an alkene is known as the Markovnikov test for unsaturation.
    (k) $CH_3CH_2CH_2^+$ is a primary carbonium ion.
    (l) A secondary carbonium ion is more stable than a primary carbonium ion and less stable than a tertiary carbonium ion.
    (m) After bromine has added to an alkene the product is no longer unsaturated.
    (n) Alkynes have the general formula C_nH_{2n-4}.
    (o) 1-butyne will yield hydrogen gas when reacted with sodium.
    (p) *Cis–trans* isomerism occurs in alkenes and alkynes.
    (q) Geometric isomers are superimposable on each other.
    (r) All six hydrogen atoms in benzene are equivalent.
    (s) The chemical behavior of benzene is similar to that of alkenes.
    (t) Toluene and benzene are isomers.
    (u) Two substituents on a benzene ring that are in the 1,2- position are *ortho* to each other.
    (v) The oxidation of toluene with hot $KMnO_4$ yields benzoic acid.

24 Alcohols, Phenols, and Ethers

After studying Chapter 24 you should be able to:

1. Understand the terms listed in Question A at the end of the chapter.
2. Name alcohols by common and IUPAC methods.
3. Write the structural formula when given the name of an alcohol.
4. Write the structural formulas for all the isomeric alcohols of a given molecular formula.
5. Recognize and identify primary, secondary, and tertiary alcohols.
6. Compare and explain the relative solubilities of alcohols and ethers in water.
7. Write equations for the oxidation of alcohols.
8. Write equations for the dehydration of alcohols to ethers and to alkenes.
9. Write equations for the synthesis of alcohols from alkyl halides and from alkenes.
10. Write equations for the reactions of alcohols with sodium metal.
11. Explain the relative reactivities of primary, secondary, and tertiary alcohols.
12. Understand the common methods of preparing methyl alcohol, ethyl alcohol, isopropyl alcohol, ethylene glycol, and glycerol, and know the properties of these alcohols.
13. Name phenols and write their formulas.
14. Understand the differences in properties of the hydroxyl group (—OH) when bonded to an aromatic ring (a phenol) and to an aliphatic group (an alcohol).
15. Be familiar with the general properties of phenols.
16. Write equations for the cumene–hydroperoxide synthesis of phenol.
17. Name ethers and write their formulas.
18. Write equations for preparing ethers by the Williamson synthesis.
19. Discuss relative boiling points and water solubilities of comparable hydrocarbons, alcohols, dihydroxy alcohols, and ethers.
20. Discuss the hazards of using ethers in the laboratory.

24.1 Functional Groups

functional group

alcohol

Organic compounds were obtained orginally from plants and animals; and these, even today, are the direct sources of many important chemicals. As a case in point, millions of tons of sucrose (table sugar) are obtained from cane and beet juices each year. As chemical knowledge developed, many naturally occurring compounds were synthesized, often at costs far less than those of the natural products, Of even greater significance than the cheaper manufacture of existing substances was the synthesis of new substances totally unlike any natural product. The synthesis of new substances was aided greatly by the realization that organic chemicals can be divided into a relatively small number of classes and studied on the basis of similar chemical properties. The various classes of compounds are identified by the presence of certain characteristic groups which are called **functional groups**. For example, if a hydroxyl group (—OH) is substituted for a hydrogen atom in an alkane molecule, the resulting compound is an **alcohol**. Thus, alcohols are a class of compounds in which the

Table 24.1 Classes of organic compounds

Class	General formula	Structure of functional group	Sample structural formula	Name (Common)	Name (IUPAC)
Alkane	RH	R—H	CH_4	Methane	Methane
			CH_3CH_3	Ethane	Ethane
Alkene	R—CH=CH_2	>C=C<	CH_2=CH_2	Ethylene	Ethene
			CH_3CH=CH_2	Propylene	Propene
Alkyne	R—C≡C—H	—C≡C—	CH≡CH	Acetylene	Ethyne
			CH_3C≡CH	Methyl acetylene	Propyne
Alkyl halide	RX	—X, X=F, Cl, Br, I	CH_3Cl	Methyl chloride	Chloromethane
			CH_3CH_2Cl	Ethyl chloride	Chloroethane
Alcohol	ROH	—OH	CH_3OH	Methyl alcohol	Methanol
			CH_3CH_2OH	Ethyl alcohol	Ethanol
Ether	R—O—R	R—O—R	CH_3—O—CH_3	Dimethyl ether	Methoxymethane
			CH_3CH_2—O—CH_2CH_3	Diethyl ether	Ethoxyethane
Aldehyde	R—CH=O	—CH=O	H—CH=O	Formaldehyde	Methanal
			CH_3—CH=O	Acetaldehyde	Ethanal

Alcohols, Phenols, and Ethers 527

Table 24.1. (Continued)

Class	General formula	Structure of functional group	Sample structural formula	Name (Common)	Name (IUPAC)
Ketone	R—C(=O)—R	R—C(=O)—R	CH$_3$—C(=O)—CH$_3$	Acetone	Propanone
			CH$_3$—C(=O)—CH$_2$CH$_3$	Methyl ethyl ketone	2-Butanone
Carboxylic acid	R—C(=O)OH	—C(=O)OH	HCOOH	Formic acid	Methanoic acid
			CH$_3$COOH	Acetic acid	Ethanoic acid
Ester	R—C(=O)OR	—C(=O)OR	HCOOCH$_3$	Methyl formate	Methyl methanoate
			CH$_3$COOCH$_3$	Methyl acetate	Methyl ethanoate
Amide	R—C(=O)NH$_2$	—C(=O)NH$_2$	HC(=O)—NH$_2$	Formamide	Methanamide
			CH$_3$C(=O)—NH$_2$	Acetamide	Ethanamide
Amine	R—NH$_2$	—NH$_2$	CH$_3$NH$_2$	Methylamine	Aminomethane
			CH$_3$CH$_2$NH$_2$	Ethylamine	Aminoethane
Nitrile	R—C≡N	—C≡N	CH$_3$C≡N	Acetonitrile	Ethanonitrile
			CH$_3$CH$_2$C≡N	Propiononitrile	Propanonitrile
Amino acid	R—CH(NH$_2$)—COOH		H$_2$NCH$_2$COOH	Glycine	2-Aminoethanoic acid
			CH$_3$CH(NH$_2$)—COOH	Alanine	2-Aminopropanoic acid

functional group is the hydroxyl group. Each class of compounds contains a different functional group, and the members comprise a homologous series. Furthermore, the members of a homologous series exhibit similar chemical properties as a result of their common functional group. Examples of the more common classes of compounds and their functional groups are shown in Table 24.1.

Through the reactions of functional groups it is possible to create or synthesize new substances. The synthesis of new and possibly useful compounds or the more economical synthesis of known compounds is mostly what modern organic chemistry is all about. The vast majority of chemicals used today do not occur in nature but are synthesized from naturally occurring raw materials. The chemical and physical properties of an organic compound depend on (1) the kind and number of functional groups present and (2) the shape and size of the molecule. These relationships are easily seen among the different alcohols.

The three classes of compounds covered in this chapter are alcohols, phenols, and ethers. Their structures may be derived from water by replacing the hydrogen atoms of water by alkyl groups (R) or an aromatic ring:

Water　　Alcohol　　Ether　　Phenol

In the case of ethers, the R groups may be the same or different and may be alkyl groups or aromatic rings.

24.2 Classification of Alcohols

From a structural point of view, an alcohol is derived from a nonaromatic hydrocarbon by the replacement of at least one hydrogen atom with a hydroxyl (—OH) group. Alcohols are represented by the general formula ROH, with methanol (CH_3OH) being the first member of the homologous series. R represents an alkyl or substituted alkyl radical. Models illustrating the structural arrangements of the atoms in methanol and ethanol are shown in Figure 24.1.

CH_3OH
Methanol

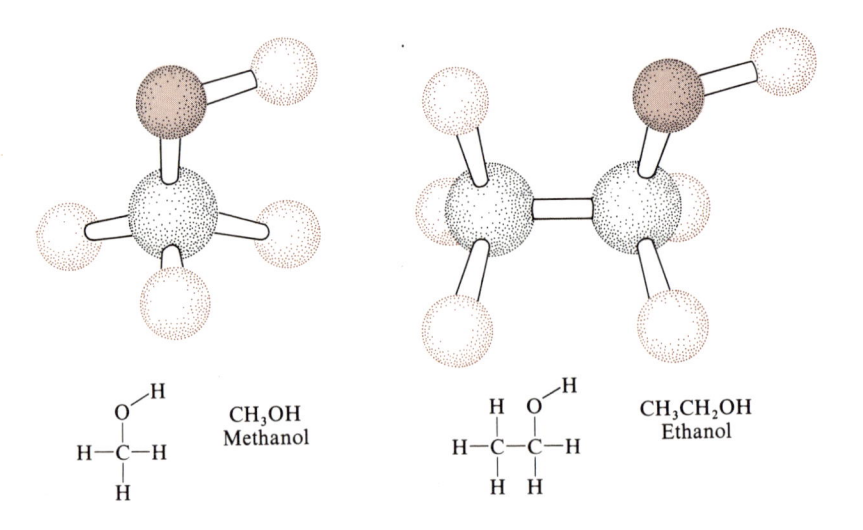

CH_3CH_2OH
Ethanol

Figure 24.1. Ball-and-stick models illustrating structural formulas of methanol and ethanol.

primary alcohol

secondary alcohol

tertiary alcohol

Alcohols are classified as **primary** (1°), **secondary** (2°), or **tertiary** (3°), depending on whether the carbon atom to which the —OH group is attached is bonded to one, two, or three other carbon atoms, respectively. Generalized formulas for 1°, 2°, and 3° alcohols are as follows.

$$\underset{\text{Primary alcohol}}{\overset{H}{\underset{H}{R-C-OH}}} \qquad \underset{\text{Secondary alcohol}}{\overset{R}{\underset{H}{R-C-OH}}} \qquad \underset{\text{Tertiary alcohol}}{\overset{R}{\underset{R}{R-C-OH}}}$$

Formulas of specific examples of these classes of alcohols are shown in Table 24.2. Methanol (CH_3OH) is grouped with the primary alcohols.

Molecular structures with more than one —OH group attached to a single carbon atom are generally not stable. But an alcohol molecule may contain two or more —OH groups if each —OH is attached to a different carbon atom. Accordingly, alcohols are further classified as monohydroxy, dihydroxy, trihydroxy, and so on, on the basis of the number of hydroxyl groups per molecule. **Polyhydroxy alcohols** and polyols are general terms for alcohols that have more than one —OH group per molecule.

polyhydroxy alcohol

Table 24.2. Naming and classification of alcohols.

Class	Formula	IUPAC name	Common name[a]	Boiling point, °C
Primary	CH_3OH	Methanol	Methyl alcohol	65.0
Primary	CH_3CH_2OH	Ethanol	Ethyl alcohol	78.5
Primary	$CH_3CH_2CH_2OH$	1-Propanol	*n*-Propyl alcohol	97.4
Primary	$CH_3CH_2CH_2CH_2OH$	1-Butanol	*n*-Butyl alcohol	118
Primary	$CH_3(CH_2)_3CH_2OH$	1-Pentanol	*n*-Amyl or *n*-pentyl alcohol	138
Primary	$CH_3(CH_2)_6CH_2OH$	1-Octanol	*n*-Octyl alcohol	195
Primary	CH_3CHCH_2OH | CH_3	2-Methyl-1-propanol	Isobutyl alcohol	108
Secondary	CH_3CHCH_3 | OH	2-Propanol	Isopropyl alcohol	82.5
Secondary	$CH_3CH_2CHCH_3$ | OH	2-Butanol	*sec*-Butyl alcohol	91.5
Tertiary	CH_3 | CH_3-C-OH | CH_3	2-Methyl-2-propanol	*t*-Butyl alcohol	82.9
Dihydroxy	$HOCH_2CH_2OH$	1,2-Ethanediol	Ethylene glycol	197
Trihydroxy	$HOCH_2CHCH_2OH$ | OH	1,2,3-Propanetriol	Glycerol or glycerine	290

[a] The abbreviations *n*, *sec*, and *t* stand for normal, secondary, and tertiary, respectively.

24.3 Naming Alcohols

If you know how to name alkanes, it is easy to name alcohols by the IUPAC System. Unfortunately, several of the alcohols are generally known by common or nonsystematic names, so it is often necessary to know more than one name for a given alcohol. The common name is usually formed from the name of the organic radical, which is attached to the —OH group, followed by the word *alcohol*. See examples given below and Table 24.2. To name an alcohol by the IUPAC System:

1. Select the longest continuous chain of carbon atoms containing the hydroxyl group.
2. Number the carbon atoms in this chain so that the one bearing the —OH group has the lowest possible number.
3. Form the alcohol name by dropping the final *-e* from the corresponding alkane name and adding *-ol*. Locate the —OH group by putting the number (hyphenated) of the carbon atom to which it is attached immediately before the alcohol name.
4. Name each alkyl side chain (or other group) and designate its position by number.

Study the application of this naming system to these examples and to those shown in Table 24.2.

$$CH_3-CH_2-CH_2OH$$
1-Propanol
(*n*-Propyl alcohol)

$$CH_3-CH-CH_3$$
$$|$$
$$OH$$
2-Propanol
(Isopropyl alcohol)

$$CH_3-CH-CH_2-CH_2OH$$
$$|$$
$$CH_3$$
3-Methyl-1-butanol
(Isoamyl alcohol or isopentyl alcohol)

$$HOCH_2-CH_2OH$$
1,2-Ethanediol
(Ethylene glycol)

Problem 24.1 Name this alcohol by the IUPAC method.

$$CH_3CH_2CHCH_2CHCH_3$$
$$||$$
$$CH_3OH$$

Step 1. There are six carbon atoms in the longest continuous carbon chain containing the —OH group.

Step 2. This carbon chain is numbered from right to left so that the —OH group has the smallest possible number:

$$\overset{6}{C}-\overset{5}{C}-\overset{4}{C}-\overset{3}{C}-\overset{2}{C}-\overset{1}{C}$$
$$||$$
$$CH_3OH$$

In this case, the —OH is on carbon number 2.

Step 3. The alcohol name is derived from the six-carbon hydrocarbon hexane by dropping the *-e* and adding *-ol*. Thus, the alcohol name is 2-hexanol since the —OH group is on carbon 2.

Step 4. A methyl group (—CH$_3$) is located on carbon 4. Therefore, the full name of the compound is 4-methyl-2-hexanol.

Problem 24.2 Write the structural formula of 3,3-dimethyl-2-hexanol.

Step 1. The 2-hexanol refers to a six-carbon chain with an —OH group on carbon number 2. Write the skeleton structure as follows:

$$\overset{1}{C}-\overset{2}{\underset{OH}{C}}-\overset{3}{C}-\overset{4}{C}-\overset{5}{C}-\overset{6}{C}$$

Step 2. Place the two methyl groups (3,3-dimethyl) on carbon number 3:

$$\overset{1}{C}-\overset{2}{\underset{HO}{C}}-\overset{3}{\underset{CH_3}{\overset{CH_3}{C}}}-\overset{4}{C}-\overset{5}{C}-\overset{6}{C}$$

Step 3. Finally, add H atoms to give each carbon atom four bonds:

$$CH_3CH-\underset{CH_3}{\overset{CH_3}{C}}-CH_2CH_2CH_3$$
$$|$$
$$OH$$

3,3-Dimethyl-2-hexanol

24.4 Physical Properties of Alcohols

The physical properties of alcohols are related to those of both water and alkane hydrocarbons. This is easily understandable if we recall certain facts about water and the alkanes. Water molecules are quite polar. The properties of water, such as high boiling point and ability to dissolve many polar substances, are largely due to the polarity of its molecules. Alkane molecules possess almost no polarity. The properties of the alkanes reflect this lack of polarity—for example, relatively low boiling point and inability to dissolve water and other polar substances. An alcohol molecule is made up of a water-like hydroxyl group joined to a hydrocarbon-like alkyl group.

$$\underset{\text{Water}}{H\overset{O}{\diagdown}H} \qquad \underset{\text{Alcohol}}{R\overset{O}{\diagdown}H}$$

One striking property of alcohols is their relatively high boiling points. The simplest alcohol, methanol, boils at 65°C. But methane, the simplest hydrocarbon, boils at −162°C. The boiling point of the normal alcohols increases in a regular fashion with increasing number of carbon atoms. Branched-chain alcohols have lower boiling points than the corresponding straight-chain alcohols.

Alcohols containing up to three carbon atoms are completely soluble in water. With some exceptions, alcohols with four or more carbon atoms have limited solubility in water. By way of contrast, all hydrocarbons are essentially insoluble in water.

The hydroxyl group on the alcohol molecule is responsible for both the water solubility and relatively high boiling points of the low-molecular-weight alcohols. Hydrogen bonding between water and alcohol molecules accounts for the solubility; and hydrogen bonding between alcohol molecules accounts for their high boiling points.

```
----H—O----H—O----H—O----          ----H—O----H—O----H—O----
     |        |        |                    |        |        |
     R        H        R                    R        R        R
        Alcohol–Water                          Alcohol–Alcohol
```

As the length of the hydrocarbon chain increases, the effect of the hydroxyl group becomes relatively less important. Alcohols with five to eleven carbons are oily liquids of limited water solubility, and in physical behavior they tend to resemble the corresponding alkane hydrocarbons. However, the effect of the —OH group is still noticeable in that their boiling points are higher than those of alkanes with similar molecular weights (see Table 24.3). Alcohols containing 12 or more carbons are waxlike solids that closely resemble solid alkanes in physical appearance.

Table 24.3. Boiling points of alkanes and monohydroxy alcohols.

Name	Boiling point (°C)	Name	Boiling point (°C)
n-Hexane	69	n-Hexanol	156
n-Octane	126	n-Octanol	195
n-Decane	174	n-Decanol	228

In alcohols with two or more hydroxyl groups, the effect of the hydroxyl groups on intermolecular attractive forces is, as we might suspect, even more striking than in monohydroxy alcohols. Ethanol (CH_3CH_2OH) boils at 78°C, but the boiling point of ethylene glycol or 1,2-ethanediol (CH_2OHCH_2OH) is 197°C. Comparison of the boiling points of ethanol, 1-propanol, and ethylene glycol shows that increased molecular weight does not account for the high boiling point of ethylene glycol (see Table 24.4). The higher boiling point is primarily a result of additional hydrogen bonding due to the two —OH groups in the ethylene glycol molecule.

Table 24.4. Comparison of the boiling points of ethanol, ethylene glycol, and 1-propanol.

Name	Formula	Molecular weight	Boiling point (°C)
Ethanol	CH_3CH_2OH	46	78
Ethylene glycol	CH_2OHCH_2OH	62	197
1-Propanol	$CH_3CH_2CH_2OH$	60	97

24.5 Reactions of Alcohols

Oxidation

We will consider only a few of the many reactions that alcohols are known to undergo. One important reaction is oxidation. Carbon atoms exist in progressively higher stages of oxidation in different functional groups in this fashion.

$$\text{Alkanes} \rightarrow \text{Alcohols} \rightarrow \begin{Bmatrix} \text{Aldehydes} \\ \text{Ketones} \end{Bmatrix} \rightarrow \text{Carboxylic acids} \rightarrow \text{Carbon dioxide}$$

The different stages shown do not necessarily indicate direct methods of synthesis. For example, some, but not all, alcohols can be converted to aldehydes by simple oxidation; but it is not practical to convert alkanes directly to alcohols.

Primary alcohols are converted to aldehydes by oxidation:

$$\underset{\text{Primary alcohol}}{RCH_2OH} + [O] \longrightarrow \underset{\text{Aldehyde}}{R\overset{H}{\underset{}{C}}=O} + H_2O$$

This reaction goes easily, but is not always a practical method for making aldehydes, because aldehydes are easily oxidized further to carboxylic acids:

$$\underset{\text{Aldehyde}}{R\overset{H}{\underset{}{C}}=O} + [O] \longrightarrow \underset{\text{Carboxylic acid}}{R\overset{O}{\underset{}{C}}-OH}$$

Secondary alcohols yield ketones upon oxidation:

$$\underset{\text{Secondary alcohol}}{R-\underset{OH}{\underset{|}{C}H}-R} + [O] \longrightarrow \underset{\text{Ketone}}{R-\underset{O}{\underset{||}{C}}-R} + H_2O$$

Unlike aldehydes, ketones cannot be oxidized further, except by very drastic procedures that break carbon–carbon bonds or by combustion. Almost any desired ketone can be made from the corresponding secondary alcohol by the action of an appropriate oxidizing agent.

The foregoing equations represent generalized oxidation reactions in which the oxidizing agent is represented by [O]. Some common oxidizing agents used for specific reactions are permanganate ions (MnO_4^-), dichromate ions ($Cr_2O_7^{2-}$ + H^+), or oxygen of the air. A complete equation for an alcohol oxidation may be fairly complicated. For example, the equation for the oxidation of ethanol to ethanal by potassium dichromate in sulfuric acid solution is

$$\underset{\text{Ethanol}}{3\,CH_3CH_2OH} + K_2Cr_2O_7 + 4\,H_2SO_4 \longrightarrow \underset{\text{Ethanal}}{3\,CH_3\overset{H}{\underset{}{C}}=O} + K_2SO_4 + Cr_2(SO_4)_3 + 7\,H_2O$$

Since our main interest is in the changes that occur in the functional groups, we can convey this information in abbreviated form:

$$CH_3CH_2OH + [O] \xrightarrow[\Delta]{K_2Cr_2O_7/H_2SO_4} CH_3\overset{\overset{\displaystyle H}{|}}{C}=O + H_2O$$

While the abbreviated equation lacks some of the details, it does show the overall reaction involving the organic compounds. Additional information is provided by notations above and below the arrow, which indicate that this reaction is carried out in heated potassium dichromate–sulfuric acid solution. Abbreviated equations of this kind will be used frequently in the remainder of this book.

Oxidation of a primary alcohol to an aldehyde or of a secondary alcohol to a ketone involves the removal of two hydrogen atoms from the alcohol molecule. One of these hydrogens is from the —OH group; the other is from the carbon atom to which the —OH group is attached.

$$CH_3CH_2OH + [O] \longrightarrow CH_3\overset{\overset{\displaystyle H}{|}}{C}=O + H_2O$$
Ethanol Ethanal (Acetaldehyde)

$$CH_3\underset{\underset{\displaystyle OH}{|}}{C}HCH_3 + [O] \longrightarrow CH_3-\underset{\underset{\displaystyle O}{\|}}{C}-CH_3 + H_2O$$
2-Propanol Propanone (Acetone)

But in a tertiary alcohol, no hydrogen atom is directly bonded to the carbon atom that is attached to the —OH group. Consequently, a tertiary alcohol cannot be oxidized, except by such drastic procedures as combustion.

$$CH_3-\overset{\overset{\displaystyle CH_3}{|}}{\underset{\underset{\displaystyle CH_3}{|}}{C}}-OH + [O] \longrightarrow \text{No reaction}$$

Dehydration

The term *dehydration* implies the elimination of water. Alcohols can be dehydrated to form alkenes or ethers. One of the more effective dehydrating reagents is sulfuric acid. Whether an ether or an alkene is formed depends on the ratio of alcohol to sulfuric acid, the reaction temperature, and the type of alcohol. With primary alcohols dehydration to an ether occurs with excess alcohol and a temperature lower than that required to dehydrate the alcohol to an alkene.

When ethanol and concentrated sulfuric acid are heated to 140°C, one molecule of water is eliminated per two molecules of alcohol to form diethyl ether. At 180°C the dehydration occurs within individual molecules and ethylene is formed. The equations are

$$\begin{array}{c} CH_3CH_2O\,|\,H \\ \raisebox{2pt}{\rule{0pt}{0pt}} \\ CH_3CH_2\,|\,OH \end{array} \xrightarrow[140°C]{96\%\ H_2SO_4} CH_3CH_2OCH_2CH_3 + H_2O$$
Diethyl ether

$$\underset{\underset{\text{H}\ \text{OH}}{}}{\text{H}-\overset{\overset{\text{H}}{|}}{\text{C}}-\overset{\overset{\text{H}}{|}}{\text{C}}-\text{H}} \xrightarrow[180°]{96\%\ H_2SO_4} CH_2=CH_2 + H_2O$$
Ethylene

The dehydration to ethers is useful for only primary alcohols. Dehydration of secondary and tertiary alcohols yields predominantly the most highly substituted alkene. Thus, the dehydration of 2-butanol gives chiefly 2-butene and the dehydration of 2-methyl-2-butanol gives mainly 2-methyl-2-butene.

$$\underset{\text{2-Butanol}}{CH_3CH_2\overset{\overset{\displaystyle}{|}}{\underset{\underset{\text{OH}}{|}}{C}}HCH_3} \xrightarrow[100°]{60\%\ H_2SO_4} \underset{\substack{\text{2-Butene}\\ \text{(chief product)}}}{CH_3CH=CHCH_3} + CH_3CH_2CH=CH_2 + H_2O$$

$$\underset{\text{2-Methyl-2-butanol}}{CH_3CH_2\overset{\overset{CH_3}{|}}{\underset{\underset{OH}{|}}{C}}CH_3} \xrightarrow[90°]{45\%\ H_2SO_4} \underset{\substack{\text{2-Methyl-2-butene}\\ \text{(chief product)}}}{CH_3CH=\overset{\overset{CH_3}{|}}{C}CH_3} + CH_3CH_2\overset{\overset{CH_3}{|}}{C}=CH_2 + H_2O$$

Industrially, dehydration reactions are often used to make relatively high-priced ethers from cheaper alcohols. But since alkenes are generally lower-priced than the corresponding alcohols, the industrial preparation of alkenes from alcohols does not make economic sense.

Reaction with Active Metals

Although alcohols are less acidic than water, they readily react with active metals such as sodium and potassium to replace the hydrogen atom of the alcohol —OH group. The products formed are a salt and hydrogen gas. The salt formed is an ionic compound and is called an alkoxide (CH_3CH_2ONa is sodium ethoxide). The equation is

$$2\,ROH + 2\,Na \longrightarrow 2\,RONa + H_2(g)$$
Alkoxide

$$2\,CH_3CH_2OH + 2\,Na \longrightarrow 2\,CH_3CH_2ONa + H_2(g)$$
Ethanol   Sodium ethoxide

The order of reactivity of alcohols with sodium is primary > secondary > tertiary. Methanol and ethanol react quite vigorously with sodium, but not as vigorously as sodium and water. Alcohols are less acidic than water. Reactivity of an alcohol also decreases as its molecular weight increases. This is because the —OH group becomes a smaller and relatively less significant part of the molecule.

Esterification

An alcohol can react with a carboxylic acid to form an ester and water. The reaction is represented as follows:

$$R-\underset{\text{Carboxylic acid}}{\underset{\parallel}{\overset{O}{C}}-OH} + \underset{\text{Alcohol}}{H-OR'} \underset{}{\overset{H^+}{\rightleftharpoons}} \underset{\text{Ester}}{R\underset{\parallel}{\overset{O}{C}}-OR'} + H_2O$$

$$\underset{\text{Acetic acid}}{CH_3\underset{\parallel}{\overset{O}{C}}-OH} + \underset{\text{Ethanol}}{HOCH_2CH_3} \overset{H^+}{\rightleftharpoons} \underset{\text{Ethyl acetate}}{CH_3\underset{\parallel}{\overset{O}{C}}-OCH_2CH_3} + H_2O$$

Esterification is an important reaction of alcohols and is discussed in greater detail in Chapter 26.

24.6 Preparation and Properties of Common Alcohols

Three general methods for making alcohols are

1. *Hydrolysis of an ester:*

$$\underset{\text{Ester}}{R\underset{\parallel}{\overset{O}{C}}-OR} + HOH \underset{\Delta}{\overset{H^+}{\longrightarrow}} \underset{\text{Carboxylic acid}}{R\underset{\parallel}{\overset{O}{C}}-OH} + \underset{\text{Alcohol}}{ROH}$$

2. *Alkaline hydrolysis of an alkyl halide* (1° and 2° alcohols only):

$$\underset{\text{Alkyl halide}}{RX} + NaOH(aq) \longrightarrow \underset{\text{Alcohol}}{ROH} + NaX$$

$$CH_3CH_2Cl + NaOH(aq) \longrightarrow CH_3CH_2OH + NaCl$$

3. *Catalytic reduction of aldehydes and ketones* to produce primary and secondary alcohols. These reactions are discussed in Chapter 25.

In theory, these general methods provide a way to make almost any alcohol. But in practice, they may be quite useless for a given alcohol because the needed ester and alkyl halide cannot be obtained at a reasonable cost. Therefore, on an industrial scale special methods that are specific for particular alcohols are used.

Methanol

When wood is heated to a high temperature in an atmosphere lacking oxygen, methanol (or wood alcohol) and other products are formed and driven off. The process is called *destructive distillation*, and until about 1925, nearly all methanol was obtained in this way. In the early 1920s the synthesis of methanol by high-pressure catalytic hydrogenation of carbon monoxide was developed in Germany. The reaction is

$$CO + 2H_2 \xrightarrow[300-400°C,\ 200\ atm]{ZnO-Cr_2O_3} CH_3OH$$

Nearly all methanol is now manufactured by this method.

Methanol is a volatile (bp 65°C), highly flammable liquid. It is poisonous and capable of causing blindness or death if taken internally. Exposure to methanol vapors for even short periods of time is dangerous. Despite these disadvantages, large quantities (over 6 billion lb annually) are manufactured and used for

1. Conversion to formaldehyde (methanal) used primarily in the manufacture of polymers
2. Manufacture of other chemicals, especially various kinds of esters
3. Denaturing or rendering ethyl alcohol unfit for use as a beverage
4. An industrial solvent
5. An inexpensive and temporary antifreeze for radiators (it is not a satisfactory permanent antifreeze, since its boiling point is lower than that of water)

The experimental use of 5 to 30% methanol in gasoline (gasahol) has shown promising results in reducing the amount of air pollutants emitted in automobile exhausts. Another important plus factor for using methanol in gasoline is that methanol can be made from nonpetroleum sources. The most economical nonpetroleum source of carbon monoxide for methanol is coal. In addition to coal, burnable materials such as wood, agricultural wastes, and sewage sludge are potential sources of methanol.

Ethanol

Ethanol is without doubt the oldest and most widely known alcohol. It is or has been known by a variety of names such as ethyl alcohol, "alcohol," grain alcohol, methyl carbinol, spirit, *aqua vitae*, to mention a few. The preparation of ethanol by fermentation is recorded in the Old Testament. Today, huge quantities of this substance are still prepared by fermentation. Starch and sugar are the raw materials. Starch must first be converted to sugar by enzyme- or acid-catalyzed hydrolysis. Conversion of simple sugars to ethanol is accomplished by the yeast enzyme zymase:

$$\underset{\text{Glucose}}{C_6H_{12}O_6} \xrightarrow{\text{Zymase}} 2\,\underset{\text{Ethanol}}{CH_3CH_2OH} + 2\,CO_2$$

For legal use in beverages, ethanol must be made by fermentation; but a large part of the alcohol intended for industrial uses (about 1.3 billion lb annually) is made from petroleum-derived ethylene. Ethylene is passed into concentrated sulfuric acid to form ethyl hydrogen sulfate and diethyl sulfate, which are then hydrolyzed with steam to obtain ethanol:

$$\underset{\text{Ethylene}}{CH_2=CH_2} + \underset{\text{Sulfuric acid}}{H_2SO_4} \longrightarrow \underset{\text{Ethyl hydrogen sulfate}}{CH_3CH_2OSO_3H} + (C_2H_5)_2SO_4$$

$$CH_3CH_2OSO_3H + (C_2H_5)_2SO_4 + HOH \longrightarrow CH_3CH_2OH + H_2SO_4$$

Some of the economically significant uses of ethanol are

1. Intermediate in the manufacture of other chemicals such as acetaldehyde, acetic acid, ethyl acetate, and diethyl ether
2. Solvent for many organic substances
3. Compounding ingredient for pharmaceuticals, perfumes, flavorings, and so on
4. Essential ingredient of alcoholic beverages

Physiologically, ethanol acts as a food, as a drug, and as a poison. It is a food in the limited sense that the body is able to metabolize small amounts of it to carbon dioxide and water with the production of energy. As a drug, ethanol is often mistakenly considered to be a stimulant, but it is in fact a depressant. In moderate quantities, ethanol causes drowsiness and depresses brain functions so that activities requiring skill and judgment (such as automobile driving) are impaired. In larger quantities, ethanol causes nausea, vomiting, impaired perception, and incoordination. If a very large amount is consumed, complete unconsciousness and, ultimately, death may occur.

Authorities maintain that the effects of ethanol on automobile drivers are a factor in about half of all fatal traffic accidents in the United States. The gravity of this problem can be grasped when you realize that traffic accidents are responsible for more than 50,000 deaths each year in the United States.

Extremely heavy taxes are imposed on alcohol in beverages. A gallon of pure alcohol costs only a dollar or two to produce but, in a distilled beverage, bears a tax of about 20 dollars!

Ethanol for industrial use that has been denatured or rendered unfit for drinking is not taxed. Denaturing is done by adding small amounts of methanol and other denaturants that are extremely difficult to remove. Denaturing is, of course, required by the federal government to protect the beverage alcohol tax source. Special tax-free use permits are issued to scientific and industrial users who require pure ethanol for nonbeverage uses.

Isopropyl Alcohol

Isopropyl alcohol (2-propanol) is made from propene derived from petroleum. This synthesis is analogous to that used for making ethanol from ethylene:

$$\underset{\text{Propene}}{CH_3CH=CH_2} + H_2SO_4 \longrightarrow CH_3\underset{OSO_3H}{\underset{|}{C}H}CH_3$$

$$\text{CH}_3\text{CHCH}_3 + \text{HOH} \longrightarrow \text{CH}_3\text{CHCH}_3 + \text{H}_2\text{SO}_4$$
$$||$$
$$\text{OSO}_3\text{H}\text{OH}$$
2-Propanol

Note that 2-propanol, not 1-propanol, is produced. This is because in the first step of the reaction one H^+ from H_2SO_4 adds to carbon number 1 of propene and $^-OSO_3H$ adds to carbon number 2 according to Markovnikov's rule. Isopropyl alcohol is a relatively low-cost alcohol which is manufactured in large quantities (about 1.7 billion lb annually). It is not a potable alcohol. Even breathing large quantities of the vapor may cause dizziness, headache, nausea, vomiting, mental depression, and coma. Isopropyl alcohol is used (1) to manufacture other chemicals, especially acetone, (2) as an industrial solvent, and (3) as the principal ingredient in rubbing alcohol formulations.

Ethylene Glycol

This is the simplest alcohol containing two —OH groups. Like most other relatively cheap, low-molecular-weight alcohols, it is derived commercially from petroleum. One industrial synthesis is from ethylene via ethylene oxide.

$$2\,CH_2{=}CH_2 + O_2 \xrightarrow[200-300°C]{\text{Ag catalyst}} 2\,\overset{\overset{\displaystyle O}{\diagup\diagdown}}{CH_2{-}CH_2}$$
Ethylene   Ethylene oxide

$$\overset{\overset{\displaystyle O}{\diagup\diagdown}}{CH_2{-}CH_2} + HOH \xrightarrow{H^+} HOCH_2CH_2OH$$
1,2-Ethanediol
(Ethylene glycol)

Major uses of ethylene glycol are (1) in the preparation of the synthetic polyester fibers Dacron and Terylene, (2) major ingredient in "permanent type" antifreeze for cooling systems, (3) solvent in the paint and plastics industries, and (4) in the formulations of printers ink and ink for ballpoint pens.

The low molecular weight, complete water solubility, and low freezing point make ethylene glycol a nearly ideal antifreeze. A 58% by weight solution of ethylene glycol freezes at $-48°C$. Its high boiling point and high heat of vaporization prevent it from being boiled away and permit higher—and therefore more efficient—engine operating temperatures than are possible with water alone. In 1977, the United States production of ethylene glycol amounts to about 4 billion lb annually. Ethylene glycol is extremely toxic when ingested.

Glycerol

This compound, also known as *glycerine* or 1,2,3-propanetriol, is an important trihydroxy alcohol. Glycerol is a syrupy liquid with a sweet warm taste. It is about 0.6 times as sweet as cane sugar. It is obtained as a by-product in processing animal and vegetable fats to make soap and other products; and it is also synthesized commercially from propene. The major uses of glycerol are (1) as a raw material in the manufacture of polymers and explosives. (2) as an emollient in cosmetic preparations, (3) as a humectant in tobacco products, and (4) as a sweetener. Each use is directly related to the three —OH groups on glycerol.

The —OH groups provide sites through which the glycerol unit may be bonded to other molecules to form a polymer (Chapter 27). The explosive, nitroglycerine or glyceryltrinitrate, is made by reacting the —OH groups with nitric acid:

$$\begin{array}{c}CH_2OH\\|\\CHOH\\|\\CH_2OH\end{array} + 3\,HONO_2 \xrightarrow{\text{Nitric acid}} \begin{array}{c}CH_2ONO_2\\|\\CHONO_2\\|\\CH_2ONO_2\end{array} + 3\,H_2O$$

Glycerol   Glyceryltrinitrate (Nitroglycerine)

The three polar —OH groups on the glycerol molecule are able to hold water molecules by hydrogen bonding. Consequently, glycerol is a hygroscopic substance; that is, it has the ability to take up water vapor from the air. It is therefore used as a skin moisturizer in cosmetic preparations. Glycerol is also used as an additive in tobacco products; by taking up moisture from the air, it prevents the tobacco from becoming excessively dry and crumbly.

24.7 Phenols

phenol

The term **phenol** is used for the class of compounds that have a hydroxy group attached to an aromatic ring. The parent compound is called *phenol* (C_6H_5OH) and is also known as carbolic acid.

Naming Phenols

Many phenols are named as derivatives of the parent compound using the general methods for naming aromatic compounds. For example:

Phenol   *m*-Bromophenol   *p*-Aminophenol   2,4,6-Trinitrophenol (Picric acid)

The *ortho-*, *meta-*, and *para-*dihydroxybenzenes have the special names catechol, resorcinol, and hydroquinone, respectively. The catechol structure occurs in many natural substances, and hydroquinone, a manufactured product, is commonly used as a photographic reducer and developer.

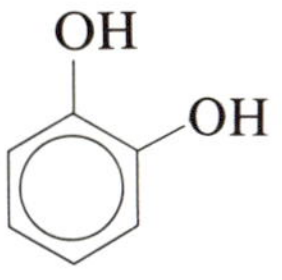

Catechol 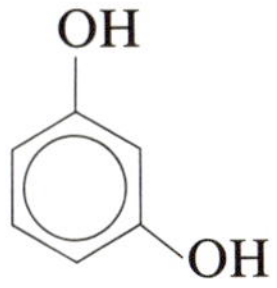   Resorcinol 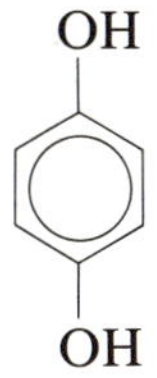   Hydroquinone

The *ortho*-, *meta*-, and *para*-methylphenols are present in coal tar and are known as cresols. They are all useful disinfectants.

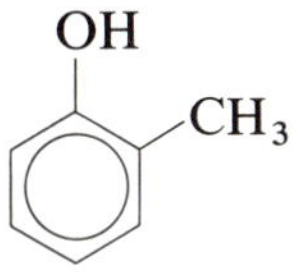

o-Cresol   m-Cresol   p-Cresol
(o-Methylphenol)   (m-Methylphenol)   (p-Methylphenol)

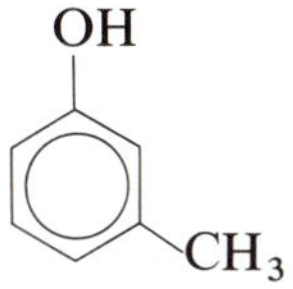

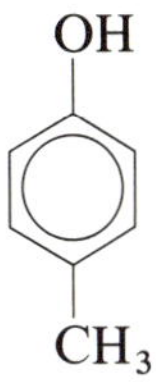

24.8 Selected Phenolic Compounds

Many well-known natural substances have phenolic groups in their structures. Several examples are described and shown below.

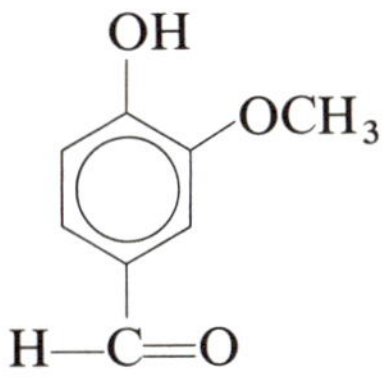

Vanillin 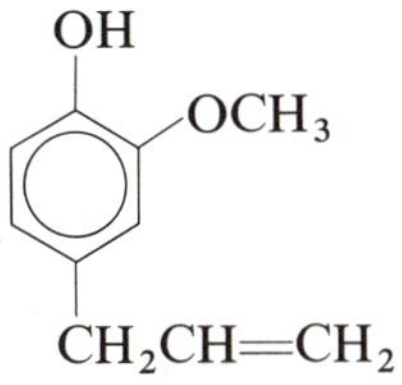   Eugenol

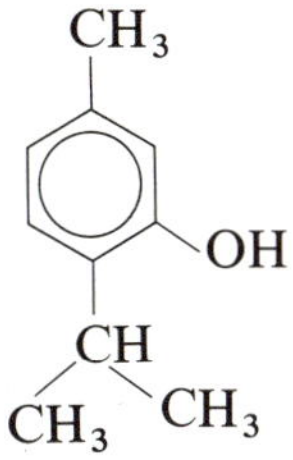

Thymol 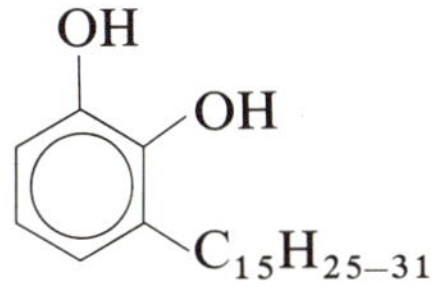   Urushiols

Vanillin is the principal odorous component of the vanilla bean. It is one of the most widely used flavorings and is also used for masking undesirable odors in many products such as paints.

Eugenol is the essence of oil of cloves. Among its uses are as a dental analgesic and for the manufacture of synthetic vanillin. Thymol occurs in the oil of thyme. It has a fairly pleasant odor and flavor and is used as an antiseptic in many preparations such as mouthwashes. Thymol is the starting material for the synthesis of menthol, the main constituent of oil of peppermint. Thymol is a widely used flavoring and pharmaceutical.

The active irritants in poison ivy and poison oak are called urushiols. They are catechol derivatives with an unbranched 15 carbon side chain in position 3 on the phenol ring.

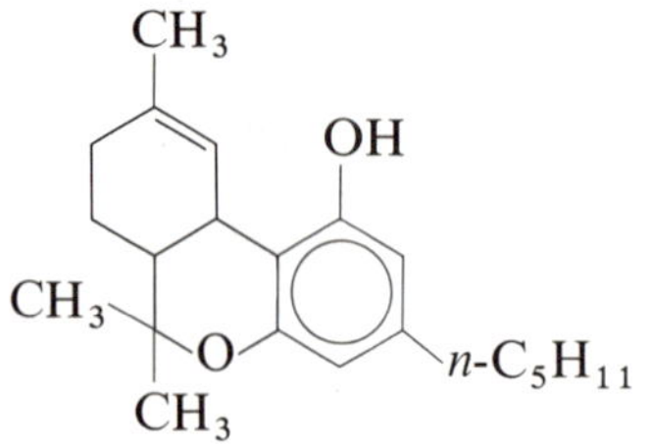

Tetrahydrocannibinol
(Marijuana)

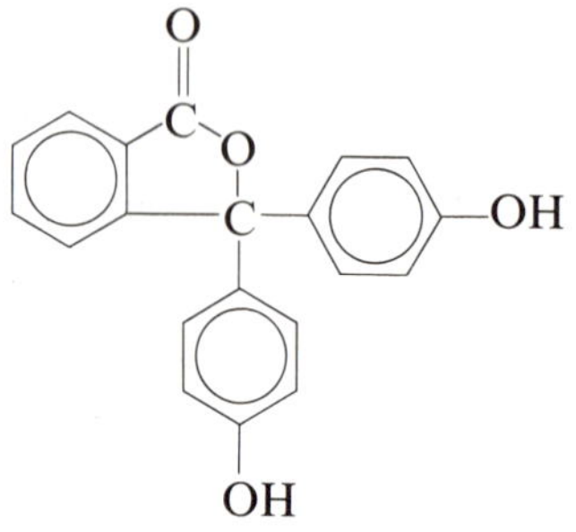

Phenolphthalein

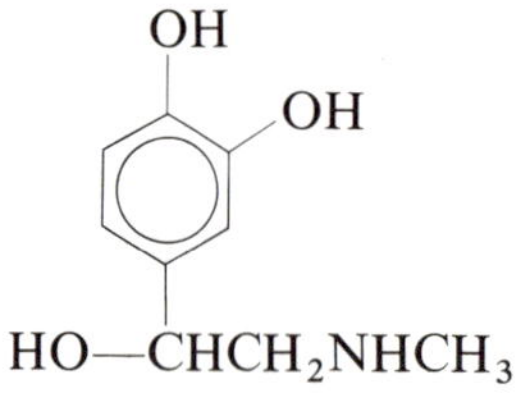

Adrenalin
(Epinephrine)

The widely discussed active principle of marijuana is tetrahydrocannibinol. It is obtained from the dried leaves and flowering tops of the hemp plant and has been used since antiquity for its physiological effects. The common acid–base indicator phenolphthalein is a phenol derivative. Phenolphthalein is also used as a medical cathartic. Epinephrine (adrenalin) is a substance that is secreted by the adrenal gland; it stimulates the conversion of glycogen to glucose in the body as a result of stress, fear, anger, or other heightened emotional states.

24.9 Properties of Phenols

In the pure state phenol is a colorless crystalline solid with a melting point at about 41°C and a characteristic odor. Phenol is highly poisonous. Ingestion of even small amounts may cause nausea, vomiting, circulatory collapse, and death from respiratory failure.

Phenol is a weak acid—more acidic than alcohols and water, but less acidic than acetic and carbonic acids. The pH values are as follows: acetic acid (0.1 M), 2.87; water, 7.0; phenol (0.1 M), 5.5. Thus, phenol will react with sodium hydroxide solution to form a salt but will not react with sodium bicarbonate. The salt formed is called sodium phenoxide or sodium phenolate. Sodium hydroxide will not remove a hydrogen atom from an alcohol, since alcohols are weaker acids than water.

$$\text{C}_6\text{H}_5\text{OH} + \text{NaOH} \longrightarrow \text{C}_6\text{H}_5\text{ONa} + \text{H}_2\text{O}$$

Sodium phenoxide
(Sodium phenolate)

In general, the phenols are toxic to microorganisms. They are widely used as antiseptics and disinfectants. Phenol was the first compound to be used extensively as an operating room disinfectant. Joseph Lister (1827–1912) first used phenol for this purpose in 1867. The antiseptic power of phenols is increased by substituting alkyl groups (up to six carbons) in the benzene ring. For example, 4-hexylresorcinol is used as an antiseptic in numerous pharmaceuticals.

4-Hexylresorcinol: benzene ring with OH, OH, and CH$_2$(CH$_2$)$_4$CH$_3$ substituents.

24.10 Production of Phenol

Phenol is obtained from coal tar. In addition, there are several commercial methods used to produce phenol synthetically. The most economical of these methods starts with benzene and propylene, which react to form cumene (isopropylbenzene). Cumene is then oxidized by air to cumene hydroperoxide, which is treated with dilute sulfuric acid to obtain phenol and acetone. The economic feasability of the process is due to the fact that two important commercial products are produced. The equations for the reactions are

$$\text{Benzene} + \text{CH}_3\text{CH}=\text{CH}_2 \xrightarrow{\text{H}_2\text{SO}_4} \text{Cumene} \xrightarrow{\text{O}_2}$$

$$\text{Cumene hydroperoxide} \xrightarrow[\text{(dilute)}]{\text{H}_2\text{SO}_4} \text{Phenol} + \text{CH}_3\overset{\text{O}}{\underset{\|}{\text{C}}}\text{CH}_3 \;\;(\text{Acetone})$$

About 2.7 billion pounds of synthetic phenol is produced annually in the United States. The chief use of phenol is for the manufacture of phenol–formaldehyde resins and plastics.

24.11 Ethers

ether

Ethers have the general formula ROR′. The radicals, R and R′, may be derived from saturated, unsaturated, or aromatic hydrocarbons; and, for a given ether, may be alike or different. Cyclic ethers are formed by joining the ends of a single hydrocarbon chain through an oxygen atom to form a ring structure. Table 24.5 shows structural formulas and names for some of the different kinds of ethers.

Table 24.5. Names and structural formulas of ethers.

Name[a]	Formula	Boiling point (°C)
Dimethyl ether (Methyoxymethane)	CH_3-O-CH_3	-25
Methyl ethyl ether (Methyoxyethane)	$CH_3CH_2-O-CH_3$	8
Diethyl ether (Ethoxyethane)	$CH_3CH_2-O-CH_2CH_3$	35
Ethyl isopropyl ether (2-Ethoxypropane)	$CH_3CH_2-O-CHCH_3$ $\quad\quad\quad\quad\quad\quad\quad\quad\;\; \|$ $\quad\quad\quad\quad\quad\quad\quad\quad CH_3$	54
Divinyl ether	$CH_2=CH-O-CH=CH_2$	39
Anisole (Methyoxybenzene)	C₆H₅—OCH₃	154
Tetrahydrofuran	CH_2-CH_2 $\|\quad\quad\quad\|$ $CH_2\;\;\;CH_2$ $\;\;\;\diagdown\;O\;\diagup$	66

[a] The IUPAC name is in parentheses when given.

Naming Ethers

Individual ethers, like alcohols, may be known by several names. The ether having the formula $CH_3CH_2-O-CH_2CH_3$ and formerly widely used as an anesthetic is called diethyl ether, ethyl ether, ethoxyethane, or simply *ether*. Common names of ethers are formed from the names of the groups attached to the oxygen atom followed by the word *ether*.

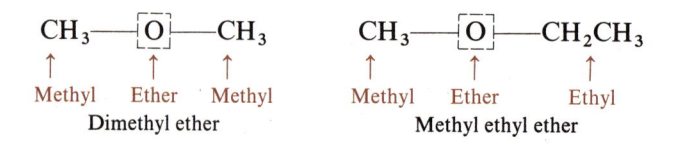

$$CH_3-\boxed{O}-CH_3 \quad\quad\quad CH_3-\boxed{O}-CH_2CH_3$$
$\;\;\;\;\;\;\uparrow\quad\;\;\;\;\uparrow\quad\;\;\;\;\uparrow\quad\quad\quad\quad\;\;\;\uparrow\quad\;\;\;\;\;\uparrow\quad\;\;\;\;\uparrow$
Methyl Ether Methyl   Methyl Ether Ethyl
     Dimethyl ether   Methyl ethyl ether

In the IUPAC System, ethers are named as alkoxy (RO—) derivatives of the alkane corresponding to the longest carbon–carbon chain in the molecule. To name an ether by this system:

1. Select the longest carbon–carbon chain and label it with the name of the corresponding alkane.
2. Change the *-yl* ending of the other hydrocarbon radical to *-oxy* to obtain the alkoxy radical name. For example, CH$_3$O— is called *methoxy*.
3. Combine the two names from Steps 1 and 2, giving the alkoxy name first, to form the ether name.

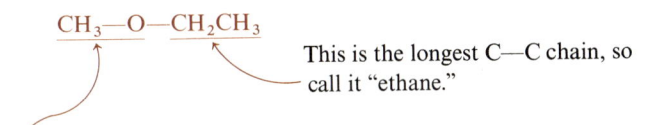

CH$_3$—O—CH$_2$CH$_3$

This is the longest C—C chain, so call it "ethane."

This is the other radical; modify its name to "methoxy" and combine with ethane to obtain the name of the ether, "methoxyethane."

Thus,

CH$_3$CH$_2$—O—CH$_2$CH$_3$ is ethoxyethane;

CH$_3$CH$_2$CH$_2$—O—CH$_2$CH$_2$CH$_2$CH$_3$ is *n*-propoxybutane.

Additional examples named by both methods are to be found in Table 24.5.

24.12 Structures and Properties of Ethers

An oxygen atom linking two carbon atoms together is the key structural feature of an ether molecule. This oxygen atom causes ether molecules to have a bent shape somewhat like that of water and alcohol molecules:

Ethers are a bit more polar than alkanes, since alkanes lack the oxygen atom with its exposed nonbonded electrons. But ethers are much less polar than alcohols, since no hydrogen is attached to the oxygen atom in an ether. The solubility and boiling point (vapor pressure) characteristics of ethers are related to the C—O—C structure. Alkanes have virtually no solubility in water or acid. But about 7.5 g of diethyl ether will dissolve in 100 g of water at 20°C. Diethyl ether will also dissolve in sulfuric acid. Hydrogen bonding between ether and water molecules and between ether and acid molecules is responsible for this solubility.

⟵ H-bond ⟶

Since no —OH group is present, hydrogen bonding does not occur between ether molecules. This lack of H-bonding can be seen by comparing the boiling points of a hydrocarbon, an ether, and an alcohol of similar molecular weight, as in Table 24.6. The boiling point of the ether is somewhat above that of the hydrocarbon but much lower than that of the relatively polar alcohol.

Table 24.6. Boiling points of ethers, alkanes, and alcohols.

Name	Formula	Molecular weight	Boiling point (°C)
Dimethyl ether	CH_3OCH_3	46	−24
Propane	$CH_3CH_2CH_3$	44	−42
Ethanol	CH_3CH_2OH	46	78
Methoxyethane	$CH_3OCH_2CH_3$	60	8
Butane	$CH_3CH_2CH_2CH_3$	58	−0.6
1-Propanol	$CH_3CH_2CH_2OH$	60	97
2-Propanol	$CH_3CH(OH)CH_3$	60	83

Ethers, especially diethyl ether, are exceptionally good solvents. Many polar compounds, including water, acids, alcohols, and other oxygenated organic compounds, will dissolve, at least to some extent, in ethers. This solubility is a result of intermolecular attractions between the slightly polar ether molecules and the molecules of the other polar substance. Nonpolar compounds such as hydrocarbons and alkyl halides will also dissolve in ethers. These substances dissolve because the ether molecules are not very polar and thus are not strongly attracted either to one another or to other kinds of molecules. Thus, ether molecules are able to intermingle freely with the molecules of a nonpolar substance and form a solution by simple mixing.

In summary, ethers are polar enough to dissolve some polar substances. On the other hand, their polarity is so slight that they act as nonpolar solvents for a great many nonpolar substances.

Ethers have little chemical reactivity, but since a great many organic substances dissolve readily in ethers, they are often used as solvents in both laboratory and manufacturing operations. Their use may be dangerous since low-molecular-weight ethers are volatile and their highly flammable vapors form explosive mixtures with air. Another hazard of ethers is that, despite their generally low chemical reactivity, oxygen of the air reacts slowly with them to form unstable peroxides which are subject to explosive decomposition.

$$CH_3CH_2-O-CH_2CH_3 + O_2 \longrightarrow \underset{\underset{O-O-H}{|}}{CH_3CH}-O-CH_2CH_3$$

Diethyl ether hydroperoxide

24.13 Preparation of Ethers

We have seen that ethers can be made by intermolecular dehydration of alcohols by heating in the presence of an acid (see Section 24.4). Ethers are also made from sodium alkoxides or sodium phenoxides and alkyl halides (Williamson synthesis).

$$\text{RONa} + \text{R'X} \longrightarrow \text{ROR'} + \text{NaX}$$

Sodium alkoxide   Alkyl halide   Ether   Sodium halide

The Williamson synthesis is especially useful in the preparation of mixed ethers (where R ≠ R′) and aromatic ethers.

$$\text{CH}_3\text{CH}_2\text{ONa} + \text{CH}_3\text{Br} \longrightarrow \text{CH}_3\text{CH}_2\text{—O—CH}_3 + \text{NaBr}$$

Sodium ethoxide   Bromomethane   Methoxyethane (Methyl ethyl ether)

$$\text{C}_6\text{H}_5\text{—ONa} + \text{CH}_3\text{Br} \longrightarrow \text{C}_6\text{H}_5\text{—OCH}_3 + \text{NaBr}$$

Sodium phenoxide   Bromomethane   Methyl phenyl ether (Anisole)

24.14 Ether as an Anesthetic

Probably the most widely known use of ethyl ether was as a general anesthetic for surgery. The introduction of ether for this purpose is one of the great landmarks of medicine. Two Americans, Crawford W. Long and William T. Morton, played important roles in this development. Long, a physician, used ether in a surgical operation as early as 1842 but did not publish his discovery until 1849. Morton, a dentist, used ether as an anesthetic for dental work in 1846. He publicly demonstrated its effectiveness by administering ether to a patient undergoing surgery at the Massachusetts General Hospital on October 16, 1846.

The word *anesthesia* is from the Greek, meaning insensibility, and was suggested to Morton by the poet and physician Oliver Wendell Holmes. A **general anesthetic** is a substance or combination of substances that produces both unconsciousness and insensitivity to pain. Many other substances, including other ethers such as divinyl ether (Vinethene) and methoxyflurane (Penthrane), are general anesthetics. These substances are superior in some respects and in recent years have replaced ethyl ether as a general anesthetic.

general anesthetic

$$\text{CH}_2\text{=CH—O—CH=CH}_2 \qquad \text{CHCl}_2\text{CF}_2\text{—O—CH}_3$$

Divinyl ether   Methoxyflurane

Questions

A. Review the meanings of the new terms introduced in this chapter.
  1. Functional group
  2. Alcohol
  3. Primary alcohol
  4. Secondary alcohol
  5. Tertiary alcohol
  6. Polyhydroxy alcohol
  7. Phenol
  8. Ether
  9. General anesthetic

B. Review questions.
  1. Write structural formulas for and give an example of:
     (a) An alkyl halide
     (b) A phenol
     (c) An ether
     (d) An aldehyde
     (e) A ketone
     (f) A carboxylic acid
     (g) An ester
  2. Although it is possible to make alkenes from alcohols, alkenes are seldom, if ever, made in this way on an industrial scale. Why not?
  3. Isopropyl alcohol is usually used in rubbing alcohol formulations. Why is this alcohol used in preference to normal propyl alcohol?
  4. What classes of compounds can be formed by the oxidation of primary alcohols? Cite examples.
  5. What is the molecular weight of an open-chain saturated alcohol containing 40 carbon atoms?
  6. Why is ethylene glycol (1,2-ethanediol) superior to methyl alcohol (methanol) as an antifreeze for automobile radiators?
  7. Briefly outline the physiological effects of:
     (a) Methanol   (b) Ethanol
  8. Explain, in terms of molecular structure, why ethanol (CH_3CH_2OH, molecular weight = 46) is a liquid at ordinary temperatures and butane ($CH_3CH_2CH_2CH_3$, molecular weight = 58) is a gas.
  9. Write structural formulas for:
     (a) Methanol
     (b) 2-Butanol
     (c) 3-Methyl-1-hexanol
     (d) 2-Methyl-2-butanol
     (e) Propylene glycol
  10. There are eight open-chain isomeric alcohols that have the formula $C_5H_{11}OH$. Write the structural formula and the IUPAC name for each of these alcohols.
  11. Which of the isomers in Question 10 are:
     (a) Primary alcohols?   (b) Secondary alcohols?   (c) Tertiary alcohols?
  12. Write the structural formula of a glycol that is both a primary and a secondary alcohol.
  13. Name the following compounds:

     (a) CH_3CH_2OH (b) $CH_3CH(OH)CH_3$ (c) $C_6H_5-CH_2CH_2OH$

     (d) $CH_3CH_2CH(OH)CH_2CH_3$ (e) $CH_3\underset{OH}{\overset{CH_2CH_3}{\underset{|}{\overset{|}{C}}}}CH_2CH_3$

     (f) $C_6H_5-CH_2\underset{CH_3}{\underset{|}{CH}}CH_2OH$ (g) $CH_2\overset{O}{\underset{}{-}}CH_2$ (h) $CH_3\underset{OH}{\underset{|}{CH}}-\underset{OH}{\underset{|}{CH_2}}$

14. Cyclic glycols show *cis–trans* isomerism because of restricted rotation about the carbon–carbon bonds in the ring. Draw and label the structures for *cis*- and *trans*-cyclopentane-1,2-diol.
15. Write the formula and the name of the chief product when the following alcohols are dehydrated to alkenes:

(a) $CH_3CHCHCH_3$ with CH_3 and OH substituents (b) $CH_3CHCH_2CH_2CH_3$ with OH substituent (c) cyclohexane ring with $-OH$ group, $C_6H_{11}OH$

16. When 1-butanol is dehydrated to an alkene, it yields mainly 2-butene rather that 1-butene. This indicates that the dehydration process is at least a two-step reaction. Suggest a mechanism to explain the reaction. [*Hint:* *n*-Butyl carbonium ion is formed initially.]
17. Alcohols can be made by reacting alkyl halides with aqueous sodium hydroxide, as follows:

$$RX + NaOH(aq) \rightarrow ROH + NaX(aq)$$

Give the names and formulas of the alkyl bromides needed to prepare the following alcohols by this method:
(a) Isopropyl alcohol   (b) 3-Methyl-1-butanol   (c) Cyclohexanol
Write the equation for the preparation of each alcohol.

18. Write the equation for the preparation of an alcohol by reacting each of the following alkenes with sulfuric acid and water:
(a) Propene   (c) 2-Butene   (e) 2-Methyl-2-butene
(b) 1-Butene   (d) 1-Pentene

19. Complete the following equations and name the principal organic product formed in each case:

(a) $CH_3CH_2OH + [O] \xrightarrow[\Delta]{Cu}$

(b) $2\,CH_3CH_2OH + H_2SO_4 \xrightarrow{140°C}$

(c) $CH_3CH_2CH_2OH + H_2SO_4 \xrightarrow{180°C}$

(d) $CH_3CH(OH)CH_2CH_3 + [O] \xrightarrow[H_2SO_4]{K_2Cr_2O_7}$

20. Benzyl alcohol ($C_6H_5CH_2OH$) is a primary alcohol. Write the formulas of two different organic compounds that can be obtained by oxidizing benzyl alcohol.
21. Name the following compounds:

(a) phenol (OH on benzene)   (b) benzene ring with two OH groups (ortho)   (c) benzene ring with OH and CH_3 (meta)

(d) [structure: benzene ring with OH, Br, NO₂ — OH top, Br on ring, NO₂ para to OH] (e) [structure: benzene ring with OH, OH, CH₂(CH₂)₄CH₃]

22. Write structural formulas for each of the following:
    (a) *o*-Cresol
    (b) *p*-Nitrophenol
    (c) Resorcinol
    (d) 2,6-Dimethylphenol
    (e) 4-Hydroxy-3-methoxybenzaldehyde (vanillin)
    (f) 2-Isopropyl-5-methylphenol (thymol)
23. Summarize the general properties of phenols.
24. Write equations for the cumene–hydroperoxide synthesis of phenol and acetone.
25. Which of the compounds given below would you expect to react with (a) sodium metal and (b) sodium hydroxide solution? Write equations for those that react.

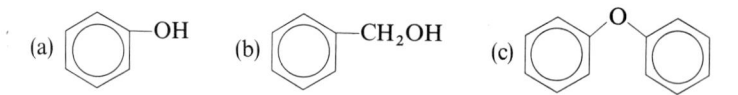

    (a) C₆H₅—OH   (b) C₆H₅—CH₂OH   (c) C₆H₅—O—C₆H₅

26. Starting with *p*-cresol and ethane show equations for the synthesis of ethyl-*p*-methylphenyl ether (*p*-CH₃C₆H₄OCH₂CH₃).
27. What two hazards may be present when working with low-molecular-weight ethers?
28. Arrange the following substances in order of increasing solubility in water:
    (a) CH₃CH₂—O—CH₂CH₂CH₃   (c) CH₃CH₂CH₂CH₂CH₃
    (b) CH₃CH(OH)CH₂CH₂CH₃   (d) CH₃CH(OH)CH(OH)CH₂CH₃
29. There are six isomeric saturated ethers that have the formula $C_5H_{12}O$. Write the structural formula and name for each of these ethers.
30. (a) Write the formula and name of the ether that is isomeric with 1-propanol.
    (b) Write formulas and names for all the possible ethers that are isomeric with butyl alcohol (C₄H₉OH).
31. Write the balanced chemical equation for the complete combustion of ethyl ether (CH₃CH₂—O—CH₂CH₃).
32. Write the formulas of all the possible combinations of RONa and RCl for making each of these ethers by the Williamson synthesis:

    (a) CH₃CH₂—O—CH₃
    (b) CH₃CH₂CH₂—O—CH₂CH₂CH₃
    (c) C₆H₅—CH₂—O—CH₂CH₃
    (d) (CH₃)₂CH—O—CH₂CH₂CH₃

33. Give a simple chemical test that will distinguish between the compounds in each of the following pairs:
    (a) Ethanol and dimethyl ether
    (b) 1-Pentanol and 1-pentene
    (c) *p*-Cresol and methyl phenyl ether

34. Which of the following statements are correct?
    (a) Another name for isopropyl alcohol is 2-propanol.
    (b) Ethanol and dimethyl ether are isomers.
    (c) Alcohols and phenols are more acid than water.
    (d) Sodium ethoxide can be prepared by reacting ethyl alcohol and sodium hydroxide solution.
    (e) Methyl alcohol is a very poisonous substance that can lead to blindness if ingested.
    (f) Tertiary alcohols are easier to oxidize than primary alcohols.
    (g) A correct name for $CH_3CH_2CH(OH)CH_3$ is secondary butyl alcohol.
    (h) $(CH_3)_3CCH_2OH$ is a primary alcohol.
    (i) When a secondary alcohol is oxidized, a ketone is formed.
    (j) Alcohols have higher boiling points than ethers with comparable molecular weights due to hydrogen bonding between the alcohol molecules.
    (k) The product formed when a molecule of water is split out between an alcohol and a carboxylic acid is called an ether.
    (l) When 1-butene is reacted with dilute H_2SO_4, the alcohol formed is 1-butanol.
    (m) Ethanol used for industrial purposes and rendered unfit for use in beverages is said to be denatured.
    (n) The common name for 1,2,3-propanetriol is ethylene glycol.
    (o) Although ethyl alcohol is used in beverages, it is still classified physiologically as a depressant and a poison.

25 Aldehydes and Ketones

After studying Chapter 25 you should be able to:

1. Understand the terms listed in Question A at the end of the chapter.
2. Recognize aldehydes and ketones from their formulas.
3. Give IUPAC and common names of aldehydes and ketones.
4. Write formulas of aldehydes and ketones when given their names.
5. Understand why aldehydes and ketones have lower boiling points than alcohols.
6. Write equations showing the oxidation of alcohols to aldehydes and ketones.
7. Write the structure of the alcohol formed when an aldehyde or a ketone is reduced.
8. Discuss the Tollens' and Fehling's tests, including the reagents used, evidence of a positive test, and the equations for the reactions that occur in positive tests.
9. Use Tollens' and Fehling's tests to differentiate between aldehydes and ketones.
10. Recognize whether an aldehyde or a ketone will undergo the aldol condensation.
11. Write equations showing the aldol condensation of aldehydes and ketones.
12. Write equations for the formation and hydrolysis of cyanohydrins.
13. Write equations for the formation and/or decomposition of hemiacetals, hemiketals, acetals, and ketals.
14. Understand what a positive iodoform test is.
15. Identify, from their structural formulas, compounds that will show a positive iodoform test.
16. Prepare a Grignard reagent.
17. Write equations for the reaction of a Grignard reagent with a given aldehyde or ketone.
18. Choose the reagents for a Grignard reaction that will produce a given alcohol.
19. Distinguish among alcohols, aldehydes, and ketones.

25.1 Structure of Aldehydes and Ketones

carbonyl group

aldehyde

ketone

The aldehydes and ketones are closely related classes of compounds. The structure of both aldehydes and ketones contains the **carbonyl group**, $>\!\!C\!=\!\!O$, a carbon double-bonded oxygen. **Aldehydes** have at least one hydrogen atom bonded to the carbonyl group whereas **ketones** have only alkyl or aryl (aromatic, Ar) groups bonded to the carbonyl group.

$$\underset{\text{Aldehydes}}{\overset{H}{\underset{}{R-C=O}} \quad \overset{H}{\underset{}{Ar-C=O}}} \quad \underset{\text{Ketones}}{\overset{}{R-\underset{\overset{\|}{O}}{C}-R} \quad \overset{}{R-\underset{\overset{\|}{O}}{C}-Ar} \quad \overset{}{Ar-\underset{\overset{\|}{O}}{C}-Ar}}$$

The aldehyde group is often written as CHO or CH=O. In a linear expression of a ketone, the carbonyl group is written as CO; for example,

$$CH_3COCH_3 \quad \text{is equivalent to} \quad CH_3\underset{\overset{\|}{O}}{C}CH_3$$

25.2 Naming Aldehydes and Ketones

The IUPAC names of aliphatic aldehydes are obtained by dropping the final -e and adding -al to the name of the parent hydrocarbon (that is, the longest carbon–carbon chain carrying the —CHO group). The aldehyde carbon is always at the end of the carbon chain and is understood to be carbon number 1. The first member of the homologous series, $H_2C\!=\!O$, is methanal. The name *methanal* is derived from the hydrocarbon methane, which contains one carbon atom. The second member of the series is ethanal; the third member of the series is propanal; and so on. In dialdehydes the suffix *-dial* is added to the corresponding hydrocarbon name; for example,

$$\overset{H}{\underset{O}{\diagdown}}\!\!C\,CH_2CH_2C\!\overset{H}{\underset{O}{\diagup}}$$

is named butanedial.

The common name of aliphatic aldehydes is obtained by dropping *-ic acid* and adding *-aldehyde* to the common name of the corresponding carboxylic acid; for example, $H_2C\!=\!O$ is formaldehyde (from formic acid, HCOOH). It must be emphasized that the common names of aldehydes are derived from the common names of the corresponding acids, not from IUPAC names. See Table 26.1 for the common names of acids. Naming by both systems is illustrated in Table 25.1.

The IUPAC name of a ketone is derived from the name of the alkane corresponding to the longest carbon chain that contains the carbonyl group.

Table 25.1. IUPAC and common names of selected aldehydes.

Formula	IUPAC name	Common name
H–C(H)=O	Methanal	Formaldehyde
CH₃C(H)=O	Ethanal	Acetaldehyde
CH₃CH₂C(H)=O	Propanal	Propionaldehyde
CH₃CH₂CH₂C(H)=O	Butanal	Butyraldehyde
CH₃CH(CH₃)C(H)=O	2-Methylpropanal	Isobutyraldehyde

The name is formed by changing the *-e* ending of the alkane to *-one*. If the chain is longer than four carbons, it is numbered so that the carbonyl carbon has the smallest number possible; and this number is prefixed to the name of the ketone. See the following examples:

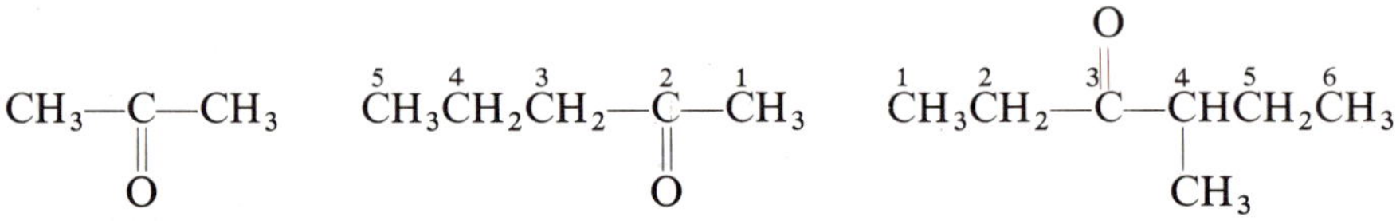

CH₃—C(=O)—CH₃   ⁵CH₃⁴CH₂³CH₂—²C(=O)—¹CH₃   ¹CH₃²CH₂—³C(=O)—⁴CH(CH₃)⁵CH₂⁶CH₃

Propanone   2-Pentanone   4-Methyl-3-hexanone

Note that in 4-methyl-3-hexanone the carbon chain is numbered from left to right to give the ketone group the lowest possible number.

An alternate non-IUPAC method commonly used to name simple ketones is to list the names of the alkyl or aromatic groups attached to the carbonyl carbon together with the word *ketone*. Thus, butanone (CH₃COCH₂CH₃) is methyl ethyl ketone:

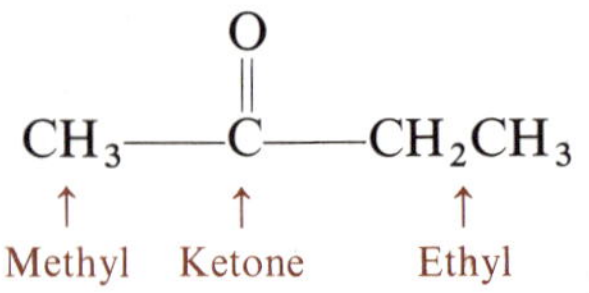

CH₃—C(=O)—CH₂CH₃
 ↑   ↑   ↑
Methyl Ketone   Ethyl

Two of the most widely used ketones have special common names: propanone is commonly called acetone; and butanone is known as methyl ethyl ketone, or MEK.

Aromatic aldehydes are commonly named after the corresponding acids; for example, benzaldehyde from benzoic acid and *p*-tolualdehyde from *p*-toluic acid.

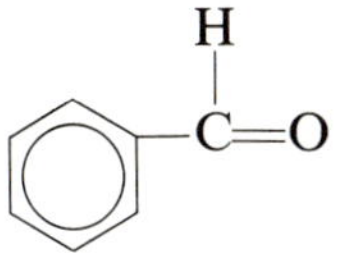

Benzaldehyde

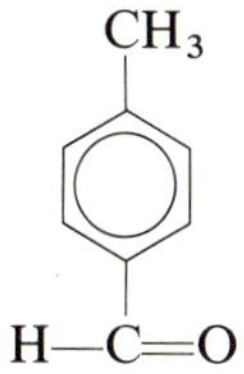

p-Tolualdehyde

Aromatic ketones are named in a similar fashion to aliphatic ketones and often have special names as well.

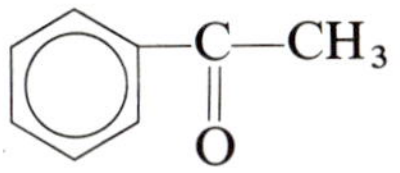

Methyl phenyl ketone
Acetophenone
1-Phenylethanone

Problem 25.1   Name and write the formulas for the straight-chain five- and six-carbon aldehydes.

The systematic names are based on the five- and six-carbon alkanes. Drop the -e of the alkane name and add the suffix -al. Pentane (C_5) becomes pentanal and hexane (C_6) becomes hexanal. The common names are derived from valeric acid and caproic acid, respectively.

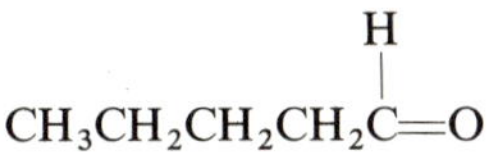

$$CH_3CH_2CH_2CH_2\overset{H}{\underset{}{C}}=O \qquad CH_3CH_2CH_2CH_2CH_2\overset{H}{\underset{}{C}}=O$$

Pentanal (Valeraldehyde) 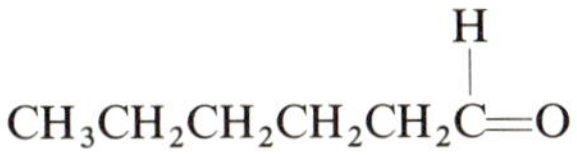   Hexanal (Caproaldehyde)

Problem 25.2   Give two names for each of the following ketones:

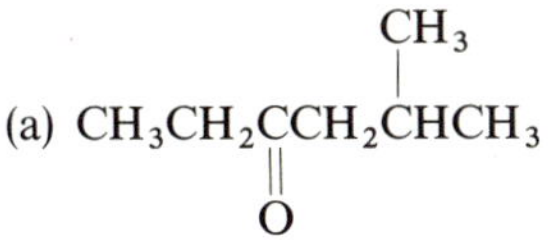

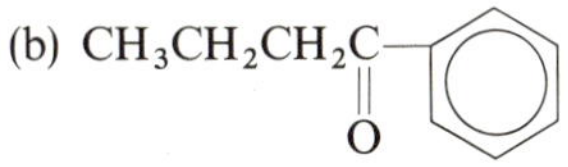

(a) $CH_3CH_2\underset{O}{\overset{CH_3}{\underset{|}{C}}}CH_2\overset{}{\underset{}{CH}}CH_3$ with the C=O indicated

(b) $CH_3CH_2CH_2\underset{O}{\overset{}{C}}$—(phenyl)

(a) Number the longest carbon chain that includes the carbonyl group to give the carbonyl group the smallest number. The longest carbon chain has six carbons. The ketone group is on carbon 3 and a methyl group is on carbon 5. Drop the -e from hexane and add -one to give hexanone. Prefix the name hexanone with 3- to locate the ketone group and with 5-methyl- to locate the methyl group. The name is 5-methyl-3-hexanone. The common name is ethyl isobutyl ketone since the C=O has an ethyl group and an isobutyl group attached to it.

(b) The longest aliphatic chain has four carbon atoms, making the parent ketone a butanone by dropping the -e of butane and adding -one. The butanone has a phenyl group attached to carbon 1. The name is therefore 1-phenyl-1-butanone. The other name for this compound is phenyl n-propyl ketone, since the C=O group has a phenyl and an n-propyl group attached to it.

25.3 Bonding and Physical Properties

The carbon atom of the carbonyl group is sp^2 hybridized and is joined to three other atoms by sigma bonds. The fourth bond is made by overlapping p electrons of carbon and oxygen to form a pi bond between the carbon and oxygen atoms.

Because the oxygen atom is considerably more electronegative than carbon, the carbonyl group is polar, with the electrons shifted toward the oxygen atom. This makes the oxygen atom partially negative (δ^-) and leaves the carbon atom partially positive (δ^+). Many of the chemical reactions of aldehydes and ketones are due to this polarity.

$$\overset{\delta^+}{>}C=\overset{\delta^-}{\ddot{\underset{..}{O}}}{:}$$

Polarity

Unlike alcohols, aldehydes and ketones cannot interact with themselves through hydrogen bonding, since there is no hydrogen atom attached to the oxygen atom of the carbonyl group. Aldehydes and ketones, therefore, have lower boiling points than alcohols of comparable molecular weight (see Table 25.2).

Table 25.2. Comparison of the boiling points of selected aldehydes and ketones with those of corresponding alcohols.

Name	Molecular weight	Boiling point (°C)
1-Propanol	60	97
Propanal	58	49
Propanone	58	56
1-Butanol	74	118
Butanal	72	76
2-Butanone	72	80
1-Pentanol	86	138
Pentanal	84	103
2-Pentanone	84	102

Low-molecular-weight aldehydes and ketones are soluble in water, but from about C_5 on, the solubility decreases markedly. Ketones are very efficient organic solvents.

The lower-molecular-weight aldehydes have a penetrating, disagreeable odor and are partially responsible for the flavor of some rancid and stale foods. As the molecular weight increases, the odor of both aldehydes and ketones, especially the aromatic ones, becomes more fragrant and some are even used in flavorings and perfumes. A few of these together with other selected aldehydes and ketones are shown in Figure 25.1.

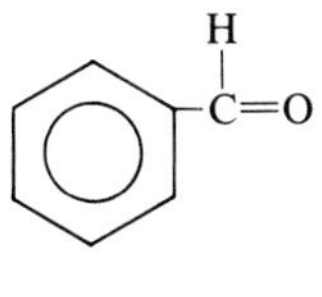

Benzaldehyde
(Oil of bitter almond)

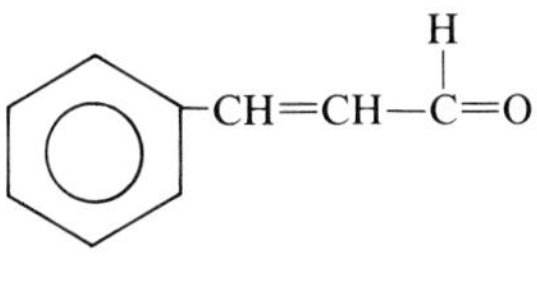

Cinnamaldehyde
(Oil of cinnamon)

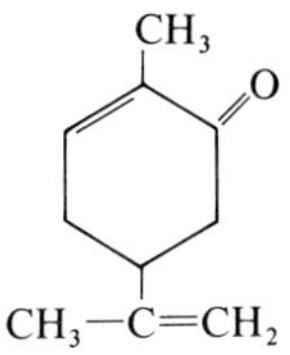

Carvone
(Chief component of spearmint oil)

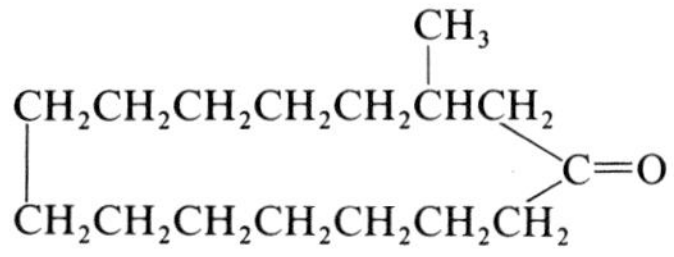

Muscone
(Gland of male musk deer, used in perfume)

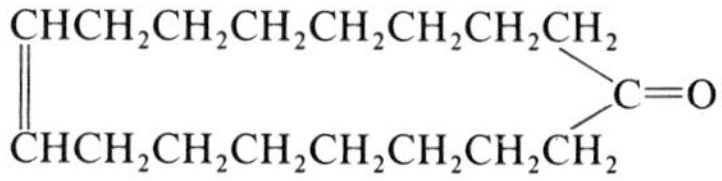

Civetone
(Secretion of the civet cat, used in perfume)

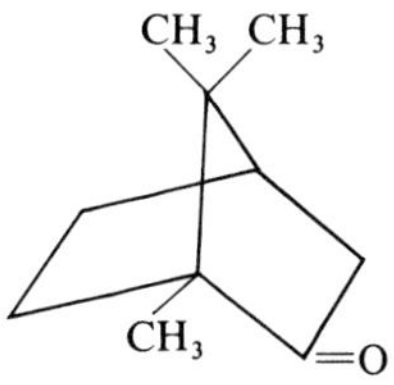

Camphor
(From the camphor tree)

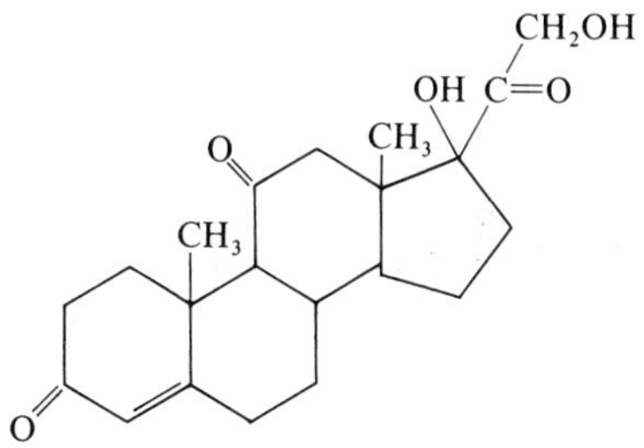

Cortisone
(Hormone; regulation of carbohydrate
and protein metabolism; used to
reduce inflammation)

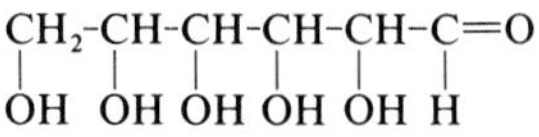

Glucose
(Sugar)

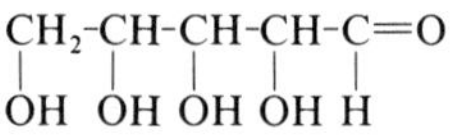

Ribose
(Sugar)

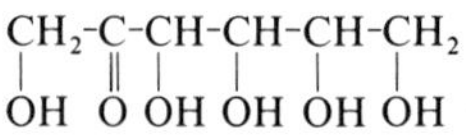

Fructose
(Sugar)

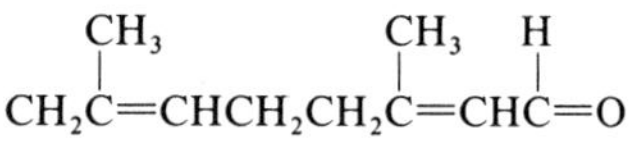

Citral
(Oil of lemon)

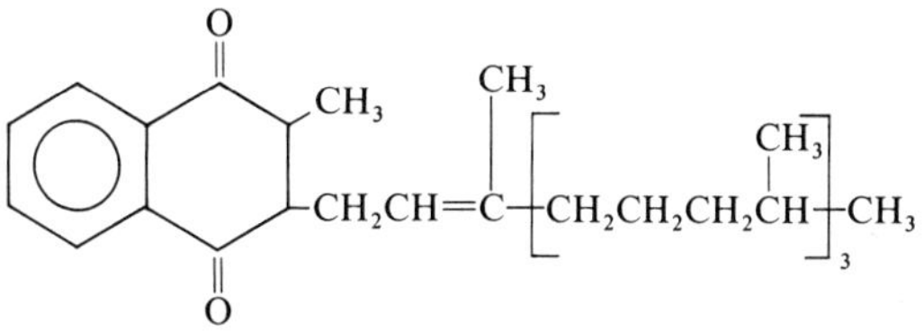

Vitamin K$_1$
(Antihemorrhagic vitamin)

Figure 25.1. Selected naturally occurring aldehydes and ketones.

25.4 Reactions of Aldehydes and Ketones

The carbonyl group undergoes a great variety of reactions. Although there are differences, aldehydes and ketones undergo many similar reactions. However, ketones are generally less reactive than aldehydes. Some typical reactions follow.

Oxidation

Aldehydes are easily oxidized to carboxylic acids by a variety of oxidizing agents, including—under some conditions—oxygen of the air. Oxidation is the reaction in which aldehydes differ most from ketones. In fact, aldehydes and ketones may be separated into classes by their relative susceptibilities to oxidation. Aldehydes are easily oxidized to carboxylic acids by $K_2Cr_2O_7$ + H_2SO_4 and by mild oxidizing agents such as Ag^+ and Cu^{2+} ions; ketones are unaffected by such reagents. Ketones can be oxidized under drastic conditions—for example, by treatment with hot potassium permanganate solution. However, under these conditions carbon–carbon bonds are broken and a variety of products are formed. Equations for the oxidation of aldehydes by dichromate are

$$RCHO + [O] \xrightarrow[H^+]{Cr_2O_7^{2-}} RCOOH \quad \text{(General reaction)}$$
$$\text{Carboxylic acid}$$

$$CH_3CHO + [O] \xrightarrow[H^+]{Cr_2O_7^{2-}} CH_3COOH$$
$$\text{Acetic acid}$$

The Tollens' test (silver mirror test) for aldehydes is based on the ability of silver ions to oxidize aldehydes. The Ag^+ ions are thereby reduced to metallic silver. In practice, a little of the suspected aldehyde is added to a solution of silver nitrate and ammonia in a clean test tube. The appearance of a silver mirror on the inner wall of the tube is a positive test for the aldehyde group. The abbreviated equation is

$$RCHO + 2\,Ag^+ \xrightarrow[H_2O]{NH_3} RCOO^-NH_4^+ + 2\,Ag\downarrow \quad \text{(General reaction)}$$

$$CH_3CHO + 2\,Ag^+ \xrightarrow[H_2O]{NH_3} CH_3COO^-NH_4^+ + 2\,Ag\downarrow$$

Fehling's and Benedict's solutions contain Cu^{2+} ions in an alkaline medium. In the Fehling's and Benedict's tests, the aldehyde group is oxidized to an acid by Cu^{2+} ions. The blue Cu^{2+} ions are reduced and form brick-red copper(I) oxide (Cu_2O), which precipitates during the reaction. These tests can be used for detecting carbohydrates that have an available aldehyde group.

The abbreviated equation is

$$\underset{\text{Blue}}{R\overset{H}{C}=O + Cu^{2+}} \xrightarrow[H_2O]{NaOH} RCOO^-Na^+ + \underset{\text{Brick-red}}{Cu_2O\downarrow}$$

Ketones do not give a positive test with Tollens', Fehling's, or Benedict's solutions. These tests are used to distinguish between aldehydes and ketones.

$$R-\overset{R}{\underset{}{C}}=O + Ag^+ \xrightarrow[H_2O]{NH_3} \text{No reaction}$$

$$R-\overset{R}{\underset{}{C}}=O + Cu^{2+} \xrightarrow[H_2O]{OH^-} \text{No reaction}$$

Aldehydes and ketones are highly combustible, yielding carbon dioxide and water when completely burned. Their vapors, like those of nearly all volatile organic substances, form explosive mixtures with air. Adequate safety precautions must be taken to guard against fire and explosions when working with aldehydes and ketones or other volatile organic compounds, especially hydrocarbons, alcohols, and ethers.

Reduction

Aldehydes and ketones are easily reduced to alcohols, either by elemental hydrogen in the presence of a catalyst or by chemical reducing agents such as aluminum hydride (LiAlH$_4$) or sodium borohydride (NaBH$_4$). Aldehydes yield primary alcohols; ketones yield secondary alcohols:

$$R-\overset{H}{\underset{}{C}}=O \xrightarrow[\Delta]{H_2/Ni} \underset{1° \text{ alcohol}}{RCH_2OH} \quad \text{(General reaction)}$$

$$R-\underset{\underset{O}{\parallel}}{C}-R \xrightarrow[\Delta]{H_2/Ni} \underset{2° \text{ alcohol}}{R-\underset{\underset{OH}{|}}{C}H-R} \quad \text{(General reaction)}$$

$$CH_3\overset{H}{\underset{}{C}}=O \xrightarrow[\Delta]{H_2/Ni} CH_3CH_2OH$$

$$CH_3\underset{\underset{O}{\parallel}}{C}CH_3 \xrightarrow[\Delta]{H_2/Ni} CH_3\underset{\underset{OH}{|}}{C}HCH_3$$

Tertiary alcohols cannot be obtained by direct reduction, but they can be obtained by reacting ketones with Grignard reagents (see Section 25.7).

Aldol Condensation

In the reaction known as the aldol condensation, two aldehyde or two ketone molecules react in the presence of dilute sodium hydroxide to form a hydroxy aldehyde or ketone. For example, acetaldehyde reacts in this fashion:

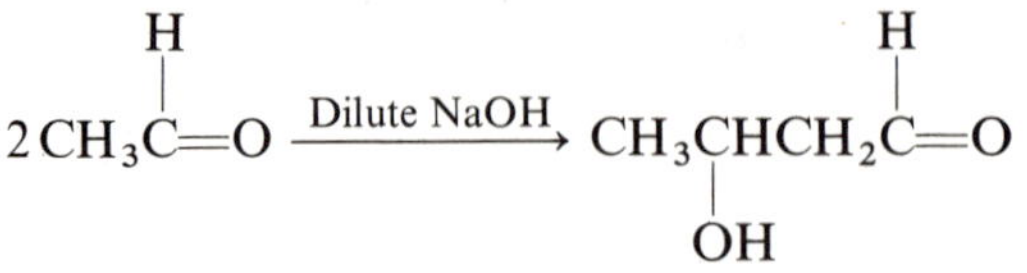

$$2\,CH_3\overset{\overset{H}{|}}{C}=O \xrightarrow{\text{Dilute NaOH}} CH_3\overset{}{C}H\,CH_2\overset{\overset{H}{|}}{C}=O$$
$$\qquad\qquad\qquad\qquad\qquad\quad |$$
$$\qquad\qquad\qquad\qquad\qquad\,\,OH$$
Aldol
(3-Hydroxybutanal)

In order to explain the aldol condensation we first need to identify some additional terms. The carbon atoms in a carbonyl compound are labeled alpha (α), beta (β), gamma (γ), delta (δ), and so on, according to their positions with respect to the carbonyl group. The alpha carbon is adjacent to the carbonyl carbon, the beta carbon is the next one, the gamma carbon the third one, and so forth. In addition, the hydrogen atoms attached to the alpha carbon atom are called alpha hydrogens, and so on as shown below.

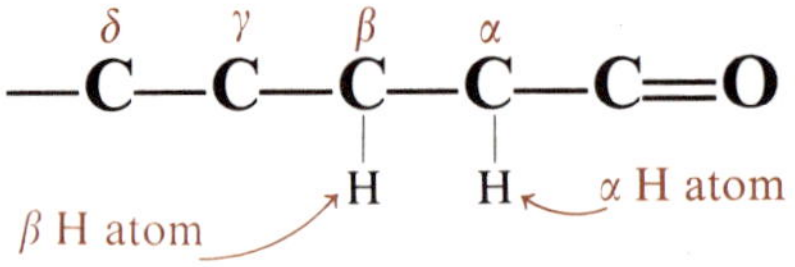

$$\overset{\delta}{-C}-\overset{\gamma}{C}-\overset{\beta}{C}-\overset{\alpha}{C}-C=O$$

β H atom → H   H ← α H atom

The aldol condensation is a fairly general-type reaction. Aldehydes that have at least one alpha hydrogen will react in a manner similar to acetaldehyde. The result is that the alpha hydrogen of one molecule transfers to the oxygen of the carbonyl group of a second molecule. The carbon atom that lost the alpha hydrogen bonds to the carbonyl carbon atom that accepted the hydrogen atom:

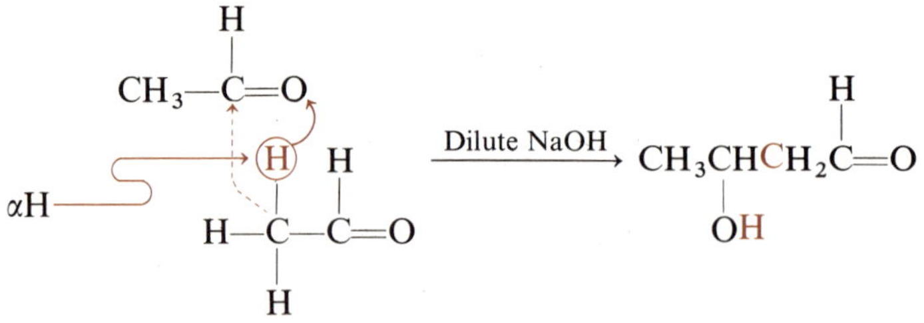

$$CH_3-\overset{\overset{H}{|}}{C}=O$$
$$\qquad\qquad H \quad H$$
$$\alpha H \longrightarrow \quad | \quad |$$
$$\qquad\qquad H-C-C=O \xrightarrow{\text{Dilute NaOH}} CH_3CHCH_2\overset{\overset{H}{|}}{C}=O$$
$$\qquad\qquad\qquad |\qquad\qquad\qquad\qquad\qquad\quad |$$
$$\qquad\qquad\qquad H\qquad\qquad\qquad\qquad\qquad\,\,OH$$

Acetone will also undergo the aldol condensation:

$$CH_3\underset{\underset{OH}{\|}}{C}CH_3 + H-CH_2\underset{\underset{O}{\|}}{C}CH_3 \xrightarrow[\text{NaOH}]{\text{Dilute}} CH_3\underset{\underset{OH}{|}}{\overset{\overset{CH_3}{|}}{C}}-CH_2\underset{\underset{O}{\|}}{C}CH_3$$

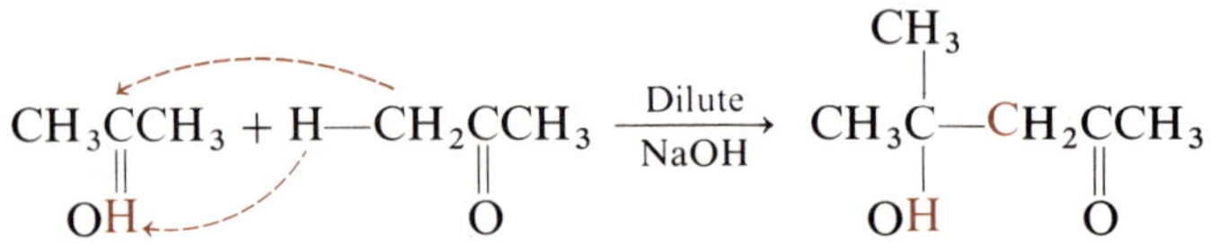

Acetone   Acetone   Diacetone alcohol
4-Hydroxy-4-methyl-2-pentanone

In the foregoing reaction, an alpha hydrogen first transfers from one molecule to the oxygen of the other molecule. This breaks the C=O pi bond, leaving a carbon atom of each molecule with three bonds. The two carbon atoms then bond to each other, forming the product diacetone alcohol.

Problem 25.3 Write the equation for the aldol condensation of propanal. First write the structure for propanal and locate the alpha hydrogen atoms:

$$CH_3CHC=O \quad \text{with} \quad H \leftarrow \alpha H$$

Now write two propanal molecules and transfer an alpha hydrogen from one molecule to the oxygen of the second molecule. After the pi bond breaks, the two carbon atoms that are bonded to only three other atoms are attached to each other to form the product.

$$CH_3CH_2C(=O)H + CH_3CH(H)C=O \xrightarrow{\text{Dilute NaOH}} CH_3CH_2\overset{H}{\underset{OH}{C}} + CH_3\overset{H}{\underset{H}{C}}C=O \longrightarrow CH_3CH_2\overset{CH_3}{\underset{OH}{C}H}CHC=O$$

3-Hydroxy-2-methylpentanal

Addition of Alcohols

hemiacetal

hemiketal

acetal

ketal

Compounds derived from aldehydes and ketones that contain an alkoxy and hydroxy group on the same carbon atom are known as **hemiacetals** and **hemiketals**. In a like manner, compounds that have two alkoxy groups on the same carbon atom are known as **acetals** and **ketals**.

$$\underset{\text{Hemiacetal}}{RCH\begin{matrix}OH\\OR'\end{matrix}} \quad \underset{\text{Hemiketal}}{R\underset{OH}{\overset{R}{C}}-OR'} \quad \underset{\text{Acetal}}{RCH\begin{matrix}OR'\\OR'\end{matrix}} \quad \underset{\text{Ketal}}{R\underset{OR'}{\overset{R}{C}}-OR'}$$

Most hemiacetals and hemiketals are so unstable that they cannot be isolated. On the other hand, acetals and ketals are stable in alkaline solutions but are unstable in acid solutions, where they are hydrolyzed back to the original aldehyde or ketone. The hemiacetal and acetal structures are important in the study of carbohydrate chemistry (Chapter 29).

In the reactions below we will show only aldehydes in the equation, but keep in mind that ketones behave in a similar fashion, although they are not as reactive.

Aldehydes react with one molecule of an alcohol in an acid medium to form hemiacetals:

$$CH_3CH_2\underset{\text{Propanal}}{\overset{H}{\underset{|}{C}}}=O + \underset{\text{Methanol}}{CH_3OH} \underset{}{\overset{H^+}{\rightleftharpoons}} CH_3CH_2\underset{\underset{\text{1-Methoxy-1-propanol}}{\text{(Propionaldehyde methyl hemiacetal)}}}{CH}\begin{smallmatrix}OH\\ \\OCH_3\end{smallmatrix}$$

In the presence of a strong acid such as dry HCl and excess alcohol, the hemiacetal reacts with another molecule of the alcohol to give an acetal:

$$CH_3CH_2CH\begin{smallmatrix}OH\\ \\OCH_3\end{smallmatrix} + CH_3OH \overset{\text{Dry HCl}}{\rightleftharpoons} CH_3CH_2CH\begin{smallmatrix}OCH_3\\ \\OCH_3\end{smallmatrix} + H_2O$$

1,1-Dimethoxypropane
(Propionaldehyde dimethyl acetal)

One molecule of ethylene glycol can furnish both of the hydroxy groups for acetal formation:

Benzaldehyde + Ethylene glycol $\overset{\text{Dry HCl}}{\rightleftharpoons}$ Ethylene glycol acetal of benzaldehyde

All hemiacetal and acetal reactions are reversible.

Addition of Hydrogen Cyanide

cyanohydrin

The addition of hydrogen cyanide, HCN, to aldehydes and ketones forms a class of compounds known as cyanohydrins. **Cyanohydrins** have a cyano (CN) group and a hydroxyl group on the same carbon atom. The reaction is catalyzed by a small amount of base:

$$\underset{\text{Acetaldehyde}}{CH_3\overset{H}{\underset{|}{C}}=O} + HCN \overset{OH^-}{\longrightarrow} \underset{\text{Acetaldehyde cyanohydrin}}{CH_3\overset{OH}{\underset{|}{C}}HCN}$$

$$\underset{\text{Acetone}}{CH_3\underset{\underset{O}{\|}}{C}CH_3} + HCN \overset{OH^-}{\longrightarrow} \underset{\text{Acetone cyanohydrin}}{CH_3\overset{CH_3}{\underset{\underset{OH}{|}}{\underset{|}{C}}}-CN}$$

In the cyanohydrin reaction, the more positive H atom of HCN adds to the oxygen of the carbonyl group and the —CN group adds to the carbon atom of the carbonyl group. In the aldehyde addition, the length of the carbon chain is increased by one carbon. The ketone addition product also contains an additional carbon atom.

Cyanohydrins are useful intermediates for the synthesis of several important compounds. For example, the hydrolyses of cyanohydrins produce α-hydroxy acids:

$$\underset{\underset{OH}{|}}{CH_3CH}-CN + H_2O \xrightarrow{H^+} \underset{\underset{OH}{|}}{CH_3CHCOOH}$$
Lactic acid

25.5 Preparation and Properties of Common Aldehydes and Ketones

Numerous methods have been devised for making aldehydes and ketones. The oxidation of alcohols is a very general method. Special methods are often used for the commercial production of individual aldehydes and ketones.

Formaldehyde (methanal). This aldehyde is made from methanol by reaction with oxygen (air) in the presence of a silver or copper catalyst:

$$2\ CH_3OH + O_2 \xrightarrow[400°C]{Ag\ or\ Cu} 2\ H_2C=O + 2\ H_2O$$
Formaldehyde

Formaldehyde is a poisonous, irritating gas, which is very soluble in water. It is marketed as a 40% aqueous solution called *formalin*. Since formaldehyde is a powerful germicide, it is used in embalming and to preserve biological specimens. Formaldehyde is also used for disinfecting dwellings, ships, and storage houses; for destroying flies; for tanning hides; and as a fungicide for plants and vegetables. But by far the largest use of this chemical is in the manufacture of polymers (Chapter 27). About 2.5 billion lb of formaldehyde are manufactured annually in the United States.

Formaldehyde vapors are intensely irritating to the mucous membranes. Ingestion may cause severe abdominal pains, leading to coma and death.

Acetaldehyde (ethanal). This aldehyde is made by vapor-phase oxidation of ethanol by a process similar to the one used for formaldehyde:

$$2\ CH_3CH_2OH + O_2 \xrightarrow[450°C]{Ag\ or\ Cu} 2\ CH_3\overset{\overset{H}{|}}{C}=O + 2\ H_2O$$
Acetaldehyde

564   Chapter 25

One of the newest commercial processes for making acetaldehyde is known as the Wacker reaction. It is a direct oxidation of ethylene to acetaldehyde using palladium chloride as a catalyst:

$$CH_2=CH_2 + O_2 \xrightarrow[HCl/H_2O]{PdCl_2/CuCl_2} CH_3\overset{H}{\underset{}{C}}=O$$

Acetaldehyde is a volatile liquid (bp 21°C) with a pungent, irritating odor. It has general narcotic action and in large doses may cause respiratory paralysis. Its principal use is as an intermediate in the manufacture of other chemicals such as acetic acid, 1-butanol, and paraldehyde. Acetic acid, for example, is made by air oxidation of acetaldehyde.

$$2\ CH_3\overset{H}{\underset{}{C}}=O + O_2 \xrightarrow[\Delta]{Mn^{2+}} 2\ CH_3\overset{O}{\underset{}{C}}-OH$$

Acetaldehyde undergoes reactions in which three or four molecules condense or polymerize to form the cyclic compounds paraldehyde and metaldehyde.

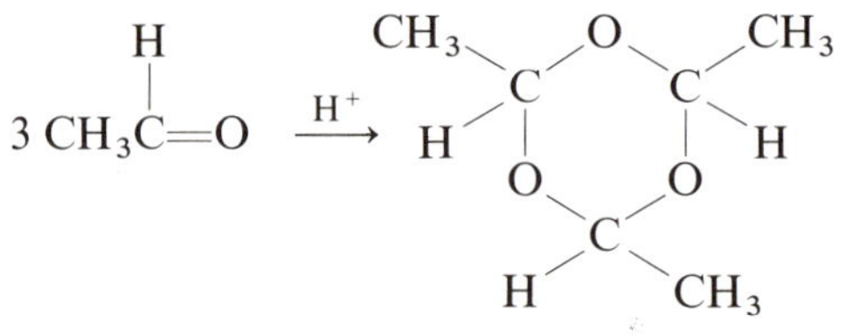

$$3\ CH_3\overset{H}{\underset{}{C}}=O \xrightarrow{H^+}$$

Paraldehyde
(bp 125°C)

Paraldehyde is used in medical practice as a hypnotic or sleep-inducing drug.

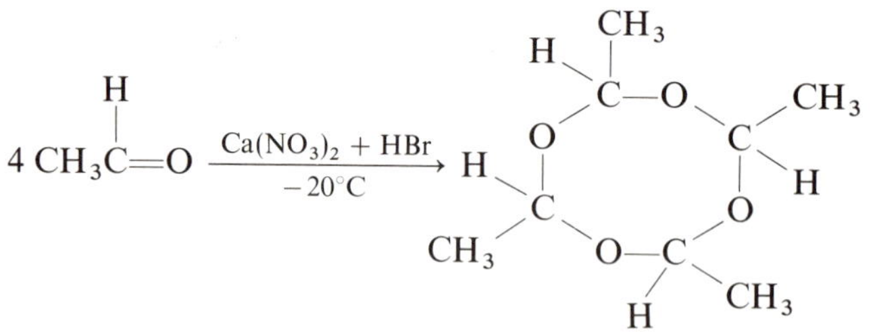

$$4\ CH_3\overset{H}{\underset{}{C}}=O \xrightarrow[-20°C]{Ca(NO_3)_2 + HBr}$$

Metaldehyde
(mp 246°C)

Metaldehyde is very attractive to slugs and snails, and it is also very poisonous to them. For this reason it is an active ingredient in some pesticides that are sold for lawn and garden use. Metaldehyde is also used as a solid fuel.

Benzaldehyde.   Benzaldehyde (C_6H_5CHO) is known as *oil of bitter almonds*. It is found in almonds and in the seeds of stone fruits—for example, apricots and peaches. Benzaldehyde is made synthetically and is used in the manufacture of artificial flavors. The synthesis begins with the free radical chlorination of toluene, as shown in the following sequence of reactions:

$$\text{Toluene (C}_6\text{H}_5\text{—CH}_3) \xrightarrow[\text{Light}]{\text{Cl}_2} \text{Benzyl chloride (C}_6\text{H}_5\text{—CH}_2\text{Cl)} \xrightarrow[\text{Light}]{\text{Cl}_2} \text{Benzal chloride (C}_6\text{H}_5\text{—CHCl}_2) \xrightarrow{\text{H}_2\text{O}} \text{Benzaldehyde (C}_6\text{H}_5\text{—CHO)} + 2\,\text{HCl}$$

Cinnamaldehyde, $C_6H_5CH=CHCHO$, is the principal substance contributing to the flavor of oil of cinnamon. Like benzaldehyde, it is made synthetically and used in the preparation of artificial flavoring agents.

$$C_6H_5\text{—CH}=\text{CH}\text{—CHO}$$
Cinnamaldehyde

Acetone and methyl ethyl ketone. Ketones are widely used organic solvents. Acetone, in particular, is used in very large quantities for this purpose. United States production of acetone is about 2 billion lb annually. It is used as a solvent in the manufacture of drugs, chemicals, and explosives; for removal of paints, varnishes, and fingernail polish; and as a solvent in the plastics industry. Methyl ethyl ketone (MEK) is also widely used as a solvent, especially for lacquers. Both acetone and MEK are made by oxidation (dehydrogenation) of secondary alcohols. Acetone is also a coproduct in the manufacture of phenol (see Section 24.10).

$$\underset{\text{2-Propanol}}{CH_3\underset{\underset{OH}{|}}{CH}CH_3} \xrightarrow[250-300°]{Cu} \underset{\underset{\text{Acetone (Propanone)}}{}}{CH_3\underset{\underset{O}{\|}}{C}CH_3} + H_2$$

$$\underset{\text{2-Butanol}}{CH_3CH_2\underset{\underset{OH}{|}}{CH}CH_3} \xrightarrow[250-300°]{Cu} \underset{\underset{\text{Methyl ethyl ketone (2-Butanone)}}{}}{CH_3CH_2\underset{\underset{O}{\|}}{C}CH_3} + H_2$$

25.6 Iodoform Test

Iodoform (CHI_3) is a canary yellow solid with a distinctive, somewhat unpleasant odor. Iodoform is formed when a methyl ketone, acetaldehyde, ethyl alcohol, or a secondary alcohol that contains at least one methyl group

iodoform test

bonded to the —C—OH group is treated with sodium hypoiodite (NaOI) solution (NaOH + I$_2$). All these substances either contain the methyl carbonyl (acetyl) group (CH$_3$C=O) or are easily oxidized to form this structure. The **iodoform test** is useful in the qualitative examination of organic compounds. A positive test is the appearance of a yellow precipitate of iodoform. The reactions involved are represented by the following equations:

$$R-\underset{\underset{O}{\|}}{C}-CH_3 + 3\ NaOI \longrightarrow R\underset{\underset{O}{\|}}{C}-CI_3 + 3\ NaOH$$

$$R\underset{\underset{O}{\|}}{C}-CI_3 + NaOH \longrightarrow R\underset{\underset{O}{\|}}{C}-ONa + CHI_3 \text{ (Iodoform)}$$

Ethyl alcohol gives a positive test because it is first oxidized to acetaldehyde, which contains the CH$_3$C=O group:

$$CH_3CH_2OH \xrightarrow{NaOI} CH_3\underset{\underset{}{|}}{\overset{H}{C}}=O \xrightarrow[NaOH]{NaOI} CHI_3$$

The reaction that occurs in the iodoform test is sometimes called the haloform reaction, since similar reactions occur when chlorine or bromine is used. However, iodine is commonly used because chloroform (CHCl$_3$) and bromoform (CHBr$_3$) are more difficult to detect than iodoform.

25.7 Grignard Reactions

Grignard reagent

A **Grignard reagent** is an organic magnesium halide. It can be either an alkyl or an aryl compound (RMgX or ArMgX). Grignard (pronounced *green-yard*) reagents were first prepared in France around 1900 by Victor Grignard (1871–1935). Grignard was awarded the Nobel Prize in organic chemistry in 1912.

Grignard reagents are usually prepared by reacting an organic halide and magnesium metal in an ether solvent:

$$RX + Mg \xrightarrow{Ether} \underset{\substack{\text{Grignard reagent} \\ (X = Cl, Br, or I)}}{RMgX}$$

$$ArX + Mg \xrightarrow{Ether} \underset{\substack{\text{Grignard reagent} \\ (X = Br)}}{ArMgX}$$

In the Grignard reagent, the bonding electrons between carbon and magnesium are shifted away from the electropositive Mg to form a strongly polar covalent bond. As a result, the charge distribution in the Grignard reagent is such that the organic group (R) is partially negative (δ^-) and the —MgX group is partially positive (δ^+). This charge distribution directs the manner in which the Grignard reacts with other compounds.

$$\overset{\delta^-\ \delta^+}{\mathbf{R\ MgX}}$$

The Grignard reagent is one of the most versatile and widely used reagents in organic chemistry. We will consider only its reactions with aldehydes and ketones at this time. Grignards react with aldehydes and ketones to give intermediate products, which form alcohols when hydrolyzed. With formaldehyde, primary alcohols are formed; with other aldehydes, secondary alcohols are formed; with ketones, tertiary alcohols are formed:

Grignard reagent + Formaldehyde ⟶ 1° ROH
Grignard reagent + Other aldehydes ⟶ 2° ROH
Grignard reagent + Ketones ⟶ 3° ROH

Specific examples of each type of reaction follow.

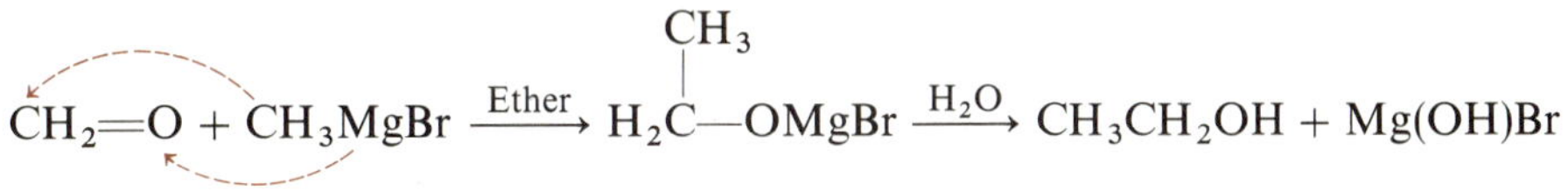

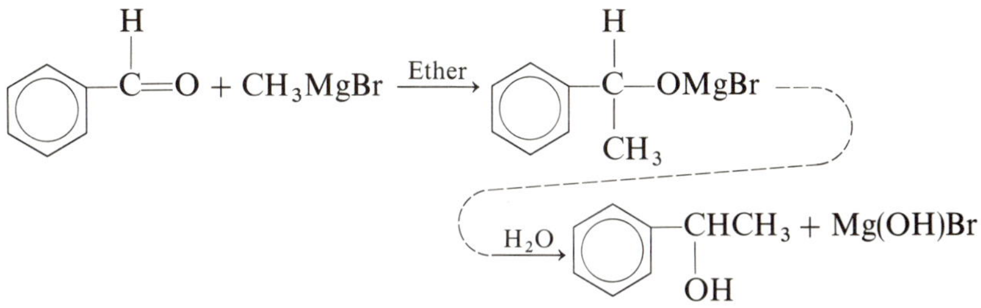

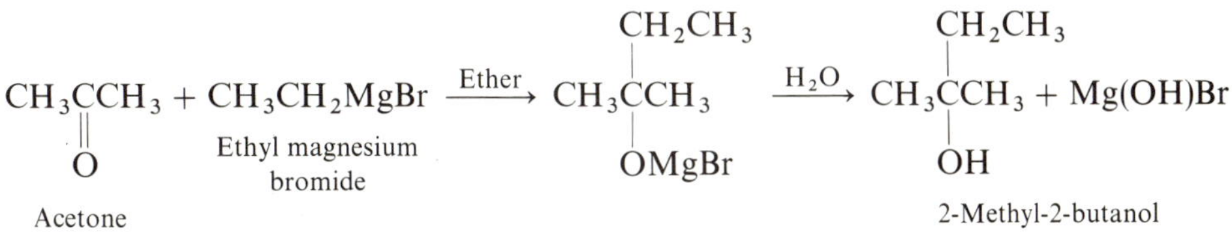

$$CH_2=O + CH_3MgBr \xrightarrow{Ether} H_2\overset{CH_3}{\underset{}{C}}-OMgBr \xrightarrow{H_2O} CH_3CH_2OH + Mg(OH)Br$$

$$C_6H_5-\overset{H}{\underset{}{C}}=O + CH_3MgBr \xrightarrow{Ether} C_6H_5-\overset{H}{\underset{CH_3}{C}}-OMgBr$$

$$\xrightarrow{H_2O} C_6H_5-\overset{}{\underset{OH}{C}}HCH_3 + Mg(OH)Br$$

$$\underset{\underset{Acetone}{O}}{CH_3\overset{\|}{C}CH_3} + \underset{\substack{Ethyl\ magnesium\\bromide}}{CH_3CH_2MgBr} \xrightarrow{Ether} \underset{OMgBr}{CH_3\overset{CH_2CH_3}{\underset{}{C}}CH_3} \xrightarrow{H_2O} \underset{\underset{2\text{-Methyl-2-butanol}}{OH}}{CH_3\overset{CH_2CH_3}{\underset{}{C}}CH_3} + Mg(OH)Br$$

The Grignard reaction with acetone may be explained in this way. In the first step of the addition of ethyl magnesium bromide, the partially positive —MgBr of the Grignard bonds to the oxygen atom and the partially negative CH₃CH₂— bonds to the carbon atom of the carbonyl group of acetone.

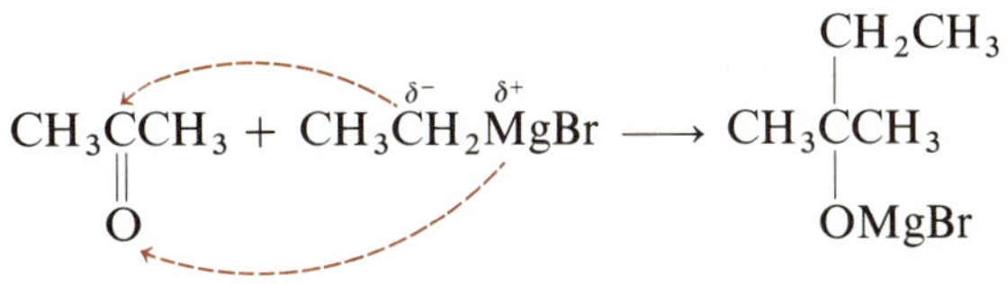

$$\underset{O}{CH_3\overset{\|}{C}CH_3} + \overset{\delta^-\quad \delta^+}{CH_3CH_2MgBr} \longrightarrow \underset{OMgBr}{CH_3\overset{CH_2CH_3}{\underset{}{C}}CH_3}$$

In the hydrolysis step, a proton (H⁺) from water combines with the —OMgBr, leaving the negative —OH to bond to the carbon that was attached to the —OMgBr. Thus, the alcohol is formed.

$$\underset{\underset{OMgBr}{|}}{\overset{\overset{CH_2CH_3}{|}}{CH_3CCH_3}} + H\text{—}OH \longrightarrow \underset{\underset{OH}{|}}{\overset{\overset{CH_2CH_3}{|}}{CH_3CCH_3}} + Mg(OH)Br$$

Problem 25.4 What combination(s) of aldehyde or ketone and Grignard reagent will react to give 2-butanol, $CH_3\underset{\underset{OH}{|}}{CH}CH_2CH_3$?

Since the alcohol is secondary, the reactants must be an aldehyde (other than formaldehyde) and a Grignard reagent. If the two organic groups attached to the >CHOH of the alcohol are the same, only one combination is possible; if they are different, two combinations are possible. Isolate the groups as follows:

$$\boxed{CH_3\text{—}\underset{\underset{OH}{|}}{CH}}\boxed{\text{—}CH_2CH_3} \qquad \boxed{CH_3\text{—}}\boxed{\text{—}\underset{\underset{OH}{|}}{CH}\text{—}CH_2CH_3}$$

This part from the aldehyde / This part from the Grignard   This part from the Grignard / This part from the aldehyde

In the first case, the aldehyde and Grignard each have two carbon atoms. Therefore, one possible combination to make 2-butanol is ethanal and ethyl magnesium bromide.

$$\overset{\overset{H}{|}}{CH_3C}=O \quad \text{and} \quad CH_3CH_2MgBr$$

In the second case, the aldehyde has three carbon atoms and the Grignard has one carbon atom. Thus, another combination to make 2-butanol is propanal and methyl magnesium bromide.

$$\overset{\overset{H}{|}}{CH_3CH_2C}=O \quad \text{and} \quad CH_3MgBr$$

Questions A. *Review the meanings of the new terms introduced in this chapter.*

1. Carbonyl group
2. Aldehyde
3. Ketone
4. Hemiacetal
5. Hemiketal
6. Acetal
7. Ketal
8. Cyanohydrin
9. Iodoform test
10. Grignard reagent

B. Review questions.
1. Name each of these aldehydes:

(a) $CH_2=O$ (two names)

(b) $CH_3CH_2CH_2\overset{H}{\underset{\|}{C}}=O$ (two names)

(c) $CH_3\underset{\underset{CH_3}{|}}{CH}CH_2\overset{H}{\underset{\|}{C}}=O$ (one name)

(d) 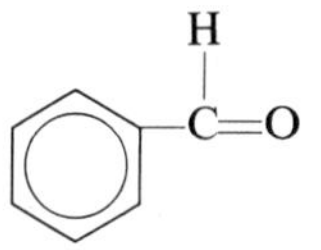 $C_6H_5\overset{H}{\underset{\|}{C}}=O$ (one name)

(e) $O=\overset{H}{\underset{\|}{C}}CH_2CH_2\overset{H}{\underset{\|}{C}}=O$ (one name)

(f) 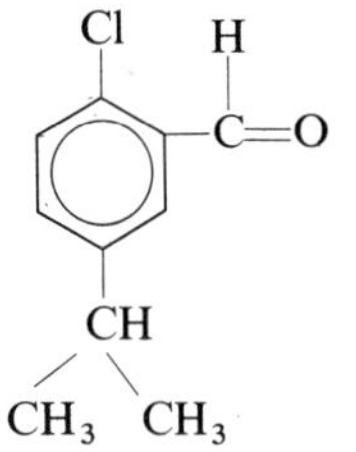 (2-chloro-4-isopropyl benzaldehyde structure) (one name)

(g) 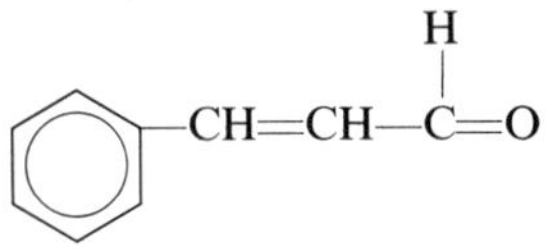 $C_6H_5-CH=CH-\overset{H}{\underset{\|}{C}}=O$ (one name)

(h) $CH_3\underset{\underset{OH}{|}}{CH}CH_2\overset{H}{\underset{\|}{C}}=O$ (two names)

2. Name each of these ketones:

(a) $CH_3\underset{\underset{O}{\|}}{C}CH_3$ (three names)

(b) $CH_3CH_2\underset{\underset{O}{\|}}{C}CH_3$ (two names)

(c) 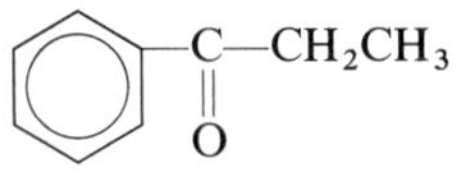 $C_6H_5-\underset{\underset{O}{\|}}{C}-CH_2CH_3$ (two names)

(d) $CH_3\underset{\underset{O}{\|}}{C}-\underset{\underset{CH_3}{|}}{\overset{CH_3}{\overset{|}{C}}}CH_3$ (two names)

(e) 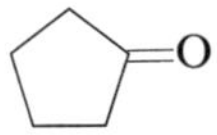 cyclopentanone (one name)

(f) $CH_3\underset{\underset{O}{\|}}{C}CH_2CH_2\underset{\underset{O}{\|}}{C}CH_3$ (one name)

(g) $CH_3\underset{\underset{OH}{|}}{\overset{CH_3}{\overset{|}{C}}}-CH_2\underset{\underset{O}{\|}}{C}CH_3$ (two names)

3. Write structural formulas for:
   (a) 1,3-Dichloropropanone
   (b) Phenylacetaldehyde
   (c) 3-Butenal
   (d) 3-Hydroxypropanal
   (e) Diisopropyl ketone
   (f) 4-Methyl-3-hexanone

4. Write structural formulas for propanal and propanone. From these formulas, do you think that aldehydes and ketones are isomeric with each other? Show evidence and substantiate your answer by testing with a four-carbon aldehyde and a four-carbon ketone.

570   Chapter 25

5. Explain, in terms of structure, why aldehydes and ketones have lower boiling points than alcohols of similar molecular weights.
6. Write equations to show how each of the following are oxidized by (1) $K_2Cr_2O_7$ + H_2SO_4 and (2) air + Cu or Ag + heat:
   (a) 1-Propanol   (b) 3-Pentanol   (c) 2,3-Dimethyl-2-butanol
7. Ketones are prepared by oxidation of secondary alcohols. Which alcohol should be used to prepare
   (a) Diethyl ketone   (b) Diisopropyl ketone   (c) 4-Phenyl-2-butanone
8. (a) What functional group is present in a compound that gives a positive Tollens' test?
   (b) What is the visible evidence for a positive Tollens' test?
   (c) Write an equation showing the reaction involved in a positive Tollens' test.
9. (a) What functional group is present in a compound that gives a positive Fehling's test?
   (b) What is the visible evidence for a positive Fehling's test?
   (c) Write an equation showing the reaction involved in a positive Fehling's test.
10. Give the products of the reaction of the following with Tollens' reagent:
    (a) Butanal   (b) Benzaldehyde   (c) Methyl ethyl ketone
11. Write equations showing the aldol condensation for the following compounds:
    (a) Acetaldehyde   (b) Propanal   (c) Butanal   (d) 2-Pentanone
12. How many aldol condensation products are possible if a mixture of acetaldehyde and propanal are reacted?
13. Complete the following equations:

   (a) $CH_3CH_2\overset{H}{\underset{}{C}}=O + CH_3CH_2CH_2OH \underset{}{\overset{Dry\ HCl}{\rightleftarrows}}$

   (b) $CH_3\underset{\underset{O}{\|}}{C}CH_3 + \underset{\underset{OH\ OH}{|\ \ |}}{CH_2CH_2} \overset{Dry\ HCl}{\rightleftarrows}$

   (c) $CH_3CH_2\overset{H}{\underset{}{C}}=O + CH_3CH_2OH \overset{H^+}{\rightleftarrows}$

   (d) ⬡=O + $CH_3OH \overset{H^+}{\rightleftarrows}$

   (e) $CH_3CH_2CH_2CH(OCH_3)_2 \overset{H_2O}{\underset{H^+}{\rightarrow}}$

14. Write equations for the following sequence of reactions:
    (a) Benzaldehyde + HCN →
    (b) Product from part (a) + H_2O →
    (c) Product from part (b) + $K_2Cr_2O_7$ + H_2SO_4 →
15. What structural grouping must a molecule have—or be easily converted to—in order to give a positive iodoform test?
16. Which of these compounds will give a positive iodoform test?
    (a) CH_3OH (b) CH_3CH_2OH (c) $CH_3CH_2CH_2OH$ (d) $CH_3\underset{\underset{OH}{|}}{CH}CH_3$

(e) $CH_3\overset{H}{\underset{|}{C}}=O$ (f) $CH_3CH_2\overset{CH_3}{\underset{|}{C}}=O$ (g) $CH_3-\overset{CH_3}{\underset{\underset{CH_3}{|}}{\overset{|}{C}}}-OH$

(h) $CH_3\overset{CH_3}{\underset{|}{CH}}-\underset{\underset{O}{\|}}{C}-CH_3$ (i) C₆H₅–CH₂OH (j) C₆H₅–C(=O)–CH₃

17. There are four isomeric butyl alcohols. Which of them will give a positive iodoform test?
18. (a) Write an equation for the preparation of the Grignard reagent ethyl magnesium bromide, starting with ethyl bromide.
    (b) Using the Grignard reagent from part (a) and whatever other reagents are needed, write equations for the preparation of the following alcohols:
       (1) 1-propanol, (2) 3-hexanol, (3) 3-methyl-3-pentanol, (4) 2-methyl-3-pentanol, (5) 1-phenyl-2-butanol.
19. What combination(s) of Grignard reagent and aldehyde or ketone are needed to prepare the following?
    (a) *t*-Butyl alcohol   (b) 2-Pentanol   (c) 1-Hexanol   (d) Benzyl alcohol
20. Give a simple chemical test that will distinguish between the compounds in each of the following pairs:

(a) $CH_3CH_2\overset{H}{\underset{|}{C}}=O$ and $CH_3\underset{\underset{O}{\|}}{C}CH_3$

(b) $CH_3\overset{H}{\underset{|}{C}}=O$ and $CH_3CH_2\overset{}{\underset{\underset{H}{|}}{C}}=O$

(c) $CH_3CH_2\overset{H}{\underset{|}{C}}=O$ and $CH_2=CH\overset{H}{\underset{|}{C}}=O$

(d) C₆H₅–$CH_2\overset{H}{\underset{|}{C}}=O$ and   C₆H₅–$\underset{\underset{O}{\|}}{C}-CH_3$

(e) C₆H₅–CH₂CH₂OH   and   C₆H₅–$\underset{\underset{OH}{|}}{CH}CH_3$

21. Which of the following statements are correct?
    (a) The functional group that characterizes aldehydes and ketones is called a carboxyl group.
    (b) The carbonyl group contains a sigma and a pi bond.
    (c) The carbonyl group is polar, with the oxygen atom being more electronegative than the carbon atom.

(d) The higher-molecular-weight aldehydes and ketones are very soluble in water.
(e) Ketones, like aldehydes, are easily oxidized to carboxylic acids.
(f) A compound of formula $C_6H_{12}O$ can be either an aliphatic aldehyde or ketone.
(g) Diethyl ketone has the same molecular formula as butyraldehyde.
(h) In aldehydes and ketones, the hydrogen atoms that are bonded to carbon atoms alpha to the carbonyl group are more reactive than other hydrogen atoms in the molecule.
(i) In order for an aldehyde or a ketone to undergo the aldol condensation it must have at least one alpha hydrogen atom.
(j) Primary, secondary, and tertiary alcohols can be prepared by reacting aldehydes with a Grignard reagent.
(k) A hemiacetal has an alcohol and an ether group bonded to the same carbon atom.
(l) Acetals are stable in acid solution, but not in alkaline solution.
(m) The reagent used for the iodoform test is NaOI (NaOH + I_2).
(n) The formula for iodoform is $CH_3\underset{\underset{O}{\|}}{C}-CI_3$.
(o) Formaldehyde is a gas, but it is usually handled in a solution.
(p) The major use for formaldehyde is for making plastics.
(q) Ethanal may be distinguished from propanal by use of Tollens' reagent.

26 Carboxylic Acids, Esters, and Amines

After studying Chapter 26 you should be able to:

1. Understand the terms listed in Question A at the end of the chapter.
2. Give the common and IUPAC names of carboxylic acids.
3. Write structural formulas for carboxylic acids.
4. Tell how the water solubility of carboxylic acids varies with increasing molecular weight.
5. Relate the boiling points of carboxylic acids to their structures.
6. Write structural formulas and names for unsaturated acids, hydroxy acids, aromatic carboxylic acids, and dicarboxylic acids.
7. Write equations for the preparation of carboxylic acids by (a) oxidation of alcohols and aldehydes, (b) hydrolysis or saponification of esters and fats, (c) oxidation of aromatic hydrocarbons, (d) hydrolysis of nitriles.
8. Write equations for reactions of carboxylic acids to form (a) salts, (b) esters, (c) amides, (d) *N*-substituted amides, (e) acid chlorides.
9. Write equations for reactions of acid chlorides to form esters and amides.
10. Name and write structural formulas for amides.
11. Write common names, IUPAC names, and formulas of esters.
12. Identify the carboxylic acid and alcohol needed to prepare an ester.
13. Describe the significant physical properties of esters.
14. Write the structure of a triglyceride when given the fatty acid composition.
15. Tell how a fat differs from an oil.
16. Write equations illustrating the (a) hydrogenation, (b) hydrogenolysis, (c) hydrolysis, and (d) saponification of a fat or oil.
17. Write the structural formulas for the three principal unsaturated carboxylic acids found in fats and oils.
18. Explain how a soap or synthetic detergent acts as a cleaning agent.
19. Explain why syndets are effective and soaps are not effective as cleaning agents in hard water.
20. Write structural formulas for a soap, a biodegradable synthetic detergent, and a nonbiodegradable detergent.
21. Explain how cationic, anionic, and nonionic detergents differ in how they act as cleaning agents.
22. Name and write structural formulas for amines.
23. Distinguish among primary, secondary, and tertiary amines.
24. Show that amines are bases in their reactions with water and with acids.

25. Write equations for reactions of amines to form substituted amides.
26. Write equations for the formation of amines from (a) alkyl halides plus ammonia, (b) reduction of amides with LiAlH$_4$, (c) reduction of nitriles with H$_2$ and Ni.
27. Write equations showing the effect of heat on malonic, succinic, and glutaric acids.

Carboxylic acids and their derivatives are among the most widely distributed substances in nature. Acetic acid, the familiar substance that gives vinegar its sour taste, has been known since antiquity. Salicylic acid is the acid from which aspirin is made. Butyric acid is one of the foulest smelling substances imaginable, yet derivatives of butyric acid are found in butter and are used as flavorings and perfumes. Some acid derivatives are important to the sustenance of life. Two of the principal classes of foods are acid derivatives— proteins consist primarily of amino acids, and fats are esters of carboxylic acids. A number of organic acids play essential roles in metabolic and energy-producing processes in plant and animal life.

26.1 Carboxylic Acids

carboxyl group

The functional group of the carboxylic acids is called a **carboxyl group** and is represented in the following ways:

$$-\overset{\overset{\displaystyle O}{\|}}{C}-OH \quad \text{or} \quad -COOH \quad \text{or} \quad -CO_2H$$

Carboxylic acids can be either aliphatic or aromatic.

$$R\overset{\overset{\displaystyle O}{\|}}{C}-OH \qquad Ar\overset{\overset{\displaystyle O}{\|}}{C}-OH$$

$$CH_3\overset{\overset{\displaystyle O}{\|}}{C}-OH \qquad C_6H_5\overset{\overset{\displaystyle O}{\|}}{C}-OH$$

Aliphatic   Aromatic

26.2 Nomenclature and Sources of Aliphatic Carboxylic acids

Aliphatic carboxylic acids form a homologous series. The carboxyl group is always at the end of a carbon chain, and the C atom in this group is understood to be carbon number 1 in naming the compound.

To name a carboxylic acid by the IUPAC System, first identify the longest carbon chain including the carboxyl group. Then form the acid name by dropping the -*e* from the corresponding parent hydrocarbon name and add -*oic acid*. Thus, the names corresponding to the one-, two-, and three-carbon acids are methanoic acid, ethanoic acid, and propanoic acid. These names are, of course, derived from methane, ethane, and propane.

CH_4	Methane	HCOOH	Methanoic acid
CH_3CH_3	Ethane	CH_3COOH	Ethanoic acid
$CH_3CH_2CH_3$	Propane	CH_3CH_2COOH	Propanoic acid

The Greek lettering system α, β, γ, δ, . . . , has traditionally been used in naming certain acid derivatives, especially hydroxy, amino, and halogen acids. When Greek letters are used, the lettering begins with a carbon atom adjacent to the —COOH group. When numbers are used, the numbers begin with the carbon in the —COOH group. Common and IUPAC systems should not be intermixed.

$$\overset{\delta}{\underset{5}{C}}-\overset{\gamma}{\underset{4}{C}}-\overset{\beta}{\underset{3}{C}}-\overset{\alpha}{\underset{2}{C}}-\overset{\overset{O}{\|}}{\underset{1}{C}}-OH$$

$$CH_3CH_2\underset{\underset{OH}{|}}{C}HCOOH \qquad CH_3\underset{\underset{NH_2}{|}}{C}HCOOH \qquad CH_2ClCH_2COOH$$

Common name: α-Hydroxybutyric acid   α-Aminopropionic acid   β-Chloropropionic acid
IUPAC name: 2-Hydroxybutanoic acid   2-Aminopropanoic acid   3-Chloropropanoic acid

Unfortunately, the IUPAC method is not the only, nor most generally used, method of naming acids. Organic acids are ordinarily known by common names. Methanoic, ethanoic, and propanoic acids are commonly called formic, acetic, and propionic acids, respectively. These names usually refer to a natural source of the acid and are not really systematic. Formic acid was named from the Latin word *formica*, meaning "ant." This acid contributes to the stinging sensation of ant bites. Acetic acid is found in vinegar and is so named from the Latin word for vinegar. The name of butyric acid is derived from the Latin term for butter, since it is a constituent of butterfat. The 6-, 8-, and 10-carbon acids are found in goat fat and have names derived from the Latin word for goat. These three acids—caproic, caprylic, and capric—along with butyric acid, have characteristic and quite disagreeable odors. In a similar way, the names of the 12-, 14-, and 16-carbon acids—lauric, myristic, and palmitic—are from plants from which the corresponding acid has been isolated. The name stearic acid is derived from a Greek word meaning beef fat or tallow, which is a good source of this acid. Many of the carboxylic acids, principally those having even numbers of carbon atoms ranging from 4 to about 20, exist in combined form in plant and animal fats. These are called *fatty acids*. Table 26.1 lists the common and IUPAC names, together with some of the physical properties of the more important saturated aliphatic acids.

Table 26.1. Physical properties of saturated carboxylic acids.

Common name (IUPAC name)	Formula	Melting point (°C)	Boiling point (°C)	Solubility in water[a]
Formic acid (methanoic acid)	HCOOH	8.4	100.8	∞
Acetic acid (ethanoic acid)	CH$_3$COOH	16.6	118	∞
Propionic acid (propanoic acid)	CH$_3$CH$_2$COOH	−21.5	141.4	∞
Butyric acid (butanoic acid)	CH$_3$(CH$_2$)$_2$COOH	−6	164	∞
Valeric acid (pentanoic acid)	CH$_3$(CH$_2$)$_3$COOH	−34.5	186.4	3.3
Caproic acid (hexanoic acid)	CH$_3$(CH$_2$)$_4$COOH	−3.4	205	1.1
Caprylic acid (octanoic acid)	CH$_3$(CH$_2$)$_6$COOH	16.3	239	0.1
Capric acid (decanoic acid)	CH$_3$(CH$_2$)$_8$COOH	31.4	269[b]	Insoluble
Lauric acid (dodecanoic acid)	CH$_3$(CH$_2$)$_{10}$COOH	44.1	225[b]	Insoluble
Myristic acid (tetradecanoic acid)	CH$_3$(CH$_2$)$_{12}$COOH	54.2	251[b]	Insoluble
Palmitic acid (hexadecanoic acid)	CH$_3$(CH$_2$)$_{14}$COOH	63	272[b]	Insoluble
Stearic acid (octadecanoic acid)	CH$_3$(CH$_2$)$_{16}$COOH	69.6	287[b]	Insoluble
Arachidic acid (eicosanoic acid)	CH$_3$(CH$_2$)$_{18}$COOH	77	298[b]	Insoluble

[a] Grams of acid per 100 g of water.
[b] Boiling points of acids beginning with lauric acid are given at 100 mm Hg pressure instead of atmospheric pressure, since thermal decomposition occurs before these acids reach their boiling points at atmospheric pressure.

Problem 26.1 Write formulas for the following:
(a) 3-Chloropentanoic acid
(b) γ-Hydroxybutyric acid
(c) Phenylacetic acid

(a) Pentanoic (from pentane) indicates a five-carbon acid. Substituted on this acid on carbon 3 is a chlorine atom. Write five carbon atoms in a row. Make carbon 1 a carboxyl group, place a Cl on carbon 3, and add hydrogens to give each carbon four bonds. The formula is

CH$_3$CH$_2$CHClCH$_2$COOH

(b) Butyric (from butane) indicates a four-carbon acid. The gamma (γ) position is three carbons removed from the carboxyl group. Therefore, the formula is

γ carbon

HO—CH$_2$CH$_2$CH$_2$COOH

(c) Acetic acid is the familiar two-carbon acid. There is only one place to substitute the phenyl group and still call the compound an acid—that is the CH_3 group. Substitute a phenyl group for one of the three H atoms to give the formula.

C₆H₅—CH₂COOH

26.3 Physical Properties of Carboxylic Acids

Each aliphatic carboxylic acid molecule is polar and consists of a carboxyl group and a hydrocarbon radical. These two unlike parts have great bearing on the physical, as well as chemical, behavior of the molecule as a whole. The first four acids, formic through butyric, are completely soluble (miscible) in water (Table 26.1). Beginning with pentanoic acid (valeric acid), the water solubility falls sharply and is only about 0.1 g acid per 100 g water for octanoic acid (caprylic acid). Acids of this series with more than eight carbons are virtually insoluble in water. The water-solubility characteristics of the first four acids are evidently determined by the highly soluble polar carboxyl group. Thereafter, the water insolubility of the nonpolar hydrocarbon chain is dominant.

The polarity due to the carboxyl group is evident in the boiling point data. Formic acid (HCOOH) boils at about 101°C. Carbon dioxide, a nonpolar substance of similar molecular weight, remains in the gaseous state until it is cooled to −78°C. In like manner, the boiling point of acetic acid (molecular weight 60) is 118°C, while nonpolar butane (molecular weight 58) boils at −0.6°C. The comparatively high boiling points for carboxylic acids are due to intermolecular attractions resulting from hydrogen bonding. In fact, molecular weight determinations on gaseous acetic acid (near its boiling point) show a value of about 120, indicating that two molecules are joined together to form a *dimer*, $(CH_3COOH)_2$.

Hydrogen bonding in carboxylic acids

Acetic acid dimer

Saturated monocarboxylic acids that have fewer than ten carbon atoms are liquids at room temperature, whereas those with more than ten carbon atoms are waxlike solids.

Carboxylic acids and phenols, like mineral acids such as HCl, ionize in water to produce hydronium ions and anions. Carboxylic acids are generally weak; that is, they are only slightly ionized in water. Phenols are, in general, even weaker acids than carboxylic acids. For example, the ionization constant for acetic acid is 1.8×10^{-5}, and that for phenol is 1.1×10^{-10}. Equations illustrating these ionizations are given below.

$$HCl + H_2O \rightleftharpoons H_3O^+ + Cl^-$$
Hydrogen chloride   Hydronium ion   Chloride ion

$$CH_3\overset{O}{\overset{\|}{C}}-OH + H_2O \rightleftharpoons H_3O^+ + CH_3\overset{O}{\overset{\|}{C}}-O^-$$
Acetic acid   Hydronium ion   Acetate ion

$$C_6H_5{-}OH + H_2O \rightleftharpoons H_3O^+ + C_6H_5{-}O^-$$
Phenol   Hydronium ion   Phenoxide ion

26.4 Classification of Carboxylic Acids

Thus far, our discussion has dealt mainly with a single type of acid—that is, saturated monocarboxylic acids. But various other kinds of carboxylic acids are known. Some of the more important ones are discussed here.

Unsaturated Acids

An unsaturated acid contains one or more carbon–carbon double bonds. The first member of the homologous series of unsaturated carboxylic acids containing one carbon–carbon double bond is acrylic acid, $CH_2{=}CHCOOH$. The IUPAC name for $CH_2{=}CHCOOH$ is propenoic acid. Derivatives of acrylic acid are used industrially to manufacture a class of synthetic polymers known as the acrylates (see Chapter 27). These polymers are widely used as textiles and as paints and lacquers. Unsaturated carboxylic acids undergo the reactions of both an unsaturated hydrocarbon and a carboxylic acid.

Even one carbon–carbon double bond in the molecule exerts a major influence on the physical and chemical properties of an acid. The effect of a double bond can be seen when comparing the two 18-carbon acids, stearic and oleic. Stearic acid $[CH_3(CH_2)_{16}COOH]$, a solid melting at 70°C, shows only the reactions of a carboxylic acid. On the other hand, oleic acid $[CH_3(CH_2)_7CH{=}CH(CH_2)_7COOH$, mp 16°C], with one double bond, is a liquid at room temperature and shows the reactions of an unsaturated hydrocarbon as well as those of a carboxylic acid.

Aromatic Carboxylic Acids

An aromatic carboxylic acid is one in which the carbon of the carboxyl group (—COOH) is bonded directly to a carbon in an aromatic nucleus. The parent compound of this series is benzoic acid. Other common examples are the three isomeric toluic acids.

Benzoic acid, o-Toluic acid, m-Toluic acid, p-Toluic acid

Dicarboxylic Acids

Acids of both the aliphatic and aromatic series containing two or more carboxyl groups are known. These are called dicarboxylic acids. The simplest member of the aliphatic series is oxalic acid. The next member in the homologous series is malonic acid. Several dicarboxylic acids and their names are listed in Table 26.2.

Table 26.2. Names and formulas of selected dicarboxylic acids.

Common name	IUPAC name	Formula
Oxalic acid	Ethanedioic acid	HOOCCOOH
Malonic acid	Propanedioic acid	HOOCCH$_2$COOH
Succinic acid	Butanedioic acid	HOOC(CH$_2$)$_2$COOH
Glutaric acid	Pentanedioic acid	HOOC(CH$_2$)$_3$COOH
Adipic acid	Hexanedioic acid	HOOC(CH$_2$)$_4$COOH
Fumaric acid	trans-2-Butenedioic acid	HOOCCH=CHCOOH
Maleic acid	cis-2-Butenedioic acid	HOOCCH=CHCOOH

The IUPAC names for dicarboxylic acids are formed by modifying the corresponding hydrocarbon names to end in -*dioic acid*. Thus, the two-carbon acid is ethanedioic acid. However, the common names for dicarboxylic acids are more frequently used.

Oxalic acid is found in various plants including spinach, cabbage, and rhubarb. Among its many uses are bleaching straw and leather, and removing rust and ink stains. Although oxalic acid is poisonous, the amounts present in the above-mentioned vegetables are generally not harmful.

Malonic acid is made synthetically but was originally prepared from malic acid, which is commonly found in apples and many fruit juices. Malonic acid is one of the major compounds used in the manufacture of the class of drugs known as barbiturates. When heated above their melting points, malonic acid and substituted malonic acids lose carbon dioxide to give monocarboxylic acids. Thus, malonic acid yields acetic acid when strongly heated:

$$\begin{array}{c} \text{COOH} \\ | \\ \text{CH}_2 \\ | \\ \text{COOH} \end{array} \xrightarrow{150°C} \text{CH}_3\text{COOH} + \text{CO}_2\uparrow$$

Malonic acid → Acetic acid

580   Chapter 26

Succinic acid has been known since the 16th century, when it was obtained as a distillation product of amber. Succinic, fumaric, and citric acids are among the important acids in the energy-producing metabolic pathway known as the citric acid cycle (see Chapter 32). Citric acid is a *tricarboxylic acid* and has this formula.

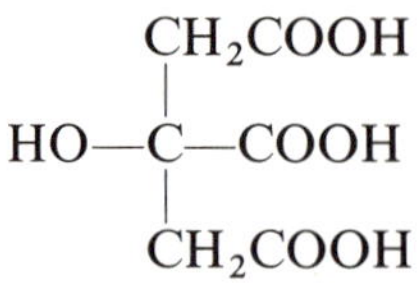

$$\begin{array}{c} CH_2COOH \\ | \\ HO-C-COOH \\ | \\ CH_2COOH \end{array}$$

Citric acid

When succinic acid is heated, it loses water, forming succinic anhydride, an acid anhydride. Glutaric acid behaves similarly.

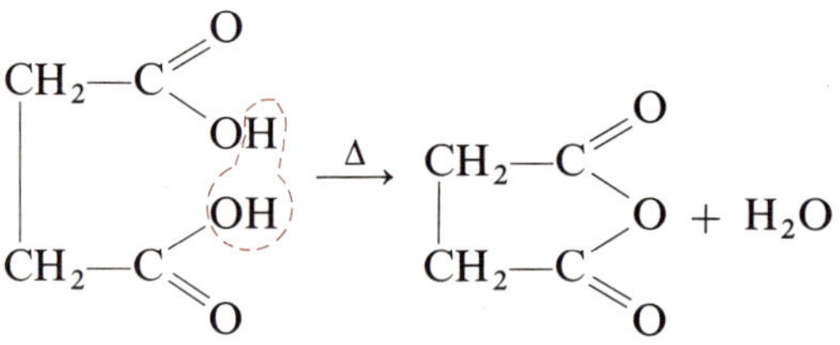

Succinic acid → Succinic anhydride + H$_2$O

Adipic acid is the most important commercial dicarboxylic acid. It is made from benzene by first converting it to cyclohexene followed by oxidation to adipic acid. About 1.7 billion lb of adipic acid are produced annually in the United States. Most of the adipic acid is used to produce nylon (Chapter 27). It is also used in polyurethane foams, plasticizers, and lubricating oil additives.

Aromatic dicarboxylic acids contain two carboxyl groups attached directly to an aromatic nucleus. Examples are the three isomeric phthalic acids, $C_6H_4(COOH)_2$.

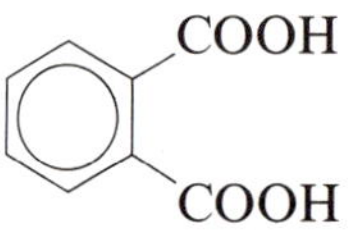

o-Phthalic acid
(Phthalic acid)

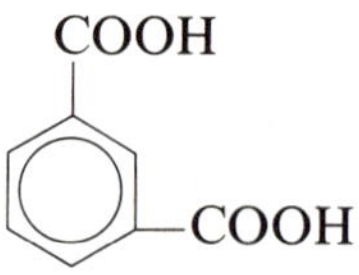

m-Phthalic acid
(Isophthalic acid)

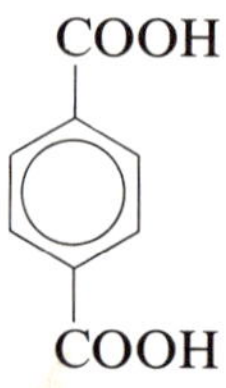

p-Phthalic acid
(Terephthalic acid)

Dicarboxylic acids are *bifunctional*; that is, they have two different sites where reactions can occur. Therefore, they are often used as monomers in the preparation of synthetic polymers such as Dacron polyester (Section 27.7).

Hydroxy Acids

Lactic acid, found in sour milk, sauerkraut, and dill pickles, has the functional groups of both a carboxylic acid and an alcohol. Salicylic acid is both a carboxylic acid and a phenol. It is of special interest since a family of useful drugs—the salicylates—are derivatives of this acid. The salicylates include aspirin and function as *analgesics* (pain relievers) and as *antipyretics* (fever reducers). The structural formulas of several hydroxy acids are shown below.

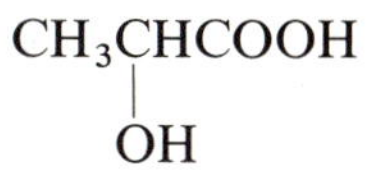

$$CH_3CHCOOH$$
$$|$$
$$OH$$

Lactic acid
(α-Hydroxypropionic acid)

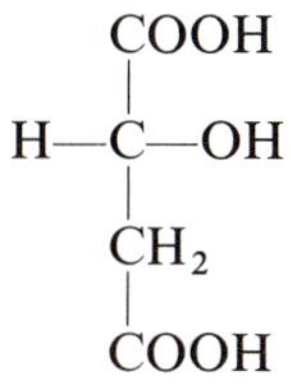

$$COOH$$
$$|$$
$$H-C-OH$$
$$|$$
$$CH_2$$
$$|$$
$$COOH$$

Malic acid
(α-Hydroxysuccinic acid)

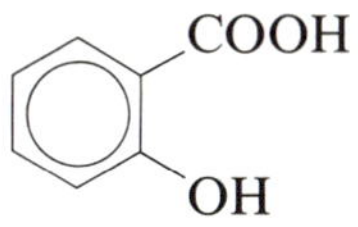

Salicylic acid (o-Hydroxybenzoic acid) — benzene ring with COOH and OH in ortho positions

$$HO-CHCOOH$$
$$|$$
$$HO-CHCOOH$$

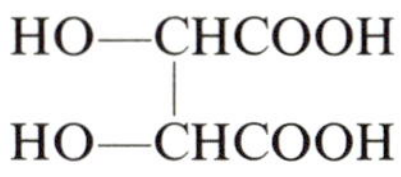

Tartaric acid
(2,3-Dihydroxybutanedioic acid)

Amino Acids

Naturally occurring amino acids have this general formula; the amino group is in the alpha position:

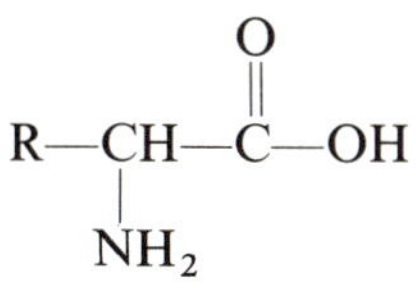

$$R-CH(NH_2)-C(=O)-OH$$

Each amino acid molecule has a carboxyl group that acts as an acid and an amino group that acts as a base. About 20 biologically important amino acids, each with a different group represented by R, have been found in nature. (In amino acids, R does not always represent an alkyl group.) The immensely complicated protein molecules, found in every form of life, are built from amino acids. Some protein molecules contain more than 10,000 amino acid units.

26.5 Preparation of Carboxylic Acids

Many different methods of preparing carboxylic acids are known. We will consider only a few examples.

Oxidation of an aldehyde or a primary alcohol. This is a general method that can be used to convert an aldehyde or primary alcohol to the corresponding carboxylic acid.

$$RCH_2OH + [O] \longrightarrow RCOOH$$

$$R\overset{H}{\underset{}{C}}=O + [O] \longrightarrow RCOOH$$

Butyric acid (butanoic acid) can be obtained by oxidizing either 1-butanol or butanal with sodium dichromate in the presence of sulfuric acid. Aromatic acids may be prepared by the same general method. For example, benzoic acid is obtained by oxidizing benzyl alcohol.

$$CH_3CH_2CH_2CH_2OH + [O] \xrightarrow[H_2SO_4]{Cr_2O_7^{2-}} CH_3CH_2CH_2COOH$$
1-Butanol → Butanoic acid

$$CH_3CH_2CH_2\overset{H}{\underset{}{C}}=O + [O] \xrightarrow[H_2SO_4]{Cr_2O_7^{2-}} CH_3CH_2CH_2COOH$$ 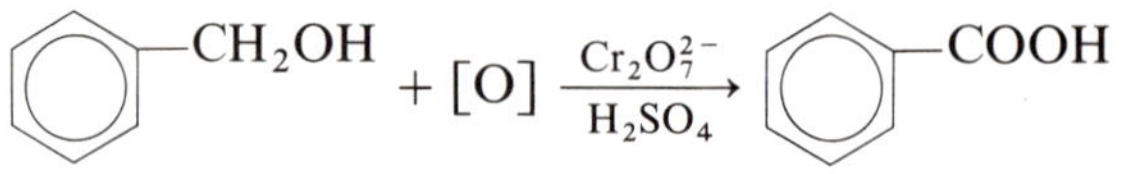
Butanal → Butanoic acid

$$C_6H_5\text{-}CH_2OH + [O] \xrightarrow[H_2SO_4]{Cr_2O_7^{2-}} C_6H_5\text{-}COOH$$
Benzyl alcohol → Benzoic acid

Hydrolysis or saponification of an ester. Esters of carboxylic acids have the general formula RCOOR′ (see Section 26.7).

$$R-\overset{O}{\underset{\|}{C}}-O-R'$$

Ester structure

Both a carboxylic acid, RCOOH, and an alcohol, R′OH, can be obtained from the ester. This preparation of an acid from an ester is practical, of course, only when an ester of the desired acid is available. As a rule, a chemist is more likely to want to make an ester from an acid than to make an acid from an ester. However, large quantities of fatty acids and their salts (soaps) are obtained commercially by hydrolysis or saponification of animal and vegetable fats and oils. These fats and oils are esters of fatty acids and glycerol (Section 26.8).

Hydrolysis of an ester involves reaction with water to form an acid and an alcohol. Hydrolysis is catalyzed by strong acids (H_2SO_4 and HCl) or by certain enzymes.

$$R\overset{O}{\underset{\|}{C}}-OR' + H_2O \xrightarrow[\text{or enzyme}]{\text{Mineral acid}} R\overset{O}{\underset{\|}{C}}-OH + R'OH$$

Ester → Acid + Alcohol

$$CH_3CH_2\overset{\overset{O}{\|}}{C}-O-CH_3 + H_2O \xrightarrow{H^+} CH_3CH_2COOH + CH_3OH$$
Methyl propanoate   Propanoic acid

$$\underset{\text{Methyl salicylate}}{\text{C}_6\text{H}_4(\text{OH})\text{COOCH}_3} + H_2O \xrightarrow{H^+} \underset{\text{Salicylic acid}}{\text{C}_6\text{H}_4(\text{OH})\text{COOH}} + CH_3OH$$

saponification   **Saponification** is the hydrolysis of an ester by a strong base (NaOH or KOH) to produce an alcohol and a salt (or soap).

$$\underset{\text{Ester}}{R\overset{\overset{O}{\|}}{C}-OR'} + NaOH \xrightarrow[\Delta]{H_2O} \underset{\substack{\text{Salt} \\ \text{(or soap)}}}{R\overset{\overset{O}{\|}}{C}-O^- Na^+} + \underset{\text{Alcohol}}{R'OH}$$

$$\underset{\text{Ethyl stearate}}{CH_3(CH_2)_{16}\overset{\overset{O}{\|}}{C}-OCH_2CH_3} + NaOH \xrightarrow[\Delta]{H_2O} \underset{\text{Sodium stearate}}{CH_3(CH_2)_{16}\overset{\overset{O}{\|}}{C}ONa} + CH_3CH_2OH$$

The carboxylic acid may be obtained by reacting the salt with a strong acid such as HCl:

$$RCOONa + HCl \longrightarrow RCOOH + NaCl$$

Oxidation of an alkyl group. When reacted with a strong oxidizing agent (alkaline permanganate solution or potassium dichromate and sulfuric acid), alkyl groups on aromatic hydrocarbons are oxidized to carboxyl groups. Regardless of the size of the alkyl group, the carboxyl group so obtained is directly attached to the aromatic ring. Thus, sodium benzoate is obtained when either methylbenzene (toluene) or ethylbenzene is heated with alkaline permanganate solution:

$$\underset{\text{Toluene}}{C_6H_5-CH_3} + [O] \xrightarrow[\text{NaOH}]{\text{NaMnO}_4} \underset{\text{Sodium benzoate}}{C_6H_5-\overset{\overset{O}{\|}}{C}-O^- Na^+}$$

$$\underset{\text{Ethylbenzene}}{C_6H_5-CH_2CH_3} + [O] \xrightarrow[\text{NaOH}]{\text{NaMnO}_4} \underset{\text{Sodium benzoate}}{C_6H_5-\overset{\overset{O}{\|}}{C}-O^- Na^+} + CO_2$$

Since the reaction was conducted in an alkaline medium, a salt of the carboxylic acid (sodium benzoate) is formed instead of the free acid. To obtain the free carboxylic acid, the reaction mixture is acidified with a strong mineral acid (HCl or H_2SO_4) in a second step.

$$\text{C}_6\text{H}_5\text{—COONa} + \text{H}^+ \longrightarrow \text{C}_6\text{H}_5\text{—COOH} + \text{Na}^+$$

Sodium benzoate → Benzoic acid

Hydrolysis of nitriles. Nitriles, RCN, which can be prepared by adding HCN to aldehydes and ketones (Section 25.4) or by reacting alkyl halides with KCN, can be hydrolyzed to carboxylic acids.

$$RCN + 2H_2O \xrightarrow{H^+} RCOOH + NH_4^+$$

$$CH_3CN + 2H_2O \xrightarrow{H^+} CH_3COOH + NH_4^+$$

26.6 Reactions of Carboxylic Acids

Acid–base reactions. Because of their ability to form hydrogen ions in solution, acids in general have the following properties:

1. Sour taste
2. Change blue litmus to red and affect other suitable indicators
3. Form water solutions with pH values of less than 7
4. Undergo neutralization reactions with bases to form water and a salt

All of the foregoing general properties of an acid are readily seen in low-molecular-weight carboxylic acids such as acetic acid. However, these general acid properties can be greatly influenced by the size of the hydrocarbon radical attached to the carboxyl group. In stearic acid, for example, taste, effect on indicators, and pH are not detectable because the large size of the hydrocarbon radical makes the acid insoluble in water. But it is an acid and it will react with a base to form water and a salt. With sodium hydroxide, the equation for the reaction is

$$C_{17}H_{35}COOH + NaOH \longrightarrow C_{17}H_{35}COONa + H_2O$$

Stearic acid → Sodium stearate

Ester formation. Carboxylic acids react with alcohols to form esters:

$$\underset{\text{Carboxylic acid}}{RC(=O)\text{—OH}} + \underset{\text{Alcohol}}{R'O\text{—H}} \xrightleftharpoons{H^+} \underset{\substack{\text{Ester} \\ (R \text{ can be } H, \\ \text{but } R' \text{ cannot be } H)}}{RC(=O)\text{—OR}'} + H_2O$$

$$\underset{\text{Formic acid}}{H\text{—C(=O)—OH}} + CH_3CH_2OH \xrightleftharpoons{H^+} \underset{\text{Ethyl formate}}{H\text{—C(=O)—OCH}_2CH_3} + H_2O$$

Many biologically important substances are esters (see Section 26.7).

Amide formation. Carboxylic acids react with ammonia to form ammonium salts.

$$RC\overset{O}{\overset{\|}{-}}OH + NH_3 \longrightarrow RC\overset{O}{\overset{\|}{-}}O^-NH_4^+$$

Carboxylic acid　　Ammonia　　　Ammonium salt

$$CH_3C\overset{O}{\overset{\|}{-}}OH + NH_3 \longrightarrow CH_3C\overset{O}{\overset{\|}{-}}O^-NH_4^+$$

Ammonium acetate

Ammonium salts of carboxylic acids are ionic substances. Ammonium acetate, for example, is ionized and exists as ammonium ions and acetate ions in both the crystalline form and when dissolved in water.

When heated, ammonium salts of carboxylic acids are converted to *amides:*

$$RC\overset{O}{\overset{\|}{-}}ONH_4 \xrightarrow{\Delta} RC\overset{O}{\overset{\|}{-}}NH_2 + H_2O$$

Ammonium salt　　　　Amide

$$CH_3C\overset{O}{\overset{\|}{-}}ONH_4 \xrightarrow{\Delta} CH_3C\overset{O}{\overset{\|}{-}}NH_2 + H_2O$$

Ammonium acetate　　　Acetamide
　　　　　　　　　　　(Ethanamide)

Amides are molecular substances and exist as molecules (not ions) in both the crystalline form and when dissolved in water. An amide contains the following characteristic structure:

$$R-\boxed{C\overset{O}{\overset{\|}{-}}N\,H_2}$$

Amide structure

In amides, the carbon atom of a carbonyl group is bonded directly to a nitrogen atom of an —NH$_2$, —NHR, or —NR$_2$ group. The amide structure occurs in numerous substances, including proteins and some synthetic polymers such as nylon.

Two systems are in use for naming amides. The common names are derived from the corresponding common acid names by dropping the *-ic* ending and adding *-amide*. For example, the one-, two-, and three-carbon amides are called formamide, acetamide, and propionamide, respectively. The IUPAC name is based on the longest continuous carbon chain that includes the amide structure. The name is formed by dropping the *-e* ending of the corresponding hydrocarbon name and adding *-amide*. Thus, the first three members of the series are methanamide, ethanamide, and propanamide (Table 26.3).

586 Chapter 26

Table 26.3. Formulas and names of selected amides.

Formula	Common name	IUPAC name
$\text{HC}(=\!\!O)-\text{NH}_2$	Formamide	Methanamide
$\text{CH}_3\text{C}(=\!\!O)-\text{NH}_2$	Acetamide	Ethanamide
$\text{CH}_3\text{CH}_2\text{C}(=\!\!O)-\text{NH}_2$	Propionamide	Propanamide
$\text{CH}_3\text{CH}(\text{CH}_3)-\text{C}(=\!\!O)-\text{NH}_2$	Isobutyramide	2-Methylpropanamide

Acid chloride formation. Thionyl chloride (SOCl$_2$) reacts with carboxylic acids to form acid chlorides, which are very reactive and can be used to synthesize other substances, such as amides and esters.

$$\underset{\text{Acid}}{\text{RC}(=\!\!O)-\text{OH}} + \underset{\substack{\text{Thionyl}\\\text{chloride}}}{\text{SOCl}_2} \longrightarrow \underset{\text{Acid chloride}}{\text{RC}(=\!\!O)-\text{Cl}} + \text{SO}_2 + \text{HCl}$$

$$\text{CH}_3\text{C}(=\!\!O)-\text{OH} + \text{SOCl}_2 \longrightarrow \underset{\text{Acetyl chloride}}{\text{CH}_3\text{C}(=\!\!O)-\text{Cl}} + \text{SO}_2 + \text{HCl}$$

Acid chlorides are extremely reactive substances. They must be kept away from moisture, or they will hydrolyze back to the acid:

$$\text{RC}(=\!\!O)-\text{Cl} + \text{H}_2\text{O} \longrightarrow \text{RC}(=\!\!O)-\text{OH} + \text{HCl}$$

Acid chlorides are more reactive than acids and can be used in place of acids to prepare esters and amides.

$$\text{CH}_3\text{C}(=\!\!O)-\text{Cl} + \text{CH}_3\text{OH} \longrightarrow \underset{\text{Methyl acetate}}{\text{CH}_3\text{C}(=\!\!O)-\text{OCH}_3} + \text{HCl}$$

$$\text{CH}_3\text{C}(=\!\!O)-\text{Cl} + 2\,\text{NH}_3 \longrightarrow \underset{\text{Acetamide}}{\text{CH}_3\text{C}(=\!\!O)-\text{NH}_2} + \text{NH}_4\text{Cl}$$

26.7 Esters

Carboxylic acids react with alcohols to form esters in an equilibrium reaction represented by this equation:

$$\underset{\text{Carboxylic acid}}{RC(=O)-OH} + \underset{\text{Alcohol}}{R'O-H} \underset{}{\overset{H^+}{\rightleftarrows}} \underset{\text{Ester}}{RC(=O)-OR'} + H_2O$$

At first glance, this looks like the familiar acid–base neutralization reaction. But this is not the case, since the alcohol does not yield OH^- ions and the ester, unlike a salt, is a molecular, not an ionic substance. The forward reaction of an acid and an alcohol is called *esterification;* the reverse reaction of an ester with water is called hydrolysis. The work of a chemist may call for manipulating reaction conditions to favor the formation of either esters or their component parts, alcohols and acids. We will not go into the details of ester preparation at this point.

Naming Esters

Esters are alcohol derivatives of carboxylic acids. They are named in much the same way as salts. The alcohol part is named first, followed by the name of the acid modified to end in *-ate*. The *-ic* ending of the organic acid name is replaced by the ending *-ate*. Thus, in the IUPAC System, ethan*oic acid* becomes ethan*oate*. In the common names, acet*ic acid* becomes acet*ate*. To name an ester it is necessary to recognize the portion of the ester molecule that comes from the acid and the portion that comes from the alcohol. In the general formula for an ester, the RC=O comes from the acid and the R'O comes from the alcohol:

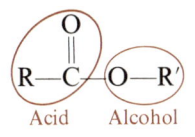

$$\underset{\text{Acid} \quad \text{Alcohol}}{R-C(=O)-O-R'}$$

The R' in R'O is named first, followed by the name of the acid modified by replacing *-ic acid* with *-ate*. The ester derived from ethyl alcohol and acetic acid is called ethyl acetate or ethyl ethanoate. Consider the ester formed from CH_3CH_2COOH and CH_3OH:

$$\underset{\substack{\text{Propanoic acid}\\\text{(Propionic acid)}}}{CH_3CH_2C(=O)-OH} + \underset{\substack{\text{Methanol}\\\text{(Methyl alcohol)}}}{H-OCH_3} \overset{H^+}{\rightleftarrows} \underset{\substack{\text{Methyl propanoate}\\\text{(Methyl propionate)}}}{CH_3CH_2C(=O)-OCH_3} + H_2O$$

Esters of aromatic acids are named in the same general way as those of aliphatic acids. For example, the ester of benzoic acid and isopropyl alcohol is

$$\text{C}_6\text{H}_5-\overset{\overset{\displaystyle O}{\|}}{\text{C}}-\text{O}-\overset{\overset{\displaystyle CH_3}{|}}{\underset{\underset{\displaystyle CH_3}{|}}{\text{CH}}}$$

Isopropyl benzoate

Formulas and names for additional esters are given in Table 26.4.

Table 26.4. Odors and flavors of selected esters.

Formula	IUPAC name	Common name	Odor or flavor
$\text{CH}_3\overset{\overset{\displaystyle O}{\|}}{\text{C}}-\text{OCH}_2\text{CH}_2\overset{\overset{\displaystyle CH_3}{|}}{\text{CH}}\text{CH}_3$	Isopentyl ethanoate	Isoamyl acetate	Banana, pear
$\text{CH}_3\text{CH}_2\text{CH}_2\overset{\overset{\displaystyle O}{\|}}{\text{C}}-\text{OCH}_2\text{CH}_3$	Ethyl butanoate	Ethyl butyrate	Pineapple
$\text{H}\overset{\overset{\displaystyle O}{\|}}{\text{C}}-\text{OCH}_2\overset{\overset{\displaystyle CH_3}{|}}{\text{CH}}\text{CH}_3$	Isobutyl methanoate	Isobutyl formate	Raspberry
$\text{CH}_3\overset{\overset{\displaystyle O}{\|}}{\text{C}}-\text{OCH}_2(\text{CH}_2)_6\text{CH}_3$	Octyl ethanoate	n-Octyl acetate	Orange
benzene ring with $-\overset{\overset{\displaystyle O}{\|}}{\text{C}}-\text{OCH}_3$ and $-\text{OH}$		Methyl salicylate	Wintergreen

Occurrence and Properties of Esters

Since many acids and many alcohols are known, the number of esters theoretically possible is very large. In fact, esters, both natural and man-made, do exist in almost endless variety. Simple esters derived from monocarboxylic acids and monohydroxy alcohols are colorless, generally nonpolar liquids or solids. The low polarity of ester molecules is substantiated by the fact that both their water solubility and boiling points are lower than those of either acids or alcohols of similar molecular weights.

Low- and intermediate-molecular-weight esters (from both acids and alcohols up to about ten carbons) are liquids with characteristic, usually fragrant or fruity odors. The distinctive odor and flavor of many fruits is caused by one or more of these esters. The difference in properties between an acid and its esters is remarkable. For example, in contrast to the extremely unpleasant odor of butyric acid, ethyl butyrate has the pleasant odor of pineapple and methyl butyrate the odor of artificial rum. Esters are used in flavoring and scenting

Carboxylic Acids, Esters, and Amines   589

agents (see Table 26.4). They are generally good solvents for organic substances, and those having relatively low molecular weights are volatile. Therefore, esters such as ethyl acetate, butyl acetate, and isoamyl acetate are extensively used in paints, varnishes, and lacquers.

High-molecular-weight esters (formed from acids and alcohols of 16 or more carbons) are waxes and are obtained from various plants. They are used in furniture wax and automobile wax preparations; for example, carnauba wax contains esters of 24- and 28-carbon fatty acids and 32- and 34-carbon alcohols. Polyesters with very high molecular weights, such as Dacron, are widely used in the textile industries (see Chapter 27).

Problem 26.2   Name the following esters:

(a) $H-\overset{\overset{O}{\|}}{C}-OCH_2CH_2CH_3$ (b) $\text{C}_6\text{H}_5-\overset{\overset{O}{\|}}{C}-OCH_2CH_3$ (c) $\begin{array}{c} O=C-OCH_2CH_3 \\ | \\ CH_2 \\ | \\ O=C-OCH_2CH_3 \end{array}$

(a) First identify the acid and alcohol components. The acid contains one carbon and is formic acid. The alcohol is propyl alcohol.

$\underbrace{H-\overset{\overset{O}{\|}}{C}}_{\text{Formic acid}}-\underbrace{O-CH_2CH_2CH_3}_{\text{Propyl alcohol}}$

Change the *-ic* ending of the acid to *-ate*, making the name formate or methanoate. The name of the ester then is propyl formate or propyl methanoate.
(b) The acid is benzoic acid; the alcohol is ethyl alcohol. Using the same procedure as in part (a), the name of the ester is ethyl benzoate.
(c) The acid is the three-carbon dicarboxylic acid malonic acid. The alcohol is ethyl alcohol. Both acid groups are in the ester form. The name, therefore, is diethyl malonate.

26.8   Glycerol Esters

fat

oil

triglyceride

Fats and **oils** are esters of glycerol and predominantly long-chain fatty acids. Fats and oils are also called **triglycerides**, since each molecule is derived from one molecule of glycerol and three molecules of fatty acid.

$$\text{Glycerol portion} \rightarrow \begin{array}{c} CH_2-O-\overset{\overset{O}{\|}}{C}-R \\ | \\ CH\ -O-\overset{\overset{O}{\|}}{C}-R' \\ | \\ CH_2-O-\overset{\overset{O}{\|}}{C}-R'' \end{array} \qquad \begin{array}{c} CH_2-O-\overset{\overset{O}{\|}}{C}-C_{17}H_{35} \\ | \\ CH\ -O-\overset{\overset{O}{\|}}{C}-C_{15}H_{31} \\ | \\ CH_2-O-\overset{\overset{O}{\|}}{C}-C_{11}H_{23} \end{array}$$

General formula for a triglyceride   Typical triglyceride containing three different fatty acids

The structural formulas of triglyceride molecules vary because:

1. The length of the fatty acid chain may vary from 4 to 20 carbons, but the number of carbon atoms in the chain is nearly always even.
2. Each fatty acid may be saturated or may be unsaturated and contain one, two, or three carbon–carbon double bonds.
3. An individual triglyceride may, and frequently does, contain three different fatty acids.

The most abundant saturated fatty acids in fats and oils are lauric, myristic, palmitic, and stearic acids (see Table 26.1). The most abundant unsaturated acids in fats and oils contain 18 carbon atoms and have one, two, or three carbon–carbon double bonds. Their formulas are:

$$CH_3(CH_2)_7CH=CH(CH_2)_7COOH$$
Oleic acid

$$CH_3(CH_2)_4CH=CHCH_2CH=CH(CH_2)_7COOH$$
Linoleic acid

$$CH_3CH_2CH=CHCH_2CH=CHCH_2CH=CH(CH_2)_7COOH$$
Linolenic acid

The major physical difference between fats and oils is that fats are solid and oils are liquid at room temperature. Since the glycerol part of the structure is the same for a fat and an oil, the difference must be due to the fatty acid end of the molecule. Fats contain a higher proportion of saturated fatty acids, whereas oils contain higher amounts of unsaturated fatty acids. The term *polyunsaturated* has been popularized in recent years; this means that the molecules of a particular product each contain several double bonds.

Table 26.5. Fatty acid composition of selected fats and oils.

Fat or oil	Myristic acid	Palmitic acid	Stearic acid	Oleic acid	Linoleic acid
Animal fat					
Butter[a]	7–10	23–26	10–13	30–40	4–5
Lard	1–2	28–30	12–18	41–48	6–7
Tallow	3–6	24–32	14–32	35–48	2–4
Vegetable oil					
Olive	0–1	5–15	1–4	49–84	4–12
Peanut	—	6–9	2–6	50–70	13–26
Corn	0–2	7–11	3–4	43–49	34–42
Cottonseed	0–2	19–24	1–2	23–33	40–48
Soybean	0–2	6–10	2–4	21–29	50–59
Linseed[b]	—	4–7	2–5	9–38	3–43

[a] Butyric acid, 3–4%.
[b] Linolenic acid, 25–58%.

Fats and oils are obtained from natural sources. In general, fats come from animal sources and oils from vegetable sources. Thus, lard is obtained from hogs and tallow from cattle and sheep. Olive, cottonseed, corn, soybean, linseed, and other oils are obtained from the fruit or seed of their respective vegetable sources. Table 26.5 shows the major constituents of several fats and oils.

Hydrogenation of glycerides. Addition of hydrogen is a characteristic reaction of the carbon–carbon pi bonds in ethylenic linkages. Industrially, low-cost vegetable oils are partially hydrogenated to obtain solid fats that are useful in baking as shortening or in making oleomargarine. In this process, hydrogen gas is bubbled through hot oil containing a finely dispersed nickel catalyst. The hydrogen adds to the carbon–carbon double bonds of the oil to saturate the double bonds and form fats:

$$H_2 + -CH=CH- \xrightarrow{Ni} -CH_2-CH_2-$$
$$\text{In oil or fat}$$

In actual practice, only some of the double bonds are allowed to become saturated. The degree of hydrogenation can be controlled to obtain fats of any desired degree of saturation. The products resulting from the partial hydrogenation of oils are marketed as solid shortening (Crisco, Spry, etc.) and are used for cooking and baking. Oils and fats are also partially hydrogenated to improve their keeping qualities. Rancidity in fats and oils results from air oxidation at points of unsaturation, producing low-molecular-weight aldehydes and acids of disagreeable odor and flavor.

Hydrogenolysis. Triglycerides can be split and reduced in a reaction called hydrogenolysis (splitting by hydrogen). Hydrogenolysis requires higher temperatures and pressures and a different catalyst (copper chromite) than for hydrogenation of double bonds. Each triglyceride molecule yields a molecule of glycerol and three primary alcohol molecules. The hydrogenolysis of glyceryl trilaurate is represented as follows:

$$\begin{array}{l} CH_2-O-\overset{\overset{O}{\|}}{C}-C_{11}H_{23} \\ | \\ CH-O-\overset{\overset{O}{\|}}{C}-C_{11}H_{23} \\ | \\ CH_2-O-\overset{\overset{O}{\|}}{C}-C_{11}H_{23} \end{array} + 6\,H_2 \xrightarrow[\text{chromite}]{\text{Copper}} 3\,CH_3(CH_2)_{10}CH_2OH + \begin{array}{l} CH_2OH \\ | \\ CHOH \\ | \\ CH_2OH \end{array}$$

Glyceryl trilaurate $\qquad\qquad$ Lauryl alcohol (1-Dodecanol) $\qquad$ Glycerol

Long-chain primary alcohols obtained by this reaction are used to manufacture other products, especially synthetic detergents, or syndets.

Hydrolysis. Triglycerides can be hydrolyzed to give three molecules of fatty acid and glycerol. The hydrolysis is catalyzed by digestive enzymes at room temperatures and by mineral acids at high temperatures.

$$\begin{array}{c} CH_2-O-\overset{O}{\underset{\|}{C}}-R \\ | \\ CH-O-\overset{O}{\underset{\|}{C}}-R' \\ | \\ CH_2-O-\overset{O}{\underset{\|}{C}}-R'' \end{array} + 3\,H_2O \xrightarrow[\text{enzymes}]{H^+ \text{ or}} \begin{array}{c} RCOOH \\ R'COOH \\ R''COOH \\ \text{Fatty acids} \end{array} + \begin{array}{c} CH_2OH \\ | \\ CHOH \\ | \\ CH_2OH \\ \text{Glycerol} \end{array}$$

Triglyceride

The enzyme-catalyzed reaction occurs in digestive reactions and in biological degradation (or decay) processes. The acid-catalyzed reaction is employed in the commercial preparation of fatty acids and glycerol.

Saponification. Saponification of a fat or oil involves the alkaline hydrolysis of a triester. Salts of fatty acids (soaps) and glycerol are formed. As a specific example, glyceryl tripalmitate reacts with sodium hydroxide to produce sodium palmitate and glycerol:

$$\begin{array}{c} CH_2-O-\overset{O}{\underset{\|}{C}}-C_{15}H_{31} \\ | \\ CH-O-\overset{O}{\underset{\|}{C}}-C_{15}H_{31} \\ | \\ CH_2-O-\overset{O}{\underset{\|}{C}}-C_{15}H_{31} \end{array} + 3\,NaOH \xrightarrow{\Delta} 3\,C_{15}H_{31}COONa + \begin{array}{c} CH_2OH \\ | \\ CHOH \\ | \\ CH_2OH \end{array}$$

Glyceryl tripalmitate   Sodium palmitate (a soap)   Glycerol

Glycerol, fatty acids, and soaps are valuable articles of commerce, and the processing of fats and oils to obtain these products is a major industry.

26.9 Soaps and Synthetic Detergents

In the broadest sense possible, a detergent is simply a cleaning agent. Soap has been used as a cleaning agent for at least 2000 years and thereby is classified as a detergent under this definition. Beginning about 1930, a number of new cleaning agents that were superior in many respects to ordinary soap began to appear on the market. Because they were both synthetic organic products and

Carboxylic Acids, Esters, and Amines   593

synthetic detergent
detergents, they were called **synthetic detergents**, or syndets. A soap is distinguished from a synthetic detergent on the basis of chemical composition and not on the basis of function or usage.

Soaps

soap
In former times, soap-making was a crude operation. Surplus fats were boiled with wood ashes or with some other alkaline material. Today, soap is generally made in large manufacturing plants under controlled conditions. Salts of long-chain fatty acids are called **soaps**. However, only the sodium and potassium salts of carboxylic acids containing 12 to 18 carbon atoms are of great value as detergents.

$$\text{Fat or Oil} + \text{NaOH} \longrightarrow \text{Soap} + \text{Glycerol}$$

In order to understand how a soap works as a cleaning agent, let us consider sodium palmitate $[CH_3(CH_2)_{14}COONa]$ as an example of a typical soap. In water, this substance exists as sodium ions (Na^+) and palmitate ions $[CH_3(CH_2)_{14}COO^-]$. The sodium ion is an ordinary hydrated metal ion. The cleansing property, then, must be centered in the palmitate ion. This ion is made up of a polar, negatively charged carboxylate group attached to a relatively large nonpolar hydrocarbon group. The hydrocarbon end of the ion is soluble in oils and greases, and the negative carboxylate end is soluble in water.

The cleansing action of a soap is explained in this fashion. When the soap comes in contact with grease on a soiled surface, the hydrocarbon end of the soap dissolves in the grease, leaving the negatively charged carboxylate end exposed on the grease surface. Since the negatively charged carboxylate groups are strongly attracted by water, small droplets are formed and the grease is literally lifted or floated away from the soiled object (see Figure 26.1).

The grease–soap emulsion is stable because the oil droplets repel each other due to the negatively charged carboxyl groups on their surfaces. Some insoluble particulate matter is carried away with the grease; the remainder is wetted and mechanically washed away in the water. Synthetic detergents function in a similar way.

Ordinary soap is a good cleaning agent in soft water, but it is not satisfactory in hard water because insoluble calcium and magnesium salts are formed. Palmitate ions, for example, are precipitated by calcium ions:

$$Ca^{2+}(aq) + 2\,CH_3(CH_2)_{14}COO^-(aq) \longrightarrow [CH_3(CH_2)_{14}COO]_2Ca\downarrow$$
Palmitate ion   Calcium palmitate

These precipitates are sticky substances and are responsible for "bathtub ring" and the sticky feel of hair after being shampooed with soap in hard water.

Soaps are ineffective in acidic solutions because water-insoluble molecular fatty acids are formed.

$$CH_3(CH_2)_{14}COO^- + H^+ \longrightarrow CH_3(CH_2)_{14}COOH$$
Palmitic acid molecule

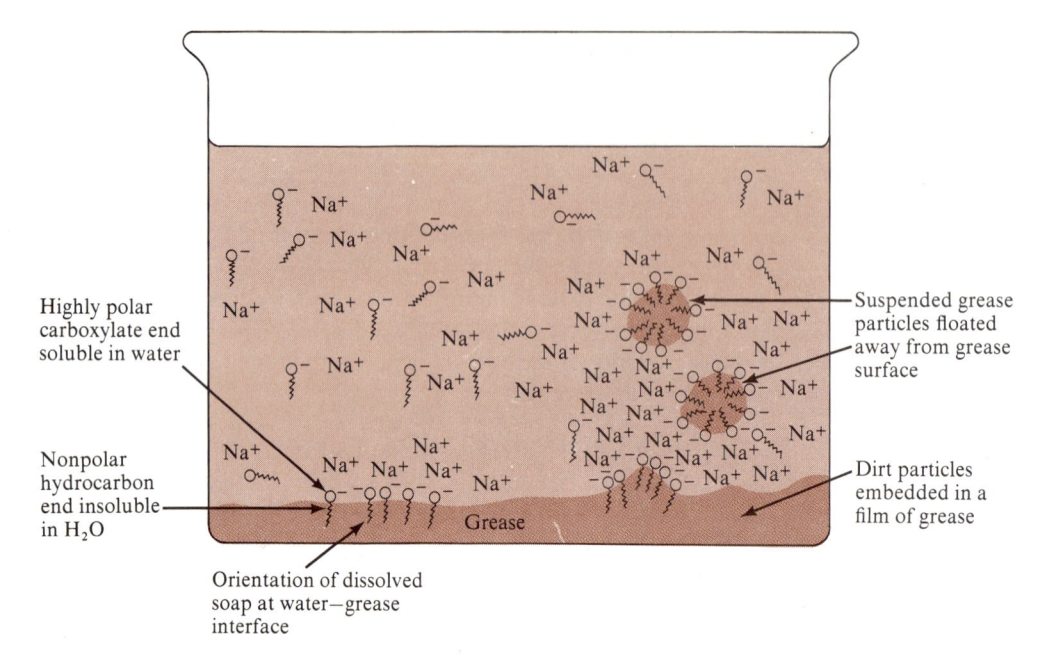

Figure 26.1. Cleansing action of soap. Dirt particles are embedded in a surface film of grease. The hydrocarbon ends of negative soap ions dissolve in the grease film, leaving exposed carboxylate groups. These carboxylate groups are attracted to water, and small droplets of grease bearing dirt are formed and floated away from the surface.

Synthetic Detergents

Once it was recognized that the insoluble hydrocarbon radical joined to a highly polar group was the key to the detergent action of soaps, chemists set out to make new substances that would have similar properties. About 1930, synthetic detergents (syndets) began to replace soaps, and at present about four pounds of syndets are sold for each pound of soap.

Although literally hundreds of substances having detergent properties are known, an idea of their general nature can be had from consideration of sodium lauryl sulfate and sodium dodecylbenzene sulfonate. Sodium lauryl sulfate is made by the following sequence of reactions, starting with lauryl alcohol (1-dodecanol) obtained from glycerides or from petroleum:

$$CH_3(CH_2)_{10}CH_2OH \xrightarrow{H_2SO_4} CH_3(CH_2)_{10}CH_2OSO_3H$$

Lauryl alcohol → Lauryl hydrogen sulfate

$$\xrightarrow{NaOH} CH_3(CH_2)_{10}CH_2OSO_3^- \; Na^+$$

Sodium lauryl sulfate

Sodium dodecylbenzene sulfonate is prepared from dodecylbenzene, which is synthesized from petroleum:

$$\underset{\underset{\text{Dodecylbenzene}}{CH_2(CH_2)_{10}CH_3}}{\bigcirc} \xrightarrow{H_2SO_4} \underset{\underset{p\text{-Dodecylbenzene sulfonic acid}}{CH_2(CH_2)_{10}CH_3}}{\overset{SO_3H}{\bigcirc}} \xrightarrow{NaOH} \underset{\underset{\text{Sodium } p\text{-dodecylbenzene sulfonate}}{CH_2(CH_2)_{10}CH_3}}{\overset{SO_3^- \ Na^+}{\bigcirc}}$$

Both sodium lauryl sulfate and sodium *p*-dodecylbenzene sulfonate act in water in much the same way as sodium palmitate. The negative ion of each substance is the detergent. For example, like the palmitate ion, the negative lauryl sulfate ion has a long hydrocarbon chain that is soluble in grease and a sulfate group that is attracted to water. The one great advantage that these synthetic detergents have over soap is that their calcium and magnesium salts, as well as their sodium salts, are soluble in water. Therefore, they are nearly as effective in hard water as in soft water.

The foregoing are anionic detergents by virtue of the fact that the detergent activity is located in negative ions. Other detergents, both cationic and nonionic, have been developed for special purposes. The detergent activity of a cationic detergent is located in a cation that has a long hydrocarbon chain and a positive charge.

Nonionic detergents are molecular substances. The molecule of a nonionic detergent contains a grease-soluble component and a water-soluble component. Some of these substances are especially useful in automatic washing machines because they have good detergent but low sudsing properties. The structure of a representative nonionic detergent is

$$\underbrace{CH_3(CH_2)_{10}CH_2}_{\text{Grease-soluble}}-O-\underbrace{(CH_2CH_2O)_7-CH_2CH_2OH}_{\text{Water-soluble}}$$

Biodegradability

biodegradable

Organic substances that are readily decomposed by microorganisms in the environment are said to be **biodegradable**. All naturally occurring organic substances must eventually be converted to simple inorganic molecules and ions, such as CO_2, H_2O, N_2, Cl^-, and SO_4^{2-}. Most of these conversions are catalyzed by enzymes produced by microorganisms. These enzymes are capable of attacking only certain specific molecular configurations that are found in substances occurring in nature.

A few years ago, a serious environmental pollution problem arose in connection with synthetic detergents. Some syndets, especially those containing branched-chain hydrocarbons, have no counterparts in nature. Therefore, enzymes capable of degrading them do not exist, and the detergents were nonbiodegradable and broke down very, very slowly. These syndets accumulated in water supplies, where they produced excessive foaming and other undesirable effects.

Happily, this particular pollution problem has been eliminated or at least greatly diminished. Detergent manufacturers, acting on the recommendations of chemists and biologists, changed from a branched-chain alkyl benzene to a straight-chain alkyl benzene raw material. Detergents containing the straight-chain alkyl groups are biodegradable.

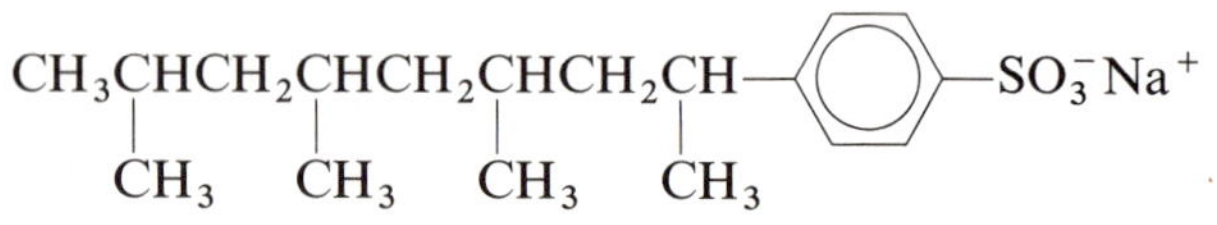
CH3CHCH2CHCH2CHCH2CH
SO3- Na+
CH3 CH3 CH3 CH3

A nonbiodegradable detergent

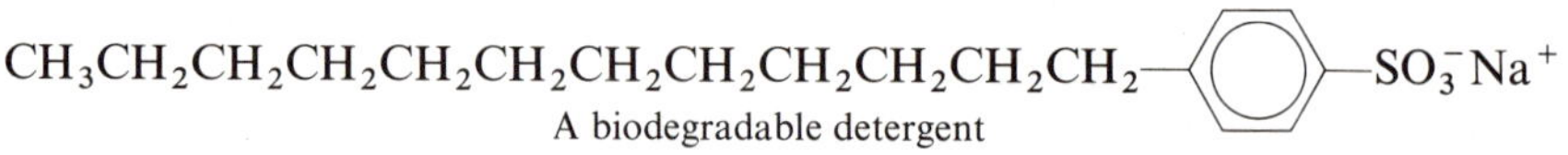
CH3CH2CH2CH2CH2CH2CH2CH2CH2CH2CH2CH2
SO3- Na+

A biodegradable detergent

26.10 Amines

amine

An **amine** is a substituted ammonia molecule and has the general formula RNH_2, R_2NH, or R_3N, where R is an alkyl or an aryl group. Amines are classified as primary (1°), secondary (2°), or tertiary (3°), depending upon the number of hydrocarbon radicals attached to the nitrogen atom. Examples are given below.

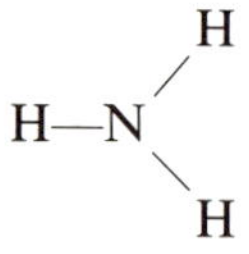
H
H—N
H

Ammonia

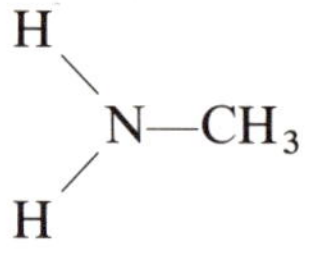
H
N—CH3
H

Methylamine
(1° amine)

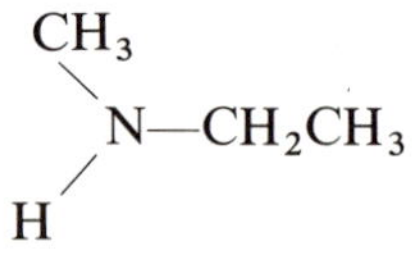
CH3
N—CH2CH3
H

Methylethylamine
(2° amine)

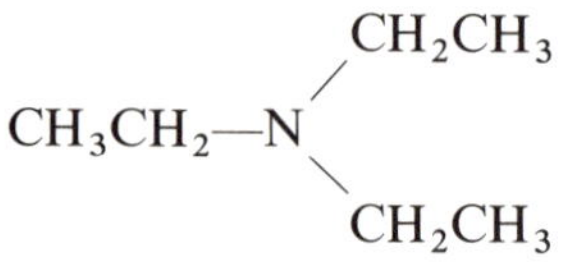
CH2CH3
CH3CH2—N
CH2CH3

Triethylamine
(3° amine)

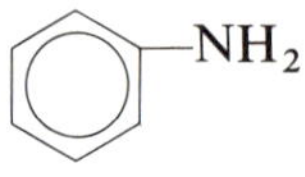
NH2

Aniline
(1° amine)

Naming Amines

Common names for aliphatic amines are formed by naming the alkyl group or groups attached to the nitrogen atom followed by the ending *-amine*.

Thus, CH_3NH_2 is methylamine, $(CH_3)_2NH$ is dimethylamine, and $(CH_3)_3N$ is trimethylamine. A few more examples follow:

$CH_3CH_2NH_2$ $(CH_3)_3N$ [cyclohexyl]—NH_2

Ethylamine   Trimethylamine   Cyclohexylamine ($C_6H_{11}NH_2$)

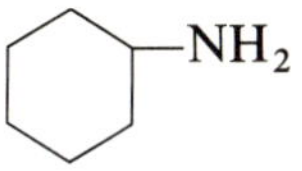

The most important aromatic amine is aniline ($C_6H_5NH_2$). Derivatives are named as substituted anilines. To identify a substituted aniline where the substituent group is attached to the nitrogen atom, an *N*- is placed before the group name. For example, this compound is called *N*-methylaniline.

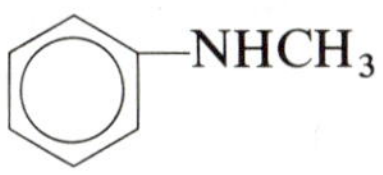

[phenyl]—$NHCH_3$

N-Methylaniline

The monomethyl ring-substituted anilines are known as toluidines: Study the names for the following substituted anilines.

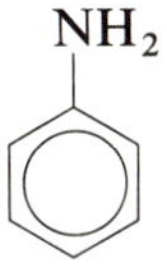

Aniline    *p*-Toluidine (*p*-Methylaniline)   *N*-Ethylaniline

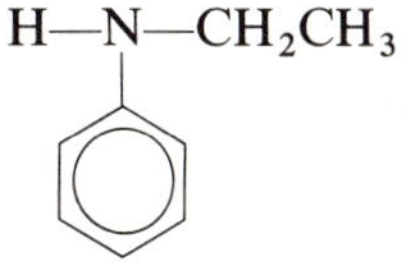

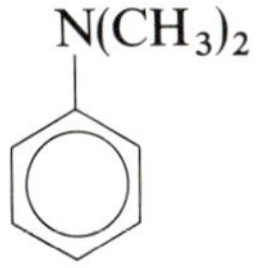

N,N-Dimethylaniline 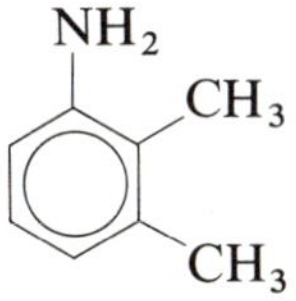   2,3-Dimethylaniline

Physiologically, aniline is a toxic substance. It is easily absorbed through the skin and affects both the blood and the nervous system. Aniline reduces the oxygen-carrying capacity of the blood by converting hemoglobin to methemoglobin. Methemoglobin is the oxidized form of hemoglobin in which the iron has gone from a $+2$ to a $+3$ oxidation state.

Ring compounds in which all the atoms in the ring are not alike are known as **heterocyclic compounds**. A number of the nitrogen-containing heterocyclic compounds are present in naturally occurring biological substances such as the genes in DNA, which control heredity. The structural formulas of several nitrogen-containing heterocyclics follow.

heterocyclic compounds

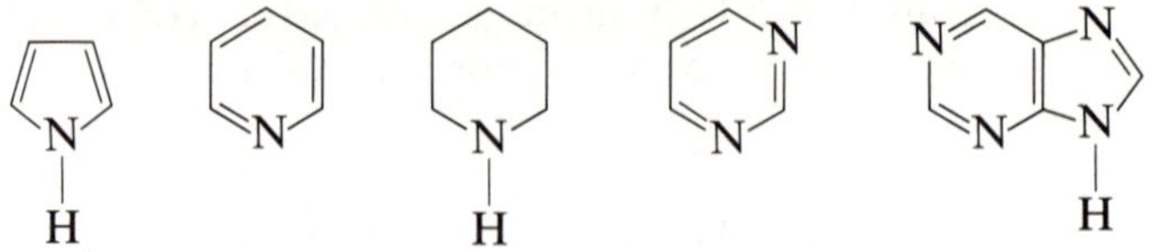

Pyrrole
C_4H_5N

Pyridine
C_5H_5N

Piperidine
$C_5H_{11}N$

Pyrimidine
$C_4H_4N_2$

Purine
$C_5H_4N_4$

Physical Properties

Amines are capable of hydrogen bonding with water. As a result, the aliphatic amines with up to six carbons are quite soluble in water. The methylamines and ethylamine are flammable gases with a strong ammoniacal odor. Trimethylamine has a "fishy" odor. Higher-molecular-weight amines have disagreeable, obnoxious odors. The foul odors arising from dead fish and decaying flesh are due to amines released by bacterial decomposition.

Simple aromatic amines are all liquids and solids. They are colorless or almost colorless when freshly prepared, but become dark brown or red when exposed to air and light.

Chemical Properties

In many respects amines resemble ammonia in their reactions. Thus, amines are bases and, like ammonia, ionize in water to produce OH^- ions:

$$NH_3 + (H)OH \rightleftharpoons NH_4^+ + OH^-$$

Ammonia molecule — Ammonium ion — Hydroxide ion

Methylamine and aniline react in the same manner:

$$CH_3NH_2 + (H)OH \rightleftharpoons CH_3NH_3^+ + OH^-$$

Methylamine molecule — Methylammonium ion — Hydroxide ion

$$C_6H_5-NH_2 + (H)OH \rightleftharpoons C_6H_5-NH_3^+ + OH^-$$

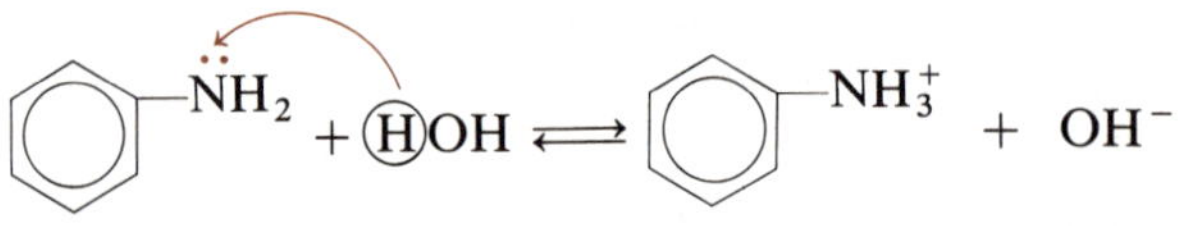

Aniline molecule — Anilinium ion — Hydroxide ion

Like ammonia, amines are weak bases. Methylamine is a slightly stronger base than ammonia and aniline is considerably weaker than ammonia. The pH values for 0.1 M solutions are: methylamine, 11.8; ammonia, 11.1; and aniline, 8.8.

Carboxylic Acids, Esters, and Amines

An amine will react with a strong acid to form a salt; for example, methylamine and hydrogen chloride react in this fashion:

$$CH_3NH_2(g) + HCl(g) \rightleftharpoons [CH_3NH_3]^+ Cl^-$$

Methylamine molecule   Hydrogen chloride molecule   Methylammonium chloride (salt)

Methylammonium chloride is made up of methylammonium ions ($CH_3NH_3^+$) and chloride ions (Cl^-). It is a white crystalline salt which in physical appearance resembles ammonium chloride very closely.

Aniline reacts in a similar manner, forming anilinium chloride:

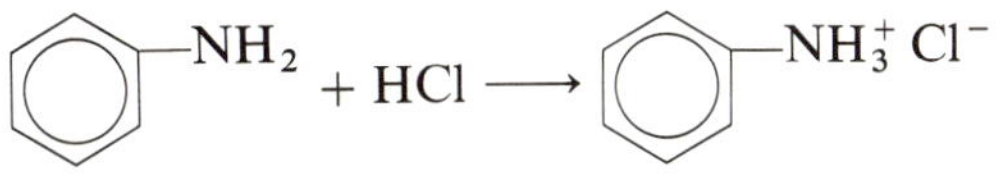

$$C_6H_5-NH_2 + HCl \longrightarrow C_6H_5-NH_3^+ Cl^-$$

Anilinium chloride

Amines react with organic acids as well as with mineral acids. For example, methylamine reacts with acetic acid according to the following equation:

$$CH_3\overset{O}{\overset{\|}{C}}-OH(aq) + CH_3NH_2(aq) \longrightarrow CH_3\overset{O}{\overset{\|}{C}}-O^-\overset{+}{N}H_3CH_3$$

N-Methylammonium acetate

Primary and secondary amines react with acid chlorides to form substituted amides. For example,

$$CH_3\overset{O}{\overset{\|}{C}}-Cl + 2(CH_3CH_2)_2NH \longrightarrow CH_3\overset{O}{\overset{\|}{C}}-N\begin{smallmatrix}CH_2CH_3\\CH_2CH_3\end{smallmatrix} + (CH_3CH_2)_2NH_2^+ Cl^-$$

N,N-Diethylacetamide

Amides can be reduced with lithium aluminum hydride to give amines. For example, when N,N-diethylacetamide is reduced, triethylamine is formed:

$$CH_3\overset{O}{\overset{\|}{C}}-N\begin{smallmatrix}CH_2CH_3\\CH_2CH_3\end{smallmatrix} \xrightarrow{LiAlH_4} (CH_3CH_2)_3N$$

Triethylamine

Nitriles, RCN, are also reducible to amines using hydrogen and a metal catalyst.

$$CH_3CH_2C\equiv N \xrightarrow{H_2/Ni} CH_3CH_2CH_2NH_2$$

Propionitrile   n-Propylamine

Sources and Uses of Amines

Nitrogen compounds are found throughout the entire plant and animal kingdoms. Amines, substituted amines, and amides occur in every living cell.

Colorful dyes; many medicinals including the alkaloids quinine and morphine, sulfa drugs, and amphetamines; and vitamins are but a few of the classes of substances where organic nitrogen compounds are found. The formulas for several well-known nitrogen compounds are shown below.

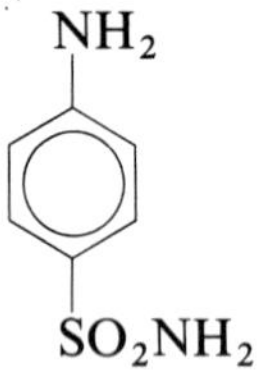

Sulfanilamide

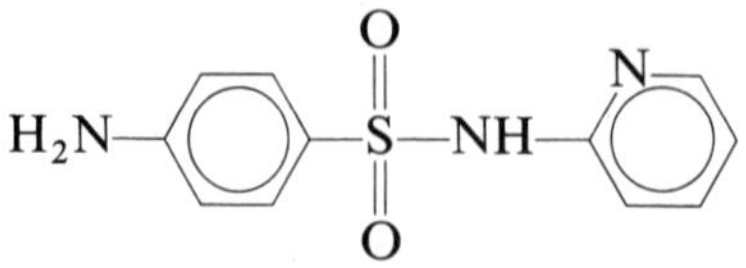

Sulfapyridine
(antibacterial agent)

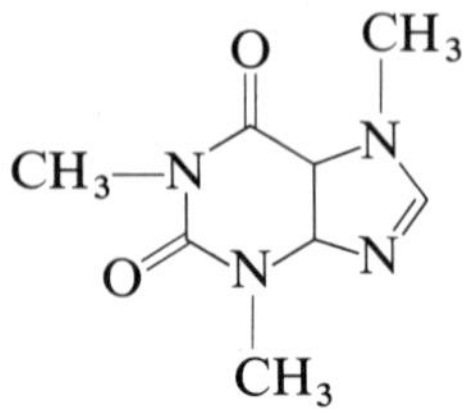

Caffeine
(tea, coffee, cola nut)

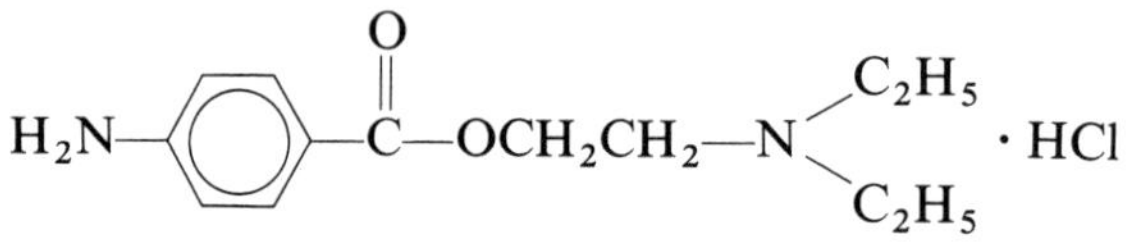

Procaine hydrochloride
(novocaine, local anesthetic)

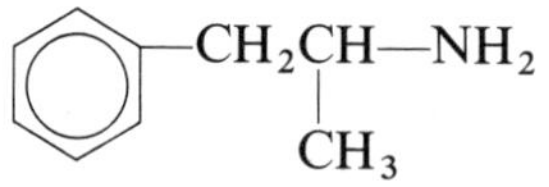

Amphetamine (Benzidrene)
(stimulant)

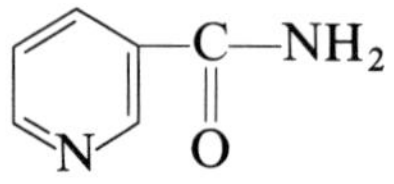

Nicotinamide
(niacin, antipellagra vitamin)

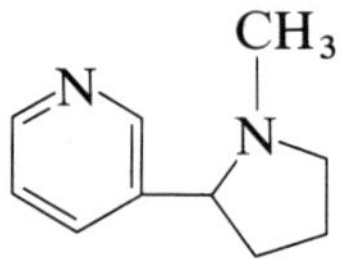

Nicotine
(tobacco leaves, used as an agricultural insecticide)

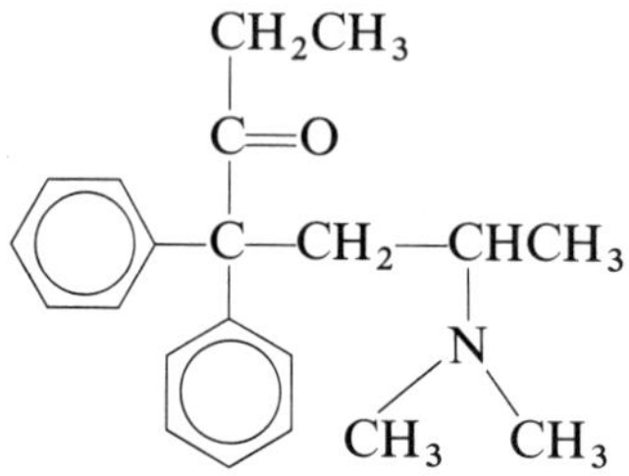

Methadone
(narcotic analgesic, substitute for heroin)

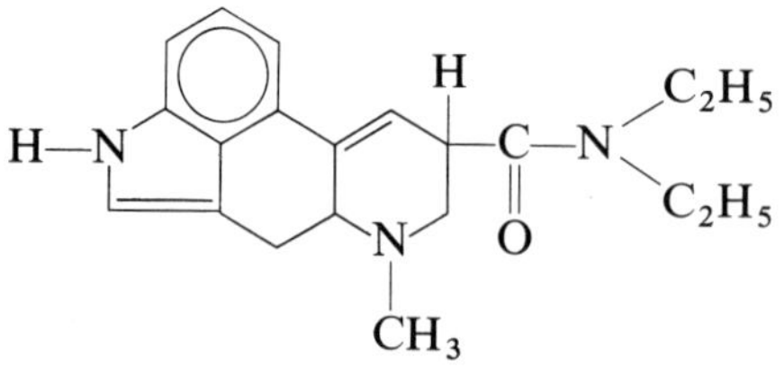

Lysergic acid diethylamide
(LSD)

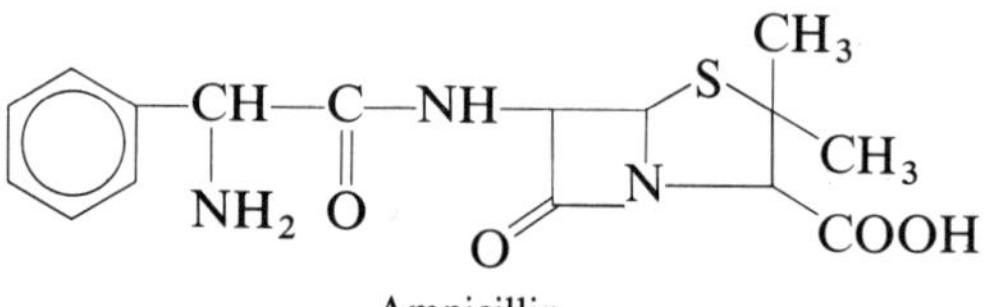

Ampicillin
(antibacterial agent)

Questions

A. Review the meanings of the new terms introduced in this chapter.
1. Carboxyl group
2. Saponification
3. Fat
4. Oil
5. Triglyceride
6. Synthetic detergent
7. Soap
8. Biodegradable
9. Amine
10. Heterocyclic compounds

B. Review questions.
 1. Name the following compounds:

  (a) $CH_3(CH_2)_5COOH$

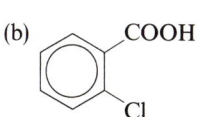

  (b) benzene ring with COOH and Cl (ortho)

  (c) $CH_3CH=CHCOOH$

  (d) $CH_3(CH_2)_7CH=CH(CH_2)_7COOH$

  (e) benzene ring with two COOH groups (ortho)

  (f) benzene ring with COOH and CH_3 (meta)

 2. Write structures for the following compounds:
  (a) Caproic acid
  (b) Adipic acid
  (c) o-Toluic acid
  (d) Oxalic acid
  (e) β-Hydroxybutyric acid
  (f) p-Phthalic acid
  (g) Sodium benzoate
  (h) Linoleic acid

 3. Which of the following would have the more objectionable odor?
  (a) a 1% solution of butyric acid (C_3H_7COOH) or
  (b) a 1% solution of sodium butyrate (C_3H_7COONa)

  Cite a satisfactory reason for your answer.

 4. Suggest a logical scheme for obtaining reasonably pure stearic acid from a solution containing 2.0% sodium stearate [$CH_3(CH_2)_{16}COONa$] dissolved in water.

 5. Assume that you have a 0.01 M solution of each of the following substances:

  (a) CH_3NH_2 (b) HCl (c) NaCl (d) NaOH

  (e) CH_3COOH (f) phenol (benzene ring with OH)

  Arrange them in order of increasing pH (list the most acidic solution first).

 6. Give at least one name—more if you can—for each of the following:

  (a) $HC(=O)-OCH_3$

  (b) benzene-C(=O)-O-CH(CH_3)_2

  (c) benzene-C(=O)-O-benzene with OH

  (d) $CH_3CH_2C(=O)-OCH_2CH_3$

(e) $CH_3CH_2O-\overset{\overset{O}{\|}}{C}-\overset{\overset{O}{\|}}{C}-OCH_2CH_3$
(f) $CH_2=CH\overset{\overset{O}{\|}}{C}-OCH_3$

7. Write structural formulas for each:
   (a) Methyl formate
   (b) n-Octyl acetate
   (c) n-Propyl propanoate
   (d) Ethyl benzoate

8. Write the structural formula and name of the principal organic product for each of the following reactions:

   (a) $CH_3(CH_2)_7CH=CH(CH_2)_7-\overset{\overset{O}{\|}}{C}-OH + H_2 \xrightarrow{Ni}$

   (b) $CH_3CH_2\overset{\overset{O}{\|}}{C}-OH + NH_3 \xrightarrow{\Delta}$

   (c) 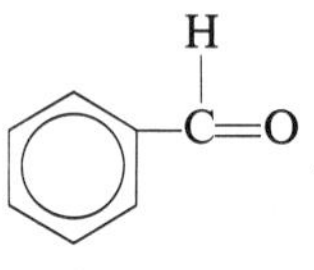 $C_6H_5-\overset{\overset{H}{|}}{\underset{}{C}}=O + [O] \xrightarrow{Na_2Cr_2O_7}{H_2SO_4}$

   (d) 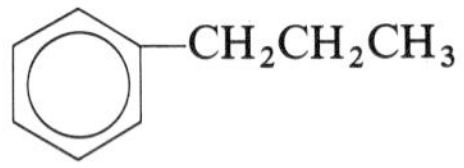 $C_6H_5-CH_2CH_2CH_3 + [O] \xrightarrow{NaMnO_4/NaOH}{\Delta}$

   (e) 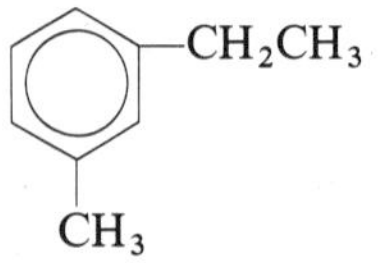 3-methylphenyl-CH_2CH_3 + [O] $\xrightarrow{NaMnO_4/NaOH}{\Delta}$

   (f) $CH_3CH_2COOH + NaOH \rightarrow$

   (g) $CH_3CH_2CH_2CH_2CH_2\overset{\overset{O}{\|}}{C}-OCH_2CH_2CH_3 + NaOH \xrightarrow{\Delta}$

9. For each of the first eight acids listed in Table 26.1, give the common and IUPAC names of the corresponding amides.

10. Write structural formulas for the products of the following reactions:

    (a) 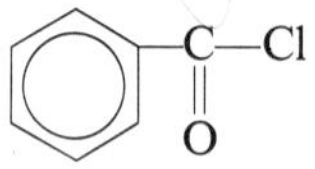 $C_6H_5-\overset{\overset{O}{\|}}{C}-Cl + CH_3NH_2 \rightarrow$

    (b) 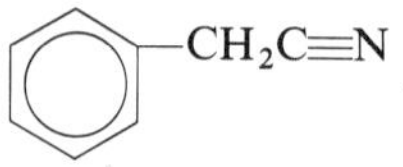 $C_6H_5-CH_2C\equiv N + H_2O \xrightarrow{H^+}$

    (c) $CH_3CH_2CH_2Br + KCN \rightarrow A \xrightarrow{LiAlH_4} B$

    (d) $CH_3CH_2\overset{\overset{O}{\|}}{C}-NHCH_3 \xrightarrow{LiAlH_4}$

    (e) $CH_3CH_2CH_2CN \xrightarrow{H_2/Ni}$

11. Write structural formulas for the products of the following reactions:

(a)
$$CH_3-\underset{\underset{COOH}{|}}{\overset{\overset{COOH}{|}}{CH}} \xrightarrow{150°C}$$

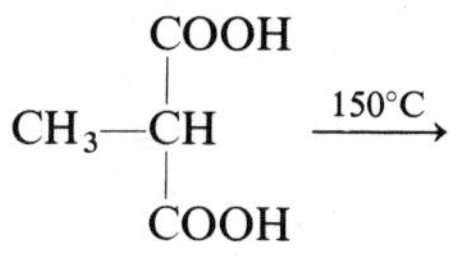

(b) 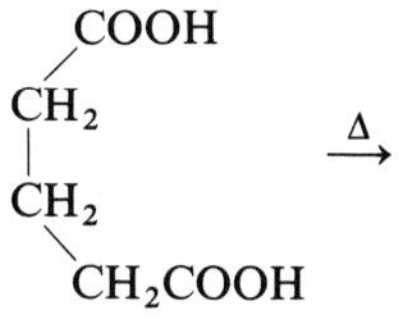
$$\begin{array}{c} CH_2\text{-COOH} \\ | \\ CH_2 \\ | \\ CH_2 \\ | \\ CH_2\text{-COOH} \end{array} \xrightarrow{\Delta}$$
Glutaric acid

(c) $C_6H_5-\overset{O}{\underset{\|}{C}}-OCH_2CH_3 + H_2O \xrightarrow{H^+}$

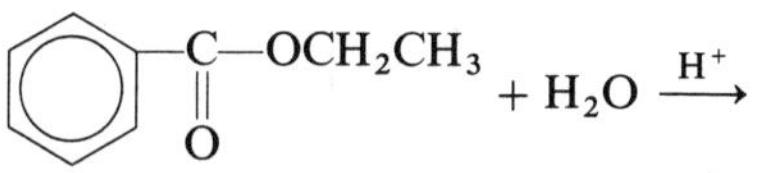

(d) $C_6H_5-COOH + CH_3\underset{\underset{OH}{|}}{CH}CH_3 \xrightarrow{H^+}$

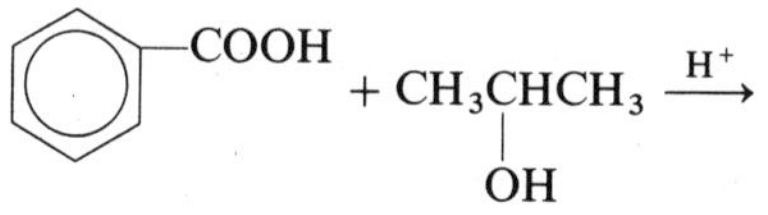

12. Write the structural formulas for the organic products formed in the following reactions:

(a) $CH_2CH_2COOH \xrightarrow[\Delta]{H_2SO_4}$
    $|$
    OH

(b) $CH_2=CHCOOH + Br_2 \rightarrow$

(c) $CH_3COOH + SOCl_2 \rightarrow$

(d) $CH_3\overset{O}{\underset{\|}{C}}-Cl + NH_3 \rightarrow$

(e) $CH_3\overset{O}{\underset{\|}{C}}-Cl + CH_3CH_2OH \rightarrow$

13. Write the structural formula of the ester that when hydrolyzed would yield:
    (a) Methanol and acetic acid
    (b) Ethanol and formic acid
    (c) 1-Octanol and acetic acid
    (d) 2-Propanol and benzoic acid
    (e) Methanol and salicylic acid

14. Explain how the amide formed from ammonia and acetic acid differs in properties from the ammonium salt formed from ammonia and acetic acid.

15. Write structural formulas for the following compounds:
    (a) Isopropyl formate
    (b) Methyl palmitate
    (c) Diethyl adipate
    (d) 4-Methylpentyl ethanoate
    (e) Phenyl propionate
    (f) Methyl-2-aminopentanoate

16. What simple tests can be used to distinguish between the following pairs of compounds?
    (a) Benzoic acid and sodium benzoate
    (b) Maleic acid and malonic acid

17. The geometric configuration of naturally occurring unsaturated 18-carbon acids is all *cis*. Draw structural formulas for:
    (a) *cis*-Oleic acid   (b) *cis, cis*-Linoleic acid
18. Write the structural formula of a triglyceride that contains one unit each of lauric acid, palmitic acid, and stearic acid. How many other triglycerides, each containing all three of these acids, are possible?
19. Triolein (glyceryl trioleate) has this structure:

$$\begin{array}{l} CH_2-O-\overset{O}{\underset{\|}{C}}(CH_2)_7CH=CH(CH_2)_7CH_3 \\ CH-O-\overset{O}{\underset{\|}{C}}(CH_2)_7CH=CH(CH_2)_7CH_3 \\ CH_2-O-\overset{O}{\underset{\|}{C}}(CH_2)_7CH=CH(CH_2)_7CH_3 \end{array}$$

Write the names and formulas of all products expected when triolein is:
(a) Reacted with hydrogen in the presence of Ni
(b) Reacted with water at high temperature and pressure in the presence of mineral acid
(c) Boiled with potassium hydroxide
(d) Reacted with hydrogen at relatively high pressure and temperature in the presence of a copper chromite catalyst

20. Which has the greater solubility in water?
    (a) Methyl propanoate *or* propanoic acid
    (b) Sodium palmitate *or* palmitic acid
    (c) Decylamine *or* decylammonium chloride
    (d) Sodium stearate *or* barium stearate
    (e) Phenol *or* sodium phenoxide
21. Write the structural formulas for each pair of compounds mentioned in Question 20.
22. Explain the difference between:
    (a) A fat and an oil   (c) Hydrolysis and saponification
    (b) A soap and a syndet
23. Cite the principal advantages synthetic detergents (syndets) have over soaps.
24. Would $CH_3(CH_2)_{12}COOH$ or $CH_3(CH_2)_{12}COONa$ be the more useful cleaning agent in soft water? Explain.
25. Would $CH_3(CH_2)_{11}OSO_3Na$ (sodium lauryl sulfate) or $CH_3CH_2CH_2OSO_3Na$ (sodium propyl sulfate) be the more effective detergent in hard water? Explain.
26. Explain the cleaning action of detergents.
27. Which one of the following substances is a good detergent in water? Is this substance a nonionic, anionic, or cationic detergent?

(a) $C_{16}H_{33}N(CH_3)_3Cl$ (b) $C_{16}H_{34}$
   Hexadecyltrimethyl   Hexadecane
   ammonium chloride

(c) $C_{15}H_{31}COOH$ (d) $C_{15}H_{31}\overset{O}{\underset{\|}{C}}-O-C_{15}H_{31}$
   Palmitic acid   Palmityl palmitate
   (Hexadecanoic acid)   (Hexadecyl hexadecanoate)

28. Oleomargarine consists principally of a relatively small amount of water emulsified in vegetable or animal fats and oils. Monoglycerides such as glycerol monooleate,

$$C_{17}H_{33}\overset{O}{\underset{\|}{C}}-O-CH_2-\underset{OH}{\underset{|}{CH}}-\underset{OH}{\underset{|}{CH_2}}$$

are often used in small quantities in making oleomargarine.
   (a) What is the function of the monoglycerides?
   (b) Explain how this function is achieved.
29. If 1.00 kg of triolein (glyceryl trioleate) is converted to tristearin (glyceryl tristearate) by hydrogenation,
   (a) How many litres of hydrogen (at STP) are required?
   (b) What is the weight of the tristearin that is produced?
30. Low-molecular-weight aliphatic amines generally have odors suggestive of ammonia and/or stale fish. Which would have the more objectionable odor?
   (a) a 1% trimethylamine solution in 1.0 M H_2SO_4 or
   (b) a 1% trimethylamine solution in 1.0 M NaOH
   Give a satisfactory reason for your answer.
31. Name the following compounds:

   (a) benzene with CH_2CH_3 and NH_2 substituents

   (b) $CH_3CH_2NH_3^+ Br^-$

   (c) ring with $\overset{O}{\underset{\|}{C}}$, CH_2, NH_2, CH_2, NH_2, $\overset{O}{\underset{\|}{C}}$

   (d) benzene with $-NHCH_2CH_3$

   (e) pyridine

   (f) $CH_3\overset{O}{\underset{\|}{C}}-NHC_2H_5$

   (g) pyridine with $-\overset{O}{\underset{\|}{C}}-NH_2$

32. Write structural formulas for the following compounds:
   (a) Methylethylamine   (d) Ethylammonium chloride
   (b) Tributylamine   (e) N-Methylanilinium chloride
   (c) Aniline   (f) 1,4-Butanediamine
33. Write structural formulas for all the isomeric amines that have the formula $C_4H_{11}N$.
34. (a) What is a heterocyclic compound?
   (b) How many heterocyclic rings are present in (1) LSD, (2) ampicillin, (3) methadone, and (4) nicotine?
35. Which acid will require more base for neutralization?
   (a) 1 g of acetic acid or   (b) 1 g of propanoic acid
   Explain your answer.
36. Starting with ethyl alcohol as the only source of organic material and using any other reagents you desire, write equations to show the synthesis of:
   (a) Acetic acid   (c) N-Ethylacetamide
   (b) Ethyl acetate   (d) β-Hydroxybutyric acid

37. Upon hydrolysis, an ester of formula $C_6H_{12}O_2$ yields an acid A and an alcohol B. When B is oxidized, it yields a product identical to A. What is the structure of the ester? Explain your answer.

38. Which of these statements are correct?
   (a) Carboxylic acids can be either aliphatic or aromatic.
   (b) The functional group —COOH is known as a carboxyl group.
   (c) The name for $CH_3CH_2CHBrCH_2COOH$ is γ-bromovaleric acid.
   (d) Acetic acid is a stronger acid than hydrochloric acid.
   (e) Benzoic acid is more soluble in water than sodium benzoate.
   (f) Carboxylic acids have relatively high boiling points because of hydrogen bonding between molecules.
   (g) The formula $C_{17}H_{33}COOH$ represents oleic acid.
   (h) Fumaric and maleic acids are *cis–trans* isomers.
   (i) If $CH_3CH_2CH_2COCl$ is allowed to come into contact with water, it will be hydrolyzed to butyric acid.
   (j) Volatile esters generally have a pleasant smell.
   (k) The common name for CH_3CONH_2 is methanamide.
   (l) The general formula for a primary amine is $R\underset{\underset{O}{\|}}{C}—NH_2$.

   (m) Glycerol is a trihydroxy alcohol.
   (n) Fatty acids in fats usually have an even number of carbon atoms.
   (o) Fats and oils are esters of glycerol.
   (p) Oils are largely of vegetable origin.
   (q) The presence of unsaturation in the acid component of a fat tends to raise its melting point compared to the corresponding saturated compound.
   (r) Alkali metal salts of long-chain fatty acids are called soaps.
   (s) The name for $[(CH_3)_2CH]_2NH$ is isopropyl amine.
   (t) Aniline is soluble in dilute HCl because it forms a soluble salt.
   (u) Most amines have pleasant odors.
   (v) Lactic acid is an α-amino acid.
   (w) Sulfanilamide is an aniline derivative.

27

Polymers—Macromolecules

After studying Chapter 27 you should be able to:

1. Understand the terms listed in Question A at the end of the chapter.
2. Write formulas for addition polymers when given the monomer(s).
3. Write formulas for condensation polymers when given the monomers.
4. Describe the properties of a thermoplastic polymer and a thermosetting polymer.
5. Explain the free radical mechanism for polymer formation.
6. Identify polymers from their trade names (for example, Dacron, nylon, and Teflon).
7. Explain the effect of cross-linking in polymers.
8. Draw a segment of the structural formula of natural rubber to illustrate the all *cis* configuration.
9. Explain how butadiene-type polymers are formed.
10. Explain vulcanization and its effect on rubber.
11. Identify polymers by type (such as vinyl, polyester, polyamide, or polyurethane).

27.1 Introduction

Up to now we have dealt mainly with rather small organic molecules that contain up to 50 atoms and some that contain up to about 150 atoms (fats). But there exist in nature some very large molecules (macromolecules) containing tens of thousands of atoms. Some of these, such as starch, glycogen, cellulose, proteins, and DNA, have molecular weights in the millions and are central to many of our life processes. Man-made macromolecules touch every phase of modern living. It is hard to imagine a world today without polymers. Textiles for clothing, carpeting, and draperies; shoes; toys; automobile parts; materials of construction; synthetic rubber; chemical equipment; medical supplies; cooking utensils; synthetic leather; recreational equipment—the list could go on and on. All these and a host of others that we consider to be essential in our daily life are wholly or partly man-made polymers. Most of these modern-day polymers were unknown 50 years ago. The vast majority of these polymeric materials are based on petroleum. Since petroleum is a nonreplaceable resource, our dependence on polymers is another good reason for not squandering our limited world supply of petroleum.

Under proper conditions, ethylene reacts with itself to form a substance known as polyethylene (or polythene). Polyethylene is a long-chain hydrocarbon made from many ethylene units:

$$n\ CH_2{=}CH_2 \longrightarrow -(CH_2-CH_2)_n-$$
Polyethylene

A typical polyethylene molecule is made up from about 2500–25,000 ethylene molecules joined in a continuous structure:

$\bullet = CH_2$

The process of forming very large, high-molecular-weight molecules from smaller units is called **polymerization**. The large molecule, or unit, is called the **polymer** and the small unit, the **monomer**. The term polymer is derived from the Greek word *polumerēs*, meaning "having many parts." Ethylene is a monomer and polyethylene is a polymer. Because of their large size, polymers are often called *macromolecules*. Another commonly used term is *plastics*. The word plastic means "to be capable of being molded, or pliable." Although all polymers are not pliable and capable of being remolded, the word plastics has gained general use and has come to mean any of a variety of polymeric substances.

polymerization
polymer
monomer

27.2 Synthetic Polymers

Some of the early commercial polymers were merely modifications of naturally occurring substances. One chemically modified natural polymer—nitrated cellulose—was first made and sold as Celluloid late in the 19th century. But the first commercially successful fully synthetic polymer, Bakelite, was made from phenol and formaldehyde by Leo Baekeland in 1909. This was the beginning of the modern plastics industry. Chemists began to create synthetic polymers in large numbers in the late 1920s. Today, dozens of different synthetic macromolecular materials are available commercially. Even greater numbers of polymers have been made and discarded as impractical for technical or economic reasons.

Although there is a great variety of synthetic polymers on the market, they can be classified into the following general groups on the basis of properties and uses:

1. Rubber materials or elastomers
2. Flexible films
3. Synthetic textiles
4. Resins (or plastics) for casting, molding, and extruding
5. Coating resins for dip, spray, or solvent-dispersed applications
6. Miscellaneous, including hydraulic fluids, foamed insulation, ion-exchange resins

27.3 Polymer Types

addition polymer

condensation polymer

thermoplastic polymer

thermosetting polymer

Two general types of polymers—addition and condensation—are known. An **addition polymer** is one that is produced by the successive addition of repeating monomer molecules. Polyethylene is an example of an addition polymer. A **condensation polymer** is one that is formed from monomers that react to split out water or some other simple substance.

Polymers are also classified as being either thermoplastic or thermosetting. Those that soften on reheating are **thermoplastic polymers**; those that set to an infusible solid and do not soften on reheating are **thermosetting polymers**. Thermoplastic polymers are formed when monomer molecules join end to end in a linear chain with little or no cross-linking between the chains. Thermosetting polymers are macromolecules in which the polymeric chains are cross-linked to form a network structure. The structures of thermoplastic and thermosetting polymers are illustrated in the schematic diagram in Figure 27.1.

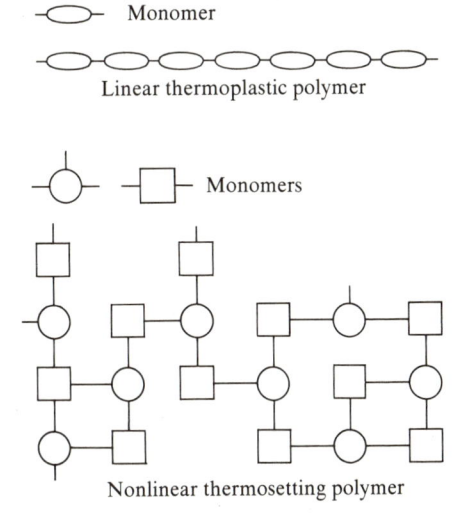

Figure 27.1. Schematic diagrams of thermoplastic and thermosetting polymer structures.

27.4 Addition Polymerization

Ethylene polymerizes by addition to form polyethylene, according to the reaction represented in Section 27.1. Polyethylene is the most important and widely used polymer on the market today. It is a tough, inert, but flexible thermoplastic material. Over ten billion pounds of polyethylene are made annually in the United States. Polyethylene is made into hundreds of different articles such as bread wrappers, toys, squeeze bottles, containers of all kinds, electrical insulation, and so on.

The double bond is the key structural feature involved in the polymerization of ethylene. Ethylene derivatives, in which one or more hydrogen atoms

Table 27.1. Polymers derived from modified ethylene monomers.

Monomer	Polymer	Uses
$CH_2{=}CH_2$ Ethylene	$-(CH_2-CH_2)_n-$ Polyethylene	Packaging material, molded articles, containers, toys
$CH_2{=}CH-CH_3$ Propylene	$(-CH_2-CH(CH_3)-)_n$ Polypropylene	Textile fibers, molded articles, lightweight ropes, autoclavable biological equipment
$CH_2{=}C(CH_3)_2$ Isobutylene	$(-CH_2-C(CH_3)_2-)_n$ Polyisobutylene	Pressure-sensitive adhesives, butyl rubber (contains some isoprene as copolymer)
$CH_2{=}CHCl$ Vinyl chloride	$(-CH_2-CHCl-)_n$ Polyvinyl chloride (PVC)	Phonograph records, garden hoses, pipes, molded articles, floor tile, electrical insulation, vinyl leather
$CH_2{=}CCl_2$ Vinylidene chloride	$-(CH_2-CCl_2)_n-$ Saran[a]	Food packaging, textile fibers, pipes, tubing
$CH_2{=}CHCN$ Acrylonitrile	$(-CH_2-CH(CN)-)_n$ Orlon, Acrilan	Textile fibers
$CF_2{=}CF_2$ Tetrafluoroethylene	$-(CF_2-CF_2)_n-$ Teflon	Gaskets, valves, insulation, heat-resistant and chemically resistant coatings, linings for pots and pans
$CH_2{=}CH{-}C_6H_5$ Styrene	$(-CH_2-CH(C_6H_5)-)_n$ Polystyrene	Molded articles, styrofoam, insulation
$CH_2{=}CH-OC(O)CH_3$ Vinyl acetate	$(-CH_2-CH(OC(O)CH_3)-)_n$ Polyvinyl acetate	Adhesives, paint and varnish
$CH_2{=}C(CH_3)-C(O)OCH_3$ Methylmethacrylate	$(-CH_2-C(CH_3)(C(O)OCH_3)-)_n$ Lucite, Plexiglas (acrylic resins)	Contact lenses, clear sheets for windows and optical uses, molded articles, automobile finishes

[a] Contains some vinyl chloride as copolymer.

have been replaced by other atoms or groups, can also be polymerized. This is often called *vinyl polymerization*. Many of our commercial synthetic polymers are made from such modified ethylene monomers. The names, structures, and uses of some of these polymers are given in Table 27.1.

Free radicals catalyze or initiate many addition polymerizations. Organic peroxides, ROOR, are frequently used for this purpose. The steps in the reaction are as follows.

Step 1. *Free radical formation*: The peroxide splits into free radicals.

$$RO:OR \longrightarrow 2\ RO\cdot$$

Step 2. *Propagation of polymeric chain*:

$$RO\cdot + CH_2{=}CH_2 \longrightarrow ROCH_2CH_2\cdot$$
$$ROCH_2CH_2\cdot + CH_2{=}CH_2 \longrightarrow ROCH_2CH_2CH_2CH_2\cdot,\ \text{etc.}$$

Step 3. *Termination*: Polymerization stops when the free radicals are used up.

$$RO(CH_2CH_2)_n\cdot + \cdot OR \longrightarrow RO(CH_2CH_2)_nOR$$
$$RO(CH_2CH_2)_n\cdot + \cdot(CH_2CH_2)_nOR \longrightarrow RO(CH_2CH_2)_n(CH_2CH_2)_nOR$$

Addition (or vinyl) polymerization of ethylene and its substituted derivatives yields saturated polymers—that is, polymer chains without double bonds. The pi bond is eliminated when a free radical adds to an ethylene molecule. One of the electrons of the pi bond pairs with the unpaired electron of the free radical, thus bonding the radical to the ethylene unit. The other pi bond electron remains unpaired, generating a new and larger free radical. This new free radical then adds another ethylene molecule, continuing the building of the polymeric chain. This process is illustrated by the following electron-dot diagram:

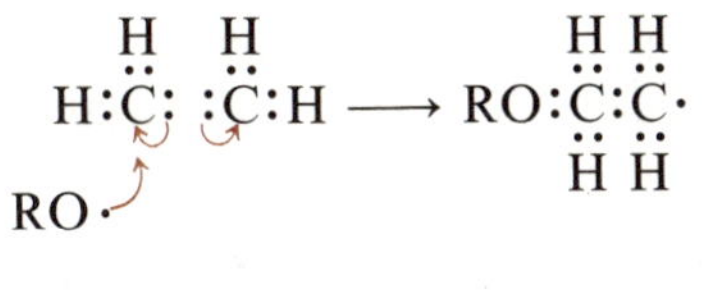

27.5 Butadiene Polymers

A diene is a compound that contains two carbon–carbon double bonds. Another type of addition polymer is based on the compound 1,3-butadiene or its derivatives.

$$\overset{1}{C}H_2{=}\overset{2}{C}H{-}\overset{3}{C}H{=}\overset{4}{C}H_2$$

1,3-Butadiene

Natural rubber is a polymer of isoprene (2-methyl-1,3-butadiene). Many synthetic elastomers or rubberlike materials are polymers of isoprene or of butadiene. Unlike the saturated ethylene polymers, these polymers are unsaturated—that is, they have double bonds in their polymeric structures.

$$n\ CH_2=\underset{\underset{CH_3}{|}}{C}-CH=CH_2 \longrightarrow -(CH_2-\underset{\underset{CH_3}{|}}{C}=CH-CH_2)_n-$$

Isoprene → Rubber polymer chain

$$n\ CH_2=CH-CH=CH_2 \longrightarrow -(CH_2-CH=CH-CH_2)_n-$$

1,3-Butadiene → Butadiene polymer chain

$$n\ CH_2=\underset{\underset{Cl}{|}}{C}-CH=CH_2 \longrightarrow -(CH_2-\underset{\underset{Cl}{|}}{C}=CH-CH_2)_n-$$

2-Chloro-1,3-butadiene (Chloroprene) → Neoprene polymer chain

In this kind of polymerization the free radical adds to the butadiene monomer at carbon 1 of the carbon–carbon double bond. At the same time, a double bond is formed between carbon 2 and carbon 3, and a new free radical is formed at carbon 4. This process is illustrated in the following diagram.

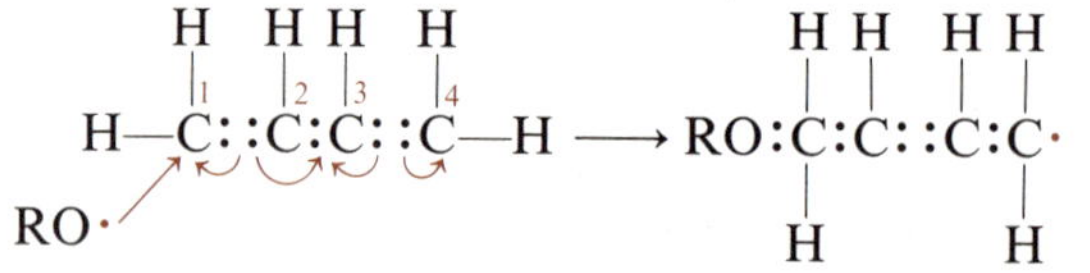

Free radical + 1,3-Butadiene → Radical chain lengthened by four carbon atoms

One of the outstanding synthetic rubbers (styrene–butadiene rubber, SBR) is made from two monomers, styrene and 1,3-butadiene. These substances form a **copolymer**—that is, a polymer containing two different kinds of monomer units. Styrene and butadiene do not necessarily have to combine in a 1:1 ratio. In the actual manufacture of SBR polymers, about 3 moles of butadiene are used per mole of styrene. Thus, the butadiene and styrene units are intermixed, but in a ratio of about 3:1.

copolymer

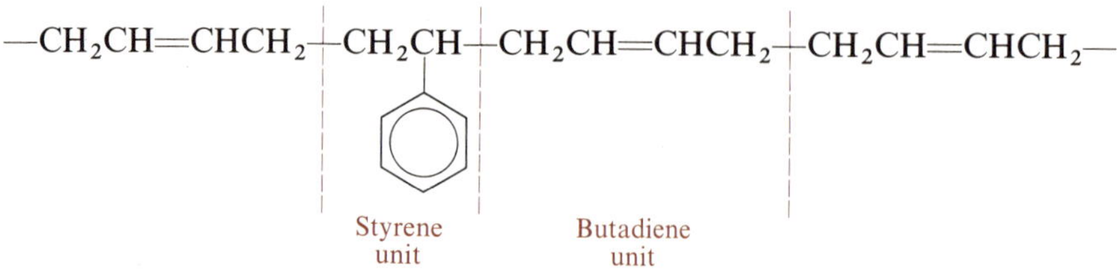

Segment of styrene–butadiene rubber (SBR)

The presence of double bonds at intervals along the chains of rubber and rubberlike synthetic polymers designed for use in tires is almost a necessity and, at the same time, a disadvantage. On the positive side, double bonds make vulcanization possible. On the negative side, double bonds afford sites where ozone, especially in smoggy atmospheres, can attack the rubber, causing "age hardening" and cracking. Vulcanization vastly extends the useful temperature range of rubber products and imparts greater abrasion resistance to them. It is usually accomplished by heating raw rubber with sulfur and other auxiliary agents. The vulcanization process consists of introducing sulfur atoms that connect or cross-link the long strands of polymeric chains. Vulcanization was devised through trial-and-error experimentation by the American inventor Charles Goodyear in 1839, long before any real understanding of the chemistry of the process was known. Goodyear's patent on "Improvement in India Rubber" was issued on June 15, 1844.

$$-CH_2\underset{\underset{\underset{S}{|}}{\overset{\overset{CH_3}{|}}{C}}}{}=CHCHCH_2\underset{\underset{\underset{S}{|}}{\overset{\overset{CH_3}{|}}{C}}}{}=CHCH_2CH\overset{\overset{CH_3}{|}}{C}=CHCH_2-$$

$$-CH_2\underset{\underset{CH_3}{|}}{\overset{\overset{S}{|}}{C}}=CHCHCH_2\underset{\underset{CH_3}{|}}{\overset{\overset{S}{|}}{C}}=CHCH_2CH\underset{\underset{CH_3}{|}}{C}=CHCH_2-$$

Segment of vulcanized rubber

27.6 Stereochemistry of Polymers

The recurring double bonds in isoprene and butadiene polymers make it possible to have polymers with specific spatial orientation as a result of *cis–trans* isomerism. Recall from Section 23.4 that two carbon atoms joined by a double bond are not free to rotate, and thus give rise to *cis–trans* isomerism. An isoprene polymer can have all *cis*, all *trans*, or a random distribution of *cis* and *trans* configurations about the double bonds.

Natural rubber is *cis*-polyisoprene with an all *cis* configuration. Gutta-percha, also obtained from plants, is an all *trans*-polyisoprene. Although these two polymers have the same composition, their properties are radically different. The *cis* natural rubber is a soft, elastic material, whereas the *trans* gutta-percha is a tough, nonelastic, hornlike substance.

All *cis* configuration of natural rubber

All *trans* configuration of gutta-percha

Chicle is another natural substance containing polyisoprenes. It is obtained by concentrating the latex from the sapodilla tree and contains about 5% *cis*-polyisoprene and 12% *trans*-polyisoprene. The chief use of chicle is in chewing gum.

Only random or nonstereospecific polymers are obtained by free radical polymerization. Synthetic polyisoprenes made by free radical polymerization are much inferior to natural rubber since they contain both the *cis* and the *trans* isomers. But in the 1950s, Karl Ziegler (1898–1973) of Germany and Guilio Natta (1903–1973) of Italy developed catalysts [for example, $(C_2H_5)_3Al/TiCl_4$] that allowed polymerization to proceed by an ionic mechanism, producing stereochemically controlled polymers. Ziegler–Natta catalysts made possible the synthesis of polyisoprene with an all *cis* configuration and with properties fully comparable to those of natural rubber. This material is known by the odd but logical name *synthetic natural rubber*. In 1963, Natta and Ziegler were jointly awarded the Nobel Prize for their work on stereochemically controlled polymerization reactions.

27.7 Condensation Polymers

Condensation polymers are formed by reactions between functional groups on adjacent monomer molecules. As a rule, a smaller molecule—usually water—is eliminated in the reaction. The monomers must be at least bifunctional, and if cross-linking is to occur, there must be more than two functional groups on some monomer molecules.

Many different condensation polymers have been synthesized. Important classes include the polyesters, polyamides, phenol–formaldehyde polymers, and polyurethanes.

Polyesters

In polyesters the monomers are joined by ester linkages between carboxylic acid and alcohol groups:

$$\underset{\text{Carboxylic acid}}{-\overset{\overset{O}{\|}}{C}-\boxed{OH}} + \underset{\text{Alcohol}}{\boxed{H}-OCH_2-} \longrightarrow \underset{\text{Ester}}{-\overset{\overset{O}{\|}}{C}-O-CH_2-} + H_2O$$

Polyesters may be linear or cross-linked. From the bifunctional monomers terephthalic acid and ethylene glycol, a linear polyester is obtained. Esterification occurs between the alcohol and acid groups on both ends of both monomers, forming long-chain macromolecules:

HOOC—⌬—COOH   HOCH₂CH₂OH
Terephthalic acid   Ethylene glycol

$$-OCH_2CH_2O\left[\overset{\overset{O}{\|}}{C}-⌬-\overset{\overset{O}{\|}}{C}-OCH_2CH_2O\right]_n \overset{\overset{O}{\|}}{C}-⌬-\overset{\overset{O}{\|}}{C}-$$

This polymer may be drawn into fibers or formed into transparent films of great strength. Dacron and Terylene synthetic textiles and Mylar films are made from this polyester. In actual practice, the dimethyl ester of terephthalic acid is used and the molecule split out is methyl alcohol instead of water.

When trifunctional acids or alcohols are used as monomers, cross-linked thermosetting polyesters are obtained (see Figure 27.1). One common example is the reaction of glycerol and *o*-phthalic acid. The polymer formed is one of a group of polymers known as alkyd resins. Glycerol has three functional —OH groups and phthalic acid has two functional —COOH groups.

$$\underset{\text{Glycerol}}{\text{HOCH}_2\overset{\overset{\displaystyle\text{OH}}{|}}{\text{CH}}\text{CH}_2\text{OH}} \qquad \underset{\textit{o}\text{-Phthalic acid}}{\text{C}_6\text{H}_4(\text{COOH})_2}$$

A cross-linked macromolecular structure is formed which, with modifications, has proved to be one of the most outstanding materials used in the coatings industry. Alkyd resins have been used as "baked-on" finishes for automobiles and household appliances. Each year more than 800 million pounds of these resins are used in paints, varnishes, lacquers, electrical insulation, and so on.

Polyamides

Although there are several nylons, one of the best known and the first commercially successful polyamide is Nylon-66. This polymer was so named because it was made from two six-carbon monomers, adipic acid [$\text{HOOC}(\text{CH}_2)_4\text{COOH}$] and hexamethylene diamine [$\text{H}_2\text{N}(\text{CH}_2)_6\text{NH}_2$]. The polymer chains of polyamides contain recurring amide linkages. The amide linkage can be made by reacting a carboxylic acid group with an amine group:

$$\underset{\substack{\text{Carboxylic}\\\text{acid group}}}{-\overset{\overset{\displaystyle\text{O}}{\|}}{\text{C}}-(\text{OH}\ +\ \text{H})}\underset{\substack{\text{Amine}\\\text{group}}}{\overset{\overset{\displaystyle\text{H}}{|}}{\text{N}}-\text{CH}_2-} \xrightarrow{\Delta} \underset{\substack{\text{Amide}\\\text{linkage}}}{-\overset{\overset{\displaystyle\text{O}}{\|}}{\text{C}}-\text{NH}-\text{CH}_2-} + \text{H}_2\text{O}$$

The repeating structural unit of the Nylon-66 chain consists of one adipic acid unit and one hexamethylene diamine unit:

$$-\text{NH}(\text{CH}_2)_6-\text{NH}\left[\overset{\overset{\displaystyle\text{O}}{\|}}{\text{C}}(\text{CH}_2)_4-\overset{\overset{\displaystyle\text{O}}{\|}}{\text{C}}-\text{NH}(\text{CH}_2)_6-\text{NH}\right]_n\overset{\overset{\displaystyle\text{O}}{\|}}{\text{C}}-(\text{CH}_2)_4-\overset{\overset{\displaystyle\text{O}}{\|}}{\text{C}}-$$

Segment of Nylon-66 polyamide

Nylon was developed as a synthetic fiber for the production of stockings and other wearing apparel. It was introduced to the public at the New York

World's Fair in 1939. Nylon is used to make fibers for clothing and carpeting, filaments for fishing lines and ropes, bristles for brushes, and molded objects such as gears and bearings. For the latter application no lubrication is required, because nylon surfaces are inherently slippery.

Phenol–formaldehyde Polymers

As noted earlier, a phenol–formaldehyde condensation polymer (Bakelite) was first marketed over 65 years ago. Polymers of this type are still widely used, especially in electrical equipment, because of their insulating and fire-resisting properties.

Each phenol molecule can react with formaldehyde to lose an H atom from the *para* position and each of the *ortho* positions (indicated by arrows).

Each formaldehyde molecule reacts with two phenol molecules to eliminate water.

Similar reactions occur at the other two reactive sites on each phenol molecule, leading to the formation of the polymer. This polymer is thermosetting because it has an extensively cross-linked network structure. A typical section of this structure is illustrated as follows.

Phenol–formaldehyde polymer

Polyurethanes

The compound urethane has structural features of both an ester and an amide. Its formula is

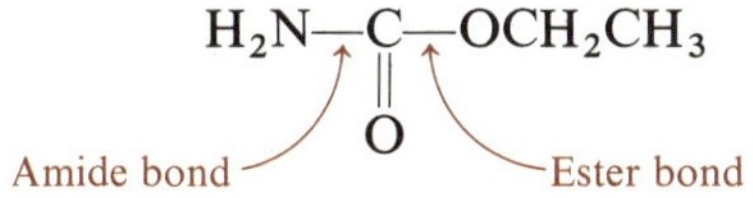

$$\text{H}_2\text{N}-\underset{\underset{\text{O}}{\|}}{\text{C}}-\text{OCH}_2\text{CH}_3$$

Amide bond — Ester bond

A substituted urethane can be made by reacting an isocyanate with an alcohol. For example,

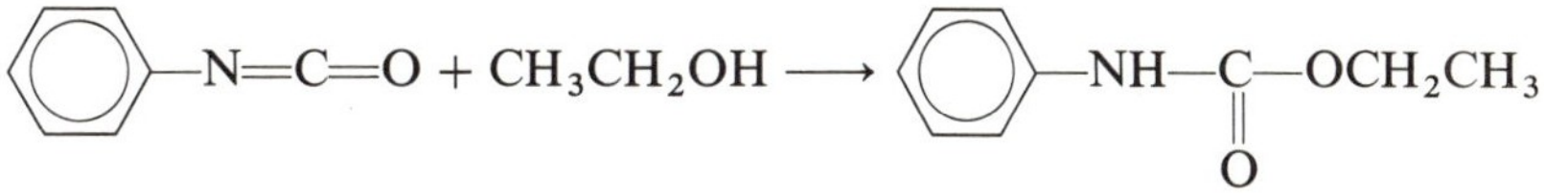

C₆H₅—N=C=O + CH₃CH₂OH ⟶ C₆H₅—NH—C(=O)—OCH₂CH₃

Phenyl isocyanate — N-Phenyl urethane

Diisocyanates and diols are both difunctional; therefore, they yield polymers called *polyurethanes*. The polyurethanes are classified as condensation polymers, although no water or other small molecule is split out when they are formed. From phenylene diisocyanate and ethylene glycol we obtain a polyurethane that has a structure as shown.

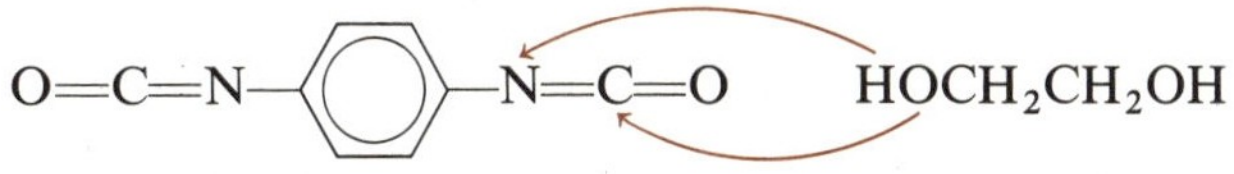

O=C=N—C₆H₄—N=C=O   HOCH₂CH₂OH

p-Phenylene diisocyanate   Ethylene glycol

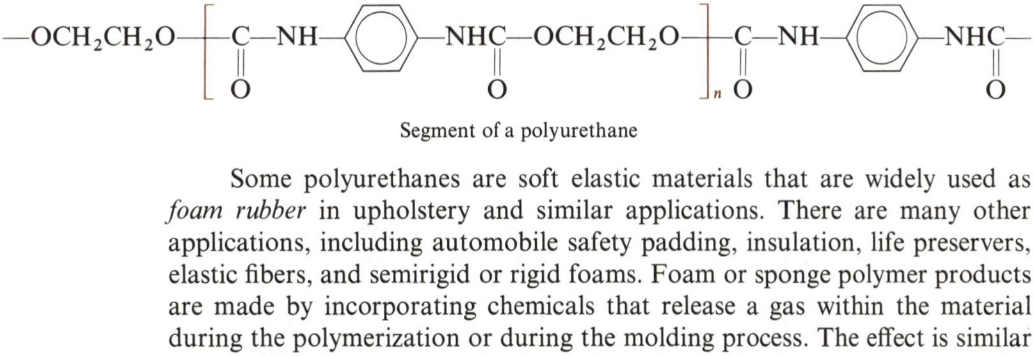

—OCH₂CH₂O—[C(=O)—NH—C₆H₄—NHC(=O)—OCH₂CH₂O—]ₙ—C(=O)—NH—C₆H₄—NHC(=O)—

Segment of a polyurethane

Some polyurethanes are soft elastic materials that are widely used as *foam rubber* in upholstery and similar applications. There are many other applications, including automobile safety padding, insulation, life preservers, elastic fibers, and semirigid or rigid foams. Foam or sponge polymer products are made by incorporating chemicals that release a gas within the material during the polymerization or during the molding process. The effect is similar to that of baking powder releasing carbon dioxide in dough, causing it to rise and become light. The result is a polymer containing innumerable tiny gas-filled cavities that give the product a spongelike quality.

27.8 Silicone Polymers

The silicones are an unusual group of polymers. They include oils and greases, molding resins, rubbers (elastomers), and Silly Putty, the latter being a remarkable material that bounces like a rubber ball when dropped but that can be shaped like putty! Silicones have properties common to both organic and inorganic compounds. The mineral quartz (found in igneous rocks and sand) has the empirical formula SiO$_2$ and is actually a mineral high polymer. Each silicon atom is bonded to four oxygen atoms, and each oxygen to two silicon atoms to form a three-dimensional structure:

$$\begin{array}{c} | \quad\quad | \\ \mathrm{O} \quad\quad \mathrm{O} \\ | \quad\quad | \\ -\mathrm{Si}-\mathrm{O}-\mathrm{Si}-\mathrm{O}-\mathrm{Si}-\mathrm{O}-\mathrm{Si}- \\ | \quad\quad | \\ \mathrm{O} \quad\quad \mathrm{O} \\ | \quad\quad | \\ -\mathrm{Si}- \quad -\mathrm{Si}- \\ | \quad\quad | \end{array}$$

The silicon–oxygen bonds are very strong, and quartz is stable at very high temperatures. But it is also very hard and brittle and difficult to form into desired shapes.

Linear silicone polymers (or polysiloxanes) consist of silicon–oxygen chains with two alkyl groups attached to each silicon atom.

$$\begin{array}{c} \mathrm{R} \quad\;\; \left[\mathrm{R}\quad\right] \;\; \mathrm{R} \\ | \quad\;\; \left[| \quad\right] \;\; | \\ -\mathrm{O}-\mathrm{Si}-\mathrm{O}-\mathrm{Si}-\mathrm{O}-\mathrm{Si}-\mathrm{O}- \\ | \quad\;\; \left[| \quad\right]_n \;\; | \\ \mathrm{R} \quad\;\; \left[\mathrm{R}\quad\right] \;\; \mathrm{R} \end{array}$$

The physical properties of silicones can be modified (1) by varying the length of the polymer chain, (2) by varying the R groups, and (3) by introducing cross-linking between the chains.

Because of their special properties, silicones have found a great variety of uses despite their high cost. Some of their useful qualities are (1) good insulating properties (used in high-temperature applications), (2) little viscosity change over a wide temperature range (therefore they are used as lubricating oils and greases at extreme temperatures), (3) excellent water repellency (used to waterproof many types of surfaces), and (4) good biological compatibility (hence their use in medical and plastic surgery applications). A few of the specific uses of silicone polymers are for coatings on printed electronic circuits, synthetic lubricants, hydraulic systems, brake fluids, electrical insulation, foam shields for nuclear power plants, solar energy, heat-transfer systems, hair sprays, body and hand lotions, automobile and furniture polishes, and in urethane foams. Because of their stability and tissue compatibility silicones are frequently used for permanent surgical implants. For example, heart pacemakers are encased in protective casings made of silicone and silicones are used to replace destroyed nose cartilage.

Questions

A. *Review the meanings of the new terms introduced in this chapter.*
 1. Polymerization
 2. Polymer
 3. Monomer
 4. Addition polymer

5. Condensation polymer
6. Thermoplastic polymer
7. Thermosetting polymer
8. Copolymer

B. Review questions.
1. How does condensation polymerization differ from addition polymerization?
2. What property distinguishes a thermoplastic polymer from a thermosetting polymer?
3. Show the free radical mechanism for the polymerization of propylene to polypropylene.
4. How many ethylene units are in a polyethylene molecule that has a molecular weight of approximately 25,000?
5. Write a structural formula showing the polymer that can be formed from:
   (a) Ethylene   (b) Propylene   (c) 1-Butene   (d) 2-Butene
6. Write structural formulas for the following polymers:
   (a) Saran   (c) Teflon   (e) Lucite
   (b) Orlon   (d) Polystyrene
7. Write structures showing two possible ways in which vinyl chloride can polymerize to form polyvinyl chloride. Show four units in each structure.
8. (a) Write the structure for the polymer that can be formed from 2,3-dimethyl-1,3-butadiene.
   (b) Can this substance form *cis* and *trans* polymers? Explain.
9. Write the chemical structures for the monomers of
   (a) Natural rubber   (c) Synthetic natural rubber
   (b) Synthetic (SBR) rubber
10. Why is the useful life of natural rubber and that of many synthetic rubbers shortened in smoggy atmospheres?
11. How are rubber molecules modified by vulcanization?
12. Natural rubber is the all *cis* polymer of isoprene and gutta-percha is the all *trans* isomer. Write the structure for each polymer, showing at least three isoprene units.
13. Nitrile rubber (Buna N) is a copolymer of 2 parts of 1,3-butadiene to 1 part of acrylonitrile (CH_2=CHCN). Write a structure for this synthetic rubber, showing at least two units of the polymer.
14. Ziegler–Natta catalysts can orient the polymerization of propylene to form isotactic polypropylene—that is, polypropylene with all the methyl groups on the same side of the long carbon chain. Write the structure for (a) isotactic polypropylene and (b) another possible geometric form of polypropylene.
15. Write formulas showing the structure of:
    (a) Dacron   (b) Nylon-66   (c) Bakelite   (d) Polyurethane
16. Write structures for the monomers of each of the polymers shown below, and classify each as polyvinyl, polyester, polyamide, or polyurethane.

(a) —$CH_2CHCH_2CHCH_2CHCH_2CH$—
        | | | |
        C=O   C=O   C=O   C=O
        | | | |
       OC_2H_5 OC_2H_5 OC_2H_5 OC_2H_5

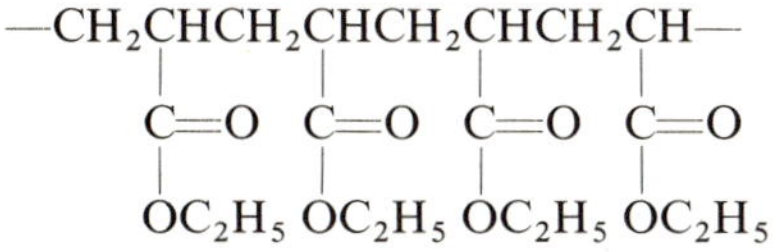

(b) —C—⟨◯⟩—C—OCH_2—⟨◯⟩—CH_2O—C—⟨◯⟩—C—OCH_2—⟨◯⟩—CH_2O—
    ‖   ‖   ‖   ‖
    O   O   O   O

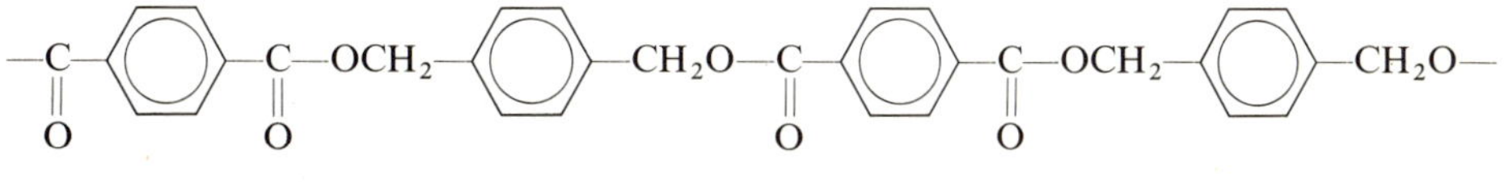

(c) [structure: 3-methylaniline with —NH— substituent, bearing —NHCOCH₂CH₂—O—C(=O)—NH— linked to methylphenyl ring, then —NH—C(=O)—OCH₂CH₂O—C(=O)—NH— linked to another methylphenyl ring with —NH—]

(d) —HN—C(=O)—(CH₂)₈—C(=O)—NH(CH₂)₆NHC(=O)—(CH₂)₈—C(=O)—NH(CH₂)₆NH—

(e)
$$-CH_2\underset{Cl}{\overset{CH_3}{\underset{|}{\overset{|}{C}}}}-CH_2-\underset{Cl}{\overset{CH_3}{\underset{|}{\overset{|}{C}}}}-CH_2-\underset{Cl}{\overset{CH_3}{\underset{|}{\overset{|}{C}}}}-$$

17. Why must at least some monomer molecules be trifunctional to form a thermosetting polyester?
18. Glyptal polyesters are made from glycerol and phthalic acid. Would this kind of polymer more likely be thermoplastic or thermosetting? Explain.
19. How is "foam" introduced into foam or sponge rubber materials?
20. Silicone polymers are more resistant to high temperatures than the usual organic polymers. Suggest an explanation for this property of the silicones.
21. Which of these statements are correct?
    (a) The process of forming macromolecules from small units is called polymerization.
    (b) The monomers in condensation polymerization must be at least bifunctional.
    (c) The monomers of Nylon-66 are a dicarboxylic acid and a diol.
    (d) Dacron and Mylar are both made from the same monomers.
    (e) Hexamethylene diamine is a secondary amine.
    (f) Teflon is an addition polymer.
    (g) Vulcanization was invented by Charles Goodrich.
    (h) The monomer for polystyrene is

    C₆H₅—CH₂CH=CH₂

    (i) Polyurethanes have both ester and amide bonds in their structure.
    (j) Bakelite is a copolymer of phenol and ethylene glycol.
    (k) The monomer of natural rubber is 2-methyl-1,3-butadiene.

28 Optical Isomerism

After studying Chapter 28 you should be able to:

1. Understand the terms listed in Question A at the end of the chapter.
2. Identify all asymmetric (chiral) carbon atoms from a given formula.
3. Explain the use of a polarimeter.
4. Explain how polarized light is obtained.
5. Calculate the specific rotation of a compound.
6. Explain the phenomenon of optical isomerism.
7. Determine whether a compound is chiral.
8. Draw projection formulas for all possible stereoisomers of a given compound. Label enantiomers, diastereomers, and *meso* compounds.
9. Calculate the maximum possible optical isomers from the formula of a compound.
10. Understand the meaning of (+) and (−) relative to the optical activity of a compound.
11. Draw the mirror image of a given structure.
12. Compare projection formulas to ascertain whether they represent identical compounds or enantiomers.
13. Define a racemic mixture.
14. Compare the physical properties of enantiomers, diastereomers, and racemic mixtures.
15. Explain why *meso* compounds are optically inactive.
16. Determine whether optical isomers are formed in simple chemical reactions.

28.1 Review of Isomerism

The phenomenon of two or more compounds having the same number and kind of atoms is isomerism. Thus, isomers have the same composition and the same molecular formula.

There are two types of isomerism. In the first type, known as structural isomerism, the difference between isomers is due to a different structural arrangement of the atoms in the molecules. For example, butane and isobutane, ethanol and dimethyl ether, and 1-chloropropane and 2-chloropropane are structural isomers.

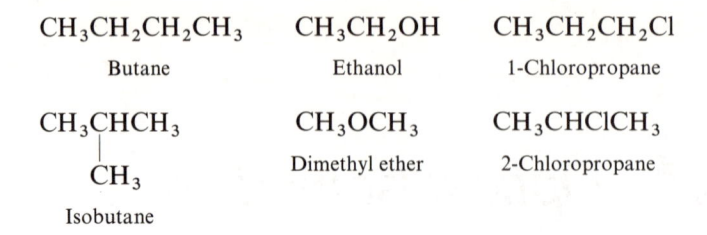

CH₃CH₂CH₂CH₃ CH₃CH₂OH CH₃CH₂CH₂Cl
Butane Ethanol 1-Chloropropane

CH₃CHCH₃ CH₃OCH₃ CH₃CHClCH₃
  | Dimethyl ether   2-Chloropropane
  CH₃
Isobutane

In the second type of isomerism, the isomers have the same structural formulas but differ in the arrangement of the atoms in space. This type of isomerism is known as **stereoisomerism**. Thus compounds that have the same structural formulas but differ in their spatial arrangement are called **stereoisomers**. There are two types of stereoisomers: *cis–trans* or geometric isomers, which we have already considered; and optical isomers, the subject of this chapter. The outstanding feature of optical isomers is that they have the ability to rotate plane-polarized light.

28.2 Plane-Polarized Light

What is plane-polarized light? **Plane-polarized light** is light that has passed through a polarizer and is vibrating in only one plane. Oridnary light consists of electromagnetic waves vibrating in all directions (planes) perpendicular to the direction in which it is traveling. When ordinary light is passed through a polarizer, the light that emerges is vibrating in only one plane and is called plane-polarized light (see Figure 28.1).

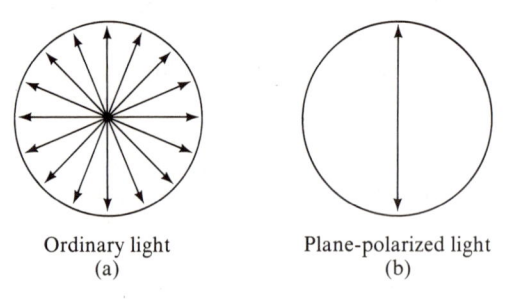

Ordinary light
(a)

Plane-polarized light
(b)

Figure 28.1. (a) Diagram of ordinary light vibrating in all possible planes and (b) diagram of plane-polarized light vibrating in a single plane.

Polarizers can be made from calcite or tourmaline crystals or from Polaroid film, which is a transparent plastic containing properly oriented imbedded crystals. Two pieces of Polaroid film with their axes parallel will allow the passage of plane-polarized light. But when one piece is placed so that its axis is at right angles to that of the other piece, the passage of light is blocked and the color appears black (see Figure 28.2).

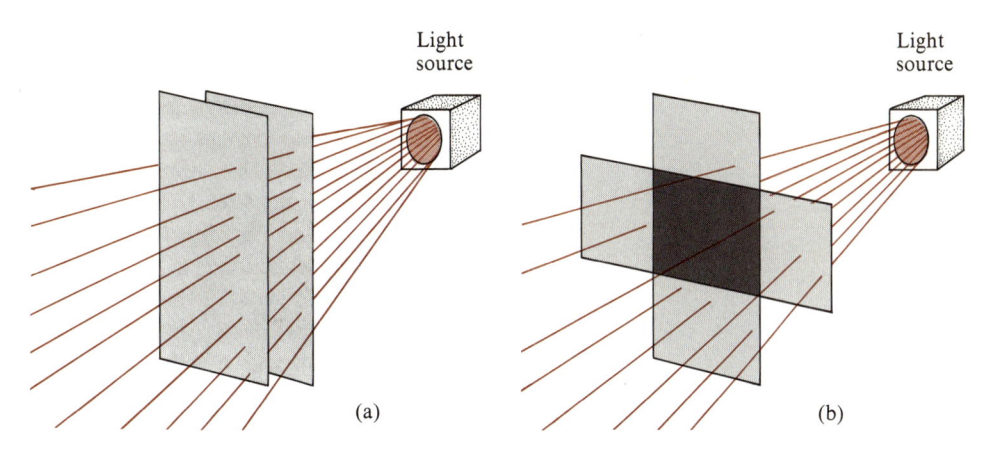

Figure 28.2. Two pieces of Polaroid film (a) with axes parallel and (b) with axes at right angles. In part (a), light passes through both pieces of film and emerges polarized. In part (b), the polarized light that emerges from one film is blocked and does not pass through the second film, which is at right angles to the first. With no light emerging, the film appears black.

Specific Rotation

The ability of a substance to rotate the plane of polarized light is measured with a polarimeter. The essential features of this instrument are (1) a light source (usually a sodium lamp), (2) a polarizer, (3) a sample tube, (4) an analyzer (which is another matched polarizer), and (5) a calibrated scale (360°) to measure the number of degrees the plane of polarized light is rotated. The calibrated scale is attached to the analyzer (see Figure 28.3). When the sample tube contains a solution of an optically inactive material, the axes of the polarizer and the analyzer will be parallel and the scale will be at zero degrees with the light passing through at a maximum. When a solution of an optically active substance is placed in the sample tube, the plane in which the polarized light is vibrating is rotated through an angle. The analyzer is then rotated to the position where the emerging light is at a maximum intensity. The number of degrees and the direction of rotation by the solution is then read from the scale as the observed rotation.

specific rotation

The **specific rotation**, $[\alpha]$, of a compound is the number of degrees that polarized light would be rotated by passage through 1 decimetre (dm) of a solution of the substance at a concentration of 1 g per millilitre. The specific rotation of optically active substances is listed in chemical handbooks along with other physical properties. The following formula is used to calculate specific rotation from polarimeter data:

$$[\alpha] = \frac{\text{Observed rotation in degrees}}{\left(\begin{array}{c}\text{Length of}\\ \text{sample tube, decimetres}\end{array}\right)\left(\begin{array}{c}\text{Sample concentration,}\\ \text{grams per millilitre}\end{array}\right)}$$

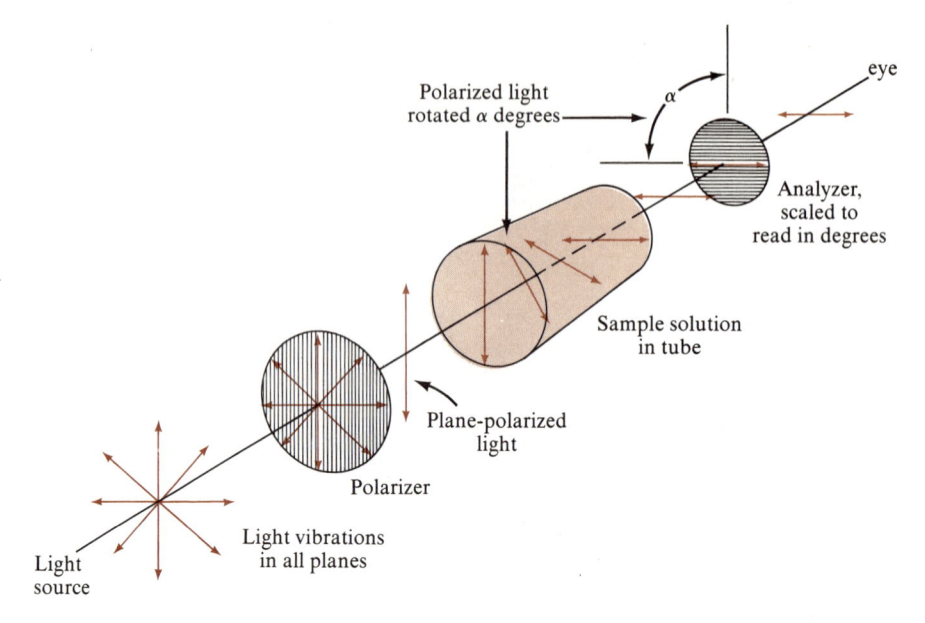

Figure 28.3. Schematic diagram of a polarimeter. This instrument is used to measure the angle, α, through which an optically active substance rotates the plane of polarized light.

28.3 Optical Activity

optical activity

dextrorotatory

levorotatory

Many naturally occurring substances are able to rotate the plane of polarized light. Such substances are said to be **optically active**. When plane-polarized light passes through an optically active substance, the plane of the polarized light is rotated. If the rotation of the plane is to the right (clockwise), the substance is said to be **dextrorotatory**; if the rotation is to the left (counterclockwise), the substance is said to be **levorotatory**.

Some minerals, notably quartz, rotate the plane of polarized light. In fact, optical activity was discovered in minerals. However, when such mineral crystals are melted, the optical activity disappears. This means that, for these substances, the optical activity must be due to an ordered arrangement within the crystals.

In 1848, Louis Pasteur (1822–1895) observed that sodium ammonium tartrate, a salt of tartaric acid, existed as a mixture of two different kinds of crystals. Pasteur carefully hand-separated the two kinds of crystals. Investigating their properties, he found that solutions made from either kind of crystal would rotate the plane of polarized light—but in opposite directions, Since this activity was present in a solution, it could not be caused by a specific arrangement within a crystal.

In 1874, J. H. van't Hoff (1852–1911) and J. A. Le Bel (1847–1930) concluded that the presence of at least one asymmetric carbon atom in the molecule

asymmetric carbon atom

of an optically active substance is a key factor in optical activity. An **asymmetric carbon atom** is one to which four different atoms or groups are attached; it is a center of dissymmetry in a molecule. The lack of symmetry is due to the tetrahedral nature of this carbon atom.

$$\begin{array}{c} D \\ | \\ A \diagup \overset{\circ}{|} \diagdown B \\ C \end{array}$$

◯ = Carbon atom

A, B, C, D = Four different atoms or groups of atoms

The first Nobel Prize in chemistry was awarded to van't Hoff in 1901.

Molecules of optically active substances must have at least one center of dissymmetry. Although optically active compounds are known that do not contain asymmetric carbon atoms (that is, the center of dissymmetry is due to some other structural feature), most optically active organic substances do contain one or more asymmetric carbon atoms.

28.4 Projection Formulas

Molecules of a compound that contain one asymmetric carbon atom occur in two optically active isomeric forms. This is because the four groups attached to the asymmetric carbon atom can be oriented in space in two different arrangements. It is important to understand how we represent such isomers on paper. Let us consider the spatial arrangement of a lactic acid molecule, which contains one asymmetric carbon atom.

$$\begin{array}{c} {}^1COOH \\ | \\ H-{}^2C^*-OH \\ | \\ {}^3CH_3 \end{array}$$

C* = Asymmetric carbon atom

Lactic acid

Three-dimensional models are the best means of representing such a molecule, but by adopting certain conventions and using our imaginations, we can formulate the needed images on paper. The geometrical arrangement of the four groups about the asymmetric carbon (carbon 2) is the key to the stereoisomerism of lactic acid. The four bonds attached to carbon 2 are separated by angles of about 109° and are directed outward to the corners of an imaginary tetrahedron, as shown by the tetrahedral diagrams I and II in Figure 28.4. Diagram III is a three-dimensional representation in which the asymmetric carbon atom is represented as a sphere, with its center in the plane of the paper. The —H and —OH groups are projected forward from the paper (toward the observer) and the —COOH and —CH₃ groups are projected back from the paper (away from the observer). Diagrams I, II, and III represent three different ways to illustrate three-dimensional tetrahedral structures.

626   Chapter 28

COOH

(Structures I, II, III, IV, V showing three-dimensional representations)

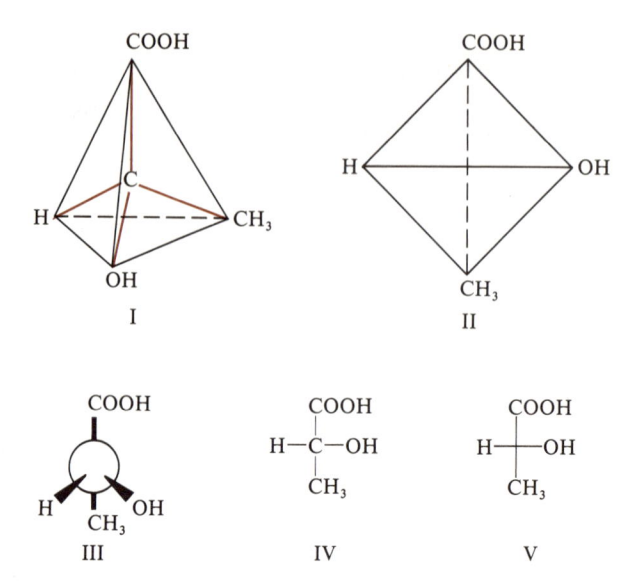

I   II   III   IV   V

Figure 28.4. Methods of representing three-dimensional formulas of a compound that contains one asymmetric carbon atom. All five structures represent the same molecule.

projection formula

For convenience of expression, simpler diagrams such as IV and V are used. These are much easier and faster to draw. In IV, it is understood that the groups (—H and —OH) attached to the horizontal bonds are coming out of the plane of the paper toward the observer, and the groups attached to the vertical bonds are projected backward away from the paper. The molecule in formula V is made by drawing a cross and attaching the four groups in their respective positions, as in formula IV. The asymmetric carbon atom is understood to be located where the lines cross. Formulas IV and V are called **projection formulas**.

It is important to be careful when comparing projection formulas such as IV and V (Figure 28.4). They must not be rotated 90° or lifted out of the plane of the paper. The formulas may, however, be rotated 180°. Consider the following formulas:

COOH	CH₃	H
H—⊢—OH	HO—⊢—H	CH₃—⊢—COOH
CH₃	COOH	OH
V	VI	VIII

Formulas III, V, VI, and VII represent identical molecules. Formula VI is a 180° rotation of formula V. Formula VII is formula VI drawn in a

```
    COOH            CH₃             H
     |               |              |
  H─C─OH         HO─C─H         CH₃─C─COOH
     |               |              |
    CH₃            COOH            OH

     III            VII             IX
```

(V rotated 180°)

three-dimensional representation. Formulas VIII and IX would be obtained by rotating formula V only 90° in the plane of the paper, but since this is not allowed, they represent the other stereoisomer of lactic acid.

28.5 Enantiomers

super-imposable

Your right and your left hands are mirror images of each other; that is, the left hand is a mirror reflection of the right hand and vice versa. Furthermore, the two hands are not superimposable on each other (see Figure 28.5). **Superimposable** means that when we lay one object upon another, all parts of both objects will coincide exactly.

(a) Left hand | Mirror image of right hand | Right hand | (b) Right and left hands are not superimposable

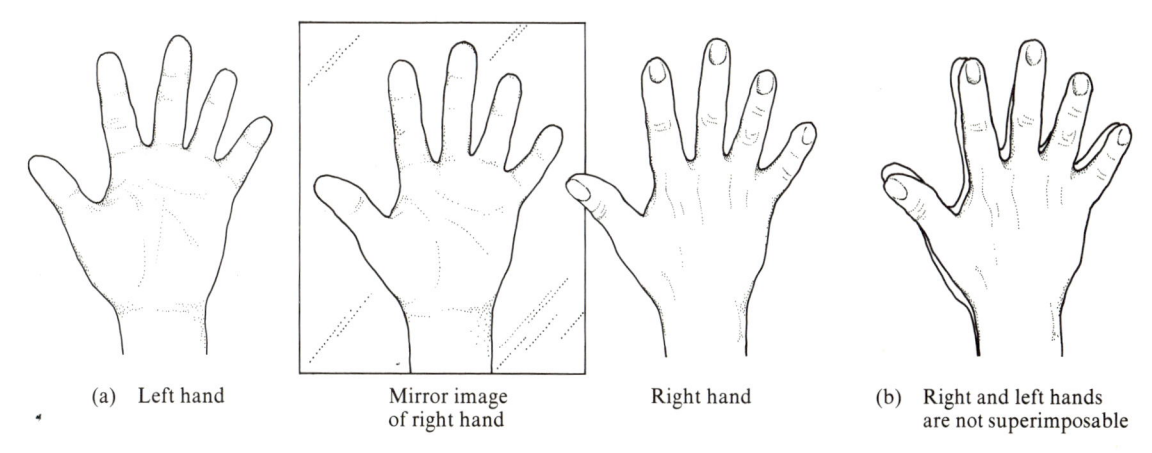

Figure 28.5. (a) Left hand shown is the same as the mirror image of the right hand. (b) Right and left hands are not superimposable, hence they are chiral.

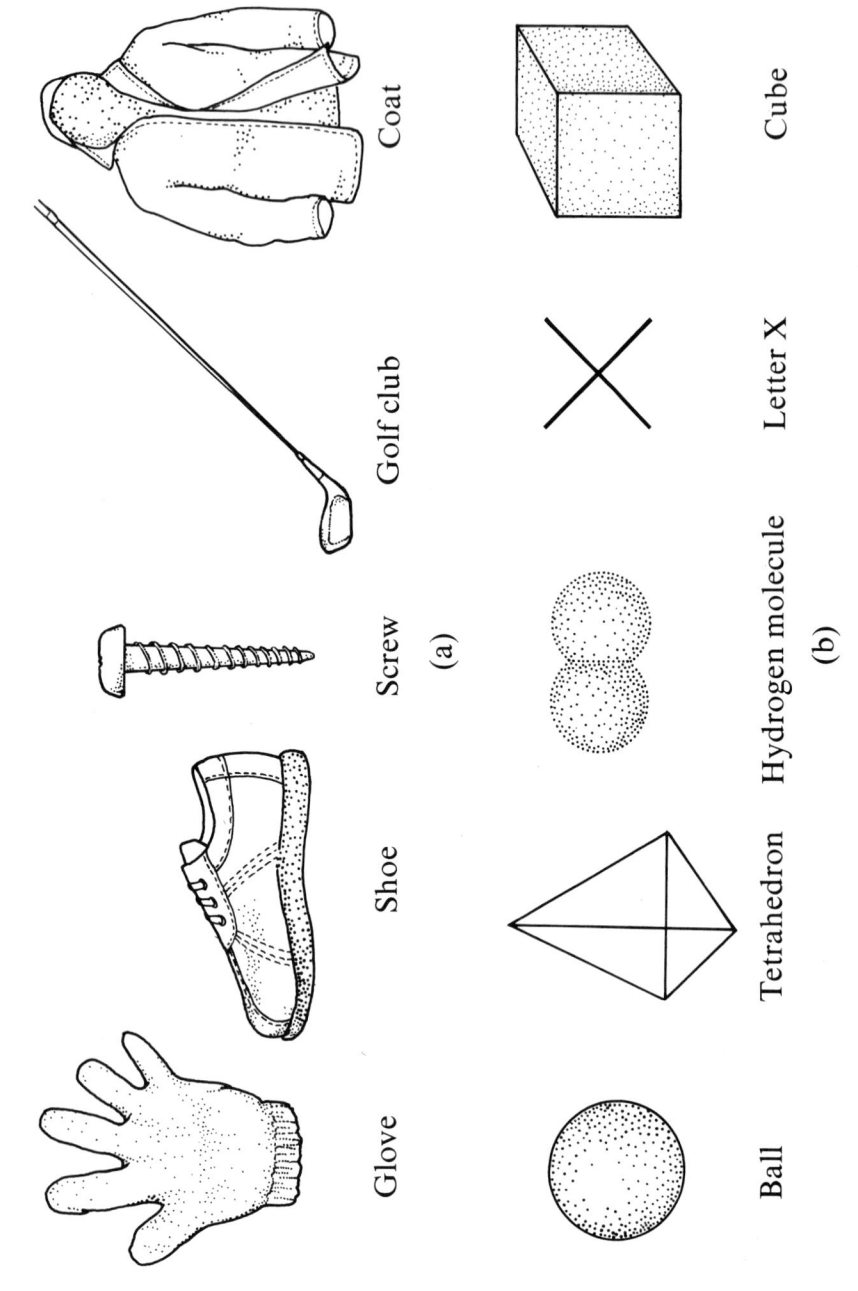

Figure 28.6. (a) Chiral and (b) achiral objects.

Optical Isomerism 629

chiral

achiral

A molecule that is not superimposable on its mirror image is said to be **chiral**. The word chiral comes from the Greek word *cheir*, meaning "hand." Chiral molecules have the property of "handedness"—that is, they are related to each other in the same manner as the right and left hand. Molecules, or objects, that are superimposable are **achiral**. Some chiral and achiral objects are shown in Figure 28.6.

The formulas developed in Section 28.4 dealt primarily with a single kind of lactic acid molecule. But two stereoisomers of lactic acid are known—one that rotates the plane of polarized light to the right and one that rotates it to the left. These two forms of lactic acid are shown below:

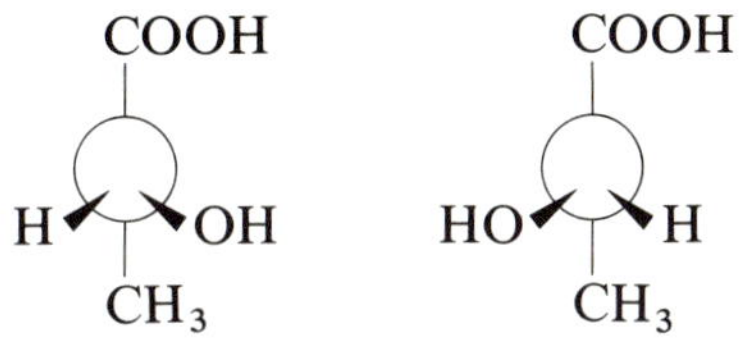

$$\begin{array}{cc}
\text{COOH} & \text{COOH} \\
| & | \\
\text{H} \diagdown\text{OH} & \text{HO}\diagup\text{H} \\
| & | \\
\text{CH}_3 & \text{CH}_3
\end{array}$$

If we examine these two structural formulas carefully, we can see that they are mirror images of each other. The reflection of either molecule in a mirror corresponds to the structure of the other molecule. Even though the two molecules have the same molecular formula and the same four groups attached to the central carbon atom, they are not superimposable. Therefore, the two molecules are not identical, but are isomers. One molecule will rotate the plane of polarized light to the left, the other will rotate it to the right. A plus (+) or a minus (−) sign written in parentheses in front of the name or formula indicates the direction of rotation of polarized light and becomes part of the name of the compound. Plus (+) indicates rotation to the right and minus (−) to the left. Using projection formulas, the two lactic acids are written:

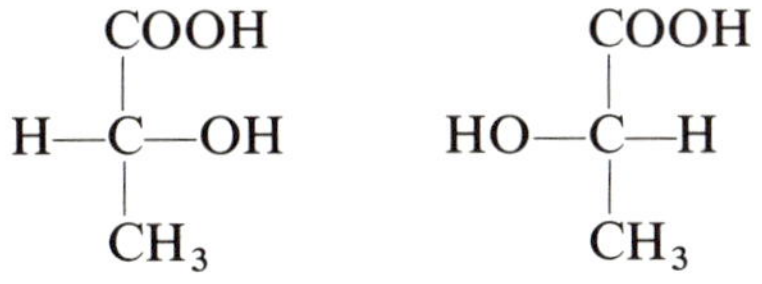

$$\begin{array}{cc}
\text{COOH} & \text{COOH} \\
| & | \\
\text{H}-\text{C}-\text{OH} & \text{HO}-\text{C}-\text{H} \\
| & | \\
\text{CH}_3 & \text{CH}_3 \\
(-)\text{-Lactic acid} & (+)\text{-Lactic acid}
\end{array}$$

enantiomer

Originally, it was not known which lactic acid structure was the (+) or the (−) compound. However, it is now known that they are as shown. Isomers that are mirror images of each other are called **enantiomers**.

Many, but not all, molecules that contain an asymmetric carbon are chiral. Most of the molecules we shall study that have an asymmetric carbon atom are chiral. To decide whether a molecule is chiral and has an enantiomer, make models of the molecule and of its mirror image and see if they are superimposable. This is the ultimate test, but instead of making models every time, first examine the formula to see if it has an asymmetric carbon atom. If an asymmetric carbon atom is found, draw a cross and attach the four groups on the asymmetric carbon to the four ends of the cross. The asymmetric carbon is understood to be located where the lines cross. Remember that an asymmetric carbon atom has four different groups attached to it. Let us test the compounds 2-butanol and 2-chloropropane:

$$CH_3CH_2CHCH_3 \atop |\ \ \ \ \ \ \ \ \ \ OH$$ $$CH_3CHCH_3 \atop |\ \ \ \ \ \ Cl$$

2-Butanol   2-Chloropropane

In 2-butanol, carbon 2 is asymmetric. The four groups attached to carbon 2 are H, OH, CH$_3$, and CH$_2$CH$_3$. Draw the structure and its mirror image:

```
        CH3                CH3                CH3
         |                  |                  |
  H——————OH         HO——————H          CH2
         |                  |                  |
        CH2                CH2         H——————OH
         |                  |                  |
        CH3                CH3                CH3
        (a)    Enantiomers  (b)         (b) rotated 180°
```

Rotating structure (b) 180° allows H and OH to coincide with (a), but CH$_3$ and CH$_2$CH$_3$ do not coincide. We conclude that the mirror images (a) and (b) are enantiomers since they are not superimposable.

In 2-chloropropane, the four groups attached to carbon 2 are H, Cl, CH$_3$, and CH$_3$. Note that two groups are the same. Draw the structure and its mirror image:

```
        CH3                CH3                CH3
         |                  |                  |
  H——————Cl         Cl——————H          H——————Cl
         |                  |                  |
        CH3                CH3                CH3
        (c)                (d)          (d) rotated 180°
```

When we rotate structure (d) 180°, the two structures are superimposable, proving that 2-chloropropane, which does not have an asymmetric carbon, does not exist in enantiomeric forms.

Problem 28.1   Draw mirror images for any of the following compounds that can exist as enantiomers:
(a) CH$_3$CH$_2$CH$_2$OH   (b) CH$_3$CH$_2$CHClCH$_2$CH$_3$
(c) CH$_3$CH$_2$CHClCH$_2$CH$_3$

First check each formula for asymmetric carbon atoms.
(a) No asymmetric carbon atoms; each carbon has at least two groups that are the same.
(b) No asymmetric carbon atoms; carbon 3 has H, Cl, and two CH$_2$CH$_3$ groups.
(c) Carbon 3 is asymmetric; the four groups on carbon 3 are H, Cl, CH$_3$CH$_2$, and CH$_3$CH$_2$CH$_2$. Draw mirror images:

```
              H                               H
              |                               |
   CH3CH2————C————CH2CH2CH3      CH3CH2CH2————C————CH2CH3
              |                               |
              Cl                              Cl
```

The relationship between enantiomers is such that if we change the positions of any two groups on a compound containing only one asymmetric carbon atom, we obtain the structure of its enantiomer. If we make a second change, the structure of the original isomer is obtained again. In both cases shown below, (+)-lactic acid is formed by interchanging the positions of two groups on (−)-lactic acid.

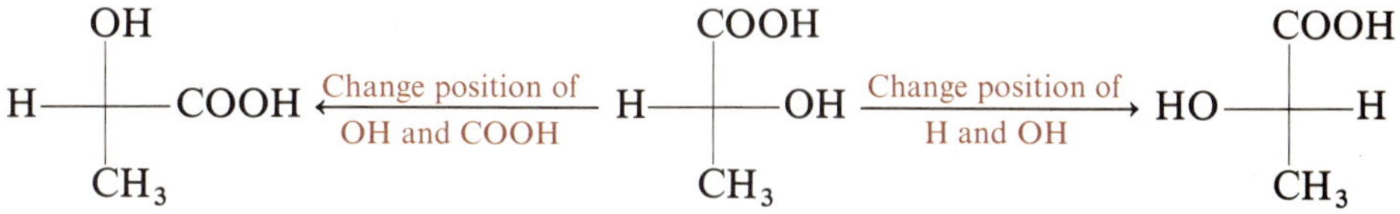

$$\underset{(+)\text{-Lactic acid}}{\overset{\text{OH}}{\underset{\text{CH}_3}{\text{H}\!\!-\!\!\!\!\!-\!\!\text{COOH}}}} \xleftarrow[\text{OH and COOH}]{\text{Change position of}} \underset{(-)\text{-Lactic acid}}{\overset{\text{COOH}}{\underset{\text{CH}_3}{\text{H}\!\!-\!\!\!\!\!-\!\!\text{OH}}}} \xrightarrow[\text{H and OH}]{\text{Change position of}} \underset{(+)\text{-Lactic acid}}{\overset{\text{COOH}}{\underset{\text{CH}_3}{\text{HO}\!\!-\!\!\!\!\!-\!\!\text{H}}}}$$

To compare two projection formulas to test whether they are the same structure or enantiomers, (1) rotate one structure 180° and compare or (2) make successive group interchanges until the formulas appear to be identical. If an odd number of interchanges was made, the two original formulas represent enantiomers; if an even number of interchanges was made, the two formulas represent the same compound. The following two examples illustrate this method:

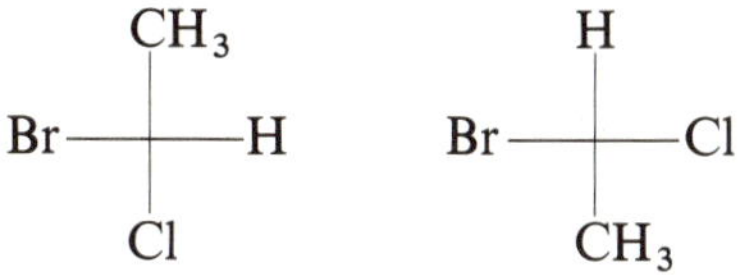

Structure I: Br—C(CH₃)(H)—Cl (with CH₃ up, Cl down, Br left, H right)

Structure II: Br—C(H)(CH₃)—Cl (with H up, CH₃ down, Br left, Cl right)

II → Interchange CH₃ and Cl → Br—C(H)(CH₃)—Cl (H up, Cl down, Br left, CH₃ right) → Interchange CH₃ and H → Br—C(CH₃)(H)—Cl (CH₃ up, Cl down, Br left, H right)

Two interchanges were needed to make structure II identical to structure I. Therefore, structures I and II represent the same compound. Do structures III and IV represent the same compound?

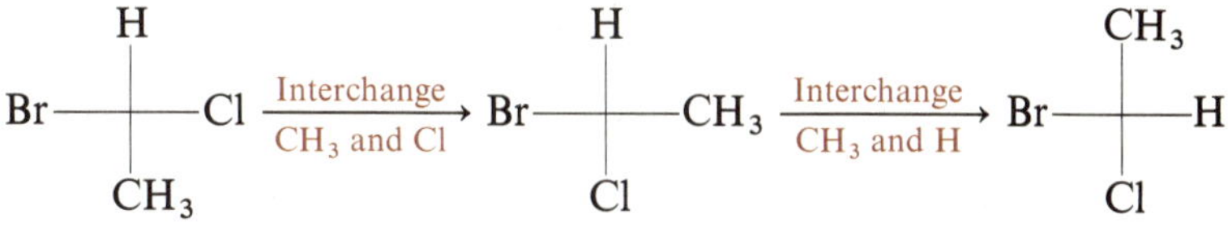

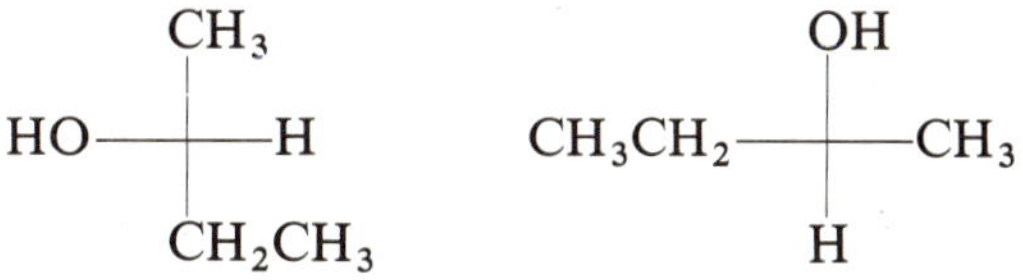

Structure III: HO—C(CH₃)(CH₂CH₃)—H (CH₃ up, CH₂CH₃ down, HO left, H right)

Structure IV: CH₃CH₂—C(OH)(H)—CH₃ (OH up, H down, CH₃CH₂ left, CH₃ right)

$$\text{CH}_3\text{CH}_2\overset{\text{OH}}{\underset{\text{H}}{\vert}}\text{CH}_3 \xrightarrow[\text{CH}_3\text{CH}_2 \text{ and OH}]{\text{Interchange}} \text{HO}\overset{\text{CH}_2\text{CH}_3}{\underset{\text{H}}{\vert}}\text{CH}_3 \xrightarrow[\text{CH}_3 \text{ and H}]{\text{Interchange}}$$

IV

$$\rightarrow \text{HO}\overset{\text{CH}_2\text{CH}_3}{\underset{\text{CH}_3}{\vert}}\text{H} \xrightarrow[\text{CH}_3 \text{ and CH}_3\text{CH}_2]{\text{Interchange}} \text{HO}\overset{\text{CH}_3}{\underset{\text{CH}_2\text{CH}_3}{\vert}}\text{H}$$

Three interchanges were needed to make structure IV identical to structure III. Therefore, structures III and IV do not represent the same compound, but are enantiomers.

Enantiomers ordinarily have the same chemical properties and the same physical properties other than optical rotation (see Table 28.1). They rotate plane-polarized light the same number of degrees but in opposite directions. Enantiomers differ in their physiological properties. Since their reactions are stereoselective, enzymes act on only one of a pair of enantiomers. For example, enzyme-catalyzed reduction of pyruvic acid in muscle tissue yields only (+)-lactic acid, but reduction catalyzed with H_2/Pt yields both (+)- and (−)-lactic acids (see Section 28.6).

28.6 Racemic Mixtures

racemic mixture

A mixture containing equal amounts of a pair of enantiomers is known as a **racemic mixture**. Such a mixture is optically inactive and will show no rotation of polarized light when tested in a polarimeter. Each of the enantiomers rotates the plane of polarized light by the same amount but in opposite directions. The (±) symbol is often used to designate racemic mixtures. For example, a racemic mixture of lactic acid is written as (±)-lactic acid because this mixture contains equal molar amounts of (+)-lactic acid and (−)-lactic acid.

Racemic mixtures are usually obtained in laboratory syntheses of compounds where an asymmetric carbon atom is formed. Thus, catalytic reduction of pyruvic acid to lactic acid produces a racemic mixture containing equal amounts of (+)- and (−)-lactic acid:

$$\underset{\text{Pyruvic acid}}{\text{CH}_3\underset{\underset{\text{O}}{\|}}{\text{C}}\text{COOH}} + H_2 \xrightarrow{\text{Ni}} \underset{(\pm)\text{-Lactic acid}}{\text{CH}_3\underset{\underset{\text{OH}}{\vert}}{\text{CH}}\text{COOH}}$$

As a general rule in the biological synthesis of potentially optically active compounds, only one of the isomers is produced. For example, (+)-lactic acid is produced by reactions occurring in muscle tissue, and (−)-lactic acid is produced by lactic acid bacteria in the souring of milk. These stereospecific reactions occur because biochemical syntheses are enzyme-catalyzed. The preferential production of one isomer over another is apparently due to the configuration (shape) of the specific enzyme involved. Returning to the hand analogy—if the "right-handed" enantiomer is produced, then the enzyme responsible for the product can be likened to a right-handed glove.

The mirror-image isomers (enantiomers) of a racemic mixture are alike in all ordinary physical properties except in their action on polarized light. However, they usually differ in their biochemical properties. It is possible to separate or resolve racemic mixtures into their optically active components. In fact, Pasteur's original work with sodium ammonium tartrate involved such a separation. But a general consideration of the methods involved in such separations is beyond the scope of our present discussion.

28.7 Diastereomers and Meso Compounds

The enantiomers (mirror-image isomers) discussed in the preceding sections are stereoisomers. That is, they differ only in the spatial arrangement of the atoms and groups within the molecule. The number of stereoisomers increases as the number of centers of dissymmetry increases. The maximum number of stereoisomers for a given compound is obtained by the formula 2^n, where n is the number of asymmetric carbon atoms in the compound. As we have seen, there are two ($2^1 = 2$) stereoisomers of lactic acid. But for a substance with two asymmetric carbon atoms, such as 2-bromo-3-chlorobutane ($CH_3CHBrCHClCH_3$), four stereoisomers are possible. These four possible stereoisomers are written as projection formulas in this way (carbons 2 and 3 are asymmetric):

$$
\begin{array}{cccc}
\text{CH}_3 & \text{CH}_3 & \text{CH}_3 & \text{CH}_3 \\
| & | & | & | \\
\text{H—C—Br} & \text{Br—C—H} & \text{H—C—Br} & \text{Br—C—H} \\
| & | & | & | \\
\text{H—C—Cl} & \text{Cl—C—H} & \text{Cl—C—H} & \text{H—C—Cl} \\
| & | & | & | \\
\text{CH}_3 & \text{CH}_3 & \text{CH}_3 & \text{CH}_3 \\
\text{I} & \text{II} & \text{III} & \text{IV} \\
\underbrace{\qquad\qquad\qquad}_{\text{Enantiomers}} & & \underbrace{\qquad\qquad\qquad}_{\text{Enantiomers}} &
\end{array}
$$

Remember that for comparison, projection formulas may be rotated 180° in the plane of the paper, but they cannot be lifted out of the plane. Formulas I and II and formulas III and IV represent two pairs of nonsuperimposable mirror-image isomers and are, therefore, two pairs of enantiomers. Stereoisomers that are not mirror images of each other are called **diastereomers**. Four different pairs of diastereomers of 2-bromo-3-chlorobutane exist. They are I and III, I and IV, II and III, and II and IV.

diastereomer

The 2^n formula indicates that four stereoisomers of tartaric acid are possible. The projection formulas of these four possible stereoisomers are written in this way (carbons 2 and 3 are asymmetric):

```
     COOH              COOH              COOH              COOH
      |                 |                 |                 |
HO—C—H             H—C—OH            H—C—OH           HO—C—H
      |                 |                 |                 |
 H—C—OH            HO—C—H            H—C—OH           HO—C—H
      |                 |                 |                 |
     COOH              COOH              COOH              COOH
       V                VI                VII              VIII
```

Formulas V and VI represent nonsuperimposable mirror-image isomers and are, therefore, *enantiomers*. Formulas VII and VIII are also mirror images. But by rotating VIII 180°, we see that it is exactly superimposable on VII. Therefore, VII and VIII represent the same compound, and only *three* stereoisomers of tartaric acid actually exist. Compound VII is achiral and does not rotate polarized light. A plane of symmetry can be passed between carbons 2 and 3 so that the top and bottom halves of the molecule are mirror images:

```
     COOH
      |
 H—C—OH
      |          ——Plane of symmetry
 H—C—OH
      |
     COOH
```

Thus, the molecule is internally compensated. The rotation of polarized light in one direction by half of the molecule is exactly compensated by an opposite rotation by the other half. Stereoisomers that contain asymmetric carbon atoms and are superimposable on their own mirror images are called **meso compounds** or **meso structures**. All *meso* compounds are optically inactive.

meso compound

meso structure

The term *meso* comes from the Greek word *mesos*, meaning "middle." It was first used by Pasteur to name a kind of tartaric acid that was optically inactive and could not be separated into different forms by any means. Pasteur called it *meso*-tartaric acid, because it seemed intermediate between the (+)- and (−)-tartaric acids. The three stereoisomers of tartaric acid are represented and designated in this fashion:

```
     COOH              COOH              COOH
      |                 |                 |
HO—C—H             H—C—OH            H—C—OH
      |                 |                 |
 H—C—OH            HO—C—H            H—C—OH
      |                 |                 |
     COOH              COOH              COOH
 (−)-Tartaric acid   (+)-Tartaric acid   meso-Tartaric acid
       V                 VI                 VII
```

The physical properties of tartaric acid stereoisomers are given in Table 28.1. Note that the properties of (+)-tartaric acid and (−)-tartaric acid, except for opposite rotation of polarized light, are identical. However, *meso*-tartaric acid has properties that are entirely different from those of the other isomers. But most surprising is the fact that the racemic mixture, although composed of equal parts of the (+) and (−) enantiomers, differs from them in specific gravity, melting point, and solubility. Why, for example, is the melting point of the racemic mixture higher than that of any of the other forms? The melting point of any substance is largely dependent on the attractive forces holding the ions or molecules together. The melting point of the racemic mixture is higher than that of either enantiomer. Therefore, we can conclude that the attraction between molecules of the (+) and (−) enantiomers in the racemic mixture is greater than the attraction between molecules of the (+) and (+) or the (−) and (−) enantiomers.

Table 28.1. Properties of tartaric acids (HOOC—CHOH—CHOH—COOH).

Name	Specific gravity	Melting point (°C)	Solubility (g/100 g H$_2$O)	Specific rotation [α]
(+)-Tartaric acid	1.760	170	147$^{20°C}$	+12°
(−)-Tartaric acid	1.760	170	147$^{20°C}$	−12°
(±)-Tartaric acid (racemic mixture)	1.687	206	20.6$^{20°C}$	0°
meso-Tartaric acid	1.666	140	125$^{15°C}$	0°

Problem 28.2

How many stereoisomers do the following compounds have? Write their structures and label any pairs of enantiomers and *meso* compounds. Point out any diastereomers.
(a) CH$_3$CHBrCHBrCH$_2$CH$_3$ (b) CH$_2$BrCHClCHClCH$_2$Br

(a) Carbons 2 and 3 are asymmetric, so there can be a maximum of four stereoisomers ($2^2 = 4$). Write structures around the asymmetric carbons:

```
      CH3           CH3           CH3           CH3
       |             |             |             |
   H—C—Br        Br—C—H        H—C—Br        Br—C—H
       |             |             |             |
   H—C—Br        Br—C—H        Br—C—H        H—C—Br
       |             |             |             |
      CH2           CH2           CH2           CH2
       |             |             |             |
      CH3           CH3           CH3           CH3
       I             II            III            IV
       Enantiomers                 Enantiomers
```

There are four stereoisomers; two pairs of enantiomers; no *meso* compounds; I and III, I and IV, II and III, and II and IV are diastereomers.

(b) Carbons 2 and 3 are asymmetric, so there can be a maximum of four stereoisomers. Write structures around the asymmetric carbons:

```
    CH₂Br          CH₂Br          CH₂Br          CH₂Br
     |              |              |              |
  H—C—Cl         Cl—C—H         Cl—C—H         H—C—Cl
     |              |              |              |
  H—C—Cl         Cl—C—H         H—C—Cl         Cl—C—H
     |              |              |              |
    CH₂Br          CH₂Br          CH₂Br          CH₂Br

      I             II             III            IV
```

              meso Compound   Enantiomers

There are three stereoisomers; one pair of enantiomers; one *meso* compound; I and II are the same compound. There is a plane of symmetry between carbons 2 and 3. Rotating II 180° makes it superimposable on I; therefore it is a *meso* compound. Structures I and III and I and IV are diastereomers.

Questions

A. Review the meanings of the new terms introduced in this chapter.
  1. Stereoisomerism
  2. Stereoisomers
  3. Plane-polarized light
  4. Specific rotation
  5. Optical activity
  6. Dextrorotatory
  7. Levorotatory
  8. Asymmetric carbon atom
  9. Projection formula
  10. Superimposable
  11. Chiral
  12. Achiral
  13. Enantiomer
  14. Racemic mixture
  15. Diastereomer
  16. *meso* compound or *meso* structure

B. Review questions.
  1. Which of these objects are chiral?
     (a) A wood screw
     (b) A pair of pliers
     (c) The letter O
     (d) The letter G
     (e) A coiled spring
     (f) Your ear
  2. What is an asymmetric carbon atom? Draw structural formulas of three different compounds that contain one asymmetric carbon atom, and mark the asymmetric carbon in each with an asterisk.
  3. How can you tell when the axes of two pieces of Polaroid film are parallel? When one piece has been rotated by 90°?
  4. How many asymmetric carbon atoms are present in each of the following?

```
          H  Cl                          Br Cl
          |  |                           |  |
  (a) Cl—C—C—Br              (b)  H—C—C—H
          |  |                           |  |
          H  H                           H  H

  (c) CH₃CH₂CH₂CHClCH₃        (d) H—C=O
                                       |
                                    H—C—OH
                                       |
                                    H—C—OH
                                       |
                                    H—C—OH
                                       |
                                       H
```

(e)
```
    H
    |
H—C—OH
    |
    C=O
    |
HO—C—H
    |
 H—C—OH
    |
 H—C—OH
    |
 H—C—OH
    |
    H
```

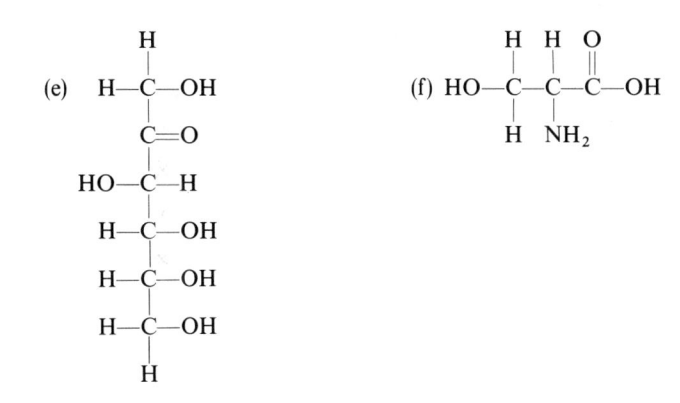

(f)
```
   H  H  O
   |  |  ||
HO—C—C—C—OH
   |  |
   H  NH₂
```

5. Using projection formulas, decide which of the following compounds will show optical activity:
   (a) 1-Chloropentane
   (b) 2-Chloropentane
   (c) 3-Chloropentane
   (d) 1-Chloro-2-methylpentane
   (e) 2-Chloro-2-methylpentane
   (f) 3-Chloro-2-methylpentane
   (g) 4-Chloro-2-methylpentane
   (h) 3-Chloro-3-methylpentane

6. Suppose a carbon atom is located at the center of a square with four different groups attached to the corners in a planar arrangement. Would the compound rotate polarized light? Explain.

7. What is a necessary and sufficient condition for a compound to show enantiomerism?

8. Glucose ($C_6H_{12}O_6$) has four asymmetric carbon atoms. How many stereoisomers of glucose are theoretically possible?

9. Structures (a) and (b) are projection formulas. Do they represent enantiomers or the same compound? Justify your answer.

```
      Cl                Br
      |                 |
   Br—C—H            H—C—Cl
      |                 |
      F                 F
     (a)               (b)
```

10. Which of these projection formulas represent (−)-lactic acid and which represent (+)-lactic acid?

(a)
```
      CH₃
       |
   HO—┼—H
       |
      COOH
```

(b)
```
      OH
       |
    H—┼—COOH
       |
      CH₃
```

(c)
```
     COOH
       |
    H—┼—CH₃
       |
      OH
```

(d)
```
      CH₃
       |
    H—┼—OH
       |
      COOH
```

(e)
```
     COOH
       |
   CH₃—┼—H
       |
      OH
```

(f)
```
       H
       |
   CH₃—┼—OH
       |
      COOH
```

638   Chapter 28

11. Draw projection formulas for all the possible stereoisomers of the following compounds. Label pairs of enantiomers and *meso* compounds.
    (a) 1,2-Dibromopropane   (d) 2,4-Dibromopentane
    (b) 2,3-Dichlorobutane   (e) 3-Chlorohexane
    (c) 2-Butanol

12. Draw projection formulas for all the stereoisomers of 1,2,3-trihydroxybutane. Point out enantiomers, *meso* compounds, and diastereomers, where present.

13. Write structures for the four stereoisomers of 3-pentene-2-ol.

14. Some substituted cycloalkanes are chiral. Draw the structures and enantiomers for any of the following that are chiral:

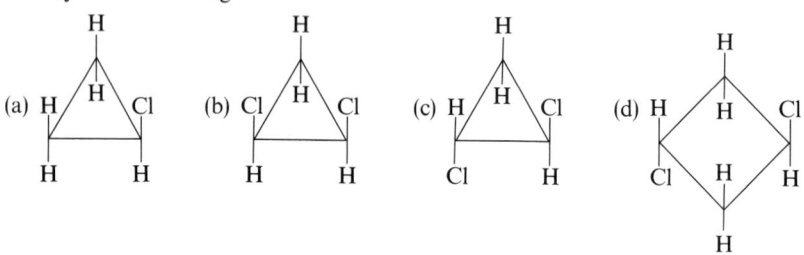

    (a) cyclopropane with H,H / H,Cl / H,H substituents
    (b) cyclopropane with Cl,H / H,Cl / H,H substituents
    (c) cyclopropane with H,H / H,Cl / Cl,H substituents
    (d) cyclobutane with H,H / H,H / Cl,H / H,H substituents

15. The physical properties for (+)-2-methyl-1-butanol are: specific rotation, +5.76°; bp, 129°C; density, 0.819 g/ml. What are these same properties for (−)-2-methyl-1-butanol?

16. Draw projection formulas for all the stereoisomers of 2,3,4-tribromopentane and point out enantiomers and *meso* compounds, where present.

17. (+)-2-Chlorobutane is further chlorinated to give dichlorobutanes ($C_4H_8Cl_2$). Write structures for all the possible isomers formed and indicate which of these isomers will be optically active. [Remember that (+)-2-chlorobutane is optically active.]

18. In the chlorination of butane, 1-chlorobutane and 2-chlorobutane are obtained as products. After separation by distillation, neither product rotates the plane of polarized light. Explain these results.

19. (a) Explain why it is not possible to separate enantiomers by ordinary chemical and physical means.
    (b) Explain why diastereomers can usually be separated by ordinary physical and chemical means.

20. The observed rotation of polarized light was 12.5° for a solution of compound X at a concentration of 10.0 g/100 ml. The measurement was made using a polarimeter sample tube 20.0 cm in length. Calculate the specific rotation, [α], for compound X.

21. Which, if any, of the following are *meso* compounds?

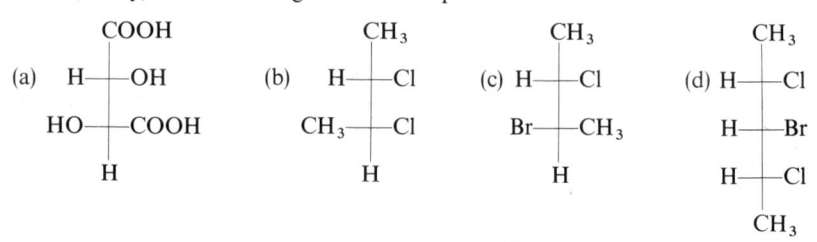

    (a) COOH / H—OH / HO—COOH / H
    (b) CH₃ / H—Cl / CH₃—Cl / H
    (c) CH₃ / H—Cl / Br—CH₃ / H
    (d) CH₃ / H—Cl / H—Br / H—Cl / CH₃

22. Certain substances of the type $R_1R_2R_3R_4N^+X^-$ are known to rotate the plane of polarized light. What inference can be drawn regarding the spatial character of the nitrogen bonds?

23. Which of these statements are correct?
    (a) The polarizer and the analyzer of a polarimeter are made of the same material.
    (b) The specific rotation of a compound is dependent on the number of molecules in the path of the plane-polarized light.
    (c) Very few natural products are optically active.
    (d) *cis–trans* isomers are not considered to be stereoisomers.
    (e) J. A. Le Bel received the first Nobel Prize in chemistry in 1901.
    (f) A compound that rotates plane-polarized light $+25°$ would be at the same position on the polarimeter as one that rotates the light $-335°$.
    (g) Molecules that contain only one asymmetric carbon atom are chiral, but all chiral molecules do not contain an asymmetric carbon atom.
    (h) The compound $CH_3CHBrCHBrCH_2OH$ will have eight optical isomers.
    (i) The compounds shown in these projection formulas are enantiomers:

    $$\begin{array}{cc} Cl & H \\ Br-\!\!\!\!+\!\!\!\!-H & Br-\!\!\!\!+\!\!\!\!-Cl \\ CH_3 & CH_3 \end{array}$$

    (j) Two diastereomers have identical melting points.
    (k) A molecule that contains two asymmetric carbon atoms may not be chiral.
    (l) A racemic mixture contains equal amounts of dextrorotatory and levorotatory molecules of a compound.

29 Carbohydrates

After studying Chapter 29 you should be able to:

1. Understand the terms listed in Question A at the end of the chapter.
2. Classify carbohydrates as mono-, di-, oligo-, or polysaccharides.
3. Explain the use of D, L, (+), and (−) in naming carbohydrates.
4. Identify and write pyranose and furanose ring structures of carbohydrates.
5. Identify and write Fischer projection, "tree", and Haworth formulas for carbohydrates.
6. Identify the structural feature of a carbohydrate that makes it a reducing sugar.
7. Explain the phenomenon of mutarotation.
8. Distinguish between hemiacetal and acetal structures in a carbohydrate.
9. Understand what a glycoside linkage is.
10. Understand monosaccharide composition and the manner in which monosaccharides are linked together in sucrose, lactose, and maltose.
11. Identify monosaccharides that are epimers.
12. List the major sources of glucose, galactose, fructose, sucrose, lactose and maltose.
13. Describe the structural differences among starch, cellulose, glycogen, and inulin.
14. Describe the Benedict's and Tollens' tests and tell what evidence must be seen to indicate a positive test for each.
15. Write chemical equations for the oxidation of monosaccharides by bromine and by nitric acid.
16. Write chemical equations for the reduction of monosaccharides.
17. Write chemical equations for the formation of osazones.
18. Identify monosaccharides that give identical osazones.
19. Identify monosaccharides that give optically inactive (*meso*) products when reduced to polyhydroxy alcohols or when oxidized to dicarboxylic acids.
20. Discuss the structural difference between amylose and amylopectin.
21. Rate the relative sweetness of the common monosaccharides and disaccharides.

29.1 Introduction

Carbohydrates are one of the three principal classes of foods (carbohydrates, fats, and proteins). The majority of all matter in plants, excepting water, is made up of carbohydrates. Carbohydrates have central roles in such vital processes as photosynthesis and the metabolism of all plants and animals. The name *carbohydrates* was given to this class of compounds many years ago by French scientists who called them *hydrates de carbone* because their empirical formulas approximated $(C \cdot H_2O)_n$. It was found later that all substances classified as carbohydrates do not conform to this formula (for example, rhamnose, $C_6H_{12}O_5$; deoxyribose, $C_5H_{10}O_4$). Carbohydrates are not simply hydrated carbon, but are complex substances containing from three to many thousands of carbon atoms. Some carbohydrates are macromolecules with molecular weights of several million. The general definition is: **Carbohydrates are polyhydroxyaldehydes or polyhydroxy ketones or substances that yield these compounds when hydrolyzed.** The simplest carbohydrates are glyceraldehyde and dihydroxyacetone.

carbohydrate

$$\begin{array}{cc}
H-C=O & H-C-OH \\
| & | \\
H-C-OH & C=O \\
| & | \\
H-C-OH & H-C-OH \\
| & | \\
H & H \\
\text{Glyceraldehyde} & \text{Dihydroxyacetone}
\end{array}$$

29.2 Classification

A carbohydrate is classified as a monosaccharide, a disaccharide, an oligosaccharide, or a polysaccharide, based on the number of monosaccharide units linked together to form a molecule. A **monosaccharide** is a carbohydrate that cannot be hydrolyzed to simpler carbohydrate units. A **disaccharide** yields two monosaccharides—either alike or different—when hydrolyzed:

monosaccharide

disaccharide

$$\text{Disaccharide} + \text{Water} \xrightarrow{H^+ \text{ or enzyme}} 2\,\text{Monosaccharides}$$

The monosaccharides and disaccharides generally have names ending in *-ose*; for example, glucose, sucrose, lactose. These water-soluble carbohydrates, which have a characteristically sweet taste, are called *sugars*.

oligosaccharide

An **oligosaccharide** has two to six monosaccharide units linked together. *Oligo-* comes from the Greek word *oligos*, which means "small" or "few."

polysaccharide

A **polysaccharide** is a macromolecular substance that can be hydrolyzed to yield many monosaccharide units:

$$\text{Polysaccharide} + \text{Water} \xrightarrow{\text{H}^+ \text{ or enzyme}} \text{Many monosaccharide units}$$

In addition to the primary divisions given above, carbohydrates are classified in a variety of other ways. A given monosaccharide might be described with respect to several of these categories.

1. As a triose, tetrose, pentose, or hexose, depending on whether the molecule contains three, four, five, or six carbon atoms.

   Trioses $C_3H_6O_3$
   Tetroses $C_4H_8O_4$
   Pentoses $C_5H_{10}O_5$
   Hexoses $C_6H_{12}O_6$

2. As an aldose or ketose, depending on whether an aldehyde group (—CHO) or keto group ($\supset$C=O) is present.
3. As a D or L isomer, depending on the spatial orientation of the —H and —OH groups attached to the carbon atom adjacent to the terminal primary alcohol group. When the —OH is written to the right of this carbon in the projection formula, the D isomer is represented. When the —OH is written to the left, the L isomer is represented. The reference compounds for this classification are the trioses, D-glyceraldehyde and L-glyceraldehyde, which are shown below. Also shown are two aldohexoses (D- and L-glucose) and a ketohexose (D-fructose):

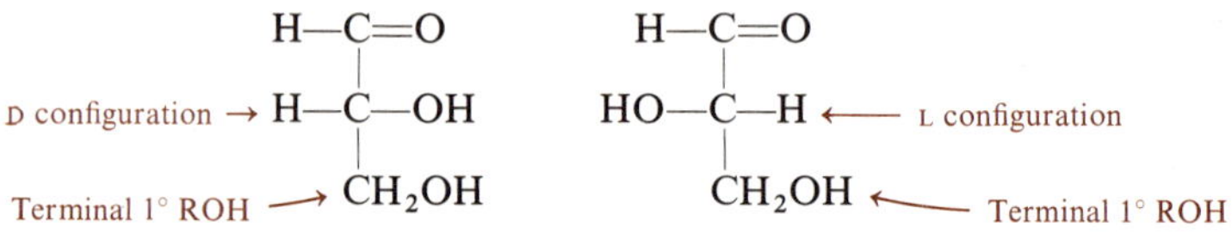

```
                    H—C=O                    H—C=O
D configuration → H—C—OH          HO—C—H ← L configuration
Terminal 1° ROH → CH₂OH               CH₂OH ← Terminal 1° ROH
                D-Glyceraldehyde      L-Glyceraldehyde


     H—C=O          H—C=O                     CH₂OH
   HO—C—H          H—C—OH                     C=O
    H—C—OH         HO—C—H                    HO—C—H
   HO—C—H          H—C—OH     D configuration  H—C—OH
   HO—C—H          H—C—OH  ←                  H—C—OH
    CH₂OH           CH₂OH  ← Terminal 1° ROH → CH₂OH
   L-Glucose       D-Glucose                  D-Fructose
```

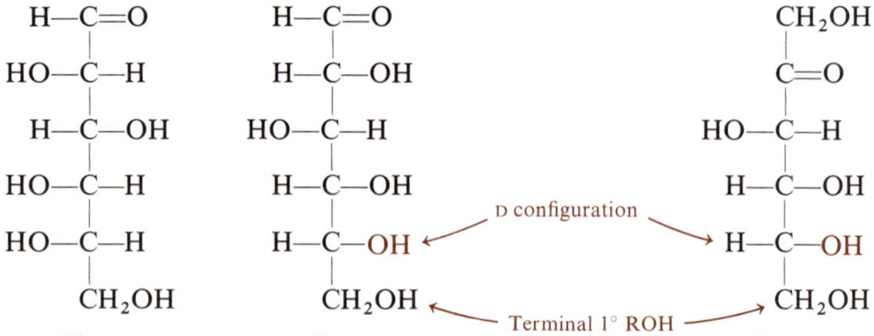

The letters D and L do not in any way refer to the direction of optical rotation of a carbohydrate. The D and L forms of any specific compound are enantiomers (for example, D- and L-glucose).

4. As a (+) or (−) isomer, depending on whether the monosaccharide rotates the plane of polarized light to the right (+) or to the left (−) (Section 28.5).
5. As a furanose or a pyranose, depending on whether the cyclic structure of the carbohydrate is related to that of the five-membered or six-membered heterocyclic

ring compound furan or pyran (a heterocyclic ring contains more than one kind of atom in the ring):

Furan, C_4H_4O
(five-membered ring containing oxygen in the ring)

Pyran, C_5H_6O
(six-membered ring containing oxygen in the ring)

6. As having an alpha (α) or beta (β) configuration, based on the orientation of the H and OH groups about a specific asymmetric carbon in the cyclic form of the monosaccharide (Section 29.5).

Problem 29.1 Write projection formulas for (a) an L-aldotriose, (b) a D-ketotetrose, and (c) a D-aldopentose.

(a) Triose indicates a three-carbon carbohydrate; aldo indicates the compound is an aldehyde; L- indicates the —OH on carbon 2 (adjacent to the terminal CH_2OH) is on the left. The aldehyde group is carbon 1.

```
H—C=O
   |
HO—C—H
   |
  CH₂OH
```
an L-Aldotriose

(b) Tetrose indicates a four-carbon carbohydrate; keto indicates a ketone group (on carbon 2); D- indicates the —OH on carbon 3 (adjacent to the terminal CH_2OH) is on the right. Carbons 1 and 4 are primary alcohols.

```
  CH₂OH
   |
   C=O
   |
H—C—OH
   |
  CH₂OH
```
a D-Ketotetrose

(c) Pentose indicates a five-carbon carbohydrate; aldo indicates an aldehyde group (on carbon 1); D- indicates the —OH on carbon 4 (adjacent to the terminal CH_2OH) is on the right. The orientation of the —OH groups on carbons 2 and 3 is not specified here and therefore can be written in either direction for this problem.

```
H—C=O
   |
H—C—OH
   |
H—C—OH
   |
H—C—OH
   |
  CH₂OH
```
a D-Aldopentose

29.3 Monosaccharides

Although a great many monosaccharides have been synthesized, only a very limited number appear to be of great biological significance. One pentose monosaccharide—ribose—and its deoxy derivative are essential components of ribonucleic acid (RNA) and of deoxyribonucleic acid (DNA) (see Chapter 31). Three hexose monosaccharides—glucose, galactose, and fructose—are of significance in nutrition. All three have the same molecular formula, $C_6H_{12}O_6$. Of these, glucose is of outstanding importance because it is a key substance in photosynthesis and in the metabolic schemes of living things. The structure of glucose is considered in detail in Section 29.4.

Glucose

Glucose is the most important of the monosaccharides. It is an aldohexose and is found in the free state in plants and animal tissue. Glucose is commonly known as *dextrose* or *grape sugar*. It is a component of the disaccharides sucrose, maltose, and lactose, and is also the monomer of the polysaccharides starch, cellulose, and glycogen. Among the common sugars, glucose is of intermediate sweetness (see Table 29.1).

Glucose is the key sugar of the body and is carried by the bloodstream to all of its parts. The concentration of glucose in the blood is normally 80 to 100 mg/100 ml blood. Glucose requires no digestion; therefore, it may be given intravenously to patients who cannot take food by mouth. The body's heat is derived primarily from the oxidation of glucose. Glucose is found in the urine of those who have diabetes mellitus (sugar diabetes). The condition in which glucose is excreted in the urine is called glycosuria.

Galactose

Galactose is also an aldohexose and occurs along with glucose in lactose and in many oligo- and polysaccharides such as pectin, gums, and mucilages. Galactose is an isomer of glucose, differing only in the spatial arrangement of the —H and —OH groups around carbon 4 (see Section 29.4). Galactose is synthesized in the mammary glands to make the lactose of milk. It is also a constituent of glycolipids and glycoproteins in nervous tissue. Galactose is less than half as sweet as glucose.

A servere inherited disease, called galactosemia, is the inability of infants to metabolize galactose. The galactose concentration increases markedly in the blood and also appears in the urine. Galactosemia causes vomiting, diarrhea, enlargement of the liver, and often mental retardation. If not recognized within a few days after birth, it can lead to death. If diagnosis is made early and lactose is excluded from the diet, the symptoms disappear and normal growth may be resumed.

Fructose

Fructose, also known as levulose, is a ketohexose and occurs in fruit juices, honey, and (along with glucose) as a constituent of sucrose. Fructose is the major constituent of the polysaccharide inulin, a starch-like substance present in many plants such as dahlia tubers, chicory roots, and Jerusalem artichokes. Fructose is the sweetest of all the sugars, being about twice as sweet as glucose. This accounts for the sweetness of honey; the enzyme invertase present in bees splits sucrose into glucose and fructose. Fructose is metabolized directly, but is also readily converted to glucose in the liver.

29.4 Structure of Glucose and Other Aldoses

In one of the classic feats of research in organic chemistry, Emil Fischer (1852–1919), working in Germany, established the structural configuration of glucose along with that of many other sugars. Emil Fischer received the Nobel Prize in chemistry in 1902. Fischer devised projection formulas that relate the structure of a sugar to one or the other of the two enantiomeric forms of glyceraldehyde. In this system, the molecule is represented with the aldehyde (or ketone) group at the top. The —H and —OH groups attached to other than end carbons are written to the right or to the left as they would appear when projected toward the observer. The two glyceraldehydes are represented thus:

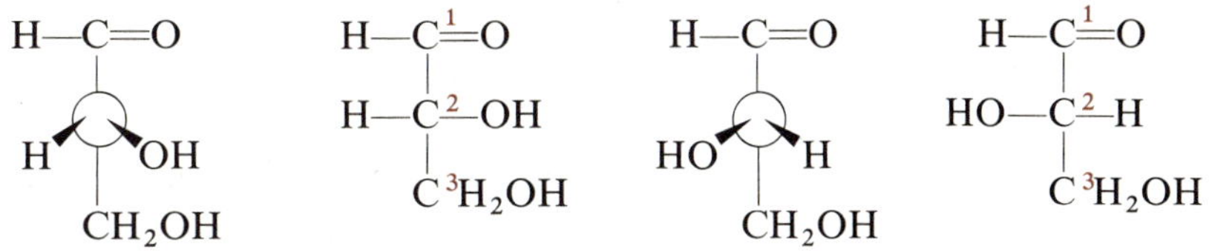

D-Glyceraldehyde (three-dimensional representation)

D-Glyceraldehyde (Projection formula)

L-Glyceraldehyde (three-dimensional representation)

L-Glyceraldehyde (Projection formula)

In the three-dimensional molecules represented by these formulas, the number 2 carbon atoms are in the plane of the paper. The —H and —OH groups project forward (toward the observer); the —CHO and —CH$_2$OH groups project backward (away from the observer). Any two monosaccharides that differ only in the configuration around a single carbon atom are called **epimers**. Thus, D- and L-glyceraldehyde are epimers.

epimer

Fischer recognized that there were two enantiomeric forms of glucose. To these forms he assigned the following structures and names:

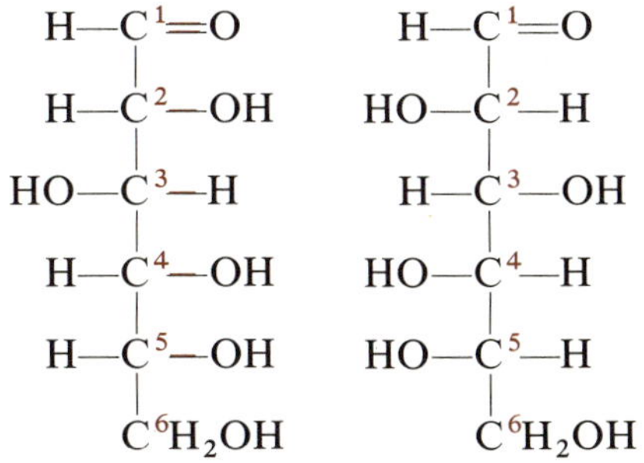

D-Glucose

L-Glucose

646 Chapter 29

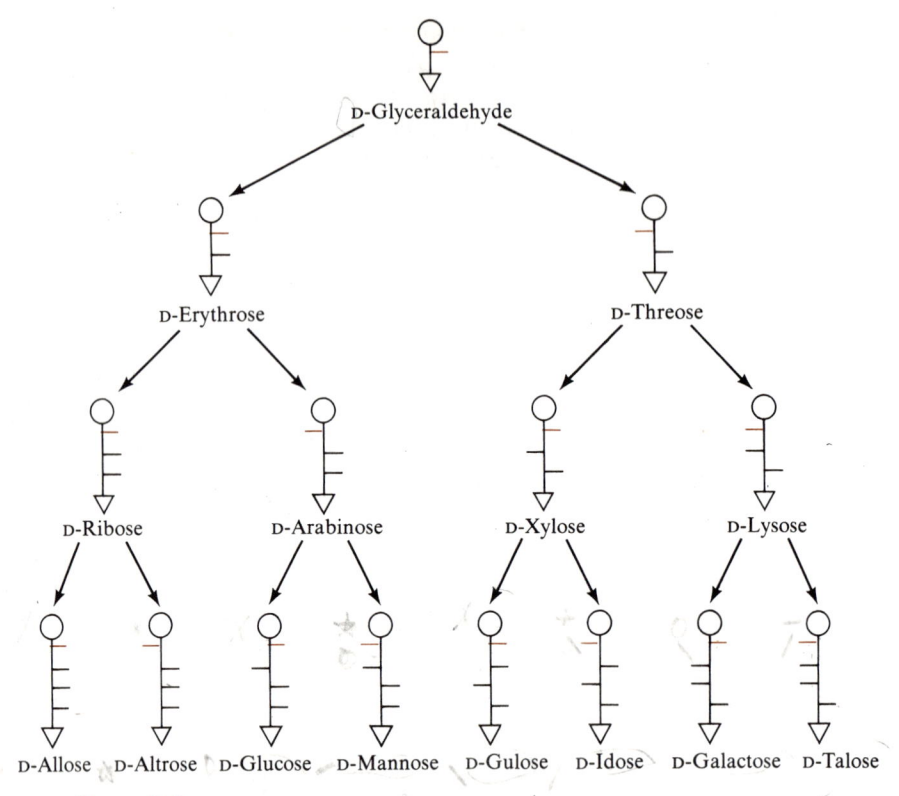

Figure 29.1. Configurations of the D family of aldoses.

The structure named D-glucose is so named because the —H and —OH on carbon 5 are in the same configuration as the —H and —OH on carbon 2 in D-glyceraldehyde. The configuration of the —H and —OH on carbon 5 in L-glucose corresponds to the —H and —OH on carbon 2 in L-glyceraldehyde.

Fischer recognized that 16 different aldohexoses, 8 with the D configuration and 8 with the L configuration, were possible. This follows our formula 2^n for optical isomers. Glucose has four asymmetric carbon atoms and should have 16 stereoisomers (2^4). The configurations of the D-aldose family are shown in Figure 29.1. The shorthand "tree" formulas in this figure use the following symbols:

$\bigcirc$ = H—C=O

$\vdash$ = H—C—OH

$\dashv$ = HO—C—H

$\bigtriangledown$ = CH$_2$OH

For example,

represents

H—C=O
H—C—OH
CH$_2$OH

In this family (Figure 29.1), new asymmetric carbon atoms are formed as we go from triose to tetrose to pentose to hexose. Each time a new asymmetric carbon is formed, a pair of epimers are formed around carbon 2, with the rest of the molecule remaining the same. This sequence continues until eight D-aldohexoses are formed. A similar series starting with L-glyceraldehyde is known, making a total of 16 aldohexoses.

All 16 aldohexoses have been synthesized, but only D-glucose and D-galactose appear to be of considerable biological importance. Since the metabolism of most living organisms revolves about D-glucose, our discussion will be centered on this substance.

The conversion of one aldose into another aldose containing one more carbon atom is known as the Kiliani–Fischer synthesis. This synthesis involves (1) the addition of HCN to form a cyanohydrin, (2) hydrolysis of the CN group to COOH, and (3) reduction with sodium amalgam, Na(Hg), to form the aldehyde. As an example, the formation of two aldotetroses from an aldotriose is shown in Figure 29.2.

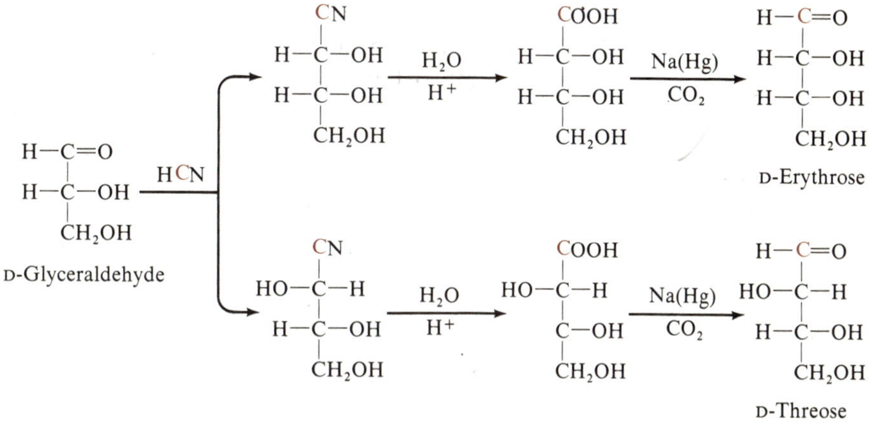

Figure 29.2. An example of the Kiliani–Fischer synthesis.

29.5 Cyclic Structure of Glucose; Mutarotation

mutarotation

Two crystalline forms of D-(+)-glucose are known. These two forms are diastereomers and differ with respect to their rotation of polarized light. One form, labeled α-D-glucose, has a specific rotation of +113°; the other, labeled β-D-glucose, has a specific rotation of +19°. A very interesting phenomenon occurs when these two forms of glucose are put into separate solutions and allowed to stand for several hours. The specific rotation of each solution changes to +52.5°. This phenomenon is known as **mutarotation**. An explanation of mutarotation is that in solution, D-glucose exists as an equilibrium mixture of two cyclic forms and the open-chain form (see Figure 29.3). When dissolved some α-D-glucose molecules are transformed into β-D-glucose molecules and vice versa until an equilibrium is reached between the α and β forms. The

648 Chapter 29

equilibrium mixture contains about 36% α molecules and 64% β molecules, with a trace of open-chain molecules.

In the cyclic forms of D-glucose, carbon 1 is linked to carbon 5 by means of an oxygen atom. As a result, an asymmetric center is formed at carbon 1 with two possible arrangements of the —H and —OH groups attached to it. When two diastereomers such as α- and β-glucose differ in their stereo arrangement only about carbon 1, they are called **anomers**.

anomer

The cyclic forms of D-glucose may be represented by either Fischer projection formulas or by Haworth perspective formulas. These structures are shown in Figure 29.3. In the projection formulas, the α form has the —OH on carbon 1 written to the right; in the β form the —OH on carbon 1 is on the left. The Haworth structure represents the molecule as a flat hexagon with the —H and —OH groups above and below the plane of the hexagon. In the α form the —OH on carbon 1 is written below the plane; in the β form the —OH on carbon 1 is above the plane. In converting the projection formula of a D-aldohexose to a

Modified Fischer Projection Formulas

α-D-(+)-Glucose
[α] + 113°

D-(+)-Glucose
(open-chain form)

β-D-(+)-Glucose
[α] + 19°

Haworth Perspective Formulas

α-D-(+)-Glucose

D-(+)-Glucose
(open-chain form)

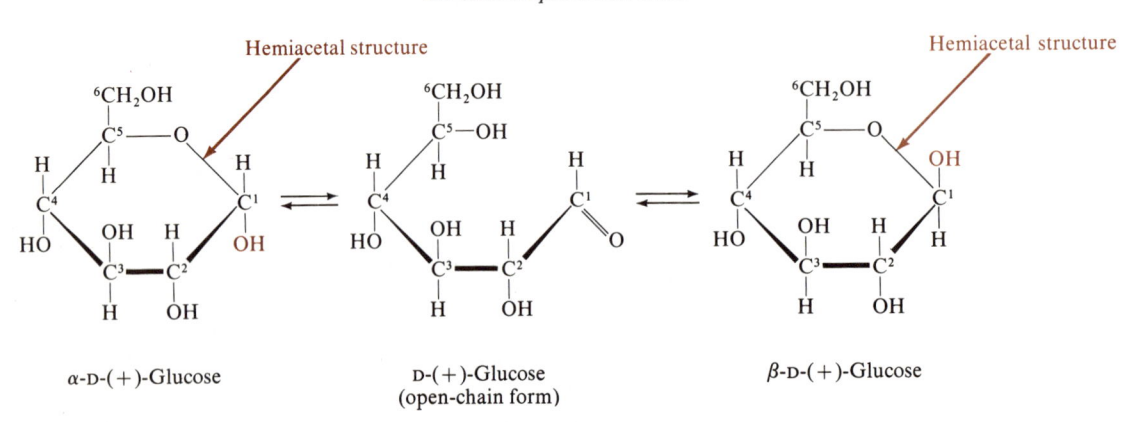

β-D-(+)-Glucose

Figure 29.3. Modified Fischer projection and Haworth perspective structures of D-glucose.

Haworth formula, the —OH groups on carbons 2, 3, and 4 are written below the plane if they project to the right and above the plane if they project to the left. Carbon 6 is written above the plane.

The two cyclic forms of D-glucose differ only in the relative positions of the —H and —OH groups attached to carbon 1. As viewed in the Haworth perspective formulas, the —OH group is on the lower side of the molecule in α-D-(+)-glucose and on the upper side in β-D-(+)-glucose. This seemingly minor structural difference in the cyclic forms of glucose has startling biochemical consequences. For example, the fundamental chemical difference between starch and cellulose is that starch is a polymer of α-D-glucose and cellulose is a polymer of β-D-glucose. Starch is a major and easily digested human food yet we are totally unable to digest cellulose.

Haworth formulas are sometimes shown in abbreviated schematic form. For example, α-D-(+)-glucose is shown in these diagrams.

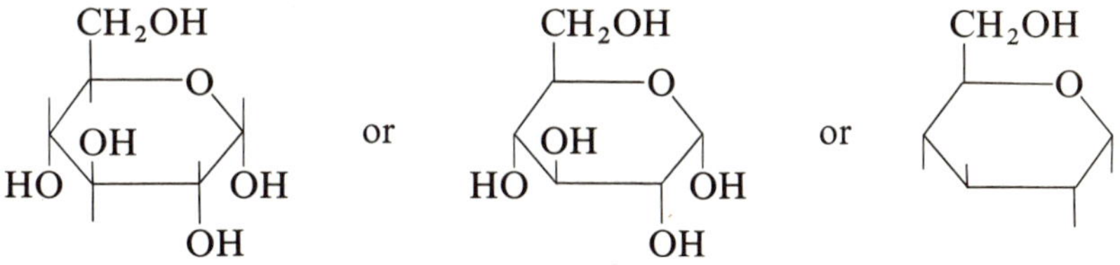

α-D-(+)-Glucose

At best, any two-dimensional representation is a compromise in portraying the three-dimensional configuration of such molecules. Models are much more effective at this job, especially if constructed by the student!

29.6 Hemiacetals and Acetals

As you may recall, in Chapter 25 we studied the reactions of aldehydes (and ketones) to form hemiacetals and acetals (see Section 25.4). The hemiacetal structure consists of an ether linkage and an alcohol linkage on the same carbon atom, whereas the acetal structure has two ether linkages to the same carbon atom.

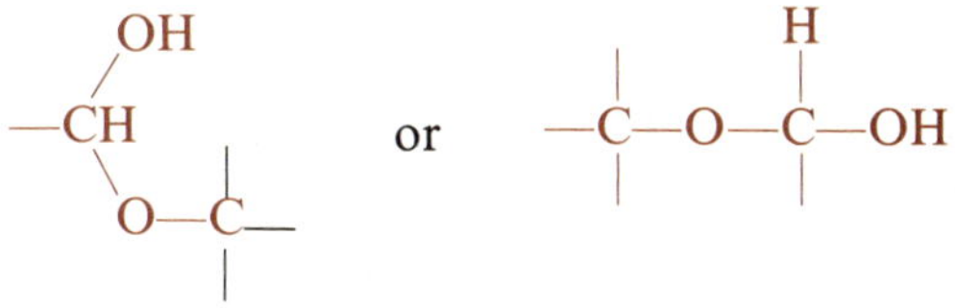

Hemiacetal

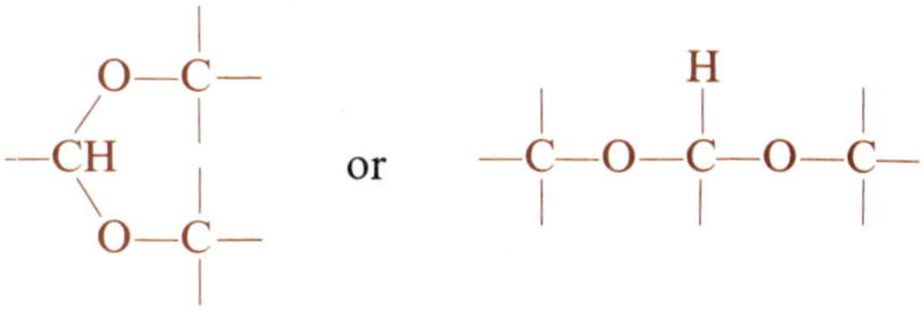

Acetal

Cyclic structures of monosaccharides are hemiacetals.

[Structure showing α-D-glucose with arrow indicating Hemiacetal structure in α-D-glucose]

Hemiacetals are unstable and are readily hydrolyzed by water. When hydrolyzed, the ring opens and the hemiacetal reverts back to the open-chain aldehyde. Thus, mutarotation results from this opening and closing of the hemiacetal ring (see Figure 29.3). When the ring closes it can form either the α or the β anomer.

When an alcohol, ROH, reacts with another alcohol, R'OH, the product formed can be an ether, ROR'. Carbohydrates are alcohols and behave in a similar manner. When a monosaccharide reacts to form an ether at carbon 1, the resulting product is an acetal. In carbohydrate terminology this acetal structure is called a **glycoside**. In the case of glucose, it would be a glucoside; if galactose, a galactoside; and so on.

glycoside

[Structure of An α-glycoside showing Glycosidic linkage at O—R]

An α-glycoside
(R = a variety of groups)

When α-D-glucose is heated with methyl alcohol and a small quantity of hydrogen chloride is added, two optically active isomers are formed, methyl-α-D-glucoside and methyl-β-D-glucoside:

[Reaction: α-D-glucose + CH₃OH/HCl ⇌ Methyl-α-D-glucoside + Methyl-β-D-glucoside]

α-D-glucose   Methyl-α-D-glucoside   Methyl-β-D-glucoside
                     mp 165°C   mp 107°C
                     $[\alpha] = +158°$ $[\alpha] = -33°$

These two glucosides do not undergo mutarotation and do not behave at all like normal monosaccharides. Because they are full acetals, they are stable and do not hydrolyze in water, but they may be hydrolyzed in acidic solutions.

The glycosidic linkage occurs in a wide variety of natural substances. All carbohydrates other than monosaccharides are glycosides. Heart stimulants

29.7 Structure of Galactose and Fructose

Galactose, like glucose, is an aldohexose and differs structurally from glucose only in the orientation of the —OH group on the fourth carbon:

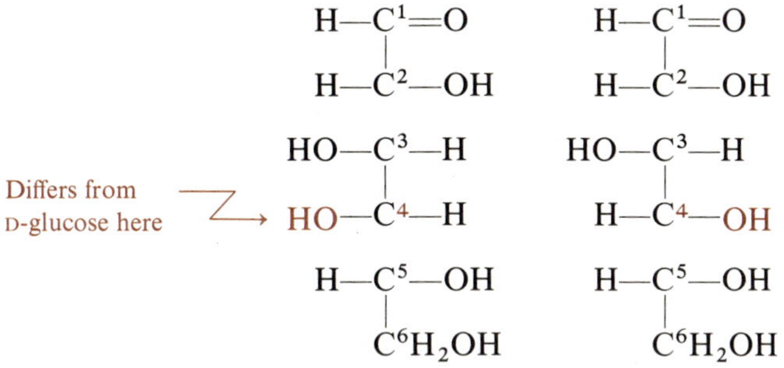

$$\begin{array}{c} H-C^1=O \\ | \\ H-C^2-OH \\ | \\ HO-C^3-H \\ | \\ HO-C^4-H \\ | \\ H-C^5-OH \\ | \\ C^6H_2OH \end{array} \qquad \begin{array}{c} H-C^1=O \\ | \\ H-C^2-OH \\ | \\ HO-C^3-H \\ | \\ H-C^4-OH \\ | \\ H-C^5-OH \\ | \\ C^6H_2OH \end{array}$$

D-Galactose   D-Glucose

Differs from D-glucose here ↗

Galactose, like glucose, also exists in two cyclic (pyranose) forms that have hemiacetal structures and undergo mutarotation.

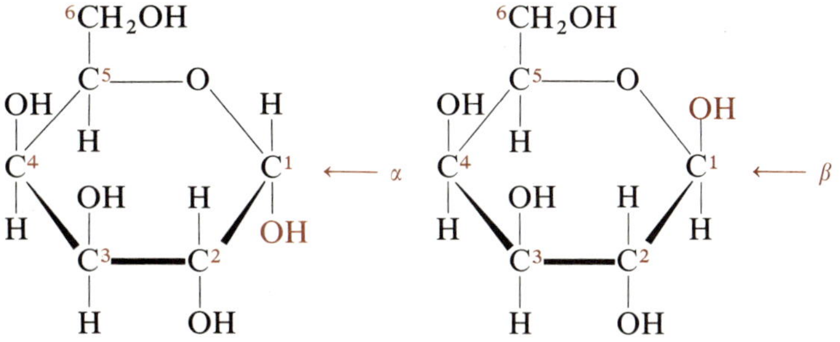

α-D-Galactose   β-D-Galactose

Fructose is a ketohexose. The open-chain form may be represented in a Fischer projection formula:

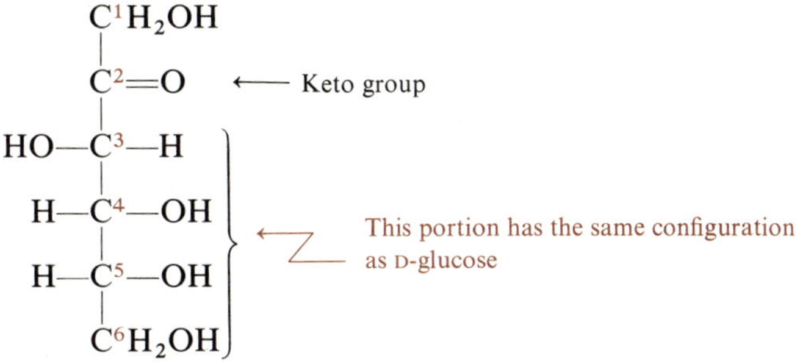

$$\begin{array}{c} C^1H_2OH \\ | \\ C^2=O \\ | \\ HO-C^3-H \\ | \\ H-C^4-OH \\ | \\ H-C^5-OH \\ | \\ C^6H_2OH \end{array}$$

← Keto group

⎱ This portion has the same configuration as D-glucose

D-Fructose

Like glucose and galactose, fructose exists in both cyclic and open-chain forms. The cyclic form is a five-membered (furan-type) ring and has the β configuration.

β-D-Fructose ← β configuration

29.8 Pentoses

An open-chain aldopentose has three asymmetric carbon atoms. Therefore, eight (2^3) isomeric aldopentoses are possible. The four possible D-pentoses, written in shorthand tree form, are shown in Figure 29.1. Arabinose and xylose occur in certain plants as polysaccharides called pentosans. Because of their relationship to nucleic acids and the genetic code (Chapter 31), D-ribose and its derivative, D-2-deoxyribose, are the most interesting pentoses. D-Ribose and D-2-deoxyribose are synthesized in the body. Note the two names D-ribose and D-2-deoxyribose. In the latter name, the 2-deoxy means that oxygen is missing from the D-ribose molecule at carbon 2. Check the formulas below to verify this difference.

D-Ribose β-D-Ribose

D-2-Deoxyribose β-D-2-Deoxyribose

D-Ribulose is a ketopentose. It forms phosphate esters that are important in the photosynthesis cycle of converting carbon dioxide and water to carbohydrates.

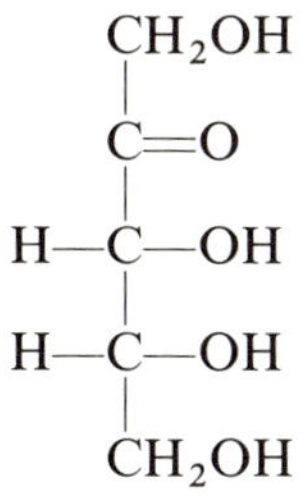

```
        CH₂OH
         |
         C=O
         |
     H—C—OH
         |
     H—C—OH
         |
        CH₂OH
```
D-Ribulose

29.9 Disaccharides

Disaccharides are carbohydrates composed of two monosaccharide residues united by a glycosidic linkage. The three disaccharides that are especially important from a biological viewpoint are sucrose, lactose, and maltose. Sucrose, which is commonly known as *table sugar*, is found in the free state throughout the plant kingdom. Sugar cane contains 15 to 20% sucrose and sugar beets 10 to 17%. Maple syrup and sorghum are also good sources of sucrose.

Lactose, also known as *milk sugar*, is found free in nature mainly in the milk of mammals. Human milk contains about 6.7% lactose and cow milk about 4.5% of this sugar.

Maltose is found in sprouting grain, but occurs much less commonly (in nature) than either sucrose or lactose. Maltose is prepared commercially by the partial hydrolysis of starch, catalyzed either by enzymes or by dilute acids.

Upon hydrolysis disaccharides yield two monosaccharide molecules. The hydrolysis is catalyzed by hydrogen ions (acids), usually at elevated temperatures, or by certain enzymes that act effectively at room or body temperatures. An enzyme is a protein that acts as a biochemical catalyst and is specific in its action; that is, a particular enzyme catalyzes a specific biochemical reaction. Thus, a different enzyme is required for the hydrolysis of each of the three disaccharides:

$$\text{Sucrose} + \text{Water} \xrightarrow{H^+ \text{ or Sucrase}} \text{Glucose} + \text{Fructose}$$

$$\text{Lactose} + \text{Water} \xrightarrow{H^+ \text{ or Lactase}} \text{Galactose} + \text{Glucose}$$

$$\text{Maltose} + \text{Water} \xrightarrow{H^+ \text{ or Maltase}} \text{Glucose} + \text{Glucose}$$

29.10 Structure and Properties of Disaccharides

Disaccharides contain an acetal structure (glycosidic linkage), and some also contain a hemiacetal structure. Maltose has the structure shown below. It may be considered as being derived from two glucose molecules by the elimination of a molecule of water between the OH group on carbon 1 on one glucose unit and the OH group on carbon 4 on the other glucose unit. This is an α-1,4-glycosidic linkage, since the glucose units have the α configuration and are joined at carbons 1 and 4.

In its usual form lactose consists of a β-D-galactose unit linked to an α-D-glucose unit. These are joined by a β-1,4-glycosidic linkage from carbon 1 on galactose to carbon 4 on glucose.

Lactose
(α-D-Glucose-4-β-D-galactoside)

Sucrose consists of an α-D-glucose unit and a β-D-fructose unit. These units are joined by an oxygen bridge from carbon 1 on glucose to carbon 2 on fructose; that is, by an α-1,2-glycosidic linkage.

α-D-Glucose unit   β-D-Fructose unit

Sucrose

In this perspective formula the fructose unit has been turned through 180° to bring its number 2 carbon close to the number 1 carbon on the glucose unit. The groups on the fructose unit are therefore shown reversed from the perspective representation in Section 29.7.

Ordinary sugar is nearly pure sucrose. There are no essential chemical differences between cane sugar and beet sugar. Astonishingly large amounts of sugar are produced—world production is on the order of 90 million tons annually.

Lactose and maltose both show mutarotation, which indicates one of the monosaccharide units has a hemiacetal ring that can open and close to give the α and β configurations. Sucrose does not show mutarotation. Its structure has only acetal linkages to both monosaccharide units.

The three disaccharides—sucrose, lactose, and maltose—have physical properties associated with large polar molecules. All three are crystalline solids and are quite soluble in water, the solubility of sucrose amounting to 179 g/100 g water at 0°C. Hydrogen bonding between the polar —OH groups on the sugar molecules and water molecules is a major factor in this high solubility. These sugars are not easily melted. In fact, lactose is the only one with a clearly defined melting point, namely 201.6°C. Sucrose and maltose begin to decompose when heated to 186°C and 102.5°C, respectively. When sucrose is heated to melting, it darkens and undergoes partial decomposition. The resulting mixture is known as caramel or burnt sugar and is used as coloring and as a flavoring agent in foods.

A scale of relative sweetness of common sugars, where fructose is assigned a value of 100, is given in Table 29.1. Note that sucrose is only 58% as sweet as fructose.

Sucrose has a tendency to crystallize from concentrated solutions or syrups. Therefore, in commercial food preparations (for example, candies, jellies, canned fruits) the sucrose is often hydrolyzed:

$$\text{Sucrose} + H_2O \xrightarrow{H^+} \text{Glucose} + \text{Fructose}$$

Table 29.1. Relative sweetness of sugars.

Fructose	100	Galactose	19
Sucrose	58	Lactose	9.2
Glucose	43	Invert sugar	75
Maltose	19		

invert sugar

The resulting mixture of glucose and fructose, usually in solution, is called **invert sugar**. Invert sugar has less tendency to crystallize than sucrose, and it has greater sweetening power than the corresponding amount of sucrose. The nutritive value of the sucrose is not affected in any way by conversion to invert sugar, since the same hydrolysis reaction occurs in normal digestion.

29.11 Reducing Sugars

reducing sugar

Some sugars are capable of reducing silver ions to free silver and copper(II) ions to copper(I) ions under prescribed conditions. Such sugars are called **reducing sugars**. This reducing ability, which is useful in classifying sugars and in certain clinical tests, is dependent on the presence of aldehyde groups, an α-hydroxyketone group (—CH$_2$COCH$_2$OH) such as is present in fructose, or a hemiacetal structure in a cyclic molecule such as maltose. These groups are easily oxidized to carboxylic acid (or carboxylate ion) groups; the metal ions are thereby reduced (Ag$^+$ → Ag^0; Cu^{2+} → Cu$^+$). Several different reagents, including Tollens', Fehling's, Benedict's, and Barfoed's reagents, are used to detect reducing sugars.

Tollens' reagent is silver nitrate solution made alkaline with ammonia. A positive Tollens' test is the formation of a metallic silver mirror on the inner surface of a clean test tube. Using an aldehyde group to illustrate, the reaction is represented by

$$\underset{\text{Aldehyde}}{RC\overset{H}{=}O} + \underset{\substack{\text{Silver}\\\text{ion}}}{2\,Ag^+} + 3\,NH_3 + H_2O \longrightarrow \underset{\substack{\text{Carboxylate}\\\text{ion group}}}{RC\overset{O}{\underset{\|}{-}}O^-} + \underset{\substack{\text{Free}\\\text{silver}}}{2\,Ag\downarrow} + 3\,NH_4^+$$

Benedict's and Fehling's reagents contain Cu^{2+} ions in an alkaline citrate or tartrate solution. A reducing sugar in sufficient concentration produces a brick-red Cu$_2$O precipitate with either reagent. For an aldehyde group the equation is

$$\underset{\substack{\text{Aldehyde}\\\text{group}}}{RC\overset{H}{=}O} + \underset{\text{Blue}}{2\,Cu^{2+}} + 5\,OH^- \longrightarrow \underset{\substack{\text{Carboxylate}\\\text{ion group}}}{RC\overset{O}{\underset{\|}{-}}O^-} + \underset{\substack{\text{Copper(I)}\\\text{oxide}\\\text{(brick-red)}}}{Cu_2O\downarrow} + 3\,H_2O$$

Benedict's and Fehling's reagents are used to detect the presence of glucose in urine. Initially, the reagents are deep blue in color. A positive test is indicated by a color change to greenish-yellow, yellowish-orange, or brick-red, depending on the glucose (reducing sugar) concentration. These tests are used for estimating the amount of glucose in the urine of diabetics in order to adjust the amount of insulin needed for proper glucose utilization.

Barfoed's reagent contains Cu^{2+} ions in the presence of acetic acid. It is used to distinguish reducing monosaccharides from reducing disaccharides. Under the same reaction conditions the reagent is reduced more rapidly by monosaccharides.

Glucose and galactose contain aldehyde groups; fructose contains an α-hydroxyketone group. Therefore, all three of these monosaccharides are reducing sugars.

A carbohydrate molecule need not have an actual free aldehyde or α-hydroxyketone group to be a reducing sugar. A hemiacetal structure (see below) is a "potential aldehyde group." Maltose and the cyclic form of glucose are examples of molecules with the hemiacetal structure:

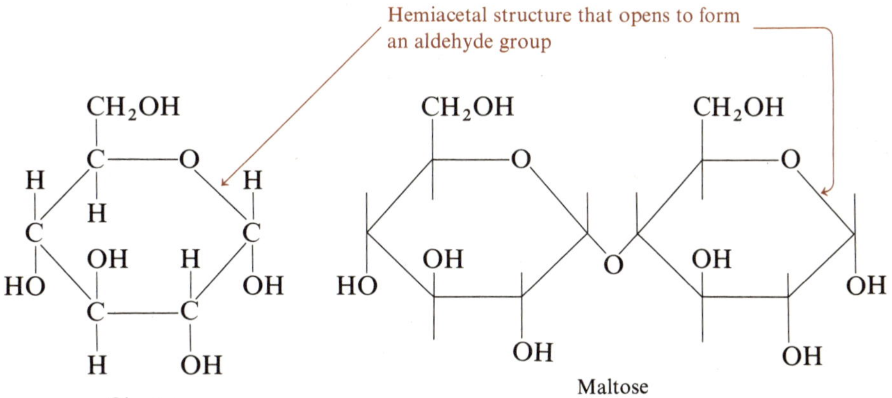

Glucose

Maltose

Under mildly alkaline conditions, the rings open at the points indicated by the arrows to form aldehyde groups:

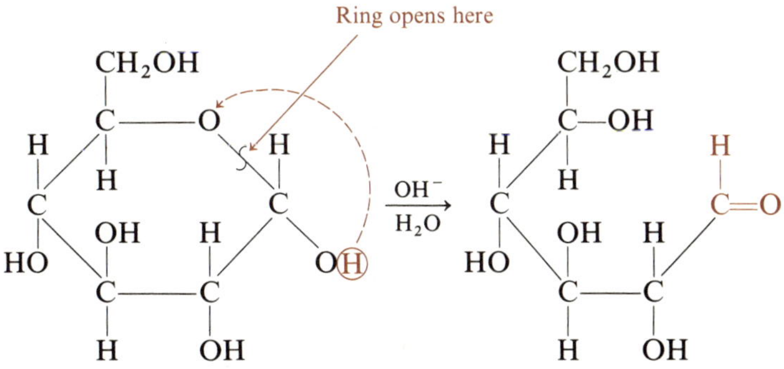

Glucose, ring structure

Glucose, open-chain structure

Since the Tollens', Fehling's, and Benedict's tests are carried out under alkaline conditions, they give positive tests with such molecules. Any sugar that has the hemiacetal structure is classified as a reducing sugar. Among the disaccharides, lactose and maltose have hemiacetal structures and are therefore reducing sugars. Sucrose is not a reducing sugar because it does not have the hemiacetal structure.

29.12 Reactions of Monosaccharides

Oxidation

The oxidation of monosaccharides by copper ions and silver ions was described in Section 29.11. The aldehyde groups in monosaccharides are also oxidized to monocarboxylic acids by other mild oxidizing agents such as bromine water. The carboxylic acid group is formed at carbon 1. The name of the resulting acid is formed by changing the *-ose* ending to *-onic acid*. Glucose yields gluconic acid; galactose, galactonic acid; and so on.

$$\text{D-Glucose} + Br_2 + H_2O \longrightarrow \text{D-Gluconic acid} + 2\,HBr$$

Dilute nitric acid, a vigorous oxidizing agent, oxidizes both carbon 1 and carbon 6 of aldohexoses to form dicarboxylic acids. Glucose yields saccharic acid; galactose, mucic acid.

$$\text{D-Glucose} \xrightarrow{\text{Warm } HNO_3} \text{Saccharic acid}$$

This reaction serves as the basis of the *mucic acid test*, which is sometimes used to distinguish glucose from galactose. Mucic acid has very limited solubility and saccharic acid is quite soluble. Hence, when oxidized with nitric acid, galactose (and lactose, which hydrolyzes to form glucose and galactose) yields a precipitate of mucic acid crystals. When glucose is oxidized under the same conditions, the saccharic acid does not precipitate. Other aldoses react similarly when heated with nitric acid. The aldehyde and the —CH$_2$OH groups are oxidized to carboxylic acids.

Problem 29.2 Two samples labeled A and B are known to be D-threose and D-erythrose. Water solutions of each sample are optically active. However, when each solution was warmed with nitric acid, the solution from sample A was optically inactive while that from sample B was still active. Identify samples A and B.

In this problem we need to examine the structures of D-threose and D-erythrose, write equations for the reaction with nitric acid, and examine the products to see why one is optically active and the other optically inactive. Start by writing the formulas for D-threose and D-erythrose:

```
   H—C=O              H—C=O
   |                  |
HO—C—H             H—C—OH
   |                  |
 H—C—OH            H—C—OH
   |                  |
  CH₂OH              CH₂OH

 D-Threose          D-Erythrose
```

The oxidation of these tetroses would yield dicarboxylic acids:

```
   H—C=O            COOH           H—C=O             COOH
   |                |              |                 |
HO—C—H   HNO₃   HO—C—H          H—C—OH   HNO₃    H—C—OH
   |     ——→       |              |       ——→       |
 H—C—OH          H—C—OH         H—C—OH            H—C—OH
   |                |              |                 |
  CH₂OH           COOH            CH₂OH             COOH
                   I                                  II
```

Product I is a chiral molecule and would be optically active. Product II is a *meso* compound and would be optically inactive. Therefore, sample A is D-erythrose, since the *meso* acid results from its oxidation. Sample B must then be D-threose.

Osazone formation

Phenylhydrazine (C$_6$H$_5$NHNH$_2$) reacts with carbons 1 and 2 of reducing sugars to form derivatives called osazones. Osazone formation is useful for comparing the structures of sugars. Glucose and fructose react as shown in Figure 29.4.

Identical osazones are obtained from D-glucose and D-fructose. This demonstrates that carbons 3 through 6 of D-glucose and D-fructose molecules are identical. The same osazone is also obtained from D-mannose. This indicates that carbons 3 through 6 of the D-mannose molecule are the same as those of D-glucose and D-fructose molecules. In fact, D-mannose differs from D-glucose only in the configuration of the —H and —OH groups on carbon 2.

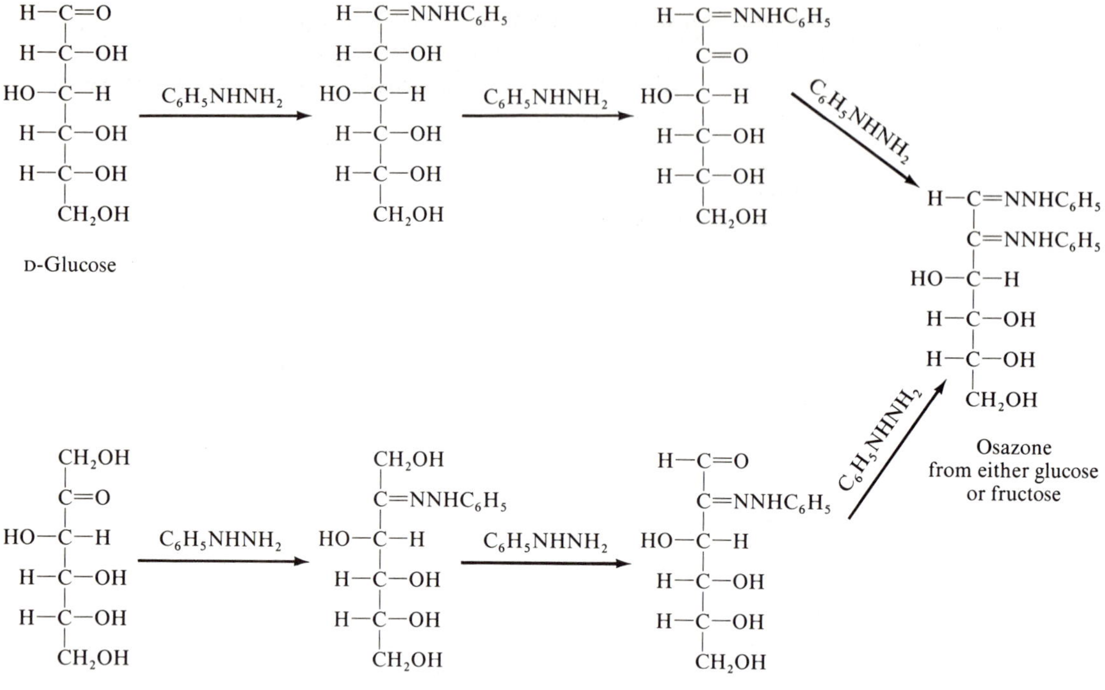

660 Chapter 29

$$
\begin{array}{c}
\text{H}-\text{C}=\text{O} \\
\text{H}-\text{C}-\text{OH} \\
\text{HO}-\text{C}-\text{H} \\
\text{H}-\text{C}-\text{OH} \\
\text{H}-\text{C}-\text{OH} \\
\text{CH}_2\text{OH}
\end{array}
\xrightarrow{\text{C}_6\text{H}_5\text{NHNH}_2}
\begin{array}{c}
\text{H}-\text{C}=\text{NNHC}_6\text{H}_5 \\
\text{H}-\text{C}-\text{OH} \\
\text{HO}-\text{C}-\text{H} \\
\text{H}-\text{C}-\text{OH} \\
\text{H}-\text{C}-\text{OH} \\
\text{CH}_2\text{OH}
\end{array}
\xrightarrow{\text{C}_6\text{H}_5\text{NHNH}_2}
\begin{array}{c}
\text{H}-\text{C}=\text{NNHC}_6\text{H}_5 \\
\text{C}=\text{O} \\
\text{HO}-\text{C}-\text{H} \\
\text{H}-\text{C}-\text{OH} \\
\text{H}-\text{C}-\text{OH} \\
\text{CH}_2\text{OH}
\end{array}
\xrightarrow{\text{C}_6\text{H}_5\text{NHNH}_2}
\begin{array}{c}
\text{H}-\text{C}=\text{NNHC}_6\text{H}_5 \\
\text{C}=\text{NNHC}_6\text{H}_5 \\
\text{HO}-\text{C}-\text{H} \\
\text{H}-\text{C}-\text{OH} \\
\text{H}-\text{C}-\text{OH} \\
\text{CH}_2\text{OH}
\end{array}
$$

D-Glucose

Osazone from either glucose or fructose

$$
\begin{array}{c}
\text{CH}_2\text{OH} \\
\text{C}=\text{O} \\
\text{HO}-\text{C}-\text{H} \\
\text{H}-\text{C}-\text{OH} \\
\text{H}-\text{C}-\text{OH} \\
\text{CH}_2\text{OH}
\end{array}
\xrightarrow{\text{C}_6\text{H}_5\text{NHNH}_2}
\begin{array}{c}
\text{CH}_2\text{OH} \\
\text{C}=\text{NNHC}_6\text{H}_5 \\
\text{HO}-\text{C}-\text{H} \\
\text{H}-\text{C}-\text{OH} \\
\text{H}-\text{C}-\text{OH} \\
\text{CH}_2\text{OH}
\end{array}
\xrightarrow{\text{C}_6\text{H}_5\text{NHNH}_2}
\begin{array}{c}
\text{H}-\text{C}=\text{O} \\
\text{C}=\text{NNHC}_6\text{H}_5 \\
\text{HO}-\text{C}-\text{H} \\
\text{H}-\text{C}-\text{OH} \\
\text{H}-\text{C}-\text{OH} \\
\text{CH}_2\text{OH}
\end{array}
$$

D-Fructose

Figure 29.4. Reaction of glucose and fructose to form osazones.

Reduction

Monosaccharides may be reduced to their corresponding polyhydroxy-alcohols by reducing agents such as H_2/Pt or sodium amalgam (Na/Hg). For example, glucose yields sorbitol (glucitol), galactose yields galactitol (dulcitol), and mannose yields mannitol; all these are hexahydric alcohols (containing six —OH groups).

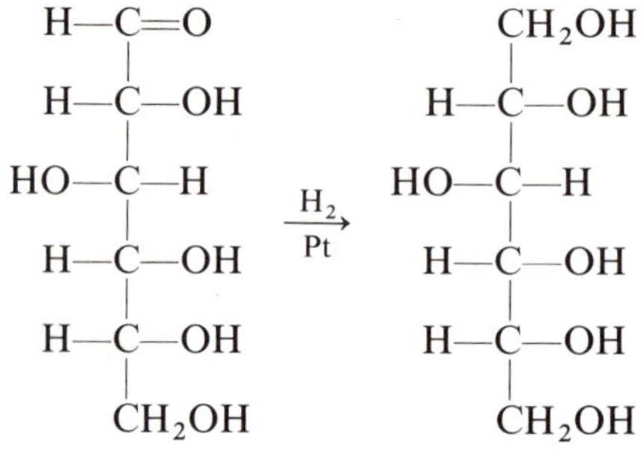

$$
\begin{array}{c}
\text{H}-\text{C}=\text{O} \\
\text{H}-\text{C}-\text{OH} \\
\text{HO}-\text{C}-\text{H} \\
\text{H}-\text{C}-\text{OH} \\
\text{H}-\text{C}-\text{OH} \\
\text{CH}_2\text{OH}
\end{array}
\xrightarrow[\text{Pt}]{H_2}
\begin{array}{c}
\text{CH}_2\text{OH} \\
\text{H}-\text{C}-\text{OH} \\
\text{HO}-\text{C}-\text{H} \\
\text{H}-\text{C}-\text{OH} \\
\text{H}-\text{C}-\text{OH} \\
\text{CH}_2\text{OH}
\end{array}
$$

D-Glucose → D-Glucitol (Sorbitol)

Hexahydric alcohols have properties resembling those of glycerol (Section 24.5). Because of their affinity for water, they are used as "moisturizing" agents in food and cosmetic products. Sorbitol is about 60% as sweet as sucrose. It is used in "sugarless" products such as "sugar-free" chewing gum. About 70% of the sorbitol ingested orally is converted to CO_2 without appearing as glucose in the blood. Sorbitol, galactitol, and mannitol occur naturally in a variety of plants.

29.13 Polysaccharides

Although several polysaccharides are known, three—starch, cellulose, and glycogen—are of outstanding importance. All three are polymers and, when hydrolyzed, yield D-glucose as the only product, according to this approximate equation.

$$(C_6H_{10}O_5)_n + nH_2O \longrightarrow nC_6H_{12}O_6$$

Polysaccharide molecule (approximate)   D-Glucose

This hydrolysis reaction establishes that all three polysaccharides are polymers made up of glucose monosaccharide units. It also means that the differences in properties among the three polysaccharides must be due to differences in the structure and/or size of their molecules.

Many years of research were required to determine the detailed structures for polysaccharide molecules. Consideration of this work is beyond the scope of our discussion, but an abbreviated summary of the results is given in the following paragraphs.

Starch

Starch is found in plants, mainly in the seeds, roots, or tubers. Corn, wheat, potatoes, rice, and cassava are the chief sources of starch. The two main components of starch are amylose and amylopectin. Amylose molecules are unbranched chains made up of from about 25 to 1300 α-D-glucose units joined by α-1,4-glycosidic linkages, as shown in Figure 29.5.

Amylopectin is a branched-chain polysaccharide with much larger molecules than amylose. Amylopectin molecules consist on the average of several thousand α-D-glucose units with molecular weights ranging up to one million or more. The main chain contains glucose units connected by α-1,4-glycosidic linkages. In addition, branch chains are linked to the main chain through α-1,6-glycosidic linkages about every 25 glucose units, as shown in Figure 29.5.

Apparently because of their very large size and despite the presence of many polar —OH groups, starch molecules are insoluble in cold water. Starch readily forms colloidal dispersions in hot water. Such starch "solutions" form an intense blue-black color in the presence of free iodine. Hence, a starch solution can be used to detect free iodine, or a dilute iodine solution can be used to detect starch.

Starch is readily converted to glucose by heating with water and a little acid (for example, hydrochloric or sulfuric acid). It is also readily hydrolyzed

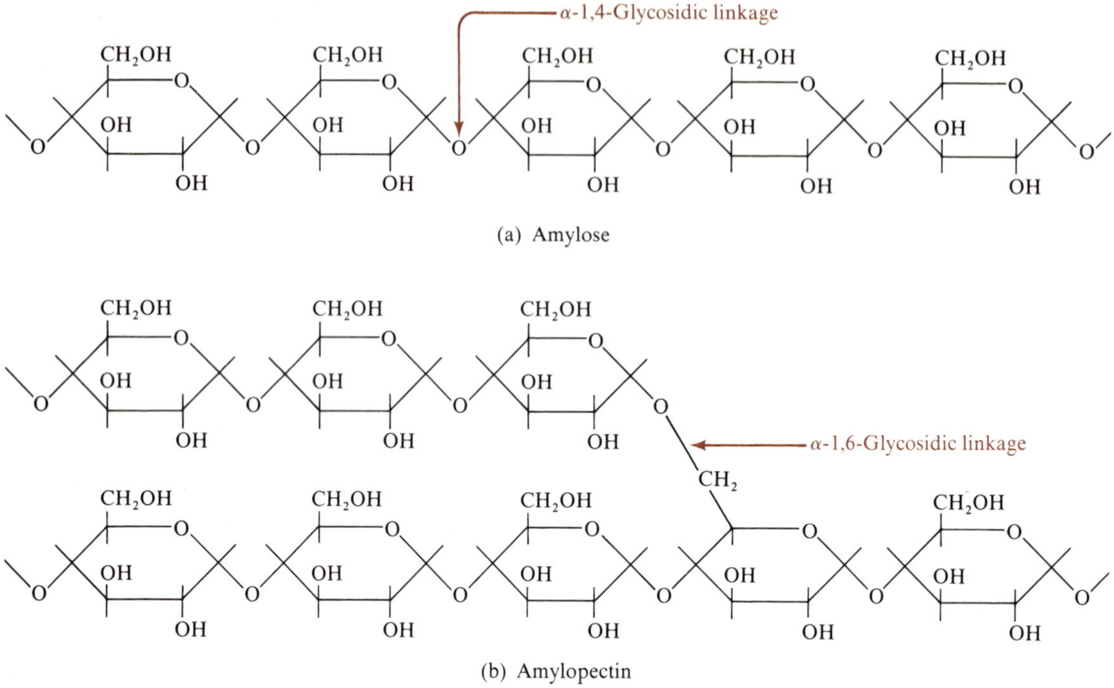

(a) Amylose — α-1,4-Glycosidic linkage

(b) Amylopectin — α-1,6-Glycosidic linkage

Figure 29.5. Structure of (a) amylose and (b) amylopectin, the two chief components of starch.

at room temperature by certain digestive enzymes. One of these enzymes, ptyalin (salivary amylase found in saliva), converts starch to maltose. The hydrolysis of starch to maltose and glucose is shown in the following equations:

$$\text{Starch} \xrightarrow[\text{or amylase}]{\text{Acid} + \Delta} \text{Dextrins} \xrightarrow[\text{or amylase}]{\text{Acid} + \Delta} \text{Maltose} \xrightarrow[\text{or maltase}]{\text{Acid} + \Delta} \text{D-Glucose}$$

Dextrins are intermediate-size polysaccharides.

The hydrolysis of starch can be followed qualitatively by periodically testing samples from a mixture of starch and saliva with very dilute iodine solution. The change of color sequence is blue-black → blue → purple → pink → colorless as the starch molecules are broken down into smaller and smaller fragments.

Starch is the reserve carbohydrate of the plant kingdom. Humans, as well as many other animals, utilize huge quantities of starch as food. It is also hydrolyzed to make glucose and various industrial products such as adhesives used on postage stamps.

Glycogen

Glycogen is the reserve carbohydrate of the animal kingdom. It is formed in the body by polymerization of glucose and stored in the liver and in muscle tissues. Structurally, it is very similar to the amylopectin fraction of starch, except that it is more highly branched. The α-1,6-glycosidic linkages occur on 1 of every 12 to 18 glucose units.

Cellulose

Cellulose is the most abundant organic substance found in nature. It is the chief structural component of plants and wood. Cotton fibers are almost pure cellulose, and wood, after removal of moisture, consists of about 50% cellulose. Cellulose is an important substance in the textile and paper industries.

Cellulose, like starch and glycogen, is a polymer of glucose. But cellulose differs from starch and glycogen because the glucose units are joined by β-1,4-glycosidic linkages instead of α-1,4-glycosidic linkages. A portion of a cellulose molecule is represented in Figure 29.6. Cellulose has greater resistance to hydrolysis than either starch or glycogen. It is not appreciably hydrolyzed when boiled in a 1% sulfuric acid solution. It does not show a color reaction with iodine. Humans cannot digest cellulose, since they have no enzymes capable of catalyzing its hydrolysis. Fortunately, some microorganisms found in soil and in the digestive tracts of certain animals produce enzymes that do catalyze the breakdown of cellulose. The presence of these microorganisms explains why cows and other herbivorous animals thrive on grass—and also why termites thrive on wood!

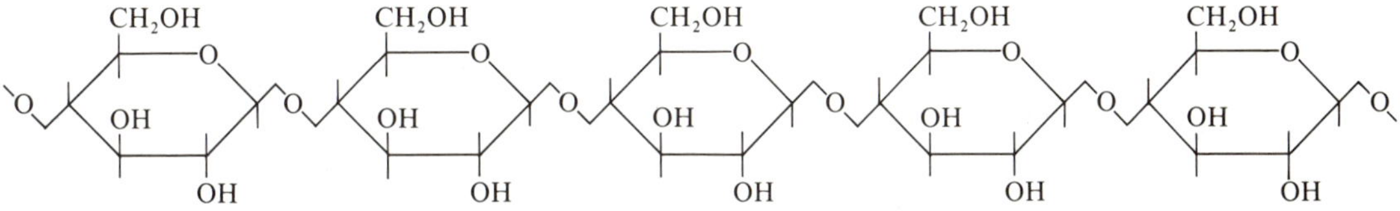

Figure 29.6. Portion of a cellulose molecule. Cellulose molecules consist of thousands of glucose anhydride units joined together by β-1,4-glycosidic linkages. Cellulose molecules are largely unbranched.

The —OH groups of starch and cellulose can be reacted without destruction of the macromolecular structures. For example, nitric acid converts an —OH group to a nitrate group in this fashion.

$$\text{Cellulose—OH} + \text{HONO}_2 \longrightarrow \text{Cellulose—ONO}_2 + \text{H}_2\text{O}$$

| Hydroxyl group on cellulose molecule | Nitric acid | Nitrate group on cellulose molecule | Water |

If only a portion of the —OH groups on the cellulose molecule are nitrated, a plastic nitrocellulose material known as Celluloid or pyroxylin is obtained. This material has been used to make such diverse articles as billiard balls, Celluloid collars, and photographic film. By nitration of nearly all —OH groups, a powerful high explosive is obtained. This highly nitrated cellulose, or "guncotton," is the basic ingredient in modern "smokeless" gunpowder.

Another modified cellulose, cellulose acetate, is made by esterification of —OH groups with acetic acid (acetic anhydride). About two-thirds of the —OH groups are esterified:

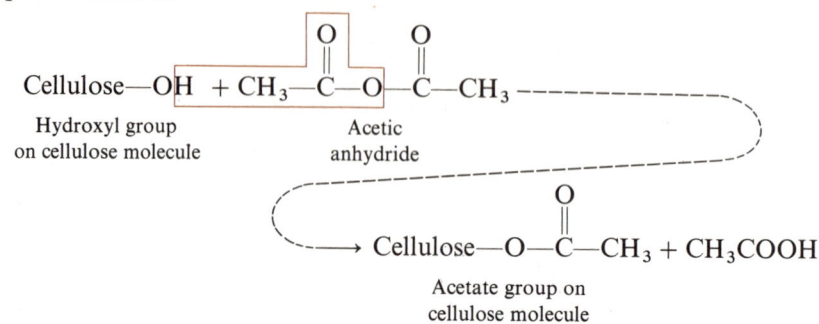

$$\text{Cellulose—OH} + \text{CH}_3-\overset{\overset{\text{O}}{\|}}{\text{C}}-\text{O}-\overset{\overset{\text{O}}{\|}}{\text{C}}-\text{CH}_3 \longrightarrow \text{Cellulose}-\text{O}-\overset{\overset{\text{O}}{\|}}{\text{C}}-\text{CH}_3 + \text{CH}_3\text{COOH}$$

Hydroxyl group on cellulose molecule
Acetic anhydride
Acetate group on cellulose molecule

Cellulose acetate, unlike the dangerously flammable cellulose nitrate, can be made to burn only with difficulty. For this reason, cellulose acetate has displaced cellulose nitrate in almost all kinds of photographic films. The textile known as acetate rayon is made from cellulose acetate. Cellulose acetate is also used as a clear, transparent packaging film. In another process, cellulose is reacted with carbon disulfide in the presence of sodium hydroxide to form a soluble cellulose derivative called cellulose xanthate, from which cellulose can be regenerated. Viscose rayon textiles and cellophane packaging materials are made of regenerated cellulose prepared by this process.

Not all polysaccharides are polymers of glucose. For example, inulin is a starchlike substance that is largely a polymer composed of fructose units joined by 1,2-glycosidic linkages. It has a molecular weight of about 5000. Inulin is found in the tubers of artichokes and dahlias and in the roots of dandelions and chicory. It is not digestable by animal enzymes. Clinically, inulin is used as a diagnostic aid in testing kidney function.

Questions

A. *Review the meanings of the new terms introduced in this chapter.*
   1. Carbohydrate
   2. Monosaccharide
   3. Disaccharide
   4. Oligosaccharide
   5. Polysaccharide
   6. Epimer
   7. Mutarotation
   8. Anomer
   9. Glycoside
   10. Invert sugar
   11. Reducing sugar

B. *Review questions.*
   1. What is the significance of the notations D and L in the name of a carbohydrate?
   2. What is the significance of the notations (+) and (−) in the name of a carbohydrate?
   3. Explain how a carbohydrate with a pyranose structure differs from one with a furanose structure.
   4. What is galactosemia and what are its effects on humans?
   5. Which of the D-aldohexoses in Figure 29.1 are epimers?
   6. Write the cyclic structures for α-D-glucose, β-D-galactose, and α-D-mannose.
   7. Explain how α-D-glucose differs from β-D-glucose.

8. Starting with the proper D-tetrose, show the steps for the synthesis of D-glucose by the Kiliani-Fischer synthesis.
9. Are the cyclic forms of monosaccharides hemiacetals or glycosides?
10. Explain the phenomenon of mutarotation.
11. Is D-2-deoxyarabinose the same as D-2-deoxyribose? Explain.
12. What is (are) the major source(s) of each?
    (a) Sucrose   (b) Lactose   (c) Maltose
13. What is the monosaccharide composition of each?
    (a) Sucrose   (c) Lactose   (e) Amylose
    (b) Maltose   (d) Glycogen
14. Cite two reasons why sucrose is sometimes changed to invert sugar before use in commercial food preparation.
15. Explain why invert sugar is sweeter than sucrose.
16. Which of the disaccharides—sucrose, maltose, and lactose—will show mutarotation? Explain why.
17. Draw structural formulas for maltose and sucrose. Point out the portion of the structure that is responsible for one of these disaccharides being classified as a reducing sugar.
18. (a) What is Tollens' reagent?
    (b) Write an equation showing the reaction of glucose with Tollens' reagent.
19. What visible evidence of reaction would you see when testing a reducing sugar with (a) Benedict's reagent?   (b) Tollens' reagent?
20. Which of the D-aldopentoses (Figure 29.1) will be optically inactive when reduced with H_2/Pt? Write structures for the reduced products and explain why they are optically inactive.
21. The simplest aldose is glyceraldehyde, $CH_2(OH)$—$CH(OH)$—CHO.
    (a) Write projection formulas for and identify the D and L forms of glyceraldehyde.
    (b) Write the structural formula of the product obtained by reacting glyceraldehyde with hydrogen in the presence of a platinum catalyst.
    (c) Would the compound obtained in part (b) be optically active? Why or why not?
22. Consider the eight aldohexoses given in Figure 29.1 and answer the following:
    (a) Which of these aldohexoses will give the same osazone?
    (b) The eight aldohexoses are oxidized by nitric acid to dicarboxylic acids. Which of these will give *meso* (optically inactive) dicarboxylic acids?
    (c) Write the structures and names for the enantiomers of D-(+)-mannose and D-(−)-idose.
23. Explain the principal differences and similarities between the members of each of the following pairs:
    (a) D-Glucose and D-fructose   (d) Amylose and amylopectin
    (b) D-Ribose and D-2-deoxyribose   (e) Cellulose and glycogen
    (c) Maltose and sucrose   (f) D-(−)-Ribose and D-(+)-glucose
24. When glucose is oxidized with nitric acid, saccharic acid is formed. Write the structural formula and the name of the dicarboxylic acid that is formed when galactose is oxidized with nitric acid.
25. When galactose is oxidized by nitric acid, will the dicarboxylic acid formed be optically active or inactive? Show evidence.
26. Write tree formulas for the four L-aldopentoses. Indicate which pair (or pairs) of the L-aldopentoses will give identical osazones with phenylhydrazine.

27. Write the structure of a disaccharide that hydrolyzes to give only D-glucose and (a) is a reducing sugar, shows mutarotation, and will form an osazone; (b) is a nonreducing sugar, does not mutarotate, and will not form an osazone.
28. What are the structural differences between starch and cellulose?
29. What are the two main components of starch? How are they alike and how do they differ?
30. Write the structure for cellobiose, a disaccharide obtained by the hydrolysis of cellulose.
31. A molecular weight value of about 325,000 was calculated for a sample of amylopectin. Approximately how many glucose anhydride units are present in an average molecule of this amylopectin?
32. Draw enough of the structural formula of cellulose acetate to show the repeating units and how they are linked.
33. Write three units of the polymeric structure of inulin.
34. Which of these statements are correct?
    (a) α-D-Glucose and β-D-glucose are enantiomers.
    (b) D-Glyceraldehyde and L-glyceraldehyde are epimers.
    (c) D-Threose and L-threose are epimers.
    (d) There are eight stereoisomers of the aldopentoses.
    (e) D-Glucose and L-glucose will form identical osazones.
    (f) Raffinose, which consists of one unit each of galactose, glucose, and fructose, is an oligosaccharide.
    (g) Two aldohexoses that react with phenylhydrazine to yield identical osazones are epimers.
    (h) Dextrose is another name for glucose.
    (i) Methyl glucosides are capable of reducing Fehling's and Tollens' reagents.
    (j) Humans are incapable of using cellulose directly as a food.
    (k) Fructose can be classified as a hexose, a monosaccharide, and an aldose.
    (l) The change in the optical rotation of a carbohydrate solution to an equilibrium value is called mutarotation.
    (m) The disaccharide found in mammalian milk is galactose.
    (n) The reserve carbohydrate of animals is glycogen.
    (o) Starch consists of two polysaccharides known as amylose and amylopectin.
    (p) Carbohydrates that are capable of reducing silver ions in Tollens' reagent are called reducing sugars.

30 Amino Acids, Polypeptides, and Proteins

After studying Chapter 30 you should be able to:

1. Understand the terms listed in Question A at the end of the chapter.
2. List five foods that are major sources of proteins.
3. List the elements that are contained in proteins.
4. Write formulas for and know the abbreviations for the common amino acids.
5. Write the zwitterion formula of an amino acid and know how it behaves in a dilute acid and in a dilute basic solution.
6. Understand why a protein in solution at its isolectric pH will not migrate in an electrolytic cell.
7. Combine amino acids into polypeptide chains.
8. Understand how peptide chains are named and numbered.
9. Understand what is meant by the *N*-terminal and *C*-terminal residues of a peptide chain.
10. Briefly explain the primary, secondary, and tertiary structure of a protein.
11. Understand the bonding in the α-helix and β-pleated sheet structure of a protein.
12. Explain the role cysteine plays in the structure of polypeptides and proteins.
13. Explain how enzymes function as biological catalysts.
14. Discuss the role of enzymes in the body.
15. List the six main classes of enzymes and their functions.
16. Tell what is meant by the specificity of an enzyme.
17. Discuss the role of specific enzymes in the hydrolysis of proteins.
18. Describe Sanger's work on determining the primary structure of proteins.
19. Reconstruct a peptide chain knowing its hydrolysis products.
20. Describe chromatographic methods of separating amino acids.
21. State the physical evidence observed in a positive reaction in the (a) xanthoproteic reaction, (b) biuret test, and (c) ninhydrin test.
22. Know what is meant by denaturation and the various methods by which proteins can be denatured.

30.1 Introduction

Proteins are present in every living cell. Their very name, derived from the Greek word *proteios*, which means "holding first place," signifies the importance of these substances. Proteins are one of the three major classes of foods. The other two, carbohydrates and fats, are needed for energy; proteins are needed for growth and/or maintenance of body tissue. Some common foods with high (over 10%) protein content are fish, beans, nuts, cheese, eggs, poultry, and meat of all kinds. These kinds of foods tend to be scarce and relatively expensive. Proteins are, therefore, the class of foods that is least available to the undernourished people of the world. Hence, the question of how to secure an adequate supply of high-quality protein for an ever-increasing population is one of the more critical problems now confronting us.

The problem of protein shortage is being partly solved by (1) increased vegetable protein production (principally soybeans) and (2) increased processing of fish and other marine organisms to obtain meal and other edible products. Improved soybean varieties have been developed; and United States production has increased to the point that soybeans are our number one agricultural export product. It seems likely that protein production can be increased greatly in other countries, especially Brazil and India, by extensive plantings of improved soybean varieties. There is also hope that the yield of food protein from the ocean can be increased by a large factor. But much research needs to be done to determine how this can be best accomplished. Unfortunately, the ocean fisheries are already responsible for international friction as various nations compete for this resource.

Proteins function as structural materials and as enzymes (catalysts) which regulate the countless chemical reactions taking place in every living organism, including the reactions involved in the decomposition and synthesis of proteins.

All proteins are polymeric substances that yield amino acids on hydrolysis. Those that yield only amino acids when hydrolyzed are classified as **simple proteins**; those that yield amino acids and one or more additional products are classified as **conjugated proteins**. There are approximately 200 different known amino acids in nature. Some are found in only one particular species of plant or animal, others are found in only a restricted number of life forms. But 20 of these amino acids are found in almost all proteins. Furthermore, these same 20 amino acids are used by all forms of life in the synthesis of proteins.

All proteins contain carbon, hydrogen, oxygen, nitrogen, and sulfur. Some proteins contain additional elements, usually phosphorus, copper, or zinc. The presence of nitrogen in proteins sets them apart from carbohydrates and lipids. The average nitrogen content is about 16%.

Proteins are highly specific in their functions. The amino acid units are arranged in a definite sequence in any given protein molecule. An amazing fact about proteins is that in some cases if just one of the hundreds or thousands of amino acid units is missing or out of place, the biological function of that protein is seriously damaged or destroyed.

30.2 The Nature of Amino Acids

Each amino acid has two functional groups, an amino group (—NH$_2$) and a carboxylic acid group (—COOH). The amino acids that are found in proteins are alpha (α) amino acids because the amino group is attached to the first or alpha carbon atom adjacent to the carboxyl group. The beta (β) position is the next adjacent carbon; the gamma (γ) position the next carbon; and so on. The following formula represents an alpha (α) amino acid:

$$\overset{\gamma}{C}H_3\overset{\beta}{C}H_2\overset{\alpha}{C}HCOOH$$
$$\qquad\qquad\;\; |$$
$$\qquad\qquad NH_2$$

α-Amino butyric acid

Alpha amino acids are represented by this general formula:

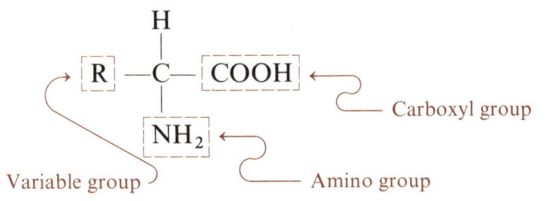

$$\text{Variable group} \rightarrow \boxed{R} - \underset{\underset{\boxed{NH_2}}{|}}{\overset{\overset{H}{|}}{C}} - \boxed{COOH} \leftarrow \text{Carboxyl group}$$

$$\qquad\qquad\qquad\qquad\qquad\uparrow$$
$$\qquad\qquad\qquad\qquad\text{Amino group}$$

The part of the amino acid molecule designated R is not restricted to the usual alkyl radical; R may represent open-chain or cyclic hydrocarbons, aromatic hydrocarbons, structures containing additional amino or carboxyl groups, or groups containing sulfur.

Amino acids are divided into three groups: neutral, acidic, and basic. They are classified as neutral amino acids when their molecules have the same number of amino and carboxyl groups; as acidic when their molecules have more carboxyl groups than amino groups; or as basic when their molecules have more amino groups than carboxyl groups.

The names, formulas, and abbreviations of the common amino acids are given in Table 30.1. Two of these, aspartic acid and glutamic acid, are classified as acidic; three, lysine, arginine, and histidine, as basic; the remainder are classified as neutral amino acids.

30.3 Essential Amino Acids

essential amino acid

Eight of the amino acids are **essential amino acids**. These amino acids—isoleucine, leucine, lysine, methionine, phenylalanine, threonine, tryptophan, and valine—are essential to the functioning of the human body. Since the body is not capable of synthesizing them, they must be supplied in our diets if we are to enjoy normal health. It is known that some other animals require amino acids in addition to those listed for humans. Rats, for example, require two additional amino acids—arginine and histidine.

On a nutritional basis, proteins are classified as complete or incomplete. A complete protein supplies all the essential amino acids; an incomplete protein

Table 30.1. Common amino acids derived from proteins.

Name	Abbreviation	Formula
Alanine	Ala	$CH_3CH(NH_2)COOH$
Arginine	Arg	$NH_2-C(=NH)-NH-CH_2CH_2CH_2CH(NH_2)COOH$
Asparagine	Asn	$NH_2C(=O)-CH_2CH(NH_2)COOH$
Aspartic acid	Asp	$HOOCCH_2CH(NH_2)COOH$
Cysteine	Cys	$HSCH_2CH(NH_2)COOH$
Glutamic acid	Glu	$HOOCCH_2CH_2CH(NH_2)COOH$
Glutamine	Gln	$NH_2C(=O)CH_2CH_2CH(NH_2)COOH$
Glycine	Gly	$HCH(NH_2)COOH$
Histidine	His	imidazole-$CH_2CH(NH_2)COOH$
Isoleucine[a]	Ile	$CH_3CH_2CH(CH_3)CH(NH_2)COOH$
Leucine[a]	Leu	$(CH_3)_2CHCH_2CH(NH_2)COOH$
Lysine[a]	Lys	$NH_2CH_2CH_2CH_2CH_2CH(NH_2)COOH$
Methionine[a]	Met	$CH_3SCH_2CH_2CH(NH_2)COOH$
Phenylalanine[a]	Phe	$C_6H_5CH_2CH(NH_2)COOH$

Table 30.1 (*Continued*)

Name	Abbreviation	Formula
Proline	Pro	![pyrrolidine]—COOH with N-H in ring
Serine	Ser	HOCH$_2$CHCOOH 　　　　　| 　　　　　NH$_2$
Threonine[a]	Thr	CH$_3$CH—CHCOOH 	| 　　　OH　NH$_2$
Tryptophan[a]	Trp	indole—C—CH$_2$CHCOOH with CH and NH in ring, NH$_2$ on α-carbon
Tyrosine	Tyr	HO—C$_6$H$_4$—CH$_2$CHCOOH 　　　　　　　　　　| 　　　　　　　　　　NH$_2$
Valine[a]	Val	(CH$_3$)$_2$CHCHCOOH 　　　　　　　| 　　　　　　　NH$_2$

[a] Amino acids essential in human nutrition.

is deficient in one or more essential amino acids. Many proteins, especially those from vegetable sources, are incomplete. For example, protein from corn (maize) is deficient in lysine. The nutritional quality of such vegetable proteins can be greatly improved by supplementing them with the essential amino acids that are lacking, if these can be synthesized at reasonable costs. Lysine, methionine, and tryptophan are being sold at present for enriching human food and livestock feed. This is another way to extend the world's limited supply of high-quality food protein. In still another approach to the problem of obtaining more high-quality protein, plant breeders have recently developed maize varieties with greatly improved lysine content.

30.4　D-Amino Acids and L-Amino Acids

All amino acids, except glycine, have at least one asymmetric carbon atom. For example, two optically active stereoisomers of alanine are possible:

```
      COOH                COOH
        |                   |
   H—C*—NH₂          H₂N—C*—H
        |                   |
       CH₃                 CH₃

  D-(−)-Alanine        L-(+)-Alanine
```

The D and L configurations of amino acids are shown in projection formulas in the same way as in the configuration of D- and L-glyceraldehyde (Section 29.2). The —COOH group is written at the top of the projection formula, and the D configuration is indicated by writing the alpha—NH_2 group to the right of carbon 2. The L configuration is indicated by writing the alpha—NH_2 group to the left of carbon 2 (see the projection formulas of alanine above). Although some D-amino acids occur in nature, only L-amino acids occur in proteins. The (+) and (−) signs in the name indicate the direction of optical rotation of the amino acid.

30.5 Amphoterism

Amino acids are *amphoteric* (or *amphiprotic*); that is, they can react either as an acid or as a base. For example, with a strong base such as sodium hydroxide, alanine reacts as an acid, as shown in equation (1); with a strong acid such as HCl, alanine reacts as a base, as shown in equation (2).

$$CH_3CHCOOH + NaOH \longrightarrow CH_3CHCOO^- Na^+ + H_2O \quad (1)$$
$$||$$
$$NH_2 NH_2$$

Alanine — Sodium alanate

$$CH_3CHCOOH + HCl \longrightarrow CH_3CHCOOH \quad (2)$$
$$||$$
$$NH_2 NH_3^+ \, Cl^-$$

Alanyl ammonium chloride

Amino acids actually do not exist to any extent in the molecular form as shown in equations (1) and (2). Instead, they exist mainly as dipolar ions called **zwitterions**. Again using alanine as an example, the proton on the carboxyl group transfers to the amino group, forming a zwitterion by an acid–base reaction within the molecule.

zwitterion

$$CH_3CHCOO(H) \longrightarrow CH_3CHCOO^\ominus$$
$$||$$
$$H_2N: NH_3^\oplus$$

Alanine molecule — Alanine zwitterion

On an ionic basis the reaction of alanine with NaOH and HCl is

$$CH_3CHCOO^- + OH^- \longrightarrow CH_3CHCOO^- + H_2O \quad (3)$$
$$||$$
$$NH_3^+ NH_2$$

Alanine zwitterion — Alanate anion

$$CH_3CHCOO^- + H^+ \longrightarrow CH_3CHCOOH \quad (4)$$
$$||$$
$$NH_3^+ NH_3^+$$

Alanine zwitterion — Alanyl ammonium cation

Other amino acids behave like alanine. They, together with protein molecules that contain —COOH and —NH$_2$ groups, help to buffer or stabilize the pH of the blood at about 7.4. The pH is maintained close to 7.4 because any excess acid or base in the blood is neutralized by reactions such as shown in equations (3) and (4).

$$\underset{\underset{\text{II}}{\text{Cation form}}}{\underset{\overset{|}{NH_3^+}}{RCHCOOH}} \underset{OH^-}{\overset{H^+}{\rightleftharpoons}} \underset{\underset{\text{I}}{\text{Zwitterion form}}}{\underset{\overset{|}{NH_3^+}}{RCHCOO^-}} \underset{H^+}{\overset{OH^-}{\rightleftharpoons}} \underset{\underset{\text{III}}{\text{Anion form}}}{\underset{\overset{|}{NH_2}}{RCHCOO^-}}$$

isoelectric point

When an amino acid in solution has equal positive and negative charges, as in formula I, it is electrically neutral and does not migrate toward either the positive or negative electrode when placed in an electrolytic cell. The pH at which there is no migration toward either electrode is called the **isoelectric point** (see Table 30.2). If acid (H$^+$) is added to an amino acid at its isoelectric point, the equilibrium is shifted toward formula II and the cation formed will migrate toward the negative electrode. When base (OH$^-$) is added, the anion formed (formula III) will migrate toward the positive electrode. Differences in isoelectric points are important in isolating and purifying amino acids and proteins, since their rates and directions of migration can be controlled in an electrolytic cell by adjusting the pH. This method of separation is called electrophoresis.

Table 30.2. Isoelectric points of selected amino acids.

Amino acid	pH at isoelectric point
Arginine	10.8
Lysine	9.7
Alanine	6.0
Glycine	6.0
Serine	5.7
Glutamic acid	3.2
Aspartic acid	2.9

Amino acids are classified as basic, neutral, or acidic depending on whether the ratio of —NH$_2$ to —COOH groups in the molecules is greater than 1:1, equal to 1:1, or less than 1:1, respectively. Isoelectric points are found at pH values ranging from 7.8 to 10.8 for basic, 4.8 to 6.3 for neutral, and 2.8 to 3.3 for acidic amino acids. It is logical that a molecule such as glutamic acid, with one amino group and two carboxyl groups, would be classified as acidic and that its isoelectric point would be at a pH lower than 7.0. It might also seem that the isoelectric point of an amino acid that is classified as neutral, such as alanine, with one amino and one carboxyl group, would have an isoelectric point of 7.0. However, the isoelectric point of alanine is 6.0, not 7.0. This is because the carboxyl group and the amino group are not equally ionized. The carboxyl group of alanine ionizes to a greater degree as an acid than the amino group ionizes as a base.

30.6 Formation of Polypeptides

peptide linkage

Proteins are polyamides consisting of amino acid units joined through amide structures. If we react two glycine molecules, with the elimination of a molecule of water, we form a compound containing the amide structure, also called the **peptide linkage** or peptide bond. The elimination of water occurs between the carboxyl group of one amino acid and the α-amino group of a second amino acid. The product formed from two glycine molecules is called glycylglycine (abbreviated Gly-Gly). Since it contains two amino acid units, it is a dipeptide.

$$CH_2-C\overset{O}{\underset{OH}{\diagup}} + CH_2-C\overset{O}{\underset{OH}{\diagup}} \longrightarrow CH_2-C\overset{O}{\underset{NH_2}{\diagup}} \quad CH_2-C-OH + H_2O$$

Peptide linkage

Glycylglycine (Gly-Gly)

polypeptide

If three amino acid units are included in a molecule, it is a tripeptide; if four, a tetrapeptide; if five, a pentapeptide; and so on. Peptides containing up to about 40–50 amino acid units in a chain are called **polypeptides**. The units making up the peptide are amino acids less the elements of water and are referred to as *amino acid residues* or simply residues. Still larger chains of amino acids are known as proteins.

In linear peptides, one end of the chain will have a free amino group and the other end a free carboxyl group. The amino group end is called the *N-terminal residue* and the other end the *C-terminal residue*:

```
       1    2    3    4    5    6    7
       Ala-Pro-Tyr-Met-Gly-Lys-Gly
       ↗                        ↖
   N-Terminal              C-Terminal
```

The sequence of amino acids in a chain is numbered starting with the *N*-terminal residue, which is usually written to the left with the *C*-terminal residue at the right. Any segment of the sequence that is not specifically known is placed in parentheses. Thus, in the heptapeptide above, if the order of tyrosine and methionine is not known, the structure would be written as:

Ala-Pro-(Met, Tyr)-Gly-Lys-Gly

Peptides are named as acyl derivatives of the *C*-terminal amino acid with the *C*-terminal unit keeping its complete name. The *-ine* ending of all but the *C*-terminal amino acids is changed to *-yl*, and these are listed in the order they appear, starting with the *N*-terminal amino acid.

$$\underset{\text{Alanyl}}{\underset{|}{\overset{\overset{O}{\|}}{CH_3CH\!-\!\!C}}\!-\!}\underset{\text{Tyrosyl}}{\underset{\underset{\underset{OH}{|}}{\underset{|}{\bigcirc}}}{\underset{|}{\overset{\overset{O}{\|}}{NHCH\!-\!\!C}}}\!-\!}\underset{\text{Glycine}}{NHCH_2COOH}$$

Ala-Tyr-Gly

Thus, Ala-Tyr-Gly is called alanyltyrosylglycine. The name of Arg-Gln-His-Ala is arginylglutamylhistidylalanine.

Alanine and glycine can form two different dipeptides, Gly-Ala and Ala-Gly, using each amino acid only once:

$$\underset{\text{Glycylalanine (Gly-Ala)}}{\underset{|}{\overset{|}{CH_2}}\overset{\overset{O}{\|}}{-C-}\underset{|}{\overset{|}{N}}-\underset{|}{\overset{|}{CH}}\overset{\overset{O}{\|}}{-C}-OH} \quad \text{and} \quad \underset{\text{Alanylglycine (Ala-Gly)}}{CH_3\underset{|}{\overset{|}{CH}}\overset{\overset{O}{\|}}{-C}-\underset{|}{\overset{|}{N}}-CH_2\overset{\overset{O}{\|}}{-C}-OH}$$

If we react three different amino acids—for example, glycine, alanine, and threonine—six tripeptides in which each amino acid appears only once are possible:

Gly-Ala-Thr   Ala-Thr-Gly   Thr-Ala-Gly
Gly-Thr-Ala   Ala-Gly-Thr   Thr-Gly-Ala

The number of peptides possible goes up very rapidly as the number of amino acid units increases. For example, there are 120 ($1 \times 2 \times 3 \times 4 \times 5 = 120$) different ways to combine five different amino acids to form a pentapeptide using each amino acid only once in each molecule. If the same constraints are applied to 15 different amino acids, the number of possible combinations is greater than 1 trillion (10^{12})! Since a protein molecule may contain several hundred amino acid units, with individual amino acids occurring several times, the number of possible combinations from 20 amino acids is simply beyond imagination.

There are a number of small, naturally occurring polypeptides with significant biochemical functions. The amino acid sequences of two of these, oxytocin and vasopressin, are shown in Figure 30.1. Oxytocin controls uterine contractions during labor in childbirth and also causes contraction of the smooth muscles of the mammary gland, resulting in milk excretion. Vasopressin in high concentration raises the blood pressure and has been used in surgical shock treatment for this purpose. Vasopressin is also an antidiuretic, regulating the excretion of fluid by the kidneys. The absence of vasopressin leads to diabetes insipidus. This condition is characterized by excretion of up to 30 litres of urine

676   Chapter 30

$$\text{Cy-Tyr-Ile-Gln-Asn-Cy-Pro-Leu-Gly-NH}_2$$
$$\overset{1}{}\quad\overset{3}{}\quad\quad\quad\quad\overset{8}{}$$
              |___S—S___|

Oxytocin

$$\text{Cy-Tyr-Phe-Gln-Asn-Cy-Pro-Arg-Gly-NH}_2$$
              |___S—S___|

Vasopressin

Figure 30.1. Amino acid sequence of oxytocin and vasopressin. The difference in only two amino acids in these two compounds results in very different physiological activity. The *C*-terminal amino acid has an amide structure instead of the free COOH (indicated as -Gly-NH$_2$).

per day, but may be controlled by administration of vasopressin or its derivatives. Oxytocin and vasopressin are similar nonapeptides, differing only at positions 3 and 8.

The isolation and synthesis of oxytocin and vasopressin was accomplished by Vincent du Vigneaud (1901–1978) and coworkers at Cornell University. Du Vigneaud was awarded the Nobel Prize in chemistry in 1955 for this work. Synthetic oxytocin is indistinguishable from the natural material. It is available commercially and is used for the induction of labor in the late stages of pregnancy.

30.7   Protein Structure

By 1940, a great deal of information concerning proteins had been assembled. Their elemental composition was known, and they had been carefully classified according to solubility in various solvents. Proteins were known to be polymers of amino acids, and the different amino acids had, for the most part, been isolated and identified. Protein molecules were known to be very large in size, with molecular weights ranging from several thousand to several million.

Knowledge of protein structure would help to answer many chemical and biological questions. But for awhile the task of determining the actual structure of molecules of such colossal size appeared to be next to impossible. Then Linus Pauling (1901–   ), at the California Institute of Technology, attacked the problem by a new approach. Using X-ray diffraction techniques, Pauling and his collaborators painstakingly determined the bond angles and dimensions of amino acids and of dipeptides and tripeptides. After building accurate scale models of the dipeptides and tripeptides, they determined how these could be fitted into likely polypeptide configurations. Based on this work, Pauling and R. B. Corey proposed in 1951 that two different configurations—the α-*helix* and the β-*pleated sheet*—were the most probable stable polypeptide chain configurations of protein molecules. These two macromolecular structures are illustrated in Figure 30.2. Within a short time it was established that many proteins do have actual structures corresponding to those predicted by Pauling and

Amino Acids, Polypeptides, and Proteins 677

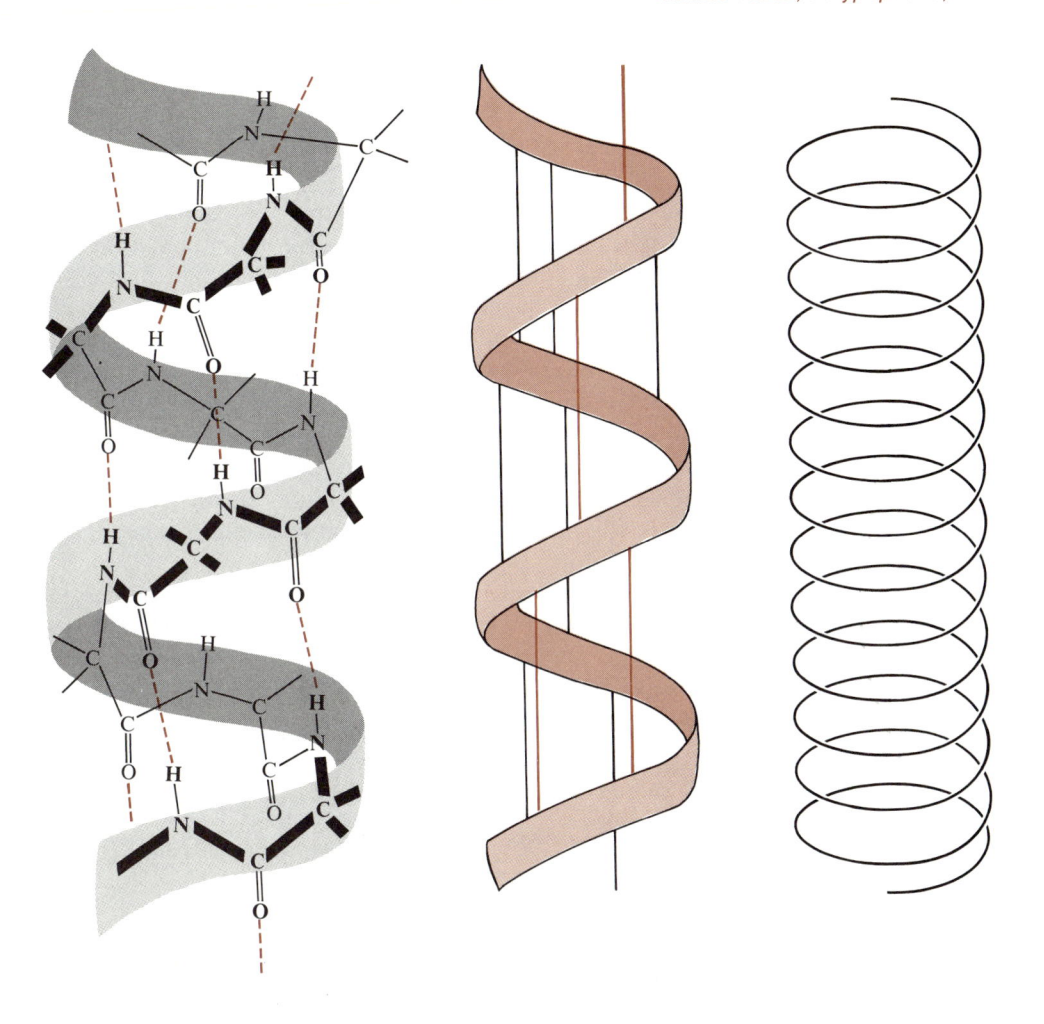

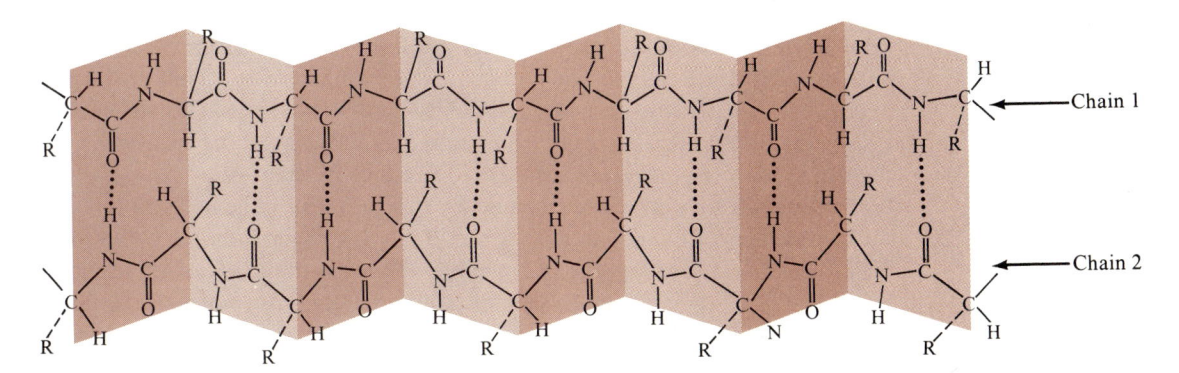

Figure 30.2. α-Helix and β-pleated sheet structures of protein molecules.

Corey. This work was in itself a very great achievement. Pauling received the 1954 Nobel Prize in chemistry for his work on protein structure. Pauling's and Corey's work provided the inspiration for another great biochemical breakthrough—the double helix structure concept for deoxyribonucleic acid, DNA (see Section 31.6).

Proteins are very large molecules. But just how many amino acid units must be present for a substance to be a protein? There is no universally agreed-upon answer to this question. Some authorities state that a protein must have a molecular weight of at least 6000, or contain about 50 amino acid units. Smaller amino acid polymers, containing from 5 to 50 amino acid units, are classified as polypeptides and are not proteins. In reality, there is no natural clearly defined lower limit to the molecular size of proteins; the distinction, based on size, is one of convenience.

In general, if a protein molecule is to perform a specific biological function, it must have a highly detailed overall conformation, or "shape." This overall conformation is due to (1) a primary structure, (2) a secondary structure, (3) a tertiary structure, and sometimes (4) a quaternary structure.

primary structure

The **primary structure** of a protein is established by the number, kind, and sequence of amino acid units comprising the polypeptide chain or chains making up the molecule.

In 1902, Emil Fischer proposed that proteins consisted of amino acid units joined by amide bonds in this way:

$$\begin{array}{c} \quad\quad H \;\; R \;\; O \quad\quad H \quad\quad H \;\; R \;\; O \quad\quad H \\ \quad\quad | \;\;\; | \;\;\; \| \quad\quad | \quad\quad | \;\;\; | \;\;\; \| \quad\quad | \\ \sim\!\!\sim\!\!\sim\!\!N\!-\!C\!-\!C\!-\!N\!-\!C\!-\!C\!-\!N\!-\!C\!-\!C\!-\!N\!-\!C\!-\!C\!\sim\!\!\sim\!\!\sim \\ \quad\quad | \;\;\; | \;\;\; \quad\quad\quad | \;\;\; \| \quad\quad | \;\;\; \quad\quad\quad | \;\;\; \| \\ \quad\quad H \;\; H \quad\quad R \;\; O \quad\quad H \quad\quad H \;\; R \;\; O \end{array}$$

Determining the *sequence* of the amino acids in even one protein molecule was a formidable task. The amino acid sequence of beef insulin was announced in 1955 by the British biochemist Frederick Sanger (1918–   ). This structure determination required several years of effort by a team under Sanger's direction. He was awarded the 1958 Nobel Prize in chemistry for this work. Insulin is a hormone that regulates the blood sugar level. A deficiency of insulin leads to the condition of diabetes. Beef insulin consists of 51 amino acid units in two polypeptide chains. The two chains are connected by disulfide linkages (—S—S—) of two cysteine residues at two different sites. The structure is shown in Figure 30.3. Insulins from other animals, including humans, differ slightly by one, two, or three amino acid residues in chain A.

secondary structure

The **secondary structure** of proteins, represented by the helix and pleated sheet structures of Pauling and Corey, is due to hydrogen bonding between the O of the C=O groups and the H of the N—H groups in the polypeptide chains. Hydrogen bonding between groups in the same chain (intramolecular bonding) is responsible for the helical structure, with each peptide bond participating in the hydrogen bonding. An α-helix forms spontaneously since this is the lowest energy configuration of a polypeptide chain. When the residues are L-amino acids as in proteins, the right-handed or α-helix is more stable than the left-

Amino Acids, Polypeptides, and Proteins 679

```
                          15                         20      21
                      ┌─Gln──Leu──Glu──Asn──Tyr──Cy──Asn
                    Tyr                              │
                     │                               S
                    Leu                              │
                     │                               S
                    Ser            10                │        20
                     └─Cy──Val─┐                    ┌Cy──Gly┐
                        │        │                  Val      Glu
                        S       Ser                  │        │
                        │        │                  Leu      Arg
                        S       Ala                  │        │
       1                │        │                  Tyr      Gly
      Gly─Ile─Val─Glu─Gln─Cy────Cy─┘                 │        │
         Chain A                                  15 Leu     Phe
                                 S                   │        │
                                 │                  Ala      Phe  25
                                 S                   │        │
       1          5              │         10       Glu      Tyr
      Phe─Val─Asn─Gln─His─Leu─Cy─Gly─Ser─His─Leu─Val          │
                                                             Thr
         Chain B                                              │
                                                             Pro
                                                              │
                                                             Lys
                                                              │
                                                             Ala  30
```

Figure 30.3. Amino acid sequence of beef insulin.

handed helix. Hydrogen bonding between groups on different chains (intermolecular bonding) is responsible for the pleated sheet structures.

When sections of a single polypeptide chain or of different polypeptide chains are parallel and are close enough so that hydrogen bonding can form between them, a pleated sheet structure will be formed. A single protein molecule may have both the helix and pleated sheet structure within it.

tertiary structure   The **tertiary structure** of a protein refers to the distinctive and characteristic conformation, or "shape," of a protein molecule. Myoglobin, a muscle protein, is described as having the appearance of a folded sausage (see Figure 30.4). The tertiary structure is due to a variety of interactions between the side-chain (R) groups on the amino acid units. These interactions include (1) hydrogen bonding, (2) ionic bonding, and (3) disulfide bonding.

1. Glutamic acid–tryosine hydrogen bonding:

$$\mathrm{-CH_2CH_2\overset{OH}{\underset{}{C}}=O\;\cdots\cdots\;HO-\!\!\!\bigcirc\!\!\!-}$$

2. Glutamic acid–lysine ionic bonding:

$$\mathrm{-CH_2CH_2\overset{O}{\overset{\|}{C}}-O^{\ominus}\quad H_3\overset{\oplus}{N}-CH_2CH_2CH_2CH_2-}$$

3. Cysteine–cysteine disulfide bonding:

$$\mathrm{-CH_2-S-S-CH_2-}$$

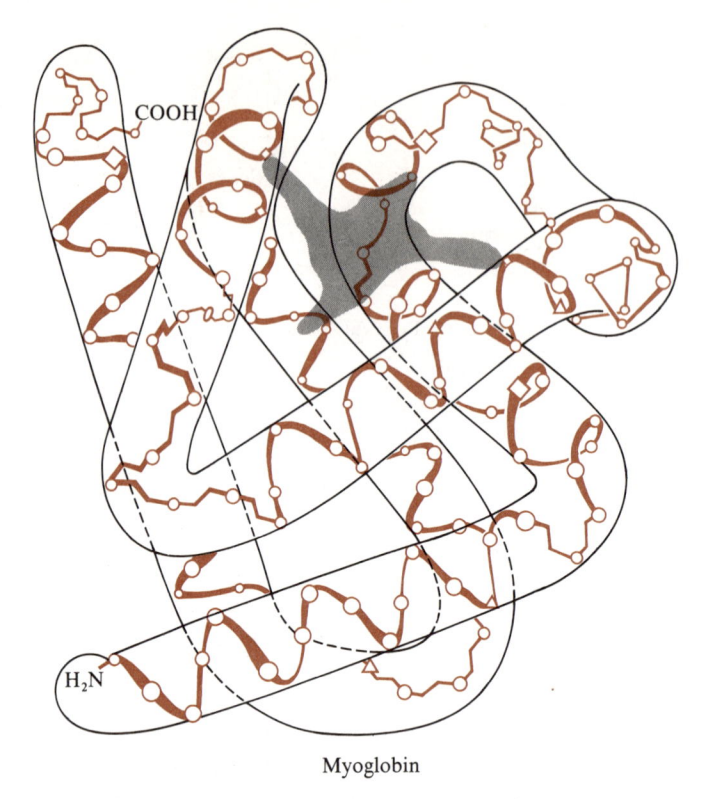

Figure 30.4. Tertiary structure of a protein represented by myoglobin.

Note that in the insulin molecule (Figure 30.3) there are three points at which disulfide bonding has an obvious bearing on the shape of the molecule.

Hair is especially rich in disulfide bonds. These can be broken by certain reducing agents and restored by an oxidizing agent. This fact is the key to "cold" permanent waving of hair. Some of the disulfide bonds are broken by applying a reducing agent to the hair. The hair is then styled with the desired curls or waves. These are then permanently set by using an oxidizing agent to reestablish the disulfide bonds at different points.

$$|\text{—CH}_2\text{—S—S—CH}_2\text{—}| \xrightarrow{\text{Reducing agent}}$$

Bonds in normal hair

$$|\text{—CH}_2\text{—S—H} \quad \text{H—S—CH}_2\text{—}| \xrightarrow{\text{Oxidizing agent}} |\text{—CH}_2\text{—S—S—CH}_2\text{—}|$$

Broken bonds of "reduced" form

Reestablished bonds of permanently waved hair

quaternary structure

A fourth type of structure, called **quaternary structure**, is found in some complex proteins. These proteins are made up of two or more smaller protein subunits (polypeptide chains). Nonprotein molecules may also be present. The quaternary structure refers to the shape of the entire complex molecule and is determined by the way in which the subunits are held together by *noncovalent* bonds—that is, by hydrogen bonding, ionic bonding, and so on. For example, collagen, which is the main protein component of connective tissue, consists of three polypeptide chains twisted together into a ropelike structure. Hemoglobin molecules are made up of four polypeptide chains folded about four nonprotein heme groups (oxygen receptor sites). Many enzymes and viruses are known to have quaternary structures.

To summarize, the four different types of protein structures are (1) the primary structure, determined by the amino acid sequence in the polypeptide chain; (2) the secondary structure, determined by the hydrogen bonding between amino acid units in the same polypeptide chain (α-helix) or in different polypeptide chains (pleated sheet); (3) the tertiary structure, determined by bonding interactions of various kinds between amino acid side chains on the same polypeptide or on different polypeptide chains; and (4) the quaternary structure, determined by noncovalent bonding between the subunits of a complex protein.

30.8 Enzymes

enzymes

Enzymes are the catalysts of biochemical reactions. All enzymes are proteins and they catalyze nearly all of the myriad reactions that occur in living cells. Uncatalyzed reactions that may require hours of boiling in the presence of a strong acid or a strong base may occur in a fraction of a second in the presence of the proper enzyme at room temperature and nearly neutral pH. This is all the more remarkable when we realize that enzymes do not actually cause chemical reactions. They act as catalysts by greatly lowering the activation energy of specific biochemical reactions. The lowered activation energy permits these reactions to proceed at high speed at body temperature.

Louis Pasteur was one of the first scientists to study enzyme-catalyzed reactions. He believed that living yeasts or bacteria were required for these reactions, which he called *fermentations*—for example, the conversion of glucose to alcohol by yeasts. In 1897, Eduard Büchner (1860–1917) made a cell-free filtrate that contained enzymes prepared by grinding yeast cells with very fine sand. The enzymes in this filtrate converted glucose to alcohol, thus proving that the presence of living cells was not required for enzyme activity. For this work Büchner received the Nobel Prize in chemistry in 1907.

Each organism contains thousands of enzymes. Some are simple proteins consisting only of amino acids units. Others are conjugated and consist of a protein part, or *apoenzyme*, and a nonprotein part, or *coenzyme*. Both parts are essential, and a functioning enzyme consisting of both the protein and nonprotein parts is called a *holoenzyme*.

Apoenzyme + Coenzyme = Holoenzyme

Often the coenzyme is a vitamin, and the same coenzyme may be associated with many different enzymes.

For some enzymes an inorganic component such as a metal ion—for example, Ca^{2+}, Mg^{2+}, or Zn^{2+}—is required. This inorganic component is an *activator*. From the standpoint of function, an activator is analogous to a coenzyme, but inorganic components are not called coenzymes.

Another remarkable property of enzymes is their specificity of reaction; that is, a certain enzyme will catalyze the reaction of a specific type of substance. For example, the enzyme maltase catalyzes the reaction of maltose and water to form glucose. Maltase has no effect on the other two common disaccharides, sucrose and lactose. Each of these sugars requires a specific enzyme; sucrase to hydrolyze sucrose; lactase to hydrolyze lactose. These reactions are indicated by the following equations:

$$C_{12}H_{22}O_{11} + H_2O \xrightarrow{\text{maltase}} C_6H_{12}O_6 + C_6H_{12}O_6$$
Maltose $\qquad\qquad\qquad\qquad$ Glucose $\quad$ Glucose

$$C_{12}H_{22}O_{11} + H_2O \xrightarrow{\text{sucrase}} C_6H_{12}O_6 + C_6H_{12}O_6$$
Sucrose $\qquad\qquad\qquad\qquad$ Glucose $\quad$ Fructose

$$C_{12}H_{22}O_{11} + H_2O \xrightarrow{\text{lactase}} C_6H_{12}O_6 + C_6H_{12}O_6$$
Lactose $\qquad\qquad\qquad\qquad$ Glucose $\quad$ Galactose

The substance acted on by an enzyme is called the *substrate*. Sucrose is the substrate of the enzyme sucrase. Enzymes have been named by adding the suffix *-ase* to the root of the substrate name. Note the derivations of maltase, sucrase, and lactase from maltose, sucrose, and lactose. Many enzymes, especially digestive enzymes, have trivial names such as pepsin, rennin, trypsin, and so on. These names have no systematic significance.

In the International Union of Biochemistry (IUB) System, enzymes are assigned to one of six classes, the names of which clearly describe the nature of the reaction they catalyze. Each of the classes has several subclasses. In this system, the name of the enzyme has two parts; the first gives the name of the substrate and the second, ending in *-ase*, indicates the type of reactions catalyzed by all enzymes in the group. The six main classes of enzymes are:

1. *Oxidoreductases.* Enzymes that catalyze the oxidation–reduction between two substrates.
2. *Transferases.* Enzymes that catalyze the transfer of a functional group between two substrates.
3. *Hydrolases.* Enzymes that catalyze the hydrolysis of esters, carbohydrates, and proteins (polypeptides).
4. *Lyases.* Enzymes that catalyze the removal of groups from substrates by mechanisms other than hydrolysis.
5. *Isomerases.* Enzymes that catalyze the interconversion of optical, geometric, and structural isomers.
6. *Ligases.* Enzymes that catalyze the linking together of two compounds with the breaking of a pyrophosphate bond in adenosine triphosphate (ATP, see Chapter 31).

Since the systematic name is usually long and often complex, a working or practical name is used for enzymes. For example, adenosine triphosphate creatine phosphotransferase is called creatine kinase; and acetylcholine acylhydrolase is called cholinesterase.

Enzymes act according to the following general sequence. Enzyme (E) and substrate (S) combine to form an enzyme–substrate intermediate (E–S). This intermediate decomposes to give the product (P) and regenerate the enzyme:

$$E + S \rightleftharpoons E\text{–}S \longrightarrow E + P$$

For the hydrolysis of maltose the sequence is

$$\underset{E}{\text{Maltase}} + \underset{S}{\text{Maltose}} \longrightarrow \underset{E\text{–}S}{\text{Maltase–Maltose}}$$

$$\underset{E\text{–}S}{\text{Maltase–Maltose}} + H_2O \longrightarrow \underset{E}{\text{Maltase}} + \underset{P}{\text{Glucose}}$$

Enzyme specificity is believed to be due to the particular shape of a small part of the enzyme which exactly fits a complementary-shaped part of the substrate (see Figure 30.5). It is analogous to a lock and key; the substrate is the lock and the enzyme the key. Just as a key will open only the lock it fits, the enzyme will act on only a molecule that fits its particular shape. When the substrate and the enzyme come together, they form a substrate–enzyme complex unit. The substrate, activated by the enzyme in the complex, reacts to form the products, regenerating the enzyme. The *lock-and-key theory* for enzyme–substrate interaction was proposed by Emil Fischer.

A more recently suggested model of the enzyme–substrate catalytic site is known as the "induced fit" model, introduced by D. E. Koshland, Jr. In this model, the enzyme site of attachment to the substrate is flexible, with the substrate inducing a change in the enzyme shape to fit the shape of the substrate.

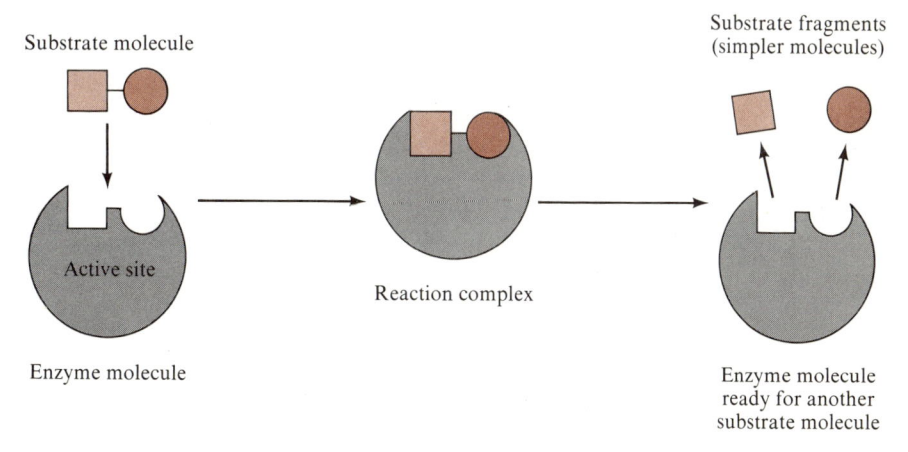

Figure 30.5. Enzyme–substrate interaction illustrating specificity of an enzyme by the lock-and-key analogy.

This theory allows for the possibility that in some cases the enzyme might wrap itself around the substrate and so form the correct shape of lock and key. Thus, the enzyme does not need to have an exact preformed catalytic site to match the substrate.

Enzymes show a high degree of specificity toward the substrates they catalyze. A given enzyme will catalyze only a few reactions and frequently only one specific reaction. The specificity of reaction is one of the most significant properties of an enzyme. Some enzymes have absolute specificity; that is, they catalyze only a single reaction and no others. For example, urease has absolute specificity for the hydrolysis of urea to form ammonia and carbon dioxide:

$$\underset{\text{Urea}}{H_2N-\underset{\underset{O}{\|}}{C}-NH_2} + H_2O \xrightarrow{\text{Urease}} 2 NH_3 + CO_2$$

Many compounds exist that contain substituents on the NH_2 group of urea. None are touched by the enzyme urease.

Some enzymes catalyze certain functional groups. Esterases catalyze the hydrolysis of esters to alcohols and carboxylic acids. Steapsin catalyzes the hydrolysis of fats to fatty acids and glycerol. Proteinases attack peptide bonds—they hydrolyze polypeptides and proteins to form free amino and carboxyl groups. Many enzymes have stereo specificity. For example, amylase hydrolyzes only starches in which the glucose units are linked by α-1,4-glycosidic linkages. Amylase will not hydrolyze cellulose, which has beta linkages. Most mammalian enzymes act on the L isomers of amino acids and have no effect on the corresponding D-amino acids.

Enzyme technology is being used for the commercial production of an increasing number of products. It holds the promise of converting such substances as cellulose to glucose, which can then be used as a source of food or in the manufacture of such products as ethanol and acetone. Enzymes have long been used in fermentation processes for the production of a myriad of compounds. More recently, enzymes have been used to convert cornstarch into syrups equivalent in sweetness and calories to ordinary table sugar. More than 1 billion pounds of such syrups are produced annually. The process uses three enzymes: the first, α-amylase, catalyzes the liquifaction of starch to dextrins; the second, a glucoamylase, catalyzes the breakdown of oligosaccharides and other carbohydrates to glucose; and the third, glucose isomerase, converts glucose to fructose:

$$\text{Starch} \xrightarrow{\alpha\text{-Amylase}} \text{Dextrins} \xrightarrow{\text{Glucoamylase}} \text{Glucose} \xrightarrow{\text{Glucose isomerase}} \text{Fructose}$$

The product is a high-fructose syrup equivalent in sweetness to sucrose. One of these syrups, sold commercially since 1968, contains about 42% fructose, 50% glucose, and 8% other saccharides (on a dry basis).

30.9 Hydrolysis of Proteins

Proteins ultimately yield amino acids when hydrolyzed. Proteins can be hydrolyzed by boiling in a solution containing a strong acid such as hydrochloric acid or in a solution containing a strong base such as sodium hydroxide. Many enzymes that will catalyze the hydrolysis of proteins are known. These enzymes, which are themselves proteins, catalyze the hydrolysis at ordinary temperatures. The essential reaction of hydrolysis is the breaking of a peptide linkage and the addition of the elements of water:

$$\sim\!\!N\!-\!\underset{\underset{H}{|}}{\overset{\overset{H}{|}}{C}}\!-\!\underset{\underset{H}{|}}{\overset{\overset{R}{|}}{C}}\!-\!\underset{\underset{R}{|}}{\overset{\overset{H}{|}}{N}}\!-\!\underset{\underset{O}{|}}{\overset{\overset{H}{|}}{C}}\!-\!C\!\sim + H_2O \longrightarrow \sim\!\!N\!-\!\underset{\underset{H}{|}}{\overset{\overset{H}{|}}{C}}\!-\!\underset{\underset{}{}}{\overset{\overset{R}{|}}{C}}\!-\!OH + H_2N\!-\!\underset{\underset{O}{|}}{\overset{\overset{H}{|}}{C}}\!-\!C\!\sim$$

(Peptide linkage in protein chain)

Any molecule with one or more peptide bonds, from the smallest dipeptide to the largest protein molecule, can be hydrolyzed. During hydrolysis, proteins are broken down into smaller and smaller fragments as follows:

Proteins ⟶ Proteoses ⟶ Peptones ⟶ Peptides ⟶ Amino Acids

Proteoses and peptones are protein degradation products containing large polypeptides which differ from proteins in that they do not coagulate with heat.

The enzymes that catalyze the hydrolysis of proteins to amino acids do so at specific peptide linkages. Some enzymes attack only N- or C-terminal amino acids, while others break the polypeptide chain of specific internal amino acids. Table 30.3 shows the enzymes involved in protein digestion and the amino acids they hydrolyze.

Sanger partially hydrolyzed insulin and then determined the amino acid sequences of a large number of the fragments. He used these data, with the aid of a computer, to determine the amino acid sequence, or primary structure, of the entire insulin molecule.

Table 30.3. Enzymes that hydrolyze peptide linkages in proteins.

Enzyme	Peptide linkage hydrolyzed
Pepsin	Hydrolyzes proteins to proteoses and peptones; attacks the amino end of tyrosine and phenylalanine.
Trypsin	Hydrolyzes the polypeptide chain on the carboxyl side of arginine and lysine.
Chymotrypsin	Hydrolyzes the polypeptide chain on the carboxyl side of tyrosine, phenylalanine, and tryptophan.
Carboxypeptidase	Hydrolyzes C-terminal amino acids.
Aminopeptidase	Hydrolyzes N-terminal amino acids.
Dipeptidase	Hydrolyzes dipeptides.

30.10 Denaturation of Proteins

denaturation

Denaturation refers to changes in the properties of a protein without hydrolysis. This means that denaturation involves alteration or disruption of the secondary, tertiary, or quaternary—but not the primary—structure of proteins (see Figure 30.6). When a protein is denatured, it loses its biological activity. Denaturation may involve changes ranging from the subtle and reversible alterations caused by a slight change of pH to the extreme alterations involved in tanning a skin to form leather. Noticeable coagulation or precipitation often, but not always, occurs when a protein is denatured. One reason for cooking meats is to denature and thereby tenderize their protein structure. The white of a fried or poached egg is an example of heat-denatured protein.

A wide variety of chemical and physical agents can denature proteins. To name a few: strong acids and strong bases; salts, especially those of heavy metals; certain specific reagents such as tannic acid and picric acid; alcohol and other organic solvents; detergents; mechanical action as in whipping; high temperature; and ultraviolet (uv) radiation.

The probable modes of action of these denaturing agents include the following: strong acids and bases disrupt the salt bridge bonds of the tertiary structures. Heavy metal ions and the alkaloidal reagents react with carboxylate and amino groups and thereby disrupt the tertiary structures. Alcohol and other polar solvents disturb hydrogen bonding patterns. High temperatures and uv radiation impart greater kinetic energy to the protein molecules, resulting in more violent molecular motion, which disrupts the structure to a greater or

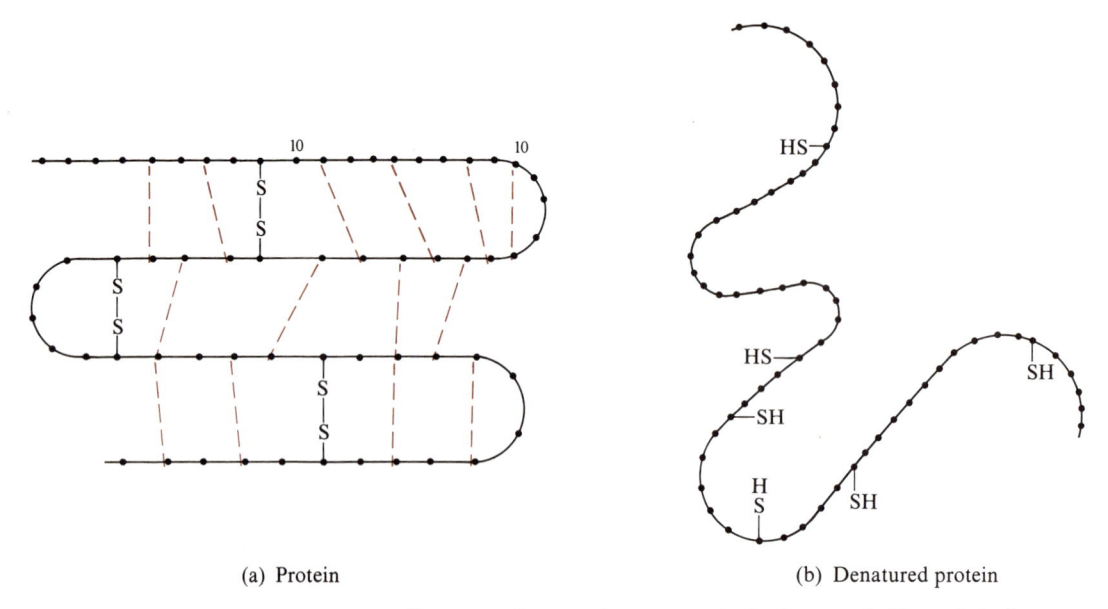

(a) Protein

(b) Denatured protein

Figure 30.6. Structure change when a protein is denatured. The relatively weak hydrogen, electrostatic, and disulfide bonds are broken, resulting in a change of structure and properties.

lesser degree. Detergents, being surface-active agents, probably disrupt the quaternary structure of protein aggregates.

Egg white contains the protein albumin. If it is stirred vigorously (as with an egg beater) or heated above 70°C, the protein structure is broken and the albumin coagulates. Denatured proteins are generally less soluble than undenatured proteins. Consequently, the denaturation of a protein will often cause it to coagulate or precipitate from solution.

In clinical laboratories, the analysis of blood serum for small molecules such as glucose and uric acid is hampered by the presence of serum protein. This problem is resolved by first treating the serum with an acid to denature and precipitate the protein. The precipitate is removed and the protein-free liquid is then analyzed.

30.11 Tests for Proteins and Amino Acids

Many tests have been devised for detecting and distinguishing among amino acids, peptides, and proteins. Some examples are described here.

Xanthoproteic reaction. Proteins containing a benzene ring—for example, phenylalanine, tryptophan, or tyrosine—react with concentrated nitric acid to give yellow reaction products. Nitric acid on skin produces a positive xanthoproteic test caused by this reaction!

Biuret test. A violet color is produced when dilute copper(II) sulfate is added to an alkaline solution of a peptide or protein. At least two peptide bonds must be present, since amino acids and dipeptides do not give a positive biuret test.

Ninhydrin test. Triketohydrindene hydrate, generally known as *ninhydrin*, is an extremely sensitive reagent for amino acids:

Ninhydrin

All amino acids, excepting proline and hydroxyproline, give a blue solution with ninhydrin. Proline and hydroxyproline produce a yellow solution. Less than one microgram (10^{-6} g) of an amino acid can be detected with ninhydrin.

Chromatographic Separation

Complex mixtures of amino acids are readily separated by thin-layer, paper, or column chromatography. In chromatographic methods, the components of a mixture are separated by differences in their distributions between two phases. Separation depends on the relative tendencies of the components to remain in one phase or the other. In *thin-layer chromatography* (TLC), for example, a liquid and a solid phase are used. The procedure is as follows: A tiny drop of a solution containing a mixture of amino acids (obtained by hydrolyzing a protein) is spotted on a strip (or sheet) coated with a thin layer of dried alumina or some other adsorbant. After the spot has dried, the bottom edge of the strip is put in a suitable solvent. The solvent ascends the strip (by diffusion), carrying the different amino acids upward at different rates. When the solvent front nears the top, the strip is removed from the solvent and dried. The locations of the different amino acids are established by spraying with ninhydrin solution and noting where colored spots appear on the chromatogram. The identities of the amino acids in an unknown mixture can then be established by comparing their migration patterns with the patterns of known amino acids. A typical chromatogram of amino acids is shown in Figure 30.7.

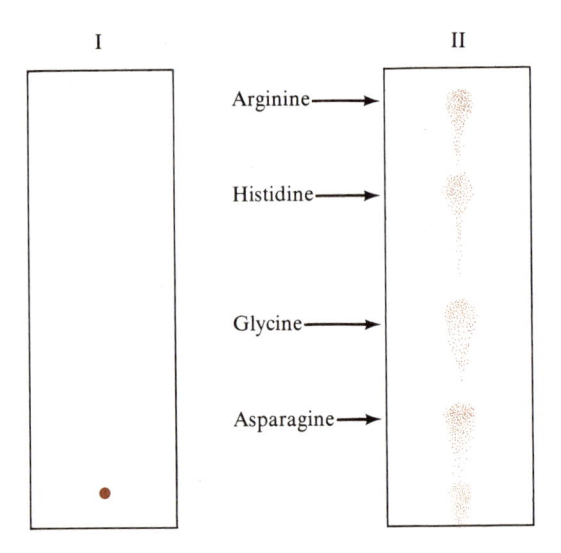

Figure 30.7. Chromatogram showing separation of selected amino acids. On the left, spotted on the chromatographic strip, is an amino acid mixture containing arginine, histidine, glycine, and asparagine. On the right is the developed chromatogram showing separated amino acids after treatment with ninhydrin. Solvent is a 250:60:250 volume ratio of *n*-butanol:acetic acid:water.

30.12 Determination of the Primary Structure of Polypeptides

Sanger's reagent. Until about 1967, the method for determining the amino acid content and the primary structure of polypeptides was to hydrolyze them into smaller peptides and free amino acids using specific enzymes. The hydrolysates were then separated and identified by various chemical reactions.

Amino Acids, Polypeptides, and Proteins   689

The pioneering work of Frederick Sanger gave us the first complete primary structure of a protein—insulin—in 1955. Sanger's reagent, 2,4-dinitrofluorobenzene (DNFB), reacts with the α-amino group of the N-terminal amino acid of a polypeptide chain. The carbon–nitrogen bond between the amino acid and the benzene ring is more resistant to hydrolysis than are the remaining peptide linkages. Thus, when the substituted polypeptide is hydrolyzed, the terminal amino acid remains with the dinitrobenzene radical and can be isolated and identified. The remaining peptide chain is hydrolyzed to free amino acids in the process. This method marked an important step in the determination of the amino acid sequence in a protein. The reaction of Sanger's reagent is illustrated in the following equations:

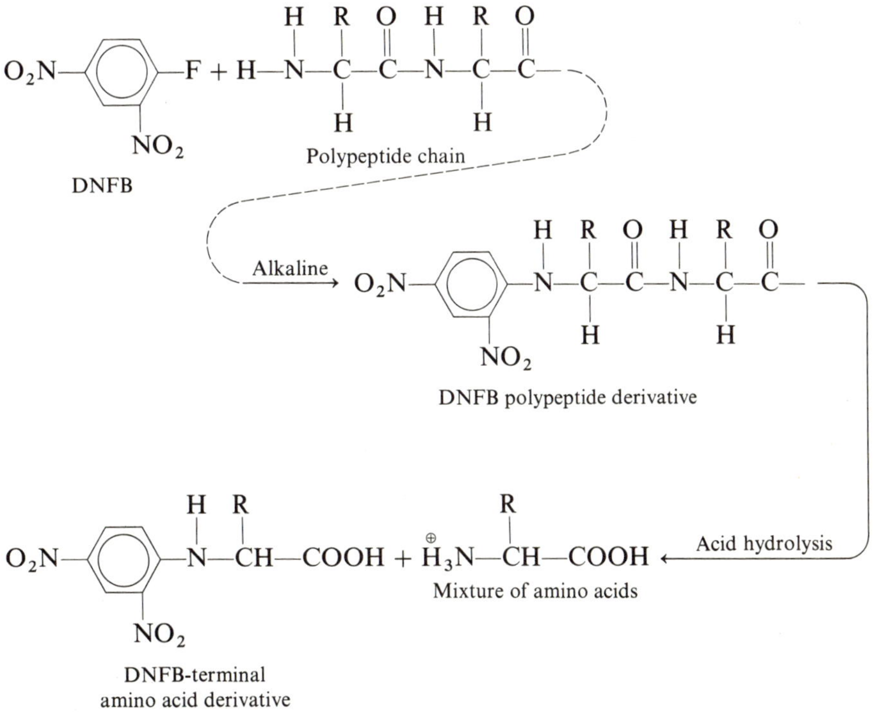

$$O_2N\text{-}\underset{NO_2}{C_6H_3}\text{-}F + H\text{-}\underset{H}{\overset{H}{N}}\text{-}\underset{}{\overset{R}{C}}\text{-}\overset{O}{\underset{}{C}}\text{-}\underset{H}{\overset{H}{N}}\text{-}\underset{}{\overset{R}{C}}\text{-}\overset{O}{\underset{}{C}}\text{-}\cdots$$

DNFB   Polypeptide chain

$$\xrightarrow{\text{Alkaline}} O_2N\text{-}\underset{NO_2}{C_6H_3}\text{-}\underset{H}{\overset{H}{N}}\text{-}\underset{}{\overset{R}{C}}\text{-}\overset{O}{\underset{}{C}}\text{-}\underset{H}{\overset{H}{N}}\text{-}\underset{}{\overset{R}{C}}\text{-}\overset{O}{\underset{}{C}}\text{-}\cdots$$

DNFB polypeptide derivative

$$\xrightarrow{\text{Acid hydrolysis}} O_2N\text{-}\underset{NO_2}{C_6H_3}\text{-}\underset{H}{\overset{H}{N}}\text{-}\overset{R}{\underset{}{CH}}\text{-}COOH + \overset{\oplus}{H_3}N\text{-}\overset{R}{\underset{}{CH}}\text{-}COOH$$

DNFB-terminal amino acid derivative   Mixture of amino acids

From the DNFB hydrolysis, Sanger learned which amino acids were present in insulin. By less drastic hydrolysis, he split the insulin molecule into peptide fragments consisting of two, three, four, or more amino acid residues. After analyzing vast numbers of fragments utilizing the N-terminal method, he pieced them together in the proper sequence by combining fragments with overlapping structures at their ends, finally elucidating the entire insulin structure. As an example, consider the overlap that occurs between the hexapeptide and heptapeptide shown below:

Gly-Glu-Arg-Gly-Phe-Phe	Hexapeptide
Gly-Phe-Phe-Tyr-Thr-Pro-Lys	Heptapeptide
Gly-Glu-Arg-Gly-Phe-Phe-Tyr-Thr-Pro-Lys	Decapeptide

690 Chapter 30

The three residues Gly-Phe-Phe- at the end of the hexapeptide match the three residues at the beginning of the heptapeptide. By using these three residues in common, the structure of the decapeptide shown, which occurs in chain B of insulin (residue 20–29), is determined.

More recently, a method has been developed to split off amino acids one at a time from the *N*-terminal end of a polypeptide chain (Edman reaction).

30.13 Synthesis of Peptides and Proteins

Since each amino acid has two functional groups, it is not difficult to form dipeptides or even fairly large polypeptide molecules. As mentioned in Section 30.6, glycine and alanine react to form two different dipeptides, glycylalanine and alanylglycine. If threonine, alanine, and glycine are reacted, six tripeptides, each made up of three different amino acids, are obtained. By reacting a mixture of several amino acids, polypeptides are produced that contain a random arrangement of the amino acid units. This fact makes it clear that even a small protein like insulin could not be produced by simply reacting a mixture of the required amino acids. If such a large polypeptide were to be synthesized, it would have to be formed by joining amino acids one by one in the proper sequence.

The *in vivo* synthesis of proteins (that is, synthesis in living tissue) proceeds under RNA-regulated enzyme control. Remarkable progress has been made in *in vitro* polypeptide synthesis (that is, synthesis in glass without the aid of living tissue). When a polypeptide chain of known primary structure is synthesized *in vitro*, the process must be controlled so that only one particular amino acid is added at each stage. This is done by using amino acids in which either the amino or carboxylic acid group has been inactivated or blocked with a suitable *blocking agent*. The blocked amino or carboxylic acid group is reactivated by removing the blocking agent in the next stage of the synthesis. For example, if the tripeptide Thr-Ala-Gly is to be made, amino-blocked threonine is reacted with carboxyl-blocked alanine to make the doubled-blocked Thr-Ala dipeptide. The carboxyl-blocked alanine end of the dipeptide can then be reactivated and reacted with carboxyl-blocked glycine to make the blocked Thr-Ala-Gly tripeptide. The free tripeptide can then be obtained by removing the blocking agents from both ends. Using B to designate the blocking agents, this synthesis is represented in schematic form as follows:

$$\boxed{B}H_2NCHCOOH + H_2NCHCOOH\boxed{B} \longrightarrow \boxed{B}H_2NCH\overset{\overset{O}{\|}}{C}-NHCHCOOH\boxed{B}$$

$$\underset{\underset{CH_3}{|}}{\underset{H-C-OH}{|}} \qquad \underset{CH_3}{|} \qquad \underset{\underset{CH_3}{|}}{\underset{H-C-OH}{|}} \quad \underset{CH_3}{|}$$

Amino-blocked threonine   Carboxyl-blocked alanine   Blocked Thr-Ala dipeptide

$$\text{B}\,H_2NCHCNHCHCOOH + H_2NCH_2COOH\,\text{B}$$

with structure showing H—C—OH, CH₃ groups on the dipeptide.

Amino-blocked Thr-Ala dipeptide + Carboxyl-blocked glycine

$$\text{B}\,H_2NCHC—NHCHCNHCH_2COOH\,\text{B}$$

Blocked Thr-Ala-Gly tripeptide

Longer polypeptide chains of specified amino acid sequence can be made by this general technique. The procedure is very tedious; several man-years of effort were required for the synthesis of insulin.

It is not necessary to prepare an entire synthetic protein polypeptide chain by starting at one end and adding amino acids one at a time. Previously prepared shorter polypeptide chains of known structure can be joined to form a longer polypeptide chain. This method was used in making synthetic insulin. The insulin molecule shown in Figure 30.3 consists of two polypeptide chains bonded together by two disulfide bonds. A third disulfide bond, between two amino acids in one chain, makes a small loop in that chain. Once the two chains had been assembled from fragments, a seemingly formidable problem remained—how to form the disulfide bonds in the correct positions. As it turned out, it was necessary only to bring the two chains together, with the cysteine side chains in the reduced condition (Section 30.7), and to treat them with an oxidizing agent. Disulfide bonds formed at the right places and biologically active insulin molecules were obtained.

In the mid-1960s, a machine capable of automatically synthesizing polypeptide chains of known amino acid sequence was designed by R. B. Merrifield of Rockefeller University. The starting amino acid is bonded to a plastic surface (polystyrene bead) in the reaction chamber of the apparatus. Various reagents needed for building a chain of predetermined structures are automatically delivered to the reaction chamber in a programmed sequence. Twelve reagents and about 100 operations are needed to lengthen the chain by a single amino acid residue. But the machine is capable of adding residues to the chain at the rate of six a day. Such a machine makes it possible to synthesize complex molecules like insulin in a few days. The first large polypeptide (pancreatic ribonuclease) containing 124 amino acid residues was synthesized by this method in 1969. More recently, a human growth hormone (HGH) containing 188 amino acid residues was synthesized using this technique.

Questions

A. Review the meanings of the new terms introduced in this chapter.
   1. Simple proteins
   2. Conjugated proteins
   3. Essential amino acid
   4. Zwitterion
   5. Isoelectric point
   6. Peptide linkage
   7. Polypeptide
   8. Primary structure
   9. Secondary structure
   10. Tertiary structure
   11. Quaternary structure
   12. Enzymes
   13. Denaturation

B. Review questions.
   1. List four foods that are major sources of proteins.
   2. Why are the amino acids of proteins called α-amino acids?
   3. What elements are present in amino acids and proteins?
   4. Why are proteins from some foods of greater nutritional value than others?
   5. Write the names of the amino acids that are essential to humans.
   6. What two general methods are now available for improving the nutritional value of corn protein?
   7. Which amino acids contain a heterocyclic ring?
   8. Write the structural formulas for D-serine and L-serine. Which form is found in proteins?
   9. Why are amino acids amphoteric? Why are they optically active? What configuration do they have in proteins?
   10. Write the structural formula representing threonine at its isoelectric point.
   11. For phenylalanine write:
       (a) The molecular formula
       (b) The zwitterion formula
       (c) The formula in 0.1 M H_2SO_4
       (d) The formula in 0.1 M NaOH
   12. Write ionic equations to show how alanine acts as a buffer toward:
       (a) H^+ ion   (b) OH^- ion
   13. (a) At what pH will arginine not migrate to either electrode in an electrolytic cell?
       (b) In what pH range will it migrate toward the positive electrode?
   14. What can you say about the number of positive and negative charges on a protein molecule at its isoelectric point?
   15. At what pH will the following amino acids be at their isoelectric points?
       (a) Histidine   (b) Phenylalanine   (c) Glutamic acid
   16. Write out the full structural formula of the two dipeptides containing glycine and phenylalanine. Indicate the location of the peptide bonds.
   17. Write structures for:
       (a) Glycylglycine   (b) Glycylglycylalanine   (c) Leucylmethionylglycylserine
   18. Using amino acid abbreviations, write all the possible tripeptides containing one unit each of glycine, phenylalanine, and leucine.
   19. Explain what is meant by:
       (a) The primary structure
       (b) The secondary structure
       (c) The tertiary structure of a protein
   20. What special role does the sulfur-containing amino acid cysteine have in protein structure?
   21. What are enzymes and what is their role in the body?
   22. Distinguish between a coenzyme and an apoenzyme.
   23. Give the names of the enzymes that catalyze the hydrolysis of:
       (a) Sucrose   (b) Lactose   (c) Maltose
   24. Explain how enzymes function as catalysts. Why is an enzyme usually specific for one particular reaction?

25. What are the six general classes of enzymes?
26. Suggest reasons why most proteins are denatured (rendered biologically inactive) by each of the following treatments:
    (a) Heating to 100°C
    (b) The addition of mercury(II) or lead(II) ions
    (c) The addition of an appreciable amount of 0.1 M sodium hydroxide
27. Explain how hydrolysis of a protein differs from denaturation.
28. What chemical change occurs when a protein is hydrolyzed to amino acids?
29. When proteins are partially hydrolyzed, what products are formed?
30. What are the specific functions of the enzymes trypsin and chymotrypsin?
31. What is the visible evidence observed in a positive reaction for the following tests?
    (a) Xanthoproteic reaction   (b) Biuret test   (c) Ninhydrin test
32. Would Thr-Ala-Gly react with each of the following?
    (a) Sanger's reagent
    (b) Concentrated HNO_3 to give a positive xanthoproteic test
    (c) Ninhydrin
33. Which amino acids give a positive xanthoproteic test?
34. (a) What is thin-layer chromatography?
    (b) Describe how amino acids are separated using this technique.
    (c) What reagent is used to locate the amino acids in the chromatogram?
35. Given the decapeptide Ala-Tyr-Gly-Gln-Lys-Phe-Gly-Arg-Ser-Gly, what peptide fragments would be found in hydrolysis using the following?
    (a) Pepsin   (d) Carboxypeptidase   (f) Sanger's reagent
    (b) Trypsin   (e) Aminopeptidase   (g) Edman's reagent
    (c) Chymotrypsin
36. Distinguish between *in vivo* and *in vitro* synthesis of proteins.
37. Threonine has two asymmetric carbon atoms. Write Fischer projection formulas for its stereoisomers.
38. Write the reaction for Ala-Gly-Phe with Sanger's reagent (DNFB). Show the full structure of the final products.
39. What is the sequence of amino acids of a heptapeptide that contains one residue each of Gly, Leu, and Tyr; contains two residues each of Ala and Phe; and is hydrolyzed to the following tripeptides?
    (a) Gly-Phe-Leu, Phe-Ala-Gly, Leu-Ala-Tyr
    (b) Phe-Gly-Tyr, Phe-Ala-Ala, Ala-Leu-Phe
40. Bradykinin is a natural nonapeptide that is obtained from plasma by treatment with the enzyme trypsin. It is a vasodilator and a potent pain-producing agent. Analysis using Sanger's reagent and carboxypeptidase showed that both terminal amino acids are arginine. Total hydrolysis of bradykinin yielded Gly, Ser, 2 Arg, 2 Phe, 3 Pro. Partial hydrolysis gave Phe-Ser, Phe-Arg, Arg-Pro, Pro-Pro, Pro-Gly-Phe, Ser-Pro-Phe. What is the amino acid sequence of bradykinin?
41. One hundred grams of a food product was analyzed and found to contain 6.0 g of nitrogen. If protein contains an average of 16% nitrogen, what percentage of the food is protein?
42. Human hemoglobin contains 0.33% iron. If each hemoglobin molecule contains four iron atoms, what is the molecular weight of hemoglobin?
43. Which of these statements are correct?
    (a) Proteins, like fats and carbohydrates, are primarily for supplying heat and energy to the body.
    (b) Proteins differ from fats and carbohydrates in that they contain nitrogen.

(c) A complete protein is one that contains all the essential amino acids.
(d) Amino acids in proteins have the L configuration.
(e) All amino acids have an asymmetric carbon atom and are therefore optically active.
(f) The amide linkages by which amino acids are joined together are called peptide linkages.
(g) Two different dipeptides can be formed from the amino acids glycine and phenylalanine.
(h) The compound Ala-Phe-Tyr has two peptide bonds and is therefore known as a dipeptide.
(i) A zwitterion is a dipolar ion form of an amino acid.
(j) The primary structure of a protein is the α-helix or pleated sheet form that it takes.
(k) The amino acid residues in a peptide chain are numbered beginning with with the *C*-terminal amino acid.
(l) Insulin contains two polypeptide chains—one with 21 amino acids and the other with 30 amino acids.
(m) Enzymes are proteins.
(n) The most outstanding property of an enzyme is its specificity of reaction.
(o) The substance acted upon by an enzyme is called the coenzyme.
(p) When a protein is denatured, the polypeptide bonds are broken, liberating the amino acids.
(q) Irreversible coagulation or precipitation of proteins is called denaturation.
(r) Sanger's reagent reacts with peptide chains, isolating the *N*-terminal amino acid.

31 Nucleic Acids and Heredity

After studying Chapter 31 you should be able to:

1. Understand the terms listed in Question A at the end of the chapter.
2. Write the structural formulas for the two purine and three pyrimidine bases found in nucleotides.
3. Distinguish between ribonucleotides and deoxyribonucleotides.
4. List the compositions, abbreviations, and structures for the ten nucleotides in DNA and RNA.
5. Write the structural formulas for ADP and ATP.
6. Identify where energy is stored in ADP and ATP.
7. Write a structural formula of a segment of a polynucleotide that contains four nucleotides.
8. Describe the double-helix structure of DNA according to Watson and Crick.
9. Explain the concept of complementary bases.
10. Describe and illustrate the replication process of DNA.
11. Understand how heredity factors are stored in DNA molecules.
12. Distinguish between mitosis and meiosis.
13. Understand how the genetic code is used in the synthesis of proteins.
14. Explain the genetic code.
15. State the functions of the three different kinds of RNA.
16. Describe the transcription of the genetic code from DNA to RNA.
17. Describe the biosynthesis of proteins.
18. Understand how mutations are caused.
19. Explain the occurrence and effects of sickle cell anemia and phenylketonuria.

31.1 Introduction

The question of how hereditary material duplicates itself was for a long time one of the most baffling problems of biology. For many years, biologists attempted in vain to solve this problem, and also to find an answer to the question "Why are the offspring of a given species undeniably of that species?" Many thought the chemical basis for heredity lay in the structure of the proteins. But no one was able to provide evidence showing how protein could reproduce itself. The answer to the heredity problem was finally found in the structure of the nucleic acids.

696   Chapter 31

nucleoprotein

nucleic acid

The unit structure of all living things is the cell. Suspended in the nuclei of cells are chromosomes, which consist largely of proteins and nucleic acids. The nucleic acids and the proteins are intimately associated within complexes called **nucleoproteins**. There are two types of **nucleic acids**—those that contain the sugar deoxyribose and those that contain the sugar ribose. Accordingly, they are called deoxyribonucleic acid (DNA) and ribonucleic acid (RNA). DNA was discovered in 1869 by the Swiss physiologist Friederich Miescher (1844–1895), who extracted it from the nuclei of cells.

31.2   Nucleosides

Nucleosides are important cellular molecules that, when linked together, form the nucleic acids. Nucleosides are derived from the two parent bases, purine and pyrimidine.

Purine, $C_5H_4N_4$ Pyrimidine, $C_4H_4N_2$

nucleoside

There are five major bases found in living matter—two purine bases (adenine and guanine) and the pyrimidine bases (cytosine, thymine, and uracil). The structures of these five bases are shown in Figure 31.1. The five purine and pyrimidine bases occur in nature primarily as nucleosides and nucleotides. A **nucleoside** is a purine or pyrimidine base linked to a sugar molecule, usually D-ribose or 2-deoxy-D-ribose. The sugar molecule is attached at carbon 1 to

Adenine
(6-Aminopurine)

Guanine
(2-Amino-6-oxypurine)

Cytosine
(2-Oxy-4-aminopyrimidine)

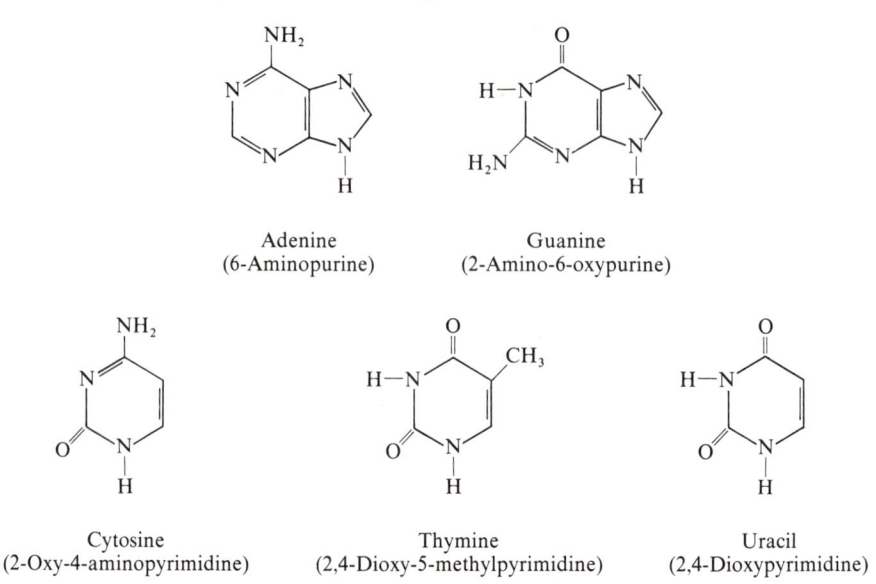

Thymine
(2,4-Dioxy-5-methylpyrimidine)

Uracil
(2,4-Dioxypyrimidine)

Figure 31.1. Purine and pyrimidine bases found in living matter.

Table 31.1. Composition of ribonucleosides and deoxyribonucleosides.

Name	Composition	Abbreviation
Adenosine	Adenine–ribose	A
Deoxyadenosine	Adenine–deoxyribose	dA
Guanosine	Guanine–ribose	G
Deoxyguanosine	Guanine–deoxyribose	dG
Cytidine	Cytosine–ribose	C
Deoxycytidine	Cytosine–deoxyribose	dC
Thymidine	Thymine–ribose	T
Deoxythymidine	Thymine–deoxyribose	dT
Uridine	Uracil–ribose	U
Deoxyuridine	Uracil–deoxyribose	dU

the nitrogen 9 position of the purine and to the nitrogen 1 position of the pyrimidine. These are sometimes called *ribonucleosides* and *deoxyribonucleosides*. Thus, the adenine ribonucleoside, adenosine, consists of adenine plus D-ribose, and the adenine deoxyribonucleoside, deoxyadenosine, consists of adenine plus 2-deoxyribose. The compositions of the ribonucleosides and deoxyribonucleosides are given in Table 31.1. Typical structures of these compounds are shown in Figure 31.2.

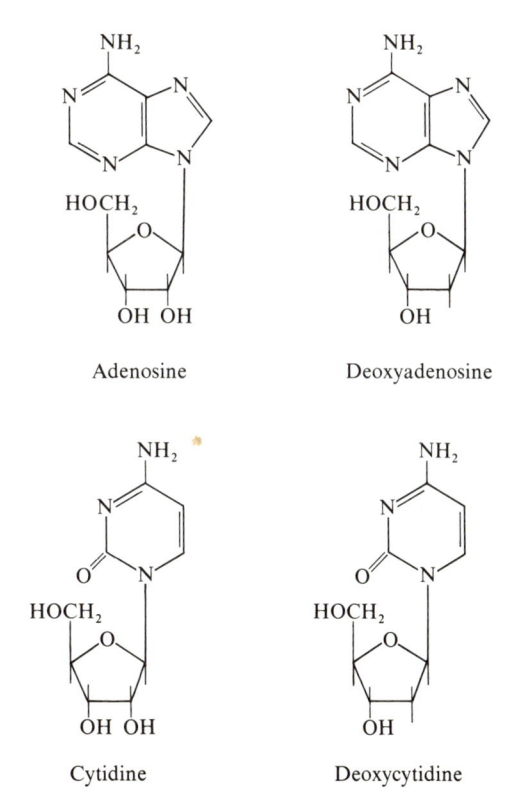

Figure 31.2. Typical structures of ribonucleosides and deoxyribonucleosides.

698 Chapter 31

31.3 Nucleotides—Phosphate Esters

nucleotide

Nucleotides are phosphate esters of nucleosides. **Nucleotides** consist of a purine or a pyrimidine base linked to a ribose sugar which in turn is linked to a phosphate group. The ester may be a monophosphate, a diphosphate, or a triphosphate. The ester linkage may be to the hydroxy group of position 2, 3, or 5 of ribose or to the 3 or 5 position of deoxyribose.

Nucleoside + Phosphoric acid = Nucleotide

Each nucleotide consists of the following sequence:

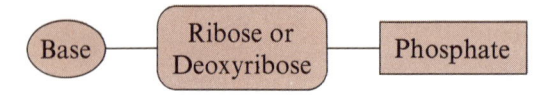

Base — Ribose or Deoxyribose — Phosphate

A typical nucleotide is adenosine-5′-monophosphate (AMP), also called adenylic acid. The abbreviations for the five nucleosides containing D-ribose are:

Adenosine = A
Guanosine = G
Cytidine   = C
Thymidine = T
Uridine   = U

The letters MP (monophosphate) can be added to any of these to designate the corresponding nucleotide. Thus, GMP is guanosine monophosphate. A lower-case d is placed in front of GMP if the nucleotide contains the deoxyribose sugar (dGMP). When the letters such as AMP or GMP are given, it is generally understood that the phosphate group is attached to the 5′ position of the ribose unit (5′-AMP). If attachment is elsewhere, it will be designated—for example, 3′-AMP. In naming these compounds, the prime designation, as in 5′, indicates

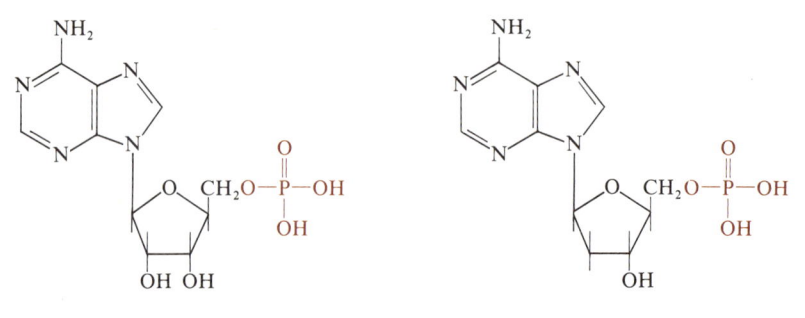

Adenosine-5′-monophosphate (AMP)

Deoxyadenosine-5′-monophosphate (dAMP)

Figure 31.3. Examples of nucleotides.

the position on the ribose or deoxyribose unit. Thus, adenosine-5′-monophosphate means that the phosphate group is attached to the 5 position of ribose. Examples of nucleotide structures are shown in Figure 31.3.

Two other important adenosine phosphate esters are adenosine diphosphate (ADP) and adenosine triphosphate (ATP). Note that the letters DP are used for diphosphate and TP for triphosphate. In these molecules the phosphate groups are linked together. The structures are similar to AMP except that they contain two or three phosphate residues, respectively (see Figure 31.4). All the nucleosides form mono-, di-, and triphosphate nucleotides.

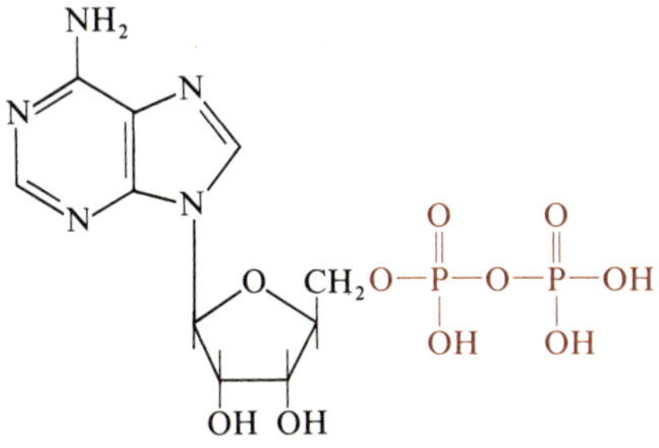

Adenosine-5′-diphosphate (ADP)

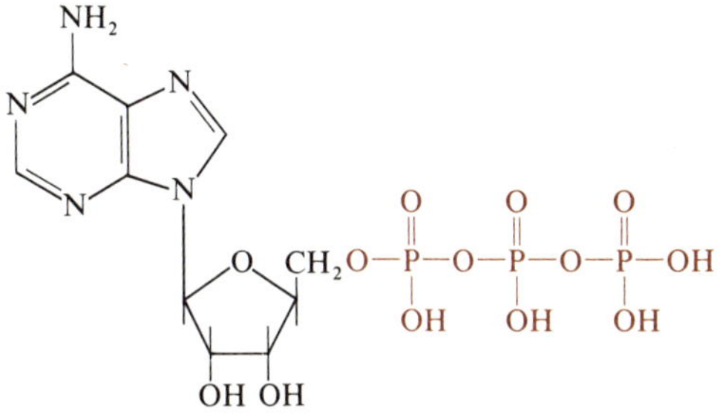

Adenosine-5′-triphosphate (ATP)

Figure 31.4. Structures of ADP and ATP.

31.4 High-Energy Nucleotides

The phosphate esters have a central role in the energy transfers involved in many metabolic processes. ATP and ADP are especially involved in these processes. The role of these two nucleotides is to store energy and to release energy to the cells and tissues. The source of energy is the foods we eat, particularly carbohydrates and fats. The energy released as these foods are metabolized is partly utilized as heat to keep our bodies warm. There is more than enough energy to maintain body temperatures, and at least part of the remainder is stored as chemical energy—principally in the phosphate bonds in such molecules as ATP. Chemical energy from energy-rich ATP is used to carry out many of the complex reactions that are essential to most of our life processes. The phosphate bonds in ATP and ADP are known as high-energy phosphate bonds. This energy is released during hydrolysis of ATP and ADP molecules.

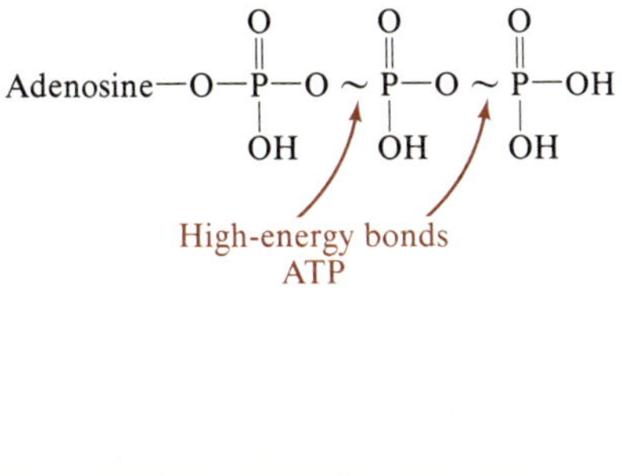

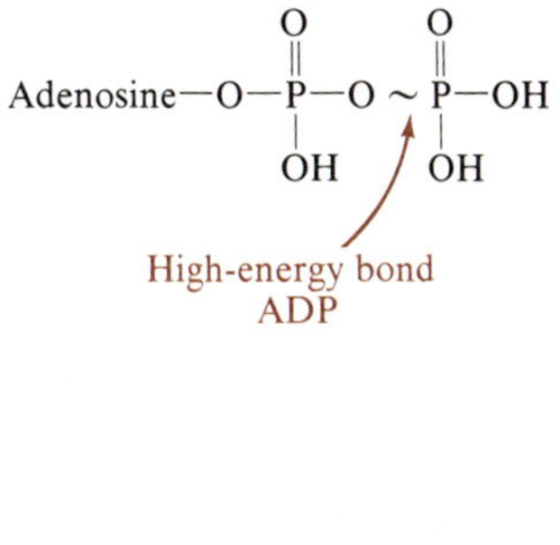

High-energy bonds
ATP

High-energy bond
ADP

In the hydrolysis, ATP forms ADP and inorganic phosphate (P_i) with the release of 7.3 kcal of energy per mole of ATP:

$$ATP + H_2O \underset{\text{Energy storage}}{\overset{\text{Energy utilization}}{\rightleftharpoons}} ADP + P_i + 7.3 \text{ kcal}$$

The hydrolysis reaction is reversible, with ADP being converted to ATP by still higher-energy phosphate molecules. In this manner, energy is supplied to the cells from ATP, and energy is stored by the synthesis of ATP from ADP and AMP. All these processes are enzyme-catalyzed. Processes such as muscle movement, nerve sensations, seeing with the eyes, even the maintenance of our heartbeats, are all dependent on energy from ATP.

31.5 Polynucleotides; Nucleic Acids

Macromolecular chains of polynucleotides are formed by the loss of a molecule of water between the —OH of the phosphate group of one nucleotide and the —OH on carbon 3 of the ribose or deoxyribose unit of a second nucleotide. This sequence can continue on both ends of the dinucleotide formed until chains of polynucleotides with very high molecular weights are produced.

Two series of polynucleotide chains are known, one containing ribose and the other 2-deoxyribose. One polymeric chain consists of the monomers AMP, GMP, CMP, and UMP and is known as a polyribonucleotide. The other chain contains the monomers dAMP, dGMP, dCMP, and dTMP and is known as a polydeoxyribonucleotide.

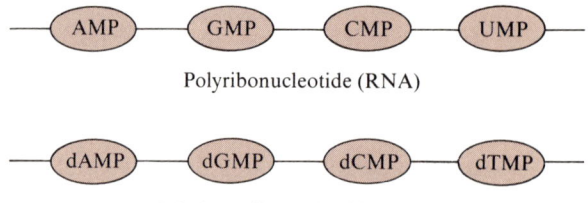

—(AMP)—(GMP)—(CMP)—(UMP)—
Polyribonucleotide (RNA)

—(dAMP)—(dGMP)—(dCMP)—(dTMP)—
Polydeoxyribonucleotide (DNA)

ribonucleic acid (RNA)

deoxyribonucleic acid (DNA)

The nucleic acids DNA and RNA are polynucleotides. **Ribonucleic acid (RNA)** is a polynucleotide which upon hydrolysis yields ribose, phosphoric acid, and the four purine and pyrimidine bases adenine, guanine, cytosine, and uracil. **Deoxyribonucleic acid (DNA)** is a polynucleotide that yields 2-deoxyribose, phosphoric acid, and the four bases, adenine, guanine, cytosine, and thymine. Note that RNA and DNA contain one different pyrimidine nucleotide. RNA contains uridine whereas DNA contains thymidine. A segment of a ribonucleic acid chain is shown in Figure 31.5.

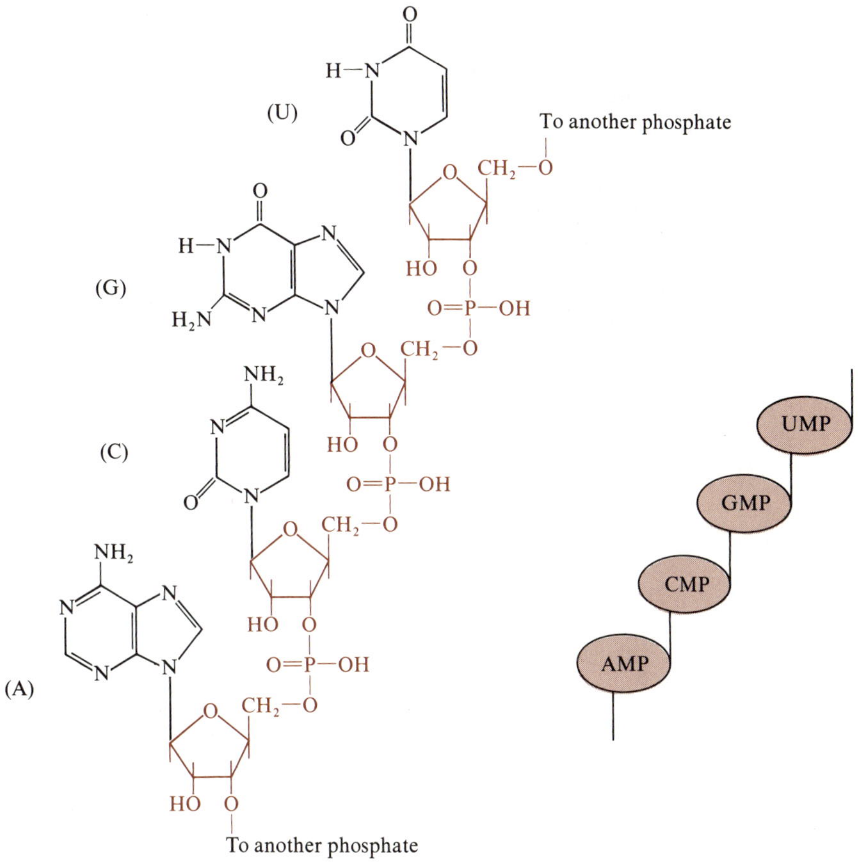

Figure 31.5. A segment of ribonucleic acid (RNA) consisting of the four nucleotides adenosine monophosphate, cytidine monophosphate, guanosine monophosphate, and uridine monophosphate.

31.6 Structure of DNA

Deoxyribonucleic acid (DNA) is a polymeric substance made up of thousands of the nucleotides dAMP, dGMP, dCMP, and dTMP. The order in which these four nucleotides occur differs in different DNA molecules, and it is this order that determines the specificity of each DNA molecule. For a long time it was thought that the four nucleotides occur in equal amounts in DNA. However, more refined methods of analysis of DNA showed that the ratios of purine and pyrimidine bases differed in different DNA samples. But in all samples, the total amount of purines equaled the total amount of pyrimidines; that is, the amount of adenine equaled the amount of thymine and the amount

Table 31.2. Relative amounts of purines and pyrimidines in samples of DNA.

Source	Adenine	Thymine	Ratio A/T	Guanine	Cytosine	Ratio G/C
Beef thymus	29.0	28.5	1.02	21.2	21.2	1.00
Beef liver	28.8	29.0	0.99	21.0	21.1	1.00
Beef sperm	28.7	27.2	1.06	22.2	22.0	1.01
Human thymus	30.9	29.4	1.05	19.9	19.8	1.00
Human liver	30.3	30.3	1.00	19.5	19.9	0.98
Human sperm	30.9	31.6	0.98	19.1	18.4	1.04
Hen red cells	28.8	29.2	0.99	20.5	21.5	0.96
Herring sperm	27.8	27.5	1.01	22.2	22.6	0.98
Wheat germ	26.5	27.0	0.98	23.5	23.0	1.02
Yeast	31.7	32.6	0.97	18.3	17.4	1.05
Vaccinia virus	29.5	29.9	0.99	20.6	20.0	1.03
Bacteriophage T_2	32.5	32.6	1.00	18.2	18.6	0.98

P = Phosphate
D = Deoxyribose
A = Adenine
T = Thymine
C = Cytosine
G = Guanine

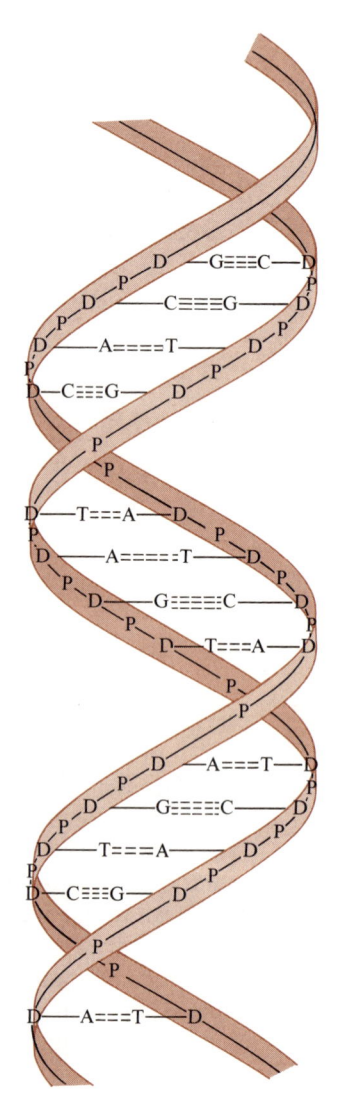

Figure 31.6. Double-stranded helix structure of DNA.

of guanine equaled the amount of cytosine. The analysis of DNA from several species is shown in Table 31.2.

The primary clues to the special configuration and structure of DNA came from X-ray diffraction studies. Most significant was the work of Maurice H. F. Wilkins (1916–) of Kings College in London. Wilkins' X-ray pictures inferred that the nucleotide bases were placed one on top of another like a stack of saucers. From this work as well as others, the American biologist James D. Watson (1928–) and British physicist Francis H. C. Crick (1916–), working at Cambridge University, designed and built a scale model of a DNA molecule. In 1953, Watson and Crick announced their now-famous double-stranded helix structure for DNA. This was a major milestone in the history of biology, and in 1962, Watson, Crick, and Wilkins were awarded the Nobel Prize in medicine and physiology for their studies of DNA.

The structure of DNA, according to Watson and Crick, consists of two polymeric strands of nucleotides in the form of a double helix, with both nucleotide strands coiled around the same axis (see Figure 31.6). Along each strand are alternate phosphate and deoxyribose units with one of the four bases adenine, guanine, cytosine, or thymine attached to deoxyribose as a side group. The double helix is held together by hydrogen bonds extending from the base

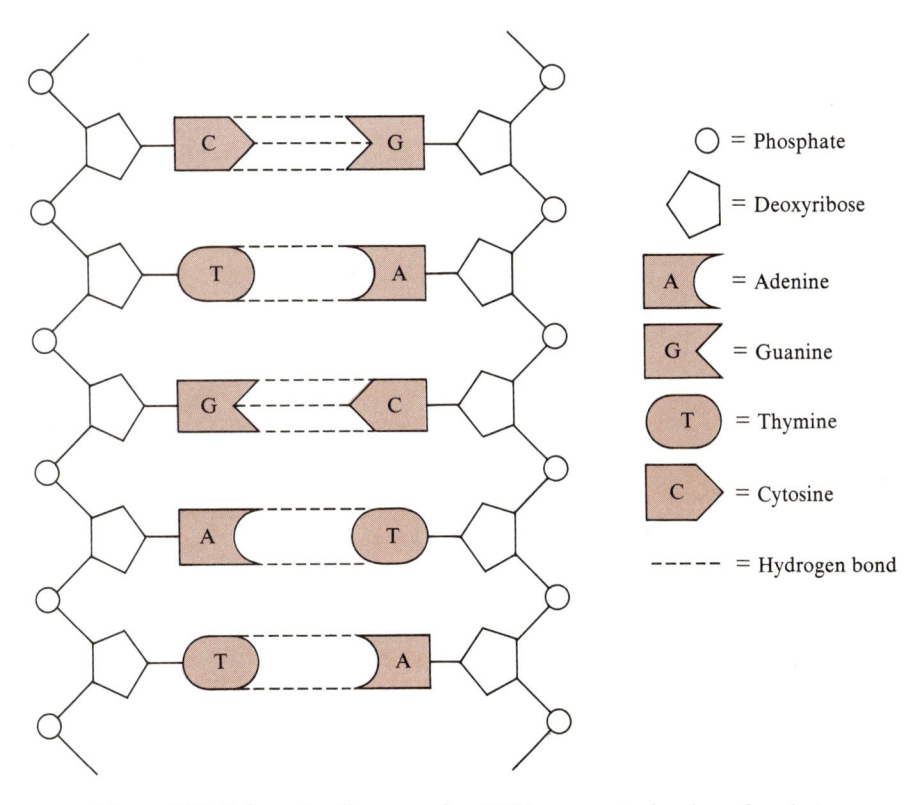

Figure 31.7. Schematic diagram of a DNA segment showing phosphate, deoxyribose, and complementary base pairings held together by hydrogen bonds.

on one strand of the double helix to a complementary base on the other strand. The structure of DNA has been likened to a ladder that has been twisted into a double helix, with the rungs of the ladder kept perpendicular to the twisted railings. The phosphate and deoxyribose units alternate along the two railings of the ladder, and two nitrogen bases form each rung of the ladder.

In the Watson–Crick model of DNA, the distance between the two nucleotide strands is such that it can accommodate one purine and one pyrimidine base. This allows four possible structures, A–T, A–C, G–C, and G–T. It was shown, however, that because of the hydrogen bonding that occurs between them, the complementary bases are adenine–thymine and guanine–cytosine. The order in which the bases occur does not matter—adenine–thymine, thymine–adenine, guanine–cytosine, cytosine–guanine. The essential requirement seems to be that adenine and thymine always be paired and that guanine and cytosine always be paired. This base arrangement is substantiated by the data in Table 31.2. A schematic diagram of a DNA segment is shown in Figure 31.7. The hydrogen bonding of complementary base pairs is shown in Figure 31.8. In this figure, note that there are two hydrogen bonds between adenine and thymine and three hydrogen bonds between guanine and cytosine.

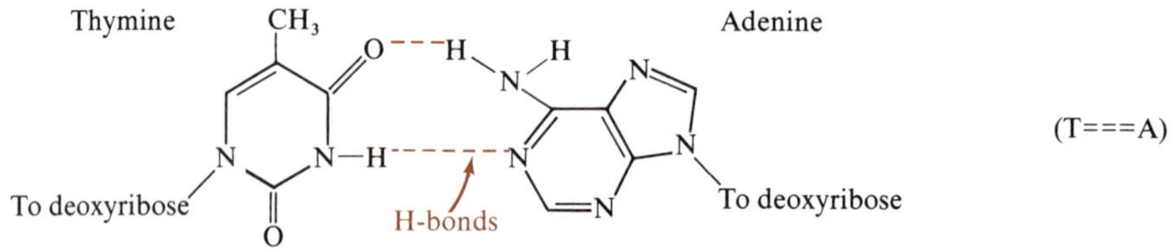

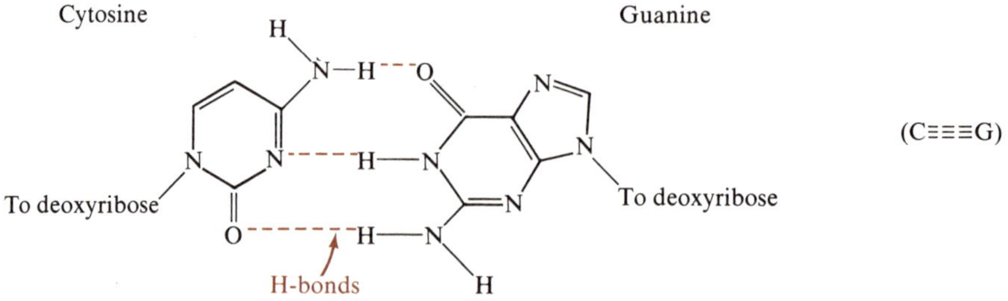

Figure 31.8. Hydrogen bonding between the complementary bases thymine and adenine (T====A) and cytosine and guanine (C====G). Note that one pair of bases has two hydrogen bonds and the other pair has three hydrogen bonds between them.

31.7 DNA—The Genetic Substance

The foundations of our present concepts of heredity and evolution were laid within the span of a decade. Charles Darwin (1809–1882), in *The Origin of Species* (1859), presented evidence supporting the concept of organic evolution and his theory of natural selection. Gregor Johann Mendel (1822–1884) discovered the basic laws of heredity in 1866, and Friederich Miescher discovered nucleic acid in 1869. Although Darwin's views were widely discussed and generally accepted by biologists within a few years, Mendel's and Miescher's work went unnoticed for many years.

Mendel's laws were rediscovered about 1900, and led to our present understanding of heredity and the science of genetics. Interest in nucleic acids lagged until nearly the 1950s, when chemical and X-ray data provided the basis for the suggestion by Watson and Crick that DNA exists in a double helix and that DNA has the possible copying mechanism for genetic material.

Heredity is the process by which the physical and mental characteristics of parents are transferred to their offspring. In order for this to occur, it is necessary for the material responsible for genetic transfer to be able to make exact copies of itself. The polymeric DNA molecule is the chemical basis for heredity. The genetic information needed for transmittal of a species' characteristics is coded along the polymeric chain. Although the chain is made from only four different nucleotides, the information content of DNA resides in the sequence of these nucleotides.

Chromosomes, which are composed of nucleic acids and proteins, are long threadlike bodies that contain the basic units of heredity, called *genes*. Genes are segments of nucleotides that reside along the DNA chain. The genes control the formation of one type of RNA, which in turn controls the sequence of amino acids in specific polypeptides or proteins. Apparently, this is a one-to-one relationship, with one gene directing the synthesis of only one polypeptide or protein molecule. Thus, each kind of gene is different from every other gene in its DNA sequence. Hundreds of genes can exist along a DNA chain. Furthermore, a gene may be repeated in various places in a DNA molecule.

Genetic information must be reproduced exactly each time a cell divides. The design for replication is built into the DNA structure of Watson and Crick, first by the nature of its double helical structure and second by the complementary nature of its nitrogen bases where adenine will bond only to thymine and guanine only to cytosine. The DNA double helix unwinds, or simply "unzips," into two separate helices at the hydrogen bonds between the bases. Each helix then serves as a template, combining only with the proper free nucleotides to produce two identical replicas of itself. The replication of DNA is illustrated in Figure 31.9. The DNA content of cells doubles just before the cell divides, and one half of the DNA goes to each daughter cell. After cell division is completed, each daughter cell contains DNA and the full genetic code that was present in the original cell. This process of ordinary cell division is known as **mitosis** and occurs in all the cells of the body except the reproductive cells.

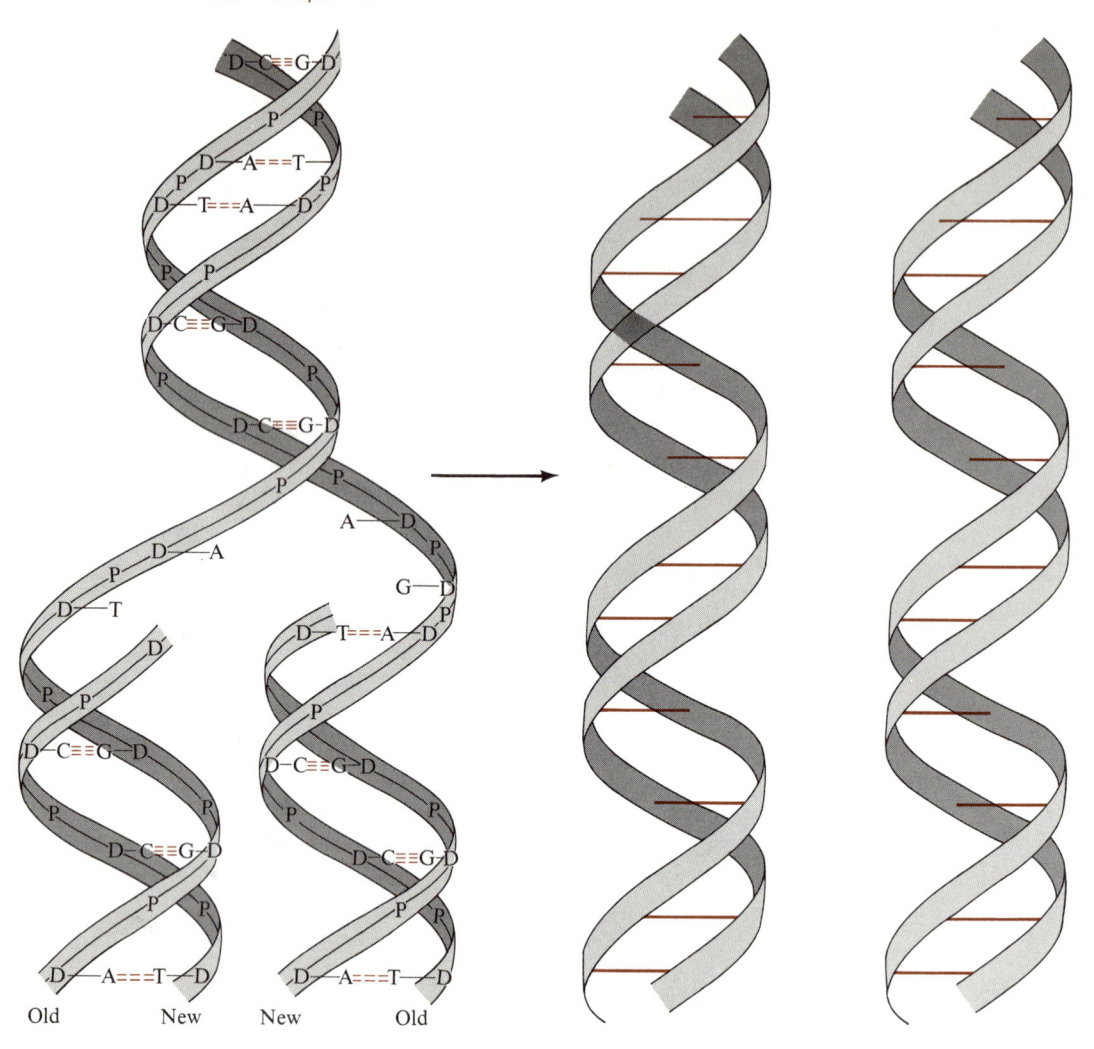

Figure 31.9. Method of replication of DNA. The two helices unwind, separating at the point of the hydrogen bonds. Each strand then serves as a template form, recombining with the proper nucleotides to duplicate itself as a double-stranded helix.

As we have indicated before, DNA is an integral part of the chromosomes. Each species carries a specific number of chromosomes in the nucleus of each of its cells. The number of chromosomes varies with different species. Humans have 23 pairs, or 46 chromosomes. The fruit fly has 4 pairs, or 8 chromosomes. Each chromosome contains DNA molecules. Mitosis produces cells with the same chromosomal content as the parent cell. However, at some point in the sexual reproductive cycle, cell division occurs by a different process known as **meiosis**.

In sexual reproduction, two cells, the sperm cell from the male and the egg cell (or ovum) from the female, unite to form the cell of the new individual. If reproduction took place with mitotic cells, the normal chromosome content

meiosis

would double each time two cells united. However, in meiosis, the cell splits in such a way as to reduce the number of chromosomes to one-half (23 in humans) the number normally present. The sperm cell carries only half of the chromosomes from its original cell, and the egg cell also carries half of the chromosomes from its original cell. When the sperm and the egg cells unite during fertilization, the cell once again contains the correct number of chromosomes and all the heredity characteristics of the species. Thus, the offspring derives half its genetic characteristics from the father and half from the mother.

31.8 The Genetic Code

For a long time after its structure was elucidated, scientists struggled with the problem of how DNA could also be a template for the synthesis of proteins. Since the backbone of the DNA molecule contains a regular structure of phosphate and deoxyribose units, the key to the code had to lie in the four bases—adenine, guanine, cytosine, and thymine.

The code, using only the four nucleotides A, G, C, and T, must be capable of coding at least the 20 amino acids that occur in proteins. If each nucleotide coded one amino acid, only four amino acids could be represented. If the code used two nucleotides to specify an amino acid, 16 (4 × 4) combinations would be possible–still not enough. Using three nucleotides, we can have 64 (4 × 4 × 4) possible combinations—more than enough to specify the 20 common amino acids in proteins. It has now been determined that each code word requires a sequence of three nucleotides. The code is therefore a triplet code. Each triplet of three nucleotides is called a **codon** and each codon specifies one amino acid. Thus, to describe a protein containing 200 amino acid units, a gene containing at least 200 codons or 600 nucleotides is required.

codon

In the sequence of biological events, the code from a gene in DNA is first transcribed to a coded RNA, which, in turn, is used to direct the synthesis of a protein. The 64 possible codons for messenger RNA (see Section 31.9) are given in Table 31.3. In this table a three letter sequence (first nucleotide–second nucleotide–third nucleotide) specifies a particular amino acid. For example, the codon CAC (cytosine nucleotide–adenine nucleotide–cytosine nucleotide) is the code for the amino acid histidine (His). You will note that three codons in the table, marked TC, do not encode any amino acids. These are called *nonsense or termination codons*. Their exact nature is not clear, but it is believed that they are signals to indicate where the synthesis of a protein molecule is to end. The other 61 codons identify 20 amino acids. Methionine and tryptophan each have only one codon. For the other amino acids, the code is redundant; that is, each amino acid is specified by at least two, and sometimes as many as six, codons.

It is believed that the genetic code is a universal code for all living organisms; that is, the same nucleotide triplet specifies a given amino acid regardless of whether that amino acid is part of a bacterial cell, a pine tree, or a human being.

Table 31.3. The genetic code for messenger RNA. The sequence of nucleotides in the triplet codons of messenger RNA which specifies a given amino acid.

First nucleotide	Second nucleotide				Third nucleotide
	U	C	A	G	
U	Phe	Ser	Tyr	Cys	U
	Phe	Ser	Tyr	Cys	C
	Leu	Ser	TC[a]	TC[a]	A
	Leu	Ser	TC[a]	Trp	G
C	Leu	Pro	His	Arg	U
	Leu	Pro	His	Arg	C
	Leu	Pro	Gln	Arg	A
	Leu	Pro	Gln	Arg	G
A	Ile	Thr	Asn	Ser	U
	Ile	Thr	Asn	Ser	C
	Ile	Thr	Lys	Arg	A
	Met	Thr	Lys	Arg	G
G	Val	Ala	Asp	Gly	U
	Val	Ala	Asp	Gly	C
	Val	Ala	Glu	Gly	A
	Val	Ala	Glu	Gly	G

[a] Terminator or nonsense codons.

31.9 Genetic Transcription, RNA

One of the main functions of DNA is in the synthesis of ribonucleic acids (RNA). RNA differs from DNA in the following ways: (1) it consists of a single polymeric strand of nucleotides rather than a double helix, (2) it contains the pentose D-ribose instead of 2-deoxy-D-ribose, and (3) it contains the pyrimidine base uracil instead of thymine.

transcription

The making of RNA from DNA is called **transcription**. The nucleotide sequence of only one strand of DNA is transcribed into a single strand of RNA. This transcription occurs in a complementary fashion. Where there is a guanine base in DNA, a cytosine base will occur in RNA. Cytosine is transcribed to guanine, thymine to adenine, and adenine to uracil (see Figure 31.10).

There are three kinds of RNA—ribosomal RNA (*r*RNA), messenger RNA (*m*RNA), and transfer RNA (*t*RNA). More than 80% of the RNA is ribosomal RNA. It is found in the ribosomes, where it is associated with protein in proportions of about 60–65% protein to 30–35% *r*RNA. Ribosomes are located in the cells and are the site where *m*RNA and *t*RNA interact to assemble the amino acids into proteins.

Messenger RNA carries genetic information from DNA to the ribosomes. It is a template made from DNA and carries the codons that direct the synthesis

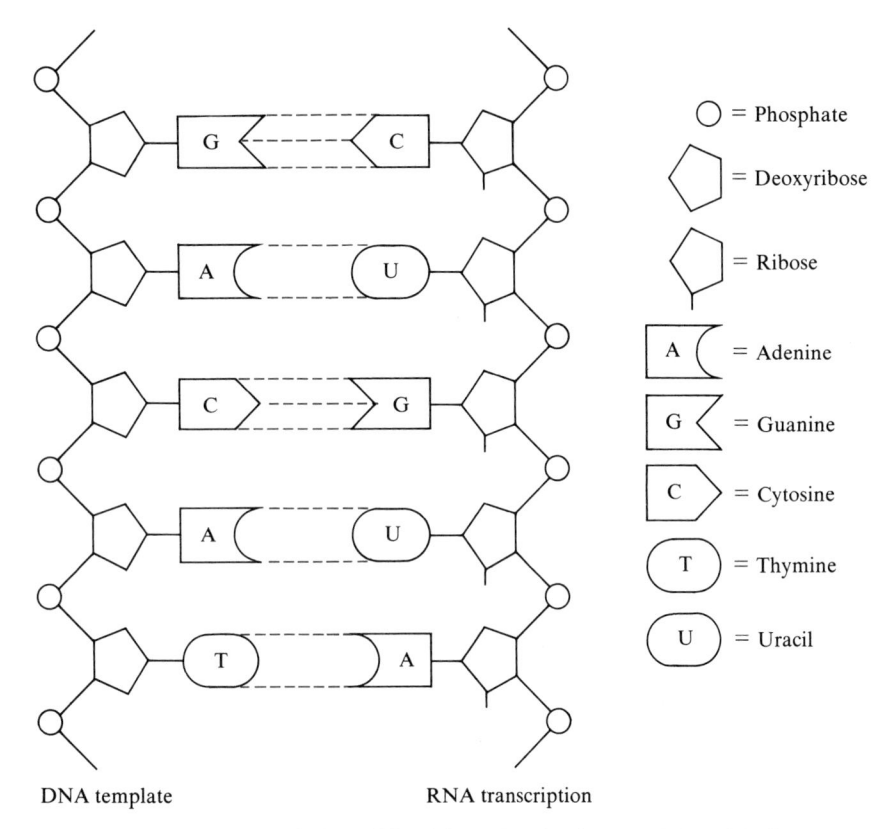

Figure 31.10. Transcription of RNA from DNA. The sugar in RNA is ribose. The complementary base of adenine is uracil. After transcription is complete, the new RNA separates from its DNA template and travels to its location for further use.

of proteins. The size of *m*RNA varies according to the length of the polypeptide chain it will encode.

The primary function of *t*RNA is to bring amino acids to the ribosomes for incorporation into protein molecules. Consequently, there exists at least one *t*RNA for each of the 20 amino acids required for proteins. Transfer RNA molecules have a number of structural features in common. The end of the chain of all *t*RNA molecules terminates in a CCA nucleotide sequence to which is attached the amino acid to be transferred to a protein chain. The primary structure of *t*RNA allows extensive folding of the molecule such that complementary bases are hydrogen bonded to each other to form a structure that appears like a cloverleaf (see Figure 31.11). The cloverleaf model of *t*RNA has an anticodon loop consisting of seven unpaired nucleotides. Three of these nucleotides make up the anticodon. The anticodon is complementary to, and hydrogen bonds to, the codon on *m*RNA. One of the other loops serves to recognize the ribosome site; the other loop serves to recognize the specific enzyme associated with that *t*RNA.

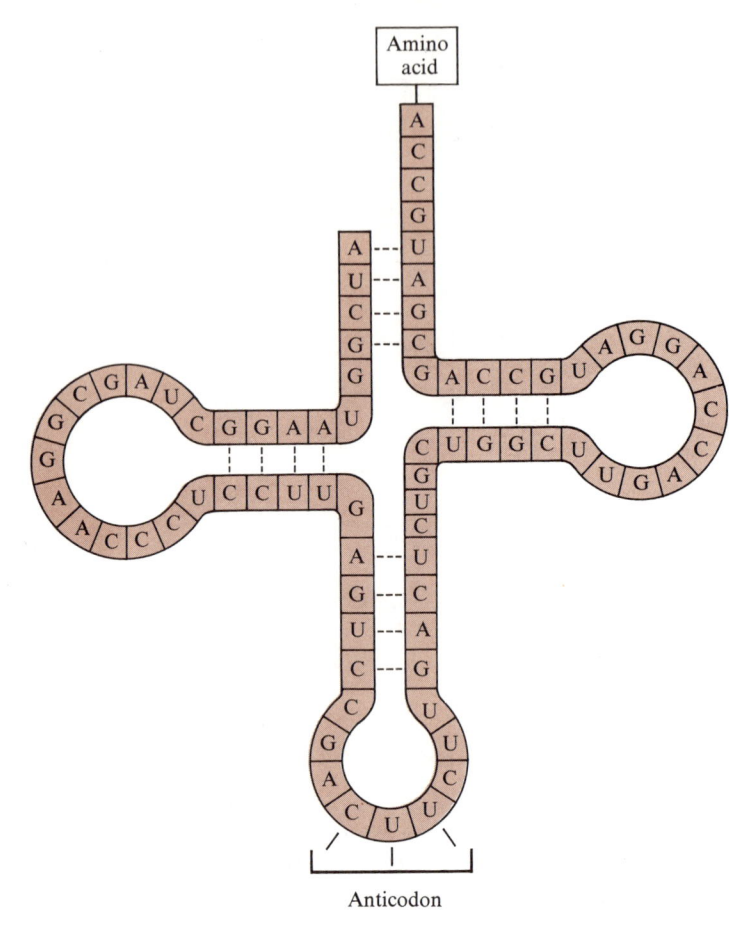

Figure 31.11. Cloverleaf model representation of *t*RNA: The anticodon located at the lower loop would be complementary to GAA (which is the code for glutamic acid) on *m*RNA.

31.10 Biosynthesis of Proteins

We present here only a cursory description of protein synthesis, since the process is very complex and the full details are beyond the scope of this book.

The biosynthesis of proteins occurs in the ribosomes. Messenger RNA leaves the nucleus of the cell and travels to the cytoplasm, where it becomes associated with a cluster of ribosomes. Each *m*RNA strand is bound to five or more ribosomes. In the next step, amino acids are activated by ATP, and with the aid of the enzyme aminoacyl-*t*RNA synthetase, bond to *t*RNA to form aminoacyl-*t*RNA complexes. The amino acid is attached by an ester linkage to the ribose unit of tRNA:

$$R-\underset{\underset{NH_2}{|}}{CH}-\underset{\underset{O}{\|}}{C}-tRNA$$

Aminoacyl-tRNA

Thus, the amino acids are brought to the synthesis site by tRNA. A different specific enzyme is utilized for binding each of the 20 amino acids to a corresponding tRNA. Although only 20 amino acids are involved, there are about 60 different tRNA molecules in the cells.

The next step in the process is the initiation of the polypeptide chain. The first amino acid in the sequence is of special importance. It has been demonstrated that the first aminoacyl-tRNA in protein synthesis is the formyl (CHO) derivative of methionine, N-formylmethionine-tRNA:

$$\begin{array}{c} H-C=O \\ | \\ NH \quad\; O \\ | \qquad \|\\ CH_3-S-CH_2CH_2CH-C-tRNA \end{array}$$

N-Formylmethionine, abbreviated as fmet-tRNA, is coded by the triplet AUG codon of mRNA and recognized by the anticodon UAC of tRNA. The mRNA codon AUG is present at the beginning of every mRNA molecule and is known as the *genetic initiation signal*. Since in protein synthesis the peptide chain always grows from the N-terminal end toward the C-terminal end, the function of fmet-tRNA is to ensure that the chain grows in this direction. In fmet-tRNA, the amino group is blocked by the formyl group, leaving the —C=O group available to react with the amino group of the next amino acid.

The next stage involves the elongation or growth of the peptide chain, which is assembled one amino acid at a time. After the fmet-tRNA is attached to the mRNA codon, the stepwise elongation of the polypeptide chain involves the following steps: (1) The next aminoacyl-tRNA enters the ribosome and becomes attached to mRNA through the hydrogen bonding of the tRNA anticodon to the mRNA codon. (2) The peptide bond between the two amino acids is formed by the transfer of the amino acid from the initial aminoacyl-tRNA to the incoming aminoacyl-tRNA. In this step, which is catalyzed by the enzyme peptidyl transferase, the free carbonyl group (—C=O) forms the peptide bond with the amino group of the incoming aminoacyl-tRNA. (3) The tRNA carrying the peptide chain (now known as a peptidyl-tRNA) moves over in the ribosome, the free tRNA is ejected, and the next aminoacyl-tRNA enters the ribosome. The peptide chain is transferred to the incoming amino acid and the sequence is repeated over and over again as the mRNA moves through the ribosome—just like a tape delivering its message. In each step, the entire peptide chain is transferred to the incoming amino acid.

The termination of the polypeptide chain occurs when a nonsense or termination codon appears and the tRNA has no amino acid at its end. Chain termination is followed by the release of the free polypeptide (protein) and the

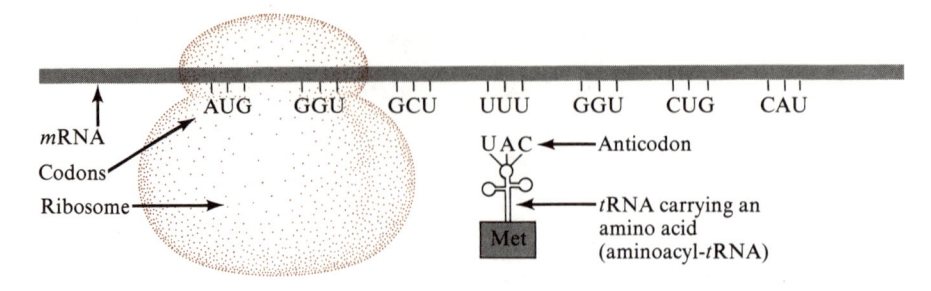

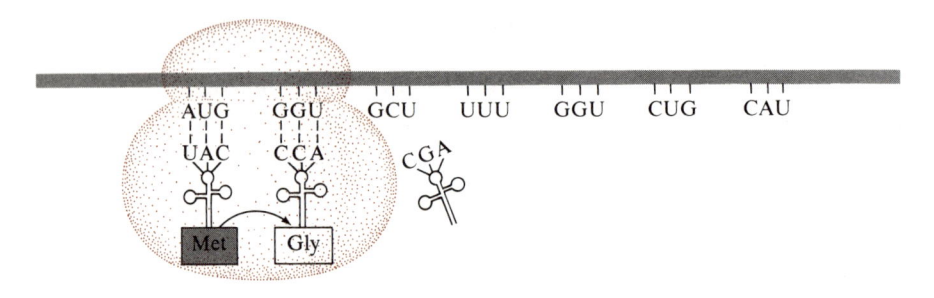

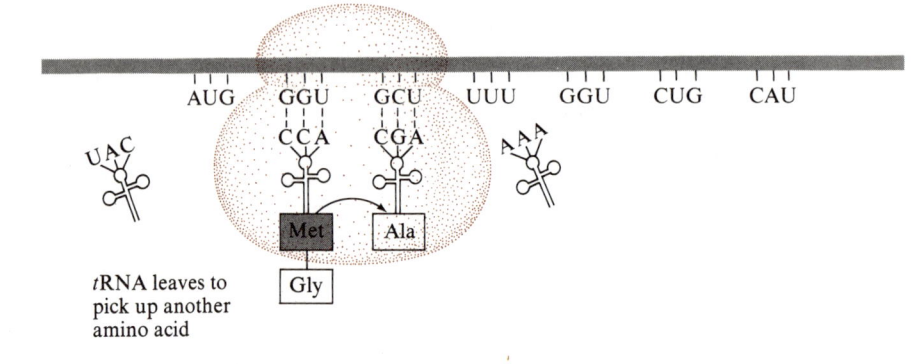

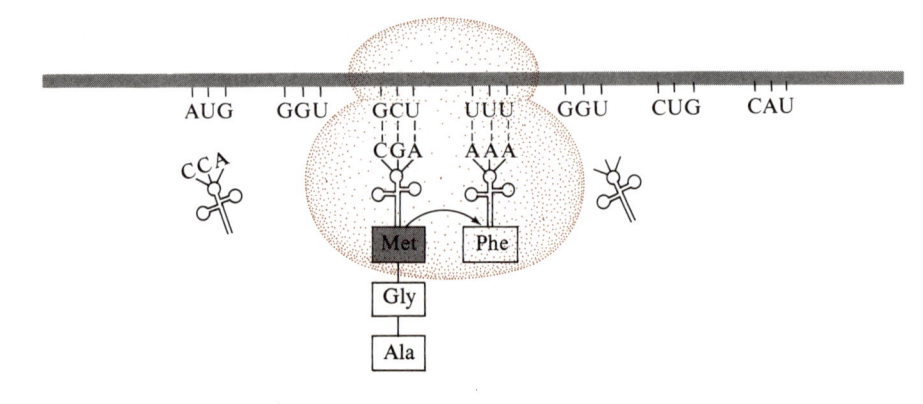

mRNA
Codons
Ribosome

AUG GGU GCU UUU GGU CUG CAU

UAC ← Anticodon

tRNA carrying an amino acid (aminoacyl-tRNA)

Met

AUG GGU GCU UUU GGU CUG CAU
UAC CCA   CGA

Met → Gly

AUG GGU GCU UUU GGU CUG CAU
     CCA CGA   AAA

UAC

Met → Ala

Gly

tRNA leaves to pick up another amino acid

AUG GGU GCU UUU GGU CUG CAU
         CGA AAA

CCA

Met → Phe

Gly

Ala

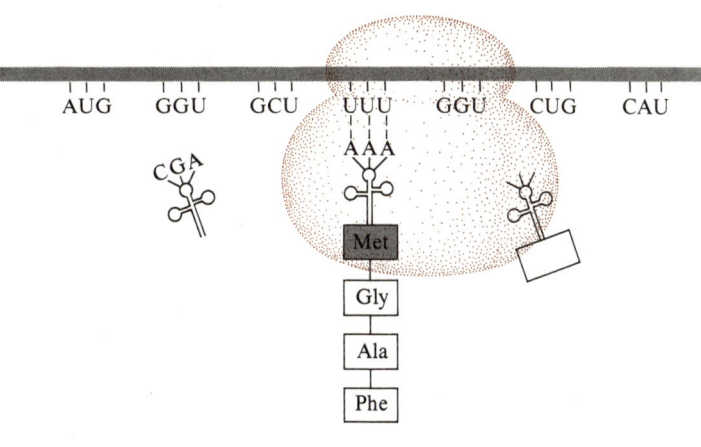

Figure 31.12. Biosynthesis of proteins: *m*RNA from DNA enters and complexes with the ribosomes. *t*RNA carrying an amino acid (aminoacyl-*t*RNA) enters the ribosome and attaches to the *m*RNA at its complementary anticodon. The peptide chain elongates when another aminoacyl-*t*RNA enters the chromosome and attaches to the *m*RNA. The peptide bond is then formed by the transfer of the peptide chain from the initial to the incoming aminoacyl-*t*RNA. The sequence is repeated until a termination codon appears in *m*RNA.

separation of fmet-*t*RNA from the protein by hydrolytic enzymes. All of these amazing, coordinated steps are accomplished at a high rate of speed—about 1 min for a 150 amino acid chain in hemoglobin and 10 to 20 seconds for a 300 to 500 amino acid chain in the species of bacteria known as *Escherichia coli* (E. coli). This mechanism of protein synthesis is illustrated in Figure 31.12.

31.11 Mutations

It is known that from time to time, a new trait appears in an individual that is not present in either parents or ancestors. These traits, which are generally the result of genetic or chromosomal changes, are called **mutations**. Some mutations are beneficial, but most are harmful and detrimental. Since mutations are genetic, they may be passed on to the next or future generations.

Mutations can occur spontaneously or may be caused by a variety of chemical agents and by various types of radiation such as X-rays, cosmic rays, and ultraviolet rays. The agent that causes the mutation is called a **mutagen**. Exposure to mutagens may produce changes in the DNA of the sperm or ova. The likelihood of such changes is increased by the intensity and length of exposure to the mutagen. Mutations may then show up as birth defects in the next generation. Common types of genetic alterations include the substitution of one purine or pyrimidine for another during DNA reproduction. Such a substitution is a change in the genetic code and causes misinformation to be transcribed by the DNA. A mutagen may alter genetic material by causing a chromosome or chromosome fragment to be added or removed.

There are hundreds of well-documented cases of mutations in bacteria, plants, and animals. In the disease known as sickle cell anemia a large number of the red blood cells form crescents (or sickle shapes) when deoxygenated, instead of the usual globular shape. This condition in the red blood cells limits the ability of the blood to transport oxygen, causes the person to be weak and unable to fight infection, and leads to a shortage of red blood cells (anemia) and, in extreme cases, to an early death.

There are 146 amino acid residues in the beta chain of hemoglobin. Sickle cell anemia is due to one misplaced amino acid in this structure. There is a valine residue in the place of the normal glutamic acid residue in position 6 from the N-terminal end:

$\overset{1}{\text{Val}}$-His-Leu-Thr-Pro-$\overset{5}{\text{Glu}}$-Glu-Lys-Ser-$\overset{10}{\text{Ala-}}$ Normal hemoglobin

$\overset{1}{\text{Val}}$-His-Leu-Thr-Pro-$\overset{5}{\text{Val}}$-Glu-Lys-Ser-$\overset{10}{\text{Ala-}}$ Sickled hemoglobin

Sickle cell anemia is an inherited disease, which is transmitted from parent to child. It affects about 10% of the Black population in the United States. Many people are carriers of the gene and are not affected by it until a situation occurs where, through exhaustion, they experience a drop in the blood level oxygen.

Phenylketonuria (PKU) is a genetic disorder that results in mental retardation. It is caused by a lack of the enzyme phenylalanine hydroxylase. Without this enzyme, the body is not able to convert the essential amino acid phenylalanine to tyrosine. This causes the concentration of phenylalanine to build up in the blood. The phenylalanine then is converted to other compounds such as phenylpyruvic acid ($C_6H_5CH_2COCOOH$), which in turn causes injury to the nervous system and inhibits brain cell development.

Phenylpyruvic acid is eliminated in the urine, where it can be easily detected. If diagnosed early enough, PKU can be prevented in children by feeding them a diet free of phenylalanine until the nervous system is developed. The diet can be terminated at about 6 years of age. Tests for phenylpyruvic acid are routinely performed on infants in most areas of the United States. Phenylketonuria affects about 1 out of 10,000 people.

Questions

A. *Review the meanings of the new terms introduced in this chapter.*
   1. Nucleoprotein
   2. Nucleic acid
   3. Nucleoside
   4. Nucleotide
   5. Ribonucleic acid (RNA)
   6. Deoxyribonucleic acid (DNA)
   7. Mitosis
   8. Meiosis
   9. Codon
   10. Transcription
   11. Mutation
   12. Mutagen

B. *Review questions.*
   1. Write the names and structural formulas for the five nitrogen bases found in nucleotides.
   2. What is the difference between a nucleoside and a nucleotide?

Nucleic Acids and Heredity 715

3. Identify the compounds represented by the following letters:
   (a) A, AMP, ADP, ATP   (b) G, GMP, GDP, GTP
4. What are the three units that make up a nucleotide?
5. Write structural formulas for the substances represented by:
   (a) A   (c) ADP   (e) dGTP
   (b) AMP   (d) ATP   (f) CDP
6. What is the major function of ATP in the body?
7. What are the principal structural differences between DNA and RNA?
8. Draw the structure for a three-nucleotide segment of RNA.
9. Show by structural formulas the hydrogen bonding between adenine and uracil.
10. Briefly describe the structure of DNA as proposed by Watson and Crick.
11. What is meant by the term *complementary bases*?
12. Give the analytical data that supported the concept of complementary bases proposed by Watson and Crick.
13. Explain why the ratio of thymine to adenine in DNA is 1:1, but the ratio of thymine to guanine is not necessarily 1:1.
14. Why is DNA considered to be the genetic substance of life?
15. Briefly describe the process of DNA replication.
16. What is the role of DNA in the genetic process?
17. What is the genetic code?
18. Why are at least three nucleotides needed for one unit of the genetic code?
19. List the three kinds of RNA and identify the role of each.
20. What is a codon? An anticodon?
21. Explain the relationship between codons and anticodons.
22. There are 146 amino acid residues in the beta chain of hemoglobin. How many nucleotides in *m*RNA are needed to designate this chain?
23. In RNA does the guanine content have to be equal to the cytosine content? Explain. Do they have to be equal in DNA? Explain.
24. Starting with DNA, briefly outline the biosynthesis of proteins.
25. Explain the role of *N*-formylmethionine in protein synthesis.
26. A segment of a DNA strand consists of GCTTAGACCTGA.
    (a) What is the nucleotide order in the complementary *m*RNA?
    (b) What is the anticodon order in *t*RNA?
    (c) What is the sequence of amino acids coded by the DNA?
27. What will the anticodon be in *t*RNA if the codon in *m*RNA is the following?
    (a) GUC   (c) CGA   (e) CCA
    (b) ACC   (d) UUU
28. Complete hydrolysis of RNA would yield what compounds?
29. What is a mutation?
30. Why do mutations occur?
31. Briefly discuss sickle cell anemia and its causes.
32. Briefly discuss phenylketonuria and its causes.
33. Why do you think that RNA has a ribose unit in it instead of a deoxyribose unit?
34. Which of these statements are correct?
    (a) Adenine and guanine are both purine bases and are found in both DNA and RNA.
    (b) The ratio of adenine to thymine and guanine to cytosine in DNA is about 1:1.
    (c) DNA and RNA are responsible for transmitting genetic information from parent to daughter cells.
    (d) DNA is a polymer made from nucleotides.
    (e) Codons are combinations of the base units in a *t*RNA molecule.

(f) The double-helix structure of DNA is held together by peptide linkages.
(g) Amino acids are linked to *t*RNA by an ester bond.
(h) The nucleotide adenosine monophosphate contains adenine, D-ribose, and a phosphate group.
(i) The letters ATP stand for adenine triphosphate.
(j) Thymine and uracil are both complementary bases to adenine.
(k) Messenger RNA is a transcribed section of DNA.
(l) The ratio of adenine to guanine and thymine to cytosine is 1:1 in DNA.
(m) Genetic information is based on the nucleotide sequence in DNA.
(n) Humans have 46 pairs of chromosomes.
(o) In mitosis, the sperm cell and the egg cell, each with 23 chromosomes, unite to give a new cell containing 46 chromosomes.
(p) The genetic code consists of triplets of nucleotides; each triplet codes an amino acid.
(q) Transfer RNA carries the code for the synthesis of proteins.

32 Digestion; Carbohydrate Metabolism; Hormones; Vitamins

After studying Chapter 32 you should be able to:

1. Understand the terms listed in Question A at the end of the chapter.
2. List the five principal digestive juices and where they originate in the body.
3. List the principal enzymes of the various digestive juices.
4. Give the main digestive functions of each of the five principal digestive juices.
5. List the classes of products formed when carbohydrates, fats, and proteins are digested.
6. Identify the role played by specific enzymes in digestion.
7. Briefly describe the absorption of nutrients from the digestive tract into the lymph and blood.
8. Discuss the function of NAD, NADP, and FAD in biological processes.
9. Explain the role of glycogen in regulating the concentration of glucose in the blood.
10. Explain the role of adrenalin, insulin, and glucagon in the control of blood glucose concentration.
11. Explain the difference between anabolic and catabolic processes.
12. Give an overall description of the Embden–Meyerhof pathway (anaerobic) for metabolism of glucose.
13. Give an overall description of the citric acid cycle (aerobic).
14. Compare the amounts of energy formed in the anaerobic and aerobic processes.
15. Describe the function of acetyl-CoA in the metabolic process.
16. Show how carbohydrates, fats, and proteins are interrelated in metabolic processes.
17. Describe the formation of glucose by gluconeogenesis.
18. Explain the function of hormones in the body.
19. Explain the function of vitamins in the body.
20. Explain how hormones and vitamins differ in origin.
21. State, in a general way, the results of vitamin deficiencies.
22. List the blood glucose levels that are considered to be normal, hyperglycemic, and hypoglycemic.
23. Explain the renal threshold.

24. Predict what might happen to the blood glucose level if a large overdose of insulin is taken.
25. Describe the glucose tolerance test and how it ties in with the condition of diabetes mellitus.

32.1 Introduction

Only simple inorganic substances are necessary to support green plants. These substances are water, carbon dioxide, nitrogen (or nitrogen compounds), and a variety of about 30 inorganic ions including Mg^{2+}, Ca^{2+}, Fe^{2+}, K^+, Na^+, PO_4^{3-}, SO_4^{2-}, NO_3^-, and Cl^-. These materials are absorbed directly by the plant from the environment. Energy is also required for all living things. Green plants absorb light energy and, through the process of photosynthesis, utilize it in making complex organic compounds. Nearly all nonphotosynthetic organisms, both plant and animal, are ultimately dependent on photosynthesis for their existence. This is because they obtain energy from organic nutrients or foods that are traceable, sometimes through a long food chain, back to photosynthetic organisms.

Cells are the functional units of living organisms, and the chemistry of life occurs almost entirely within cells. Animals require nutrients or foods that must enter the cells in order to be utilized. Foods, classified as carbohydrates, fats, and proteins, consist of large complex molecules. These molecules are generally too large to pass through the membranes enclosing the cells. They must be broken down to smaller molecules by the digestive process in order to be absorbed into the cells.

32.2 Human Digestion

digestion

The human digestive tract is shown diagrammatically in Figure 32.1. Although food is broken up mechanically by chewing in the mouth and by a kind of churning action in the stomach, digestion is a chemical process. **Digestion** is a series of enzymatically catalyzed reactions by which large molecules are hydrolyzed to molecules small enough to be absorbed through the intestinal walls. Foods are digested to smaller molecules according to this general scheme:

Carbohydrates ⟶ Monosaccharides

Fats ⟶ {Fatty acids, Glycerol, Mono- and diesters of glycerol}

Proteins ⟶ Amino acids

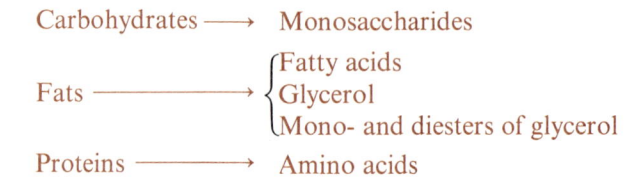

Food passes through the human digestive tract in this sequence: mouth → esophagus → stomach → small intestine (duodenum, jejunum, and ileum) → large intestine (see Figure 32.1). Five principal digestive juices (or fluids) enter

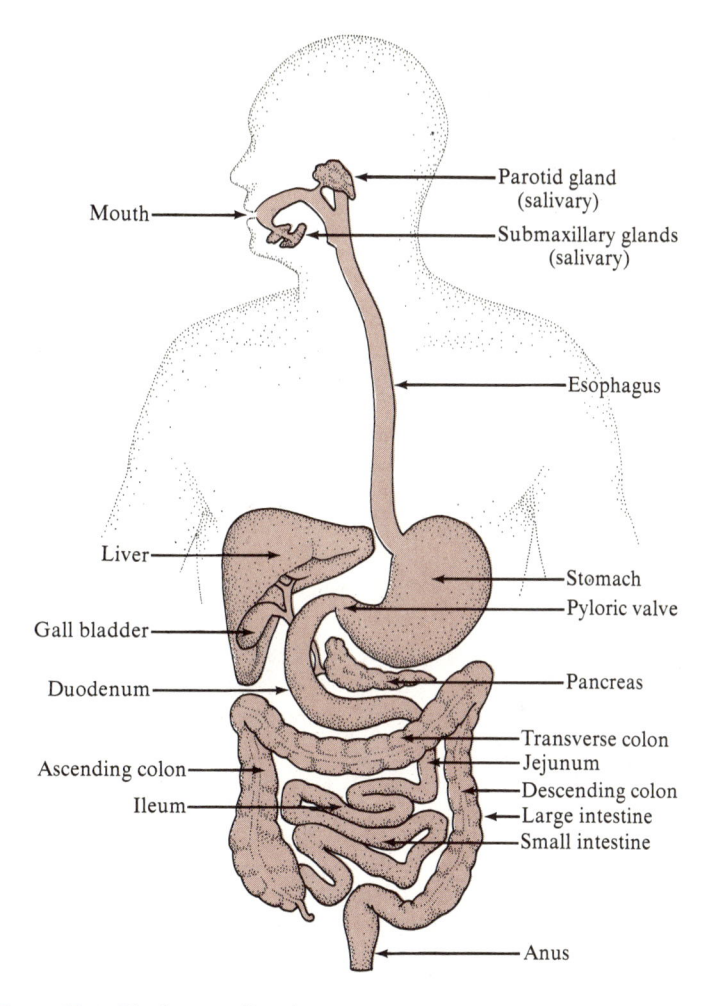

Figure 32.1. The human digestive tract.

the digestive tract at various points:

1. Saliva from three pairs of salivary glands in the mouth
2. Gastric juice from glands in the walls of the stomach
3. Pancreatic juice, which is secreted by the pancreas and enters the duodenum through the pancreatic duct
4. Bile, which is secreted by the liver and enters the duodenum via a duct from the gall bladder
5. Intestinal juice from glands in the duodenum

The main functions and principal enzymes found in each of these fluids are summarized in Table 32.1. The important digestive enzymes occur in gastric, pancreatic, and intestinal juices. An outline of the digestive process follows. Detailed accounts of the various stages of digestion are to be found in biochemistry and physiology texts.

Table 32.1. Digestive fluids.

Fluid (volume produced daily)	Source	Principal enzymes and/or function
Saliva (1000–1500 ml)	Salivary glands	Lubricant, aids chewing and swallowing; also contains salivary amylase (ptyalin), which begins the digestion of starch
Gastric juice (2000–3000 ml)	Glands in stomach wall	Pepsin, rennin, gastric lipase; pepsin catalyzes partial hydrolysis of proteins to proteoses and peptones in the stomach
Pancreatic juice (500–800 ml)	Pancreas	Trypsinogen, chymotrypsinogen, procarboxypeptidase (converted after secretion to trypsin, chymotrypsin, and carboxypeptidase, respectively); amylopsin (α-amylase), steapsin (a lipase)
Bile (500–1000 ml)	Liver	Contains no enzymes, but contains bile salts, which aid digestion by emulsifying lipids; serves to excrete cholesterol and bile pigments derived from hemoglobin
Intestinal juice	Glands in duodenum	Contains a variety of finishing enzymes—sucrase, maltase, and lactase for carbohydrates; aminopolypeptidase and dipeptidase for final protein breakdown; intestinal lipase; nucleases and phosphatase for hydrolysis of nucleic acids

Salivary digestion. Food is chewed (masticated) and mixed with saliva in the mouth and the hydrolysis of starch begins. The composition of saliva depends on many factors—age, diet, condition of teeth, time of day, and so on. Normal saliva is about 99.5% water. Saliva also contains mucin (a glycoprotein); a number of mineral ions such as K^+, Ca^{2+}, Cl^-, PO_4^{3-}, SCN^-; and one enzyme, ptyalin. The pH of saliva ranges from slightly acid to slightly basic with the optimum pH about 6.6 to 6.8.

Mucin acts as a lubricant and facilitates the chewing and swallowing of food. The enzyme ptyalin catalyzes the hydrolysis of starch to maltose:

$$\text{Starch} + \text{Water} \xrightarrow{\text{Ptyalin}} \text{Maltose}$$

Ptyalin is inactivated at a pH of 4.0, so it has very little time to act before the food reaches the highly acid stomach juices.

Saliva is secreted continuously, but the rate of secretion is greatly increased by the sight and odor, or even the thought, of many foods. The mouth-watering effect of the sight or thought of pickles is familiar to most of us. This is an example of a *conditioned reflex*.

Gastric digestion. When food is swallowed, it passes through the esophagus to the stomach. In the stomach, mechanical action continues;

food particles are reduced in size and are mixed with gastric juices until a material of liquid consistency, known as *chyme*, is obtained.

Gastric juice is a clear, pale yellow, acid fluid having a pH of about 1.5–2.5. It contains hydrochloric acid; the mineral ions Na^+, K^+, Cl^-, and some phosphates; and the digestive enzymes pepsin, rennin, and lipase. The flow of gastric juice is accelerated by conditioned reflexes and by the presence of food in the stomach. The secretion of the hormone *gastrin* is triggered by food entering the stomach. This hormone, which is produced by the gastric glands, is absorbed into the bloodstream and returned to the stomach wall where it stimulates the secretion of additional gastric juice. Control of gastric secretion by this hormone is an example of one of the many chemical control systems that exist in the body.

The chief digestive function of the stomach is the partial digestion of protein. The principal enzyme of gastric juice is pepsin, which digests protein. The enzyme is secreted in an inactive form called pepsinogen, which is activated by hydrochloric acid to pepsin. Pepsin catalyzes the hydrolysis of proteins to fragments called proteoses and peptones, which are still fairly large molecules. Pepsin splits the peptide bonds adjacent to only a few amino acid residues, particularly tyrosine and phenylalanine.

$$\text{Protein} + \text{Water} \xrightarrow{\text{Pepsin}} \text{Proteoses} + \text{Peptones}$$

Rennin is present in the gastric juice of children and is known as a milk-curdling, or protein-coagulation, enzyme. It coagulates casein, the protein of milk. In the presence of calcium ions, also in milk, an insoluble precipitate is formed. This curdling action of rennin is beneficial, since it prevents casein from leaving the stomach too rapidly, allowing time for digestion by proteolytic enzymes.

The third enzyme in the stomach, gastric lipase, is a fat-digesting enzyme. Its action in the stomach is only slight and not important because the acidity is too high for lipase activity. Food may be retained in the stomach for as long as six hours. It then passes through the pyloric valve into the duodenum.

Intestinal digestion. The next section of the digestive tract, the small intestine, is where most of the digestion occurs. The stomach contents are first made alkaline by secretions from the pancreatic and bile ducts. The pH of the pancreatic juice is 7.5 to 8.0 and the pH of bile is 7.1 to 7.7. The shift in pH is necessary because the enzymes of the pancreatic and intestinal juices are active only in an alkaline medium. Enzymes that digest all three kinds of food—carbohydrates, fats, and proteins—are secreted by the pancreas. Pancreatic secretion is stimulated by hormones that are secreted into the bloodstream by the duodenum and the jejunum.

The enzymes occurring in the small intestine include pancreatic amylases (diastase), which hydrolyze most of the starch to maltose; and carbohydrases (α-amylase, maltase, sucrase, and lactase), which complete the hydrolysis of disaccharides to monosaccharides. The proteolytic enzymes trypsin and chymotrypsin attack proteins, proteoses, and peptones, hydrolyzing them to dipeptides. Then the peptidases, carboxypeptidase, aminopeptidase, and dipeptidase complete the hydrolysis of proteins to amino acids. Pancreatic lipases catalyze the hydrolysis of almost all fats. Fats are split into fatty acids, glycerol, and mono- and diesters of glycerol by these enzymes.

The liver is another important organ in the digestive system. A fluid known as bile is produced by the liver and stored in the gall bladder, a small organ located on the surface of the liver. When food enters the duodenum, the gall bladder contracts and the bile enters the duodenum through a duct which is also used by the pancreatic juice. In addition to water, the major constituents of the bile are bile acids (as salts), bile pigments, inorganic salts, and cholesterol. The bile acids are steroid monocarboxylic acids, two of which are shown below:

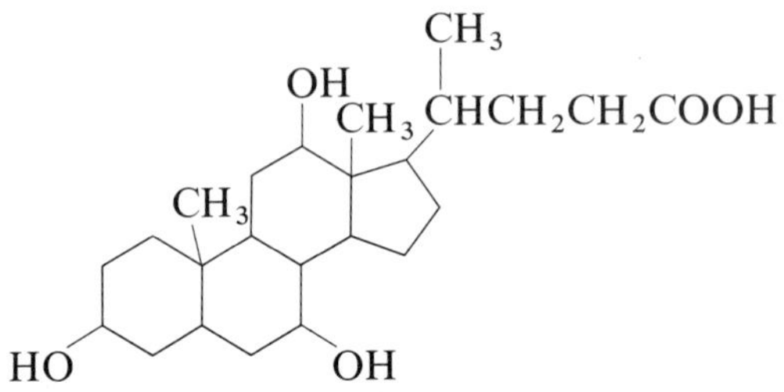

Cholic acid

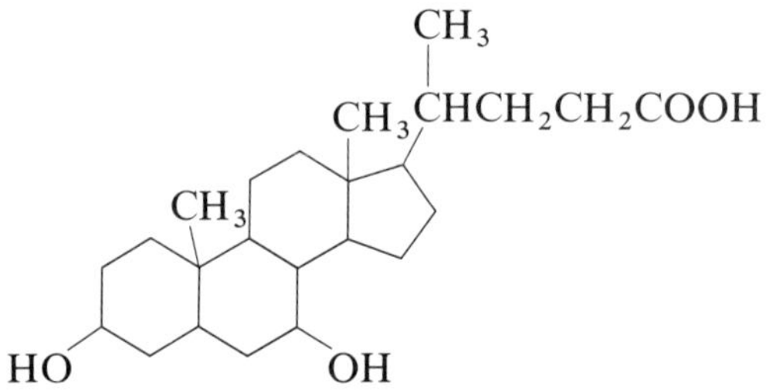

Chenodeoxycholic acid

The bile acids are synthesized in the liver from cholesterol, which is also synthesized in the liver. The presence of bile in the intestine is important for the digestion and absorption of fats. When released into the duodenum, the bile acids emulsify the fats, allowing them to be hydrolyzed by the pancreatic lipases. About 90% of the bile salts are readsorbed in the lower part of the small intestine and are transported back to the liver and used again.

Most of the digested food is absorbed while in the small intestine. Undigested and indigestible material passes from the small intestine to the large intestine, where it is retained for varying periods of time before final elimination as feces. Additional chemical breakdown, sometimes with the production of considerable amounts of gases, is brought about by bacteria (or rather by bacterial enzymes) in the large intestine. For a healthy person this additional breakdown is not important from the standpoint of nutrition since absorption of nutrients does not occur from the large intestine. However, large amounts of water, partly from digestive juices, are absorbed from the large intestine so that the contents become more solid before elimination.

32.3 Absorption

absorption

In order for digested food to be utilized in the body, it must pass from the intestine into the blood and lymph systems. The process by which digested foods pass through the membrane linings of the small intestine and enter the blood and lymph is called **absorption**. Absorption is a very complicated process and we will consider only an overview of it.

After the food you have eaten is digested, the body must absorb billions upon billions of nutrient molecules into the bloodstream. The absorption system is in the walls of the small intestine, which, upon microscopic inspection, is seen to be wrinkled into hundreds of folds. These folds are covered with thousands of small projections called *villi*. Nutrient molecules are absorbed through the membranes of the villi and pass into the blood or lymph circulatory systems of the body. A nutrient molecule may enter either the lymph or the blood, but all finally end up in the bloodstream. Water-soluble nutrients such as monosaccharides, glycerol, short-chain fatty acids, amino acids, and minerals enter directly into the bloodstream. Fat-soluble nutrients such as long-chain fatty acids and monoglycerides first enter the lymph fluid and then enter the bloodstream where the two fluids come together.

An important factor in the absorption process is that the walls of the intestine are selectively permeable; that is, they prevent the passage of most large molecules but allow the passage of smaller molecules. For example, polysaccharides, disaccharides, and proteins are not ordinarily absorbed, but in general monosaccharides and amino acids are readily absorbed.

Simple diffusion probably has a role in absorption, but it cannot account for the entire process. In diffusion, the ions or molecules move from a region of high concentration to one of lower concentration. Diffusion thus is a spontaneous process and requires no energy input.

In the overall process of absorption at least some substances move against a concentration gradient; that is, from a region of lower concentration to one of higher concentration. For example, it has been shown that both glucose and L-amino acids are absorbed through membranes from a lower to a higher concentration level. Such absorption from lower to higher concentration cannot occur by simple diffusion but does occur by a process called active transport.

active transport

Active transport is the movement of ions or molecules through a membrane against a concentration gradient. This kind of movement requires input of metabolic energy; therefore, unlike simple diffusion, it is not a spontaneous process. Active transport occurs only in living tissue, but it is not limited to a role in the absorption of digested food. Active transport is quite generally involved in establishing and maintaining the necessary concentration gradients across the various membranes found in all kinds of living organisms.

32.4 NAD, NADP, and FAD

Many biological processes involve an enzyme-catalyzed oxidation or reduction of a substrate. In these reactions coenzymes that either remove hydrogen atoms (oxidation) or donate hydrogen atoms (reduction) to the

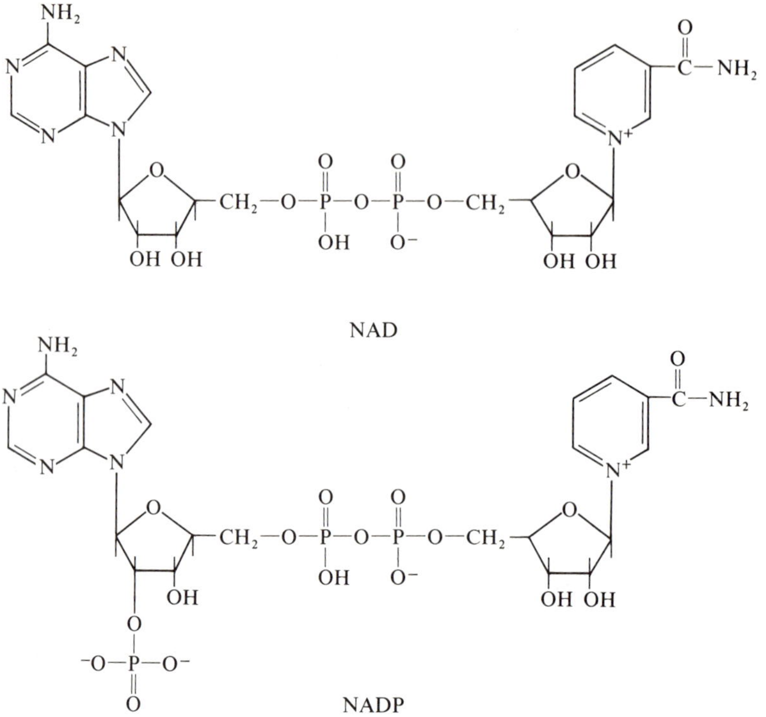

Figure 32.2. Structures of nicotinamide adenine dinucleotide (NAD) and nicotinamide adenine dinucleotide phosphate (NADP).

substrate are involved. These are classified as *dehydrogenases*. Two such coenzymes are *nicotinamide adenine dinucleotide* (*NAD*) and *nicotinamide adenine dinucleotide phosphate* (*NADP*). Their structures are shown in Figure 32.2. The difference between these two compounds is that NADP has a phosphate group on carbon 2 of the ribose attached to the adenine.

Two forms of NAD and NADP exist—the oxidized and the reduced forms. The oxidized structures are normally written as NAD$^+$ and NADP$^+$; the reduced forms as NADH and NADPH. The change in structure from oxidized to reduced form involves only the nicotinamide part of the molecules.

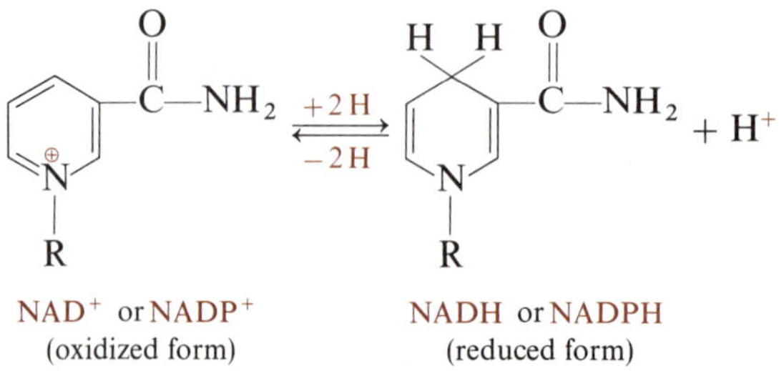

NAD$^+$ or NADP$^+$ (oxidized form) ⇌ NADH or NADPH (reduced form) + H$^+$

The main source of the nicotinamide part of these molecules is from niacin or nicotinic acid (one of the B vitamins), which can be converted to nicotinamide.

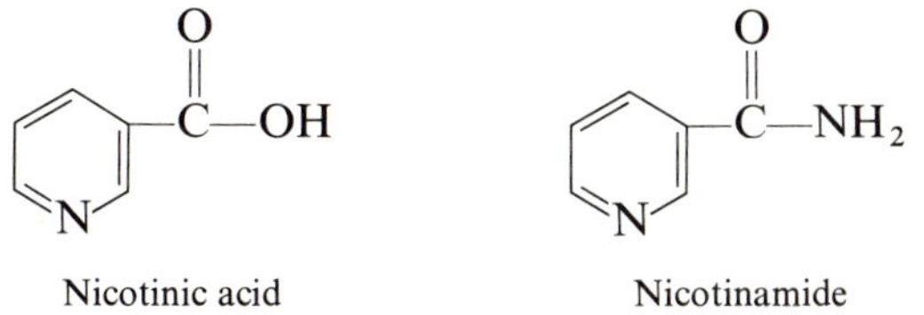

Nicotinic acid   Nicotinamide

Since the body cannot synthesize nicotinic acid, it must be included as part of the regular diet. A small amount of nicotinic acid can be formed from tryptophan, but tryptophan is also essential to the diet.

Another dehydrogenase coenzyme involved in biological oxidation–reduction reactions is *flavin adenine dinucleotide* (*FAD*). The formula for FAD, shown in Figure 32.3, is somewhat different from NAD, but its function is similar—to accept two hydrogen atoms from a substrate.

$$\text{FAD} \underset{-2\,H}{\overset{+2\,H}{\rightleftharpoons}} \text{FADH}_2$$

The structure modification from oxidized to reduced form that occurs only in the flavin part of FAD is also shown in Figure 32.3.

All of these dehydrogenases are key substances in biochemistry; their functions in specific metabolic processes will be shown in the sections that follow.

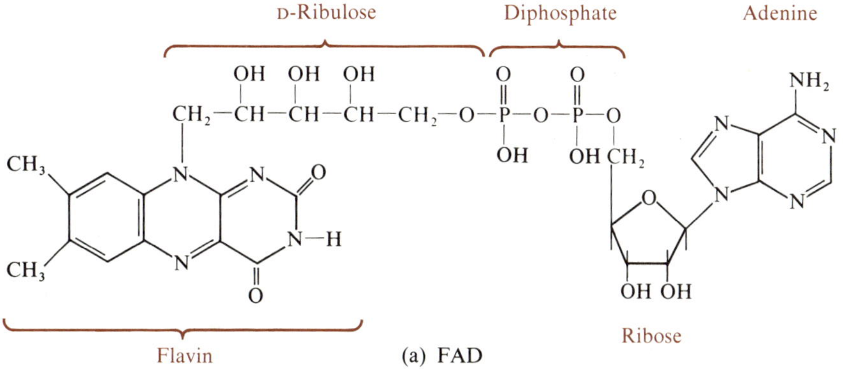

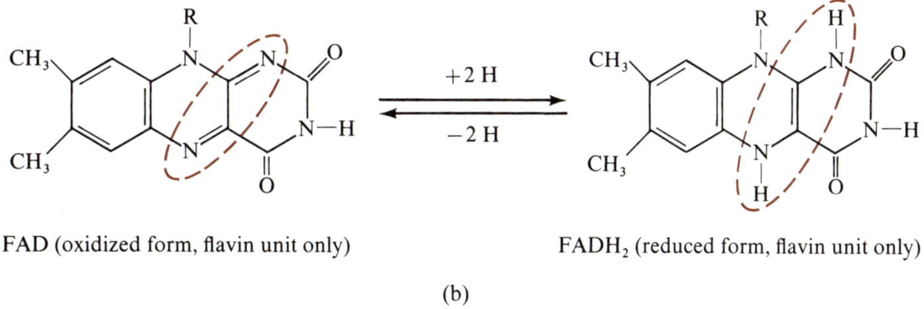

Figure 32.3. (a) Structure of flavin adenine diribonucleotide (FAD) and (b) structure changes from oxidized to reduced form of FAD.

32.5 Metabolism

metabolism

anabolism

catabolism

Metabolism is defined as the sum of all the chemical changes that occur in a living organism. Metabolism has two contrasting aspects—anabolism and catabolism. **Anabolism** is constructive metabolism or the process by which simple substances are synthesized into the complex materials found in living tissue. **Catabolism** is destructive metabolism and refers to the breaking down of complex substances to simpler substances. Anabolic reactions involve reduction and require energy (are endergonic), whereas catabolic reactions involve oxidation and are energy-releasing (exergonic).

Although we study metabolic processes separately, it is important to remember that they are all interrelated and that the simple units of carbohydrates, fats, and proteins are generally interconvertible.

32.6 Carbohydrate Metabolism

What is the source of the energy needed by living organisms to carry on the multitude of activities that distinguish living from inanimate matter? The answer is "the sun." But how does an organism, such as a human being, utilize energy from sunlight? This question is not easily answered.

Our utilization of solar energy is closely tied to the anabolism and catabolism of glucose as represented by these two generalized equations.

$$6\,CO_2 + 6\,H_2O + 673 \text{ kcal} \xrightarrow[\text{Light}]{\text{Chlorophyll}} C_6H_{12}O_6 + 6\,O_2 \qquad (1)$$

$$C_6H_{12}O_6 + 6\,O_2 \xrightarrow{\text{Enzymes}} 6\,CO_2 + 6\,H_2O + 673 \text{ kcal} \qquad (2)$$

Equation (1) summarizes the production of glucose and oxygen by an endothermic process, photosynthesis. Photosynthesis is the process whereby plants utilize energy from the sun to synthesize carbohydrates from carbon dioxide and water. The prefix *photo-* comes from the Greek word *photos*, meaning "light." Energy from sunlight is absorbed by the green pigment chlorophyll, converted to potential chemical energy, and stored in the chemical bonds of glucose and oxygen. Chlorophyll and specific enzymes are required, and many intermediate reactions occur. Since glucose is made from simpler substances, photosynthesis is an example of anabolism or constructive metabolism.

The 1961 Nobel Prize in chemistry was awarded to the American chemist Melvin Calvin (1911–   ) of the University of California at Berkeley for his work in photosynthesis. Calvin and his coworkers used radioactive carbon tracer techniques to establish the detailed and complicated sequence of chemical reactions that occur in the overall process of photosynthesis.

Equation (2) represents the oxidation of glucose and corresponds to the reversal of the overall photosynthesis reaction. Glucose actually can be burned in oxygen to produce carbon dioxide, water, and heat energy. But in the living cell, the oxidation of glucose does not proceed directly to carbon dioxide and water. Instead, like photosynthesis, the overall process proceeds by a series of enzyme-catalyzed intermediate reactions. These intermediate steps channel some of the liberated energy into uses other than heat production. Specifically, a portion of the energy is stored in the chemical bonds of adenosine triphosphate (ATP). This chemical energy is subsequently converted to other forms needed by the living organism.

32.7 Overview of Human Carbohydrate Metabolism

This section is an overview of carbohydrate metabolism, details of which are considered in the following sections. Monosaccharides are absorbed into the blood and transported to the liver. The key monosaccharide is glucose; other monosaccharides are converted to glucose in the liver. Glucose circulates in the blood and, in a healthy person, is maintained within certain well-defined concentration limits. Excess blood glucose is converted to glycogen in the liver and in muscle tissue. Glycogen is a reserve or storage carbohydrate; that is, it can be hydrolyzed to replace depleted glucose supplies in the blood. The synthesis of glycogen from glucose is called **glycogenesis**; the hydrolysis or breakdown of glycogen to glucose is known as **glycogenolysis**.

glycogenesis

glycogenolysis

In muscle, glucose is converted to lactic acid in an *anaerobic* (oxygen not required) sequence of reactions known as the Embden–Meyerhof metabolic pathway. Part of the lactic acid produced is sent to the liver and is converted to glycogen. The balance is converted to carbon dioxide and water via the citric acid cycle, an *aerobic* (free or respiratory oxygen required) sequence. The citric acid cycle, also known as the Krebs cycle and tricarboxylic acid cycle, is much more efficient at ATP production than the Embden–Meyerhof pathway. Consequently, the citric acid cycle is the major vehicle for obtaining metabolic energy from carbohydrates.

The diagram in Figure 32.4 represents a general overview of carbohydrate metabolism. The entire process requires specific enzymes at each reaction stage and is controlled by chemical regulatory compounds called hormones. Other special essential substances called vitamins are required for the function of some of the enzymes involved.

Figure 32.5 is a simplified diagram showing the interrelationships of carbohydrates, fats, and proteins in the body. The hub of this cycle is *acetyl-CoA* (CoA is an abbreviation for coenzyme A, which is described later in the chapter). Acetyl-CoA undergoes transformations leading to (1) the synthesis of amino acids; (2) the synthesis of fatty acids; and (3) irreversible oxidation to carbon dioxide and water with the release of energy, which is stored in the form of ATP. According to the diagram, your body can synthesize body fat from glucose, but you cannot manufacture glucose from fat. This is because CoA is specific and reacts in only one direction to form acetyl-CoA. All three nutrients are convertible to body fat, which means that even if your diet excludes fat, you can still deposit body fat by eating too much carbohydrate and protein.

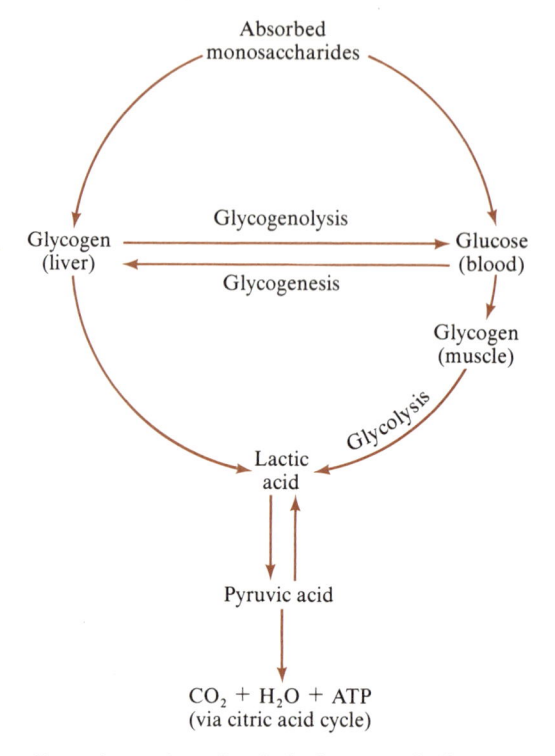

Figure 32.4. General overview of carbohydrate metabolism.

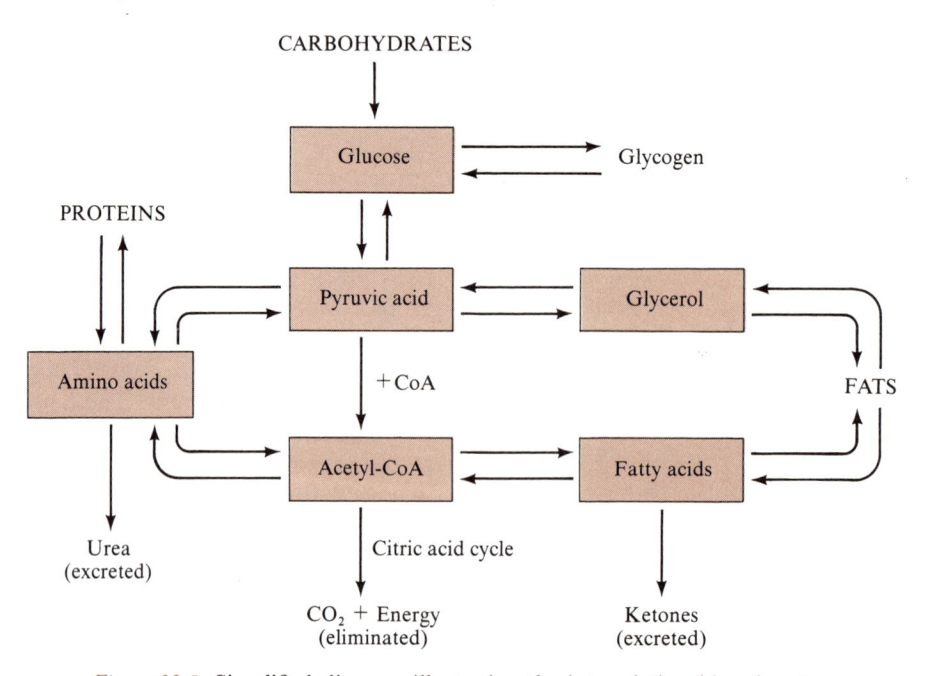

Figure 32.5. Simplified diagram illustrating the interrelationship of carbohydrates, fats, and proteins in the body.

32.8 Anaerobic Sequence

In the absence of oxygen, glucose in living cells may be converted to lactic acid (in muscle) or to alcohol (in yeast). The sequence of reactions involved is similar in different kinds of cells. At least a dozen reactions, several different enzymes, ATP, and inorganic phosphate (P_i) are required. Such a sequence of reactions from a particular reactant to end products is called a *metabolic pathway*. The key role of glucose in metabolism was established by tracing metabolic pathways.

The anaerobic conversion of glucose to pyruvic acid is a primary metabolic reaction sequence that occurs in nearly all living cells. It is known as the Embden–Meyerhof pathway. There are two different end products in this metabolic pathway, depending upon whether the process occurs in muscle tissue or in yeast cells. In muscle tissue glucose is converted to lactic acid by a process called **glycolysis**. In yeast cells glucose is converted to ethyl alcohol and carbon dioxide by a process called *fermentation*. These sequences are identical up to the formation of pyruvic acid (see Figure 32.6).

margin note: glycolysis

Glucose + ATP ⟶ Glucose-1-PO$_4$ ⇌ Glucose-6-PO$_4$

Fructose-1,6-diPO$_4$ ←—(+ATP)— Fructose-6-PO$_4$

Glyceraldehyde-3-PO$_4$ ⇌ Dihydroxyacetone-PO$_4$

⟶ (with 2 P_i) (2) 1,3-Diphosphoglyceric acid

+ 2 ADP ⟶ (2) 3-Phosphoglyceric acid + 2 ATP

⇌ (2) 2-Phosphoglyceric acid

⇌ Phosphoenol pyruvic acid

+ 2 ADP ⟶ (2) Pyruvic acid + 2 ATP

Muscle tissue: NADH + H+ ⟶ 2 Lactic acid
Yeast cells: ⟶ 2 Ethanol + 2 CO$_2$

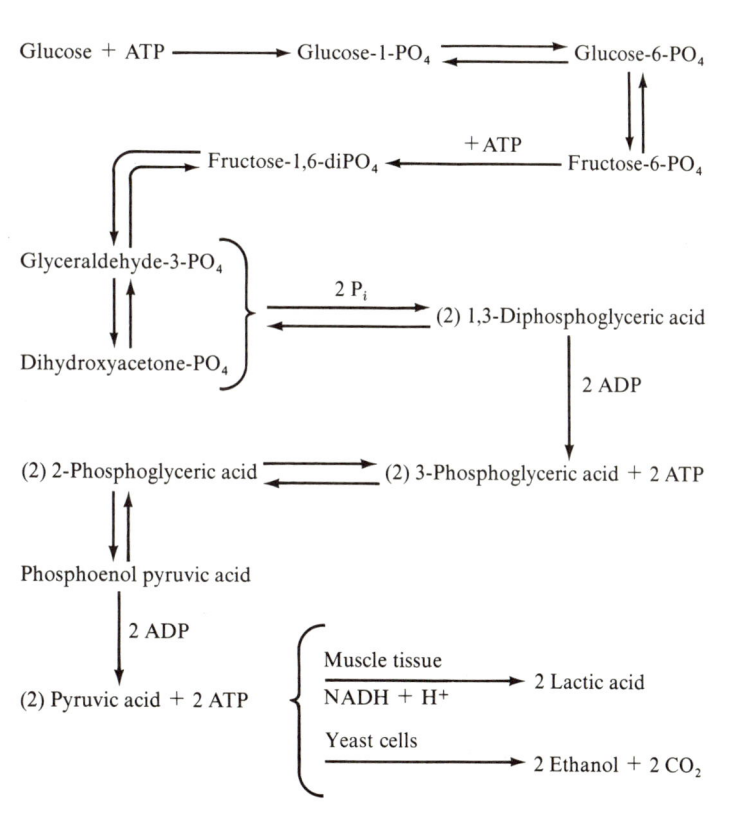

Figure 32.6. Conversion of glucose to lactic acid via the Embden–Meyerhof pathway (anaerobic sequence).

Cells obtain chemical energy in the form of ATP via the Embden–Meyerhof metabolic pathway. For each molecule of glucose converted to lactic acid or to ethyl alcohol, two molecules of ATP are consumed and four molecules of ATP are produced. Therefore, a net gain of two molecules of ATP is realized. ATP contains potential energy in a form that cells are able to utilize. Thus, the Embden–Meyerhof pathway provides a means whereby cells can obtain the energy needed to sustain life in the absence of free oxygen.

One theory of biological evolution holds that the Embden–Meyerhof reaction sequence evolved very early in the development of life on earth—even before there was free oxygen in the atmosphere and, therefore, before photosynthetic organisms existed.

The net chemical reaction for the conversion of glucose to lactic acid is shown by the following equation:

$$C_6H_{12}O_6 + 2\,ADP + 2\,P_i \longrightarrow 2\,CH_3CH(OH)COOH + 2\,ATP$$

In the anaerobic process only 2 moles of ATP are obtained per mole of glucose. However, a great deal more ATP is produced by aerobic oxidation of pyruvic acid.

32.9 Citric Acid Cycle (Aerobic Sequence)

Only a small fraction of the energy potentially available from glucose is liberated during the anaerobic conversion to pyruvic acid and lactic acid (Embden–Meyerhof pathway). The lactic acid formed may be (1) circulated back to the liver and converted to glycogen at the expense of some ATP or (2) converted back to pyruvic acid and oxidized to carbon dioxide and water via the citric acid cycle.

$$\underset{\text{Pyruvic acid}}{CH_3\underset{\underset{O}{\|}}{C}-COOH} \underset{-2H}{\overset{+2H}{\rightleftarrows}} \underset{\text{Lactic acid}}{CH_3\underset{\underset{OH}{|}}{CH}COOH}$$

The citric acid cycle is a series of reactions in the respiratory chain where the acetyl group ($CH_3\overset{|}{C}{=}O$) of acetyl-CoA is oxidized to carbon dioxide and water and where much metabolic energy is liberated and stored in ATP. The citric acid cycle takes place in the mitochondria of the cells. The mitochondria are relatively large granular particles and are often called the powerhouses of the cell because they are the major place where energy is extracted.

The citric acid cycle is aerobic because free oxygen (from respiration) is required for the reaction sequence. Thirty additional ATP molecules are produced when two pyruvic acid molecules (obtained from a single glucose molecule) are oxidized to carbon dioxide and water. This fact demonstrates that the aerobic sequence is far more efficient than the anaerobic sequence in converting the energy available from glucose to forms that are useful to cells.

Pyruvic acid is the compound that serves as the link between the anaerobic sequence (Embden–Meyerhof pathway) and the aerobic sequence (citric acid cycle). Pyruvic acid itself does not actually enter into the citric acid cycle. Instead, it is converted to acetyl coenzyme A, which is a very complex substance, and like ATP is of great importance in metabolism. Acetyl coenzyme A consists of an acetyl group bonded to a coenzyme A group. Coenzyme A contains the following units: adenine, ribose, diphosphate, pantothenic acid, and thio-ethanol amine. Coenzyme A is abbreviated as CoA. Acetyl coenzyme A is abbreviated as acetyl-CoA or acetyl-SCoA.

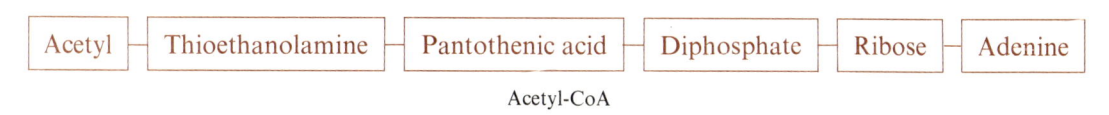

| Acetyl | Thioethanolamine | Pantothenic acid | Diphosphate | Ribose | Adenine |

Acetyl-CoA

The acetyl group is the group that is actually oxidized in the citric acid cycle. This group is attached to the large carrier molecule as a thio ester—that is, by an ester linkage in which oxygen is replaced by sulfur:

$$CoA-S-\underset{\substack{\uparrow \\ \text{Thioester linkage}}}{\overset{\overset{O}{\|}}{C}}-CH_3$$

Carrier group, Acetyl group

Acetyl-CoA not only is a key component of carbohydrate metabolism but, as we have seen, also serves to tie the metabolism of fats and that of certain amino acids to the citric acid cycle.

The sequence of reactions involved in the citric acid cycle is shown in Figure 32.7. Although a rather complicated series of reactions is represented, the net effect is the complete oxidation of the acetyl group from acetyl-CoA with the production of ATP. The overall process is represented by these two general equations:

$$CH_3\overset{\overset{O}{\|}}{C}-COOH + CoASH + [O] \longrightarrow \underset{\text{Acetyl-CoA}}{CH_3\overset{\overset{O}{\|}}{C}-SCoA} + CO_2 + H_2O + 3\,ATP$$

$$\underset{\text{Acetyl-CoA}}{CH_3\overset{\overset{O}{\|}}{C}-SCoA} + 4\,[O] \longrightarrow 2\,CO_2 + H_2O + CoASH + 12\,ATP$$

The citric acid cycle was elucidated by Hans A. Krebs (1900–   ), a British biochemist; thus, it is also called the Krebs cycle. For his studies in intermediary metabolism, Krebs shared the 1953 Nobel Prize in medicine and physiology with Fritz A. Lipmann (1899–   ), an American biochemist, who discovered coenzyme A.

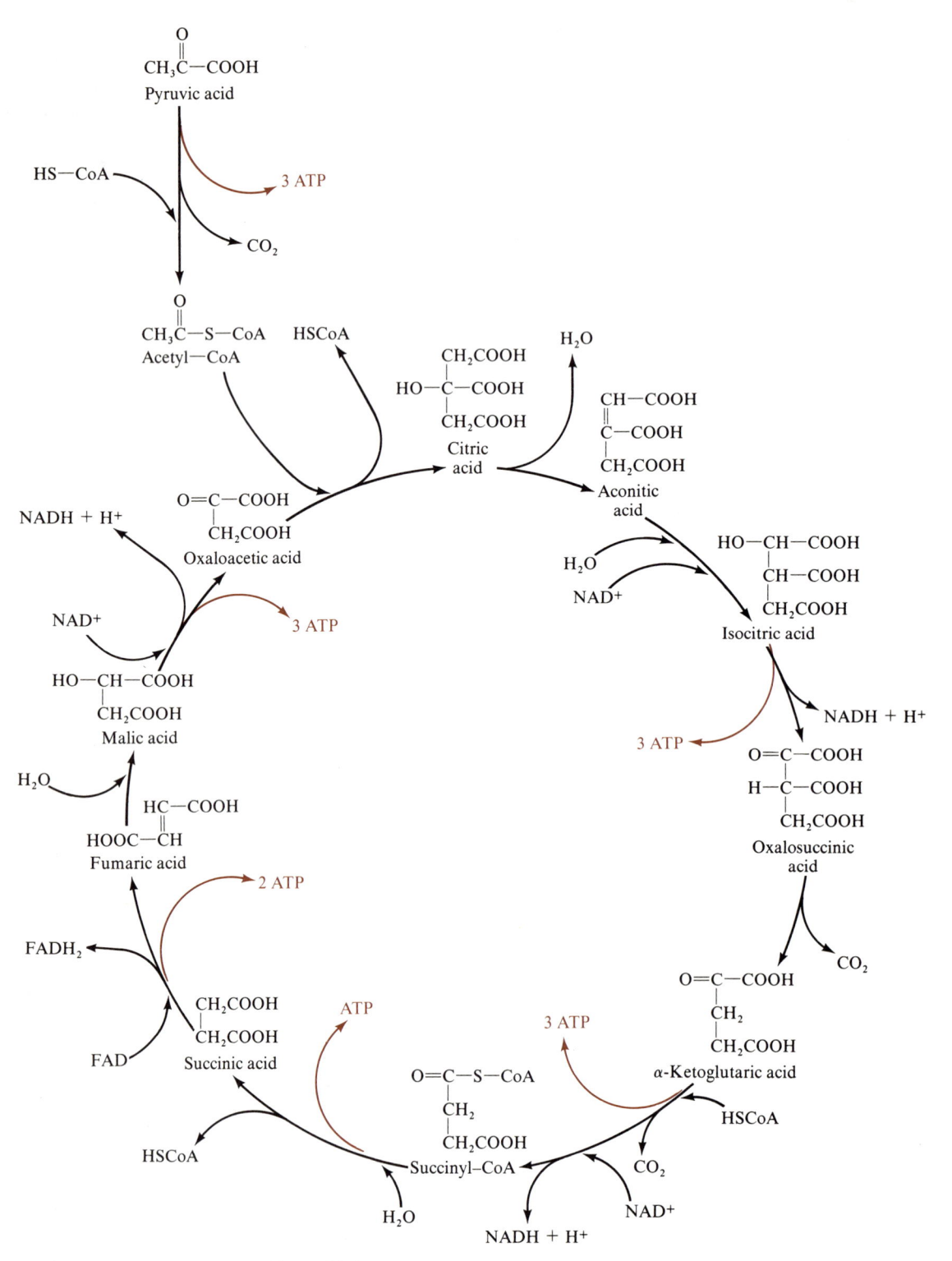

Figure 32.7. The citric acid cycle (Krebs cycle).

32.10 Gluconeogenesis

gluconeogenesis

A supply of glucose is always needed by the body, especially for the brain and the nervous system. But the amount of glucose and glycogen present in the body is sufficient to last for only about 4 hr at normal metabolic rates. When glucose is not being absorbed in sufficient quantities, the needed glucose must be obtained from noncarbohydrate sources within the body. The formation of glucose from noncarbohydrate sources is called **gluconeogenesis**.

The glucose formed during gluconeogenesis comes from certain amino acids and from the glycerol of fats. About half of the amino acids are capable of being converted to glucose. The amino acids first undergo *deamination* (loss of amino groups) and then are broken down to pyruvic acid. The pyruvic acid enters the citric acid cycle as oxaloacetic acid, which is converted to phosphoenol–pyruvic acid. From this point the process follows the Embden–Meyerhof pathway (Figure 32.6) in reverse, leading to the formation of glucose.

Gluconeogenesis takes place primarily in the liver and also in the kidneys. These organs have special enzymes that catalyze the reverse reaction in the Embden–Meyerhof pathway.

$$\text{Amino acids} \rightarrow \text{Pyruvic acid} \rightarrow \text{Phosphoenol-pyruvic acid} \xrightarrow[\text{Embden–Meyerhof pathway}]{\text{Reverse}} \text{Glucose}$$

32.11 Hormones

hormone

Hormones are chemical substances that act as control or regulatory agents in the body. They help to regulate overall physiological processes such as digestion, metabolism, growth, and reproduction. For example, the concentration of glucose in the blood is maintained within definite limits by the action of hormones. Hormones are secreted by the endocrine, or ductless, glands directly into the bloodstream and are transported to various parts of the body to exert specific control functions. The endocrine glands include the thyroid, parathyroid, pancreas, adrenal, pituitary, ovaries, testes, placenta, and certain portions of the gastrointestinal tract. A hormone produced by one species is usually active in some other species. For example, the insulin used to treat diabetes mellitus in humans is obtained from the pancreas of animals slaughtered in meat packing plants. Hormones are often referred to as the chemical messengers of the body. They do not fit into any single chemical structural classification. Many are proteins or polypeptides, some are steroids, some are phenol or amino acid derivatives; examples are shown in Figure 32.8. Since a lack of any hormone may produce serious physiological disorders, many of them are produced synthetically or are extracted from their natural sources and made available for medical use.

Like the vitamins, hormones are generally needed in only minute amounts. Concentrations range from 10^{-6} to 10^{-12} M. Unlike vitamins, which must be supplied in the diet, the necessary hormones are produced in the body of a healthy person. A number of hormones and their functions are listed in Table 32.2.

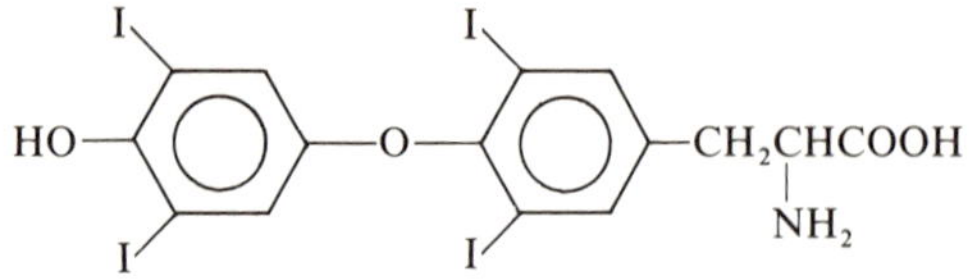

734 Chapter 32

Thyroxin: HO-C6H2I2-O-C6H2I2-CH2CH(NH2)COOH

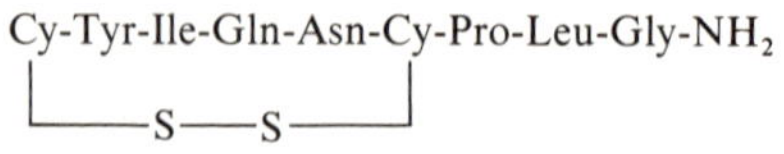

Oxytocin: Cy-Tyr-Ile-Gln-Asn-Cy-Pro-Leu-Gly-NH2
              |___S—S___|

Glucagon:
1   10   15   20   29
His-Ser-Gln-Gly-Thr-Phe-Thr-Ser-Asp-Tyr-Ser-Lys-Tyr-Leu-Asp-Ser-Arg-Arg-Ala-Gln-Asp-Phe-Val-Gln-Tyr-Leu-Met-Asn-Thr

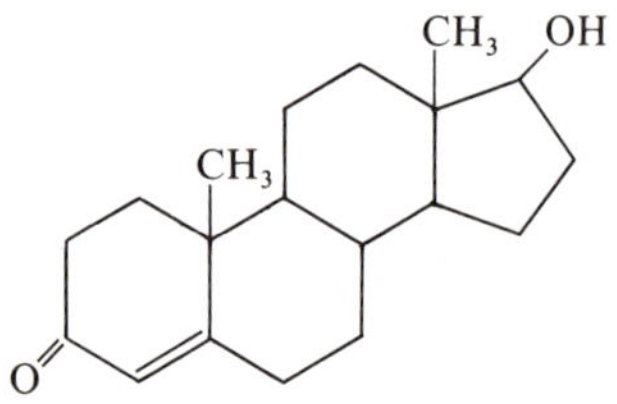

Testosterone

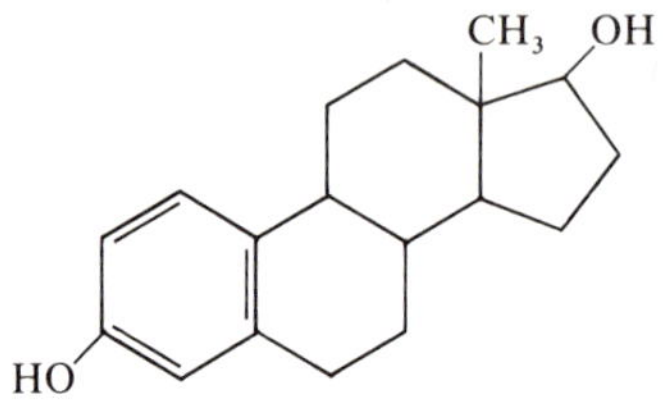

Estradiol (Estrogen)

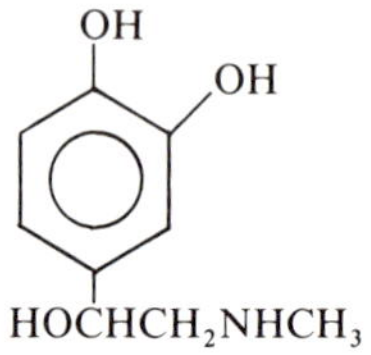

Epinephrine (Adrenalin): HOC6H3(OH)-CH(OH)CH2NHCH3

Figure 32.8. Structure of selected hormones: Thyroxin is produced in the thyroid gland; oxytocin is a polypeptide produced in the posterior lobe of the pituitary gland; glucagon is a polypeptide and is produced in the pancreas; testosterone and estradiol are steroid hormones—testosterone is produced in the testes and estradiol in the ovaries; adrenalin is produced in the adrenal glands.

Table 32.2. Selected list of hormones and their functions.

Hormone	Source	Principal functions
Insulin	Pancreas	Controls blood sugar level and storage of glycogen
Glucagon	Pancreas	Stimulates conversion of glycogen to glucose; increases blood sugar level
Oxytocin	Pituitary gland	Stimulates contraction of the uterine muscles and secretion of milk by the mammary glands
Vasopressin	Pituitary gland	Controls water excretion by the kidneys; stimulates constriction of the blood vessels
Growth hormone	Pituitary gland	Stimulates growth
Adrenocorticotrophic (ACTH)	Pituitary gland	Stimulates the adrenal cortex, which, in turn, releases several steroid hormones
Prolactin	Pituitary gland	Stimulates milk production by mammary glands shortly after birth of a baby

Table 32.2 (Continued)

Hormone	Source	Principal functions
Epinephrine (Adrenalin)	Adrenal glands	Stimulates rise in blood pressure, acceleration of heartbeat, decreased secretion of insulin, and increased blood sugar
Cortisone	Adrenal glands	Helps control carbohydrate metabolism, salt and water balance, formation and storage of glycogen
Thyroxine and triiodothyronine	Thyroid gland	Increase the metabolic rate of carbohydrates and proteins
Calcitonin	Thyroid gland	Prevents the rise of calcium in the blood above the required level
Parathormone	Parathyroid gland	Regulates the metabolism of calcium and phosphate in the body
Gastrin	Stomach	Stimulates secretion of gastric juices
Secretin	Duodenum	Stimulates secretion of pancreatic juice
Estrogen	Ovaries	Stimulates development and maintenance of female sexual characteristics
Progesterone	Ovaries	Stimulates female sexual characteristics and maintains pregnancy
Testosterone	Testes	Stimulates development and maintenance of male sexual characteristics

32.12 Vitamins

vitamin

Vitamins are a group of naturally occurring organic compounds that are essential for normal nutrition. Animals maintained on a diet consisting of proteins, fats, carbohydrates, and the necessary minerals and water are not able to sustain life. Although vitamins are required in only minute amounts, normal nutrition, growth, and development are not possible without them. The minimum daily adult requirement may be as much as 100 milligrams for some vitamins and as little as 0.1 microgram for others. The human body is unable to synthesize these substances and is dependent on vitamins being supplied in the diet. A substance that functions as a vitamin for one species does not necessarily function as a vitamin for another species.

A prolonged lack of vitamins in the diet leads to vitamin deficiency diseases such as beriberi, pellagra, pernicious anemia, rickets, and scurvy. Some of these vitamin deficiency diseases may be corrected by feeding supplementary amounts of vitamins. However, it is especially important for young children to have sufficient vitamins for proper growth and development. For example, it is difficult to correct distorted bone structures that have developed due to a lack of vitamin D. Most vitamins are manufactured synthetically and are available as dietary supplements, although a balanced diet should supply all the necessary vitamins.

Table 32.3. Some of the most important vitamins. Alternate names are given in parentheses.

Vitamin	Important dietary sources	Some deficiency symptoms
Vitamin A (Retinol)	Green and yellow vegetables, butter, eggs, nuts, cheese, fish liver oil	Poor teeth and gums, night blindness
Vitamin B_1 (Thiamin)	Meat, whole-grain cereals, liver, yeast, nuts	Beriberi (nervous system disorders, heart disease, fatigue)
Vitamin B_2 (Riboflavin)	Meat, cheese, eggs, fish, meat products, liver	Sores on the tongue and lips, bloodshot eyes, anemia
Vitamin B_6 Pyridoxine)	Cereals, liver, meat, fresh vegetables	Skin disorders (dermatitis)
Vitamin B_{12} (Cyanocobalamin)	Meat, eggs, liver, milk	Pernicious anemia
Vitamin C (Ascorbic acid)	Citrus fruits, tomatoes, green vegetables	Scurvy (bleeding gums, loose teeth, swollen joints, slow healing of wounds, weight loss)
Vitamin D (Calciferol)	Egg yolk, milk, fish liver oils; formed from provitamin in the skin when exposed to sunlight	Rickets (low blood calcium level, soft bones, distorted skeletal structure)
Vitamin E (Tocopherol)	Widely distributed in foods, meat, egg yolk, wheat germ oil, green vegetables	Not definitely known in humans
Vitamin K (Phylloquinone)	Eggs, liver, green vegetables; produced in the intestines by bacterial reactions	Blood is slow to clot (antihemorrhagic vitamin)
Niacin (Nicotinic acid and amide)	Meat, yeast, whole wheat	Pellagra (dermatitis, diarrhea, mental disorders)
Biotin (Vitamin H)	Liver, yeast, egg yolk	Skin disorders (dermatitis)
Folic acid	Liver extract, wheat germ, yeast, green leaves	Macrocytic anemia, gastrointestinal disorders

The chemical composition of vitamins varies greatly; some are relatively simple substances while others are extremely complex. On the basis of solubility characteristics, vitamins are generally divided into two groups: fat-soluble and water-soluble. The fat-soluble vitamins include A, D, E, and K. Those classified as water-soluble are vitamin C and the B complex, which includes about a dozen different compounds. One of the B vitamins, pantothenic acid, is a component of coenzyme A and therefore takes part in a great many metabolic reactions. Table 32.3 lists some of the important vitamins, their main food sources, and deficiency symptoms. The structural formulas of several vitamins are shown in Figure 32.9.

Most of the vitamins are known to function as coenzymes. For example, the B complex vitamins riboflavin, niacin, and pantothenic acid have coenzyme functions in glucose metabolism.

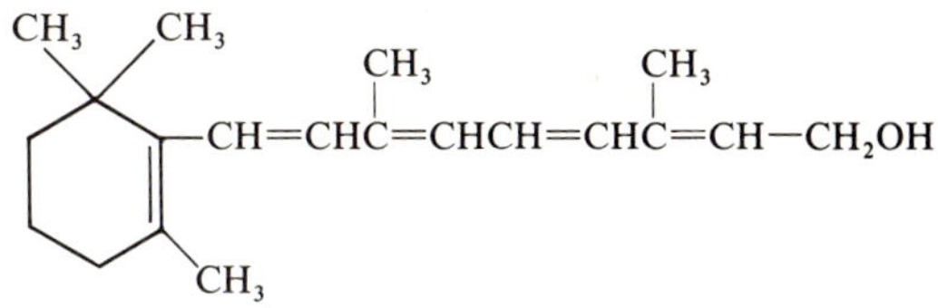

Vitamin A
(Retinol)

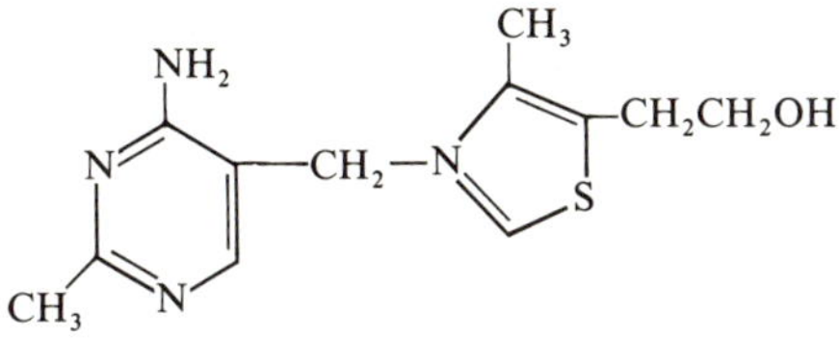

Vitamin B_1
(Thiamin)

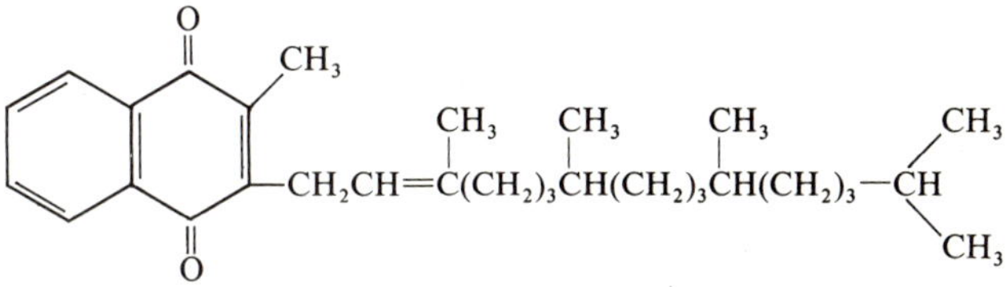

Vitamin K_1
(Phylloquinone)

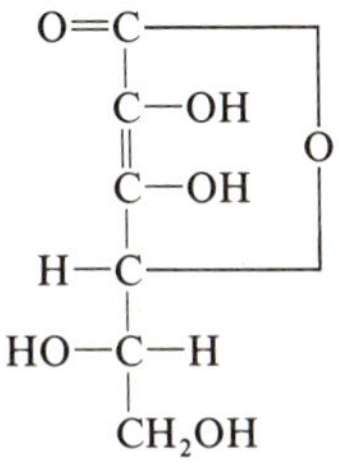

Vitamin C
(Ascorbic acid)

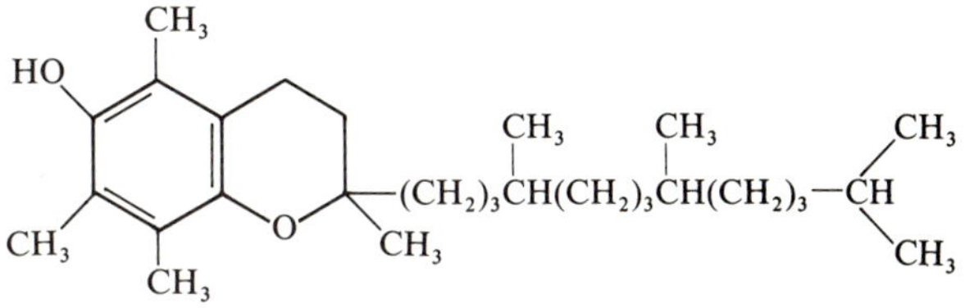

Vitamin E
(α-Tocopherol)

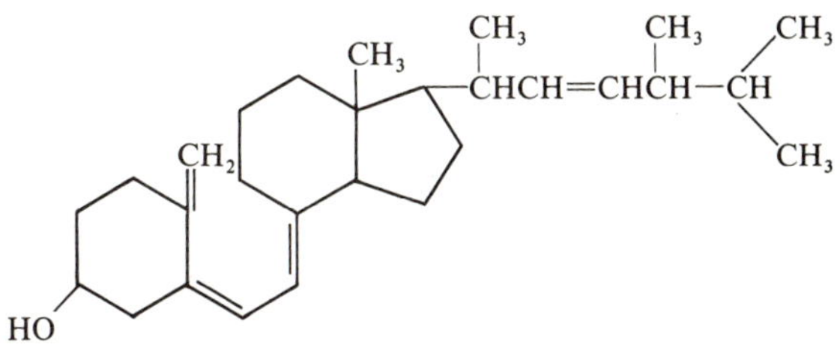

Vitamin D_2
(Calciferol)

Figure 32.9. The structures of selected vitamins.

32.13 Glucose Concentration in the Blood

Glucose concentrations average about 70 to 90 milligrams per 100 millilitres of blood under normal fasting conditions; that is, when no nourishment has been taken for several hours. For most people a normal fasting condition exists before eating breakfast. After a meal or ingestion of carbohydrates, the glucose concentration rises above the normal level and a condition of **hyperglycemia** exists. The **renal threshold** is the concentration of a substance in the blood above which the kidneys begin to excrete that substance in the urine. The renal threshold for glucose is about 140 to 170 milligrams per 100 millilitres of blood. Glucose excreted by the kidneys can be detected in the urine by a test for reducing sugars (for example, Benedict's test; see Section 29.11). When the glucose concentration of the blood is below the normal fasting level, **hypoglycemia** exists (see Figure 32.10).

hyperglycemia

renal threshold

hypoglycemia

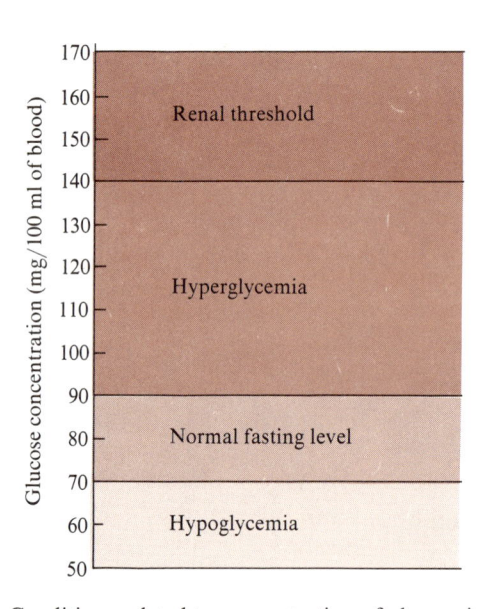

Figure 32.10. Conditions related to concentration of glucose in the blood.

Glucose concentration in the blood is under the control of various hormones. These hormones act as checks on one another and establish a condition of **homeostasis**; that is, of self-regulated equilibrium. Three hormones—insulin, epinephrine (adrenalin), and glucagon—are of special significance in maintaining glucose concentration within the proper limits. Insulin, secreted by the islands of Langerhans in the pancreas, acts to reduce blood glucose levels by increasing the rate of glycogen formation. Epinephrine from the adrenal glands and glucagon from the pancreas act to increase the rate of glycogen breakdown (glycogenolysis) and thereby to increase blood glucose levels. These opposing effects are summarized in this fashion.

homeostasis

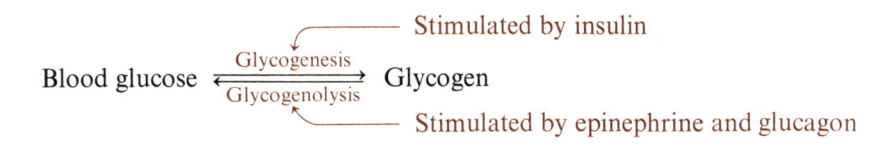

$$\text{Blood glucose} \underset{\text{Glycogenolysis}}{\overset{\text{Glycogenesis}}{\rightleftarrows}} \text{Glycogen}$$

— Stimulated by insulin (glycogenesis)
— Stimulated by epinephrine and glucagon (glycogenolysis)

During the digestion of a meal rich in carbohydrates, the blood glucose level of a healthy person rises into the hyperglycemic range. This stimulates insulin secretion, and the excess glucose is converted to glycogen, thereby returning the glucose level to normal. A large amount of ingested carbohydrates can overstimulate insulin production and thereby produce a condition of mild hypoglycemia. This in turn triggers the secretion of additional epinephrine and glucagon, and the blood glucose levels are again restored to normal. The body is able to maintain the normal fasting level of blood glucose for long periods of time without food by drawing on liver glycogen, muscle glycogen, and finally on body fat as glucose replacement sources. Thus, in a normal person neither hyperglycemia nor mild hypoglycemia has serious consequences, since the body is able to correct these conditions. However, either condition, if not corrected, can have very serious consequences. Since the brain is heavily dependent upon blood glucose for energy, hypoglycemia affects the brain and the central nervous system. Mild hypoglycemia may result in impaired vision, dizziness, and fainting spells. Severe hypoglycemia produces convulsions and unconsciousness; if prolonged, it may result in permanent brain damage and death.

Hyperglycemia may be induced by fear or anger, because the rate of epinephrine secretion is increased under emotional stress. Glycogen hydrolysis is thereby speeded up and glucose concentration levels rise sharply. This whole sequence readies the individual for the strenuous effort of either fighting or fleeing as the exigencies of the situation demand!

Diabetes mellitus is a serious metabolic disorder characterized by hyperglycemia, glucosuria (glucose in the urine), frequent urination, thirst, weakness, and loss of weight. Prior to 1921, diabetes often resulted in death. In that year Frederick Banting and Charles Best, working at the University of Toronto, discovered insulin and devised methods for extracting the hormone from animal pancreases. For his work on insulin, Banting, with J. J. MacLeod, received the Nobel Prize in medicine and physiology in 1923. Insulin is very effective in controlling diabetes. It must be given by injection because, like any other protein, it would be hydrolyzed to amino acids in the gastrointestinal tract.

People with mild or borderline diabetes may show normal fasting blood glucose levels, but they are unable to produce sufficient insulin for prompt control of ingested carbohydrates. As a result, their blood glucose rises to an abnormally high level and does not return to normal for a long period of time. Such a person has a decreased tolerance for glucose and the condition may be diagnosed by a glucose-tolerance test. After not eating for at least 12 hr, a fasting blood and urine specimen is taken for a beginning reference level. The person then drinks a solution containing 100 g of glucose (amount for adults). Blood and urine specimens are then collected at 0.5, 1, 2, and 3 hour intervals and tested for glucose content. In a normal situation, the blood glucose level

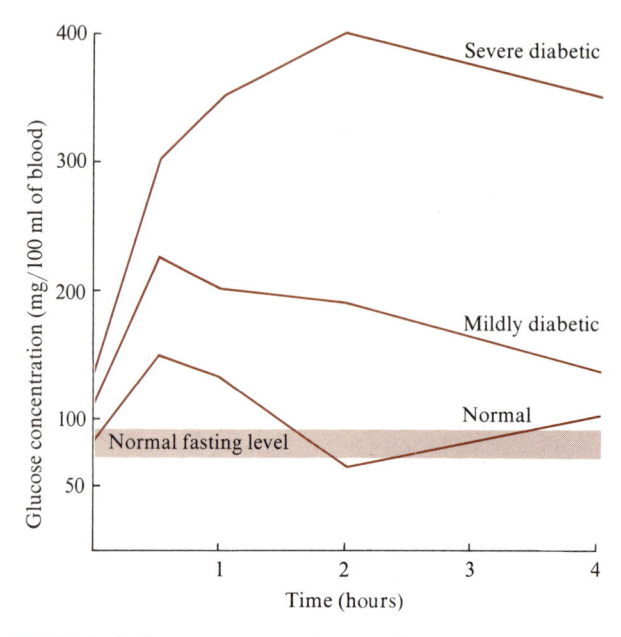

Figure 32.11. Typical responses to a glucose-tolerance test.

returns to normal in about 3 hours. Individuals with mild diabetes show a slower drop in glucose levels, while in a severe diabetic the glucose level remains high for the entire 3 hours. Responses to a glucose-tolerance test are shown in Figure 32.11.

The chemical structure of insulin has been determined (see Figure 30.3), and biologically active insulin has actually been synthesized in the laboratory. To date, no practical method has been devised whereby insulin can be synthesized in quantities large enough for medical use. In 1978, scientists at the City of Hope Medical Center in Duarte, California, announced the production of insulin identical in structure to that made in the human pancreas. Genes containing the codons required to produce the A and B polypeptide chains of insulin were made and attached to the bacteria *Escherichia coli*, which then synthesized the two chains. The A and B chains were extracted, brought together, and insulin was formed when the two chains linked together through the two disulfide groups. Commercially available insulin produced by this method is still several years away, but when accomplished it will be a tremendous boon for the millions of diabetics throughout the world. For now, insulin is still obtained as a by-product of meat-packing operations, but there is need for more insulin that can be obtained from this source.

Questions

A. *Review the meanings of the new terms introduced in this chapter.*
1. Digestion
2. Absorption
3. Active transport
4. Metabolism
5. Anabolism
6. Catabolism
7. Glycogenesis
8. Glycogenolysis

Digestion; Carbohydrate Metabolism; Hormones; Vitamins   741

9. Glycolysis
10. Glugoneogenesis
11. Hormone
12. Vitamin
13. Hyperglycemia
14. Renal threshold
15. Hypoglycemia
16. Homeostasis

B. Review questions.
1. Why must the food of higher animals be digested before it can be utilized?
2. What are the five principal digestive juices?
3. What is chyme?
4. What is the approximate pH of each of the following?
   (a) Saliva   (c) Pancreatic juice
   (b) Gastric juice   (d) Bile
5. What enzymes are present in each of the digestive juices?
6. In what parts of the digestive system are each of the following digested?
   (a) Carbohydrates   (b) Fats   (c) Proteins
7. In what parts of the digestive tract does the absorption of food occur?
8. What are the end products of carbohydrate digestion? List the specific compounds that are formed.
9. What is the digestive function of the liver?
10. List the principal digestive enzymes that act on proteins.
11. How does active transport differ from simple diffusion?
12. What is meant by the expression "moving against a concentration gradient"?
13. What changes in carbohydrate composition occur in liver glycogenesis?
14. What are the end products of the anaerobic catabolism of glucose in (a) muscle tissue and (b) yeast cells?
15. Briefly describe how energy is obtained from glucose and in what form it is stored.
16. Compare the amounts of metabolic energy (in the form of ATP) produced per glucose unit via the anaerobic sequence and the aerobic sequence.
17. Explain why it is believed that, in an evolutionary sense, the Embden–Meyerhof pathway probably developed before the citric acid cycle made its appearance.
18. Why was the citric acid cycle not involved in the metabolism of life forms in existence before the evolutionary development of photosynthesis?
19. Briefly explain what happens in the citric acid cycle.
20. What is acetyl-CoA and what is its function in metabolism?
21. What are the functions of NAD, NADP, and FAD?
22. What are the functions of hormones?
23. How does the function of hormones in the body differ from that of enzymes?
24. How do hormones differ from vitamins from the standpoints of origin and function?
25. Where in the body are the hormones produced?
26. What is the main function of vitamins?
27. Which vitamins are fat-soluble? Which are water-soluble?
28. (a) What is the range of glucose concentration in blood under normal fasting conditions?
    (b) What blood glucose concentrations are considered to be hyperglycemic? Hypoglycemic?
29. What is meant by the renal threshold?
30. What is meant by normal fasting blood sugar level?
31. Does the presence of glucose in the urine establish that the condition of diabetes mellitus is present? Explain.

32. Explain how the body maintains blood glucose concentrations within certain definite limits despite wide variations in the rates of glucose intake and utilization.
33. Why is adrenalin sometimes called the emergency or crisis hormone?
34. Explain how adrenalin and insulin maintain blood glucose within a definite concentration range.
35. Predict what might happen to blood glucose concentrations if a large overdose of insulin were taken by accident.
36. Why is insulin not effective when given orally?
37. Describe the glucose-tolerance test.
38. Which of these statements are correct?
    (a) The main purpose of digestion is to hydrolyze large molecules to small ones capable of being absorbed through the intestinal walls.
    (b) Most of the digestion of food occurs in the stomach.
    (c) Gastric juice contains hydrochloric acid and has a pH of 1.5–2.5.
    (d) Digestion in the small intestine occurs in an alkaline medium.
    (e) The function of bile acids is to emulsify carbohydrates; this allows them to be hydrolyzed to monosaccharides.
    (f) NAD and NADP are coenzymes that react as oxidizing agents in some metabolic processes.
    (g) Anabolic processes are those in which complex biological substances are broken down to simpler substances.
    (h) The Embden–Meyerhof metabolic pathway is an anaerobic sequence of reactions.
    (i) In the Embden–Meyerhof metabolic pathway, 1 mole of glucose is converted to 2 moles of lactic acid in muscle tissue.
    (j) The citric acid cycle is much more efficient in energy production than is the Embden–Meyerhof pathway.
    (k) Hormones are regulatory agents that are secreted into the stomach and intestine to control metabolism.
    (l) Vitamins must be included in the diet because they are not synthesized in the human body.
    (m) Most vitamins function as coenzymes.
    (n) A person with diabetes mellitus suffers from hypoglycemia.
    (o) Hypoglycemia can affect the brain due to low blood sugar level.
    (p) When the blood glucose level exceeds the renal threshold, glucose is eliminated through the kidneys into the urine.

33 Lipids; Metabolism of Lipids and Proteins

After studying Chapter 33 you should be able to:

1. Understand the terms listed in Question A at the end of the chapter.
2. Describe the different classes of lipids and their functions.
3. State which fatty acids commonly occur in fats and oils.
4. State which fatty acids are essential to human diets.
5. Describe the general structural make-up of phospholipids.
6. Draw the structural feature common to all steroids.
7. Draw the structures for cholesterol and several of the other common steroids.
8. Discuss atherosclerosis and the factors that affect it.
9. Discuss the digestion, absorption, and reconversion of fats to triglycerides.
10. Briefly describe Knoop's experiments on fatty acid oxidation and degradation and the conclusions derived from them.
11. Explain what is meant by beta oxidation and beta cleavage in relation to the metabolism of fatty acids.
12. Explain why fats supply more energy than carbohydrates.
13. Briefly describe the biosynthesis of fatty acids using palmitic acid as an example.
14. List the possible metabolic fates of amino acids in humans.
15. Write equations to illustrate the oxidative deamination of amino acids and the transamination of amino acids.
16. Explain how metabolism of proteins (amino acids) is tied into that of carbohydrates and fats.
17. Explain how a lack of essential amino acids in the diet affects the nitrogen balance.

33.1 Classification of Lipids

lipids

Lipids are compounds that are found in living organisms. They are arbitrarily classified on the basis of their solubility in fat solvents such as diethyl ether, benzene, chloroform, and carbon tetrachloride, and on the basis of their insolubility in water. Lipids are classified as follows:

1. Simple lipids
   a. *Fats and oils*: esters of fatty acids and glycerol

b. *Waxes*: esters of high-molecular-weight fatty acids and high-molecular-weight alcohols
2. Compound lipids
   a. *Phospholipids*: substances that yield glycerol, phosphoric acid, fatty acids, and nitrogen-containing bases upon hydrolysis
   b. *Glycolipids*: substances that yield an alcohol (other than glycerol), fatty acids, a nitrogen-containing base, and a carbohydrate upon hydrolysis
3. Steroids
   Substances possessing a 17-carbon unit structure containing four fused rings known as the steroid nucleus. Cholesterol and several hormones are in this class.
4. Miscellaneous lipids
   Substances that do not fit into the preceding classifications; these include the fat-soluble vitamins A, D, E, and K

The most abundant lipids are the fats and oils. These substances are one of the three important classes of foods. The discussion that follows is centered on fats and oils. A more complete consideration of the properties and composition of various fats and oils is given in Section 26.8.

33.2 Fats and Oils

Chemically, fats and oils are esters of glycerol and the higher-molecular-weight fatty acids. They have the general formula

$$\begin{array}{l} CH_2-O-\overset{\displaystyle \underset{\displaystyle O}{\|}}{C}-R \\ \\ CH-O-\overset{\displaystyle \underset{\displaystyle O}{\|}}{C}-R' \\ \\ CH_2-O-\overset{\displaystyle \underset{\displaystyle O}{\|}}{C}-R'' \end{array}$$

where the R's can be either long-chain saturated or unsaturated hydrocarbon radicals.

The alcohol of these esters is glycerol. Most of the fatty acids in these esters have 14 to 18 carbon atoms; the remainder have carbon atoms ranging down to 4. Fats may be considered to be formed from the alcohol, glycerol, and three molecules of fatty acids. Because there are three ester groups per molecule, fats are called triglycerides. The three R groups are generally different.

The formulas for some of the most common fatty acids found in fats are shown in Table 33.1. All these acids are straight-chain carbon compounds. Five of these fatty acids—palmitoleic, oleic, linoleic, arachidonic, and linolenic—are unsaturated, having carbon–carbon double bonds in their structures.

Table 33.1. Some naturally occurring fatty acids in fats.

Fatty acid	Number of C atoms	Formula
Saturated acids		
Butyric acid	4	$CH_3CH_2CH_2COOH$
Caproic acid	6	$CH_3(CH_2)_4COOH$
Caprylic acid	8	$CH_3(CH_2)_6COOH$
Capric acid	10	$CH_3(CH_2)_8COOH$
Lauric acid	12	$CH_3(CH_2)_{10}COOH$
Myristic acid	14	$CH_3(CH_2)_{12}COOH$
Palmitic acid	16	$CH_3(CH_2)_{14}COOH$
Stearic acid	18	$CH_3(CH_2)_{16}COOH$
Arachidic acid	20	$CH_3(CH_2)_{18}COOH$
Unsaturated acids		
Palmitoleic acid	16	$CH_3(CH_2)_5CH=CH(CH_2)_7COOH$
Oleic acid	18	$CH_3(CH_2)_7CH=CH(CH_2)_7COOH$
Linoleic acid	18	$CH_3(CH_2)_4CH=CHCH_2CH=CH(CH_2)_7COOH$
Linolenic acid	18	$CH_3CH_2CH=CHCH_2CH=CHCH_2CH=CH(CH_2)_7COOH$
Arachidonic acid	20	$CH_3(CH_2)_4(CH=CHCH_2)_4CH_2CH_2COOH$

$$
\begin{array}{c}
CH_2\text{—}O\text{—}H \quad H\text{—}O\text{—}\underset{O}{\overset{\parallel}{C}}\text{—}R \\
| \\
CH\text{—}O\text{—}H \quad H\text{—}O\text{—}\underset{O}{\overset{\parallel}{C}}\text{—}R' \\
| \\
CH_2\text{—}O\text{—}H \quad H\text{—}O\text{—}\underset{O}{\overset{\parallel}{C}}\text{—}R''
\end{array}
\longrightarrow
\begin{array}{c}
CH_2\text{—}O\text{—}\underset{O}{\overset{\parallel}{C}}\text{—}R \\
| \\
CH\text{—}O\text{—}\underset{O}{\overset{\parallel}{C}}\text{—}R' \\
| \\
CH_2\text{—}O\text{—}\underset{O}{\overset{\parallel}{C}}\text{—}R''
\end{array}
+ 3H_2O
$$

Glycerol   Fatty acids   A triglyceride

Three unsaturated fatty acids—linoleic, linolenic, and arachidonic—are essential for animal nutrition and must be supplied in the diet. Diets lacking these fatty acids lead to impaired growth and reproduction, and skin disorders such as eczema and dermatitis. But all dermatitis disorders are not due to unsaturated fatty acid deficiencies. A dermatitis disorder can be attributed to an unsaturated fatty acid deficiency if the symptoms clear up when that fatty acid is supplied in the diet.

Fats are an important food source for humans and normally account for about 25 to 50% of their caloric intake. When oxidized to carbon dioxide and water, fats supply about 9 kcal of energy per gram, which is more than twice the amount obtained from carbohydrates and proteins.

33.3 Compound Lipids

phospholipids

The **phospholipids** are a group of compounds that yield one or more fatty acid molecules, a phosphate group, and usually a nitrogenous base upon hydrolysis. Phospholipids are believed to be present in all animal and vegetable cells. In most cells the phospholipids play an important integral part in the internal and external cell membranes. They are also involved in the metabolism of other lipids and nonlipids. Most of the phospholipids that enter the blood are formed in the liver, although they are also formed to some extent in practically all cells. Representative phospholipids are described below.

Phosphatidic Acids

Phosphatidic acids are glyceryl esters of fatty acids and phosphoric acid. The phosphatidic acids are important intermediates in the synthesis of triglycerides and other phospholipids.

$$\begin{array}{c} CH_2-O-\underset{\underset{O}{\|}}{C}-R \\ | \\ R'-\underset{\underset{O}{\|}}{C}-O-CH \\ | \\ CH_2-O-\underset{\underset{OH}{|}}{\overset{\overset{O}{\|}}{P}}-OH \end{array}$$

A phosphatidic acid

Lecithins

Lecithins (phosphatidyl choline) are glyceryl esters of fatty acids, phosphoric acid, and choline. Lecithins are synthesized in the liver and are present in considerable amounts in nerve tissue and brain substance. Most commercial lecithin is obtained from soybean oil and contains palmitic, stearic, palmitoleic, oleic, linoleic, linolenic, and arachidonic acids. Lecithin is an edible and digestible emulsifying agent that is used extensively in the food industry. For example, chocolate and margarine are generally emulsified with lecithin. Lecithin is also used as an emulsifier in many pharmaceutical preparations.

$$\begin{array}{c} CH_2-O-\underset{\underset{O}{\|}}{C}-R \\ | \\ R'-\underset{\underset{O}{\|}}{C}-O-CH \qquad \overbrace{\qquad\qquad\qquad}^{\text{Choline}} \\ | \qquad\qquad\qquad CH_3 \\ CH_2-O-\underset{\underset{OH}{|}}{\overset{\overset{O}{\|}}{P}}-OCH_2CH_2\underset{\underset{OH}{|}}{N}\begin{matrix}\diagup CH_3 \\ \diagdown CH_3\end{matrix} \end{array}$$

A lecithin molecule

Cephalins

Cephalins (phosphatidyl ethanolamines) are glyceryl esters of fatty acids, phosphoric acid, and ethanolamine ($HOCH_2CH_2NH_2$). The cephalins are found in all living organisms. They are a significant constituent of the nervous tissue and brain substance. Cephalin is an essential substance in the clotting of blood. It is the predominate substance in thromboplastin, which initiates the clotting process.

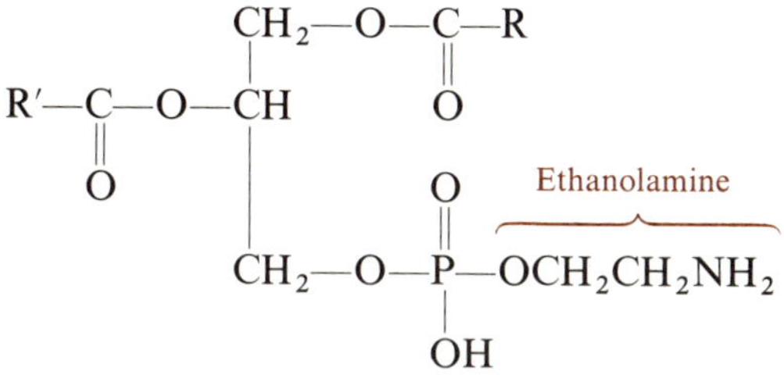

$$\begin{array}{c} CH_2-O-\overset{\overset{\displaystyle O}{\|}}{C}-R \\ R'-\underset{\underset{\displaystyle O}{\|}}{C}-O-CH \\ CH_2-O-\underset{\underset{\displaystyle OH}{|}}{\overset{\overset{\displaystyle O}{\|}}{P}}-OCH_2CH_2NH_2 \end{array}$$

Ethanolamine: $OCH_2CH_2NH_2$

A cephalin molecule

Glycolipids

Glycolipids (cerebrosides) are compounds that when hydrolyzed yield a long-chain fatty acid (18 to 26 carbons), a monosaccharide (usually galactose), and sphingosine (an unsaturated amino alcohol). Glycolipids are not esters of glycerol. They are found in many different tissues, but, as the name cerebroside indicates, they occur mainly in the brain tissue. The exact biological function of cerebrosides is unknown. But due to their location in the brain it has been suggested that their structural function is that of insulating the nerves in the myelin area. In paralyzing diseases such as multiple sclerosis there is a notable loss of cerebroside. Gaucher's disease is characterized by an enlarged liver and spleen, mental retardation, and a higher-than-normal amount of glucocerebroside in these organs and in the bone marrow.

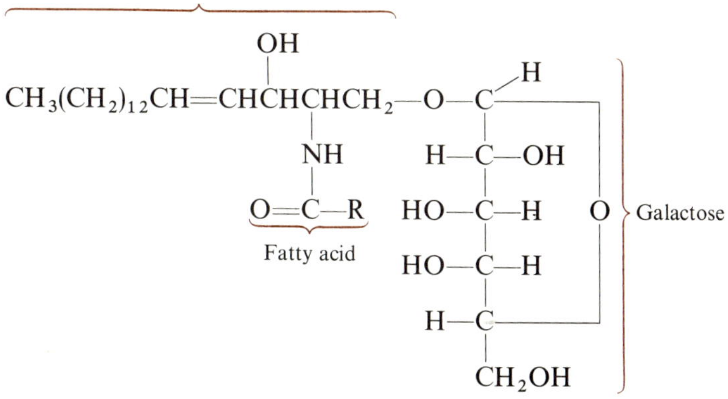

Sphingosine (an amino alcohol): $CH_3(CH_2)_{12}CH=CHCHCHCH_2-$
with OH and NH substituents; $O=C-R$ (Fatty acid) attached to NH.

Galactose portion attached via $-O-C$.

A glycolipid (galactocerebroside)

33.4 Steroids

steroids

Steroids are compounds having the steroid nucleus, which consists of four fused carbocyclic rings. This nucleus contains 17 carbon atoms in one five-membered and three six-membered rings. Modifications of this nucleus in the various steroid compounds include side chains, hydroxyl groups, carbonyl groups, ring double bonds, and so on.

Steroid ring nucleus

Steroids are closely related in structure but are highly diverse in functions. Examples of steroids and steroid-containing materials are (1) cholesterol, which is widely distributed in all cells of the body and is the most abundant steroid in the body; (2) bile salts, which aid in the digestion of fats; (3) ergosterol, a steroid present in the skin, which is converted to vitamin D by the action of ultraviolet radiation; (4) digitalis and related substances called cardiac glycosides, which are potent heart drugs; (5) the adrenal cortex hormones, which are involved in metabolism; and (6) male and female sex hormones, which control sexual characteristics and reproduction. The formulas for several steroids are given in Figure 33.1.

Cholesterol is a steroid of special interest not only because it is the precursor of many other steroids, but because of its association with atherosclerosis, a disease of the large arteries. The body synthesizes about 1 g of cholesterol per day, whereas about 0.3 g per day is ingested in the average diet. The major sources of cholesterol in the diet are meat, liver, and eggs. Cholesterol does not need to be digested to be absorbed from the intestines. A large part of the ingested cholesterol is esterified with long-chain fatty acids before being absorbed.

The biosynthesis of cholesterol begins with acetyl-CoA. Through the use of radioactive carbon-14 tracer atoms it has been shown that all 27 carbon atoms in cholesterol orginate from acetyl-CoA. The main site of cholesterol synthesis is the liver. Cholesterol is also produced in many other places such as the adrenal glands and the reproductive organs. Large concentrations of cholesterol are found in the brain, the spinal chord, and the liver.

The most abundant use of cholesterol in the body is to form cholic acid, which is further modified into the bile salts associated with the digestion of fats. A small amount of cholesterol is used by the adrenal glands to produce adrenocortical hormone (ACTH), by the ovaries to produce estrogen and progesterone, and by the testes to produce testosterone. A rather large amount of cholesterol is deposited under the skin, making the skin resistant to the absorption of water-soluble substances and helping, along with other lipids, to prevent excessive evaporation of water through the skin.

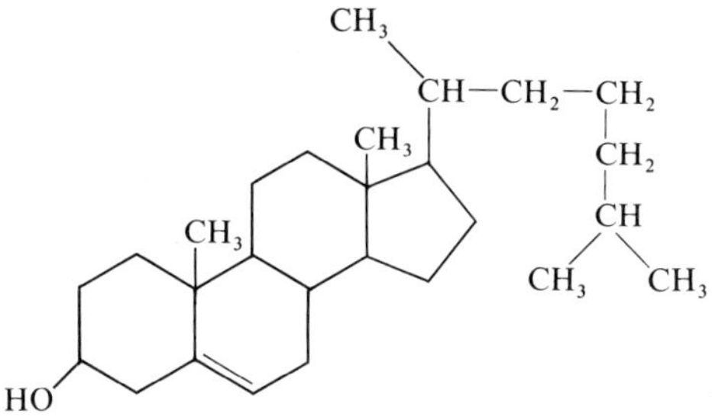

Cholesterol
(parent steroid compound—precursor for steroid hormones and bile acids)

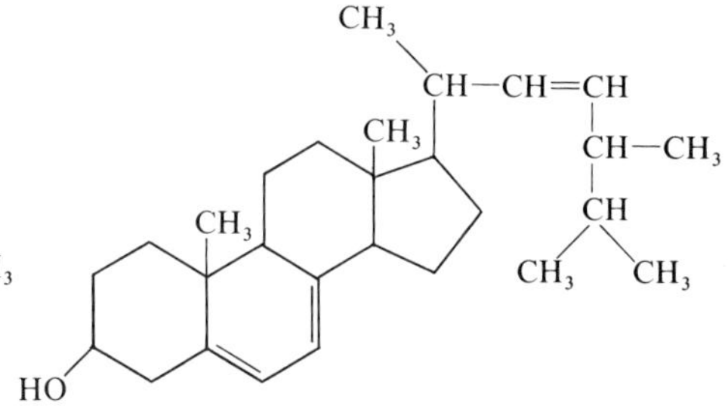

Ergosterol
(precursor of vitamin D_2)

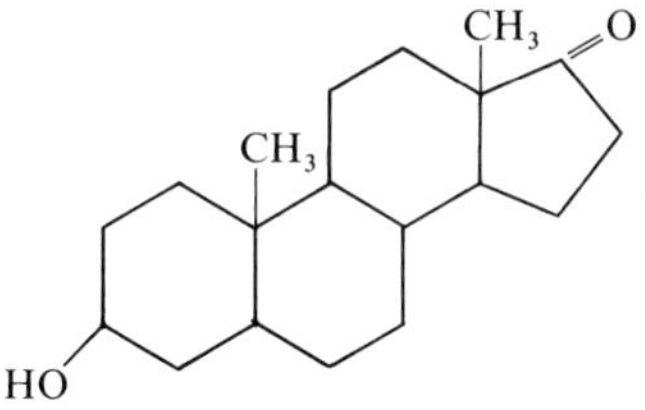

Androsterone
(male sex hormone)

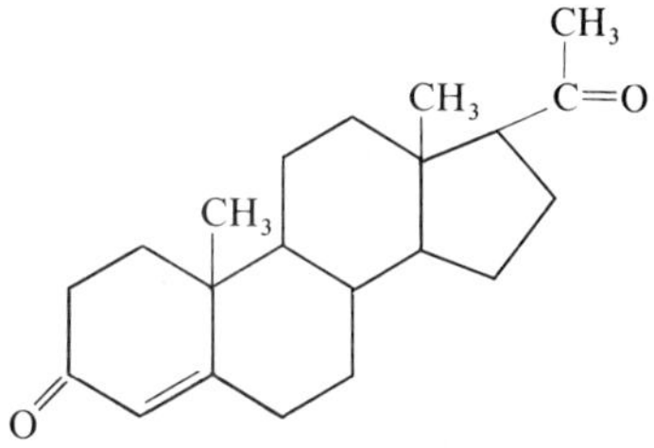

Progesterone
(progestational hormone)

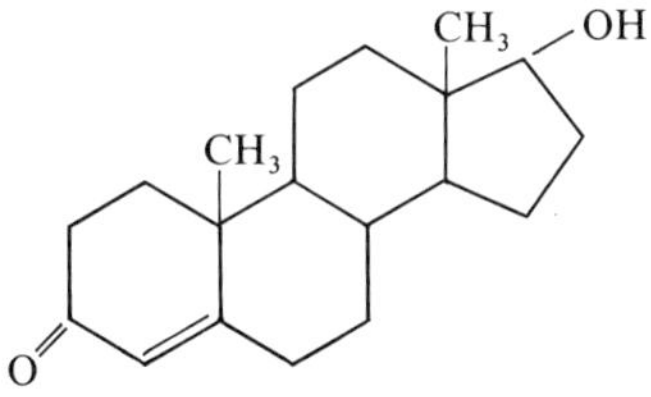

Testosterone
(male sex hormone)

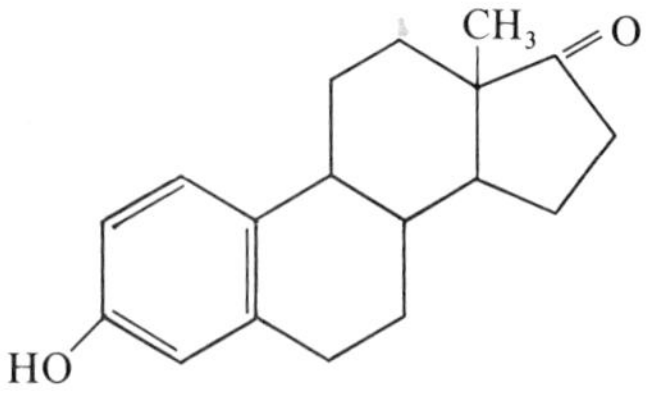

Estrone
(female sex hormone)

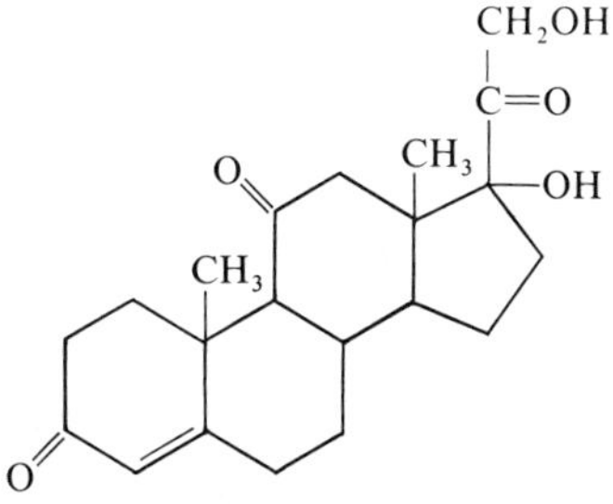

Cortisone
(adrenal corticosteroid hormone)

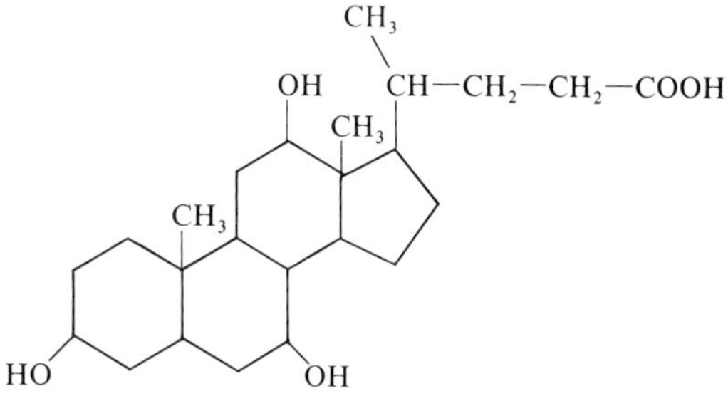

Cholic acid
(a bile acid)

Figure 33.1. Structures of selected steroids.

Cholesterol has been associated with atherosclerosis, the deposition of cholesterol and other lipids on the inner walls of the large arteries. These deposits are called *plaque*. With the accumulation of plaque, the arterial passages become progressively narrower and narrower. The walls of the arteries also lose their elasticity and are not able to expand to accommodate the volume of blood pumped by the heart. Blood pressure increases as the heart works to pump sufficient blood through the narrowed passages; this may eventually lead to a heart attack. The build-up of plaque also results in the inner walls having a rough, rather than the usual smooth surface. This condition is favorable to coronary thrombosis—heart attack due to blood clots.

A great deal remains to be learned about atherosclerosis, but certain facts relating to this disease have been established. Atherosclerosis is mainly a disease of old age, but young adults may also have plaque deposits in their arteries. This may indicate that plaque deposition is a lifetime process rather than just occurring for a few years in later life. Heredity may be a factor in atherosclerosis, and sex may also be a factor—far more men die of the disease than do women. A high-fat diet, especially one containing cholesterol, increases the chance of developing atherosclerosis. And finally, overweight people have a higher incidence of atherosclerosis than do people of normal weight.

33.5 Fat Absorption and Distribution

In the digestive process, triglycerides are hydrolyzed to diglycerides, monoglycerides, fatty acids, and glycerol. These digestion products are small enough to enter the intestinal wall where diglycerides are first hydrolyzed to monoglycerides and fatty acids. In the cells of the intestinal wall the long-chain fatty acids (greater than ten carbon atoms) are converted to an acyl-CoA by reaction with Co-ASH and ATP:

$$\underset{\text{Fatty acid}}{\text{RCOOH}} \xrightarrow[\text{Thiokinase}]{\text{Co-ASH + ATP}, \text{Mg}^{2+}} \underset{\text{Acyl-CoA}}{\text{R}\overset{\overset{\text{O}}{\|}}{\text{C}}-\text{CoA}}$$

Acyl-CoA then reacts with monoglycerides and with glycerol-3-phosphate to form triglycerides again.

The bulk of the glycerol released in digestion is converted to glycerol-3-phosphate and reconverted to triglycerides. Some free glycerol passes directly into the portal vein and is distributed to other parts of the body. The lower-molecular-weight fatty acids go directly to the portal vein and to the liver. The liver is a principal site of lipid metabolism. The digestion and resynthesis of triglycerides is summarized in Figure 33.2.

Once absorbed, the resynthesized fat forms aggregates called chylomicrons, which form water-soluble complexes with plasma proteins. These complexes are called lipoproteins. The lipoproteins are then transported by the lymphatic vessels of the abdominal cavity to the bloodstream for distribution to the rest of the body.

Lipids; Metabolism of Lipids and Proteins 751

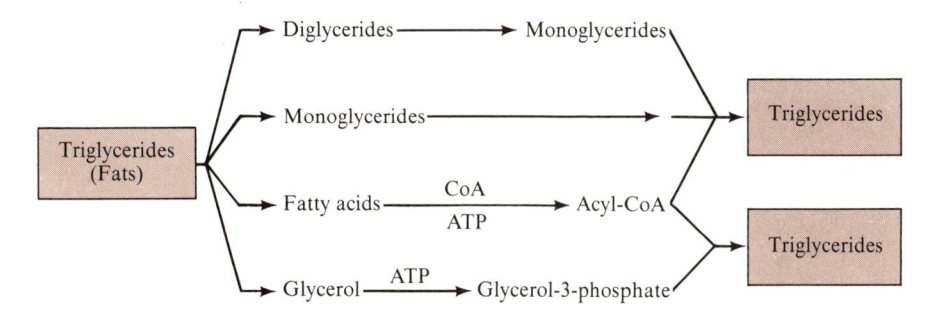

Figure 33.2. Summary of digestion and reconversion of fats to triglycerides.

Absorbed lipids are utilized mainly in these ways:

1. Oxidation for energy (heat and ATP production)
2. Synthesis of other lipids
3. Storage as reserve lipids (body fat)
4. As materials in the synthesis of other kinds of tissue (often after partial degradation)

33.6 Fatty Acid Oxidation

Fats are the most energy-rich class of foods. Palmitic acid derived from fat yields 9.36 kcal per gram when burned to form carbon dioxide and water. By way of contrast, glucose yields only 3.74 kcal per gram. From these energy values it is evident that fats are a more concentrated energy source than carbohydrates. Of course, fats are not actually burned in the body simply to produce heat. Instead, they are broken down in a series of enzyme-catalyzed reactions that also produce useful potential chemical energy stored in the form of ATP. In complete biochemical oxidation, the carbon and hydrogen of a fat are ultimately combined with oxygen (from respiration) to form carbon dioxide and water.

In 1904, Franz Knoop, a German biochemist, established that the catabolism of fatty acids involved a process whereby their carbon chains are shortened by two carbon atoms at a time. Knoop knew that animals do not metabolize benzene groups to carbon dioxide and water. Instead, the benzene nucleus remains attached to at least one carbon atom and is eliminated in the urine as a derivative of either benzoic acid or phenylacetic acid.

Benzoic acid

Phenylacetic acid

Accordingly, Knoop prepared a homologous series of straight-chain fatty acids with a phenyl group at one end and a carboxyl group at the other end. He then fed these benzene-tagged acids to test animals. Phenylaceturic acid was identified in the urine of the animals that had eaten acids with an even number of carbon atoms; hippuric acid was present in the urine of the animals that had consumed acids with an odd number of carbon atoms:

$$\text{C}_6\text{H}_5\text{-CH}_2\text{(CH}_2)_n\text{COOH}$$

→ Phenylaceturic acid: $\text{C}_6\text{H}_5\text{-CH}_2\text{-C(=O)-NHCH}_2\text{COOH}$
(metabolic end product when *n* is even)

→ Hippuric acid: $\text{C}_6\text{H}_5\text{-C(=O)-NHCH}_2\text{COOH}$
(metabolic end product when *n* is odd)

These results indicated a metabolic pathway for fatty acids in which the carbon chain is shortened by two carbon atoms at each stage.

Knoop's experiments were remarkable for their time. They involved the use of tagged molecules and served as prototypes for a great deal of the modern research that utilizes isotopes to tag molecules.

Knoop postulated that the carbon chain of a fatty acid is shortened by successive removals of acetic acid units. The process involves the oxidation of the beta carbon atom and cleavage of the carbon chain. A six-carbon fatty acid would produce three molecules of acetic acid, thus:

This C is oxidized (β), chain is cleaved:

$$\overset{\beta}{\text{CH}_3\text{CH}_2\text{CH}_2}\overset{\alpha}{\text{CH}_2\text{CH}_2\text{C}}\text{-OH} \rightarrow \text{CH}_3\text{CH}_2\text{CH}_2\text{C(=O)-OH} + \text{CH}_3\text{C(=O)-OH}$$

Caproic acid → Butyric acid + Acetic acid

$$\overset{\beta}{\text{CH}_3\text{CH}_2}\overset{\alpha}{\text{CH}_2\text{C}}\text{(=O)-OH} \rightarrow 2\,\text{CH}_3\text{C(=O)-OH}$$

The general correctness of Knoop's beta carbon atom oxidation theory has been confirmed. However, the detailed pathway for fatty acid oxidation was not established until about 50 years after his original work. The sequence of reactions involved, like those of the Embden–Meyerhof and citric acid pathways, is another fundamental metabolic pathway. The "two-carbon chop" is accomplished in a series of reactions whereby the first two carbon atoms of the fatty acid chain become the acetyl group in a molecule of acetyl-CoA.

The catabolism proceeds in this manner: a fatty acid reacts with coenzyme A (CoASH) to form an activated thioester. The energy needed for this step of the catabolism is obtained from the high-energy phosphate compound ATP. The lower-energy AMP and inorganic phosphate (P_i) are formed along with the thioester. The activated thioester next undergoes a four-step reaction sequence of *oxidation, hydration, oxidation,* and *cleavage* to produce acetyl-CoA and an activated thioester shortened by two carbon atoms. The cleavage reaction requires an additional molecule of CoA. The shortened-chain thioester again undergoes the reaction sequence of oxidation, hydration, oxidation, and cleavage to shorten the carbon chain further and produce another molecule of acetyl-CoA. Thus, eight molecules of acetyl-CoA can be produced from one molecule of palmitic acid. The reaction sequence is shown in condensed equation form in this fashion:

Step 1. Activation—formation of thioester with CoA:

$$\underset{\text{Fatty acid}}{RCH_2CH_2CH_2\overset{O}{\underset{\|}{C}}OH} + CoASH + ATP \longrightarrow$$

$$\underset{\text{CoA thioester of a fatty acid}}{RCH_2CH_2CH_2\overset{O}{\underset{\|}{C}}-SCoA} + AMP + 2P_i$$

Step 2. Oxidation—dehydrogenation at carbons 2 and 3 (α and β carbons):

$$RCH_2CH_2CH_2\overset{O}{\underset{\|}{C}}-SCoA \xrightarrow[FAD]{-2H} RCH_2CH=CH\overset{O}{\underset{\|}{C}}-SCoA + FADH_2$$

Step 3. Hydration—conversion to secondary alcohol:

$$RCH_2CH=CH\overset{O}{\underset{\|}{C}}-SCoA \xrightarrow{H_2O} RCH_2\overset{OH}{\underset{|}{C}}HCH_2\overset{O}{\underset{\|}{C}}-SCoA$$

Step 4. Oxidation—dehydrogenation of carbon 3 (β carbon) to a keto group:

$$RCH_2\overset{OH}{\underset{|}{C}}HCH_2\overset{O}{\underset{\|}{C}}-SCoA \xrightarrow[NAD^+]{-2H} RCH_2\overset{O}{\underset{\|}{C}}CH_2\overset{O}{\underset{\|}{C}}-SCoA + NADH + H^+$$

Step 5. Carbon chain cleavage—reaction with CoA to produce acetyl-CoA and activated thioester of a fatty acid shortened by two carbons:

$$RCH_2\overset{O}{\underset{\|}{C}}CH_2\overset{O}{\underset{\|}{C}}-SCoA + CoASH \longrightarrow RCH_2\overset{O}{\underset{\|}{C}}-SCoA + \underset{\text{Acetyl-CoA}}{CH_3\overset{O}{\underset{\|}{C}}-SCoA}$$

We have seen that acetyl-CoA is produced from both fatty acid and glucose metabolism. Acetyl-CoA from either source enters the citric acid cycle

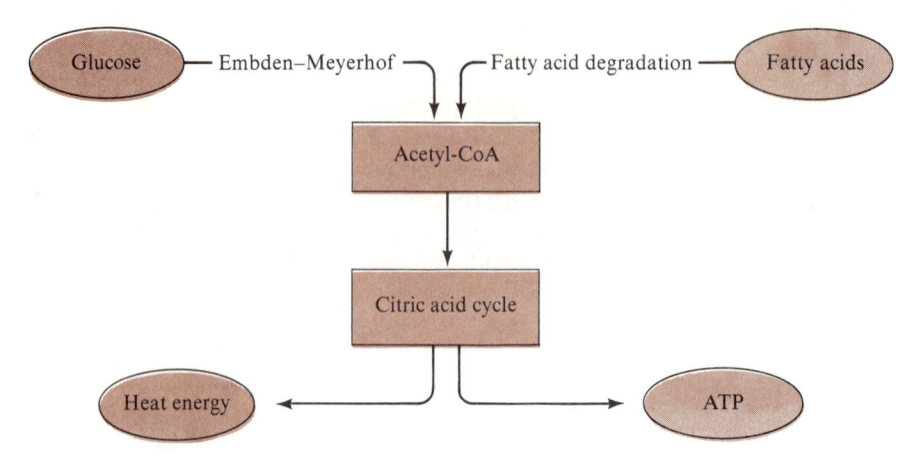

Figure 33.3. Schematic diagram showing both glucose and fatty acids related to the citric acid cycle.

to yield energy-rich ATP (see Figure 33.3). On a comparative weight basis, it has been shown that palmitic acid yields about 2.4 times as much ATP as glucose. Thus, fats supply the body with somewhat more than twice as much energy as carbohydrates.

As in the metabolic pathways for glucose, each reaction in the fatty acid oxidation pathway is enzyme-catalyzed. In addition to the pantothenic acid in the CoA, two other B vitamins—riboflavin and nicotinamide—are required for the conversion of a fatty acid to acetyl-CoA. Riboflavin is a component of FAD (flavin adenine dinucleotide), and nicotinamide is a component of NAD (nicotinamide adenine dinucleotide). FAD and NAD are required for the first and second oxidation reactions, respectively, of an activated fatty acid.

33.7 Fat Storage and Utilization

The primary storage area for fats is adipose tissue, which is widely distributed in the body. Fat tends to accumulate under the skin (subcutaneous fat), in the abdominal region, and about some internal organs, especially the kidneys. Fat is deposited about internal organs as a shock absorber or cushion. Subcutaneous fat acts as an insulating blanket. It is developed to an extreme degree in mammals that live in cold water, such as seals, walruses, and whales.

Fat is the major body reserve of potential energy. It is depleted to supply energy and to help maintain carbohydrate levels in times of starvation. However, body fat is metabolized continuously under both normal and starving conditions. Stored fat does not remain in the body unchanged; there is a rapid exchange between the triglycerides of the lipoproteins in the plasma and the triglycerides in the adipose tissue. Thus, the triglycerides in the adipose tissue are continually renewed. When there is more energy available from the food in the diet than the body needs, the excess energy is used to make extra body fat, which is stored in the adipose tissue. Continued eating of more food than the body utilizes ultimately results in obesity (excess body fat).

There is some relationship between the diet and the character of body fats. For example, when hogs are fed large amounts of highly unsaturated fats or fatty acids, they produce a relatively unsaturated lard. Conversely, hogs fed large amounts of saturated fats or starch produce a relatively saturated lard.

33.8 Biosynthesis of Fatty Acids (Lipogenesis)

The catabolism reactions of glucose—namely, those of the Embden–Meyerhof and citric acid pathways—may, with a few exceptions, be written as reversible reactions. This implies that glucose can be synthesized by simply reversing the reactions involved in its breakdown or catabolism. This is true only in the sense that the substances involved in the biological synthesis of glucose are generally the same ones that are encountered in its breakdown. But the synthesis of glucose requires a set of enzymes that are different from the set required for its catabolism. The synthesis (anabolic) pathways must therefore be different from the catabolic pathways.

At this point we might predict that the biosynthesis of fatty acids would be a reversal of the "two-carbon chop" degradation, but that a different set of enzymes would be required for synthesis. Such a prediction would be correct except on one major point: there is a preliminary set of reactions for each cycle of the synthesis that has no counterpart in the degradation. These reactions involve malonyl-CoA. Malonyl is a three-carbon group and malonyl-CoA is synthesized from acetyl-CoA and carbon dioxide in the presence of the enzyme acetyl-CoA carboxylase, ATP, and the vitamin biotin. The equation representing the formation of malonyl-CoA is

$$CH_3\overset{O}{\underset{\|}{C}}-SCoA + CO_2 \xrightarrow[\text{Biotin}]{\text{ATP}} HO\overset{O}{\underset{\|}{C}}CH_2\overset{O}{\underset{\|}{C}}-SCoA$$
$$\text{Malonyl-CoA}$$

The preliminary set of reactions is represented by the condensed equations below.

Step 1. Acetyl-CoA undergoes a condensation reaction with malonyl-CoA:

$$CH_3\overset{O}{\underset{\|}{C}}-SCoA + HO\overset{O}{\underset{\|}{C}}CH_2\overset{O}{\underset{\|}{C}}-SCoA \rightarrow CH_3\overset{O}{\underset{\|}{C}}\underset{\underset{COOH}{|}}{CH}\overset{O}{\underset{\|}{C}}-SCoA + CoASH$$

Acetyl-CoA   Malonyl-CoA   Condensation product

Step 2. The condensation product undergoes decarboxylation (loss of CO_2) to form a CoA thioester of a β-keto acid (acetoacetyl-CoA):

$$CH_3\overset{O}{\overset{\|}{C}}\underset{\underset{COOH}{|}}{CH}\overset{O}{\overset{\|}{C}}-SCoA \longrightarrow CH_3\overset{O}{\overset{\|}{C}}CH_2\overset{O}{\overset{\|}{C}}-SCoA$$
Acetoacetyl-CoA

The preliminary reactions are completed by formation of this β-keto thioester.

In the next step the β-keto group is reduced by reversing three steps of the "two-carbon chop" degradation.

$$CH_3\overset{O}{\overset{\|}{C}}CH_2\overset{O}{\overset{\|}{C}}-SCoA \xrightarrow[\text{Hydrogenation}]{NADPH + H^+} CH_3\overset{OH}{\overset{|}{C}H}CH_2\overset{O}{\overset{\|}{C}}-SCoA$$
β-Hydroxybutyryl-CoA

$$CH_3\overset{OH}{\overset{|}{C}H}CH_2\overset{O}{\overset{\|}{C}}-SCoA \xrightarrow[\text{Dehydration}]{-H_2O} CH_3CH=CH\overset{O}{\overset{\|}{C}}-SCoA$$
Crotonyl-CoA

$$CH_3CH=CH\overset{O}{\overset{\|}{C}}-SCoA \xrightarrow[\text{Hydrogenation}]{NADPH + H^+} CH_3CH_2CH_2\overset{O}{\overset{\|}{C}}-SCoA$$
Butyryl-CoA

This completes the first cycle of the synthesis; the chain has been lengthened by two carbon atoms. Biosynthesis of longer-chain fatty acids proceeds by a series of such cycles, each lengthening the carbon chain by an increment of two carbon atoms. The next cycle would begin with the reaction of butyryl-CoA and malonyl-CoA, leading to a six-carbon chain, and so on. This synthesis commonly produces palmitic acid (16 carbons) as its end product. The synthesis of palmitic acid from acetyl-CoA and malonyl-CoA requires cycling through the series of steps seven times. The condensed equation for the formation of palmitic acid is

$$CH_3CO-SCoA + 7\,HO\overset{O}{\overset{\|}{C}}CH_2CO-SCoA + 14(NADPH + H^+) \longrightarrow$$
Acetyl-CoA Malonyl-CoA

$$CH_3(CH_2)_{14}COOH + 7\,CO_2 + 6\,H_2O + 8\,CoASH + 14\,NADP^+$$
Palmitic acid

Nearly all naturally occurring fatty acids have even numbers of carbon atoms. A sound reason for this fact is that both the degradation and synthesis proceed by two-carbon increments.

In conclusion, the metabolism of fats has many features in common with that of carbohydrates. The acetyl-CoA produced in the degradation of both carbohydrates and fatty acids may be utilized as a raw material for making other substances and as an energy source. When acetyl-CoA is oxidized via the citric acid cycle, its potential energy is, for the most part, released as heat or used in making ATP. The ATP in turn serves as the source of energy needed for the production of other substances, including the synthesis of carbohydrates and fats from acetyl-CoA.

33.9 Amino Acid Absorption and Distribution

Proteins are for the most part digested to alpha amino acids, which are absorbed through the intestinal walls into the bloodstream. Large molecules, such as intact proteins, apparently do not ordinarily pass through the intestinal wall. However, some proteins may be absorbed by very young infants; and certain allergic reactions may be caused by absorption of protein molecules or fragments. Absorption of at least some of the amino acids involves active transport and enzymes. Pyridoxine (vitamin B_6) deficiency results in poor absorption of amino acids.

Soon after a person eats a meal containing protein, the amino acid concentration in the blood rises, but by only a small amount. This is mainly because protein digestion and absorption take place over a 2- to 3-hour period. In addition, excess amino acids are absorbed within 5 to 10 minutes by the cells in all parts of the body. Therefore, rarely do high concentrations of amino acids build up in the blood. Once in the cells, amino acids are rapidly converted to proteins (see Section 31.10).

Each amino acid is maintained at a particular concentration in the blood plasma. When the plasma amino acid level drops below its normal concentration, proteins in the cells are rapidly hydrolyzed and amino acids are transported to the plasma to replenish the loss. However, if the concentration of a particular amino acid in the plasma becomes too high, the excess is eliminated in the urine. Thus, there is a renal threshold for amino acids. However, under normal circumstances, the loss of amino acids in the urine is insignificant.

Amino acids vary greatly in their composition, and there seems to be no one general pathway of amino acid metabolism. Once absorbed, an amino acid may meet one of the following fates:

1. Be incorporated into a protein
2. Be utilized in the synthesis of other nitrogenous compounds such as nucleic acids
3. Be deaminated to a keto acid, which can (a) be utilized to synthesize other compounds, or (b) be oxidized for energy to carbon dioxide and water

amino acid pool

Absorbed amino acids enter the **amino acid pool**. This pool includes the total supply of amino acids available for use throughout the body. The amount of amino acids in the pool is maintained approximately in equilibrium by the various input and output factors, as shown in Figure 33.4.

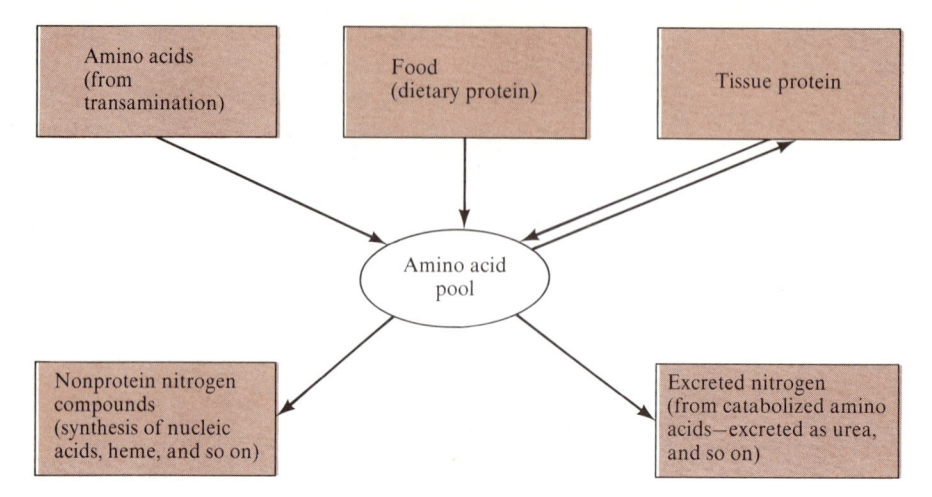

Figure 33.4. Amino acid pool showing nitrogen intake and output factors.

Dynamic equilibrium also exists between the amino acids in the amino acid pool and tissue protein. Amino acids are continually moving back and forth between the amino acid pool and tissue proteins. In other words, our body proteins are constantly being broken down and resynthesized. The rate of turnover varies greatly with different proteins. Research with tagged (isotopically labeled) amino acids has shown that some proteins from liver and other active tissues have a half-life of less that a week, whereas the half-life of some muscle proteins is about six months.

Carbohydrates are stored in the body as liver and muscle glycogen; fats are stored as adipose tissue. But there is no clearly identified bank for storage of amino acids. However, despite the lack of an obvious amino acid reserve, a healthy person can survive a fairly long period of fasting without harm. Needed amino acids are obtained from tissue proteins, and metabolic functions continue more or less normally, provided sufficient water is taken in. Obviously, one cannot fast indefinitely; serious physiological consequences, and ultimately death, result from prolonged starvation.

Nitrogen Balance

nitrogen balance

positive nitrogen balance

negative nitrogen balance

In a healthy, well-nourished adult the amount of nitrogen excreted is equal to the amount of nitrogen ingested. Such a person is said to be in nitrogen equilibrium, or **nitrogen balance**. In a growing child the amount of nitrogen excreted is less than that consumed, and the child is in **positive nitrogen balance**. A person undergoing fasting or starvation, or suffering from certain diseases, will excrete more nitrogen than is ingested; such a person is in **negative nitrogen balance.** A person on a diet lacking one or more essential amino acids will be in negative nitrogen balance. This is because proteins are ordinarily being broken down and resynthesized. Resynthesis is blocked if an amino acid, which the body cannot make, is missing from the diet. The effect is much like starvation, even though the diet may be adequate in all other respects.

33.10 Amino Acid Utilization

One obvious pathway by which dietary amino acids are utilized is by direct incorporation into tissue proteins, as described in Section 31.10. The supply of essential amino acids needed for protein synthesis must follow this pathway, since essential amino acids cannot be made in the body. Absorbed amino acids that do not go directly into protein structure may be utilized in one of the following ways.

Oxidative Deamination

Deamination means the removal of an amino group from a compound. In this reaction sequence, the removed amino group is replaced by oxygen and an alpha keto acid is formed:

$$\underset{\alpha\text{-Amino acid}}{R-\underset{\underset{NH_2}{|}}{CH}-COOH} + H_2O + O_2 \longrightarrow \underset{\alpha\text{-Keto acid}}{R-\underset{\underset{O}{\|}}{C}-COOH} + NH_3 + H_2O_2$$

This equation represents only the overall net reaction. In mammals, some of the ammonia is converted by liver enzymes to other amino acids, but the bulk of it is converted to urea and eliminated in the urine. The biosynthesis of urea is a complicated process involving ammonia, carbon dioxide, ATP, Mg^{2+}, six amino acids, and several enzymes. The urea, which is formed in the liver, is released into the blood and excreted through the kidneys into the urine (see Figure 33.5).

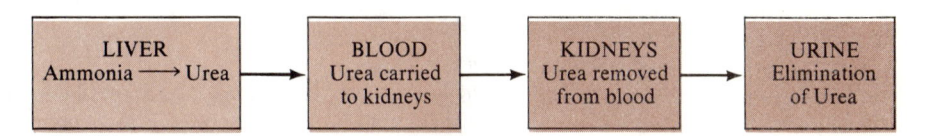

LIVER	BLOOD	KIDNEYS	URINE
Ammonia ⟶ Urea	Urea carried to kidneys	Urea removed from blood	Elimination of Urea

Figure 33.5. Path for elimination of ammonia as urea. (Recall that it was urea that Wöhler synthesized in 1828, an event that was the turning point in organic chemistry.)

Urea accounts for 80 to 90% of the nitrogen excreted by humans. The net equation is

$$2\,NH_3 + CO_2 \longrightarrow \underset{\text{Urea}}{H_2N-\underset{\underset{O}{\|}}{C}-NH_2} + H_2O$$

The hydrogen peroxide produced in the deamination is decomposed to water and oxygen. The keto acid may be catabolized or utilized in some other fashion. Oxidative deamination serves to tie the metabolism of amino acids to that of carbohydrates and lipids. As a specific example, alanine [$CH_3CH(NH_2)COOH$] is converted to pyruvic acid ($CH_3COCOOH$), which can then form glucose by gluconeogenesis or form acetyl-CoA. This in turn can form fats or enter the citric acid cycle for the production of ATP (see Figure 33.3).

Transamination

By this reaction amino acids can be converted to α-keto acids and vice versa. The effect is to transfer an amino group from an amino acid to an α-keto acid. The reaction is reversible, so it can function in either the catabolism or in the biosynthesis of amino acids. The reaction sequence is summarized by this general equation:

$$\underset{RCHCOOH}{\overset{NH_2}{|}} + \underset{R'-CCOOH}{\overset{O}{\|}} \xrightarrow{\text{Transaminase}} \underset{RCCOOH}{\overset{O}{\|}} + \underset{R'CHCOOH}{\overset{NH_2}{|}}$$

Vitamin B_6 (pyridoxine) in the form of pyridoxal phosphate acts as a coenzyme with the transaminase enzymes involved in this reaction sequence. Transamination is another way in which amino acid metabolism is linked with the metabolism of carbohydrates and lipids. It provides a way to make nonessential amino acids from intermediates obtained in carbohydrate and lipid metabolism. Transamination can also convert amino acids to substances that enter the metabolic pathways that are common to carbohydrates and lipids.

Questions

A. Review the meanings of the new terms introduced in this chapter.
   1. lipids
   2. Phospholipids
   3. Steroids
   4. Amino acid pool
   5. Nitrogen balance
   6. Positive nitrogen balance
   7. Negative nitrogen balance

B. Review questions.
   1. Why are the lipids, which are dissimilar substances, classified as a group?
   2. Write the structural formula of a triglyceride that contains one unit each of palmitic, stearic, and oleic acids. How many other triglycerides are possible, each containing one unit of each of these acids?
   3. What products are formed when fats are metabolized?
   4. What are the three essential fatty acids? What are the consequences of their being absent from the diet?
   5. In what organ in the body are phospholipids mainly produced?
   6. Lecithins and cephalins are both derivatives of phosphatidic acid. Indicate how they differ from each other.
   7. What common structural feature is possessed by all steroids? Write the structural formulas of two steroids.
   8. Where is cholesterol found in the body? Where is it primarily produced?
   9. Cholesterol can be synthesized from carbohydrates, fats, and proteins. Explain how this can occur.
   10. What is atherosclerosis? How is it produced and what are its symptoms?
   11. Explain how proteins, carbohydrates, and fats can contribute to atherosclerosis.
   12. How is a triglyceride molecule "solubilized" for transportation in the bloodstream?

13. Show how fat digestion products are resynthesized to triglycerides.
14. How is it possible to become obese even though very little fat is included in the diet?
15. Briefly describe Knoop's experiments on fatty acid oxidation and degradation.
16. What is meant by beta oxidation and beta cleavage in relation to the biochemistry of fatty acids?
17. By means of a diagram, outline how caproic acid is oxidized to butyric acid.
18. Aside from being the source of certain fatty acids, are fats essential in our diet? Explain your answer.
19. Aside from being a food reserve, what are the two principal functions of body fat?
20. In what way is the citric acid cycle involved in obtaining energy from fats?
21. Why is the ATP yield from a six-carbon fatty acid greater than the ATP yield from a six-carbon hexose (glucose) when each is catabolized to CO_2?
22. Outline the parts played by malonyl-CoA and acetyl-CoA in the biosynthesis of fatty acids.
23. Are carbohydrates essential in our diets? Explain your answer.
24. Prepare a diagram showing the principal steps in the conversion of starch to body fat.
25. What are the possible metabolic fates of amino acids?
26. Compare oxidative deamination with transamination in the utilization of amino acids.
27. Write the structural formulas of the compounds produced by the oxidative deamination of the following amino acids:
    (a) Alanine   (c) Phenylalanine
    (b) Aspartic acid   (d) Glycine
28. Distinguish between positive nitrogen balance and negative nitrogen balance.
29. How does the lack of an essential amino acid in the diet affect the nitrogen balance? Explain.
30. Discuss the renal threshold for amino acids.

Appendix I
Mathematical Review

1. **Multiplication.** Multiplication is a process of adding any given number or quantity a certain number of times. Thus, 4 times 2 means 4 added two times, or 2 added together four times, to give the product 8. Various ways of expressing multiplication are

$$ab \quad a \times b \quad a \cdot b \quad a(b) \quad (a)(b)$$

All mean *a* times *b*, or *a* multiplied by *b*, or *b* times *a*.

When $a = 16$ and $b = 24$, we have $16 \times 24 = 384$.

The expression $°F = (1.8 \times °C) + 32$ means that we are to multiply 1.8 times $°C$ and add 32 to the product. When $°C$ equal 50,

$$°F = (1.8 \times 50) + 32 = 90 + 32 = 122°F$$

The result of multiplying two or more numbers together is known as the *product*.

2. **Division.** The word *division* has several meanings. As a mathematical expression, it is the process of finding how many times one number or quantity is contained in another. Various ways of expressing division are

$$a \div b \quad \frac{a}{b} \quad a/b$$

All mean *a* divided by *b*.

When $a = 15$ and $b = 3$, $\frac{15}{3} = 5$.

The number above the line is called the *numerator*; the number below the line is the *denominator*. Both the horizontal and the slanted (/) division signs also mean "per." For example, in the expression for density, the mass per unit volume:

$$\text{Density} = \text{Mass/Volume} = \frac{\text{Mass}}{\text{Volume}} = \text{g/ml}$$

The diagonal line still refers to a division of grams by the number of millilitres occupied by that weight.

The result of dividing one number into another is called the *quotient*.

A-1

3. **Fractions and decimals.** A fraction is an expression of division, showing that the numerator is divided by the denominator. A *proper fraction* is one in which the numerator is smaller than the denominator. In an *improper fraction*, the numerator is the larger number. A decimal or a decimal fraction is a proper fraction in which the denominator is some power of 10. The decimal fraction is determined by carrying out the division of the proper fraction. Examples of proper fractions and their decimal fraction equivalents are shown in the table.

Proper fraction		Decimal fraction		Proper fraction
$\frac{1}{8}$	=	0.125	=	$\frac{125}{1000}$
$\frac{1}{10}$	=	0.1	=	$\frac{1}{10}$
$\frac{3}{4}$	=	0.75	=	$\frac{75}{100}$
$\frac{1}{100}$	=	0.01	=	$\frac{1}{100}$
$\frac{1}{4}$	=	0.25	=	$\frac{25}{100}$

4. **Adding or subtracting fractions.** We cannot directly add $\frac{1}{2} + \frac{1}{4}$, but we can add these two fractions if they both have the same denominator. A *common denominator* is a number which, when divided by each denominator in a series of fractions, gives whole-number quotients. We usually use the smallest possible common denominator. Thus, $\frac{1}{2} = \frac{2}{4}$, and $\frac{2}{4} + \frac{1}{4} = \frac{3}{4}$. Therefore, to add or subtract fractions, first change the denominator of each fraction to be the same number; that is, make each fraction have a common denominator. Then add the numerators of each fraction and place this sum over the common denominator. It may then be possible to reduce the final fraction to a simpler fraction or to a decimal number.

$$\frac{3}{8} + \frac{4}{8} = \frac{3+4}{8} = \frac{7}{8}$$

$$\frac{13}{18} - \frac{3}{18} = \frac{13-3}{18} = \frac{10}{18} = \frac{5}{9}$$

$$\frac{1}{4} + \frac{2}{3} = \frac{3}{12} + \frac{8}{12} = \frac{3+8}{12} = \frac{11}{12}$$

$$\frac{2}{3} + \frac{5}{6} - \frac{1}{5} = \frac{20}{30} + \frac{25}{30} - \frac{6}{30} = \frac{20+25-6}{30} = \frac{39}{30} = 1\frac{9}{30} = 1.3$$

$$\frac{3}{2} + \frac{3}{4} + \frac{3}{5} = \frac{30}{20} + \frac{15}{20} + \frac{12}{20} = \frac{57}{20} = 2\frac{17}{20} = 2.85$$

5. Multiplication of fractions. A fraction may be multiplied by another fraction by first multiplying the numerators together, and then multiplying the denominators together and placing the product of the numerators over the product of the denominators. This fraction may then be reduced to its lowest terms.

$$\frac{4}{5} \times \frac{1}{3} = \frac{4 \times 1}{5 \times 3} = \frac{4}{15}$$

$$\frac{3}{8} \times 4 = \frac{12}{8} = 1\frac{4}{8} = 1\frac{1}{2} = 1.5$$

$$\frac{1}{2} \times \frac{5}{3} \times \frac{4}{7} = \frac{1 \times 5 \times 4}{2 \times 3 \times 7} = \frac{20}{42} = \frac{10}{21}$$

6. Division by a fraction. Dividing a number or a fraction by a fraction can be accomplished in this manner: Invert the denominator (which is a fraction) and multiply this inverted expression by the numerator, using the usual multiplication methods. For example, divide 6 by $\frac{3}{4}$.

$$\frac{6}{\frac{3}{4}}$$

Invert the denominator $\frac{3}{4}$ to give $\frac{4}{3}$. Then multiply:

$$6 \times \frac{4}{3} = \frac{24}{3} = 8$$

Divide $\frac{2}{3}$ by $\frac{3}{4}$.

$$\frac{\frac{2}{3}}{\frac{3}{4}}$$

Invert $\frac{3}{4}$ to give $\frac{4}{3}$. Then multiply:

$$\frac{2}{3} \times \frac{4}{3} = \frac{8}{9}$$

7. Addition of numbers with decimals. To add numbers with decimals, we use the same procedure as that used when adding whole numbers, but always line up the decimal points in the same column. For example, add 8.21 + 143.1 + 0.325.

```
    8.21
+ 143.1
+   0.325
---------
  151.635
```

When adding numbers expressing units of measurement, always be certain that the numbers added together represent the same units. For example, what is the total length of these three pieces of glass tubing: 10.0 cm, 125 mm, 8.4 cm? If we add these directly, we obtain a value of 143.4, but we are not certain what the unit of measurement is. To add these lengths correctly, first change 125 mm to 12.5 cm. Now all the lengths are expressed in the same units and can be added.

```
 10.0 cm
 12.5 cm
  8.4 cm
───────
 30.9 cm
```

8. **Subtraction of numbers with decimals.** To subtract numbers containing decimals, we use the same procedure as for subtracting whole numbers, but always line up the decimal points in the same column. For example, subtract 20.60 from 182.49.

```
   182.49
 −  20.60
 ────────
   161.89
```

9. **Multiplication of numbers with decimals.** To multiply two or more numbers together that contain decimals, we first multiply as if they were whole numbers. To locate the decimal point in the product, we add together the number of digits to the right of the decimal in all the numbers multiplied together. The product should contain this total number of digits to the right of the decimal point.

Multiply 2.05×2.05 (total of four digits to the right of the decimal):

```
    2.05
 ×  2.05
 ───────
    1025
   4100
 ───────
  4.2025   (Four digits to the right of the decimal)
```

$14.25 \times 6.01 \times 0.75 = 64.231875$ (Six digits to the right of the decimal)

$39.26 \times 60 = 2355.60$ (Two digits to the right of the decimal)

[*Note*: When at least one of the numbers that is multiplied is a measurement, the answer must be adjusted to contain the correct number of significant figures. (See Section 12 on significant figures.)]

10. **Division of numbers with decimals.** To divide numbers containing decimals, we first relocate the decimal points of the numerator and denominator by moving them to the right as many places as needed to make the denominator a whole number. (Move the decimal of both the numerator and the denominator the same amount and in the same direction.) For example,

$$\frac{136.94}{4.1} = \frac{1369.4}{41}$$

The decimal point adjustment in this example is equivalent to multiplying both numerator and denominator by 10. Now we carry out the division normally, locating the decimal point immediately above its position in the dividend.

$$41\overline{)1369.4} \frac{0.441}{26.25} = \frac{44.1}{2625} = 2625\overline{)44.1000}$$

$$\begin{array}{r} 33.4 \\ \hline 123 \\ \hline 139 \\ 123 \\ \hline 164 \\ 164 \\ \hline \end{array} \qquad \begin{array}{r} 0.0168 \\ \hline 2625 \\ \hline 17850 \\ 15750 \\ \hline 21000 \\ 21000 \\ \hline \end{array}$$

[*Note:* When at least one of the numbers in the division is a measurement, the answer must be adjusted to contain the correct number of significant figures. (See Section 12 on significant figures.)]

The examples above are merely guides to the principles used in performing the various mathematical operations illustrated. There are, no doubt, shortcuts and other methods, and the student will discover these with experience. Every student of chemistry should use either an electronic calculator or a slide rule for solving problems. The use of these devices will save many hours of time that would otherwise be spent in doing tedious longhand calculations. After solving a problem, the student should check for errors and evaluate the answer to see if it is logical and consistent with the data given.

11. Algebraic equations. Many mathematical problems that are first encountered in chemistry fall into the following algebraic forms. Solutions to these problems are simplified by first isolating the desired term on one side of the equation. This is accomplished by treating both sides of the equation in an identical manner (so as not to destroy the equality) until the desired term is isolated.

(a) $$a = \frac{b}{c}$$

To solve for a, simply divide b by c.
To solve for b, multiply both sides of the equation by c.

$$a \times c = \frac{b}{\ell} \times \ell$$

$$b = a \times c$$

To solve for c, multiply both sides of the equation by $\frac{c}{a}$.

$$\not{a} \times \frac{c}{\not{a}} = \frac{b}{\ell} \times \frac{\ell}{a}$$

$$c = \frac{b}{a}$$

(b) $$\frac{a}{b} = \frac{c}{d}$$

To solve for a, multiply both sides of the equation by b.

$$\frac{a}{\cancel{b}} \times \cancel{b} = \frac{c}{d} \times b$$

$$a = \frac{c \times b}{d}$$

To solve for b, multiply both sides of the equation by $\frac{b \times d}{c}$.

$$\frac{a}{\cancel{b}} \times \frac{\cancel{b} \times d}{c} = \frac{\cancel{c}}{\cancel{d}} \times \frac{b \times \cancel{d}}{\cancel{c}}$$

$$b = \frac{a \times d}{c}$$

(c) $a \times b = c \times d$

To solve for a, divide both sides of the equation by b.

$$\frac{a \times \cancel{b}}{\cancel{b}} = \frac{c \times d}{b}$$

$$a = \frac{c \times d}{b}$$

(d) $$\frac{(b - c)}{a} = d$$

To solve for b, first multiply both sides of the equation by a.

$$\frac{\cancel{a}(b - c)}{\cancel{a}} = d \times a$$

$$b - c = d \times a$$

Then add c to both sides of the equation.

$b - \cancel{c} + \cancel{c} = d \times a + c$

$b = (d \times a) + c$

When $a = 1.8$, $c = 32$, and $d = 35$,

$b = (35 \times 1.8) + 32 = 63 + 32 = 95$

12. Significant figures. Every measurement that we make has some inherent error due to the limitations of the measuring instrument and the experimenter. The numerical value recorded for a measurement should give some indication of the reliability (precision) of that measurement. In measuring a temperature using a thermometer calibrated at one-degree intervals, we can easily read the thermometer to the nearest one degree, but we normally estimate and record the temperature to the nearest tenth of a degree (0.1°C). For example, a temperature falling between 23°C and 24°C might be estimated at 23.4°C. There is some uncertainty about the last digit, 4, but an estimate of it is better information than simply reporting 23°C or 24°C. If we read the thermometer as "exactly" twenty-three degrees, the temperature should be reported as 23.0°C, not 23°C, because 23.0°C indicates our estimate to the nearest 0.1°C. Thus, in recording any measurement, we retain one uncertain digit. The digits retained in a physical measurement are said to be significant, and are called **significant figures** or **digits**.

Some numbers are exact and therefore have an infinite number of significant figures. Exact numbers occur in simple counting operations, such as 5 bricks, and in defined relationships, as 100 cm = 1 m, 24 hr = 1 day, and so on. Because of their infinite number of significant figures, exact numbers do not limit the number of significant figures in a calculation.

Counting significant figures. Digits other than zero are always significant. Depending on their position in the number, zeros may or may not be significant. There are several possible situations.

1. All zeros between other digits in a number are significant. For example: 3.076, 4002, 790.2. Each of these numbers has four significant figures.
2. Zeros to the left of the first nonzero digit are used to locate the decimal point and are not significant. Thus, 0.013 has only two significant figures (1 and 3).
3. Zeros to the right of the last nonzero digit and to the right of the decimal point are significant, for they would not have been included except to express precision. For example, 3.070 has four significant figures; 0.070 has two significant figures.
4. Zeros to the right of the last nonzero digit, but to the left of the decimal, as in the numbers 100, 580, 37,000, may or may not be significant. For example, in 37,000 the measurement might be good to the nearest 1000, 100, 10, or 1. There are two conventions that may be used to show the intended precision. If all the zeros are significant, then an expressed decimal may be added, as 580., or 37,000. But a better system, and one which is applicable to the case when some but not all of the zeros are significant, is to express the number in exponential notation, including only the significant zeros. Thus, for 300, if the zero following 3 is significant, we would write 3.0×10^2. For 17,000, if two zeros are significant, we would write 1.700×10^4. The number we correctly expressed as 580. can also be correctly expressed as 5.80×10^2. With exponential notation there is no doubt as to the number of significant figures.

The mass of an object given as 28.2 grams (g) is expressed to the nearest 0.1 g. This indicates that the weighing was done on a balance having a precision of 0.1 g. It also means that the true value actually lies between 28.15 and 28.25 g. The value 28.2 has three significant figures (2, 8, 2). This same mass, weighed on the same balance, but expressed in milligrams (mg) is 28,200 mg. This value, 28,200 mg, also has three significant figures (2, 8, 2); the zeros are needed to express the magnitude of the number in milligrams. This same mass expressed as 0.0282 kilograms (kg) still contains three significant figures; the zeros are used to locate the decimal point. Better expressions showing three significant figures for these masses are 2.82×10^4 mg and 2.82×10^{-2} kg.

Additional examples illustrating the significant figures in a number are given in this table.

Number	Number of significant figures
2.45	3
2.450	4
2.045	4
0.245	3
0.0245	3
245.0	4
245	3

13. **Exponents; powers of 10; expression of large and small numbers.** In scientific measurements and calculations, we often encounter very large and very small numbers; for example, 0.00000384 and 602,000,000,000,000,000,000,000. These numbers are troublesome to write and awkward to work with, especially in calculations. A convenient method of expressing these large and small numbers in a simplified form is by means of exponents or powers of 10. This method of expressing numbers is known as **scientific** or **exponential notation**.

An *exponent* is a number written as a superscript following another number; it is also called a *power* of that number, and it indicates how many times the number is used as a factor. In the number 10^2, 2 is the exponent and the number means 10 squared, or 10 to the second power, or $10 \times 10 = 100$. Three other examples are

$3^2 = 3 \times 3 = 9$

$3^4 = 3 \times 3 \times 3 \times 3 = 81$

$10^3 = 10 \times 10 \times 10 = 1000$

For ease of handling, large and small numbers are expressed in powers of 10. Powers of 10 are used because multiplying or dividing by 10 coincides with moving the decimal point in a number by one place. Thus, a number multiplied by 10^1 would move the decimal point one place to the right; 10^2, two places to the right; 10^{-2}, two places to the left. To express a number in powers of 10, we move the decimal point in the original number to a new position, placing it so that the number is a value between 1 and 10. This new decimal number is multiplied by 10 raised to the proper power. For example, to write the number 42,389 in exponential form (powers of 10), the decimal point is placed between the 4 and the 2 (4.2389) and the number is multiplied by 10^4; thus, the number is 4.2389×10^4. The power of 10 (4) tells us the number of places that the decimal point must be moved to restore it to its original position. The exponent of 10 is determined by counting the number of places that the decimal point is moved from its original position. If the decimal point is moved to the left, the exponent is a positive number; if it is moved to the right, the exponent is a negative number. To express the number 0.00248 in exponential notation (as a power of 10), the decimal point is moved three places to the right; the exponent of 10 is -3, and the number is 2.48×10^{-3}. Study the examples below.

$$1237 = 1.237 \times 10^3$$
$$988 = 9.88 \times 10^2$$
$$147.2 = 1.472 \times 10^2$$
$$2{,}200{,}000 = 2.2 \times 10^6$$
$$0.0123 = 1.23 \times 10^{-2}$$
$$0.00005 = 5 \times 10^{-5}$$
$$0.000368 = 3.68 \times 10^{-4}$$

The use of powers of 10 in multiplication and division greatly simplifies locating the decimal point in the answer. In multiplication, first change all numbers to powers of 10, then multiply the numerical portion in the usual manner, and finally add the exponents of 10 algebraically, expressing them as a power of 10 in the product. In multiplication, the exponents (powers of 10) are added algebraically.

$$10^2 \times 10^3 = 10^{(2+3)} = 10^5$$
$$10^2 \times 10^2 \times 10^{-1} = 10^{(2+2-1)} = 10^3$$

Multiply: $40{,}000 \times 4200$

Change to powers of ten: $4 \times 10^4 \times 4.2 \times 10^3$

Rearrange: $4 \times 4.2 \times 10^4 \times 10^3$

$$16.8 \times 10^{(4+3)}$$
$$16.8 \times 10^7 \quad \text{or} \quad 1.68 \times 10^8 \quad \text{(Answer)}$$

Multiply: 380×0.00020

$$3.80 \times 10^2 \times 2.0 \times 10^{-4}$$
$$3.80 \times 2.0 \times 10^2 \times 10^{-4}$$
$$7.6 \times 10^{(2-4)}$$
$$7.6 \times 10^{-2} \quad \text{or} \quad 0.076 \quad \text{(Answer)}$$

Multiply: $125 \times 284 \times 0.150$

$$1.25 \times 10^2 \times 2.84 \times 10^2 \times 1.50 \times 10^{-1}$$
$$1.25 \times 2.84 \times 1.50 \times 10^2 \times 10^2 \times 10^{-1}$$
$$5.325 \times 10^{(2+2-1)}$$
$$5.32 \times 10^3 \quad \text{(Answer)}$$

In division, after changing the numbers to powers of 10, move the 10 and its exponent from the denominator to the numerator, changing the sign of the exponent. Carry out the division in the usual manner and evaluate the power of 10. The following is a proof of the equality of moving the power of 10 from the denominator to the numerator.

$$1 \times 10^{-2} = 0.01 = \frac{1}{100} = \frac{1}{10^2} = 1 \times 10^{-2}$$

In division, the exponents in the denominator are subtracted algebraically from the exponents in the numerator.

$$\frac{10^5}{10^3} = 10^5 \times 10^{-3} = 10^{(5-3)} = 10^2$$

$$\frac{10^3 \times 10^4}{10^{-2}} = 10^3 \times 10^4 \times 10^2 = 10^{(3+4+2)} = 10^9$$

Divide: $\dfrac{2871}{0.0165}$

Change to powers of 10: $\dfrac{2.871 \times 10^3}{1.65 \times 10^{-2}}$

Move 10^{-2} to the numerator, changing the sign of the exponent. This is mathematically equivalent to multiplying both numerator and denominator by 10^2.

$$\frac{2.87 \times 10^3 \times 10^2}{1.65}$$

$$\frac{2.87 \times 10^{(3+2)}}{1.65} = 1.74 \times 10^5 \quad \text{(Answer)}$$

Divide: $\dfrac{0.000585}{0.00300}$

$$\frac{5.85 \times 10^{-4}}{3.00 \times 10^{-3}}$$

$$\frac{5.85 \times 10^{-4} \times 10^3}{3.00} = \frac{5.85 \times 10^{(-4+3)}}{3.00}$$

$$1.95 \times 10^{-1} \quad \text{or} \quad 0.195 \quad \text{(Answer)}$$

Calculate: $\dfrac{760 \times 300 \times 40.0}{700 \times 273}$

$$\frac{7.60 \times 10^2 \times 3.00 \times 10^2 \times 4.00 \times 10^1}{7.00 \times 10^2 \times 2.73 \times 10^2}$$

$$\frac{7.60 \times 3.00 \times 4.00 \times 10^2 \times 10^2 \times 10^1}{7.00 \times 2.73 \times 10^2 \times 10^2}$$

$$4.77 \times 10^1 \quad \text{or} \quad 47.7 \quad \text{(Answer)}$$

14. **Rounding off numbers.** When numbers are added, subtracted, multiplied, or divided, we often obtain answers with more figures than we are justified in using. Numbers are rounded off so that we can retain a specific number of digits consistent with the accuracy that they represent.

We round off a number by dropping digits from the end of the number, adjusting the last digit retained either to remain the same or be increased by one number.

Rules for rounding off numbers:

1. When the first digit after those being retained is less than 5, all digits retained remain the same.
2. When the first digit after those being retained is larger than 5, the last digit retained is rounded off by increasing it one number.
3. When the first digit after those being retained is 5 and all others beyond it are zeros, the last digit retained remains the same if it is an even number or is increased by one number if it is an odd number.

In the examples illustrating these rules, all numbers are rounded off to four digits.

Rule 1.	1.0263	Round off to 1.026
	23.04193	Round off to 23.04
Rule 2.	1.0268	Round off to 1.027
	23.04728	Round off to 23.05
	18.998	Round off to 19.00
Rule 3.	140.25	Round off to 140.2
	63.3750	Round off to 63.38

15. **Significant figures in calculations.** The result of a calculation based on experimental measurements cannot be more precise than the measurement that has the greatest uncertainty.

Addition and subtraction. The result of an addition or subtraction should contain no more digits to the right of the decimal point than are contained in that quantity which has the least number of digits to the right of the decimal point.

Perform the operation indicated and then round off the number to the proper significant figures.

```
 142.8              93.45
  18.843           -18.0
  36.42            ─────
─────               75.45
 198.063
 198.1  (Answer)    75.4  (Answer)
```

Multiplication and division. In calculations involving multiplication or division, the answer should contain the same number of significant figures as the measurement that has the least number of significant figures. In multiplication or division the position of the decimal point has nothing to do with the number of significant figures in the answer. Study the following examples:

	Round off to
$2.05 \times 2.05 = 4.2025$	4.20
$18.48 \times 5.2 = 96.096$	96
$0.0126 \times 0.020 = 0.000252$ or	
$1.26 \times 10^{-2} \times 2.0 \times 10^{-2} = 2.52 \times 10^{-4}$	2.5×10^{-4}
$\dfrac{1369.4}{41} = 33.4$	33
$\dfrac{2268}{4.20} = 540$	540

16. Dimensional analysis. Many problems of chemistry can be solved readily by dimensional analysis using the factor-label or conversion factor method. Dimensional analysis involves the use of proper units of dimension on all factors that are multiplied, divided, added, or subtracted in setting up and solving a problem. Dimensions are physical quantities such as length, mass, and time, which are expressed in such units as centimetres, grams, and seconds, respectively. In solving a problem, these units are treated mathematically just as though they were numbers, giving us an answer that contains the correct dimensional units.

A measurement or quantity given in one kind of unit can be converted to any other kind of unit having the same dimension. To convert from one kind of unit to another, the original quantity or measurement is multiplied or divided by a conversion factor. The key to success lies in choosing the correct conversion factor. This general method of calculation is illustrated in the following examples.

Suppose we want to change 24 ft to inches. We need to multiply 24 ft by a conversion factor containing feet and inches. Two such conversion factors can be written relating inches and feet.

$$\frac{12 \text{ in.}}{1 \text{ ft}} \quad \text{or} \quad \frac{1 \text{ ft}}{12 \text{ in.}}$$

We choose the factor that will mathematically cancel feet and leave the answer in inches. Note that the units are treated in the same way we treat numbers, multiplying or dividing as required. Two possibilities then arise to change 24 ft to inches:

$$24 \text{ ft} \times \frac{12 \text{ in.}}{1 \text{ ft}} \quad \text{or} \quad 24 \text{ ft} \times \frac{1 \text{ ft}}{12 \text{ in.}}$$

In the first case (the correct method), feet in the numerator and the denominator cancel, giving us an answer of 288 in. In the second case, the units of the answer are ft²/in., the answer being 2.0 ft²/in. In the first case, the answer is reasonable since it is expressed in units having the proper dimensions. That is, the dimension of length expressed in feet has been converted to length in inches according to the mathematical expression

$$\text{ft} \times \frac{\text{in.}}{\text{ft}} = \text{in.}$$

In the second case, the answer is not reasonable since the units (ft²/in.) do not correspond to units of length. The answer is therefore incorrect. The units are the guiding factor for the proper conversion.

The reason we can multiply 24 ft times 12 in./ft and not change the value of the measurement is because the conversion factor is derived from two equivalent quantities. Therefore, the conversion factor 12 in./ft is equal to unity. And when you multiply any factor by 1, it does not change the value.

$$12 \text{ in.} = 1 \text{ ft} \quad \text{and} \quad \frac{12 \text{ in.}}{1 \text{ ft}} = 1$$

Convert 16 kg to milligrams. In this problem it is best to proceed in this fashion:

$$\text{kg} \rightarrow \text{g} \rightarrow \text{mg}$$

The possible conversion factors are

$$\frac{1000 \text{ g}}{1 \text{ kg}} \quad \text{or} \quad \frac{1 \text{ kg}}{1000 \text{ g}} \qquad \frac{1000 \text{ mg}}{1 \text{ g}} \quad \text{or} \quad \frac{1 \text{ g}}{1000 \text{ mg}}$$

We use the conversion factor that leaves the proper unit at each step for the next conversion. The calculation is

$$16 \text{ kg} \times \frac{1000 \text{ g}}{1 \text{ kg}} \times \frac{1000 \text{ mg}}{1 \text{ g}} = 1.6 \times 10^7 \text{ mg}$$

Many problems may be solved by a sequence of steps involving unit conversion factors. This sound, basic approach to problem solving, together with neat and orderly setting up of data, will lead to correct answers having the right units, fewer errors, and considerable saving of time.

17. **Graphical representation of data.** A graph is often the most convenient way to present or display a set of data. Various kinds of graphs have been devised, but the most common type uses a set of horizontal and vertical coordinates to show the relationship of two variables. It is called an *x–y* graph because the data of one variable are represented on the horizontal or *x* axis (abscissa) and the data of the other variable are represented on the vertical or *y* axis (ordinate). (See Figure I.1.)

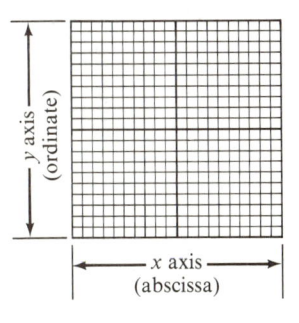

Figure I.1

A-14 Appendix I

As a specific example of a simple graph, let us graph the relationship between Celsius and Fahrenheit temperature scales. Assume that initially we have only the information in the table.

°C	°F
0	32
50	122
100	212

On a set of horizontal and vertical coordinates (graph paper), scale off at least 100 Celsius degrees on the x axis and at least 212 Fahrenheit degrees on the y axis. Locate and mark the three points corresponding to the three temperatures given and draw a line connecting these points (see Figure I.2). Here is how a point is located on the graph: Using the 50°C–122°F data, trace a vertical line up from 50°C on the x axis and a horizontal line across from 122°F on the y axis and mark the point where the two lines intersect. This process is called *plotting*. The other two points are plotted on the graph in the same way. [*Note*: The number of degrees per scale division was chosen to give a graph of convenient size. In this case there are 5 Fahrenheit degrees per scale division and 2 Celsius degrees per scale division.]

The graph in Figure I.2 shows that the relationship between Celsius and Fahrenheit temperature is that of a straight line. The Fahrenheit temperature

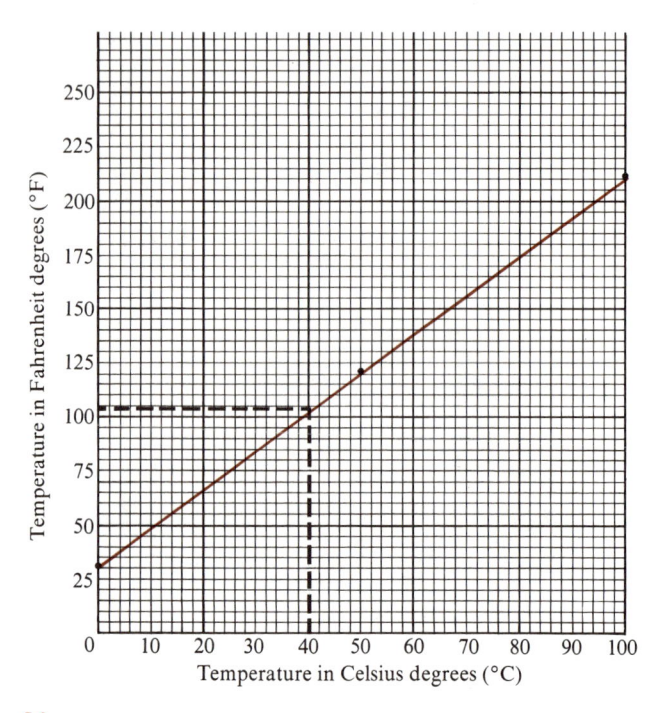

Figure I.2

corresponding to any given Celsius temperature between 0° and 100° can be determined from the graph. For example, to find the Fahrenheit temperature corresponding to 40°C, trace a perpendicular line from 40°C on the x axis to the line plotted on the graph. Now trace a horizontal line from this point on the plotted line to the y axis and read the corresponding Fahrenheit temperature (104°F). See the dotted lines on Figure I.2. In turn, the Celsius temperature corresponding to any Fahrenheit temperature between 32° and 212° can be determined from the graph. This is accomplished by tracing a horizontal line from the Fahrenheit temperature to the plotted line and reading the corresponding temperature on the Celsius scale directly below the point of intersection.

The mathematical relationship of Fahrenheit and Celsius temperatures is expressed by the equation °F = 1.8°C + 32. Figure I.2 is a graph of this equation. Since the graph is a straight line, it can be extended indefinitely at either end. Any desired Celsius temperature can be plotted against the corresponding Fahrenheit temperature by extending the scales along both axes as necessary. Negative, as well as positive, values can be plotted on the graph (see Figure I.3).

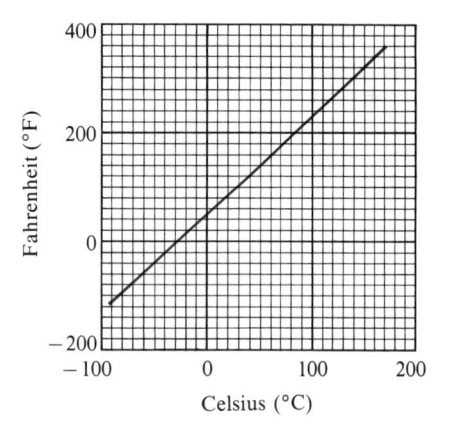

Figure I.3

Figure I.4 is a graph showing the solubility of potassium chlorate in water at various temperatures. The solubility curve on this graph was plotted from the data in the following table.

Temperature (°C)	Solubility (g KClO$_3$/100 g water)
10	5.0
20	7.4
30	10.5
50	19.3
60	24.5
80	38.5

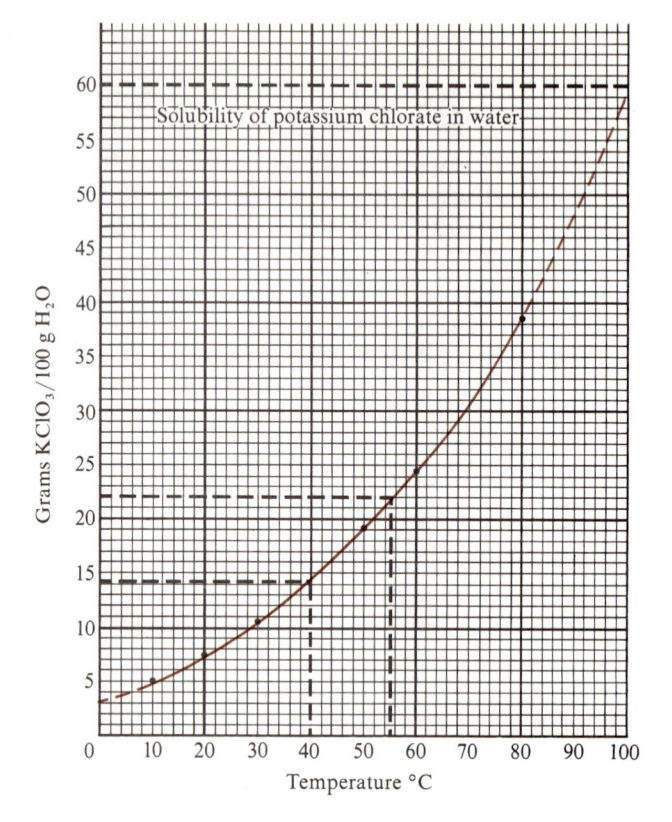

Figure I.4

In contrast to the Celsius–Fahrenheit temperature relationship, there is no known mathematical equation that describes the exact relationship between temperature and the solubility of potassium chlorate. The graph in Figure I.4 was constructed from experimentally determined solubilities at the six temperatures shown. These experimentally determined solubilities are all located on the smooth curve traced by the unbroken line portion of the graph. We are therefore confident that the unbroken line represents a very good approximation of the solubility data for potassium chlorate covering the temperature range from 10 to 80°C. All points on the plotted curve represent the composition of saturated solutions. Any point below the curve represents an unsaturated solution.

The dotted line portions of the curve are *extrapolations*; that is, they extend the curve above and below the temperature range actually covered by the plotted solubility data. Curves such as this are often extrapolated a short distance beyond the range of the known data, although the extrapolated portions may not be highly accurate. Extrapolation is justified only in the absence of more reliable information.

The graph in Figure I.4 can be used with confidence to obtain the solubility of $KClO_3$ at any temperature between 10°C and 80°C, but the solubilities between 0°C and 10°C and between 80°C and 100°C are less reliable. For example, what is the solubility of $KClO_3$ at 55°C, at 40°C, and at 100°C?

First draw a perpendicular line from each temperature to the plotted solubility curve. Now trace a horizontal line to the solubility axis from each point on the curve and read the corresponding solubilities. The values that we read from the graph are

    55°   22.0 g $KClO_3$/100 g water
    40°   14.2 g $KClO_3$/100 g water
    100° 60 g $KClO_3$/100 g water

Of these solubilities, the one at 55°C is probably the most reliable because experimental points are plotted at 50°C and at 60°C. The 40°C solubility value is a bit less reliable because the nearest plotted points are at 30°C and 50°C. The 100°C solubility is the least reliable of the three values because it was taken from the extrapolated part of the curve, and the nearest plotted point is at 80°C. Actual handbook solubility values are 14.0 and 57.0 g of $KClO_3$/100 g of water at 40°C and 100°C, respectively.

The graph in Figure I.4 can also be used to determine whether a solution is saturated or unsaturated. For example, a solution contains 15 g of $KClO_3$/100 g of water and is at a temperature of 55°C. Is the solution saturated or unsaturated? *Answer*: The solution is unsaturated because the point corresponding to 15 g and 55°C on the graph is below the solubility curve—and all points below the curve represent unsaturated solutions.

Appendix II
Vapor Pressure of Water at Various Temperatures

Temperature (°C)	Vapor Pressure (mm Hg)	Temperature (°C)	Vapor Pressure (mm Hg)
0	4.6	26	25.2
5	6.5	27	26.7
10	9.2	28	28.3
15	12.8	29	30.0
16	13.6	30	31.8
17	14.5	40	55.3
18	15.5	50	92.5
19	16.5	60	149.4
20	17.5	70	233.7
21	18.6	80	355.1
22	19.8	90	525.8
23	21.2	100	760.0
24	22.4	110	1074.6
25	23.8		

Appendix III
Units of Measurements

Numerical Value of Prefixes Used with Units

tera	1,000,000,000,000	10^{12}
giga	1,000,000,000	10^{9}
mega	1,000,000	10^{6}
kilo	1,000	10^{3}
hecto	100	10^{2}
deca	10	10^{1}
deci	0.1	10^{-1}
centi	0.01	10^{-2}
milli	0.001	10^{-3}
micro	0.000001	10^{-6}
nano	0.000000001	10^{-9}
pico	0.000000000001	10^{-12}
femto	0.000000000000001	10^{-15}
atto	0.000000000000000001	10^{-18}

Length

1 in. = 2.54 cm
10 mm = 1 cm
100 cm = 1 m
1000 mm = 1 m
1000 m = 1 km
1 mile = 1.61 km
1 Å = 10^{-8} cm

Mass

1 lb = 453.6 g
1000 mg = 1 g
1000 g = 1 kg
1 ounce = 28.3 g
2.20 lb = 1 kg

Volume

1 ml = 1 cm³
1000 ml = 1 litre
1 fluid ounce = 29.6 ml
1 qt = 0.946 litre
1 gal = 3.785 litres

Temperature

°F = 1.8°C + 32

$$°C = \frac{(°F - 32)}{1.8}$$

K = °C + 273
°F = 1.8(°C + 40) − 40
Absolute zero = −273.18°C or −459.72°F

Appendix IV
Solubility Table

	F⁻	Cl⁻	Br⁻	I⁻	O²⁻	S²⁻	OH⁻	NO₃⁻	CO₃²⁻	SO₄²⁻	C₂H₃O₂⁻
Na⁺	S	S	S	S	S	S	S	S	S	S	S
K⁺	S	S	S	S	S	S	S	S	S	S	S
NH₄⁺	S	S	S	S	—	S	S	S	S	S	S
Ag⁺	S	I	I	I	I	I	—	S	I	I	I
Mg²⁺	I	S	S	S	I	d	I	S	I	S	S
Ca²⁺	I	S	S	S	I	d	I	S	I	I	S
Ba²⁺	I	S	S	S	s	d	s	S	I	I	S
Fe²⁺	s	S	S	S	I	I	I	S	s	S	S
Fe³⁺	I	S	S	—	I	I	I	S	I	S	I
Co²⁺	S	S	S	S	I	I	I	S	I	S	S
Ni²⁺	s	S	S	S	I	I	I	S	I	S	S
Cu²⁺	s	S	S	—	I	I	I	S	I	S	S
Zn²⁺	s	S	S	S	I	I	I	S	I	S	S
Hg²⁺	d	S	I	I	I	I	I	S	I	d	S
Cd²⁺	s	S	S	S	I	I	I	S	I	S	S
Sn²⁺	S	S	S	s	I	I	I	S	I	S	S
Pb²⁺	I	I	I	I	I	I	I	S	I	I	S
Mn²⁺	s	S	S	S	I	I	I	S	I	S	S
Al³⁺	I	S	S	S	I	d	I	S	—	S	S

Key: S = soluble in water
s = slightly soluble in water
I = insoluble in water (less than 1 g/100 g H₂O)
d = decomposes in water

Appendix V
Answers to Problems

Chapter 2

C.9 The following statements are correct: b, c, e, f, g, i, j, l, p, r.
D.1 (a) 1 (b) 3 (c) 4 (d) 4 (e) 6 (f) 5 (g) 3 (h) 3
D.2 (a) 3.001 (b) 9.378 (c) 41.13 (d) 25.56 (e) 2144 (f) 82.36
 (g) 20.00
D.3 (a) 8.47×10^2 (b) 5.86×10^{-4} (c) 2.24×10^4 (d) 8.8×10^{-2}
 (e) 6.11×10^{-5} (f) 4.286×10^3 (g) 6.50×10^{-2}
D.4 (a) 38.2 (b) 148.8 (c) 104 (d) 2.44×10^4 (e) $\frac{5}{9}$ (0.556)
 (f) 0.429, 0.733, 0.667, 0.784 (g) 0.889 (h) $7.5X + 30$ (i) 93
 (j) 19.4 (k) 0.9357
D.5 (a) 0.12 m (b) 0.142 km (c) 2.5×10^8 Å (d) 424 mm (e) 30.5 cm
 (f) 8.05 km (g) 2.10×10^2 cm (h) 3.0×10^3 m (i) 10^{-7} cm
 (j) 40 cm (k) 8.66 in. (l) 43.5 miles
D.6 89 km/hr   D.7 1.86×10^5 miles/s
D.8 4.9×10^2 s   D.9 2×10^5 m^2
D.10 (a) 1.200×10^3 mg (b) 0.454 kg (c) 1×10^{-3} kg (d) 1.16×10^3 g
 (e) 5.0×10^{-2} g (f) 2.2×10^3 g (g) 3.50×10^5 mg (h) 5.64×10^{-2} lb
D.11 77.3 kg   D.12 0.32 g
D.13 $2863   D.14 4.83×10^3 cm^3; 4.83 litres
D.15 $19.87   D.16 39 miles/gal
D.17 3.0×10^3 times as heavy   D.18 3×10^4 mg
D.19 8.8×10^4 tons/day; 8.0×10^7 kg/day
D.20 1×10^8 g/day
D.21 (a) 0.145 litre (b) 81.9 cm^3 (c) 568 litres (d) 2.50×10^3 ml
 (e) 6.00×10^3 ml (f) 3.54 in.^{3} (g) 0.661 gal (h) 2.24×10^4 ml
D.22 16 litres   D.23 $15.20
D.24 74 litres; 19 gal
D.25 (a) 60°C (b) −18°C (c) 255 K (d) −24°C (e) 77°F (f) 10°F
 (g) 546 K (h) 273 K
D.26 98.6°F   D.27 −135°F is colder than −90°C
D.28 (a) −40°F = −40°C (b) 11.4°F and −11.4°C
D.29 3.0×10^3 cal   D.30 339 cal
D.31 0.0920 cal/g °C   D.32 16.7°C
D.33 1.039 g/ml   D.34 3.12 g/ml
D.35 7.1 g/ml   D.36 1.19×10^3 g
D.37 72 g   D.38 680 g Hg
D.39 A is Mg; B is Al; C is Ag   D.40 1.94×10^3 g Ag
D.41 0.965 g/ml at 90°C   D.42 (a) 3.20 g/ml (b) 3.20
D.43 100 g ethyl alcohol   D.44 (a) 1.16 (b) 77.7 ml H$_2$SO$_4$

A-21

A-22   Appendix V

Chapter 3   C.17 The following statements are correct: b, c, d, e, g, j.
D.1 −30.3°F   D.2 69.4 g mercury
D.3 (a) 6.25 g oxygen   (b) 60.3% magnesium
D.4 (a) 1.1×10^{14} cal   (b) 3.6×10^{8} gal

Chapter 4   D.1 9.82 g sodium   D.2 78% Cu; 22% Zn
D.3 16.8 g CaO   D.4 2.67 g/ml
D.5 (a) 44.4% S   (b) An atom of Ca has a greater mass.   (c) 16.0 g S
D.6 18 carat gold   D.7 25% H in methane

Chapter 5   C.16 The following statements are correct: b, c, e.
C.23 The following statements are correct: a, d.
C.29 (a) 6.02×10^{23} (b) 6.02×10^{23} (c) 12.04×10^{23} (d) 16.0   (e) 32.0
C.31 The following statements are correct: a, b, e, h, i, l, m, o, p, q, r, s.
D.1 (a) 6.4×10^{5} (b) 5.68×10^{-4} (c) 1×10^{-4} (d) 1.25×10^{2}
D.2 (a) 4,200,000   (b) 0.00009   (c) 0.0001   (d) 35,000
D.3 $K = 2; L = 8; M = 18; N = 32; O = 50; P = 72$
D.4 Ca, 20; Ni, 31; Sn, 69; Pb, 125; U, 143; No, 152
D.5 (a) 3.14×10^{23} atoms Na   (b) 1.5×10^{22} atoms P   (c) 8.03×10^{24} atoms C
      (d) 2.37×10^{23} atoms Cu   (e) 3.01×10^{23} atoms Cd   (f) 6×10^{19} atoms H
D.6 Atomic number 9: F; 9; 19.0; 19.0 g; 3.16×10^{-23} g/atom
      Atomic number 33: As; 33; 74.9; 74.9 g; 1.24×10^{-22} g/atom
      Atomic number 82: Pb; 82; 207.2; 207.2 g; 3.44×10^{-22} g/atom
D.7 (a) 5.33×10^{-23} g/atom S   (b) 1.97×10^{-22} g/atom Sn
      (c) 3.33×10^{-22} g/atom Hg   (d) 6.64×10^{-24} g/atom He
D.8 (a) 7×10^{-17} g Ar   (b) 122 g Al   (c) 106 g Cl   (d) 0.108 g Ag
D.9 (a) 31.4 g C   (b) 9.52 g Cu   (c) 2.43 kg Ag   (d) 2.1×10^{2} g Cl_2
      (e) 2.24 moles Fe   (f) 1.68×10^{-2} mole Sn   (g) 1.00 mole N_2
      (h) 1.69 moles Hg
D.10 (a) 6.02×10^{23} molecules H_2O (b) 6.02×10^{23} atoms O
      (c) 1.20×10^{24} atoms H
D.11 1.2×10^{24} atoms P   D.12 3.64 g Mg
D.13 137 g/mole   D.14 1.5×10^{14} dollars/person
D.15 1.5×10^{19} miles   E.6 2.87 g Na
E.7 58.7 g/mole

Chapter 6   C.28 The following statements are correct: a, d, e, f, g, k, l.

Chapter 7   C.25 The following statements are correct: a, d, e, f, h, j, k, l, m, n, p, q, t, u.

Chapter 9   B.6 The following statements are correct: a, b, e, g, h.
C.1 (a) 149.9   (b) 106.8   (c) 174.3   (d) 228.7   (e) 342.3   (f) 64.5
      (g) 180.0   (h) 159.8   (i) 136.1
C.2 (a) 60.0 g/mole   (b) 331.2 g/mole   (c) 123.0 g/mole   (d) 63.0 g/mole
      (e) 162.1 g/mole   (f) 136.4 g/mole   (g) 352.0 g/mole   (h) 329.1 g/mole
      (i) 237.9 g/mole
C.3 (a) 0.800 mole NaOH   (b) 0.643 mole N_2 (c) 1.56 moles CH_3OH
      (d) 0.244 mole $Ca(NO_3)_2$ (e) 0.0100 mole $MgCl_2$ (f) 6.1 moles KCl
C.4 (a) 0.30 mole Mg atoms   (b) 1.00 mole Ar atoms   (c) 0.338 mole Cl atoms
      (d) 0.500 mole F atoms

C.5 (a) 900 g H_2O (b) 19.0 g $SnCl_2$ (c) 119 g H_3PO_4 (d) 16.0 g O_2
  (e) 4.16×10^{-2} g NH_4Br (f) 72.0 g CH_4
C.6 (a) 6.0×10^{23} molecules F_2 (b) 2.1×10^{23} molecules N_2
  (c) 6.0×10^{22} molecules C_2H_6 (d) 3.76×10^{23} molecules SO_3
C.7 (a) 3.33×10^{-22} g/Hg atom (b) 2.99×10^{-23} g/H_2O molecule
  (c) 6.64×10^{-24} g/He atom (d) 7.31×10^{-23} g/CO_2 molecule
C.8 (a) 1.66×10^{-21} mole C_6H_6 (b) 2×10^{-12} mole Zn
  (c) 1.66×10^{-21} mole CH_4 (d) 9.97×10^{-21} mole NO_2
  (e) 2×10^{-24} mole Mg (f) 3×10^{-18} mole H_2O
C.9 (a) 2.41×10^{24} atoms C (b) 6.92×10^{22} atoms C (c) 1.80×10^{21} atoms C
  (d) 2.41×10^{24} atoms C (e) 7.5×10^{22} atoms C (f) 3.00×10^{10} atoms C
C.10 (a) 30.1 g Ag (b) 7.99 g Br (c) 144 g S (d) 6.69 g Cr
C.11 0.0618 mole K_2CrO_4; 11.1 moles H_2O
C.12 10.3 moles H_2SO_4
C.13 1.77 moles HNO_3
C.14 (a) 60.3% Mg; 39.7% O (b) 7.79% C; 92.2% Cl
  (c) 29.4% Ca; 23.6% S; 47.0% O (d) 38.7% K; 13.8% N; 47.5% O
  (e) 65.9% Al; 34.1% N (f) 2.7% H; 97.3% Cl
  (g) 63.5% Ag; 8.24% N; 28.3% O (h) 52.2% Fe; 44.9% O; 2.8% H
C.15 Mg^{2+}, C^{4+}, Ca^{2+}, K^+, Al^{3+}, H^+, Ag^+, Fe^{3+}
C.16 (a) 77.7% Fe (b) 69.9% Fe (c) 72.3% Fe (d) 15.2% Fe
C.17 (a) Li_2O, 53.7% O (b) MgO, 39.7% O (c) Bi_2O_3, 10.3% O
  (d) TiO_2, 40.1% O
C.18 93.1% Ag, 6.9% O
C.19 Empirical formulas: (a) CuS (b) Cu_2S (c) CaC_2 (d) N_2O_3
  (e) Cl_2O_7 (f) K_2MnO_4 (g) Na_2SO_4 (h) $ZnCO_3$ (i) HClO
  (j) C_3H_8O
C.20 (a) 65.2% Cd (b) 56.3% C (c) 34.7% Mn (d) 35.0% N
C.21 (a) H_2O (b) Na_2MnO_4 (c) $K_2Cr_2O_7$ (d) Both the same
  (e) Na_2SO_4
C.22 Ga_2O_3
C.23 No, not enough Mg to react with all the N_2
C.24 $C_6H_6O_2$
C.25 $C_6H_{12}O_6$
C.26 (a) CCl_4 (b) C_2Cl_6 (c) C_6Cl_6 (d) C_3Cl_8

Chapter 10 C.12 The following statements are correct: a, d, e, f, h, i, j.

Chapter 11 B.1 (a) 0.288 mole MnO_2 (b) 3.57 moles H_2SO_4 (c) 0.282 mole Br_2
  (d) 6.49×10^{-3} mole CCl_4 (e) 17.1 moles NaCl (f) 0.250 mole C_2H_6O
  (g) 9.32 moles CO_2 (h) 1.88 moles O_2 (i) 0.100 mole HNO_3
  (j) 3.0 moles HCl
B.2 (a) 17.6 g C_3H_8 (b) 40.5 g Al (c) 15.0 g H_2 (d) 17.0 g $AgNO_3$
  (e) 152 g $FeSO_4$ (f) 33.4 g $AlCl_3$ (g) 118 g Au (h) 6.6 g $Ni(NO_3)_2$
B.3 (a) $\dfrac{2\text{ Mg}}{O_2}$, $\dfrac{2\text{ Mg}}{2\text{ MgO}}$, $\dfrac{O_2}{2\text{ MgO}}$, $\dfrac{O_2}{2\text{ Mg}}$, $\dfrac{2\text{ MgO}}{2\text{ Mg}}$, $\dfrac{2\text{ MgO}}{O_2}$

(b) $\dfrac{2\text{ Al}}{6\text{ HCl}}$, $\dfrac{2\text{ Al}}{2\text{ AlCl}_3}$, $\dfrac{2\text{ Al}}{3\text{ H}_2}$, $\dfrac{6\text{ HCl}}{2\text{ Al}}$, $\dfrac{6\text{ HCl}}{2\text{ AlCl}_3}$, $\dfrac{6\text{ HCl}}{3\text{ H}_2}$, $\dfrac{2\text{ AlCl}_3}{2\text{ Al}}$, $\dfrac{2\text{ AlCl}_3}{6\text{ HCl}}$,

$\dfrac{2\text{ AlCl}_3}{3\text{ H}_2}$, $\dfrac{3\text{ H}_2}{2\text{ Al}}$, $\dfrac{3\text{ H}_2}{6\text{ HCl}}$, $\dfrac{3\text{ H}_2}{2\text{ AlCl}_3}$

A-24 Appendix V

(c) $\dfrac{3\,Zn}{N_2},\ \dfrac{3\,Zn}{Zn_3N_2},\ \dfrac{N_2}{3\,Zn},\ \dfrac{N_2}{Zn_3N_2},\ \dfrac{Zn_3N_2}{3\,Zn},\ \dfrac{Zn_3N_2}{N_2}$

(d) $\dfrac{2\,C_2H_6}{7\,O_2},\ \dfrac{2\,C_2H_6}{4\,CO_2},\ \dfrac{2\,C_2H_6}{6\,H_2O},\ \dfrac{7\,O_2}{2\,C_2H_6},\ \dfrac{7\,O_2}{4\,CO_2},\ \dfrac{7\,O_2}{6\,H_2O},\ \dfrac{4\,CO_2}{2\,C_2H_6},\ \dfrac{4\,CO_2}{7\,O_2},$

$\dfrac{4\,CO_2}{6\,H_2O},\ \dfrac{6\,H_2O}{2\,C_2H_6},\ \dfrac{6\,H_2O}{7\,O_2},\ \dfrac{6\,H_2O}{4\,CO_2}$

B.4 (a) 8.0 g CH_4 (b) 4.0 g CO   B.5 3.4 moles HCl
B.6 4.0 moles CO_2, 8 moles SO_2 B.7 16 moles O_2
B.8 0.80 mole $Na_2S_2O_3$, 2.1 moles $KMnO_4$, 0.27 mole H_2O
B.9 21.0 g HNO_3 B.10 55.6 moles O_2
B.11 (a) 1.0 mole $FeCl_3$ (b) 5.8 moles HCl (c) 25 g $K_2Cr_2O_7$
    (d) 1.08 g $CrCl_3$ (e) 2.3 moles H_2O (f) 7.68 g $FeCl_3$
B.12 560 g CaO   B.13 323 kg C
B.14 77.8 g H_2O, 181 g Fe   B.15 3.54 g weight loss (as O_2)
B.16 (a) 8.96 g H_3PO_4 (b) 9.98 g Fe_2O_3 (c) 15.2 g SiF_4 (d) 2.13 g B_2O_3
B.17 $MgCl_2$
B.18 (a) KOH is the limiting reagent; HCl is in excess
    (b) $Bi(NO_3)_3$ is the limiting reagent; H_2S is in excess
    (c) H_2O is the limiting reagent; Fe is in excess
    (d) C_2H_6 is the limiting reagent; O_2 is in excess
B.19 571 g CH_3OH; CO is the limiting reagent
B.20 94.0% yield   B.21 890 g CaC_2
B.22 372 kg Li_2O
B.23 The following statements are correct: a, c, e.

Chapter 12

B.4 6.6×10^3 g   B.8 56.0 litres
C.12 The following statements are correct: a, d, g, i, j, l, m, n, o.
D.1 (a) 0.829 atm (b) 24.8 in. Hg (c) 12.2 lb/in.^{2} (d) 630 torr
    (e) 840 mbar
D.2 0.691 atm   D.3 (a) 150 ml (b) 600 ml
D.4 (a) 514 mm Hg (b) 1.03×10^3 mm Hg
D.5 80 atm
D.6 (a) 3.24 litres (b) 2.29 litres (c) 2.37 litres (d) 3.56 litres
D.7 317°C   D.8 27 lb/in.^{2}
D.9 1.54×10^3 ml   D.10 2.11×10^3 mm Hg
D.11 1.75×10^5 litres   D.12 36.9 litres
D.13 112 litres   D.14 2.37 moles
D.15 0.112 mole H_2 D.16 46.5 g/mole
D.17 87.8 g/mole   D.18 0.255 litre
D.19 (a) 1.34 g/litre (b) 3.58 g/litre (c) 3.17 g/litre (d) 0.179 g/litre
D.20 (a) 2.32 g/litre (b) −78°C   D.21 79.5 g/mole
D.22 (a) 22.4 litres (b) 0.267 litre (c) 11.2 litres
    (d) 3.68 litres
D.23 6.99 litres   D.24 1350 mm Hg
D.25 1.07×10^{24} molecules   D.26 719 mm Hg
D.27 383 ml   D.28 0.365 g CH_4
D.29 He = 380 mm Hg; Ne = 228 mm Hg; Ar = 152 mm Hg
D.30 121 litres H_2
D.31 (a) 5 moles NH_3 (b) 6.2 moles O_2 (c) 100 litres NH_3 and 125 litres O_2
    (d) 165 litres O_2 (STP) (e) 179 litres (STP) of NO

Answers to Problems A-25

D.32 (a) 513 litres O_2 (b) 373 litres SO_2
D.33 No CO; 2.5 moles O_2; 15 moles CO_2
D.34 3 cu ft H_2; 1 cu ft CO   D.35 47.6 litres air
D.36 65.6% $KClO_3$
D.37 (a) 21.3 litres O_2 (b) 798 g C_2H_6 (c) 9.94 g N_2 (d) 53.5 litres CO_2

Chapter 13

C.42 The following statements are correct: a, b, c, f, h, l, m, o.
D.1 (a) 2.00 moles H_2O (b) 0.819 mole H_2O
D.2 62.89% H_2O D.3 9.00×10^4 cal
D.4 7.88×10^3 cal
D.5 (a) 9.72×10^3 cal/mole (b) 4.43×10^4 cal
D.6 Yes, there is sufficient ice   D.7 80.4 g H_2O
D.8 (a) 2.8 moles O_2 (b) 79.2 litres O_2
D.9 9.88 litres O_2 D.10 $CdBr_2 \cdot 4H_2O$
D.11 61.0 g $CuSO_4 \cdot 5H_2O$ D.12 1.24×10^3 litres H_2
D.13 (a) 18.0 g H_2O (b) 18.0 g H_2O (c) 0.783 g H_2O (d) 0.460 g H_2O
D.14 0.851 atm, or 647 mm Hg
D.15 7×10^{18} molecules/sec (7 billion billion molecules/sec)
D.16 92 g H_2SO_4

Chapter 14

C.25 The following statements are correct: a, c, d, e, h, j, k, m
D.1 17% $CuSO_4$
D.2 (a) 9.1% KCl (b) 17% KCl (c) 17% $MgCl_2$ (d) 6.2% $KMnO_4$
D.3 (a) 3.0 g NaCl (b) 37.5 g KCl
D.4 3.6 g NaCl
D.5 88.6 g H_2O must evaporate
D.6 Yes, there is sufficient KOH solution
D.7 240 g sugar
D.8 208 g H_2SO_4 solution; 178 ml H_2SO_4 solution
D.9 (a) 0.700 M $CaBr_2$ (b) 2.24 M NH_4Cl (c) 1.50 M NaOH
    (d) 0.0682 M $BaCl_2 \cdot 2H_2O$
D.10 (a) 6.00 moles $CaCl_2$ (b) 0.338 mole $KC_2H_3O_2$ (c) 0.500 mole $AgNO_3$
     (d) 0.180 mole HNO_3 (e) 0.00800 mole NaOH (f) 3.75 moles KF
D.11 (a) 19 g KCl (b) 26.6 g Na_2SO_4 (c) 98 g H_3PO_4
     (d) 0.455 g $Zn(NO_3)_2$
D.12 8.90 M HBr
D.13 (a) 33 ml 12 M HCl (b) 75 ml 16 M HNO_3 (c) 167 ml 18 M H_2SO_4
     (d) 4.4 ml 17 M $HC_2H_3O_2$
D.14 91.5 ml
D.15 (a) 0.25 M (b) 0.65 M (c) 0.68 M
D.16 (a) 1.0 mole $FeCl_3$ (b) 0.33 mole $CrCl_3$ (c) 6.7×10^{-3} mole $K_2Cr_2O_7$
     (d) 83 ml 0.080 M $K_2Cr_2O_7$ (e) 16 ml 6 M HCl
D.17 (a) 1.0 mole $MnCl_2$ (b) 2.5 moles Cl_2 (c) 8.0 moles HCl
     (d) 1000 ml 0.100 M HCl (e) 2.10 litres Cl_2
D.18 (a) 3.17 g $BaCrO_4$ (b) 12 ml 1.0 M K_2CrO_4
D.19 (a) 9.52 g Cu (b) 2.24 litres NO
D.20 0.150 mole H_2 D.21 2.14 M HCl
D.22 $Al(OH)_3$ will neutralize more acid
D.23 (a) 37.0 g/eq wt (b) 26.0 g/eq wt (c) 29.2 g/eq wt (d) 23.9 g/eq wt
     (e) 56.1 g/eq wt

A-26 Appendix V

D.24 (a) 85.5 g/eq wt   (b) 12.2 g/eq wt   (c) 32.7 g/eq wt
    (d) 103.6 g Pb/eq wt; 31.8 g Cu/eq wt   (e) 18.6 g Fe/eq wt;
    (f) 23.2 g/eq wt
D.25 (a) $-3.72°C$ (b) 2.00 m   D.26 (a) $-12.0°C$ (b) 103.4°C
D.27 (a) $-0.6°C$ (b) 83.1°C   D.28 258 g/mole
D.29 $C_6H_{12}O_6$
D.30 (a) 6.67×10^3 g ethylene glycol   (b) 6.01×10^3 ml

Chapter 15

C.37 The following statements are correct: a, b, c, d, e, f, i, k, l, m, o, p
D.1 (a) 0.10 M Na^+, 0.10 M Cl^- (b) 0.32 M K^+, 0.32 M NO_3^-
    (c) 2.50 M Na^+, 1.25 M SO_4^{2-} (d) 0.68 M Ca^{2+}, 1.36 M Cl^-
    (e) 0.22 M Fe^{3+}, 0.66 M Cl^- (f) 0.75 M Mg^{2+}, 0.75 M SO_4^{2-}
    (g) 0.15 M NH_4^+, 0.050 M PO_4^{3-} (h) 0.100 M Al^{3+}, 0.150 M SO_4^{2-}
D.2 0.263 M $MgBr_2$
D.3 6×10^{18} Al^{3+} ions; 2×10^{19} Cl^- ions
D.4 (a) 0.210 M HCl   (b) 0.243 M HCl   (c) 1.19 M HCl   (d) 0.257 M NaOH
    (e) 0.0637 M NaOH
D.5 1.68 M Mg^{2+}; 3.36 M Cl; 0.840 mole AgCl
D.6 0.131 M $Ba(OH)_2$
D.7 2.69 moles C_6H_6; 6.49 moles CCl_4
D.8 239 ml C_6H_6; 627 ml CCl_4
D.9 0.118 M HCl
D.10 0.468 litre H_2
D.11 (a) 3   (b) 1.0   (c) 7   (d) 9.07   (e) 0.30
D.12 (a) 3.7   (b) 2.80   (c) 5.20   (d) 10.49
D.13 126 ml
D.14 20.0 weight percent $BaCl_2$; 1.15 M $BaCl_2$
D.15 3.33 litres 18.0 M H_2SO_4
D.16 2.23 g AgI

Chapter 16

C.23 The following statements are correct: b, c, d, e, f, g, i, l, m, n
D.1 2.40 moles HI
D.2 (a) 2.4 moles HI   (b) 0.22 mole H_2; 0.42 mole I_2; 2.6 moles HI
D.3 1.38 moles H_2; 0.269 mole I_2; 0.250 mole HI
D.4 0.52 mole H_2; 0.52 mole I_2; 4.0 (3.95) moles HI
D.5 256 times faster at 100 than at 20°C.
D.6 (a) 0.422 mole Cl_2 (b) 16.4 litres Cl_2
D.7 $K_{eq} = 57$
D.8 Propanoic 1.4×10^{-5}; hydrofluoric, 7.9×10^{-4}; hydrocyanic, 4.0×10^{-10}
D.9 $[H^+] = 1.9 \times 10^{-3}$; pH = 2.72   D.10 1.1×10^{-2} (1.06×10^{-2}) M NO_2^-
D.11 1.2×10^{-3} MH^+ D.12 4.2% ionized
D.13 3.38
D.14 (a) 2.50×10^{-10} (b) 1.2×10^{-23} (c) 7.9×10^{-26} (d) 8.19×10^{-12}
    (e) 1.0×10^{-15} (f) 2.4×10^{-5} (g) 5.13×10^{-17} (h) 1.80×10^{-18}
D.15 $\sim 4 \times 10^{-2}$ (4.5×10^{-2}) mole/litre; ~ 0.7 (0.75) g/100 ml
D.16 (a) 3.9×10^{-8} mole/litre   (b) 8.7×10^{-4} mole/litre
    (c) 1.6×10^{-2} mole/litre   (d) 1.3×10^{-4} mole/litre
D.17 (a) $CaCO_3$ (b) Ag_2CO_3
D.18 (a) $[H^+] = 3.6 \times 10^{-5}$; pH = 4.4 (4.44)   (b) $[H^+] = 1.8 \times 10^{-5}$;
    pH = 4.7 (4.74)

Answers to Problems   A-27

Chapter 17 C.11 The following statements are correct: a, c, e, g, j, k, m
D.1 25.3 g Cl$_2$; 7.99 litres Cl$_2$ (STP)   D.2 3.0 moles SO$_2$
D.3 1,000 ml KMnO$_4$ solution; 5.60 litres O$_2$
D.4 68.0 g Cu   D.5 24.7 g Br$_2$
D.6 275 g I$_2$ D.7 1.5 moles Fe^{2+}

Chapter 18 C.22 The following statements are correct: c, e, f, g, i, l, m, n
D.1 6836   D.2 $190,000
D.3 0.390 g Sn-119 left   D.4 ~22,700 years old
D.5 24 minutes/half-life
D.6 (a) mass defect = 0.0305 g   (b) binding energy = 6.7 × 10^{11} cal

Chapter 19 C.1 (a) 12.7 kg NaCl   (b) 109 moles Cl$_2$
C.2 (a) 664 g NaCl   (b) 127 litres Cl$_2$
C.3 2.4 × 10^8 gals.
C.4 The mineral was magnetite
C.5 329 tons bauxite ore

Chapter 20 C.1 105 litres SO$_2$ C.2 29.7 litres Cl$_2$
C.3 60 g Cl$_2$ per hour   C.4 28.1 kg HNO$_3$
C.5 0.536 mole Br$_2$
C.6 (a) 18 M H$_2$SO$_4$ (b) 16 M HNO$_3$ (c) 12 M HCl

Chapter 21 B.23 The following statements are correct: a, b, c, e, f, i, k, m, n, r

Chapter 22 B.23 The following statements are correct: b, c, d, g, i, j, l, m, n

Chapter 23 B.28 The following statements are correct: a, e, f, g, h, i, k, l, m, o, r, u, v

Chapter 24 B.34 The following statements are correct: a, b, e, g, h, i, j, m, o

Chapter 25 B.21 The following statements are correct: b, c, f, h, i, k, m, o, p

Chapter 26 B.38 The following statements are correct: a, b, f, g, h, i, j, m, n, o, p, r, s, t, w

Chapter 27 B.21 The following statements are correct: a, b, d, f, g, i, k

Chapter 28 B.20 62.5°
B.23 The following statements are correct: a, f, g, h, i, k, l

Chapter 29 B.31 About 2000 glucose anhydride units
B.34 The following statements are correct: b, d, f, g, h, j, l, n, o, p

Chapter 30 B.41 38% protein
B.42 6.8 × 10^4
B.43 The following statements are correct: b, c, d, f, g, i, l, m, n, r

Chapter 31 B.34 The following statements are correct: a, b, d, g, h, j, m, p

Chapter 32 B.38 The following statements are correct: a, c, d, f, h, i, j, l, m, o, p

Index

Absolute temperature scale, 19–20
  graphic representation, 201
Absolute zero, 19, 201
Absorption, 723
Acetal, 561–562
  in carbohydrates, 649, 654
Acetaldehyde (ethanol), 563–564
Acetic acid, 282
  equilibrium equation, 291, 293
  equilibrium solutions of, 293
  ionization of, 293
  ionization constant of, 325–326
  properties of, 38
Acetone, 565
Acetyl-Co Enzyme A, 663, 727, 731, 754
  biosynthesis of cholesterol, 748
  fat formation, 750–751
  from fats and fatty acids, 750, 753
Acetylene, 138, 508–510
  preparation from calcium carbide, 509
  reactions of, 229, 509–510
Achiral, 629
Acid anhydride, 231
  reaction with water, 231
Acid chloride, 586
Acids:
  binary, 142
  definitions of, 282–285
  hydrohalic, 426
  naming of, 142–145
  organic, 574–586
  of phosphorus, 444
  properties of, 283
  as a proton donor, 283–284
  reactions of, 285
  ternary, 144
Acid-base theories:
  Arhhenius, 283
  Brønsted-Lowry, 283
  Lewis, 284–285
Active transport, 723
  Activity Series of Metals, 350–351
  principles for use of, 351
  table of, 351
ADP (adenosine diphosphate), 698–700
Adrenalin, 735, 738–739
Aerobic sequence (*see* Citric acid cycle)

Aerosol, 452
Air, composition of, 428, 450
Air pollutants, 449–450
  carbon monoxide, 454–456
  classification, 451–452
  effect on visibility, 452, 460–461
  hydrocarbons, 454
  major sources, 452
  nitrogen oxides, 456–457
  ozone, 457–459
  sulfur dioxide, 452–453
Air pollution, 448–468
  major episodes, 450–451
  from motor vehicles, 452, 453, 454, 456, 457, 465–466
  particulate matter in, 451–452
  school smog and health warnings, 452
  smog, 451, 459–460
  temperature inversion effect, 461–463
Air pollution control, 463–466
Air quality standards, 464–466
Alchemy, 3
Alcohols, 526–540
  classification, 528–529
  dehydration, 534
  esterification, 536
  ethyl, 537–538
  isopropyl, 263, 538–539
  methyl, 537
  naming of, 530–531
  olefins from, 534–535
  oxidation of, 533–534, 582
  physical properties, 531–532
  polyhydroxy, 529
  preparation of, 536
  reactions of, 533–535
  table of, 529
Aldehydes, 552–568
  from alcohols, 563
  naming of, 553–555
  physical properties, 556
  preparation, 563–565
  reactions of, 558–563
  table of, 554
Aldol condensation, 560–561
Aldose, 642
Algebraic equations, A-5–A-6
Alkali, definition of, 385

A-29

Alkali metals, 385–390
  chemical properties of, 387–389
  compounds and uses, 389–390
  family in the periodic table, 102
  hydroxides, 283
  physical properties, 385–386
  reaction with halogens, 388
  reaction with oxygen, 389
  reaction with water, 387–388
Alkaline earth metals, 391–398
  biological properties, 398
  chemical and physical properties, 394–396
  compounds of, 396–398
Alkanes, 473
  bonding in, 474–476
  general formula for, 474
  names of, 474
  reactions of, 485
  rules for naming, 482
  structural formulas of, 472–480
Alkenes, 495
  bond formation in, 496
  general formula for, 495
  naming, 499–501
  reactions of, 503–507
Alkyl group, 481
  table of, 481
Alkyl halides, 486
Alkynes, 495
  bond formation, 496
  general formula for, 495
  naming, 499
  preparation of, 508–509
  reactions of, 509–510
Allotropes:
  of carbon, 235, 462–463
  of oxygen, 235
  of phosphorus, 443–444
  of sulfur, 431
Allotropic forms of elements, 235
Alloys, 390–391
  composition of common, 391
Alpha rays or particles, 363–364
  properties of, 365
Alum, 409
Aluminum, 405–409
  chemical properties of, 408
  production of, 406–407
Amalgam, 390
Amide:
  formation of, 585
  naming, 585
  table of, 586
Amines, 596–600
  chemical properties, 598
  heterocyclic, 597–598
  naming, 596
  physical properties, 598
  salt formation, 599
Amino acids, 581, 669–673
  absorption and distribution, 757–758
  amphoterism of, 672
  chromatogram of, 688

Amino acids *(continued)*
  classification, 669, 673
  essential, 669–671
  isoelectric point of, 673
  oxidative deamination, 759
  stereoisomerism of, 671–672
  table of, 670–671
  transamination, 760
  utilization of, 759
Amino acid pool, 757–758
Ammonia:
  equilibrum with hydrogen and nitrogen, 321
  as a Lewis base, 284
  synthesis by Haber process, 321
  water solution of, 440
Ammonium hydroxide, 148
  decomposition of, 440
  ionization of, 440
  weak base, 293
Ammonium ion:
  acid properties of, 284–285
  structure of, 126, 434
Amorphous, 34
Amphoterism, 287
  in amino acids, 672
Amylopectin, 661–662
Anaerobic pathway *(see* Embden-Meyerhof)
Analgesic, 581
Anesthetic, 547
Angstrom unit, 13, 66
Aniline, 514, 596, 597, 598
Anion, 113, 290
Anode, 289, 352–356
Anomers, 648
Answers to problems, A-21–A-27
Anti-codons, 709
Antipyretic, 581
Apoenzyme, 681
Aqua regia, 442
Arhhenius, Svante, 283
Aromatic compounds, 510
  naming, 513
  fused ring compounds, 517
Aromatic hydrocarbons, 510, 517
  halogenation, 518–519
  properties of, 518
  reactions of, 518–520
  sources, 518
Aromatic ring systems, 510–513, 517
Artificial radioactivity, 368
Aspirin, 581
Asymmetric carbon atom, 625
Atherosclerosis, 750
Atmosphere, composition of, 428
Atmospheric pressure, 195–197
Atom:
  definition of, 49
  kernel of, 84
  relation to an element, 49
  size of, 66–67
Atomic bomb, 372–374
  U-235 and Pu-239 in, 373
Atomic mass, and atomic weight, 87–88

Atomic mass unit, 87
Atomic number:
 definition of, 75
 and protons in the nucleus, 75, 79
 and radioactivity, 363
Atomic radii of elements, 101–102
Atomic structure, 64–91
 diagrams of, 75, 76, 78, 81–83
 electron arrangement in, 72–75, 77–82
 nucleus of, 67, 68
Atomic theory:
 Bohr's, 69–70
 Dalton's, 65
 quantum mechanics, 70–72
Atomic weight:
 and atomic mass, 87–88
 definition of, 88
 reference standard for, 87
 table of (see back endpapers)
ATP (Adenosine triphosphate), 727–730, 751, 754
 in Embden-Meyerhof pathway, 729
 from citric acid cycle, 730–732
 structure, 699, 754
Automobile exhaust standards, 466
Avogadro's Hypothesis, 210
Avogadro's number, 89–90
 relationship to gram-atomic weight, 89, 179
 relationship to mole, 90, 179

Baekland, Leo, 608
Baeyer test, 507
Balances for mass measurement, 23–24
Banting, Frederick, 739
Barfoed's reagent, 656
Barometer, mercury, 195–196
Bartlett, N., 86
Bases:
 definition of, 283–284
 naming of, 148
 properties of, 283
 as proton acceptors, 283–284
 reactions of, 286–287
Basic anhydride, 230
 reaction with water, 230–231
Becquerel, A. H., 361
Benedict's reagent, 656
Benzaldehyde, 514, 532, 564–565
Benzene, 510
 bonding in, 511–513
 disubstituted derivatives of, 514–515
 monosubstituted derivatives of, 513–514
 polysubstituted derivatives of, 515–516
 properties of, 511–512, 518–519
 reactions of, 518–519
 structure of, 510–513
Benzene ring, naming and numbering, 516
Berzelius, J. J., 52
Bessemer, Henry, 402
Best, Charles, 739
Beta rays or particles, 362, 363–364, 365
Bile acids, 722
Bile salts, 722

Binary compounds, 59, 139–143
 naming of, 59, 139–143
Biochemistry, 4
Biodegradability, 595–596
Bisulfate, 147, 435
Bisulfites, 147, 433
Blast furnace, 400–401
Bohr, Niels, 69
Bohr's atomic theory, 69–70
Boiling point, 38
 definition of, 242
 elevation constant, 272
 normal, 242
Bond (see also Chemical bond):
 double, 472, 495
 length, 226, 512
 Pi (π), 486
 Sigma, 472, 476
 single, 472
 triple, 472, 495
Bond dissociation energy, 122
 and electronegativity, 124
Boyle, Robert, 3, 198
Boyle's Law, 198–201
Bromine, 421–422
 electron structure of, 58, 118
 laboratory preparation, 421
 physical properties of, 414
 from seawater, 421–422
Bromine water, 230
Brønsted, J. N., 283
Brønsted-Lowry theory of acids and bases, 283–285
Brown, Robert, 305
Brownian movement, 305
Büchner, Eduard, 681
Buffer solution, 332
 changes in pH, 333

Calcium carbide, 509
Calculations from chemical equations, 178–187
 limiting reagent in, 185
 mole ratio method, 179–181
 theoretical yield, 186
Calculations from chemical equations involving gases, 214–217
Calibrated glassware for measuring liquid volume, 24
Calorie, definition of, 21
Calvin, Melvin, 726
Carbides, metal, 466–467
Carbohydrates, 641–664
 classification, 641
 definition, 641
 digestion of, 718
 metabolism, 644, 726
  anaerobic sequence, 729
Carbon:
 allotropic forms of, 235
 different forms of, 471
 isotopes of, 380
 tetrahedral structure of, 471, 475
Carbon-12, atomic weight reference standard, 87

Carbon atom, 471
Carbon-carbon bonds, 472
   bond lengths, 512
Carbon cation, 505
Carbon dioxide:
   from carbonates and acids, 172, 301
   structure of, 120, 124
Carbon disulfide, 432
Carbon monoxide:
   effect on health, 454-456
   from cigarettes, 456
   from incomplete combustion, 454
   from motor vehicles, 454
   properties of, 454
Carbon tetrachloride, 120, 126
Carbonic acid, 293
Carbonium ion, 505
Carboxyl group, 574
Carboxylic acids, 574-587
   classification of, 578
   formation of esters from, 587
   naming of, 574-575
   nomenclature of, 57
   physical properties of, 577
   preparation of, 581-584
   reaction with bases, 584
   reactions of, 584-587
   table of, 576, 745
Catabolism of fatty acids, 751-754
Catalyst, 323
   definition, 323
   effect on equilibrium, 323, 335
   in preparation of oxygen from hydrogen
      peroxide, 234
Cathode, 289, 352-356
Cation, 113, 269
Cavendish, H., 438
Celsius temperature scale, 19-20
Cellulose, 663-664
   compounds of, 663-664
   structure, 663
Centigrade temperature scale (*see* Celsius scale)
Cephalins, 747
Chain reaction, 370-371
Charcoal, 471
Charles J. A. C., 201
Charles' Law, 201-204
Chemical bond, 121-125
   coordinate covalent, 124
   covalent, 121-122
      formation in a hydrogen molecule, 122
   definition of, 121
   electrovalent, 121
   ionic, 121
   nonpolar covalent, 123
   polar covalent, 123
Chemical change, definition of, 39
Chemical engineering, 4
Chemical equation(s):
   balancing, 165-169
   definition of, 40, 59, 164
   format for writing, 60, 166
   information available from, 169

Chemical equation(s) *(continued)*
   introduction to, 59-60
   ionic, 301-302
   molecular, 301
   oxidation-reduction, balancing, 341-349
   rules for writing and balancing, 166
   symbols used in, 165
   types of, 170-172
Chemical equilibrium, 313-336
   effect of catalysts on, 323
   effect of change in concentration, 318-320
   effect of pressure on, 320-322
   effect of temperature on, 322-323
   equilibrium constants, 324
   of hydrogen and iodine, 334-335
   ionization constants, 325
   reversible reactions, 314
   solubility product constants, 329
Chemical formulas:
   characteristics of, 54-55
   explanation of symbols in, 55
   oxidation number tables, 128-129
   from oxidation numbers, 129-131
Chemical kinetics, 315
Chemical properties, 37-38
Chemistry:
   branches of, 4
   definition of, 1
   history of, 2-4
   how to study, 7-8
   relationship to other sciences, 4
Chiral, 629
Chloride ion:
   hydration of, 255, 290
   size of, 114-115
   structure of, 113
Chlorine:
   commercial preparation of, 417
   compounds of, 417-420
   diatomic molecule, 58, 118
   electron structure of, 118, 413
   laboratory preparation of, 417-418
   oxides of, 418
   as an oxidizing agent, 420
   oxy-acids of, 145
   properties of, 37-38, 414
   purification of water, 420
Chlorine atom, size of, 114-115
Chlorine water, 230, 420
Chlorophyll, 726
Cholesterol, 748, 750
Cholic acids, 722
Citric acid cycle (Krebs), 730-732, 754
   ATP from, 730
Classes of organic compounds (table), 526-527
Classification of matter, 33-37
Coal, 471, 518
Coal tar, 400, 473, 518
Codon, 707, 708
Coke, 473, 518
Coenzyme, 681
   vitamins as, 682, 736
Coenzyme A(CoA), 752

Colligative properties, 271–274
　boiling point elevation, 272
　freezing point depression, 272
　osmotic pressure, 272
　vapor pressure and, 272–273
Colloids, 302–308
　preparation of, 305
　properties of, 303
　size of particles, 303, 305
　stability of, 306
Combination reaction, 170
Combustion, 429
Common names of compounds, table of, 138
Compound(s):
　comparison with mixtures, 56
　definite composition of, 54
　definition of, 52
　formulas of, 54–56
　number of, 53
　percentage composition of, 155
　relationship to a molecule, 52
Concentration of solutions, 259–271
　molar, 263
　normal, 269
　weight-percent, 261
Conductivity apparatus, 288
Conservation of energy, law of, 43
Conservation of mass, law of, 3, 41
Contact process for sulfuric acid, 435
Coordinate covalent bond, 124
Copernicus, Nicolaus, 3
Copolymer, 612
Copper, reaction with oxygen, 39–40
Copper (II) sulfate pentahydrate, 138, 232
Corey, R. B., 676
Cottrell, Friederick, 307
　process, 307
Courtois, Bernard, 414
Covalent bond, 121–122
Crick, F. H. C., 703, 705
Curie, I. J., 368
Curie, M. S., 361
　charge of radioactive particles, 365
Curie, P., 361
Curie, unit of radioactivity, 369
Cycloalkanes, 489–491

Dalton, John, 3, 65
Dalton's atomic theory, 65
Dalton's law of partial pressures, 208
Davy, Sir Humphrey, 414
de Broglie, Louis, 70
DDT, 138, 476
Decomposition reaction, 171
Definite composition:
　and hydrates, 231
　law of, 54
Deliquescence, 233
Democritus, 64
Density, 24–28
　of gases, 25
　of liquids and solids, 24–25
Dextrins, 662

Dextrorotatory, 624
Diabetes mellitis, 739
　and hyperglycemia, 739
Dialysis, 307
Diastereomers, 633–634
Diatomic molecules, 58
Dicarboxylic acids, 579
Dichlorobenzenes, 515
Digestion (human), 718–722
Digestive fluids, 718–720
　enzymes, 719–720
Digestive tract (human), 719
Dimensional analysis, 13–16, A-12
Dipole, 123
Disaccharides, 641, 653
　composition of, 653
　glycosidic linkage in, 653–655
　structure of, 653–655
Dissociation and ionization, 289–292
Distillation, 235–236
Division of numbers, A-4, A-10
DNA (deoxyribonucleic acid), 700
　and genetic code of life, 707
　composition of, 700
　hydrogen bonding in, 703–704
　replication of, 705
　structure of, 701–704
　table of purines and pyrimidines, 702
Doebereiner, J. W., 96–97
Doebereiner's triads, 96
Dorn, F. E., 86
Double replacement reaction, 172–173
Double stranded helix, 702–704

Efflorescence, 233
Einstein, Albert, 44–45
　mass-energy relationship equation, 44, 375
Electrolysis, 352
　of hydrochloric acid, 352–353
　and oxidation-reduction, 348, 355
　of sodium chloride brines, 352–353
　of water, 43–44
Electrolyte(s):
　classes of compounds, 288–289
　definition of, 289
　equilibrium of, weak, 317, 325
　freezing point of solutions, 292
　ionization of, 289–292
　as ions in solution, 292
　strong, 292
　table of strong and weak, 293
　weak, 292
Electrolytic cell, 288, 352
Electron(s):
　discovery of, 66
　energy levels of, 72–74, 77–82
　orbitals of, 70, 71, 74
　order of filling orbitals, 78, 83
　properties of, 66–67, 71
　quantum numbers of, 71
　rules for atomic structure, 71
　stable outer shell structure, 85–86
　valence, 112, 121

Electron-dot structure:
  of atoms, 84–85
  definition of, 84
  to illustrate sharing of electrons, 118–119
  use in illustrating compound formation, 114–120
  of oxy-acids of chlorine, 145
Electron orbital:
  definition of, 70
  hybrid types:
    sp, 498–499
    sp^2, 496
    sp^3, 475
  shape of s and p orbitals, 74
  sublevel energy designation, 73–74
Electron shells, 72
  maximum number of electrons in, 72
Electron structure of elements, table of, 81–82
Electron transfer in compound formation, 114–120
Electronegativity:
  and the covalent bond, 118–120, 121–123
  definition of, 119
  table of, 119
Electroplating of metals, 353
Element(s):
  alchemical symbols for, 52
  allotropic forms, 235
  atomic structure of, 65–86
  atomic weights of, (*inside back cover*)
  classification of, 57–58
  composition in the human body, 50
  definition of, 48
  distribution of, 48
  electron structure of, 71–86
  metals, 57, 58
  names of, 50–51
  nonmetals, 57–58
  number of, 48
  occurring as diatomic molecules, 58
  periodic table of, 99 (*see also back endpapers*)
  radioactive, 363
  rules governing symbols of, 50
  symbols for, 50–52 (*see also back endpapers*)
  transuranium, 375–377
Embden-Meyerhof pathway, 727, 729, 733
Empirical formula:
  calculation of, 157–161
  definition of, 156
Enantiomers, 627–632
Endothermic reactions, 173
  and equilibrium, 322–323
Energy:
  in chemical changes, 43, 173, 322–323
  definition of, 42
  kinetic, 42
  mass relationship, 43–44, 375
  potential, 42
Energy levels, of electrons, 71–75, 78
Environmental pollution, 449, 595
Enzymes, 681–685
  as catalysts, 681
  classes of, 682
  cornstarch to fructose by, 684

Enzymes (*continued*)
  "induced fit" model, 683
  lock and key theory, 683
  mechanism and action of, 683
  specificity of, 682, 683
  table of, 685
Epimers
Epinephrine (*see* Adrenalin)
Equations (*see also* Chemical equations):
  algebraic, A-5
  nuclear, 364
Equilibrium (*see also* Chemical equilibrium), 316
  chemical, 313–336
  constants, 324–332
Equivalent weight, 269–271
  and Avogadro's number, 271
Esters, 582, 584, 587–592
  formation of, 536, 586
  naming of, 587–588
  odors and flavors, 588
  properties, 588–589
  table of, 588
  waxes, 589
Ethanol (ethyl alcohol), 138, 537–538
  physiological effect, 538
  synthesis of, 538
  uses, 520
Ethers, 544–547
  anesthetic, 547
  hydroperoxide formation, 546
  naming of, 544–545
  preparation, 547
  properties, 545–546
  table of, 544
Ethylene, 495
  polymerization of, 608, 609–611
  reactions of, 504–507
Ethylene glycol, 506–507, 539
Evaporation, 239
Evylyn, John, 450
Exothermic reactions, 173
  and equilibrium, 322–323
Exponents, A-8–A-10

Factor-label method of calculation, 13–16, A-12
  (*also see* Dimensional analysis)
FAD, 723, 725, 754
Fahrenheit temperature scale, 19–20
Families of elements, 100, 102
  comparison of A and B Groups, 103
Faraday, Michael, 305, 497
Fats, 589, 744
  absorption and distribution, 750
  composition of, 744–745
  digestion of, 718
  energy content, 751
  storage and utilization, 754
  structure of, 645
  table of, 590
Fatty acids, 590, 592, 744
  biosynthesis of, 755–757
  in fats, 590

Index A-35

Fatty acids *(continued)*
    oxidation of, 751
    relation to the citric acid cycle, 753–754
    saturated, 590
    unsaturated, 590
Fehling's solution, 558–559, 656
Fission, nuclear, 370–371
Flame tests, 398–399
Fluorine, properties of, 414–415, 417
Formaldehyde, 554, 563
    from methyl alcohol, 563
Formula(s), 54–56
    of common classes of organic compounds, 526–528
    condensed structural, 478–479
    electrovalent compounds, 129
    empirical, 156
    of familiar substances, 138
    molecular, 156
    molecular weight of, 152
    from oxidation numbers, 131–133
    structural, 476–480
    tables, 128–129
Formula weight, 151, 178
    definition of, 151
    determination from formula, 152
Fractions and decimals, A-2–A-4
Frasch process, mining sulfur, 429, 430
Free radical reaction, 486–487
Freezing point, 243
    solid-liquid equilibrium, 242
Freezing point depression constant, 272
Fructose, 651–652
    osazone of, 660
    from starch, 684
    structure, 651
    sweetness of, 656
Functional groups in organic compounds, 526–528
Fusion, nuclear, 374–375

Galactose, 644
    structure of, 651
Galoctosemia, 644
Galilei, Galileo, 3
Gamma rays, 364–365, 367
Gas(es):
    Avogadro's hypothesis, 209
    Boyle's Law (pressure-volume relationship), 198–201
    calculations from chemical equations, 214–217
    Charles' Law (volume-temperature relationship) 201–205
    collection over water, 208
    Dalton's law, 207–209
    density of, 25, 212
    diffusion of, 193
    general properties of, 34, 191
    ideal, 192
    ideal gas equation, 217
    kinetic energy of, 194
    measurement of pressure of, 194–197
    relationship of pressure and number of molecules, 197
    standard pressure, 205
    weight-volume relationship, 210–212
Gasoline, 473, 507–508

Gay-Lussac, J. L., 209
Geiger counter, 369, 378
Genetic code, 707
    table of codons, 708
Genetics, 695
Genetic transcription, 708–710
Ghiorso, Albert, 376
Glucagon, 734, 739
Gluconeogenesis, 733
Glucose:
    Benedict's test for, 656
    concentration in blood, 738
    from glycogen, 738–739
    from photosynthesis, 726
    lactic acid from, 729
    mutarotation of, 647–648
    occurrence, 644
    osazone of, 660
    pyruvic acid from, 729
    relation to *(citric acid)* cycle, 730
    renal threshold for, 738
    structure of, 645, 647
Glucose tolerance test, 739–740
Glycerol (glycerine), 529, 539, 592, 593, 615, 656
    from hydrolysis of fats, 592
Glycogen, 662–663, 727
    in carbohydrate metabolism, 738–739
    from glucose, 738–739
Glycogenesis, 651, 727
Glycogenolysis, 651, 727
Glycolipids, 659, 744, 747
    structure of, 747
Glycolysis, 729
Glycosidic linkage, 653–655
Glycosuria, 644
Goodyear, Charles, 613
Graham, Thomas, 194
Graham's law of diffusion, 194
Gram, 16
Gram-atomic weight:
    definition of, 88
    relationship to atomic weight, 88
    relationship to Avogadro's number, 89, 179
Gram-formula weight, 153
    relationship to mole, 179
Gram-molecular volume, 210
Gram-molecular weight:
    relationship to Avogadro's number, 179
    relationship to molecular and formula weight, 153
Graphical representation of data, A-13–A-17
Graphite, 235
Grignard, Victor, 566
Grignard reagent, 566
    reaction with aldehydes, 567
    reaction with ketones, 567
Group VIA elements, 427–437

Haber process for ammonia, 321
Hahn, O., 370, 376
Half-life, 363
    of radium isotopes, 363
Hall, Charles M., 407

Halogenation of alkanes, 486–487
Halogens, 413–427
　biological effects, 427
　binary acids of, 142–143
　discovery of, 414
　electron-dot structure of, 118, 413
　interhalogen compounds, 425
　oxidizing strength, 425–426
　oxy-acids of, 145, 426
　physical characteristics of, 414
　reducing strength, 426
Hard water, 235
Heat:
　capacity, 22
　in chemical reactions, 164
　definition of, 21
　of fusion, 224
　of reaction, 174
　of vaporization, 224
α-Helix, 676–677
Hemiacetal, 561, 649
　in carbohydrates, 650
Hemiketal, 561
Heroult, Paul L., 407
Heterogeneous, 36
Holmes, Oliver Wendell, 547
Holoenzyme, 681
Homogeneous, 36
Homologous series, 474
Hormones, 691, 733–735, 738
　control of blood glucose, 738
　　adrenalin, 738, 739
　　insulin, 738, 739
Hydrates, 231–232
　dehydration of, 232
　table of, 232
　water of hydration, 231
Hydrides, 388
Hydrocarbons, 473
　in air pollution, 454
　alkanes, 473–474
　alkenes, 495
　alkynes, 495
　aromatic, 510
Hydrochloric acid, 283, 284
　in aqua regia, 442
　properties of, 419–420
　reaction with metals, 286
Hydrogen:
　atomic structure of, 75
　from electrolysis of hydrochloric acid, 352
　from electrolysis of water, 43, 229
　isotopes of, 76
　molecule, 58, 122
　preparation from metals and acids, 171, 286
　preparation from metals and bases, 287
　preparation from metals and water, 171
Hydrogen bond:
　definition of, 227
　in DNA, 703–704
　in hydrogen fluoride, 228, 417
　in water, 228
Hydrogen bromide, preparation of, 422

Hydrogen chloride:
　boiling point of, 437
　chemical bond in, 119
　preparation of, 437
　reaction with water, 283
Hydrogen fluoride:
　etching glass with, 417
　and hydrogen bonding, 228, 409
Hydrogen halides, reducing ability of, 426
Hydrogen iodide, preparation of, 424
Hydrogen ion:
　from acids, 283–284
　concentration in water, 294, 328
　and pH, 294–298
　in water as hydronium ion, 294
Hydrogen peroxide:
　properties of, 233–234
　structure of, 233
Hydrogen sulfide, 433–435
Hydrohalic acids, acidity of, 426
Hydrolysis:
　of esters, 536
　of proteins, 685
　of triglycerides, 592
Hydrometer, 27
Hydronium ion:
　formation of, 283–284, 291
　structure of, 284
Hydroxide(s):
　amphoteric, 287
　reaction with acids, 286, 298–299
Hydroxide ion:
　concentration in water, 294, 328
　from bases, 283
　electron-dot structure, 126
　and pOH, 329
Hygroscopic substances, 232
Hyperglycemia (see also Diabetes mellitis) 738, 739
Hypertonic, 276
Hypochlorous acid, in chlorine water, 230, 420
Hypoglycemia, 738, 739
Hypothesis, 6–7
Hypotonic, 276

Ice:
　density of, 226
　equilibrium with water, 225
　heat of fusion, 224
Immiscible, 252
Insulin:
　beef, structure of, 678–679
　in carbohydrate metabolism, 738
　principal function of, 734
　synthesis, 691
Inulin, 664
Invert sugar, 656
Iodine:
　dietary need of, 427
　laboratory preparation of, 423
　reactions of, 424
　source of, 423
Iodine solubility:
　in alcohol, 424

Iodine solubility *(continued)*
   in carbon tetrachloride, 424
   in potassium iodide solution, 424
   in water, 424
Iodoform test, 565–566
Ion(s):
   definition of, 52
   formation of, 113–117, 289
   hydrated in solution, 290–292
   polyatomic, 125
   size of, 114–115
   in solution, 290–292
   tables of, 128, 129
Ionic bonds, 121
Ionic equations, 299
   net ionic, 301
   rules for writing, 302
   total ionic, 301
Ionization, 289–292
   constants, 325–326
      of weak acids (table), 326
   demonstration of (apparatus), 288
   theories of, 283–285
Ionization energy, 110–112
   definition of, 110
   table of the elements, 111
Iron, 399–405
   cathodic protection of, 405
   chemical tests for, 403–405
   pig, 402
Isoelectric point, 673
   and pH, 673
Isomerism, 476–480
   optical, 621–636
Isomers, 478
   geometric, 501–503
Isopropyl alcohol, 262, 529, 530
   from propene, 538–539
   oxidation of, 534
   uses, 539
Isotonic, 276
Isotopes, 76–77
   definition of, 76
   of hydrogen, 76
   radioactive, 362, 363, 368
   symbolism for, 77, 362
IUPAC, 140
IUPAC rules for naming organic compounds:
   alcohols, 530
   aldehydes, 553
   alkanes, 482–485
   alkenes, 499
   alkyl halides, 488
   alkynes, 499
   carboxylic acids, 575
   dicarboxylic acids, 579
   esters, 587
   ethers, 545
   ketones, 553–554

Kekule, August, 511
   structure for benzene, 511
Kelvin temperature scale, 19–21
Kepler, Johannes, 3

Kernel of an atom, 84
Kerosene, 385, 473
Ketals, 561
Ketones, 553
   from alcohols, 534
   iodoform test, 565
   naming, 553–555
   reaction with Grignards, 567
   reactions, 559–563
   uses, 565
Ketose, 642, 643
Kilocalorie (Cal), 21
Kinetic energy, 42, 193–194
Kinetic-molecular theory, 192
Knoop, Franz, 751
   beta carbon oxidation theory, 751–752
Koshland, Jr., D. E., 683
Kreber, H. A., 731
Kurchatov, I., 376
K_w, 328

Lactic acid, 581
   from glucose metabolism, 728, 729
   stereoisomerism of, 625–627
Lactose, 653
   structure of, 654
Lavoisier, Antoine, 3
Law, 7
   Boyle's, 198–201
   Charles', 201–204
   Dalton's, 207
   Gay-Lussacs, 209
   Graham's, 194
   of combining volumes of gases, **209**
   of conservation of energy, 43
   of conservation of mass, 3, 41
   of definite composition, 54
   of mass action, 324
   of octaves, 97
   of partial pressures (Dalton's),
      207–209
   Periodic, 97–98
Lawrence, E. O., 367
Lead storage battery, 356
Le Bel, J. A., 625
Le Chatelier, H., 317
   principle of, 317
Lecithin, 746
Length, measurement of, 12
Levorotatory, 624
Lewis, G. N., 284
   theory of acids and bases, 284–285
Libby, W. F., 380
Lipids, 743–750
   classification of, 743
   uses of, 751
Lipmann, F. A., 731
Lipogenesis, 755–757
Lipoproteins, 750
Liquids, boiling point of, 241–242
   characteristics of, 34
   evaporation of, 239–241
   vapor pressure of, 240–241
Lithium compounds and uses, 388

Litre, 17–18
Logarithm scale, 296
Long, Crawford, W., 547
Lowry, T. M., 283

McMillan, E.M M., 376
Mac Leod, J. J., 652
Macromolecules, 607
Magnesium:
    compounds of, 397
    preparation from seawater, 394
    properties of, 395
Malonyl-CoA, 755, 756
Maltose, 653, 720
    structure of, 654, 657
Manganese dioxide, as oxygen preparation catalyst, 234
Markovnikov's rule, 506
Mass:
    defect, 375
    definition of, 9
    measurement of, 16–17
Mathematical Review, A-1–A-17
Matter:
    definition of, 32
    physical states of, 33–34
Measurements, 9–28
    International System, 11
Mechanism of reactions, 333–336
Meiosis, 706
Melting point, 243
Mendeleev, D. I., 97
    and the periodic table, 97–98
Mercuric oxide:
    composition of, 54
    decomposition of, 40, 164
Mercury, 137, 138
    in amalgams, 390
    barometers, 195–196
Merrifield, R. B., 691
Meso structure, 634
Metabolism:
    anabolism, 726
    carbohydrate, 726, 727
        human, 727
    catabolism, 726
    definition of, 726
Metal nitrides, 439
Metalloids, 57, 58
Metallurgy, definition of, 390
Metals:
    action on acids, 171, 286
    action on water, 171, 229
    alkali, 385–390
    alkaline earth, 391–398
    aluminum, 405–409
    flame tests for, 398–399
    general properties of, 57
    iron and steel, 399–405
    reaction with sodium and potassium hydroxide, 287
Methane:
    combustion of, 485

Methane *(continued)*
    reaction with chlorine, 486–488
    structure of, 120, 477, 478
Methanol (methyl alcohol), 138, 518, 519
    synthesis of, 537
    uses, 537
Metre, 12
Metric system of measurements, 10
    prefixes used in, 10–11, A-19
    units of, 10–11, 13, 17, 19, A-19
Meyer, Lothar, 97
Miescher, Friederich, 696
Millibar pressure, 196
Mirror images, 627, 629
Miscible, 252
Mitosis, 705
Mixture(s):
    comparison with compounds, 56
    definition of, 36, 56
Moissan, Henri, 414
Molality,
Molar solution:
    definition of, 263
    preparation of, 264
Molarity, 263–269
    of concentrated acid solutions, 265
Mole(s), 88–91, 153, 169, 178–179
    and Avogadro's number, 89–90, 153, 179
    definition of, 89
    equations for calculating, 179
    and molecular weight, 153
    relationship to volume of a gas, 210–211, 216
Mole ratio, 180
    use in calculations, 180, 187
Mole volume relationship of gases, 211, 214
Molecular equation, 301–302
Molecular formula:
    calculation from empirical formula, 160
    definition of, 156
Molecular orbital, 476
Molecular weight:
    definition of, 151
    determination from formula, 151
    relationship to mole, 153
Molecule, relation to a compound, 52–53
    definition of, 52
Monomer, 608
Monosaccharides, 641, 644
    reactions of, 658–661
Morton, William T., 547
Moseley, H. G. J., 98
Multiplication of numbers, A-1, A-3, A-4
    use of powers of ten in, A-8, A-9
Muriatic acid, 138, 420
Mutagen, 713
Mutarotation, 647–648
Mutations, 713–714

NAD, 723, 724, 732
Naming inorganic compounds, 137–149
Naming organic compounds, 480
    alcohols, 530

Naming organic compounds *(continued)*
  aldehydes, 553
  alkanes, 482–485
  alkenes, 499
  alkyl halides, 488
  alkynes, 499
  amides, 585
  amines, 596
  aromatic hydrocarbons, 513
  carboxylic acids, 574
  esters, 587
  ethers, 544
  functional groups, 526–527
  IUPAC system, 482
  ketones, 553
  phenols, 540
  radicals, 481
Naphthalene, 473, 517
Natta, Guilio, 614
Neutralization reaction, 172, 229, 286, 298
Neutron(s):
  determination from atomic weight and atomic number, 88
  and isotopes, 76
  location in the atom, 69
  in the nucleus, 76–77, 82
  properties of, 66–67
Newlands, J. A. R., 97
Newton, Isaac, 3
Nitrate ion, electron-dot structure, 126
Nitric acid, 440
  concentrated, molarity, 265
  as an oxidizing agent, 441
  Ostwald process, 441
  reactions of, 441
  structure of, 441
Nitric oxide, formation of, 439
Nitrogen, 438
  concentration in the atmosphere, 195, 438
  family of elements, 437–445
  laboratory preparation of, 438
  liquid, 438
  structure of, 118, 438
Nitrogen balance, 758
Nitrogen cycle, 442
Nitrogen dioxide-dinitrogen tetroxide equilibrium, 314–315
Nitrogen oxides, 440
  and air pollution, 456
  effect on health, 456
  formation in combustion of fuels, 456–457
  in photochemical smog, 457, 459
Nitroglycerin, 540
Nitrous acid:
  as oxidizing agent, 426
  weak acid, 293
Noble gases, 85–86
  compounds of, 86
  electron structure of, 86
  in periodic table, 102
Noble gas structure in compounds, 113, 115
Nomenclature:
  of bases, 148
  of binary acids, 142

Nomenclature *(continued)*
  of binary compounds, 59, 139–143
    containing metals of varying oxidation numbers, 140
    containing two nonmetals, 141
  common or trivial names, 137–138
  of inorganic compound, 137–149
  of salts, 139–141, 145–147
    with more than one positive ion, 147
  of ternary compounds, 143–145
  prefixes used in, 142
  Stock System, 140
Nonelectrolyte, 288–289
Nonmetals, general properties of, 57
Nonpolar covalent bond, 123
Normality, 269–271
Nuclear binding energy, 375
Nuclear equation, 364, 368
Nuclear fission, 370–371
  chain reaction, 370
Nuclear fusion, 374
Nuclear power, 371–372
Nuclear reactions, mass-energy relationship, 44–45, 369, 375
Nuclear reactor, 372
Nucleic acids, 696, 700
  DNA, 700
  RNA, 700
Nucleoproteins, 696
Nucleosides, 696, 697
Nucleotide, 698–699
  high energy, 699
  polynucleotides, 700
Nucleus:
  atomic, 67–69
  charge of, 67–69, 75
  mass of, 67–69, 77
Numerical prefixes of units, table of, 11, A-19
Nylon, 615

Octane number, 507
Oersted, H. C., 406
Oils, 589, 590, 591, 743
  composition of, 590
  hydrogenation of, 591
  polyunsaturated, 590
  structure of, 589
Oligosaccharides, 641
Optical activity, 624
Optical isomerism, 621–636
  projection formulas, 625–627
Orbital(s):
  electron, 70–72
  hybrid, 475
    sp, 497–499
    sp^2, 496
    sp^3, 475–476
  molecular, 476
  shape of s and p, 73–74
Organic acids, 574
  (*see also* carboxylic acids)
Organic bases, 598
Organic chemistry, 4
  definition, 470

Organic chemistry *(continued)*
  history, 469
Organic compounds:
  classification, 470
  naming, 480
  table of, 526–527
Osazones, 659–660
Osmosis, 274–276
Osmotic pressure, 275
Oxidation, definition of, 133, 343
Oxidation number(s), 126–133, 341
  in covalent compounds, 126, 342
  determination from formulas, 131
  and electronegativity, 127, 342
  of Group A elements, 127
  in ionic compounds, 126, 341
  of ions in ionic compounds, 128–133
  rules for, 127, 342
  writing formulas from, 129–131
Oxidation number table(s), common elements and ions, 128, 129 *(also see front endpapers)*
Oxidation-reduction:
  definition of, 133, 343–344
  equations, balancing, 345–349
    ionic, 348–349
Oxidative deamination of amino acids, 759
Oxides:
  of antimony, 445
  of arsenic, 445
  carbon dioxide, 120, 124
  carbon monoxide, 454–456
  of chlorine, 418
  of nitrogen, 410, 440, 454–456
  of phosphorus, 444
  of sulfur, 432, 435, 452, 453
Oxidizing agent, 344
Oxy-acids:
  of chlorine, electron-dot structure, 145, 426
  of halogens, 426
Oxygen:
  abundance in nature, 49–50
  and combustion, 429
    of alkali metals, 389
    of carbon compounds, 485
    of hydrogen, 43–44
    of magnesium, 42
    of methane, 485
  commercial source, 428
  composition in air, 195, 427, 428
  electron-dot structure of, 118, 234, 427
  from hydrogen peroxide, 233, 428
  isotopes of, 77, 427
  from mercuric oxide, 39–40, 41, 164, 422
  from metal oxides, 428
  physical properties, 428
  reaction with phosphorus, 443
Ozone:
  and air pollution, 234, 457–459
  danger to humans, 458
  danger to vegetation, 458
  electron-dot structure of, 234
  in photochemical smog, 234, 459

Ozone *(continued)*
  preparation of, 234
  in the stratosphere, 458

Paracelsus, 3
Pasteur, Louis, 625, 681
Pauling, Linus, 119, 676–678
Pentoses, 652
Peptide linkage, 674
Percentage composition of compounds, 155
Periodic Law, 97–98
Periodic table of the elements, 99 *(see also back endpapers)*
  general arrangement, 98, 100
  groups or families of elements, 102–103
  and new elements, 105
  periods of elements, 101
    atomic radii within, 101–102
  predicting formulas from, 104
  according to sublevel electrons, 100, 101
  transition elements, 105
Peroxide structure, 233
Petroleum, 473
pH, 294–298
  comparison of hydrochloric and acetic acid solutions, 295
  meter, 298
Phase, 36
Phenol, 504, 540–543, 616
Phenylketonuria, 714
Philosopher's stone, 2
Phosphate ion, electron-dot structure, 126
Phosphatidic acids, 746
Phospholipids, 744, 746
  structure of, 746
Phosphoric acid, structure of, 444
Phosphorous acid, structure of, 444
Phosphorus:
  acids of, 444
  allotropic forms of, 443
  oxides of, 444
  properties of, 438
Photochemical smog, 459–461
  aldehydes in, 460
  nitrogen oxides in, 459, 460
  ozone in, 459
  PAN in, 460
  role of sunlight, 459
Photosynthesis, 641, 718, 726
  glucose from, 726
Physical change, definition of, 38
Physical properties, 37
Physiological saline solution, 276
Pi (π) bond, 497
Planck, Max, 70
Plaster of Paris, 138, 232
Plutonium, 373, 376
pOH, 328, 329
Polarimeter, 623, 624
Polarized light (plane), 622, 623
Polyamides, 615
  nylon, 615
Polyesters, 614
  alkyd, 615

Polyesters *(continued)*
  Dacron, 615
  Mylar, 615
Polyethylene, 608, 609, 610
Polymer, 607–618
  butadiene, 611, 612
  classification, 608
  phenol-formaldehyde, 616
  silicone, 617
  table of, 610
  thermoplastic, 609
  thermosetting, 609
Polymerization, 608
  addition, 609
  condensation, 614, 616
  copolymer, 612
  free radical, 611
  polyamides, 615
  polyurethane, 617
  stereochemistry of, 613
Polypeptides, 674–676
  α-helix, 676, 677
  β-pleated sheet, 676, 677
  C-terminal, 674
  N-terminal, 674
  primary structure of, 688–690
  synthesis of, 690–691
  vinyl, 611
Polysaccharides, 642, 661–663
Polyurethanes, 617
  foam or sponge, 617
Potassium:
  chemical properties of, 387–389
  compounds and uses, 390
  preparation of, 387
Potassium chlorate, preparation of oxygen, 428
Potential energy, 42
Powers of ten, A-8–A-10
Pressure:
  atmospheric, 194–197
  definition of, 194
  dependence on molecules, 197
  units of, 196
  variation with altitude, 197
Priestley, Joseph, 3, 438
Problem solving, methods of, 13–16, A-12–A-13
Projection formula, 625–627
Properties, definition of, 37
Proteins, 668, 681
  amino acids in, 670–671
  analysis of, 688–690
  biosynthesis of, 710–713
  chemical bonds in, 679
  common foods containing, 668
  conjugated, 668
  denaturation of, 686
  digestion of, 638
  as enzymes, 668, 681
  hydrolysis of, 685
  molecular weight of, 676, 678
  nucleoproteins, 696
  RNA control of synthesis, 709, 710–713

Proteins *(continued)*
  structure of, 676–681
  simple, 668
  synthesis of, 690
  tests for, 687
Proton(s):
  acceptor, 283–284
  and atomic structure, 69, 76, 79
  donor, 283–284
  hydrogen ion, 283, 294
  properties of, 66–67
  relationship to atomic number, 75
Purines in DNA, 696, 702
Pyrimidines in DNA and RNA, 696, 702

Quantum Mechanics Theory, 70
Quantum numbers, 71–72
Quicksilver (mercury), 137, 138

Racemic mixture, 632, 635
Radiation, biological effects of, 377
Radioactive decay, 362
Radioactive "fallout," 373
Radioactivity, 362
  artificial or induced, 368
  discovery of, 361–362
  disintegration series, 366, 367
  half-life, 363
  measurement of, 369
  natural, 362
  particles from, 363–366
  units of, 369
Radiocarbon dating, 380
Radioisotope applications, 378–380
Raleigh, Lord, 85
Ramsey, Sir William, 86
Rate of reaction, 315
  effect of catalysts on, 323
  effect of concentration on, 318
  effect of pressure on, 320
  effect of temperature on, 322
Reaction(s):
  combination, 170
  decomposition, 171
  double replacement, 172–173
  heat in, 173–174
  neutralization, 172
  oxidation-reduction, 343
  reversible, 313–315
  single replacement, 171
  in solutions, 258
Redox, 343
Reducing agent, 344
Reducing sugar, tests for, 656–658
Reduction, definition of, 133, 343, 344
Renal threshold, 738
Reversible reactions, 313–315
Riboflavin, 754
RNA (ribonucleic acid), 700, 701
  and genetic transcription, 708
  messenger, 708, 709
  in protein synthesis, 710–713

RNA *(continued)*
  ribosomal, 708
  structure of, 701
  transfer, 708, 709
Roentgen, W. K., 361
Rounding off numbers, A-11
Rubber, 607, 608, 611–613
"Rule of eight," 85
Rutherford, Daniel, 438
Rutherford, Ernest, 67, 68, 69, 362
  determination of the charge of the nucleus, 67

Salts, 145–147
  as electrolytes, 288
  formation from acids and bases, 286–287
  naming of, 145–147
  from neutralization reactions, 172, 173
  solubility curves of, 256
  solubility, table of, 253, A-20
Sanger, Frederick, 678
Saponification, 583, 592
  of esters, 583
  of fats or oils, 592
Scheele, K. W., 414, 438
Schrodinger, E., 71
Scientific law, 7
Scientific method, 6–7
Semipermeable membrane, 275
Semipolar bond *(see* Coordinate covalent bonds)
Sex hormones, 735, 748
Sharing of electrons between atoms, 118–120
Sickle cell anemia, 714
Significant figures, 10, A-6
  in calculations, A-11
Silicone polymers, 617
Silver halides, solubility of, 253, 427
Simplest formula *(see* Empirical formula)
Single replacement reaction, 171
Smog, 451
  major episodes, 450–451
  photochemical, 459
Soap, 592–594
Sodium:
  atom, size of, 115, 385
  compounds and uses, 389
  ion:
    hydration of, 254–255, 290
    preparation of, 353, 386
    properties of, 385
  reaction with chlorine, 114, 388
  reaction with hydrogen, 388
  reaction with oxygen, 389
  reaction with water, 171, 229, 387
Sodium chloride:
  composition of, 155
  crystal lattice structure, 114
  dissociation (ionization) of, 290, 292
  formation from sodium and chlorine, 114, 388
  uses of, 389
Sodium hydroxide, 148, 283, 287, 389
Sodium hypochlorite, bleach, 420
Sodium sulfate, structure of, 125, 432

Sodium thiosulfate, structure of, 432
Solid, characteristics of, 33, 36
Solubility, 252–258
  of alkali metal halides (table), 256
  effect of concentration on, 257–258
  effect of particle size on, 257
  effect of pressure on, 257
  effect of temperature on, 257
  and the nature of the solute and solvent, 254
  rate of dissolving, 257
  rules for mineral substances, 253
  table, A-20
  of various compounds in water, 256, A-20
Solubility product constant, 329
  table of, 331
Solute, 251
  size of in solution, 251
Solutions, 250–276
  alkaline or basic, 283
  concentrated, 259
  concentration of, 259–271
  definition of, 251
  dilute, 259
  dissolving process, 254–255
  molal, 273
  molar, 263
    preparation of, 264
  normal, 269–271
  properties of, 251
  as a reaction zone, 258–259
  saturated, 260
  supersaturated, 261
  types of, 251
  unsaturated, 260
  volume-percent, 263
  weight-percent, 261
Solvent, 251
Specific gravity, 27–28
  of gases, 212
Specific heat, 22
Specific rotation, 623
Spectator ions, 295
Spectroscopy, 398–399
Standard conditions, 205
Standard temperature and pressure (STP), 205
Starch, 661–662
  degradation to fructose, 684
  hydrolysis of, 662
  iodine test for, 662
  structure of, 662
States of matter, 33, 36
Steel, 402–403
  alloys in, 403
  basic oxygen process, 402
  Bessemer process, 402
  open-hearth process, 402
Stereoisomers, 622
  number of, 633
Steroids, 744, 748
Stoichiometry, 179
Strassman, F., 370
Strong electrolytes, 292–293

Subatomic parts of the atom, 65–67
    arrangement of, 69
Subenergy level of electrons, 73–75
    maximum number of electrons in, 73
    order of placement of, 73–75, 78
    rules for electron placement in, 73–75
Sublimation, 240
Subscripts in formulas, 55
Substance, 36
Sugar (sucrose); (see also carbohydrates)
    dehydration by sulfuric acid, 436
    formula of, 53
    hydrolysis of, 653, 655
    properties of, 38, 655
    reducing, 656–658
    structure of, 655
    sweetness table, 656
Sulfate ion, electron-dot structure, 125
Sulfites, 143, 146, 433
Sulfur, 429–431
    allotropic forms of, 431
    chemical properties of, 431–434
    effects of heating, 431
    history of, 429
    isotopes of, 429
    mining of, 429, 430
    physical properties of, 429, 431
Sulfur dioxide, 432–433
    and air pollution, 433, 452, 453
    effects on health, 453
    uses of, 433
Sulfur family of elements, 427–437
Sulfuric acid, 434–437
    in air pollution, 453
    care in diluting with water, 435
    common name of, 138
    contact process, 435
    as a dehydrating agent, 436
    electron-dot structure, 125
    formula of, 55, 145
    ionization of, 435
    as an oxidizing agent, 436
    physical properties of, 434
    preparation of other acids with, 437
    reactions as an acid, 435
    salts of, 146, 147
    uses of, 434
Sulfurous acid, 433
    electron-dot structure, 125
    formula of, 144, 145
    salts of, 146, 147
Synthetic detergent, 592, 593, 594
    anionic, 595
    biodegradable, 595–596
    cationic, 595
    nonionic, 595

Teflon, 417, 610
Temperature:
    conversion formulas, 20, A-19
    definition of, 21

Temperature (continued)
    scales, 19–20, A-19
Temperature inversion (atmospheric), 461–463
Ternary compounds, 143
    naming of, 143–145
Tetraethyl lead, 508
Theory, 7–8
Thio-, meaning of, 432
Thomson, J. J., 66
Titration, 299
TNT (trinitrotoluene), 520
Tollens' reagent, 558, 559, 656
Toluene, 473, 514, 518, 520
Transamination of amino acids, 760
Transition elements, 105
Transmutation of elements, 367–368
Transuranium elements, 375–377
Travers, M. W., 86
Triglycerides, 589, 744, 745
    composition of (table), 590
    digestion of, 750
    hydrogenation, 591
    hydrogenolysis, 591
    hydrolysis, 592
    saponification of, 592
    structure of, 589–590
Tyndall effect, 305
Tyndall, John, 305

Units of measurement, tables of:
    length, 13, A-19
    mass, 17, A-19
    volume, 18, A-19
Uranium-235, 370–373
Uranium-238, 358, 359, 362, 368
    disintegration series of, 366–367
Urea:
    decomposition to ammonia, 684
    from protein metabolism, 759
    Wöhler's synthesis of, 469–470

Valence electrons, 112
Van't Hoff, J. H. 625
Vapor pressure, 240–241
    curves, 242–243
    of ethyl alcohol, 241, 243
    of ethyl chloride, 243
    of ethyl ether, 241, 243
    liquid-vapor equilibrium, 238
    of water, 241, 243
        table of, A-18
Villard, P., 362
Vitamins, 735–737
    table of, 736
Volume, measurement of, 17–18
    and Avogadro's hypothesis, 210
    gases versus liquids, 196
    and gas laws, 198–205
    gram-molecular, 210–211
Voltaic cell, 348, 351
Vulcanization, 613

Water: 223–244
 atomic composition of, 53, 54, 226
 boiling point of, 19–20, 226
 chemical properties of, 171–172, 229–231
 density of, 24–25, 226
 distillation of, 235–236
 electrolysis of, 44, 229
 equilibrium with ice, 225
 equilibrium with steam, 225
 fluoridation, 235
 freezing point of, 19, 20, 226
 hard, 235
 ionization of, 294
 K_w, ion product constant, 327–328
 natural, 235
 occurrence of, 224
 pH of, 294, 329
 physical properties of, 38, 53, 224
 pOH, 329
 pollution of, 237–239
 purification of, 235–237
 reaction with metal oxides, 230–231

Water *(continued)*
 reaction with metals, 229–230
 reaction with nonmetal oxides, 230–231
 reaction with nonmetals, 229–230
 softening of, 235–237
 structure of molecule of, 120, 226
 vapor pressure of, 241, 243
 vapor pressure, table of, A-18
Water of crystallization, 231
Water of hydration, 231
Watson, J. D., 703, 705
Wax, 589, 744
Weak electrolytes, 292–293
Weight, definition of, 10
Winkler, L., 98
Wöhler, F., 469–470, 759

X rays, 361

Zeolite, 236
Ziegler, Karl, 614
Zwitterion, 672–673

Table of Atomic Weights (Based on Carbon—12)

Name	Symbol	Atomic No.	Atomic Weight	Name	Symbol	Atomic No.	Atomic Weight
Actinium	Ac	89	(227)[a]	Mendelevium	Md	101	(256)[a]
Aluminum	Al	13	26.98154	Mercury	Hg	80	200.59
Americium	Am	95	(243)[a]	Molybdenum	Mo	42	95.94
Antimony	Sb	51	121.75	Neodymium	Nd	60	144.24
Argon	Ar	18	39.948	Neon	Ne	10	20.179
Arsenic	As	33	74.9216	Neptunium	Np	93	237.0482[b]
Astatine	At	85	(210)[a]	Nickel	Ni	28	58.71
Barium	Ba	56	137.34	Niobium	Nb	41	92.9064
Berkelium	Bk	97	(249)[a]	Nitrogen	N	7	14.0067
Beryllium	Be	4	9.01218	Nobelium	No	102	(254)[a]
Bismuth	Bi	83	208.9084	Osmium	Os	76	190.2
Boron	B	5	10.81	Oxygen	O	8	15.9994
Bromine	Br	35	79.904	Palladium	Pd	46	106.4
Cadmium	Cd	48	112.40	Phosphorus	P	15	30.97376
Calcium	Ca	20	40.08	Platinum	Pt	78	195.09
Californium	Cf	98	(251)[a]	Plutonium	Pu	94	(242)[a]
Carbon	C	6	12.011	Polonium	Po	84	(210)[a]
Cerium	Ce	58	140.12	Potassium	K	19	39.098
Cesium	Cs	55	132.9054	Praseodymium	Pr	59	140.9077
Chlorine	Cl	17	35.453	Promethium	Pm	61	(145)[a]
Chromium	Cr	24	51.996	Protactinium	Pa	91	231.0359[b]
Cobalt	Co	27	58.9332	Radium	Ra	88	226.0254[b]
Copper	Cu	29	63.546	Radon	Rn	86	(222)[a]
Curium	Cm	96	(247)[a]	Rhenium	Re	75	186.2
Dysprosium	Dy	66	162.50	Rhodium	Rh	45	102.9055
Einsteinium	Es	99	(254)[a]	Rubidium	Rb	37	85.4678
Erbium	Er	68	167.26	Ruthenium	Ru	44	101.07
Europium	Eu	63	151.96	Samarium	Sm	62	150.4
Fermium	Fm	100	(253)[a]	Scandium	Sc	21	44.9559
Fluorine	F	9	18.99840	Selenium	Se	34	78.96
Francium	Fr	87	(223)[a]	Silicon	Si	14	28.086
Gadolinium	Gd	64	157.25	Silver	Ag	47	107.868
Gallium	Ga	31	69.72	Sodium	Na	11	22.98977
Germanium	Ge	32	72.59	Strontium	Sr	38	87.62
Gold	Au	79	196.9665	Sulfur	S	16	32.06
Hafnium	Hf	72	178.49	Tantalum	Ta	73	180.9479
*Hahnium	Ha	105	(260)[a]	Technetium	Tc	43	98.9062[b]
Helium	He	2	4.00260	Tellurium	Te	52	127.60
Holmium	Ho	67	164.9304	Terbium	Tb	65	158.9254
Hydrogen	H	1	1.0079	Thallium	Tl	81	204.37
Indium	In	49	114.82	Thorium	Th	90	232.0381[b]
Iodine	I	53	126.9045	Thulium	Tm	69	168.9342
Iridium	Ir	77	192.2	Tin	Sn	50	118.69
Iron	Fe	26	55.847	Titanium	Ti	22	47.90
Krypton	Kr	36	83.80	Tungsten	W	74	183.85
*Kurchatovium	Ku	104	(260)[a]	Uranium	U	92	238.029
Lanthanum	La	57	138.9055	Vanadium	V	23	50.9414
Lawrencium	Lr	103	(257)[a]	Xenon	Xe	54	131.30
Lead	Pb	82	207.2	Ytterbium	Yb	70	173.04
Lithium	Li	3	6.941	Yttrium	Y	39	88.9059
Lutetium	Lu	71	174.97	Zinc	Zn	30	65.38
Magnesium	Mg	12	24.305	Zirconium	Zr	40	91.22
Manganese	Mn	25	54.938				

*Unofficial name and symbol
[a] Mass number of most stable or best known isotope
[b] Mass number of most commonly available long-lived isotope